Molecular and Cellular Approaches to Neural Development

Molecular and Cellular Approaches to Neural Development

Edited by
W. MAXWELL COWAN
Vice President and Chief Scientific Officer
Howard Hughes Medical Institute
THOMAS M. JESSELL
Professor of Biochemistry and Molecular Biophysics
Columbia University
S. LAWRENCE ZIPURSKY
Professor of Biological Chemistry
University of California, Los Angeles

New York Oxford
OXFORD UNIVERSITY PRESS
1997

Oxford University Press

Oxford New York
Athens Auckland Bangkok Bogota
Bombay Buenos Aires Calcutta Cape Town
Dar es Salaam Delhi Florence Hong Kong Istanbul
Karachi Kuala Lumpur Madras Madrid
Melbourne Mexico City Nairobi Paris
Singapore Taipei Tokyo Toronto

and associated companies in
Berlin Ibadan

Published by Oxford University Press, Inc.
198 Madison Avenue, New York, New York 10016

Library of Congress Cataloging-in Publication Data
Molecular and cellular approaches to neural development/
[edited by] W. Maxwell Cowan, Thomas M. Jessell, S., Lawrence Zipursky.
p. cm. Includes bibliographical references and index.
ISBN 0-19-511166-4
1. Developmental neurobiology.
2. Molecular neurobiology.
I. Cowan, W. Maxwell.
II. Jessell, Thomas M.
III. Zipursky, Stephen Lawrence.
[DNLM: 1. Nervous System—growth & development.
2. Neurons—physiology.
3. Developmental Biology.
WL 102 M7175 1997] QP363.5.M648 1997 591.1'88.—dc20
DNLM/DLC for Library of Congress 96-38648

9 8 7 6 5 4 3 2 1
Printed in the United States of America
on acid-free paper

Preface

Nearly a century ago, Ramon y Cajal initiated an extensive series of anatomical studies which revealed the extraordinary diversity, organizational complexity, and precision of connections between cells in the nervous system of both vertebrates and invertebrates. Since that time, the goal of developmental neurobiologists has been to understand the mechanisms by which this most complex of biological structures emerges during development. Over the past decade, striking advances in dissecting the molecular and cellular bases of neural development have been made, including the identification of chemotropic molecules that guide axonal outgrowth, and candidate neural inducers. The chapters in this volume review work in many areas which has shaped our current understanding of neural development and has provided a foundation for future studies.

Recent progress in developmental neurobiology largely reflects the enormous impact of molecular biology on biochemical and classical genetic approaches to neural development. For instance, molecular cloning and sequencing of genes encoding extracellular signals, key determinants of nervous system organization, have facilitated the identification of their receptors and the isolation of structurally related factors controlling other developmental processes. Assay systems for the purification of many of these factors may not be feasible; however, with the gene in hand, knockout technology provides a critical test for gene function in the developing organism. In *Drosophila* and *Caenorhabditis elegans,* organisms readily accessible to genetic analysis, methods of positional cloning have led to the isolation of key developmental determinants. Through DNA homology searches, vertebrate relatives of many of these molecules have been isolated. Their in vivo function can be assessed in the mouse through gene knockouts, and an array of molecular biological techniques provides efficient ways to assess their biochemical properties. Through these studies, it has become increasingly clear that the molecules which sculpt the nervous system have been conserved through evolution. This knowledge has spawned a powerful approach to neural development in which molecular similarities allow one to "hop" between organisms, to use different systems for which they are well suited and in so doing, establish the conserved molecular principles of neural development.

Despite the enormous progress made over the past decade, it is clear that we are still in the early days of defining the logic of neural development, and identifying the players required in different developmental processes. The pace of identifying developmental regulators will undoubtedly accelerate as additional genomic and cDNA sequences from *C. elegans, Drosophila,* and mammals become available through large-scale sequencing projects. Whereas the identification of the relevant players in specific processes will continue to be a primary goal of developmental neurobiologists, it is now possible to envision a time when many, perhaps even most, of these components will have been identified. The challenge in the future will be to determine the molecular mechanisms driving individual steps in neural development and the regulatory circuitry by which these steps are integrated into a program that leads to the orderly construction of the nervous system.

This volume was conceived when a small meeting in neural development was being organized at the Howard Hughes Medical Institute. Although the book covers a wide array of topics, it is by no means comprehensive. In the first seven chapters, specific steps in development are considered, including neural induction, determination of neuronal phenotype, neuron–glial interactions, mechanisms of axonal guidance, synaptogenesis, neurotrophic factors and receptors, and cell death. The remaining chapters focus on development of specific regions of the nervous system, including the *Drosophila* visual system and regions of the vertebrate central nervous system. Authors describe the lineage relationships between cells in the vertebrate central nervous system, patterning along both the dorsal/ventral and the anterior/posterior axes, hindbrain and forebrain development, and the formation of the cerebral cortex. The last two chapters focus on neuronal plasticity and the formation of local circuitry in the cortex. Many of the reviews represent collaborative efforts between leaders in the field including, in some cases, investigators studying vertebrate and invertebrate systems. This has provided a rather broad perspective on a number of developmental issues. We would like to thank the authors for producing an excellent series of thoughtful reviews and Mr. Jeffrey House and Oxford University Press for publishing the volume.

W. M. C.
Chevy Chase

T. M. J.
New York

S. L. Z.
Los Angeles

Contents

Contributors

AGAPITE, JULIE
Howard Hughes Medical Institute
Department of Brain and Cognitive Sciences and Department of Biology
Massachusetts Institute of Technology
Cambridge, MA 02139

ANDERSON, DAVID
Howard Hughes Medical Institute
Division of Biology 216-76
California Institute of Technology
Pasadena, CA 91125

BARRES, BARBARA A.
Department of Neurobiology
Fairchild Science Building
Stanford University School of Medicine
Stanford, CA 94305

BRAISTED, JANET E.
The Salk Institute
La Jolla, CA 92037

CAPECCHI, MARIO R.
Howard Hughes Medical Institute
Department of Human Genetics
University of Utah School of Medicine
Salt Lake City, UT 84112

CEPKO, CONSTANCE L.
Howard Hughes Medical Institute
Department of Genetics
Harvard Medical School
Boston, MA 02115

CHENN, ANJEN
Department of Biological Sciences
Stanford University
Stanford, CA 94305

FARIÑAS, ISABEL
Program in Neuroscience
Department of Physiology
University of California, San Francisco
San Francisco, CA 94143

GOLDEN, JEFFREY A.
Howard Hughes Medical Institute
Department of Genetics
Harvard Medical School
Boston, MA 02115

GOODMAN, COREY S.
Howard Hughes Medical Institute
Department of Molecular and Cell Biology
University of California, Berkeley
Berkeley, CA 94720

HARLAND, RICHARD M.
Division of Biochemistry and Molecular Biology
Department of Molecular and Cell Biology
University of California, Berkeley
Berkeley, CA 94720

JAN, YUH NUNG
Howard Hughes Medical Institute
Department of Physiology
University of California, San Francisco
San Francisco, CA 94143

JESSELL, THOMAS M.
Howard Hughes Medical Institute
Department of Biochemistry and Molecular Biophysics
Center for Neurobiology and Behavior
Columbia University
New York, NY 10032

KATZ, LAWRENCE C.
Howard Hughes Medical Institute
Department of Neurobiology
Duke University School of Medicine
Durham, NC 27710

LIN, JOHN C.
Howard Hughes Medical Institute
Department of Genetics

Harvard Medical School
Boston, MA 02115

LUMSDEN, ANDREW
MRC Brain Development Programme
Department of Developmental Neurobiology
United Medical and Dental Schools of Guy's and St. Thomas' Hospitals
London, SE1 9RT, UK

MARTIN, KATHLEEN A.
Howard Hughes Medical Institute
Department of Biological Chemistry
University of California, Los Angeles
School of Medicine
Los Angeles, CA 90095

MCCONNELL, SUSAN K.
Department of Biological Sciences
Stanford University
Stanford, CA 94305

O'LEARY, DENNIS D.M.
Molecular Neurobiology Laboratory
The Salk Institute
La Jolla, CA 92037

REICHARDT, LOUIS F.
Howard Hughes Medical Institute
Program in Neuroscience
Department of Physiology
University of California, San Francisco
San Francisco, CA 94143

RUBENSTEIN, JOHN L.R.
Nina Ireland Laboratory of Developmental Neurobiology
Programs in Developmental Biology and Neuroscience
Center for Neurobiology and Psychiatry
Department of Psychiatry and Langley Porter Psychiatric Institute
University of California, San Francisco
San Francisco, CA 94143

RUBIN, GERALD M.
Howard Hughes Medical Institute
Department of Molecular and Cell Biology
University of California, Berkeley
Berkeley, CA 94720

SANES, JOSHUA R.
Department of Anatomy and Neurobiology
Washington University School of Medicine
St. Louis, MO 63110

SCHELLER, RICHARD H.
Howard Hughes Medical Institute
Department of Molecular and Cellular Physiology
Stanford University School of Medicine
Stanford, CA 94305

SHATZ, CARLA J.
Howard Hughes Medical Institute
Department of Molecular and Cell Biology
University of California, Berkeley
Berkeley, CA 94720

SHIMAMURA, KENJI
Nina Ireland Laboratory of Developmental Neurobiology
Programs in Developmental Biology and Neuroscience
Center for Neurobiology and Psychiatry
Department of Psychiatry and Langley Porter Psychiatric Institute
University of California, San Francisco
San Francisco, CA 94143

STELLER, HERMANN
Howard Hughes Medical Institute
Department of Brain and Cognitive Sciences and
Department of Biology
Massachusetts Institute of Technology
Cambridge, MA 02139

SZELE, FRANCIS G.
Department of Genetics
Harvard Medical School
Boston, MA 02115

TESSIER-LAVIGNE, MARC
Howard Hughes Medical Institute
Department of Anatomy
University of California, San Francisco
San Francisco, CA 94143

WOLFF, TANYA
Howard Hughes Medical Institute
Department of Molecular and Cell Biology
University of California, Berkeley
Berkeley, CA 94720

ZIPURSKY, S. LAWRENCE
Howard Hughes Medical Institute
Department of Biological Chemistry
University of California, Los Angeles
School of Medicine
Los Angeles, CA 90095

Molecular and Cellular Approaches to Neural Development

1

Neural induction in *Xenopus*

RICHARD M. HARLAND

The first step in making the nervous system of a vertebrate is the segregation of the neural plate. The neural plate becomes morphologically distinct as a thickened layer of epithelial cells after gastrulation. After gastrulation and during neurulation, the neural plate rolls up into the dorsal hollow neural tube characteristic of the chordate phylum. However, before the neural plate can initiate differentiation into the diverse cell types of the functioning nervous system, it must be set aside from the rest of the ectoderm. There is evidence from a variety of transplantation studies that an active signal specifies the neural plate during gastrulation. This event, called neural induction, changes the fate of prospective epidermis so that it no longer will form a part of the skin but instead will form the nervous system. Most experiments on neural induction have been done with amphibians, which are particularly suited to the "cut and paste" experiments which define responsive tissues and signaling centers. More recently, explants of amphibian ectoderm have been used to assay candidate molecules which induce neural tissue. Because of this wealth of experimental background, the focus in this chapter is on work with amphibians, particularly the molecular biology of neural induction in *Xenopus laevis.* Background information on neural induction can be found in developmental biology textbooks and the experimental embryology has been admirably reviewed elsewhere (Gerhart et al., 1991; Doniach, 1993).

In the last year or two several apparently diverse mechanisms that have been proposed to account for neural induction have been simplified into one. This chapter will review the evidence that supports the proposed mechanism, which hinges on the function of the bone morphogenetic protein family (BMPs). Signaling by BMPs biases ectoderm toward an epidermal fate. BMP mRNAs are present in the ectoderm, and blockage of their function by cell dissociation, use of dominant negative BMP receptors, or dominant negative BMP ligands results in neural development. A number of neural-inducing molecules have been identified, such as noggin, chordin, and follistatin; these are now thought to act as antagonists of BMP action. In the case of noggin, the antagonism is direct, with noggin binding directly to BMP2 and 4 and preventing them from binding their receptor.

Blocking BMP signaling results only in anterior neural induction, leaving the question of how posterior fates are established. Different secreted proteins such as fibroblast growth factor (FGF) can induce posterior neural fates, probably by acting on tissues that have received other neuralizing signals.

Finally, progress in understanding the formation of differentiated neurons has benefited from molecular parallels to *Drosophila* neurogenesis, where activation of the Notch receptor suppresses differentiation of neurons. Similarly, proteins of the basic helix-loop-helix family related to proteins encoded by the achaete–scute complex can influence neural induction and differentiation.

THE ORGANIZER INDUCES THE NEURAL PLATE

Transplantation experiments have shown that the dorsal mesoderm (Spemann's organizer) can recruit ectodermal cells from a host embryo to make a neural tube. In these experiments, host and graft are marked by pigment differences or lineage tracers, allowing identification of the signaling and responding tissues. If the transplant is done at the early gastrula stage, with a complete organizer, the secondary axis and neural tube can contain the full complement of anterior structures. Therefore the dorsal mesoderm has neural inducing and patterning properties. If the organizer is ablated surgically in the late bastula, no neural plate forms, demonstrating that the organizer is necessary for neural induction (Stewart and Gerhart, 1990). Experimental embryology has thus posed a fascinating problem to be solved at the molecular level. What are the signals from the mesoderm that induce the neural plate and how are they coordinated to produce the full range of neural structures?

What Is a Neural Inducer?

Embryological experiments with responsive ectoderm and inducing mesoderm have determined which tissues produce signals that induce neural tissue and which cells can respond to these signals. These characteristics have defined two criteria for the activity expected of a neural inducing molecule. These criteria are illustrated by differences in the ability of *Xenopus* ectoderm to respond to Hensen's node (the chick organizer) and activin-induced mesoderm (Kintner and Dodd, 1991). Since mesoderm induces neural tissue, molecules that increase the amount of dorsal mesoderm, such as activin, indirectly produce neural tissue. Therefore, if neural induction is accompanied by the formation of mesoderm, the mesoderm may have induced the neural tissue and the induction need not be direct. Thus one test for a direct neural inducer is that it should induce neural fates without inducing mesoderm. The first criterion is straightforward and has often been used to provide evidence that a molecule or a pathway causes neural induction. The evidence is made possible by the availability of reliable markers for different tissues, which allows inductions to be scored quickly and unambiguously. Reliable molecular markers have also been essential for determining the regional nature of neural inductions since most explant assays provide few histological clues as to the nature of neural tissue.

The second criterion follows from the finding that ectodermal cells can respond to organizer tissue, the endogenous neural inducer, until the end of gastrulation; by contrast, the same cells lose their competence to respond to mesoderm inducers at the beginning of gastrulation (Sive et al., 1989; Sharpe and Gurdon, 1990; Kintner and Dodd, 1991; Servetnick and Grainger, 1991). Therefore a direct neural inducer is expected to be able to turn on neural-specific genes when applied to gastrula-stage ectoderm, and to do so without inducing mesoderm. The second criterion is more difficult to apply since it can be difficult to introduce molecules at the gastrula stage, except when purified protein is available. It is particularly difficult to study the time receptor molecules are required because there is currently no good way to inactivate a receptor at a particular time.

Attenuation of Signaling by TGF-β Superfamily Receptors Reveals Latent Neural Fates

Evidence that intercellular signaling has an important role in inhibiting neural fates came from work with dissociated embryos (Godsave and Slack, 1989; Grunz and Tacke, 1989). When dissociated for long periods, animal cap ectoderm and even presumptive mesoderm could develop autonomously into neural tissue. The timing and extent of disaggregation was crucial, with only long periods of dissociation at the late blastula or gastrula stage leading to neuralization. Both sets of authors noted that extended disaggregation might allow the loss of extracellular matrix components that could act as neural inhibitors.

Experiments with disaggregated whole embryos also resulted in neural induction in the absence of muscle differentiation, and this has been taken as evidence that latent maternal inducers must be present and activated during gastrulation (Sato and Sargent, 1989). However, an alternative explanation is provided by the experiments of Lemaire and Gurdon (1994); when embryos were disaggregated at the blastula stage, which is the crucial period for muscle induction, they found that organizer-specific genes like goosecoid can still be expressed. Although goosecoid was the only organizer marker examined, the result implies that the organizer can form despite the failure of mesoderm to differentiate. Therefore, after reaggregation, neural induction could occur in the usual way via the action of organizer on ectoderm.

The results on neuralization of animal caps at the gastrula stage (Godsave and Slack, 1989; Grunz and Tacke, 1989) suggested that neural fate is the default fate in the absence of any embryonic cell–cell signals. The idea that embryonic cells contain a latent ability to be neuralized goes back a long way, particularly to the work of Holtfreter (recently reviewed in Gerhart, 1996). However, the emergence of a molecular candidate for the signal that suppresses neural fates initially came quite by surprise from studies on mesoderm induction. Members of the TGF-β superfamily have been implicated in mesoderm induction and patterning for some time, and it has only recently been found that neural fates may be specified in part by preventing signaling by members of the TGF-β superfamily.

The TGF-β Superfamily

This family is named for its founder member, transforming growth factor–β. Subsequently a large number of secreted signaling proteins with diverse biological effects were found to be related in sequence to TGF-β (Kingsley, 1994). Although the superfamily of TGF-β–related ligands is large, this review focuses on the BMP subfamily, which has ventralizing effects in the mesoderm and epidermalizing effects in the ectoderm of the early embryo. Activins and Vg-1 are also relevant, but they have mesoderm-inducing effects. The family is defined by a characteristic spacing of seven cysteines in the active domain. All these proteins are presumed to be related structurally to TGF-β, which is a disulfide-bridged dimer with each subunit folded into a cysteine knot. A restricted set of heterodimers between family members can form, and their role will be discussed below, as will the binding of different ligands to a large family of structurally similar receptors.

The synthesis and processing of the functional domain are presumed to be similar between family members although processing has only been studied in a few. The proteins are secreted into the endoplasmic reticulum, where the signal sequence is removed. The subunits then dimerize prior to an internal cleavage that removes the mature C-terminal ligand from the pro domain. It is usually assumed that the protein is secreted and acts on other cells, although in some cases ligands are not secreted efficiently (Kessler and Melton, 1995). Thus, some TGF-β family ligands may act on receptors in the endoplasmic reticulum (ER), and since they are not permitted to diffuse, they could act with greater potency at those sites (Thomsen and Melton, 1993) and even behave as cell-autonomous cytoplasmic determinants (Lemaire and Gurdon, 1994).

Mechanism of TGF-β Signaling

Early experiments where iodinated ligand was cross-linked to the surface of responsive cells showed three classes of receptor. The largest, the type III chain, is thought to be involved in presentation of ligand and is not discussed further. The important signaling complex is a heteromeric complex formed from the type I and type II chains (reviewed by Massagué, 1996). For TGF-β and activin, the type II receptor is a high-affinity transmembrane protein with a serine threonine kinase domain. The smaller type I receptor was shown by mutant analysis to be required for signaling, and it also has a serine threonine kinase domain.

A large number of type II and type I receptors for TGF-β–related ligands have been isolated. The initial paradigm for the mode of signaling came from studies with TGF-β and activin receptors, and the paradigm has only been modified, in its details, as a result of further studies. Type II receptors encode transmembrane proteins with high affinity for ligand, even in the absence of type I receptors (Mathews and Vale, 1991; Lin et al., 1992). As diagrammed in Figure 1–1, when the type I receptor is recruited to the complex, a conserved region just inside the membrane-spanning domain, rich in glycine and serine (the GS domain), is phosphorylated (Wrana et al., 1994). The type II receptor cytoplasmic

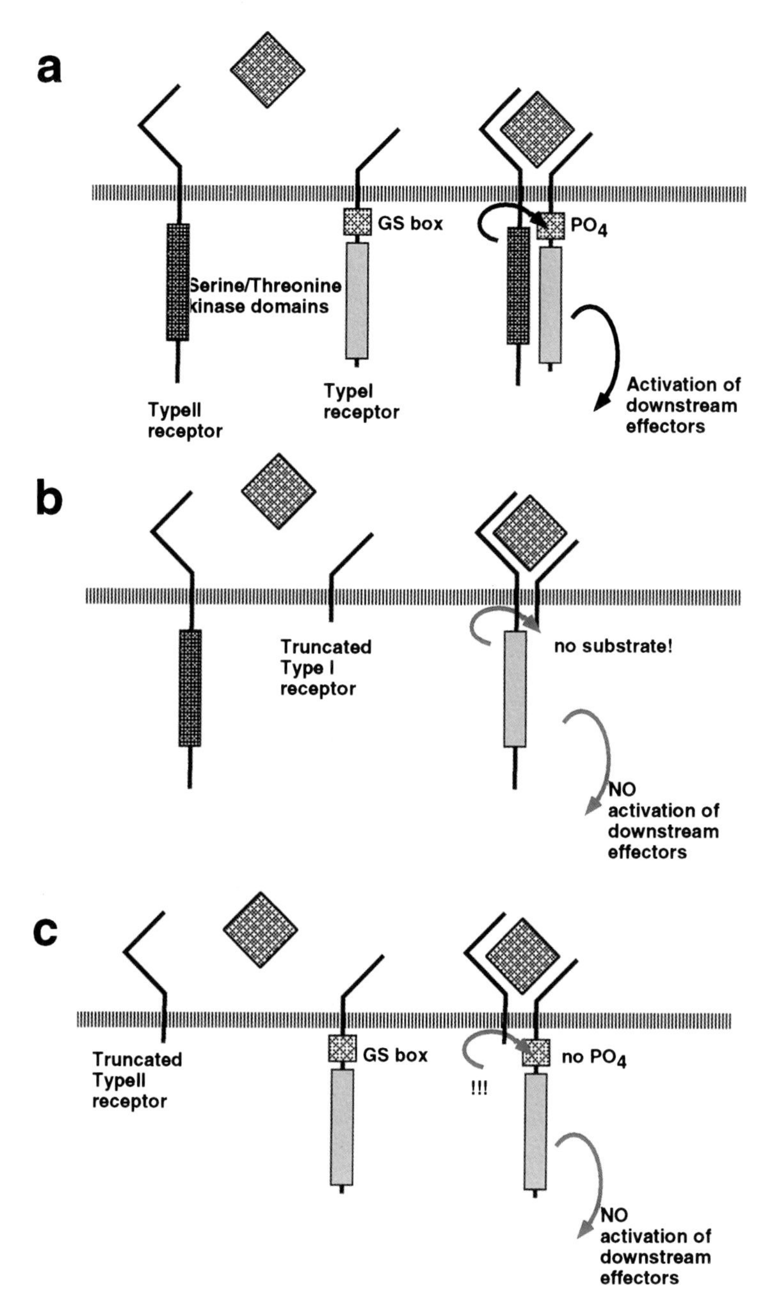

FIGURE 1–1 Models for the action of TGF-β superfamily of receptors. (A) A heteromeric complex of type II and type I receptors is induced upon ligand binding. Once formed, a constitutive activity of the type II receptor phosphorylates the GS domain of the type I receptor, and this causes it to be active in downstream signaling. The function of the ligand is to induce complex formation. If the receptors are overexpressed, they will form complexes, and signal in the absence of ligand (Ventura et al., 1994). (B) A dominant negative type I receptor cannot be phosphorylated and has no domain to confer downstream signaling. Thus, when it is expressed in sufficient excess, it effectively outcompetes the wild type for formation of the complex, and no signal is transmitted. (C) A dominant negative (truncated) type II receptor can still participate in complex formation, and if expressed in sufficient excess, will prevent wild-type receptors from binding ligand. Although a complex is formed, the GS domain of the type I remains unphosphorylated, and so no signal is transduced.

domain encodes a serine/threonine kinase domain which is necessary to activate the type I receptor by phosphorylation of the GS domain (Carcamo et al., 1995). The type I receptor has its own serine/threonine kinase domain which is presumed to phosphorylate downstream components.

Activated mutants of the type I receptor can be made by altering the GS region (Wieser et al., 1995). These show that the type I receptor is itself sufficient for signaling, and thus the role of the type II receptor and ligand is to form a stable complex which phosphorylates the GS domain. The type II receptor can therefore be considered a primary receptor and the type I receptor a transducer of the signal (Massagué, 1996).

Role of the Activin Type II Receptor in Mesoderm and Neural Induction

In experiments testing the requirement of the activin receptor for mesoderm induction, animal caps (prospective ectoderm cut from the animal pole of the blastula) were injected with a large amount of a truncated type II receptor. Since this protein lacks a kinase domain, it can dimerize with the type I receptor without activating it (Fig. 1–1C). Injection of the truncated receptor was sufficient to block any activin-mediated induction of mesoderm. However, the injected caps developed substantial amounts of neural tissue, as assayed by expression of the neural-specific marker NCAM (neural cell adhesion molecule; Hemmati-Brivanlou and Melton, 1992). Since uninjected caps normally differentiate as epidermis, this observation suggested that activin signaling normally prevents cells from adopting neural fates and instead directs them to become epidermis. The idea was further supported by the finding that follistatin, a molecule that was known to block activin signaling in other assays, could mimic the effect of the truncated activin receptor (Hemmati-Brivanlou et al., 1994). Follistatin was also found to be expressed at the right place and the right time to be active as a neural inducer, namely in the dorsal mesoderm of the gastrula. For a brief time, activin was the most attractive molecule known to mediate the blockage of neural fates in the embryo. However, the authors had pointed out that it was not clear whether the truncated activin receptor was specific in blocking only activin signaling. It subsequently emerged that the truncated activin receptor is pleiotropic and blocks the effects of activin, Vg-1, and BMPs. It also emerged subsequently that follistatin may not be specific for activin (see below).

Epidermalizing Effect of BMPs

Wilson and Hemmati-Brivanlou (1995) directly tested the role of activin and BMP4 in suppressing the ability of dissociated cells to adopt neural fates and the possible roles of activin and BMPs in epidermal induction. In the direct test, activin was found only to induce mesoderm, and at no dose did it restore epidermal gene expression. Instead, BMP4 protein was extremely potent in restoring epidermal fates; it was effective at about 15 pM. This effect of BMP4 could be blocked by the truncated activin receptor; therefore, the effect of the truncated activin receptor must indeed be pleiotropic, blocking the effects of not only activin but also BMP4.

Crosstalk Between Different TGF-β Family Receptors

The problem for embryologists would have been enormously simplified if each ligand had a specific type II receptor, and each receptor could only interact with a cognate type I receptor to specify a particular flavor of signal. However, in the presence of overexpressed TGF-β type II receptor, iodinated TGF-β could be cross-linked to all of the type I receptors (Ebner et al., 1993; ten Dijke et al., 1994a). Even though many of these complexes are not productive for signaling (ten Dijke et al., 1994a), receptor interactions must be taken into account in experiments where a large amount of truncated receptor is expressed. Therefore, a considerable amount needs to be understood about receptor biochemistry before experiments with truncated receptors can be interpreted unambiguously, since an overexpressed truncated type II receptor may associate with and prevent the activity of several type I receptors.

It is not enough to know about the interactions of receptors to interpret dominant negative experiments. Whenever a molecule is hugely overexpressed it is essential to carry out controls to show that there is no general pleiotropic effect on general cellular processes such as translation, secretion, and signal transduction. Such controls include at a minimum the observation that injection of wild-type receptor RNA can overcome the effect of the dominant negative; if the dominant negative protein is competing with the wild type, then additional wild-type protein should reverse the effects. Alternatively, if a candidate dominant negative protein is generally toxic, then the effects of overexpressing the toxic protein would not be overcome by expression of more protein, even wild type. Applying this test, the truncated activin receptor does indeed behave as a dominant negative, with efficient rescue by wild-type receptor (Hemmati-Brivanlou and Melton, 1992).

It is also useful to demonstrate that a dominant negative protein is specific for a certain pathway. In the case of the activin receptor, the blockage of mesoderm induction by truncated receptor could be very effectively overcome by activation of the FGF signaling pathway. Although such experiments showed some specificity of interference with signaling, subsequent experiments demonstrated that the truncated activin receptor blocks the effects of the Vg-1 ligand (Schulte-Merker et al., 1994; Kessler and Melton, 1995), even though Vg-1 does not bind to the receptor (Kessler and Melton, 1995). The truncated activin receptor also blocks the effects of BMPs as discussed above (Wilson and Hemmati-Brivanlou, 1995).

Although there are many caveats to interpreting experiments with dominant negative receptors, it is important not to lose sight of the positive conclusions. From results with the truncated activin type II receptor, it is fair to conclude that mesoderm formation requires signaling by members of the TGF-β superfamily, and that the suppression of neural fates is also dependent on such signals. Further precision in interpretation requires more specific reagents and a better understanding of the association of the receptors with ligand.

Promiscuity of Ligand Binding?

The lack of specificity of the truncated activin receptor may reflect not only the association of overexpressed type II receptor with different transducing type I receptors but also its ability to soak up different ligands. It is therefore important to understand the specificity of ligand binding to receptors. Receptors have been named for one ligand that binds and activates signaling, but subsequently the same receptor may be found to bind other ligands. The type I receptors particularly exemplify this problem, in particular because they tend to have low affinity for ligand and only show specific and high-affinity binding in conjunction with a type II receptor. While the type II receptors are still named for the first ligand that was found to bind with high affinity, there are many alternate names for the type I receptors. Below I use the nonjudgmental ALK (activin-like kinase) nomenclature of ten Dijke et al. (1994a) which describes sequences obtained by degenerate polymerase chain reaction (PCR) from conserved receptor domains (Table 1–1).

Which Type I Receptors Transduce BMP Signals?

ALK-3 and ALK-6 are type I receptors that can specify high-affinity binding to BMP ligands even when expressed alone (Graff et al., 1994; Koenig et al., 1994b; Suzuki et al., 1994; ten Dijke et al., 1994b). They nonetheless require type II receptors; thus the final affinity of a complex for ligand, as well as whether the signal is propagated, depends on the combination of receptor types and the concentration of ligand (Liu et al., 1995; Nohno et al., 1995). Since the main interest here is BMP signaling, the most relevant observations for neural induction are that BMP-7 (also known as osteogenic protein 1, OP-1) can bind not only the BMP receptors, but also the activin type II receptor (Yamashita et al., 1995). Indeed, the affinity of the activin type II receptor for activin is only threefold higher than for BMP-7. BMP-7 can also initiate a signal through the activin type II receptor by complexing to either ALK-3 or ALK-6 or ALK-2 (which was initially thought to be specific for activin signaling). However, it does not bind efficiently to the ac-

TABLE 1–1. The TGF-β type I Receptor Family

ALK-1	SKR3		TSR-1
ALK-2	SKR1	TSk7L	ActRI
ALK-3	SKR5	BRK-1	BMPR1A
ALK-4	SKR2		ActR1B
ALK-5			TGFβ1R
	SKR4	ESK2	TβR-I
ALK-6	SKR6	BRK-2	BMPR-1B

This table was compiled using the entrez browser service of the National Center for Biotechnology Information (http://www3.ncbi.nlm.nih.gov/Entrez/). The sequences in a line have a similarity score of at least 254 (for mammalian sequences). Relatives of ALK-5, a TGF-btransducer, include SKR4, ESK2, and TbRI, although these score only 106. For *Xenopus*, the homologous gene is more divergent; for ALK-3 the *Xenopus* gene has a similarity score of 91.

tivin type II/ALK-4 combination. This is significant because if activated type I receptors are compared, the activated ALK-4 substitutes for activin in the induction of dorsal mesoderm, but activated ALK-2 does not (R. McKendry and R.M.H., unpublished results). Since ALK-4 is likely to be responsible for mesoderm-inducing signals, that would account for why BMP-7 does not act efficiently as a mesoderm inducer (Yamashita et al., 1995).

Specificity of Signaling by Type I Receptors: Further Evidence

Comparison of activin and BMP signaling in cultured cells has used either growth inhibition of a mink lung cell line, or induction of the PAI-1 gene (plasminogen activator inhibitor), or induction of the synthetic reporter 3TP-Lux (which contains an element of the PAI-1 gene with three [TPA]/phorbol-ester–inducible elements). However, in more physiological settings, active receptors can induce a rather diverse set of responses, and it may well be that cultured cell assays do not reveal the full specificity of receptors. Thus, when BMP7 and activin were compared (Yamashita et al., 1995), both could induce erythroid differentiation through activin type II and ALK-2 (but not ALK-4) receptors; in contrast, only activin induced follicle-stimulating hormone (FSH) release from pituitary cells, which may be explained by the function of ALK-4 in conjunction with the activin type II receptor and the absence of ALK-2.

All these experiments show that, despite the ability of activin or BMPs to bind to different type II receptors, the recruitment and activation of type I receptors may have considerable specificity, so the ultimate signal may be much more specific than the affinity of ligand for the type II receptors would lead us to expect.

Use of Dominant Negative BMP Receptors in Xenopus

At the time that experiments were done with the dominant negative activin type II receptor, the heteromeric nature of receptor complexes was not fully appreciated, although the assumption that a truncated, kinase-negative version of the receptor might interfere with signaling was justified by the results. With the isolation of type I receptor cDNAs, the same kinds of dominant negative experiments could be done with type I receptors (Fig. 1–1B).

To date ALK-3, a BMP-binding receptor, is the only truncated type I receptor for which results have been reported (Graff et al., 1994; Maeno et al., 1994; Suzuki et al., 1994). In animal cap assays, BMP4 is an efficient inducer of epidermis but a poor inducer of mesoderm, although it does induce ventral mesoderm at high concentration (1 μg/ml) (Wilson and Hemmati-Brivanlou, 1995). It is not yet clear whether this induction is physiologically relevant or whether it only occurs through artifactual binding to and activation of mesoderm-inducing type I receptors at very high concentration. Nonetheless, the induction of mesoderm by high concentrations of injected BMP4 mRNA in the animal cap provides a useful assay for the specificity of the truncated ALK-3. In animal caps, truncated ALK-3 blocked the effects of BMP4 mRNA, but not activin mRNA; effects could also be reversed by excess wild-type receptor (Graff et al., 1994). Thus, it appears that this truncated BMP type I receptor inhibits BMP signaling quite specifically.

At lower concentrations, BMP4 only reveals its epidermis-inducing effects on animal caps when the caps are dissociated. The effects of BMP4 are more easily scored in the context of the mesoderm, rather than the ectoderm. BMP4 is a potent ventralizer of mesoderm (Dale et al., 1992; Jones et al., 1992; Fainsod et al., 1994; Clement et al., 1995; Gawantka et al., 1995; Hemmati-Brivanlou and Thomsen, 1995; Northrop et al., 1995; Schmidt et al., 1995; also reviewed in Harland, 1994). To assess the role of BMPs during development, the truncated receptor was injected into embryos, and the first reports focused on the mesoderm. Here, the truncated ALK-3 induced dorsal gene expression and ectopic axis formation (Graff et al., 1994; Maeno et al., 1994; Suzuki et al., 1994). The interpretation is that BMP mRNAs are normally expressed in embryos and confer ventral fates on mesoderm. However, where BMP expression or function is blocked, dorsal fates result. The interesting conclusion is that all of the marginal zone has a latent dorsal fate which is antagonized by the constitutive expression of BMPs, and normal dorsal fates result because of blockage of BMP transcription or protein function. Noggin, follistatin, and chordin, three agents that are made in dorsal mesoderm and antagonize BMP activity, are discussed below.

What effect does blocking BMP signaling have on the neural plate? In whole embryos the neural plate is expanded (Schmidt et al., 1995). However, since the domain of neural inducing tissue is expanded (as revealed by an increase in somitic mesoderm), this result does not distinguish between a direct effect of the dominant negative receptor on neural induction and an indirect effect through the change in neural-inducing mesoderm.

Blocking BMP Signals Reveals Latent Neural Fates

The experiments with dissociated animal cap ectoderm showed that BMP4 could suppress neural development, instead directing cells to differentiate as epidermis (Wilson and Hemmati-Brivanlou, 1995); in a complementary approach, the dominant negative BMP receptor was expressed in animal cap ectoderm and found to cause neuralization. In contrast to the experiments in whole embryos, neuralization was found in the absence of mesoderm, and by this criterion was counted as direct neuralization (Sasai et al., 1995; Xu et al., 1995).

Specificity of BMP Signaling Demonstrated by Dominant Negative Ligands

Another way of examining the function of TGF-β ligands is to prevent their processing. This approach takes advantage of the pathway by which functional ligands are processed. Normally, the ligand is produced in a precursor form, and after the signal sequence is removed, the subunits dimerize, after which the mature C-terminus is released by cleavage at sites that are often specified by two basic amino acids. If the cleavage site is mutated, no productive ligand can be made (Lopez et al., 1992). Such mutants can be used experimentally to prevent processing of the wild-type ligand; if expressed in sufficient excess, all new translation products will contain at least one mutant subunit, and this uncleavable subunit prevents secretion of any functional ligand from the wild-type chain.

The dominant negative ligand is only effective in embryo experiments if the ligand is being produced zygotically. Preexisting mature ligand or even unprocessed dimers are not affected (Lopez et al., 1992; Wittbrodt and Rosa, 1994). The specificity of the experimental intervention lies in the specificity of dimer formation. Thus the strategy becomes extremely useful for BMPS which are zygotically produced, either from maternal mRNA or new transcription. BMPs can heterodimerize (Aono et al., 1995; Hazama et al., 1995) so that expression of a single kind of mutant subunit will prevent secretion of several BMPs. This approach was successfully used to study mesoderm patterning and neural induction (Hawley et al., 1995). Use of either mutant BMP4 or BMP7 was effective in blocking BMP signaling, and once again, preventing BMP signaling dorsalizes mesoderm and neuralizes the ectoderm.

Where Are the BMPs Expressed?

So far the discussion has focused on neural development resulting from blockage of BMP signals, but it is important to consider where the BMP signals are made. Control of transcription of the BMPs is likely to contribute to the difference between epidermis and neural plate.

BMP4 transcripts are present at low maternal levels and accumulate during gastrula stages (Dale et al., 1992; Nishimatsu et al., 1992). Recent experiments, using in situ hybridization, have shown that from the mid–gastrula stage, BMP4 transcription is excluded from both dorsal mesoderm and prospective neural plate (Fainsod et al., 1994; Hemmati-Brivanlou and Thomsen, 1995; Schmidt et al., 1995). Nonetheless, at the onset of transcription, the RNA is present throughout the ectoderm (Hemmati-Brivanlou and Thomsen, 1995). Dissection experiments which used goosecoid and Xwnt-8 as internal controls for dissection accuracy showed that BMP4 is expressed in the organizer at the onset of gastrulation (Re'em-Kalma et al., 1995). Therefore BMP4 expression is not localized initially but is quickly downregulated in the organizer and neural plate. This level of regulation is likely to be an important part of formation of the neural plate, although the mechanism of control is not yet known. In the marginal zone, injected *goosecoid* RNA has been shown to turn off BMP4 (Fainsod et al., 1994), but *goosecoid* does not affect BMP4 expression in the neural plate (and *goosecoid* is not expressed there in any case). Experiments in mouse ES cells have suggested that BMP4 may control its own transcription (Johansson and Wiles, 1995) so that antagonists of BMP activity may effectively turn off BMP genes in the dorsal mesoderm and animal cap during gastrulation. This appears to be true for tissues analyzed at the end of neurulation (Re'em-Kalma et al., 1995), but the signals that turn off BMP4 in the neural plate of the gastrula have yet to be elucidated. It is to be expected that organizer signals induce transcription factors that both repress BMP transcription and determine neural fates. An example of a factor that is induced in explants during neural induction is X1POU2, which can cause neural development when ectopically expressed (Witta et al., 1995).

BMP7 is also expressed in early embryos (Nishimatsu et al., 1992; Hawley et al., 1995). It also is present in the ectoderm and mesoderm of the gastrula, although unlike BMP4, BMP7 transcripts are not cleared from either the organizer or neural plate at gastrula stages.

BMP2 is expressed maternally and transcripts decline at gastrulation without showing any strong dorsal/ventral localization (Shoda et al., 1994; Clement et al., 1995; Hemmati-Brivanlou and Thomsen, 1995). Maternal BMP2 protein may therefore be an important player that must be suppressed in order for dorsal mesoderm and neural plate to form.

Finally, Anti–dorsalizing morphogenetic protein (ADMP), a BMP gene related to BMP3, is expressed specifically in organizer and later in the midline of the sensorial (inner) layer of the neural plate (Moos et al., 1995). This transcript has ventralizing activities similar to the other BMPs.

Although further BMP activities may remain to be discovered, there are already enough to contend with. Their localization, with the exception of ADMP, is consistent with a role in the specification of epidermis. Among them, BMP2 turns off everywhere, and BMP4 turns off in the neural plate. In assays for bone morphogenetic activity, heterodimers of BMP2 and BMP7 are more active than either homodimer (Aono et al., 1995), leading to the possibility that the most potent activities in the embryo are specified by heterodimers of BMP2 and 7 or BMP4 and 7. If this holds true, the absence of BMP heterodimers would occur in the neural plate because of the absence of BMP4 transcripts. It remains to be determined how BMP4 transcription is repressed in the neural plate. Candidate factors that may repress BMP transcription might be expected to have weak neural inducing, or cement-gland-inducing effects.

Direct Neural Inducers Are Antagonists of BMP Activity

Noggin

Noggin was isolated in a screen for molecules which could dorsalize embryos (Smith and Harland, 1992). Injection of noggin mRNA can mimic the early signals that induce the organizer, although noggin has dorsalizing effects on the mesoderm when presented either early as mRNA or late as expression plasmids or protein (Smith et al., 1993). Since noggin is a secreted protein, and is expressed in the organizer, it was a prime candidate for other activities of the organizer, such as neural induction (Smith and Harland, 1992). When tested with purified protein, or by expression plasmid injection, noggin protein passes the various tests as a direct neural inducer. Neural tissue is induced in the absence of mesoderm, and noggin can act at gastrula stages, when mesoderm induction by activin is no longer effective (Lamb et al., 1993).

Could noggin be a BMP antagonist? In embryos, noggin has dorsalizing effects in the marginal zone that are the opposite of the ventralizing effects of BMPs (Dale et al., 1985; Dale et al., 1992; Jones et al., 1992; Smith et al., 1993), and depending on the dose of each molecule, either noggin or BMP activity can predominate (Re'em-Kalma et al., 1995). In animal caps, the neural inducing effects could also be interpreted as being antagonistic to BMPs. However, the mechanism by which antagonism occurs is not suggested by such experiments; noggin could be acting as an antagonist outside the cell, or at the level of signal transduction or transcription factor activity.

The mechanism of noggin activity has been addressed by recent experiments using injected fly embryos. While injected noggin antagonizes the activity of the fly BMP4 (decapentaplegic, dpp), it cannot inhibit the activity of the activated BMP receptor (activated *tkv*, Holley et al., 1996). This suggests that noggin works somewhere between the BMP ligand and receptor activation. The obvious implication is that noggin could interact directly with BMPs.

Indeed, noggin does block BMP4 activity directly by binding to the protein with high affinity and preventing it from binding to the BMP receptor (Zimmerman and Harland, 1996). Noggin has somewhat lower affinity for BMP7, and none detectable for TGF-β or activin. Thus the simplest explanation of how noggin neuralizes ectoderm is that it blocks BMP signaling, just like the dominant negative BMP receptors and ligands. The measured affinity of noggin for BMP4 (kD of 19 pM) is in the range where noggin would be expected to sequester BMP4 away from its receptors, which have a lower affinity for the BMP4 (0.1–1 nM kD; Koenig et al., 1994a; Iwasaki et al., 1995). What happens to the complex is then uncertain, but the difficulty of detecting endogenous noggin protein with cocktails of monoclonal antibodies (J.M. de Jesus and R.M.H., unpublished) might suggest that the noggin, and perhaps the BMP4, has a fairly short half-life.

The mechanism of noggin action therefore appears quite simple and direct, and in part accounts for the fairly high dose of noggin that is necessary to neuralize ectoderm (10 nM; Lamb et al., 1993). BMPs are made in cells of the animal cap, and presumably presented immediately to their neighbors; thus noggin must penetrate the explant and effectively block BMP signaling in a mass of tissue. However, it is also worth noting that in simplified systems, effective doses of a single recombinant protein may be much higher than those necessary in vivo, where proteins often cooperate to signal. Although the high dose of noggin is often remarked upon, it is still the only recombinant protein that has been shown to induce neural tissue, and the doses are in the same range of some other recombinant proteins in in vitro bioassays (e.g., Sonic hedgehog induction of motor neurons; Fan et al., 1995; Marti et al., 1995; Roelink et al., 1995). Furthermore, among the neural inducers that have been tested as injected mRNA, noggin has much higher activity per mole of RNA injected than follistatin or chordin (F. Mariani, M. Dionne, R.M.H., unpublished).

Ultimately, the role of noggin in neural induction must be assessed by removing its function. To date, knockout experiments in mice have shown that noggin is not required for neural induction (A.P. McMahon, R.M.H., others, unpublished). Experience has shown that many genes are functionally redundant, so this no longer comes as a surprise. The difficulty now is to understand how significant BMP antagonism is for neural induction. We need to know what factors contribute to the signal from the organizer and whether there may be separate or parallel routes to induction of neural tissue.

Follistatin

Follistatin was initially suggested as a possible neural inducer following experiments with a truncated activin receptor (Hemmati-Brivanlou and Melton, 1992) because it is a potent antagonist of activin. Follistatin interacts directly with activin and is presumed to block the interaction of activin with its receptors (Schneyer et al., 1994). Follistatin mRNA has been demonstrated to neuralize (Hemmati-Brivanlou et al., 1994), but it is not clear how this is occurring, especially in light of the experiments showing that BMPs, not activin, are the effective signal that suppresses neural fates in favor of epidermis. However, in an assay for growth inhibition, a tenfold excess of follistatin over BMP-7 antagonized BMP activity (Yamashita et al., 1995); follistatin had no effect on the growth-inhibitory effect of TGF-β, suggesting it is not acting in a parallel pathway on the cells, but rather must be inactivating BMP7 directly, although relatively inefficiently. Indeed, in the animal cap epidermis-inducing assay, follistatin had no inhibitory effect on added BMP4 (Wilson and Hemmati-Brivanlou, 1995). In line with the latter observation, the effects of follistatin in neural induction could be reversed by BMP4 expression (Sasai et al., 1995). Perhaps follistatin has a greater affinity for BMP7 than BMP4, and inactivation of BMP7 in embryos may be enough to block epidermal induction. While it is attractive to speculate that follistatin may selectively bind and inactivate members of the BMP family, to date no direct binding has been reported. Perhaps it is not surprising, in light of the weak activity of follistatin in blocking BMP4 signals, that it is a relatively weak neural inducer. Although it passes the test of inducing neural tissue in the absence of mesoderm (Hemmati-Brivanlou et al., 1994), the protein has not been reported to neuralize.

In another assay for whether a neural inducer passes the test of activity during gastrulation, a stage-specific promoter can be used to drive expression of the gene. Although most injected promoters are activated at the mid–blastula stage (Harland and Misher, 1988), some are activated later, and the *Xenopus borealis* cytoskeletal actin promoter has been used to drive late expression of neural inducing candidates. Noggin is extremely potent in this assay (Lamb et al., 1993), but follistatin fails to turn on neural-specific genes, although it is capable of activating cement gland genes (F. Mariani, P. Wilson, R.M.H., unpublished). It seems therefore that follistatin may play a subsidiary role in neural induction, despite its expression in organizer tissue (Hemmati-Brivanlou et al., 1994).

Chordin

Chordin was isolated in a screen for organizer-specific transcripts; this transcript also dorsalizes embryos and induces neural tissue (Sasai et al., 1994, 1995). Chordin resembles the *Drosophila* gene *short gastrulation (sog)* which genetically acts as an antagonist of signaling by the BMP4 homologue, *dpp* (Ferguson and Anderson, 1992). This led to tests of the relative effects of injecting BMP4 and chordin mRNAs or DNA expression vectors; again they show strong antagonism in their activities in both the marginal zone and in the ectoderm, with chordin promoting neural fates in the ectoderm (Sasai et al., 1995). It will be interesting to see how well chordin acts at the level of protein applied to the ectoderm and whether it binds various BMP family members with high affinity.

All of the factors that repress BMP activity induce anterior neural markers like Otx-2, supporting the idea that anterior is the initial state of specification of neural tissue (Doniach, 1993). Whether repression of BMP signaling is the main neural inducing pathway is not yet known, and experiments with a dominant negative FGF receptor suggest there may be yet other pathways (see below).

Other Neural Inducers: FGF

Fibroblast growth factor (FGF) has shown neural inducing activity in a number of experimental assays, which differ in detail but nonetheless are in agreement that FGF signaling can induce posterior neural fates (reviewed by Doniach, 1995). In one experimental assay, animal cap cells are dissociated and then treated with FGF (Kengaku and Okamoto, 1993, 1995). These authors found that region-specific markers were activated in response to increasing doses of FGF. In a somewhat different assay, caps were kept open in a low calcium and magnesium medium to prevent irreversible specification as epidermis. In contrast to the dissociated cells, FGF did not induce different fates in a concentration-dependent way, but rather the type of neural tissue induced depended on the age of the target tissue. Early treatments resulted in more posterior fates, whereas later treatments resulted in midbrain or hindbrain specification. Finally Cox and Hemmati-Brivanlou (1995) used explants of anterior neural tissue and caps neuralized by either follistatin or truncated *Act*RII to show that FGF can caudalize anterior neural tissue.

All of these experiments are manipulations of embryonic material and can at best illustrate what activities a factor might have. Thus it may not be important that the different assays revealed qualitative differences in the activity of FGF (such as dose-dependent anterior/posterior [A/P] fates or age-dependent A/P fates); rather, these experiments illustrate what activities may be used in the embryo and some of the mechanisms that may be used in the normal process of spinal cord specification. While it would be desirable to go back and carry out experiments in an environment that is much closer to the neural plate as it is induced in vivo, it is extremely difficult to interpret these experiments, because the cells are receiving and integrating so many signals that it is hard for the experimenter to know what other influences may be at work. These experiments also illustrate the semantic distinctions imposed by the experimenter on the embryo. For example, FGF can be defined as a direct neural inducer because it induces definitive neural markers in explants that would not normally turn on those markers. However, for FGF to work, the tissue must be manipulated in ways that will reduce BMP signaling (by partial or complete dissociation of explants). Cement gland markers turn on in the explants that can be neuralized by FGF, and some have taken cement gland to be a neural tissue, since it is an induced ectodermal organ. While this may be stretching the definition of neural tissue, cement gland can be regarded as the result of partial neuralizing treatments (Sive and Bradley, 1996). Thus FGF could be inducing neural tissue by caudalizing the extreme anterior induced cells; in such a definition, FGF is not a direct neural inducer, but rather a caudalizing molecule.

The best candidate for an endogenous FGF signal is embryonic FGF (eFGF), which is expressed at the blastopore lip and subsequently in the tailbud (Isaacs et al., 1992). eFGF would act on ectoderm that lies next to the organizer and is already neuralized by other signals, so the explant models are probably relevant to the in vivo context. Experiments suggesting that FGF signaling may be relevant to the normal mode of neural induction involve inhibition of FGF receptor function with a dominant negative construct (XFD) (Launay et al., 1996). This in turn was suggested by experiments that showed FGF receptor function is required for proper formation of the dorsal mesoderm, including muscle and notochord (Amaya et al., 1991, 1993). One of the remarkable features of the XFD-injected embryos was that the head mesoderm and anterior neural tissue developed fairly well. This implies that anterior neural inducers and the ability of ectoderm to respond were unaffected by XFD. However, to interpret such experiments clearly, one needs to know that the XFD is expressed in the tissue of interest, in this case the prospective nervous system. A continuing problem with injection of mRNAs is that they have limited diffusion, making lineage tracing essential for a cell-autonomous function such as the dominant negative FGF receptor. Tracing RNAs is often done with a co-injection of a reporter mRNA, such as that encoding lacZ (Vize et al., 1991), but mRNAs that bind to the ER diffuse even less than other RNAs, making the tracing of XFD more problematic (Lamb T.M., R.M.H., unpublished).

An alternate approach is to use explants, which is persuasive if a positive result is obtained. Launay et al. used explant assays with XFD-injected animal caps; neural induction by noggin was blocked in such explants. The interpretation of these experiments is complicated by the possible presence of mesoderm in the elongating explants. (Noggin and the truncated activin receptor do not normally cause elongation of animal caps; Hemmati-Brivanlou and Melton, 1992; Lamb et al., 1993; Cunliffe and Smith, 1994). Others have not seen a suppression of neural fates by XFD (Schulte-Merker and Smith, 1995.) If one accepts the conclusions of Launay et al., (1996) at face value, then FGF signaling is required for neural induction. Furthermore, the observation that XFD increases the amount of cement gland suggests that neural induction mediated by BMP antagonists is a result of cement gland induction followed by FGF-mediated caudalization.

However, explant experiments do not address directly whether XFD blocks neural induction in the normal embryo. Since XFD prevents chordal and paraxial mesoderm induction, the source of spinal cord inducers is eliminated in XFD-injected embryos. To address the need for FGF signals at the single-cell level, our own experiments used antibodies to the FGF receptor (FGFR) (Amaya et al., 1993) to trace which cells received XFD. In such a mosaic analysis, it was found that neither anterior nor posterior neural induction was blocked by XFD (T.M. Lamb, R.M.H.). Thus there may be differences in the sensitivity of different neural inducing pathways to XFD; the neural induction that results from inhibition of BMP signaling may require FGF signaling in parallel, even though other embryonic neural inducing pathways do not. It is attractive to speculate that this is another example of redundant signaling in the embryo, and it suggests there are other neural inducers to be found.

wnts and Their Role in A/P Pattern

In the mouse, wnt genes have been implicated in the processes of mesoderm induction and neural patterning (Parr and McMahon, 1994). The difficulty of obtaining soluble wnts has made it difficult to add wnt proteins to *Xenopus* explants at defined times, but wnts have been overexpressed by injection of mRNA. Such overexpression results in suppression of anterior markers induced by noggin and in the expression of midbrain and hindbrain markers (McGrew et al., 1995). Wnt also caudalizes the neural region of Keller explants (explants of dorsal mesoderm and ectoderm). It would be interesting to know whether wnt expression from expression plasmids also caudalizes, since the biological effects of the wnt 1 subfamily depends very strongly on the time of their expression (Christian and Moon, 1993). From wnt mRNA injection it is impossible to know whether the wnt itself causes the caudal transformation of neuralized caps or whether something induced by early wnt expression is responsible for the effect. However, it is difficult to do the experiment with late wnt expression since late expression can induce ventral mesoderm (Christian et al., 1992), itself a source of caudalizing signals. Whether wnts have a direct effect on caudalization will therefore have to await more sophisticated assays.

Neural Differentiation: A Close Parallel With Drosophila

In *Drosophila,* the neurogenic ectoderm forms in a region that is far from the source of dpp/BMP signals. However, within this territory, cells make a choice between becoming neuroblasts or epidermis in a decision analogous to the vertebrate neural plate. The analogy extends only as far as the decision to differentiate, because cells in the vertebrate neural plate that do not differentiate become neural stem cells, whereas in the neurogenic ectoderm of *Drosophila,* non-neuronal cells become epidermis. Cloning of homologues of *Drosophila* genes that play a role in this choice has provided the reagents that show the molecular machinery to be quite similar in vertebrates. These genes include homologues of the neurogenic genes Notch and Delta, and the proneural genes of the basic helix-loop-helix (bHLH) family. This area is reviewed in Chapter 2, so only the *Xenopus* experiments are discussed here.

Notch and Delta

To study the consequences of Notch signaling in vertebrates, Kintner and colleagues constructed a dominant, activated form of vertebrate Notch by deleting most of the extracellular domain (Coffman et al., 1993). Although the interpretation that the mutant Notch is an activated form was not independently supported by other experiments at the time, the interpretation has held up well; this mutant protein behaves similarly to other dominant forms of Notch (Kopan et al., 1994; Nye et al., 1994), whose invertebrate counterparts have been shown to be constitutively active (Lieber et al., 1993; Rebay et al., 1993; Roehl and Kimble, 1993; Struhl et al., 1993). The initial experiments showed that activated Notch could increase the amount of muscle and nervous system formed in the region of injection. In the case of the neural plate, the increased neural field was generated at the expense of neural crest and some neighboring epidermis. Experiments with animal caps showed that activated Notch did not cause induction of muscle or neural tissue per se but suggested that active Notch prolonged the period over which cells remained competent to respond to inducer. Thus, it was reasonable to conclude that Notch might maintain cells in an uncommitted state as long as it is expressed (until the injected mRNA and its protein product declined) and that the extended time over which cells could respond to other signals resulted in expansion of the neural plate in embryos (Coffman et al., 1993).

A second set of experiments suggested that Notch was also involved within the neural plate in the decision of neuroblasts to terminally differentiate. This relied on the finding in *Drosophila* that Delta is a ligand that activates Notch (Artavanis-Tsakonas et al., 1995). A vertebrate Delta homologue was expressed in a pattern similar to but prefiguring a pattern of neural-specific β-tubulin expression (N-tubulin; Chitnis et al., 1995) which is expressed in differentiating neurons; in the spinal cord N-tubulin is expressed in the three stripes corresponding to the primary neurons (dorsal sensory neurons, intermediate interneurons, and ventral motor neurons). While it is difficult to document that Delta expression is predictive of differentiation, its expression certainly correlates very well with the sites where differentiation is known to occur (Chitnis et al., 1995; Henrique et al., 1995). The evolutionary analogy would predict that Delta expression activates Notch and therefore should suppress differentiation of neurons. In line with such predictions, global expression of Delta resulted in inhibition of differentiated neuronal fate. The same result was obtained with the activated Notch construct. The final support for the similarity of the pathway through evolution came from truncated Delta mutants. One of these behaved as a dominant negative, and activated neuronal differentiation. Again, there is no independent proof that the truncated Delta interfered with normal Delta function, but the weight of evidence suggests that the interpretation is correct.

These experiments provide yet more evidence for conservation of molecular mechanisms in diverse groups of animals and provide further justification for using genetically tractable organisms to provide directly relevant information to the study of vertebrate development.

Proneural Genes

The high expectations for a role of bHLH genes in vertebrate neural development come from the parallel with *Drosophila* genetics, where mutations in bHLH genes of the achaete–scute complex lead to excess neurons. The expectation was also high from results with the myogenic bHLH gene family of MyoD relatives, which are key regulators of muscle development (Weintraub, 1993). Members of the bHLH family have been cloned by sequence similarity, and by their interaction with the ubiquitously expressed partners of functional bHLH heterodimers (daughterless in *Drosophila,* and E12 and its relatives in vertebrates). In *Xenopus,* two achaete–scute relatives have been isolated (Ferreiro et al., 1993; Zimmerman et al., 1993), and their possible function has been analyzed by overexpression (Ferreiro et al., 1994; Turner and Weintraub, 1994). The effects of these transcripts in embryos are twofold. First, there is an expansion of the neural field, rather like that induced by activated Notch. The neural plate expands at the expense of neural crest and some neighboring epidermis. Animal caps are not efficiently directed into neural differentiation by the achaete–scute transcripts, suggesting that in injected embryos these proteins only induce neural differentiation in regions which receive other neuralizing signals. Supporting this idea, the combination of achaete–scute transcript and noggin resulted in considerable differentiation of neurons, as measured by expression of N-tubulin.

A more powerful neural-differentiating cDNA, NeuroD, was isolated in a yeast two-hybrid screen by its interaction with daughterless (Lee et al., 1995). NeuroD transcripts not only increase the size of the neural plate but can redirect ventral epidermal cells or animal cap cells to become differentiated neurons. NeuroD is expressed fairly early in the neural plate in precursors to differentiated neurons. Thus NeuroD has the kind of attributes expected for a master regulator of neural differentiation. It is not yet known whether the activities of overexpressed NeuroD and XASH3 are qualitatively similar and only quantitatively different. While each gene may have a fairly restricted role in normal development, overexpressed genes may act promiscuously and mimic one another.

NeuroD was not isolated by sequence similarity to other bHLH genes, demonstrating the problem that searches for genes by sequence similarity alone are not comprehensive. Functional assays using yeast interaction, or screens for neural differentiating molecules, are likely to increase the size of the known bHLH family and concomitantly our understanding of the differentiation of the neuron.

Future Developments

Perhaps the main question that remains is whether the main neural inducing molecules have been identified. All of the inducing activities have been found in the last few years, and it is not unreasonable to expect that other molecules, and perhaps other signal transduction pathways that lead to neural induction are yet to be discovered. For the molecules that have been identified, it remains to be seen how these are integrated into the activities of the organizer, not only in neural induction, but in the specification of anterior-to-posterior cell identities.

ACKNOWLEDGMENTS

I thank Scott Dougan for deconvoluting many parts of the manuscript, my colleagues for valuable discussions, and the NIH and Hormone Research Foundation for support. Since there was no size constraint, any omissions are entirely due to my own ignorance and those not cited should vent their annoyance on harland@mendel.berkeley.edu.

REFERENCES

Amaya, E., Musci, T.J., Kirshner, M.W. (1991). Expression of a dominant negative mutant of the FGF receptor disrupts mesoderm formation in Xenopus embryos. *Cell* 66:257–270.

Amaya, E., Stein, P.A., Musci, T.J., Kirschner, M.W. (1993). FGF signalling in the early specification of mesoderm in *Xenopus. Development* 118:477–487.

Aono, A., Hazama M., Notoya, K., Taketomi, S., Yamasaki, H., Tsukuda, R., Sasaki, S., Fujisawa, Y. (1995). Potent ectopic bone-inducing activity of bone morphogenetic protein-4/7 heterodimer. *Biochem. Biophys. Res. Commun.* 210:670–677.

Artavanis-Tsakonas, S., Matsuno, K., Fortini, M.E. (1995). Notch signaling. *Science* 268: 225–232.

Carcamo, J., Zentella, A., Massagué, J. (1995). Disruption of transforming growth factor beta signaling by a mutation that prevents transphosphorylation within the receptor complex. *Mol. Cell. Biol.* 15:1573–1581.

Chitnis, A., Henrique, D., Lewis, J., Ish-Horowicz, D., Kintner, C. (1995). Primary neurogenesis in Xenopus embryos regulated by a homolog of the Drosophila neurogenic gene Delta [see comments]. *Nature* 375:761–766.

Christian, J.L., Moon, R.T. (1993). Interactions between Xwnt-8 and Spemann organizer signaling pathways generate dorsoventral pattern in the embryonic mesoderm of Xenopus. *Genes Dev.* 7:13–28.

Christian, J.L., Olson, D.J., Moon, R.T. (1992). Xwnt-8 modifies the character of mesoderm induced by bFGF in isolated Xenopus ectoderm. *EMBO J.* 11:33–41.

Clement, J.H., Fettes, P., Knochel, S., Lef, J., Knochel, W. (1995). Bone morphogenetic protein 2 in the early development of Xenopus laevis. *Mech. Dev.* 52:357–370.

Coffman, C.R., Skoglund, P., Harris, W.A., Kintner, C.R. (1993). Expression of an extracellular deletion of Xotch diverts cell fate in Xenopus embryos. *Cell* 73:659–671.

Cox, W.G., Hemmati-Brivanlou, A. (1995). Caudalization of neural fate by tissue recombination and bFGF. *Development* 121:4349–4358.

Cunliffe, V., Smith, J.C. (1994). Specification of mesodermal pattern in Xenopus laevis by interactions between Brachyury, noggin and Xwnt-8. *EMBO J.* 13:349–359.

Dale, L., Smith, J.C., Slack, J.M. (1985). Mesoderm induction in Xenopus laevis: a quantitative study using a cell lineage label and tissue-specific antibodies. *J. Embryol. Exp. Morphol.* 89:289–312.

Dale, L., Howes, G., Price, B.M., Smith, J.C. (1992). Bone morphogenetic protein 4: a ventralizing factor in early Xenopus development. *Development* 115:573–585.

Doniach, T. (1993). Planar and vertical induction of anteroposterior pattern during the development of the amphibian central nervous system. *J. Neurobiol.* 24:1256–1275.

Doniach, T. (1995). Basic FGF as an inducer of anteroposterior neural pattern. *Cell* 83: 1067–1670.

Ebner, R., Chen, R.H., Lawler, S., Zioncheck, T., Derynck, R. (1993). Determination of type

I receptor specificity by the type II receptors for TGF-beta or activin. *Science* 262:900–902.

Fainsod, A., Steinbeisser, H., De Robertis, E.M. (1994). On the function of BMP-4 in patterning the marginal zone of the Xenopus embryo. *EMBO J.* 13:5015–5025.

Fan, C.M., Porter, J.A., Chiang, C., Chang, D.T., Beachy, P.A., Tessier-Lavigne, M. (1995). Long-range sclerotome induction by sonic hedgehog: direct role of the amino-terminal cleavage product and modulation by the cyclic AMP signaling pathway. *Cell* 81:457–465.

Ferguson, E.L., Anderson, K.V. (1992). Localized enhancement and repression of the activity of the TGF-beta family member, decapentaplegic, is necessary for dorsal-ventral pattern formation in the Drosophila embryo. *Development* 114:583–597.

Ferreiro, B., Skoglund, P., Bailey, A., Dorsky, R., Harris, W.A. (1993). XASH1, a Xenopus homolog of achaete-scute: a proneural gene in anterior regions of the vertebrate CNS. *Mech. Dev.* 40:25–36.

Ferreiro, B., Kintner, C., Zimmerman, K., Anderson, D., Harris, W.A. (1994). XASH genes promote neurogenesis in Xenopus embryos. *Development* 120:3649–3655.

Gawantka, V., Delius, H., Hirschfeld, K., Blumenstock, C., Niehrs, C. (1995). Antagonizing the Spemann organizer: role of the homeobox gene Xvent-1. *EMBO J.* 14:6268–6279.

Gerhart, J.C. (1996). Johannes Holtfreter's contributions to ongoing studies of the organizer. *Dev. Dyn.* 205:245–256.

Gerhart, J., Doniach, T., Stewart, R. (1991). Organizing the Xenopus organizer. In *Gastrulation,* Keller, R. E., ed. *New York: Plenum Press,* pp. 57–77.

Godsave, S.F., Slack, J.M. (1989). Clonal analysis of mesoderm induction in Xenopus laevis. *Dev. Biol.* 134:486–490.

Graff, J.M., Thies, R.S., Song, J.J., Celeste, A.J., Melton, D.A. (1994). Studies with a Xenopus BMP receptor suggest that ventral mesoderm-inducing signals override dorsal signals in vivo. *Cell* 79:169–179.

Grunz, H., Tache, L. (1989). Neural differentiation of Xenopus laevis ectoderm takes place after disaggregation and delayed reaggregation without inducer. *Cell. Differ. Dev.* 28: 211–217.

Harland, R.M. (1994). Commentary: the transforming growth factor beta family and induction of the vertebrate mesoderm: bone morphogenetic proteins are ventral inducers. *Proc. Natl. Acad. Sci. U.S.A.* 91:10243–10246.

Harland, R., Misher, L. (1988). Stability of RNA in developing Xenopus embryos and identification of a destabilizing sequence in TFIIIA messenger RNA. *Development* 102: 837–852.

Hawley, S.H., Wunnenberg-Stapleton, K., Hashimoto, C., Laurent, M.N., Watabe, T., Blumberg, B.W., Cho, K.W. (1995). Disruption of BMP signals in embryonic Xenopus ectoderm leads to direct neural induction. *Genes Dev.* 9:2923–2935.

Hazama, M., Aono, A., Ueno, N., Fujisawa, Y. (1995). Efficient expression of a heterodimer of bone morphogenetic protein subunits using a baculovirus expression system. *Biochem. Biophys. Res. Commun.* 209:859–866.

Hemmati-Brivanlou, A., Melton, D.A. (1992). A truncated activin receptor inhibits mesoderm induction and formation of axial structures in Xenopus embryos [see comments]. *Nature* 359:609–614.

Hemmati-Brivanlou, A., Kelly, O.G., Melton, D.A. (1994). Follistatin, an antagonist of activin, is expressed in the Spemann organizer and displays direct neuralizing activity. *Cell* 77:283–295.

Hemmati-Brivanlou, A., Thomsen, G.H. (1995). Ventral mesodermal patterning in Xenopus embryos: expression patterns and activities of BMP-2 and BMP-4. *Dev. Genet.* 17:78–89.

Henrique, D., Adam, J., Myat, A., Chitnis, A., Lewis, J., Ish-Horowicz, D. (1995). Expression of a Delta homolog in prospective neurons in the chick. *Nature* 375:787–790.

Holley, S.A., Neul, J.L., Attisano, L., Wrana, J.L., Sasai, Y., O'Connor, M.B., De Robertis, E.M., Ferguson, E.L. (1996). The Xenopus dorsalizing factor noggin ventralizes Drosophila embryos by preventing DPP from activating its receptor. *Cell 86*, 607–17.

Isaacs, H.V., Tannahill, D., Slack, J.M. (1992). Expression of a novel FGF in the Xenopus embryo. A new candidate inducing factor for mesoderm formation and anteroposterior specification. *Development* 114:711–720.

Iwasaki, S., Tsuruoka, N., Nattori, A., Sato, M., Tsujimoto M., Kohno, M. (1995). Distribution and characterization of specific cellular binding proteins for bone morphogenetic protein-2. *J. Biol. Chem.* 270:5476–5482.

Johansson, B.M., Wiles, M.V. (1995). Evidence for involvement of activin A and bone morphogenetic protein 4 in mammalian mesoderm and hematopoietic development. *Mol. Cell. Biol.* 15:141–151.

Jones, C.M., Lyons, K.M., Lapan, P.M., Wright, C.V., Hogan, B.L. (1992). DVR-4 (bone morphogenetic protein-4) as a posterior-ventralizing factor in Xenopus mesoderm induction. *Development* 115:639–647.

Kengaku, M., Okamoto, H. (1993). Basic fibroblast growth factor induces differentiation of neural tube and neural cast lineages of cultured ectoderm cells from Xenopus gastrula. *Development* 119:1067–1078.

Kengaku, M., Okamoto, H. (1995). bFGF as a possible morphogen for the anteroposterior axis of the central nervous system in Xenopus. *Development* 121:3121–3130.

Kessler, D.S., Melton, D.A. (1995). Induction of dorsal mesoderm by soluble, mature Vg1 protein. *Development* 121:2155–2164.

Kingsley, D.M. (1994). The TGF-β superfamily: new members, new receptors, and new genetic tests of function in different organisms. *Genes Dev.* 8:133–146.

Kintner, C.R., Dodd, J. (1991). Hensen's node induces neural tissue in *Xenopus* ectoderm. Implications for the action of the organizer in neural induction. *Development* 113:1495–1506.

Koenig, B.B., Cook, J.S., Wolsing, D.H., Ting, J., Tiesman, J.P., Correa, P.E., Olson, C.A., Pecquet, A.L., Ventura, F., Grant, R.A., Chen, G.-X., Wrana, J.L., Massagué, J., Rosenbaum, J.S. (1994a). Characterization and cloning of a receptor for BMP-2 and BMP-4 from NIH3T3 cells. *Mol. Cell Biol.* 14:5961–5974.

Koenig, B.B., Cook, J.S., Wolsing, D.H., Ting, J., Tiesman, J.P., Correa, P.E., Olson, C.A., Pecquet, A.L., Ventura, F., Grant, R.A., et al. (1994b). Characterization and cloning of a receptor for BMP-2 and BMP-4 from NIH 3T3 cells. *Mol. Cell. Biol.* 14:5961–5974.

Kopan, R., Nye, J.S., Weintraub, H. (1994). The intracellular domain of mouse Notch: a constitutively activated repressor of myogenesis directed at the basic helix-loop-helix region of MyoD. *Development* 120:2385–2396.

Lamb, T.M., Knecht, A.K., Smith, W.C., Stachel, S.E., Economides, A.N., Stahl, N., Yancopolous, G.D., Harland, R.M. (1993). Neural induction by the secreted polypeptide noggin. *Science* 262:713–718.

Launay, C., Fromentoux, V., Shi, D.L., Boucaut, J.C. (1996). A truncated FGF receptor blocks neural induction by endogenous Xenopus inducers. *Development* 122:869–880.

Lee, J.E., Hollenberg, S.M., Snider, L., Turner, D.L., Lipnick, N., Weintraub, H. (1995). Conversion of Xenopus ectoderm into neurons by NeuroD, a basic helix-loop-helix protein. *Science* 268:836–844.

Lemaire, P., Gurdon, J.B. (1994). A role for cytoplasmic determinants in mesoderm patterning: cell-autonomouse activation of the goosecoid and Xwnt-8 genes along the dorsoventral axis of early Xenopus embryos. *Development* 120:1191–1199.

Lieber, T., Kidd, S., Alcamo, E., Corbin, V., Young, M.W. (1993). Antineurogenic phenotypes induced by truncated Notch proteins indicate a role in signal transduction and may point to a novel function for Notch in nuclei. *Genes Dev.* 7:1949–1965.

Lin, H.Y., Wang, X.F., Ng-Eaton, E., Weinberg, R.A., Lodish, H.F. (1992). Expression cloning of the TGF-beta type II receptor, a functional transmembrane serine/threonine kinase [published erratum appears in *Cell* 1992 Sep 18;70(6):following 1068]. *Cell* 68: 775–785.

Liu, F., Ventura, F., Doody, J., Massagué, J. (1995). Human type II receptor for bone morphogenic proteins (BMPs): extension of the two-kinase receptor model to the BMPs. *Mol. Cell. Biol.* 15:3479–3486.

Lopez, A.R., Cook, J., Deininger, P.L., Derynck, R. (1992). Dominant negative mutants of transforming growth factor-β1 inhibit the secretion of different transforming growth factor-β isoforms. *Mol. Cell. Biol.* 12:1674–1679.

Maeno, M., Ong, R.C., Suzuki, A., Ueno, N., Kung, H.F. (1994). A truncated bone morphogenetic protein 4 receptor alters the fate of ventral mesoderm to dorsal mesoderm: roles of animal pole tissue in the development of ventral mesoderm [see comments]. *Proc. Natl. Acad. Sci. U.S.A.* 91:10260–10264.

Marti, E., Bumcrot, D.A., Takada, R., McMahon, A.P. (1995). Requirement of 19K form of Sonic hedgehog for induction of distinct ventral cell types in CNS explants [see comments]. *Nature* 375:322–325.

Massagué, J. (1996). TGFβ Signaling: receptors, transducers and mad proteins. *Cell 85:947–950.*

Mathews, L.S., Vale, W.W. (1991). Expression cloning of an activin receptor, a predicted transmembrane serine kinase. *Cell* 65:973–982.

McGrew, L.L., Lai, C.J., and Moon, R.T. (1995). Specification of the anteroposterior neural axis through synergistic interaction of the Wnt signaling cascade with noggin and follistatin. *Dev Biol 172,* 337–42.

Moos, M., Jr., Wang, S., Krinks, M. (1995). Anti-dorsalizing morphogenetic protein is a novel TGF-beta homolog expressed in the Spemann organizer. *Development* 121:4293–4301.

Nishimatsu, S., Suzuki, A., Shoda, A., Murakami, K., Ueno, N. (1992). Genes for bone morphogenetic proteins are differentially transcribed in early amphibian embryos. *Biochem. Biophys. Res. Commun.* 186:1487–1495.

Nohno, T., Ishikawa, T., Saito, T., Hosokawa, K., Noji, S., Wolsing, D.H., Rosenbaum, J.S. (1995). Identification of a human type II receptor for bone morphogenetic protein-4 that forms differential heteromeric complexes with bone morphogenetic protein type I receptors. *J. Biol. Chem.* 270:22522–22526.

Northrop, J., Woods, A., Seger, R., Suzuki, A., Ueno, N., Krebs, E., Kimelman, D. (1995). BMP-4 regulates the dorsal-ventral differences in FGF/MAPKK-mediated mesoderm induction in Xenopus. *Dev. Biol.* 172:242–252.

Nye, J.S., Kopan, R. Axel, R. (1994). An activated Notch suppresses neurogenesis and myogenesis but not gliogenesis in mammalian cells. *Development* 120:2421–2430.

Parr, B.A., McMahon, A.P. (1994). Wnt genes and vertebrate development. *Curr. Opin. Genet. Dev.* 4:523–528.

Piccolo, S., Sasai, Y., Lu, B., and De Robertis, E.M. (1996). Dorsoventral patterning in Xenopus: inhibition of ventral signals by direct binding of chordin to BMP-4. *Cell 86,* 589–98.

Re'em-Kalma, Y., Lamb, T., Frank, D. (1995). Competition between noggin and bone morphogenetic prein 4 activities may regulate dorsalization during Xenopus development. *Proc. Natl. Acad. Sci. U.S.A.* 92:12141–12145.

Rebay, I., Fehon, R.G., Artavanis-Tsakonas, S. (1993). Specific truncations of Drosophila Notch define dominant activated and dominant negative forms of the receptor. *Cell* 74:319–329.

Roehl, H., Kimble, J. (1993). Control of cell fate in C. elegans by a GLP-1 peptide consisting primarily of ankyrin repeats. *Nature* 364:632–635.

Roelink, H., Porter, J.A., Chiang, C., Tanabe, Y., Chang, D.T., Beachy, P.A., Jessell, T.M. (1995). Floor plate and motor neuron induction by different concentrations of the amino-terminal cleavage product of sonic hedgehog autoproteolysis. *Cell* 81:445–455.

Sasai, Y., Lu, B., Steinbeisser, H., Geissert, D., Gont, L.K., De Robertis, E.M. (1994). Xenopus chordin: a novel dorsalizing factor activated by organizer-specific homeobox genes. *Cell* 79:779–790.

Sasai, Y., Lu, B., Steinbeisser, H., De Robertis, E.M. (1995). Regulation of neural induction

by the Chd and Bmp-4 antagonistic patterning signals in Xenopus. *Nature* 376:333–336.
Sato, S.M., Sargent, T.D. (1989). Development of neural inducing capacity in dissociated Xenopus embryos. *Dev. Biol.* 134:263–266.
Schmidt, J.E., Suzuki, A., Ueno, N., Kimelman, D. (1995). Localized BMP-4 mediates dorsal/ventral patterning in the early Xenopus embryo. *Dev. Biol.* 169:37–50.
Schneyer, A.L., Rzucidlo, D.A., Sluss, P.M., Crowley, W.F., Jr. (1994). Characterization of unique binding kinetics of follistatin and activin or inhibin in serum. *Endocrinology* 135:667–674.
Schulte-Merker, S., Smith, J.C. (1995). Mesoderm formation in response to Brachyury requires FGF signalling. *Curr. Biol.* 5:62–67.
Schulte-Merker, S., Smith, J.C., Dale, L. (1994). Effects of truncated activin and FGF receptors and of follistatin on the inducing activities of BVg1 and activin: does activin play a role in mesoderm induction? *EMBO J.* 13:3533–3541.
Servetnick, M., Grainger, R.M. (1991). Changes in neural and lens competence in Xenopus ectoderm: evidence for an autonomous developmental timer. *Development* 112:177–188.
Sharpe, C.R., Gurdon, J.B. (1990). The induction of anterior and posterior neural genes in Xenopus laevis. *Development* 109:765–774.
Shoda, A., Murakami, K., Ueno, N. (1994). Biologically active BMP-2 in early Xenopus laevis embryos. *Biochem. Biophys. Res. Commun.* 198:1267–1274.
Sive, H., Bradley, L. (1996). A sticky problem: the Xenopus cement gland as a paradigm for anteroposterior patterning. *Dev. Dyn.* 205:265–280.
Sive, H.L., Hattori, K., Weintraub, H. (1989). Progressive determination during formation of the anteroposterior axis in Xenopus laevis. *Cell* 58:171–180.
Smith, W.C., Harland, R.M. (1992). Expression cloning of noggin, a new dorsalizing factor localized to the Spemann organizer in Xenopus embryos. *Cell* 70:829–840.
Smith, W.C., Knecht, A.K., Wu, M., Harland, R.M. (1993). Secreted noggin protein mimics the Spemann organizer in dorsalizing Xenopus mesoderm. *Nature* 361:547–549.
Stewart, R.M., Gerhart, J.C. (1990). The anterior extent of dorsal development of the *Xenopus* embryonic axis depends on the quantity of organizer in the late blastula. *Development* 109:363–372.
Struhl, G., Fitzgerald, K., Greenwald, I. (1993). Intrinsic activity of the Lin-12 and Notch intracellular domains in vivo. *Cell* 74:331–345.
Suzuki, A., Thies, R.S., Yamaji, N., Song, J.J., Wozney, J.M., Murakami, K., Ueno, N. (1994). A truncated bone morphogenetic protein receptor affects dorsal-ventral patterning in the early Xenopus embryo [see comments]. *Proc. Natl. Acad. Sci. U.S.A.* 91:10255–10259.
ten Dijke, P., Yamashita, H., Ichijo, H., Franzen, P., Laiho, M., Miyazono, K., Heldin, C.H. (1994a). Characterization of type I receptors for transforming growth factor-beta and activin. *Science* 264:101–104.
ten Dijke, P., Yamashita, H., Sampath, T.K., Reddi, A.H., Estevez, M., Riddle, D.L., Ichijo, H., Heldin, C.H., Miyazono, K. (1994b). Identification of type I receptors for osteogenic protein-1 and bone morphogenetic protein-4. *J. Biol. Chem.* 269:16985–16988.
Thomsen, G.H., Melton, D.A. (1993). Processed Vg1 protein is an axial mesoderm inducer in Xenopus. *Cell* 74:433–441.
Turner, D.L., Weintraub, H. (1994). Expression of achaete-scute homolog 3 in Xenopus embryos converts ectodermal cells to a neural fate. *Genes Dev.* 8:1434–1447.
Ventura, F., Doody, J., Liu, F., Wrana, J.L., Massagué, J. (1994). Reconstitution and transphosphorylation of TGF-beta receptor complexes. *EMBO J.* 13:5581–5589.
Vize, P.D., Melton, D.A., Hemmati-Brivanlou, A., Harland, R.M. (1991). Assays for gene function in developing Xenopus embryos. *Methods Cell Biol.* 36:367–387.
Weintraub, H. (1993). The MyoD family and myogenesis: redundancy, networks, and thresholds. *Cell* 75:1241–1244.

Wieser, R., Wrana, J.L., Massagué, J. (1995). GS domain mutations that constitutively activate T beta R-I, the downstream signaling component in the TGF-beta receptor complex. *EMBO J.* 14:2199–2208.
Wilson, P.A., Hemmati-Brivanlou, A. (1995). Induction of epidermis and inhibition of neural fate by Bmp-4. *Nature* 376:331–333.
Witta, S.E., Agarwal, V.R., Sato, S.M. (1995). XIPOU 2, a noggin-inducible gene, has direct neuralizing activity. *Development* 121:721–730.
Wittbrodt, J., Rosa, F.M. (1994). Disruption of mesoderm and axis formation in fish by ectopic expression of activin variants: the role of maternal activin. *Genes Dev.* 8:1448–1462.
Wrana, J.L., Attisano, L., Wieser, R., Ventura, F., Massagué, J. (1994). Mechanism of activation of the TGF-beta receptor. *Nature* 370:341–347.
Xu, R.H., Kim, J., Taira, M., Zhan, S., Sredni, D., Kung, H.F. (1995). A dominant negative bone morphogenetic protein 4 receptor causes neuralization in Xenopus ectoderm. *Biochem. Biophys. Res. Commun.* 212:212–219.
Yamashita, H., ten Dijke, P., Huylebroeck, D., Sampath, T.K., Andries, M., Smith, J.C., Heldin, C.H., Miyazono, K. (1995). Osteogenic protein-1 binds to activin type II receptors and induces certain activin-like effects. *J. Cell Biol.* 130:217–226.
Zimmerman, K., Shih, J., Bars, J., Collazo, A., Anderson, D.J. (1993). XASH-3, a novel Xenopus achaete-scute homolog, provides an early marker of planar neural induction and position along the mediolateral axis of the neural plate. *Development* 119:221–232.
Zimmerman, L.B., De Jesus-Escobar, J.M., and Harland, R.M. (1996). The Spemann organizer signal noggin binds and inactivates bone morphogenetic protein 4. *Cell 86,* 599–606.

2

The determination of the neuronal phenotype

DAVID J. ANDERSON AND YUH NUNG JAN

What genes are involved in the determination of the neuronal phenotype, and what do they do? Are there "master" regulatory genes that can turn any cell into a neuron, in the way that MyoD is able to turn almost any cell into muscle? If so, how are different kinds of neurons generated? Are there different master regulatory genes for each neuronal type, or are different types of neurons built up "piecemeal," by the combined action of multiple regulatory genes that control different aspects of the final phenotype? If the latter is the case, do these different regulatory genes act independently or coordinately, and in series or in parallel?

These are just some of the questions one can pose concerning the determination of the neuronal phenotype. That most of them are as yet unanswered, or answered only incompletely, illustrates the relatively primitive nature of our understanding of neuronal determination. However, genetic screens in *Drosophila* have identified several genes that control basic binary decisions in neurogenesis. Moreover, vertebrate homologues of these genes have provided an increasing number of "points of entry" into the regulatory circuits controlling vertebrate neurogenesis. Yet it has proven challenging to find out exactly what these genes do and to assess whether their cellular functions as well as their sequences have been conserved in evolution. That means we are still at an early stage in first defining the important players in the process of neuronal determination and then of establishing their roles.

It would be wrong to imply that neuronal determination will be understood once all the fly genes controlling neurogenesis, and their vertebrate homologues, are cloned. *Drosophila* genetics has been very good at giving us transcriptional regulators that control early stages in neurogenesis and molecules that in turn control the activity of these genes by local cell–cell interactions or, more recently, by asymmetrically distributed cytoplasmic factors. However, the discipline has not yet connected this genetic regulatory network to the expression of the genes that actually build a neuron: the ion channels, receptors, neurotransmitter-synthesizing enzymes, synaptic vesicle proteins, and cytoskeletal proteins that determine the form and function of a particular neuronal subtype. Furthermore, relatively little information is yet available about the control of neuronal phenotype by diffusible extracellular signals, e.g., neurotrophic factors.

In contrast, both of these aspects of neurogenesis have been intensively and successfully studied in vertebrate systems. The transcriptional regulation of many neuron-specific genes has been studied in detail, and in some cases the proteins that regulate this neuron-specific transcription have been identified. And a very large number of diffusible, secreted factors that control different aspects of neuronal differentiation and survival have been identified and cloned, and their cognate receptors have been identified as well. The challenge for the future will be to link networks of transcription factors to the diffusible (and cell-associated) signals that control the neuronal phenotype from outside the cell and to the diverse structural genes that actually constitute this phenotype from within.

This chapter is chiefly concerned with the following questions: What are the key regulatory genes that specify a neuronal fate and a particular neuronal identity? What do they do? How are they controlled? In addition to identifying key regulatory genes, some attempt is made to address the "logic" of neuronal determination and differentiation. Are different aspects of the neuronal phenotype independently controlled by different regulatory genes, or do individual genes simultaneously and coordinately control multiple aspects of the final phenotype? If the former is the case, is the determination of different aspects of the phenotype coupled in some way, or does it simply reflect the superposition of several parallel pathways or circuits? We are a long way from answering these questions, but the available data at least give us a hint of what kind of solution we are likely to obtain. Not surprisingly, it is a complex one.

Determination of the Neuronal Phenotype in Drosophila

One of the best-studied cases of neuronal cell fate determination in *Drosophila* is the development of a particular type of sense organ called the external sensory (es) organ. Development of the es organ is a progressive process (Fig. 2–1) (reviewed in Ghysen and Dambly-Chaudiere, 1989; Jan and Jan, 1994).

Briefly, the first step of es organ development is the turning on of proneural genes in clusters of cells (proneural clusters). The proneural genes are so named because they endow the cells that express them with the potential to form neural precursors. The position of proneural clusters prefigures where es organs will form. Within each proneural cluster, the cells compete with each other such that only a subset of cells (often just one) is singled out to develop into an es organ precursor. This singling out process is mediated by cell–cell interaction through the action of neurogenic genes. Once a cell is singled out to become an es precursor, it starts to express two groups of genes: the pan-neural precursor genes, which are shared by most or all neural precursors and may function in controlling neural differentiation, and the neuronal-type selector genes, which are expressed in more restricted patterns (e.g., *cut* is expressed in es organ precursors

Neural competence:

Proneural genes:
achaete, scute, etc

Singling out of precursor:

Neurogenic genes:
Notch, Delta, big brain, etc

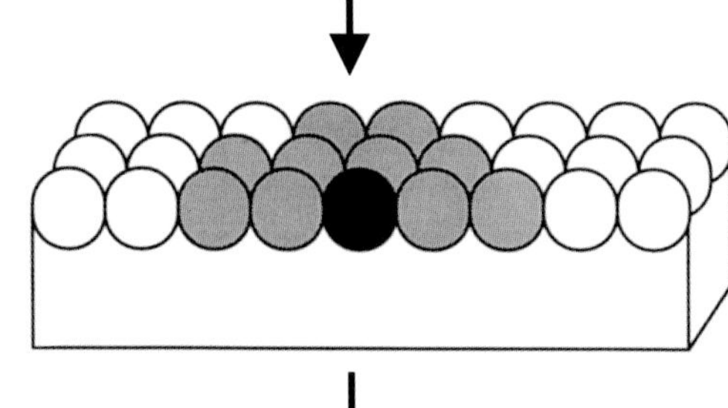

Differentiation:

Pan-neural precursor genes:
asense, deadpan, etc
and
Neuronal-type selector genes:
cut, poxneuro, etc

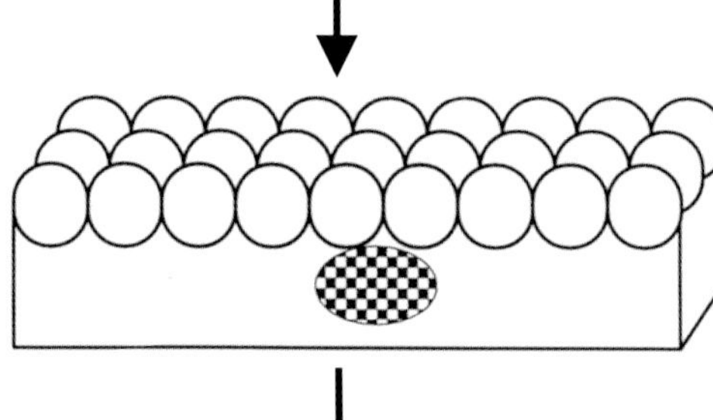

Cell division:

string, cyclins, etc

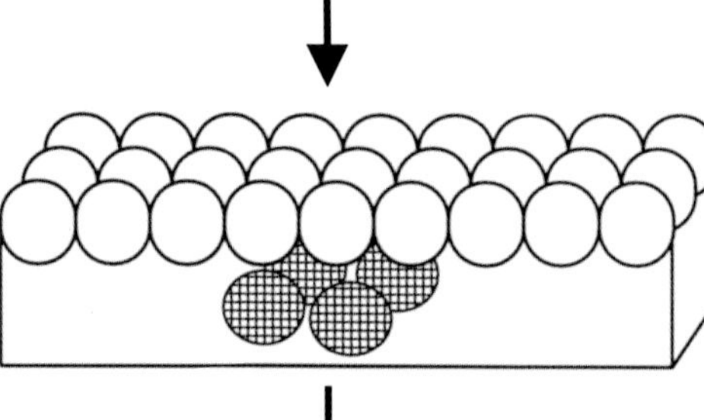

es organ cell fate specification:

Notch, Delta, numb,
Suppressor of Hairless, etc

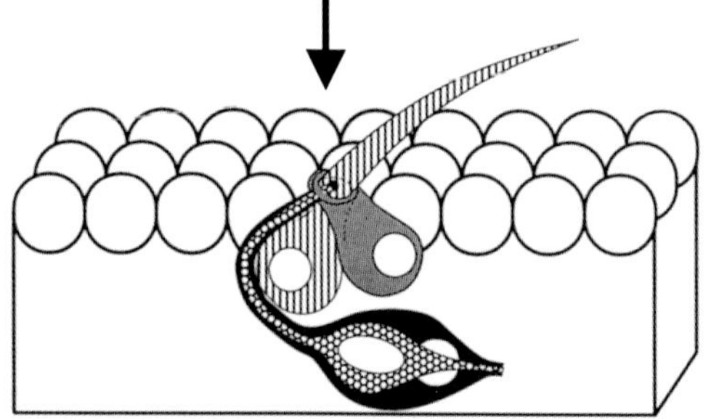

FIGURE 2–1 Progressive specifications of external sensory (es) organ cell fate. This model is a slight modification of the one proposed by Ghysen and Dambly-Chaudiere (1989). First, proneural genes are expressed in clusters of ectodermal cells (proneural cluster cells, lightly shaded) and confer those cells with the potential to develop into neuronal precursors. Those cells then compete with each other. The result is that only one cell (darkly shaded) is singled out to express a high level of one or another proneural gene and develops into a neuronal precursor; the rest adopt the alternative epidermal fate. This singled-out cell then expresses neuronal precursor genes, which probably control neuronal precursor-specific differentiation, and neuronal-type selector genes, which confer neuronal identity. The precursor cell then goes through stereotyped cell divisions and produces a fixed number of progeny cells (usually four). Using both cell–cell interactions and intrinsic mechanisms, these four cells assume four different fates. Together, they form an es organ.

but not chordotonal [ch] organs whereas *pox-neural* [*poxn*] is expressed in a subset of es organs) and specify neuronal type. The es organ precursor then goes through a stereotyped pattern of cell divisions and produces a fixed number of progeny cells which constitute an es organ.

This progressive determination involves multiple binary decisions, each controlled by groups of positive and negative regulators. In the following, we will first discuss these positive and negative regulators in the development of the es organ. We will then discuss the generality of those mechanisms.

Neural Precursor Formation

The first decision in es organ development is made by an undifferentiated ectodermal cell, which must choose between a sensory organ precursor (SOP) fate and an epidermal cell fate.

Positive Regulators for SOP Fate: achaete and scute

The genes *achaete* and *scute* belong to the *achaete–scute* complex (AS-C), which consists of *achaete* (*ac*), *scute* (*sc*), *lethal of scute* (*l'sc*), and *asense* (reviewed in Ghysen and Dambly-Chaudiere, 1988; Campuzano and Modolell, 1992). All four genes encode transcriptional regulators of the basic helix-loop-helix (bHLH) type. The *achaete–scute* complex has important roles in the development of parts of the peripheral nervous system (PNS), including es organs, and parts of the central nervous system (CNS). Of the four genes, *ac* and *sc* are the major proneural genes for es organ development. In loss-of-function *ac*/*sc* double mutants, the cells at the position of would-be *ac*/*sc* proneural clusters fail to express *ac*/*sc* and so they adopt epidermal rather than SOP fates. Conversely, in gain-of-function *ac* or *sc* mutants in which *ac* or *sc* is misexpressed in ectopic locations (where normally *ac* and *sc* are not expressed), ectopic es organs form. Based on such experiments, one can conclude that the function of *ac* and *sc* is to cause ectodermal cells to adopt on SOP fate rather than an epidermal cell fate. Therefore these genes are positive regulators of the SOP fate.

Regulators of ac/sc Expression Pattern

Where *ac* or *sc* is expressed determines where es organs form. Thus, the problem of how a fly determines where to form its es organs becomes the problem of how *ac* and *sc* expression patterns and activities are regulated during development. In the embryo, AC and SC are expressed in cell clusters at stereotyped anterior/posterior (A/P) and dorsal/ventral (D/V) coordinates in a tartan-like pattern (Skeath et al., 1992; Ruiz-Gómez and Ghysen, 1993). The A/P and D/V position of these proneural clusters is controlled by pair-rule and D/V polarity genes, and is later maintained by a subset of segment polarity genes (Skeath et al., 1992). The expression of AC and SC overlaps temporally with the expression of pair-rule and DV polarity genes, suggesting that the latter may control the expression of *ac* and *sc* directly. Indeed, *hairy*, a pair-rule gene encoding a bHLH protein, functions as a transcriptional repressor of *ac* (VanDoren et al., 1994; Ohsako et al., 1994).

In *Drosophila*, altering the A/P or D/V coordinates (for example, by mutating A/P or D/V patterning genes) leads to an altered position and/or size of the nervous system. This effect on neural development is indirect and is caused by an alteration of proneural gene expression pattern. Thus, in *Drosophila* A/P and D/V patterning can be separated from the initiation of neural development, as the latter corresponds to the turning on of proneural gene. Is there an analogous separation in vertebrates? In vertebrate, neural development starts with primary neural induction (see Chapter 1). A number of molecules have been identified as neural inducers or inhibitors of neural inducers. At least some of those molecules appear to be functioning in the patterning of the ectoderm rather than directly in initiating neural development. For example, the neuralizing effect of chordin is probably the result of its dorsalizing activity. This effect is antagonized by BMP4, which functions to induce epidermis and thus to inhibit the neural fate (Sasai et al., 1995; Wilson and Hemmati-Brivanlou, 1995). There is a striking similarity between this pair of secreted molecules with antagonistic functions in dorsal/ventral patterning and their counterparts in *Drosophila: decapentaplegic* (*dpp*) and *short-gastrulation* (*sog*). DPP is a homologue of BMP-4 and its function is to dorsalize the fly embryo. SOG is a homologue of CHORDIN protein and its function is to antagonize the actions of DPP and thereby ventralize the embryo. It therefore appears that some basic aspect of the dorsal/ventral patterning mechanism is highly conserved between vertebrates and *Drosophila*, although dorsal in vertebrates corresponds to ventral in *Drosophila* (Sasai et al., 1995; Francois and Bier, 1995; DeRobertis and Sasai, 1996). It will be interesting to learn whether initiation of neural development in vertebrates is a distinct step, perhaps carried out by genes analogous to proneural genes in fly, and whether various neural inducers or inhibitors such as NOGGIN/FOLLISTATIN act on dorsal/ventral patterning in general or formation of neural tissue per se (see Chapter 1 and below).

Negative Regulators of SOP Fate

ac and *sc* endow the ectodermal cells that express these proneural genes with the potential to develop into SOPs. However, in normal development only a subset (often one) of the cells in the proneural clusters actually develops into a SOP. The rest of the cells fail to realize their potential to develop into SOP and adopt the epidermal cell fate instead. This "singling out" of a subset of cells from the proneural clusters for an SOP fate is accomplished by the action of multiple negative regulators of SOP fate.

extramacrochaete (emc)

The ACHAETE and SCUTE bHLH proteins function as transcription regulators by forming DNA-binding homodimers or heterodimers with the ubiquitous bHLH protein DAUGHTERLESS so as to control downstream genes (Murre et al., 1989b; Cabrera and Alonso, 1991; Van Doren et al., 1991). A negative regulator of *ac* and *sc* is the gene *extramacrochaete* (*emc*), which encodes an HLH protein without a DNA-binding basic domain. EMC can dimerize with ACHAETE, SCUTE, or DAUGHTERLESS, resulting in heterodimers incapable of binding DNA, and thereby preventing the formation of active heterodimers or homodimers (Van Doren et al., 1991). The expression pattern of *emc* is approximately complementary to those of *ac* and *sc* although it is controlled independently of *ac* and *sc* (Cubas and Modelell, 1992; Van Doren et al., 1992). It appears that the relative level of *ac/sc* and *emc* determines the competence of the cells to form neural precursors. Thus, the expression pattern of *emc* contributes to refining the final pattern of neural precursor distribution. This type of negative regulation by titrating positive-acting bHLH proteins appears to be a general mechanism for regulating bHLH protein transcriptional regulator activities. A mammalian homologue of EMC, ID, functions as a negative regulator of myogenic bHLH factors by titrating myogenic factors and E12/E47 (vertebrate DAUGHTERLESS homologue).

Notch, Delta, and Suppressor of Hairless (Su[H])

Within each *ac/sc* proneural cluster, cells compete with each other through the action of neurogenic genes such that a subset of cells is singled out to become an es precursor. In loss-of-function mutants for any of the neurogenic genes (including *Notch, Delta, big brain, neuralized, mastermind,* and others), supernumerary cells in the proneural clusters adopt the es precursor fate rather than an epidermal cell fate (Hartenstein and Posakony, 1990; Goriely et al., 1991; Heitzler and Simpson, 1991). Of the neurogenic genes, *Notch* and *Delta* are by far the best studied. Within a proneural cluster, activation of NOTCH protein (a receptor) by DELTA (a ligand) leads to suppression of SOP fate. NOTCH/DELTA is a universal cell–cell interaction system used in most (possibly all) multicellular organisms in a variety of developmental processes (including vertebrate neurogenesis Chitnis et al., 1995). The general theme appears to be that whenever a subset of cells needs to be singled out from an equivalence group to adopt a particular fate, the NOTCH DELTA system is used (for a recent review see Artavanis-Tsakonas et al., 1995). SUPPRESSOR OF HAIRLESS (SU[H]) functions in the NOTCH signaling pathway by relaying a signal from the NOTCH receptor to the nucleus. SU(H) also appears to be an evolutionarily conserved component of the NOTCH/DELTA signaling pathway used in a variety of multicellular organisms (including the nematode worm, fly, and mammals). The modes of action of the other neurogenic genes *neuralized, mastermind,* and *big brain* are less well understood (for review, see Campos-Ortega, 1993).

Enhancer of Split Complex

The Enhancer of split complex (E[Spl]-C) contains at least seven genes (*ms, mr, mb, m3, m5, m7,* and *m8*). All of these genes encode bHLH proteins which are structurally related to the HAIRY protein and which appear to provide partially redundant functions (reviewed in Campos-Ortega, 1993). E(Spl)-C genes are downstream elements in the NOTCH signaling pathway. E(SPL)-C proteins are expressed in cells within proneural clusters except for the SOPs. E(SPL)-C proteins accumulate in response to NOTCH signaling activity (Jenning et al., 1994, 1995). The function of E(SPL)-C proteins is to suppress SOP fate. Deletion of E(SPL)-C leads to *Notch*-like phenotype, i.e., formation of supernumerary SOPs in the proneural clusters (Heitzler et al., 1996). Conversely, overexpression of E(SPL)-C proteins prevents SOP formation (Tata and Hartley, 1995; Nakao and Campos-Ortega, 1996). The E(Spl)-C genes were thought to promote epidermal cell fate (Campos-Ortega, 1993). However, further analysis led to an alternative view that genes of E(Spl)-C do not promote epidermal cell fate per se. Instead, they function to keep the cells that express them from differentiating into neural fate; these cells await a subsequent signal which directs them into the epidermal cell fate (Jennings et al., 1994).

Differentiation and Specification of the SOP

SOP differentiation is not an all-or-none event but rather a progressive process involving at least two stages. In the first stage, presumptive SOPs from proneural clusters are singled out by the positive and negative regulators (for SOP fate) described in the previous sections, and they express an elevated level of proneural genes *ac* and *sc*. This elevated *ac/sc* expression defines a transient state called a "nascent SOP." In a normal embryo, the nascent SOPs then express a whole set of pan-neural precursor genes and proceed with a stereotyped pattern of cell divisions to generate neurons and their non-neuronal sister cells. This group of pan-neural precursor genes (including *asense, deadpan, prospero, scratch,* and others) is expressed in most and possibly all neural precursors cells. Within the ectodermal layer, their expression is restricted to neural precursors and not found in any other cells. Temporally, their expression is transient. Some are only expressed in the neural precursors but not in their progeny, whereas others persist in the progeny for a period and then fade out. Outside of the ectoderm, some of these genes are also expressed in subsets of mesodermal or endoderm cells (Brand et al., 1993). They are hypothesized to confer "neural properties" to neural precursors, with each pan-neural precursor gene controlling a subset of neuronal properties (Vaessin et al., 1991; Bier et al., 1992). In *da*$^-$ mutants, all the pan-neural precursor genes tested (including *deadpan, asense, prospero, scratch, cyclin A,* and *couch potato*) failed to be expressed and the nascent SOPs fade away; they never divide to give rise to neuronal tissue. It is not known whether the failed nascent SOPs die or revert to an epidermal cell fate. Thus, in the embryonic PNS, *da* is not required for the formation of nascent neural precursors but is essential for subsequent neural precursor differentiation (Vaessin et al., 1994).

asense

asense (*ase*) is part of the *achaete–scute* complex and encodes a bHLH protein. *ase* functions as a proneural gene for a small subset of es organs (Dambly-Chaudiere and Ghysen, 1987). In contrast to *ac* and *sc, ase* has an additional role as a neural precursor gene in controlling neuronal differentiation later in es organ development (Brand et al., 1993; Dominguez and Campuzano, 1993).

deadpan

deadpan encodes a Hairy-related bHLH protein. Like *scute* and *daughterless, deadpan* also is an important regulator in *Drosophila* sex determination (Younger-Shepherd et al., 1992). In loss-of-function mutants the neuronal morphology appears normal (Bier et al., 1992).

prospero

prospero encodes a homeobox-containing protein. In *prospero* loss-of-function mutants, many of the neurons have axon growth and/or guidance defects (Doe et al., 1991; Vaessin et al., 1991). *prospero* is made by the neural precursors. However, when the neural precursors enter mitosis, PROSPERO protein together with another protein called NUMB (see later) becomes asymmetrically localized to one pole of the cell. Upon division, PROSPERO and NUMB are preferentially segregated into one of the two daughters (Spana and Doe, 1995; Hirata et al., 1995; Knoblich et al., 1995). The asymmetric localization of PROSPERO and NUMB provides an important clue about the mechanisms of asymmetric cell division (see later).

scratch

scratch encodes a Zn finger protein closely related to the *Drosophila* mesoderm determination gene *snail*. The morphology of loss-of-function *scratch* or *deadpan* mutants appears to be normal. However, double *scratch*$^-$/*deadpan*$^-$ mutants exhibit significant loss of neurons, suggesting that *scratch* and *deadpan* have redundant functions in neural development (Roark et al., 1995).

Inscuteable

Inscuteable encodes a novel protein with weak homology to Ankyrin (Kraut and Campos-Ortega, 1996). *Inscuteable* has an important function in specifying the apical/basal polarity and mitotic division plane of neural precursors (Kraut et al., 1996).

With the discovery of *neuroD* as a differentiation factor for neurogenesis in vertebrates (Lee et al., 1995), it was proposed that vertebrate neuronal determination and neuronal differentiation are two distinct processes. It was further suggested that in *Drosophila,* the distinction between a determination and a differentiation step is not apparent during neurogenesis (Lee et al., 1995). It might be premature to draw this conclusion because the above-mentioned pan-neuronal precursor genes appear to control aspects of neuronal differentiation. Furthermore, there is a cascade of pan-neural genes expressed in successive stages of *Drosophila* neural development. Thus, the pan-neural precursor genes are transiently expressed in *Drosophila* neural precursors. They are followed by the transient expressions of a pan-ganglion mother cell gene which encodes a novel bHLH protein and is expressed in ganglion mother cells which are daughters of neuroblasts (Frise et al., in preparation). This is then followed by the expression of pan-neuronal genes such as *elav* in the daughters of ganglion mother cells (Campos et al., 1987).

This cascade of pan-neural genes thus provides good candidates for regulators of neuronal differentiation, although the exact function of those genes remains to be clarified. One reason that relatively little is known about the function of those pan-neural genes is that many of them were not identified by classical mutant screens. Instead, they were discovered relatively recently by virtue of their pan-neural expression pattern through methods such as enhancer trap screens (O'Kane and Gehring, 1987; Bier et al., 1989). This illustrates the potential limitation of classical mutant screens (i.e., genes are knocked out randomly one at a time; only mutants that give a relevant phenotype are identified). Genes with important but redundant function (for example, *scratch* and *deadpan*) may escape detection with classical mutant screen. (However, once redundancy is suspected, one can design a mutant screen starting with a sensitized background, e.g., with one of the redundant genes already knocked out.)

An important task for the future is to understand how the various pan-neural genes function. They are likely to be key regulators of neuronal differentiations.

The Fate of the Progeny of SOPs: What Is Cell Lineage?

Each es organ is derived from a single SOP via strict lineage. The SOP divides to give rise to two secondary precursors, IIa and IIb. For a simple es organ, such as a sensory bristle, IIa divides once to give rise to hair and socket, and IIb divides once to produce neuron and sheath. What controls the fate of the cells that make up a sensory bristle? This involves a series of asymmetric cell divisions controlled by many of the same genes used in distinguishing SOP vs. epidermal cell fates as well as some additional molecular mechanisms. Two types of mechanisms may generate asymmetric cell divisions: (1) extrinsic: Daughter cells are initially equivalent, and the asymmetry is the result of the daughter cells interacting with each other or with their environment; (2) intrinsic: Daughter cells inherit unequal amounts of determinant(s). Both extrinsic and intrinsic mechanisms are used in generating different cells of a sensory bristle (Fig. 2–2) (reviewed in Posakony, 1994; Jan and Jan, 1994).

Extrinsic mechanisms

We know that extrinsic mechanisms are used because genes in the Notch signaling pathway have been found to be required for the four cells to assume their proper fate. The involvement of *Notch* and *Delta* in generating asymmetric SOP cell divisions was revealed by studying temperature-sensitive mutants of *Notch* or *Delta* (Hartenstein and Posakony, 1990; Parks and Muskavitch, 1993). Shifting these mutants to the restrictive temperature during the time periods encompassing the asymmetric cell divisions of SOP, IIa and IIb, rendered those divisions symmetric. These experiments suggest that *Notch*- and *Delta*-mediated cell–cell interactions are required to distinguish the two daughter-cell fates during multiple stages of sensory bristle development. Recent studies of loss-of-function and gain-of-function mutants of *Su(H)* and E(SPL)-C showed that these genes in the

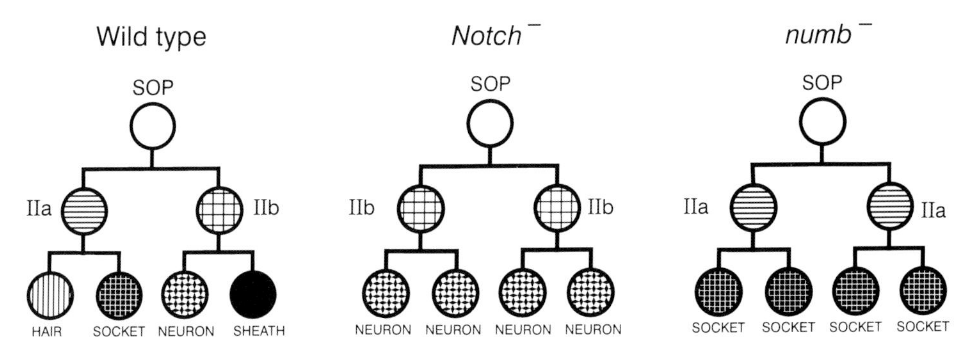

FIGURE 2–2 *Notch* and *numb* are key genes for cell–cell interaction and intrinsic mechanisms used in generating asymmetry of sensory organ precursor (SOP) cell divisions. In *Notch*⁻ and *numb*⁻ mutants, the four progeny cells of SOP develop into four neurons or four socket cells, respectively (based on Hartenstein and Posakony, 1990; and Rhyu et al., 1994).

Notch signaling pathways are also involved in sensory bristle cell fate specification (Tata and Hartley, 1995; Nakao and Campos-Ortega, 1996; Schweisguth and Posakony, 1994; Schweissguth, 1995; Bang et al., 1995).

Intrinsic mechanisms

In addition to cell–cell interactions, intrinsic mechanisms also play an important role in generating asymmetry in SOP and secondary precursor cell divisions. The gene *numb* appears to be a key gene (Uemura et al., 1989; Rhyu et al., 1994).

Immunocytochemical experiments reveal that NUMB protein is membrane associated and is asymmetrically localized in the shape of a crescent in the SOP cell prior to cell division. Upon division, NUMB protein is preferentially segregated into one of the two secondary precursors, IIb (Rhyu et al., 1994). The level of NUMB controls cell fate. In *numb*⁻ mutants, neither of the secondary precursors has NUMB protein and they both develop into IIa's. Conversely, if the level of NUMB is kept high in both secondary precursors by giving a heat pulse at around the time of SOP division to a transgenic fly carrying *numb* under the control of a heat-shock promoter, IIa is transformed into IIb. Consequently, supernumerary neurons and sheath cells are produced at the expense of hair cells and socket cells. Taken together, these experiments show that the NUMB protein is segregated unequally into the two SOP daughters and that the different NUMB levels in the two daughter cells determine their fate (Rhyu et al., 1994). *numb* also functions to generate asymmetry in the subsequent divisions of IIa and IIb. The unequal segregation of NUMB therefore, appears to be a general feature of asymmetric cell divisions in the fly nervous system. It occurs not only in divisions of sensory bristle precursors but also in other PNS precursors such as chordotonal organ precursors. It also occurs in CNS precursors (the neuroblasts) (Rhyu et al., 1994; Spana et al., 1995). However, in *numb* loss-of-function mutant, cell transformation is observed in some but not all neuroblasts (Spana et al., 1995).

How Is NUMB Asymmetrically Localized in an SOP Prior to its Division?

Studies of NUMB and PROSPERO localization (Knoblich et al., 1995; Hirata et al., 1995; Spana and Doe, 1995) have led to the hypothesis that during asymmetric cell division, there exists an "asymmetry organizer" (Knoblich et al., 1995). This hypothetical asymmetry organizer coordinates the localization of the NUMB crescent, the PROSPERO crescent, and one of the two centrosomes, presumably to make sure the cell division plane is parallel to the NUMB and PROSPERO crescents so that NUMB and PROSPERO will be segregated predominantly into one of the two daughter cells. NUMB may be an evolutionarily conserved intrinsic "determinant." A mouse *numb* gene was found to share a high degree of sequence homology with fly *numb* (Zhong et al., 1996). Mouse *numb* when introduced into *Drosophila* is as effective as fly *numb* in rescuing the loss-of-function *numb* phenotype. Further, mouse NUMB protein expressed in flies is localized in the same way as fly NUMB in both SOPs and CNS neuroblasts, suggesting that the *Drosophila* "asymmetry organizer" can recognize mouse NUMB.

During mouse cortical neurogenesis mouse NUMB is asymmetrically localized in a "crescent" in neural progenitor cells before their division in the ventricular zone. There is an interesting difference between *Drosophila* NUMB and mouse NUMB localization during neurogenesis. In fly neural precursors, the NUMB crescent is always tightly correlated with the division plane. In mouse, NUMB is always localized to the apical (ventricular) side of the neural progenitors. Depending upon the orientation of the cell division plane, mouse NUMB may be inherited by one or by both daughter cells. It is evenly divided into two daughter cells when the division plane is perpendicular to the ventricular surface, presumably leading to symmetric division (Chenn and McConnell, 1995). In contrast, mouse NUMB would be preferentially localized to one of the two daughters (the apical daughter) when the division plane is parallel to the ventricular surface and the division is asymmetric. This difference may reflect different strategies fly and mouse have adopted during evolution to exploit the asymmetric NUMB localization in order to fulfill different requirements for neurogenesis. In the fly, except in the eye, essentially all the neural precursors undergo asymmetric cell divisions to generate neurons. In the mouse, neural progenitors undergo symmetric cell divisions (presumably) to expand the neural progenitor pool and asymmetric cell divisions (presumably) to generate neurons. The mouse NUMB crescent is on the apical (ventricular) side of neural progenitors, i.e., the opposite of the basal localization of NOTCH 1 crescent observed in ferret neural progenitors (Chenn and McConnell, 1995). This observation is in accordance with the finding in *Drosophila* that one major function of NUMB is to antagonize NOTCH signaling activity (see later).

How do the Intrinsic and Extrinsic Mechanisms Interface?

Given that both cell–cell interactions mediated by the NOTCH signaling pathway and intrinsic mechanisms involving NUMB are used in specifying sensory organ cell fate, how do the extrinsic and intrinsic signaling pathways interface?

We envision the NOTCH/DELTA system as the fundamental and universal cell–cell interaction system. Consider two daughter cells: In the absence of NUMB, the two cells are initially equivalent. According to the current model of the NOTCH/DELTA system, the two cells will start with equivalent amounts of NOTCH and DELTA (Heitzler and Simpson, 1991). Due to stochastic fluctuations, one cell (say A) may have a slightly higher amount of NOTCH and would receive a slightly higher amount of inhibitory signal, which will lead to a decreased expression of DELTA and hence to decreased ability to inhibit its neighbor. Because cell B receives less inhibitory signal, it will produce more DELTA and will increase its ability to inhibit cell A. Such a feedback loop would amplify the initial slight differences in NOTCH or DELTA levels and result in the inhibition of one cell by the other so that the two cells will assume different fates. The outcome of this competition mediated by NOTCH/DELTA is stochastic; either daughter cell can become an A cell. There is now considerable genetic evidence indicating that NUMB provides a strong internal bias by suppressing NOTCH signaling activity (Guo et al., 1996; Spana and Doe, 1996). The daughter that inherited the NUMB protein will have lower NOTCH activity with respect to its sibling and will therefore always develop into a B cell. There are a number of important unresolved issues. One issue concerns how NUMB affects NOTCH signaling activity. Another issue relates to the identification of downstream effectors which actually confer the different cell fate. One such effector identified is a gene called *tramtrack* (*ttk*). *ttk* encodes a Zn finger protein which is known as a transcriptional repressor (Harrison and Travers, 1990; Brown et al., 1991; Read et al., 1992). In SOP development, *ttk* functions downstream from both *numb* and *Notch* and acts as a suppresser of neural fate (Guo et al., 1995). Formally, its function appears to be analogous to the vertebrate negative regulator of the neuronal phenotype NRSF/REST (Chong et al., 1995; Schoenherr and Anderson, 1995a; see later).

The Generality of Mechanisms Used in es Organ Development

Most of the molecular mechanisms utilized in controlling es organ development are used in *Drosophila* neural development in general.

The concept of the proneural gene is generally applicable in *Drosophila* neurogenesis

PNS. Of the four major types of sensory elements, two (es organs and the majority of multiple dendritic [md] neurons) require *ac* and *sc* for their formation, whereas the other two, chordotonal (ch) organs and photoreceptors, require a different proneural gene *atonal* (*ato*) for their formation. *ato* also encodes a bHLH protein.

ato and AS-C account for the origin of almost the entire PNS. Deletion of both AS-C and *ato* removes all but two md neurons in each hemisegment; presumably one or more proneural genes are yet to be identified (Jarman et al., 1993a). Thus, it appears that the involvement of proneural genes is a general mechanism utilized to initiate neural development in the *Drosophila* PNS.

Proneural genes also contain neuronal-type information. When expressed at an ectopic position, only *ato* but not *ac/sc* can promote the formation of ectopic chordotonal organs. From domain swap experiments in which various chimeras between *ato* and *sc* were assayed for their ability to promote chordotonal organ formation, it was found that the neuronal-type information of *ato* and *sc* resides in their basic domain (Chien et al., 1996). It will be very interesting to find out whether the basic domains of bHLH proteins involved in vertebrate neurogenesis (see below) also contain neuronal-type information. While proneural genes influence neuronal type, they do not completely specify it. For example, *ato* is required for the formation of both ch organs and photoreceptors. However, misexpression of *ato* produces only ch organs and not photoreceptors outside of the eye. In this case, there must be other factors which ensure that *ato* promotes the formation of photoreceptor rather than ch organ in the eye (Jarman et al., 1994). The genes *eyeless* and *glass* are good candidates (Halder et al., 1995; Moses et al., 1989).

CNS. Whereas the proneural genes of AS-C and ato can account for the initiation of essentially the entire PNS, the situation is more complicated in the CNS. In the embryo, *ato* is not expressed in regions that give rise to CNS and appears to play no role in initiating CNS development (Jarman et al., 1993). In contrast, genes of AS-C are expressed extensively in CNS neurogenic regions (Cabrera and Alonso, 1987; Martin-Bermundo et al., 1991; Skeath et al., 1992). However, removal of AS-C prevents the formation of only a subset of neuroblasts (Jiménez and Campos-Ortega, 1990). Therefore, there must be genes in addition to *ato* and AS-C to account for CNS formation. One such gene is *single minded,* a bHLH gene required for the formation of midline neuroblasts (Nambu et al., 1991).

Although all the proneural genes discovered so far encode bHLH proteins, it is conceivable that other types of molecules could function as proneural genes. The expression pattern of a number of genes that encode Zn finger proteins (including *hunchback* and *snail;* Roark et al., 1995) and the phenotype of a homeobox-containing protein *vnd* (Jiménez et al., 1995) suggest that they may act as proneural genes for the embryonic CNS. However, their exact function in CNS development remains to be elucidated.

In summary, proneural genes clearly are important in controlling CNS development. There are probably additional proneural genes to be discovered. Whether mechanisms other than those involving proneural genes operate in initiating CNS development is unknown.

The many binary decisions

As discussed earlier, the NOTCH signaling pathway is a general mechanism for generating asymmetry in cell fate (i.e., making two cells different). In neural development, it is used in both the PNS and CNS and both in situations where cell lineage plays no role in cell fate specification (e.g., the singling out of R8 photoreceptors in *Drosophila* eye) and in situations where there is strict cell lineage (e.g., the specification of sensory bristle cell fate). In the latter case, the NUMB system is added to ensure a reproducible outcome of cell fates within different branches of the lineage. Both NOTCH signaling and NUMB are used in multiple binary decisions. Therefore, their roles are to generate asymmetry in cell fate decisions rather than to specify a particular cell fate.

Besides the "symmetry-breaking" role of *Notch* and *numb,* a number of genes control binary decisions between two alternative cell fates. For example, *cut,* a homeobox-containing gene, controls whether a bipotential SOP will develop into an es organ or a ch organ (Blochlinger et al., 1988). *poxn,* a pair-box–containing gene, controls whether a SOP will become a polyinnervated es organ or a monoinnervated es organ (Dambly-Chaudiere et al., 1992), *glial cells missing,* which encodes a novel nuclear protein and controls many neuronal vs. glial fate decisions in both PNS and CNS (Jones et al., 1995; Hosoya et al., 1995). Whether there are homologues of these genes that carry out analogous functions in vertebrate neural development remains to be determined. Nevertheless, the use of binary genetic switches to control cell fate is probably a widely used strategy in both invertebrate and vertebrate neurogenesis.

Determination of the Neuronal Phenotype in Vertebrates

bHLH Proteins Involved in Neuronal Determination

As described in the preceding section, genes encoding transcription factors in the basic helix-loop-helix (bHLH) family play an important role in neuronal (as well as myogenic) determination (Jan and Jan, 1993). In *Drosophila,* the "proneural" genes *achaete* and *scute* are required for neural determination, defined as making uncommitted ectoderm competent to generate neural cells (Campuzano and Modolell, 1992). In mammals, *MyoD* and *myf5* are required (redundantly) for the determination of a myogenic fate in uncommitted mesodermal cells (Weintraub, 1993). However, not all bHLH proteins are necessarily involved in determination. In *Drosophila* neurogenesis, as well as in vertebrate myogenesis, different bHLH proteins can function in a cascade to control different developmental steps (such as *achaete-scute* and *asense,* which control sequential stages of es organ development (Campuzano and Modolell, 1992; Brand et al., 1993; Domínguez and Campuzano, 1993)).

Diversity among bHLH proteins also underlies diversity between different neural subtypes, as well as different developmental steps within a given lineage. For example, as mentioned in the preceding section, *achaete-scute* and *atonal* appear to carry out similar proneural functions, but for the es and chordotonal subtypes of sensory organ, respectively (Jarman et al., 1993b). These precedents suggest that there should be a similar diversity of bHLH proteins involved in vertebrate neurogenesis. The available data have borne this out. The challenge is to understand the precise developmental operations carried out by each of these genes.

XASH genes

Xenopus has been a favored system in which to search for genes involved in neuronal determination because the earliest stages of neural induction are experimentally accessible. Two *Xenopus achaete–scute* homologues thus far have been identified: *Xash1* and *Xash3*. XASH1 protein is the *Xenopus* orthologue of MASH1 (Ferreiro et al., 1992) (see below), while XASH3 is closely related to both proteins within the bHLH domain but diverges outside this region (Zimmerman et al., 1993; Turner and Weintraub, 1994). (Mammalian homologues of XASH3 have not yet been identified.) XASH3 is first expressed at stage 10.5–11, right around the time of neural induction (Zimmerman et al., 1993; Turner and Weintraub, 1994). By contrast, expression of XASH1 is not detected until much later (St. 24–25), long after primary neural induction has occurred (Ferreiro et al., 1992).

Ectopic expression of XASH3 mRNA in *Xenopus* embryos causes a dramatic expansion of the neural plate, as detected by expression of NCAM, at the apparent expense of surrounding epidermis (Ferreiro et al., 1994; Turner and Weintraub, 1994) (Fig. 2–3, lower, a). Expression of Xtwist in neural-crest–derived mesectodermal precursors of the branchial arches is also suppressed. Because in *Drosophila* ectopic expression of the proneural genes confers competence to make neural tissue on non-neurogenic regions of the ectoderm, these data (taken together with the sequence homology to *achaete–scute*) have been interpreted to indicate that *Xash3* is a vertebrate proneural gene.

This conclusion, however, is based exclusively on gain-of-function (GOF) experiments. In *Drosophila* it is clear that some genes of the AS-C, such as *asense*, can exhibit proneural activity in GOF experiments (Brand et al., 1993; Domínguez and Campuzano, 1993) even though their expression pattern strongly suggests that they do not normally perform this function (see preceding section). Furthermore XASH3 mRNA is expressed in only a very narrow region of the neural plate that appears fated to generate the sulcus limitans (a boundary zone that separates the dorsal [alar] region of the spinal cord from the ventral [basal] region; Zimmerman et al., 1993). Thus, if XASH3 normally does function as a proneural gene, it plays this role in a very restricted population of neural plate cells, implying that there must be other proneural genes for other regions of the neural plate.

A further difficulty is that the interpretation of the XASH3 GOF phenotype as "proneural" is based on the notion that the entire *Xenopus* neural plate is analogous to a proneural cluster in *Drosophila* (Ferreiro et al., 1994; Turner and Weintraub, 1994). However, the more appropriate analog of the *Drosophila* proneural cluster may be the specialized regions *within* the *Xenopus* neural plate from which different populations of primary neurons are generated. At the open neural plate stage, there are three such regions, located in roughly parallel longitudinal stripes on either side of the midline (Fig. 2–3, A). From each of these longitudinal domains a stripe of primary neurons differentiates; between these stripes there are regions of undifferentiated neuroepithelium that do not produce neurons until after neurulation has occurred. The differentiation of stripes of primary sensory, motor, and interneurons from within the longitudinal domains may, therefore, be most analogous to the selection of neuronal precursors from within each proneural cluster in the fly.

In support of this interpretation, interference with the lateral inhibition process mediated by *Xenopus* homologues of Notch and Delta causes an increased

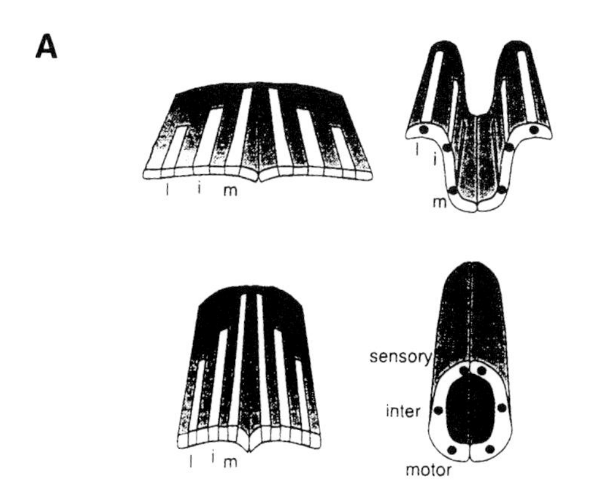

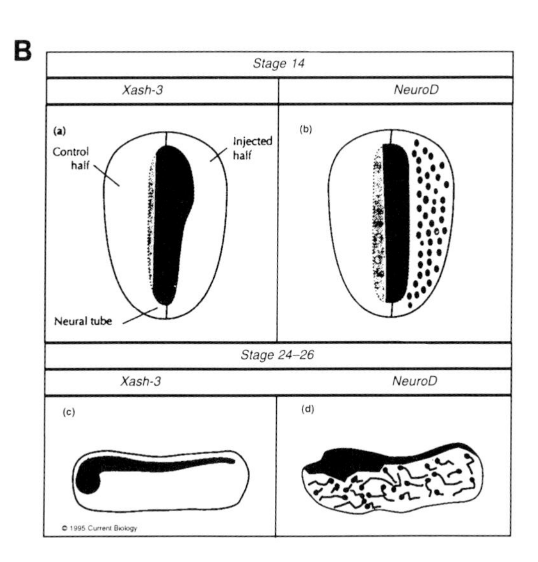

FIGURE 2–3 Primary neurogenesis in *Xenopus* and the effect of bHLH gene overexpression. (A) Three domains of primary neuron differentiation develop as parallel stripes in the neural plate. "l," lateral; "i," intermediate; "m," medial. These three stripes differentiate to form sensory neurons, interneurons and motor neurons, respectively. (B) Contrasting effects of XASH3 and NeuroD overexpression in *Xenopus* embryos at the neural plate stage. The dark shading represents the domains of neuronal differentiation as revealed by NCAM expression. Note that XASH3 expands the neural plate, but does not cause ectopic neurogenesis in surrounding ectoderm as does NeuroD. From Anderson (1995).

number of neurons to differentiate within each of the three longitudinal stripes (Chitnis et al., 1995), rather than an expansion of the entire neural plate as is observed with XASH3 overexpression (Fig. 2–3). This is precisely analogous to what happens in neurogenic mutants in *Drosophila:* An increased number of SOPs appear within each proneural cluster (Campos-Ortega, 1993). The fact that XASH3 overexpression does not produce such a neurogenic phenotype seems, therefore, inconsistent with the assignment of a "proneural" function to this bHLH protein. However, neurogenic phenotypes within the neural plate are observed following overexpression of two other, related genes that encode bHLH proteins, called NeuroD (Lee et al., 1995) and neurogenin (Fig. 2–3; see below). Of these, neurogenin shows an expression pattern most consistent with a role in neuronal determination (Ma et al., 1996). Interestingly, XASH3 has also recently been shown to produce a neurogenic phenotype like NeuroD, but only when ectopically expressed at relatively low levels (Chitnis and Kintner, 1996). The exact function of *Xash3* in neurogenesis therefore remains unclear and awaits the results of loss-of-function (LOF) perturbations.

Neurogenin

Perhaps a more attractive candidate for a vertebrate neuronal determination gene is *neurogenin* (Ma et al., 1996). NEUROGENIN is a bHLH protein that is closely related to NeuroD (Lee et al., 1995) (see below) and more closely related to *Drosophila* ATONAL than to ACHAETE-SCUTE (although it is not the mammalian orthologue of ATONAL, of which there are several; Akazawa et al., 1995; Shimizu et al., 1995). Expression of a *Xenopus neurogenin*-related gene (*XNGR1*) begins around late gastrulation (St. 10.5), in three broad patches that prefigure the domains of the neural plate in which primary neurogenesis will occur (see above) (Ma et al., 1996). Especially important is the fact that expression of XNGR1 occurs even earlier than that of XDelta, a ligand for the Notch lateral inhibition receptor (see below). Since Notch and Delta control fate choices by initially equivalent neuroectodermal cells, XNGR1 is by definition expressed before such choices are made. Such early expression is a sine qua non for a determination gene: A gene cannot determine cell fate if it is not expressed until the cells are already detectably different from their neighbors.

The gain-of-function (GOF) phenotype of XNGR1 is quite different from that of XASH3: it causes precocious and more extensive neuronal differentiation within the neural plate, as well as the induction of neuronal differentiation in surrounding non-neurogenic ectoderm (Ma et al., 1996). This is similar to the phenotype observed following ectopic expression of NeuroD (Lee et al., 1995) (Fig. 2–3, lower, b) and is indicative of a determination function. While *Xenopus* NeuroD (XNeuroD) overexpression generates a similar phenotype, endogenous XNeuroD is expressed much later than XNGR1 (Lee et al., 1995). Indeed, overexpression of XNGR1 causes ectopic expression of endogenous XNeuroD, whereas the converse is not true (Ma et al., 1996). Moreover, XNGR1 both activates expression of Delta and is itself inhibited by signaling through Notch (Ma et al., 1996), as is the case for the proneural genes in *Drosophila* (Ghysen et al., 1993). Taken together, these data suggest that XNGR1 and XNeuroD function in a cascade, controlling sequential stages in neurogenesis, with the former performing a function analogous to that of the proneural genes. Such cascades of related bHLH molecules have previously been described in vertebrate myogenesis as well as in fly neurogenesis (Jan and Jan, 1993).

Thus far, the only marker used to identify the neurons generated by ectopic expression of XNGR1 is a neuron-specific isoform of β-tubulin. Thus it is not clear whether XNGR1 is sufficient to promote expression of all aspects of a neuronal phenotype or only some. XNGR1 is expressed in all three presumptive domains of primary neurogenesis (medial, intermediate, and lateral) within the neural plate (Ma et al., 1996), suggesting that it may function in the development of at least three distinct neuronal subtypes: sensory, motor, and interneurons. However, in rodents, neurogenin is expressed only in some regions or lineages of the nervous system (Ma et al., 1996). This suggests that, like *Drosophila atonal, neurogenin* may perform its determination function only in some parts of the nervous system, and that different bHLH genes may play this role in others.

bHLH Proteins Involved in Neuronal Differentiation

NeuroD

NeuroD was isolated on the basis of its ability to heterodimerize with *Drosophila* DAUGHTERLESS, using the yeast two-hybrid system (Lee et al., 1995). Ectopic expression of XNeuroD in *Xenopus* embryos, like that of XNGR1, causes premature and ectopic neurogenesis (Lee et al., 1995). The phenotype appears distinct from that observed with ectopic expression of XASH3 (Fig. 2–3, lower, cf. a vs. b) in that the neural plate does not expand. Rather, more extensive and precocious neuronal differentiation occurs within the neural plate as well as in surrounding ectoderm. These data indicate that *XNeuroD* can act as a neuronal determination gene when it is ectopically expressed. However, the expression pattern of endogenous *NeuroD,* both in *Xenopus* and in mouse, suggests that it is more likely to be involved in neuronal differentiation than in determination. For example, NeuroD is not expressed in the ventricular zone where uncommitted stem cells are located, but rather in the intermediate zone where postmitotic neuronal precursors are present en route to the pial surface (Lee et al., 1995; Ma et al., 1996).

The results obtained with NeuroD in vertebrates, as well as with ASENSE in *Drosophila* (Brand et al., 1993; Domínguez and Campuzano, 1993), indicate how conclusions about a gene's normal function can be potentially misleading if they are based exclusively on GOF phenotypes. bHLH proteins are particularly treacherous to analyze in this way, as they can heterodimerize and bind to common enhancer elements in DNA rather promiscuously (Murre et al., 1989a, b). Ideally, interpretations of gene function should be based on both GOF and LOF manipulations, as well as on the timing of the gene's expression. However, LOF manipulations are currently difficult to accomplish in *Xenopus,* and GOF manipulations are not as easy to perform in the mouse as they are in *Xenopus.*

MASH1

Mash1 is a mammalian homologue of the *Drosophila* proneural genes *achaete-scute* and was the first such gene to be cloned in vertebrates (Johnson et al., 1990). Whether it is the *orthologue* of the proneural genes is not yet clear. MASH1 diverges considerably from ACHAETE-SCUTE in the loop region, whereas mammalian MYOD and its fly orthologue NAUTILUS are almost identical in the corresponding region (Michelson et al., 1990). Thus, it is possible that there exist other vertebrate genes more closely related to the proneural genes than *Mash1,* which have not yet been identified.

Like its *Drosophila* counterparts, MASH1 is expressed early in neurogenesis, in restricted regions of the nervous system. For example, within the PNS

MASH1 is expressed by precursors of autonomic, but not sensory, neurons (Lo et al., 1991; Guillemot and Joyner, 1993). Similarly, in *Drosophila, achaete* and *scute* are expressed in SOPs that give rise to es organs, but not in those that generate chordotonal organs (see preceding section). Expression of *Mash1,* like that of *achaete-scute,* is transient: Transcription is rapidly downregulated following overt neuronal differentiation (Lo et al., 1991).

The function of MASH1 has been investigated by both LOF and gain-of-function GOF manipulations in mammals. Thus far, no GOF phenotype has been observed in any mammalian cell line examined (Johnson et al., 1992). However, ectopic expression of XASH1, the *Xenopus* orthologue of MASH1, causes some neuronal differentiation in explanted animal caps (Ferreiro et al., 1994). Targeted deletion of the *Mash1* gene in mice prevents the development of autonomic neurons in the PNS as well as the development of primary olfactory sensory neurons (Guillemot et al., 1993). In contrast, regions of the CNS that express MASH1 appear unaffected by criteria of both morphology and expression of molecular markers.

These data indicate that *Mash1,* like the *Drosophila* proneural genes, is required for the genesis of particular subclasses of peripheral neurons; but does it control the same operation as its counterparts in the fly? A more extensive analysis of the *Mash1* mutant phenotype in vitro indicates that MASH1 function is not essential until after commitment to neuronal differentiation (Sommer et al., 1995). In the absence of MASH1 function, arrested neuronal precursors form and then die. Progression of these cells to fully differentiated neurons is dependent upon MASH1 function.

The cellular function of MASH1 is therefore distinct from that of ACHAETE-SCUTE, which is required at a relatively earlier stage of development, prior to the segregation of neuronal and glial fates (Ghysen and O'Kane, 1989; Campuzano and Modolell, 1992). This difference in function is not due to structural differences between the vertebrate and *Drosophila* genes, as *Mash1* is able to efficiently complement the *achaete-scute* LOF phenotype when expressed from a heat-shock promoter in the fly (A. Singson, D.J. Anderson, J. Posakony, unpublished). Rather, the functional differences between these genes are likely to reflect the different cellular contexts in which they are expressed. Does this difference mean that the functions of the fly proneural genes and *Mash1* have diverged in evolution even though their sequences are conserved (at least within the bHLH domain), or that *Mash1* is not the true orthologue of *achaete-scute?* This question is difficult to answer at present; however, another gene in the AS-C, *asense,* is as closely related to *Mash1* as are *achaete* and *scute,* yet in at least some lineages *asense* seems to function as a neural precursor gene rather than as a proneural gene (Brand et al., 1993; Domínguez and Campuzano, 1993). Perhaps the function of MASH1 is more analogous to that of ASENSE (although even ASENSE functions in SOPs before the segregation of neuronal and glial fates).

These data illustrate that although vertebrate homologues of *Drosophila* proneural and neural precursor genes are indeed expressed in the developing nervous system, and functionally important in neurogenesis, their exact cellular roles may have diverged in evolution. This may reflect the fact that the most significant evolutionary changes have occurred in the regulatory regions of these genes rather than in their coding sequences. In support of this idea, XASH1 (which is virtually identical to MASH1 throughout its coding region) has a very different pattern of expression within the developing *Xenopus* nervous system although, like MASH1, it is specific to that tissue (Lo et al., 1991; Ferreiro et al., 1992). The fact that different bHLH proteins act in cascades within a given lineage raises the possibility that the hierarchical positions within such cascades of bHLH proteins may have been reversed during evolution. Indeed, such differences may even occur between different neuronal lineages within the same species.

Other bHLH Proteins Expressed in Neural Development

Numerous other bHLH proteins specifically expressed in the developing nervous system have been recently isolated. These include mammalian homologues of the *Drosophila* proneural gene *atonal* (Akazawa et al., 1995; Shimizu et al., 1995), neuronal relatives of lymphoid-specific bHLH genes such as *NSCL1* and *NSCL2* (Begley et al., 1992; Göbel et al., 1992), and novel subfamilies of bHLH genes such as *eHAND/Th1* (Cserjesi et al., 1995; Hollenberg et al., 1995) and *dHAND* (Cserjesi et al., 1995). These genes show highly restricted expression patterns in the embryonic nervous system, but the phenotypes of LOF and GOF manipulations have not yet been reported.

Lineage-Specific Cascades of bHLH Proteins

As mentioned earlier, diversity among bHLH proteins appears to subserve both different stages of development within a given lineage as well as similar developmental events in different lineages. In vertebrates, this dual aspect of diversity is illustrated in the PNS. Peripheral neurons can be divided into two major sublineages: sensory and autonomic. Within each of these sublineages, different bHLH proteins are sequentially expressed, and they probably function in cascades (Fig. 2–4). Several of these proteins, moreover, show mutually exclusive expression between the sensory and autonomic lineages. For example, MASH1 is expressed in autonomic but not sensory precursors (Lo et al., 1991), whereas neurogenin is expressed at the same time of development in sensory but not autonomic precursors (Ma et al., 1996). This complementarity is reminiscent of the situation in *Drosophila,* where *achaete-scute* is expressed in precursors of es peripheral sense organs, while *atonal* is expressed in precursors of chordotonal organs (Jarman et al., 1993a). As described in the first section of this chapter, *achaete-scute* and *atonal* do not only act as proneural genes for their respective lineages but they also help to determine what type of sensory organ will develop (Jarman et al., 1993b). This illustrates the important concept that individual regulatory genes may control more than one developmental operation within a given lineage.

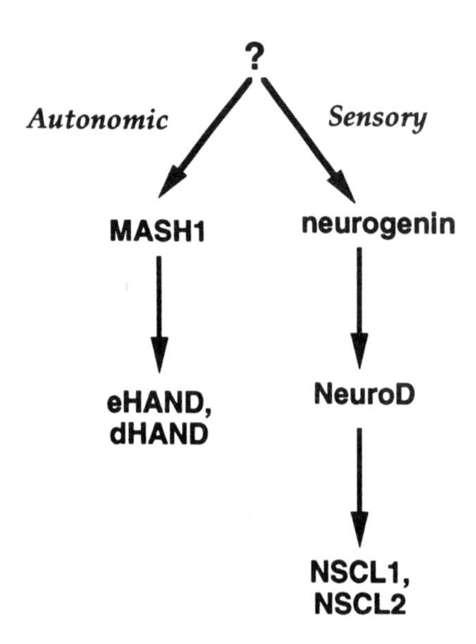

FIGURE 2–4 Cascades of different bHLH proteins function in the sensory and autonomic lineages of the neural crest. The evidence that eHAND and dHAND (Cserjesi et al., 1995) function downstream of MASH1 is based on their lack of expression in *Mash1* −/− mutants (L. Sommer, D.J.A., unpublished). Neurogenin (Ma et al., 1996), NeuroD (Lee et al., 1995), and NSCL1 and 2 (Begley et al., 1992; Göbel et al., 1992) are sequentially expressed during sensory neurogenesis in mammals, but the sequential activation implied by the diagram is based on studies in *Xenopus* (Ma et al., 1996; J. Lee, personal communication).

Negative-Acting HLH and bHLH Proteins in Neurogenesis

In *Drosophila,* a number of HLH genes have been identified that behave genetically as negative regulators of neurogenesis; that is, LOF mutations in these genes cause production of excess neurons, while GOF perturbations suppress neurogenesis. These genes include *extramachrochaete* (*emc*), *hairy,* and *enhancer of split* (*E[spl]*). These genes appear to inhibit neurogenesis by interfering with the expression and/or function of the proneural genes (Campos-Ortega, 1993; Ghysen et al., 1993). Several mammalian homologues of these genes have been identified. For example, there are multiple *Id* genes which are homologues of *emc* (Benezra et al., 1990). *Id* mRNA is expressed within the ventricular zone of the spinal cord (Duncan et al., 1992), consistent with the idea that it delays or inhibits differentiation in progenitor cells from a number of different lineages. Similarly, a family of genes that are related to *hairy* and *E(spl),* termed *HES* genes, has been cloned in mouse (Akazawa et al., 1992; Sasai et al., 1992; Ishibashi et al., 1993).

In the case of *HES-1,* the phenotypes of both LOF and GOF perturbations are consistent with the idea that this gene functions to inhibit or delay neurogenesis.

Specifically, targeted disruption of *HES-1* causes premature neurogenesis in the CNS (Ishibashi et al., 1995), whereas forced expression of *HES-1* from a retroviral vector inhibits neurogenesis (Ishibashi et al., 1994). Moreover, in the case of the LOF mutation the premature neurogenesis is accompanied by a precocious induction of MASH1 (Ishibashi et al., 1995), supporting the notion that *HES* genes function in mammals as they do in *Drosophila,* to repress or delay expression of positive-acting bHLH genes. Whether these negative-acting HLH genes are, in turn, regulated by lateral inhibition (as in *Drosophila*), by other factors, or both remains to be determined, although experiments in *Xenopus* promise to shed light on this question (Turner and Weintraub, 1994).

Genes That Directly Regulate Neuronal Terminal Differentiation Genes

Positive regulators of the neuronal phenotype

In the absence of large-scale mutagenic screens, a systematic approach to identifying regulatory factors that control the neuronal phenotype in vertebrates has involved dissecting the *cis*-acting sequences controlling the transcription of genes specifically expressed in neurons or in classes of neurons and then identifying the proteins that interact with these sequences. A detailed review of such experiments (Mandel and McKinnon, 1993) is beyond the scope of this chapter, but a few exemplary cases illustrate some of the general principles that have emerged.

POU- and LIM-domain proteins interact to regulate neuropeptide expression

One of the best-studied cases involves the regulation of the *prolactin* and *growth hormone* genes, which are specifically transcribed in lactotroph and somatotroph neuroendocrine cells of the anterior pituitary, respectively. A homeodomain protein, PIT-1/GHF-1, controls the cell-specific expression of these genes (Bodner et al., 1998; Ingraham et al., 1988). PIT-1 also contains a domain found in the *Caenorhabditis elegans* homeodomain gene *Unc-86* as well as the lymphoid transcription factor *Oct-1;* this domain is called the POU domain (Ingraham et al., 1990). Mutations in *Pit-1* in mice prevent the development of the appropriate pituitary cell types (Li et al., 1990), indicating that this gene plays an essential role in the expression of these neuroendocrine phenotypes.

The regulation and mechanism of action of PIT-1 have been studied at a level of detail beyond the scope of this chapter, but review is available elsewhere (Andersen and Rosenfeld, 1994). A recent important finding, however, is the identification of an LIM homeodomain protein, P-LIM, with which PIT-1 interacts synergistically to enhance transcription of pituitary-specific genes (Bach et al., 1995). The family of LIM homeodomains includes the *C. elegans* gene *mec-3,* which is essential for the development of a subset of mechanosensory neurons (Way and Chalfie, 1988). The MEC-3 protein has been shown to interact synergistically with UNC-86, which like PIT-1 is a member of the POU class of homeodomain proteins (Xue et al., 1993). Together these genes help establish a MECH-3 autoregulatory loop, and they may in addition control the expression of terminal differentiation genes such as the mechanosensory neuron-specific tubulin gene, *mec-7* (Hamelin et al., 1992). Thus, in the vertebrate pituitary as in the *C. elegans* mechanosensory lineage, LIM and POU homeodomain proteins interact synergistically to control the transcriptional regulation of terminal differentiation genes (Fig. 2–5). In *Drosophila,* the *Cf1a* gene encodes a POU-domain protein that binds an enhancer element in the *dopa decarboxylase* gene (Johnson and Hirsh, 1990), but interactions of this protein with LIM homeodomain proteins have not yet been reported.

The finding of a functional interaction between LIM and POU proteins that is evolutionary conserved in neural development is particularly interesting in light of the enormous diversity of these proteins that has been documented in the vertebrate nervous system. It has been estimated that there may be upward of hundreds of different POU-domain genes, each expressed in different subclasses of neurons (He and Rosenfeld, 1991). Similarly, numerous LIM-domain genes have been identified, and the expression of some of these defines subsets of neurons such as different classes of motoneurons in the spinal cord (Ericson et al., 1992; Tsuchida et al., 1994). It is difficult to escape the conclusion that the combinatorial usage of members of these two diverse families of interacting transcription factors may play a key role in the establishment of neuronal phenotypic diversity. Direct evidence for combinatorial control of neuronal identity has been provided by genetic experiments in *C. elegans* (Mitani et al., 1993).

Regulation of neurotransmitter enzymes by paired homeodomain genes

Another diverse family of transcription factors implicated in neurogenesis is that comprising the paired homeodomain (PHD) genes. Originally defined by their homology to the *Drosophila* pair-rule gene *paired,* these proteins contain a

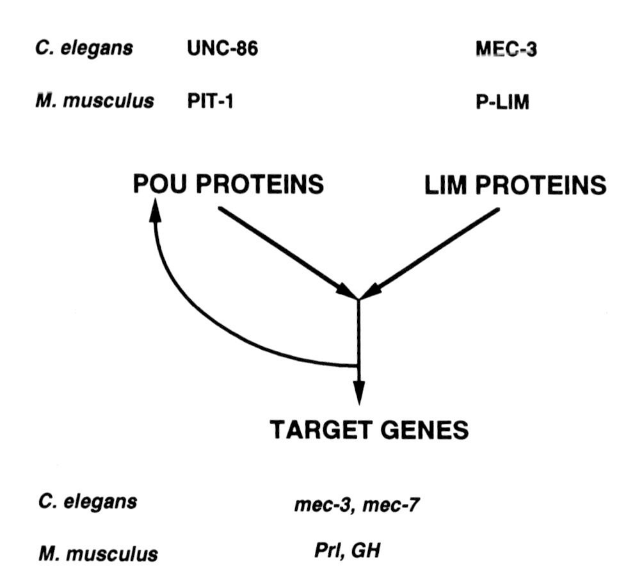

FIGURE 2–5 Cooperative interactions between POU and LIM transcription factors in neurogenesis. In mouse, PIT-1 and P-LIM interact to control the expression of the *prolactin (Prl)* and *growth hormone (GH)* genes, while in *C. elegans* UNC-86 and MEC-3 cooperate to control expression of *mec-3* and *mec-7.*

distinctive homeodomain as well as a separate DNA-binding domain called the paired region (Chalepakis et al., 1993). Binding sites for such genes have been identified in the promoter of the dopamine-β-hydroxylase (DBH) gene, an enzyme expressed in all neurons that synthesize the neurotransmitters norepinephrine and epinephrine (Tissier-Seta et al., 1993). Screening of expression libraries with a similar site from the NCAM gene identified a novel PHD gene, *Phox2,* that is expressed in precursors of all neurons that transcribe the *DBH* gene (Valarché et al., 1993; Zellmer et al., 1995). Whether *Phox2* is essential for the expression of *DBH* in vivo has not yet been reported, but this gene is able to enhance transcription from the DBH promoter in cotransfection assays (Tissier-Seta et al., 1993; Zellmer et al., 1995).

The identification of a vertebrate PHD gene as a potential regulator of the neuronal phenotype is of interest in light of the fact that genetic screens have provided evidence of a similar role for such genes in invertebrates. In *C. elegans,* for example, the gene *unc-4* encodes a PHD protein essential for the specification of VA-motoneurons (Miller et al., 1992). In *Drosophila,* the gene *pox-neuro* is required for the choice between polyinnervated and mono-innervated external sensory organs (Dambly-Chaudiere et al., 1992). (The POX-NEURO protein, however contains a paired domain but not a paired homeodomain.) Interestingly, *pox-neuro* behaves like a downstream target of *achaete-scute* (Vervoort et al., 1995), while *unc-4* may function downstream of the *C. elegans* bHLH gene *lin-32* (Zhao and Emmons, 1995) (a relative of *atonal*); similarly, in mammals *Phox2* appears to function downstream of *Mash1* (see below). Taken together, these data suggest that PHD genes may function downstream of bHLH genes to control the expression of aspects of the final neuronal phenotype (Fig. 2–6).

Pax genes comprise a large subfamily of PHD genes (Gruss and Walther, 1992). Like bHLH genes, *Pax* genes are expressed in restricted domains throughout the developing CNS and PNS, as well as sequentially within particular domains (suggesting that, like bHLH genes, they may function in cascades). LOF mutations in *Pax-3* and *Pax-6* produce spina bifida and a small eye phenotype, respectively (Chalepakis et al., 1993). *Pax-6* is the vertebrate orthologue of *Drosophila eyeless,* which is necessary and sufficient for the specification of compound eye development (Halder et al., 1995). While these data indicate that *Pax* genes play crucial roles in neurogenesis, their downstream targets, and therefore the precise aspects of the neuronal phenotype that they regulate, for the most part remain to be elucidated. However, there is evidence that some *Pax* genes may regulate expression of the neural cell adhesion molecule, NCAM, at least in transient transfection assays (Edelman and Jones, 1995).

NRSF/REST: a negative regulator of the neuronal phenotype

The consensus view of tissue-specific gene expression is that it is achieved primarily by specifically expressed, positive-acting factors (Johnson and McKnight, 1989; Mitchell and Tjian, 1989). An increasing number of neuronal genes, however, have been reported to contain negative regulatory elements, or silencers, that repress their transcription in non-neuronal cells (Mandel and McKinnon, 1993). A protein that binds to one such element, the neuron-restrictive silencer element (NRSE) (Kraner et al., 1992; Mori et al., 1992), has been isolated and is

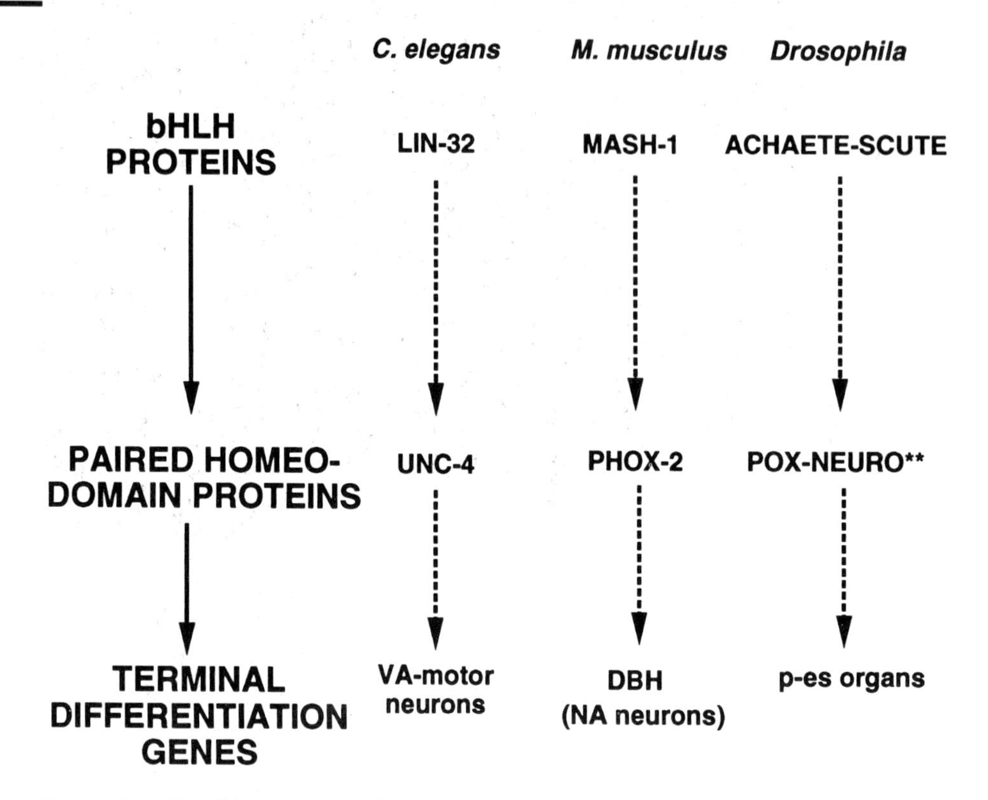

FIGURE 2–6 Possible sequential function of bHLH and paired homeodomain proteins in neuronal differentiation. The dashed arrows indicate hypothetical interactions based on genetic data. No evidence of direct transcriptional activation is yet available. **Pox-Neuro contains a paired domain but not a paired-like homeodomain. Note that different paired homeodomain genes (e.g., Pax genes) are also sequentially expressed in neurogenesis and may act in parallel as well as in series with bHLH cascades (Fig. 2–4).

called NRSF/REST (Chong et al., 1995; Schoenherr and Anderson, 1995a). NRSF/REST is a novel, large polypeptide containing nine zinc fingers and is expressed in many non-neuronal cells and tissues, but not in most neuronal cell types (Chong et al., 1995; Schoenherr and Anderson, 1995a). Computer database searches have indicated that consensus NRSEs are present five times more frequently in neuronal than in non-neuronal genes (Schoenherr et al., 1996). They are found in genes encoding proteins that contribute to virtually all aspects of the neuronal phenotype, including ion channels, receptor subunits, cytoskeletal proteins, synaptic vesicle components, neurotransmitter-synthesizing enzymes, and neuropeptides. Taken together, these data suggest that NRSF/REST may function as a "master" negative regulator that prevents ectopic expression of neuronal genes in non-neuronal cells (Schoenherr and Anderson, 1995b).

NRSF/REST mRNA is also detected in the ventricular zone of the embryonic CNS (Chong et al., 1995; Schoenherr and Anderson, 1995a), suggesting that relief from NRSF/REST-imposed repression may be important in the initial differentiation of neurons. This repression may occur directly, as suggested by the presence of NRSEs in numerous neuronal terminal differentiation genes. However, it may also occur indirectly, as suggested by the presence of an NRSE in at least one positive transcriptional activator, P-LIM (Schoenherr et al., 1996) (see above). Thus, NRSF/REST may inhibit neurogenesis by repressing positive regulators of neuronal genes. The function of NRSF/REST in neurogenesis awaits the analysis of GOF and LOF perturbations in this gene. In *Drosophila,* the *tramtrak* gene encodes a zinc finger protein that behaves genetically as a negative regulator of neuronal differentiation (Guo et al., 1995). However, there is no sequence similarity between TRAMTRAK and NRSF/REST.

Distinct Subprograms Specify Different Aspects of Neuronal Identity: Lessons from Sympathetic Neurons

As mentioned in the beginning of this chapter, the diversity of neuronal identities suggests that different aspects or components of these identities (e.g., neurotransmitter repertoire, receptor repertoire, etc.) may be assembled piecemeal through the action of distinct regulatory subprograms rather than being coordinately controlled by a unitary "master regulator" gene. Evidence consistent with this idea has come from a saturation mutagenesis analysis of one type of neuron, the HSN serotonergic motor neuron, in *C. elegans* (Desai et al., 1988). There is also evidence in support of this idea from the study of sympathetic neuron development in vertebrates.

Specifically, the expression of different aspects of the sympathetic neuron identity can be experimentally uncoupled by manipulating the environment of these neurons or their precursors. The first example of this was the demonstration that the neurotransmitter phenotype of morphologically differentiated, postmitotic neurons can be switched from noradrenergic to cholinergic by target-derived instructive factors (Patterson, 1978). In vivo, this switch normally occurs in a subpopulation of sympathetic neurons that innervates the eccrine sweat glands (Schotzinger and Landis, 1988, 1990). The exact signal that controls this switch has not yet been identified, but the available evidence suggests it is structurally related to ciliary neurotrophic factor (CNTF) (Rao et al., 1992).

The initial expression of the noradrenergic phenotype can also be uncoupled from the expression of pan-neuronal properties in developing neural crest cells. Surgical ablation of the notochord in chick embryos, which prevents formation of the floor plate (Chapter 8), prevents the initial expression of catecholamine fluorescence (Stern et al., 1991) and of noradrenergic biosynthetic enzymes such as tyrosine hydroxylase (TH) and DBH in neural crest cells aggregating in sympathetic ganglion primordia adjacent to the dorsal aorta (Groves et al., 1995). The expression of the PHD gene *Phox2,* which, as mentioned above, is thought to be a direct transcriptional regulator of DBH expression, is also prevented by this manipulation (Groves et al., 1995). Surprisingly, expression of CASH1 (the chick homologue of MASH1) is unaffected by this perturbation, as is expression of SCG10, a pan-neuronal marker (Groves et al., 1995). In a complementary experiment, neural crest cells cultured in the absence of inducing tissues (dorsal aorta and neural tube) differentiate into autonomic neurons that are MASH1 dependent, but do not express TH, DBH, or PHOX2 unless cocultured with these other tissues (A. Groves, D.J. Anderson, unpublished).

Taken together, these data suggest that the expression of the neurotransmitter identity and of other pan-neuronal aspects of sympathetic identity are separately controlled, both from inside and outside the cell. Pan-neuronal markers, such as SCG10, are controlled (directly or indirectly) by MASH1, whose expression is independent of signals derived from the notochord or floor plate. By contrast, expression of the genes that determine the neurotransmitter identity of the neurons is controlled by a different transcription factor (PHOX2) and is dependent upon signals from the notochord and/or floor plate (Fig. 2–7). Such a scheme makes sense in light of the fact that neurons throughout the autonomic system (sympathetic, parasympathetic, and a subset of enteric neurons) express and are dependent upon MASH1, whereas only subsets of these cells synthesize catecholamines.

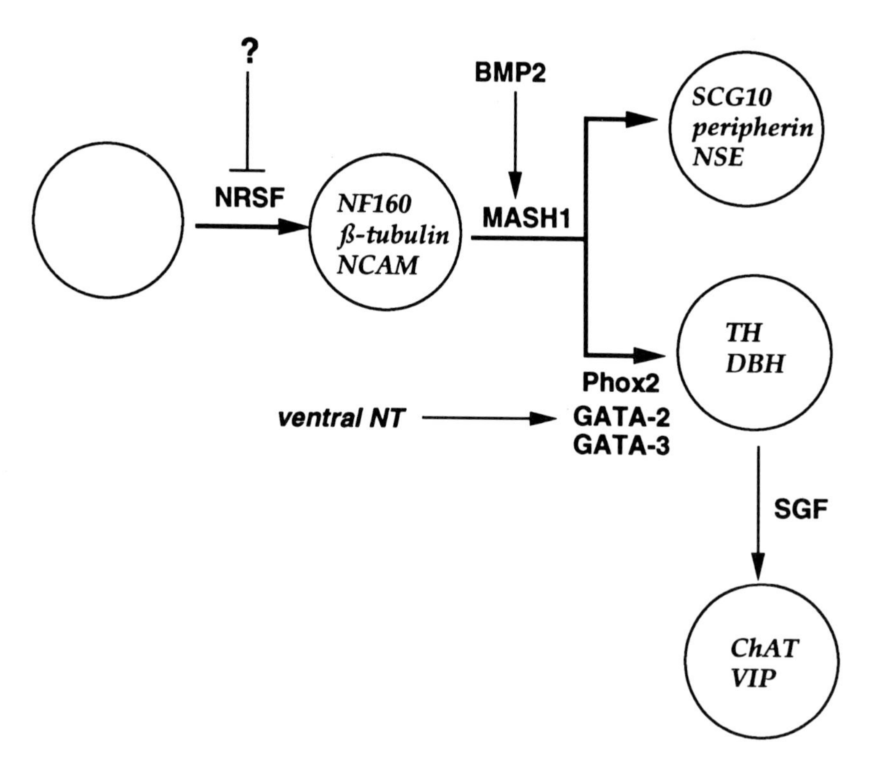

FIGURE 2–7 Regulatory circuits controlling expression of the sympathetic neuronal phenotype. The expression of pan-neuronal genes in sympathetic precursors occurs in at least two steps, one (leading to expression of NF160, β-tubulin, and NCAM) MASH1 independent and possibly dependent on repression of NRSF (Chong et al., 1995; Schoenherr and Anderson, 1995a), and the other (leading to expression of SCG10, peripherin, and NSE) MASH1 dependent (Sommer et al., 1995). Induction of MASH1 in turn is controlled by BMP2 (Shah et al., 1996), which is synthesized in the dorsal aorta. Expression of the transcription factors Phox2, GATA-2, and GATA-3 is dependent upon signals from the ventral neural tube (Groves et al., 1995) (as well as on MASH1) and is correlated with expression of the catecholamine biosynthetic enzymes TH and DBH, which contain binding sites for Phox2 (Tissier-Seta et al., 1993; Zellmer et al., 1995). SGF, sweat gland–derived factor which induces cholinergic properties in sympathetic neurons (Rao et al., 1992).

Subprograms controlling components of neuronal identity

component	character of each component	
a (e.g., neurotransmitter synthesis)	$a_1, a_2, a_3 \ldots a_i$	(ACh, NE, 5HT, etc.)
b (e.g., receptor expression)	$b_1, b_2, b_3 \ldots b_i$	(NMDA-R, Gly-R, etc.)
c	$c_1, c_2, c_3 \ldots c_i$	
. . .		
n	$n_1, n_2, n_3 \ldots n_i$	
k (pan-neuronal)	(constant)	(e.g., synapsin I, MAP2, NF160)

e.g.:

Neuron type 'A' = $a_2\, b_4\, c_5 \ldots n_i\,;\, k$

Neuron type 'B' = $a_1\, b_3\, c_2 \ldots n_j\,;\, k$

Alternative ways of combining different subprograms

(*x, y, z, w* = transcriptional regulators)

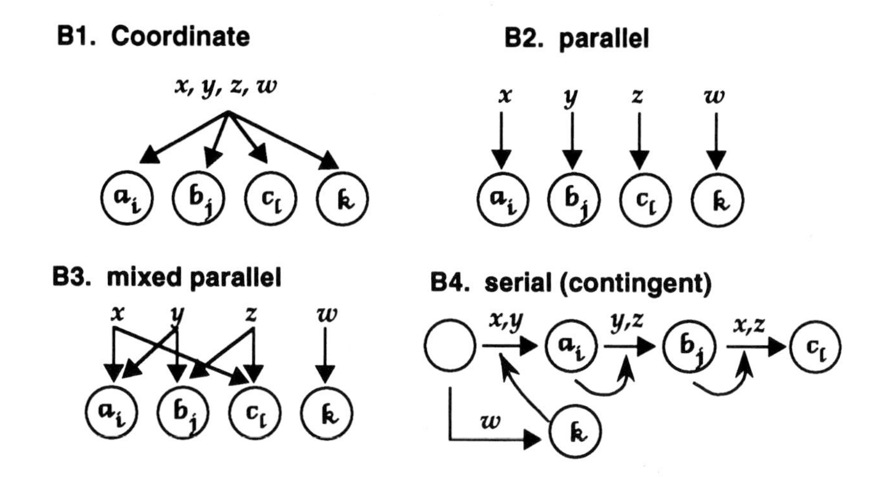

FIGURE 2–8 Schema illustrating the idea that different components of the neuronal phenotype may be specified "piecemeal" during neurogenesis. The nature of the components is arbitrary and chosen simply to illustrate the idea that each component may be regulated by a different set or subset of transcription factors (*x, y, z, w* in lower part). These transcription factors may function simultaneously or sequentially, and in series or in parallel. The extent to which the specification of different components of the phenotype is obligatorily coupled (e.g., B4) is not yet clear.

The determination of the pan-neuronal and neurotransmitter components of the sympathetic identity are not simply parallel, independent processes, however (Fig. 2–8, B2). In *Mash1* −/− mutants, sympathetic neurons do not express TH, DBH (Guillemot et al., 1993), or PHOX2 (L. Sommer, D.J. Anderson, unpublished). Thus the expression of the neurotransmitter phenotype requires both the action of MASH1 and of signals from the floor plate/notochord, presumably acting in part via induction of PHOX2 (Fig. 2–7). Whether MASH1 acts coordinately or sequentially with PHOX2 in determining expression of DBH and TH has not yet been determined.

Although induction of MASH1 in sympathetic precursors appears to be independent of signals from the floor plate/notochord, that does not mean it occurs cell-autonomously. In vitro, MASH1 expression is rapidly induced in undifferentiated neural crest cells by BMP2, which also induces autonomic neuronal differentiation (although not expression of DBH) (Shah et al., 1996). In vivo, BMP2 mRNA is expressed by endothelial cells lining the dorsal aorta at the time when migrating neural crest cells arrive at this structure (Bitgood and McMahon, 1995; Shah et al., 1996). Taken together, these data suggest that induction of MASH1 in neural crest cells may be promoted by BMP2, derived from the dorsal aorta, while induction of PHOX2, DBH, and TH occurs in response to signals from both the dorsal aorta and the floor plate/notochord (Groves et al., 1995).

The foregoing data illustrate how different aspects of the neuronal identity are controlled by converging parallel processes that may also involve feed-forward regulation (Fig. 2–8, B4). It has also become clear that although MASH1 is required for the expression of some genes in the pan-neuronal component of the sympathetic identity (e.g., SCG10), it is not required for the expression of all such genes. In *Mash1* −/− embryos, sympathetic precursors apparently are arrested in development after a stage at which they have already expressed numerous pan-neuronal markers, such as NF160, β-tubulin, NCAM, and tetanus toxin receptor (Sommer et al., 1995). Progression of these precursors to morphologically differentiated neurons expressing other genes such as SCG10, peripherin, and neuron-specific enolase requires MASH1 function. Thus, the expression of even the pan-neuronal component of the neuronal identity does not appear to involve a single genetic program but is itself assembled from different subprograms. The sequential activation of different subprograms that determine neuronal identity (Fig. 2–8, B4) is undoubtedly related to the cascades of transcription factors (e.g., bHLH proteins, PHD proteins) that are expressed during neurogenesis, but the molecular connections between these sequential processes have not yet been established.

PERSPECTIVE

Although a dizzying array of transcription factors and extracellular signals appear to be involved in neurogenesis, the recognition that different components of the neuronal identity may be specified by distinct but interconnected subprograms provides a framework for thinking about both the development and evolution of neuronal identity. Some components of neuronal identity will be shared by all classes of neurons (Fig. 2–8, "*k*"), such as expression of synaptic vesicle and cytoskeletal proteins. Others will be restricted to particular classes of neurons, such as neurotransmitter and receptor repertoire (Fig. 2–8, "*a, b, c . . . n*"). The specification of these components by distinct genetic subprograms means that different classes of neurons can be generated in development by a "mix-and-match" process in which the subprograms are implemented in different combinations. The fact that these subprograms may be independently regulated by distinct environmental signals provides a relatively easy way for such combinatorial manipulations to be achieved. However, the serial and possibly contingent implementation of some of these subprograms (Fig. 2–7 and Fig. 2–8, B4) suggests that either not all combinations of subprograms may be possible or at least that the subprograms have to fit together in a very particular way.

The determination of the neuronal identity by subprograms also provides an explanation for how new classes of neurons may have appeared during evolution. Specifically, mutations in the regulatory sequences controlling the genes that determine particular subprograms may have allowed these subprograms to be implemented in new combinations, leading to new neuronal identities. However, the apparent evolutionary conservation of certain regulatory relationships between genes controlling the neuronal phenotype (see Figs. 2-5 and 2-6) suggests that some powerful and ancient constraints may exist on the functional interactions between these genes. Whether this conservation actually sets constraints on the combinatorial usage of different subprograms depends on the functional relationships of these different classes of regulatory genes (POU domain, PHD, bHLH, LIM, etc.) to the phenotypic subprograms, something that remains to be defined.

In summary, our knowledge of the details of the underlying molecular genetic circuitry that determines the identity of any kind of neuron, in any organism, remains sketchy at best. However, a few general principles are beginning to emerge, particularly the idea that neuronal identity is implemented in series and in parallel as a collection of subprograms, rather than as one coordinately activated "master" program. Such notions will remain relatively trivial until the actual details of which genes regulate which targets and which extracellular signals in turn regulate these genes become defined. With such information in hand, it may be possible to more clearly define the repertoire of genes that comprise each "component" of neuronal identity and to clarify the logic whereby the subprograms that specify expression of these components are implemented.

References

Akazawa, C., Sasai, Y., Nakanishi, S., Kageyama, R. (1992). Molecular characterization of a rat negative regulator with a basic helix-loop-helix structure predominantly expressed in the developing nervous system. *J. Biol. Chem.* 21879–21885.

Akazawa, C., Ishibashi, M., Shimizu, C., Nakanishi, S., Kageyama, R. (1995). A mammalian helix-loop-helix factor structurally related to the product of *Drosophila* proneural gene *atonal* is a positive transcriptional regulator expressed in the developing nervous system. *J. Biol. Chem.* 270:8730–8738.

Andersen, B., Rosenfeld, M.G. (1994). Pit-1 determines cell types during development of the anterior pituitary gland—a model for transcriptional regulation of cell phenotypes in mammalian organogenesis. *J. Biol. Chem.* 269:29335–29338.

Anderson, D.J. (1995). Spinning skin into neurons. *Curr. Biol.* 5:1235–1238.

Artavanis-Tsakonas, S., Matsuno, K., Fortini, M.E. (1995). Notch signaling. *Science* 268: 225–232.

Bach, I., Rhodes, S.J., Pearse, R.V., Heinzel, T., Gloss, B., Scully, K.M., Swachenko, P.E., Rosenfeld, M.G. (1995). P-Lim, a LIM homeodomain factor, is expressed during pituitary organ and cell commitment and synergizes with Pit-1. *Proc. Natl. Acad. Sci. U.S.A.* 92:2720–2724.

Bang, A.G., Bailey, A.M., Posakony, J.W. (1995). *Hairless* promotes stable commitment to the Sensory Organ Precursor cell fate by negatively regulating the activity of the *Notch* signaling pathway. *Dev. Biol.* 172:479–494.

Begley, C.G., Lipkowitz, S., Göbel, V., Mahon, K.A., Bertness, V., Green, A.R., Gough, N.M., Kirsch, I.R. (1992). Molecular characterization of NSCL, a gene encoding a helix-loop-helix protein expressed in the developing nervous system. *Proc. Natl. Acad. Sci. U.S.A.* 89:38–42.

Benezra, R., Davis, R.L., Lockshon, D., Turner, D.L., Weintraub, H. (1990). The protein Id: a negative regulator of helix-loop-helix DNA binding proteins. *Cell* 61:49–59.

Bier, E., Vaessin, H., Shepherd, S., Lee, K., McCall, K., Barbel, S., Ackerman, L., Carretto, R., Uemura, T., Grell, E., Jan, L.Y., Jan, Y.N. (1989). Searching for pattern and mutation in the *Drosophila* genome with a P-lacZ vector. *Genes Dev.* 3:1273–1287.

Bier, E., Vaessin, H., Younger-Shepherd, S., Jan, L.Y., Jan, Y.N. (1992). *deadpan*, an essential pan-neural gene in *Drosophila*, encodes a helix-loop-helix protein similar to the *hairy* gene product. *Genes Dev.* 6:2137–2151.

Bitgood, M.J., McMahon, A.P. (1995). *Hedgehog* and *Bmp* genes are coexpressed at many diverse sites of cell-cell interaction in the mouse embryo. *Dev. Biol.* 172:126–138.

Blochlinger, K., Bodmer, R., Jack, J., Jan, L.Y., Jan, Y.N. (1988). Primary structure and expression of a product from *cut*, a locus involved in specifying sensory organ identity in *Drosophila*. *Nature* 333:629–635.

Bodner, M., Castrillo, J.-L., Theill, L.E., Deerinck, T., Ellisman, M., Karin, M. (1988). The pituitary-specific transcription factor GHF-1 is a homeobox-containing protein. *Cell* 55:505–518.

Brand, M., Jarman, A.P., Jan, L.Y., Jan, Y.-N. (1993). *asense* is a *Drosophila* neural precursor gene and is capable of initiating sense organ formation. *Development* 119:1–17.

Brown, J.L., Sonoda, S., Ueda, H., Scott, M., Wu, C. (1991). Repression of the *Drosophila fushi tarazu* (*ftz*) segmentation gene. *EMBO J.* 10:665–674.

Cabrera, C.V., Alonso, M.C. (1987). The expression of three members of the *achaete-scute* gene complex correlates with neuroblast segregation in *Drosophila*. *Cell* 50:425–433.

Cabrera, C.V., Alonso, M.C. (1991). Transcriptional activation by heterodimers of the *achaete-scute* and *daughterless* gene products of *Drosophila*. *EMBO J.* 10:2965–2973.

Campos, A.R., Rosen, D.R., Robinow, S.N., White, K. (1987). Molecular analysis of the locus *elav* in *Drosophila melanogaster:* a gene whose embryonic expression is neural specific. *EMBO J.* 6:425–431.

Campos-Ortega, J. (1993). Early neurogenesis in *Drosophila melanogaster.* In *The Development of Drosophila melanogaster,* Bate, M., Martinez-Arias, A., eds. Cold Spring Harbor, NY: Cold Spring Harbor Laboratory Press, pp. 1091–1129.

Campuzano, S., Modolell, J. (1992). Patterning of the *Drosophila* nervous system: the *achaete-scute* gene complex. *Trends Genet.* 8:202–208.

Chalepakis, G., Stoykova, A., Wijnholds, J., Tremblay, P., Gruss, P. (1993). Pax: gene regulators in the developing nervous system. *J. Neurobiol.* 24:1367–1384.
Chenn, A., McConnell, S.K. (1995). Cleavage orientation and the asymmetric inheritance of Notch1 immunoreactivity in mammalian neurogenesis. *Cell* 82:631–641.
Chien, C.T., Hsiao, C.D., Jan, L.Y., Jan, Y.N. (1996). Neuronal type information encoded in the basic-helix-loop-helix domain of proneural genes. *Proc. Natl. Acad. Sci. USA* 93: 13239–13244.
Chitnis, A., Henrique, D., Lewis, J., Ish-Horowicz, D., Kintner, C. (1995). Primary neurogenesis in *Xenopus* embryos regulated by a homologue of the *Drosophila* neurogenic gene *Delta*. *Nature* 375:761–766.
Chitnis, A. and Kintner, C. (1996). Sensitivity of proneural genes to lateral inhibition affects the pattern of primary neurons in xenopus embryos. *Development* 122:2295–2301.
Chong, J.A., Tapia-Ramirez, J., Kim, S., Toledo-Aral, J.J., Zheng, Y., Boutros, M.C., Altshuller, Y.M., Frohman, M.A., Kraner, S.D., Mandel, G. (1995). REST: a mammalian silencer protein that restricts sodium channel expression to neurons. *Cell* 80:949–957.
Cserjesi, P., Brown, D., Lyons, G.E., Olson, E.N. (1995). Expression of the novel basic helix-loop-helix gene *eHAND* in neural crest derivatives and extraembryonic membranes during mouse development. *Dev. Biol.* 170:664–678.
Cubas, P., Modolell, J. (1992). The *extramacrochaetae* gene provides information for sensory organ patterning. *EMBO J.* 11:3385–3395.
Dambly-Chaudiere, C., Ghysen, A. (1987). Independent subpatterns of sense organs require independent genes of the *achaete-scute* complex in *Drosophila* larvae. *Genes Dev.* 1:297–306.
Dambly-Chaudiere, C., Jamet, E., Burri, M., Bopp, D., Basler, K., Hafen, E., Dumont, N., Spielmann, P., Ghysen, A., Noll, M. (1992). The paired box gene *pox neuro:* a determinant of poly-innervated sense organ in *Drosophila*. *Cell* 69:159–172.
DeRobertis, E.M., Sasai, Y. (1996). A common plan for dorsoventral patterning in Bilateria. *Nature* 380:37–40.
Desai, C., Garriga, G., McIntire, S.L., Horvitz, H.R. (1988). A genetic pathway for the development of the *Caenorhabditis elegans* HSN motor neurons. *Nature* 336:638–646.
Doe, C.Q., Chu-LaGraff, Q., Wright, D.M., Scott, M.P. (1991). The *prospero* gene specifies cell fates in the *Drosophila* central nervous system. *Cell* 65:451–464.
Domínguez, M., Campuzano, S. (1993). *asense,* a member of the *Drosophila achaete-scute* complex, is a proneural and neural differentiation gene. *EMBO J.* 12:2049–2060.
Duncan, M., DiCicco-Bloom, E.M., Xiang, X., Benezra, R., Chada, K. (1992). The gene for the helix-loop-helix protein, Id, is specifically expressed in neural precursors. *Dev. Biol.* 154:1–10.
Edelman, G.M., Jones, F.S. (1995). Developmental control of N-CAM expression by Hox and Pax gene-products. *Philos. Trans. R. Soc. Lond. Biol.* 349:305–312.
Ericson, J., Thor, S., Edlund, T., Jessell, T.M., Yamada, T. (1992). Early stages of motor neuron differentiation revealed by expression of homeobox gene *Islet-1*. *Science* 256:1555–1560.
Ferreiro, B., Skoglund, P., Bailey, A., Dorsky, R., Harris, W. (1992). *XASH1,* a *Xenopus* homolog of *achaete-scute;* a proneural gene in anterior regions of the vertebrate CNS. *Mech. Dev.* 40:25–36.
Ferreiro, B., Kintner, C., Zimmerman, K., Anderson, D., Harris, W.A. (1994). *XASH* genes promote neurogenesis in *Xenopus* embryos. *Development* 120:3649–3655.
François, V., Bier, E. (1995). *Xenopus chordin* and *Drosophila short gastrulation* genes encode homologous proteins functioning in Dorsal-Ventral axis formation. *Cell* 80:19–20.
Ghysen, A., Dambly-Chaudiere, C. (1988). From DNA to form: the *achaete-scute* complex. *Genes Dev.* 7:723–733.
Ghysen, A., Dambly-Chaudiere, C., Jan, L.Y., Jan, Y.-N. (1993). Cell interactions and gene interactions in peripheral neurogenesis. *Genes Dev.* 7:723–733.

Ghysen, A., Dambly-Chaudiere, C. (1989). Genesis of the *Drosophila* peripheral nervous system. *Trends Genet.* 5:251–255.

Ghysen, A., O'Kane, C. (1989). Neural enhancer-like elements as specific cell markers in *Drosophila. Development* 105:35–52.

Göbel, V., Lipkowitz, S., Kozak, C.A., Kirsch, I.R. (1992). *NSCL-2:* a basic domain helix-loop-helix gene expressed in early neurogenesis. *Cell Growth Diff.* 3:143–148.

Góriely, A., Dumont, N., Dambly-Chaudiere, C., Ghysen, A. (1991). The determination of sense organs in *Drosophila:* effect of the neurogenic mutations in the embryo. *Development* 113:1395–1404.

Groves, A.K., George, K.M., Tissier-Seta, J.-P., Engel, J.D., Brunet, J.-F., Anderson, D.J. (1995). Differential regulation of transcription factor gene expression and phenotypic markers in developing sympathetic neurons. *Development* 121:887–901.

Gruss, P., Walther, C. (1992). Pax in development. *Cell* 69:719–722.

Guillemot, F., Joyner, A.L. (1993). Dynamic expression of the murine *Achaete-Scute* homolog *Mash-1* in the developing nervous system. *Mech. Dev.* 42:171–185.

Guillemot, F., Lo, L.-C., Johnson, J.E., Auerbach, A., Anderson, D.J., Joyner, A.L. (1993). Mammalian achaete-scute homolog-1 is required for the early development of olfactory and autonomic neurons. *Cell* 75:463–476.

Guo, M., Jan, L.Y., Jan, Y.N. (1996). Control of daughter cell fate during asymmetric division: interaction of Numb and Notch. *Neuron* 17–41.

Guo, M., Bier, E., Jan, L.Y., Jan, Y.N. (1995). *tramtrack* acts downstream of *numb* to specify distinct daughter cell fates during asymmetric cell divisions in the Drosophila PNS. *Neuron* 14:913–925.

Halder, G., Callaerts, P., Gehring, W.J. (1995). Induction of ectopic eyes by targeted expression of the eyeless gene in *Drosophila. Science* 267:1788–1792.

Hamelin, M., Scott, I.M., Way, J.C., Culotti, J.G. (1992). The *mec-7* β-tubulin gene of *C. elegans* is expressed primarily in the touch receptor neurons. *EMBO J.* 11:2885–2893.

Harrison, S., Travers, A. (1990). The *tramtrack* gene encodes a *Drosophila* finger protein that interacts with the *ftz* transcriptional regulatory region and shows a novel embryonic expression pattern. *EMBO J.* 9:207–216.

Hartenstein, V., Posakony, J.W. (1990). A dual function of the *Notch* gene in *Drosophila* sensillum development. *Dev. Biol.* 142:13–30.

He, X., Rosenfeld, M.G. (1991). Mechanisms of complex transcriptional regulation: implications for brain development. *Neuron* 7:183–196.

Heitzler, P., Simpson, P. (1991). The choice of cell fate in the epidermis of *Drosophila. Cell* 64:1083–1092.

Heitzler, P., Bourouis, M., Ruel, L., Carteret, C., Simpson, P. (1996). Genes of the *Enhancer of split* and *achaete-scute* complexes are required for a regulatory loop between *Notch* and *Delta* during lateral signalling in *Drosophila. Development* 122:161–171.

Hirata, J., Nakagoshi, H., Nabeshima, Y.-I., Matsuzaki, F. (1995). Asymmetric segregation of the homeodomain protein Prospero during *Drosophila* development. *Nature* 377:627–630.

Hollenberg, S.M., Sternglanz, R., Cheng, P.F., Weintraub, H. (1995). Identification of a new family of tissue-specific basic helix-loop-helix proteins with a two-hybrid system. *Mol. Cell. Biol.* 15:3813–3822.

Hosoya, T., Takizawa, K., Nitta, K., Hotta, Y. (1995). *glial cells missing:* a binary switch between neuronal and glial determination in *Drosophila. Cell* 82:1025–1036.

Ingraham, H.A., Chen, R., Mangalam, H.J., Elsholtz, H.P., Flynn, S.E., Lin, C.R., Simmons, D.M., Swanson, L., Rosenfeld, M.G. (1988). A tissue-specific transcription factor containing a homeodomain specifies a pituitary phenotype. *Cell* 55:519–529.

Ingraham, H.A., Albert, V.R., Chen, R., Crenshaw, E.B.I., Elsholtz, H.P., He, X., Kapiloff, M.S., Mangalam, H.J., Swanson, L.W., Treacy, M.N., Rosenfeld, M.G. (1990). A family of POU-domain and pit-1 tissue-specific transcription factors in pituitary and neuroendocrine development. *Annu. Rev. Physiol.* 52:773–791.

Ishibashi M., Sasai, Y., Nakanishi, S., Kageyama, R. (1993). Molecular characterization of HES-2, a mammalian helix-loop-helix factor structurally related to *Drosophila hairy* and *Enhancer of split*. *Eur. J. Biochem.* 215:645–652.

Ishibashi, M., Moriyoshi, K., Sasai, Y., Shiota, K., Nakanishi, S., Kageyama, R. (1994). Persistent expression of helix-loop-helix factor HES-1 prevents mammalian neural differentiation in the central nervous system. *EMBO J.* 13:1799–1805.

Ishibashi, M., Ang, S.-L., Shiota, K., Nakanishi, S., Kageyama, R., Guillemot, F. (1995). Targeted disruption of mammalian *hairy* and *Enhancer of split* homolog-1 (*HES-1*) leads to up-regulation of neural helix-loop-helix factors, premature neurogenesis, and severe neural tube defects. *Genes Dev.* 9:3136–3148.

Jan, Y.N., Jan, L.Y. (1993). HLH proteins, fly neurogenesis and vertebrate myogenesis. *Cell* 75:827–830.

Jan, Y.N., Jan, L.Y. (1994). Genetic control of cell fate specification in *Drosophila* in peripheral nervous system. *Annu. Rev. Genet.* 28:373–393.

Jarman, A.P., Brand, M., Jan, L.Y., Jan, Y.N. (1993a). The regulation and function of the helix-loop-helix gene, *asense*, in *Drosophila* neural precursors. *Development* 119:19–29.

Jarman, A.P., Grau, Y., Jan, L.Y., Jan, Y.-N. (1993b). *atonal* is a proneural gene that directs chordotonal organ formation in the Drosophila peripheral nervous system. *Cell* 73:1307–1321.

Jarman, A.P., Grell, E.H., Ackerman, L., Jan, L.Y., Jan, Y.N. (1994). *atonal* is the proneural gene for *Drosophila* photoreceptors. *Nature* 369:398–400.

Jennings, B., Preiss, A., Delidakis, C., Bray, S. (1994). The Notch signalling pathway is required for *Enhancer of split* bHLH protein expression during neurogenesis in the *Drosophila* embryo. *Development* 120:3537–3548.

Jennings, B., deCelis, J., Delidakis, C., Preiss, A., Bray, S. (1995). Role of *Notch* and *achaete-scute* complex in the expression of *Enhancer of split* bHLH proteins. *Development* 121: 3745–3752.

Jiménez, F., Campos-Ortega, J.A. (1990). Defective neuroblast commitment in mutants of the *achaete-scute* complex and adjacent genes of *Drosophila melanogaster. Neuron* 5:81–89.

Jiménez, F., Martin-Morris, L.E., Velasco, L., Chu, H., Sierra, J., Rosen, D.R., White, K. (1995). *vnd,* a gene required for early neurogenesis of *Drosophila,* encodes a homeodomain protein. *EMBO J.* 14:3487–3495.

Johnson, J.E., Birren, S.J., Anderson, D.J. (1990). Two rat homologs of *Drosophila achaete-scute* specifically expressed in neuronal precursors. *Nature* 346:858–861.

Johnson, J.E., Zimmerman, K., Saito, T., Anderson, D.J. (1992). Induction and repression of mammalian achaete-scute homolog (MASH) gene expression during neuronal differentiation of P19 embryonal carcinoma cells. *Development* 114:75–87.

Johnson, P.F., McKnight, S.L. (1989). Eukaryotic transcriptional regulatory proteins. *Annu. Rev. Biochem.* 58:799–839.

Johnson, W.A., Hirsh, J. (1990). Binding of a *Drosophila* POU-domain protein to a sequence element regulating gene expression in specific dopaminergic neurons. *Nature* 343: 467–470.

Jones, B.W., Fetter, R.D., Tear, G., Goodman, C.S. (1995). *glial cells missing:* a genetic switch that controls glial versus neuronal fate. *Cell* 82:1013–1023.

Knoblich, J.A., Jan, L.Y., Jan, Y.N. (1995). Asymmetric segregation of Numb and Prospero during cell division. *Nature* 377:624–627.

Kraner, S.D., Chong, J.A., Tsay, H.J., Mandel, G. (1992). Silencing the type II sodium channel gene: a model for neural-specific gene regulation. *Neuron* 9:37–44.

Kraut, R., Campos-Ortega, J.A. (1996). *inscuteable,* a neural precursor gene of *Drosophila,* encodes a candidate for a cytoskeleton adaptor protein. *Dev. Biol.* 174:65–81.

Kraut, R., Chia, W., Jan, L.Y., Jan, Y.N., Knoblich J. (1996). Role of *inscuteable* in orienting asymmetric cell divisions in *Drosophila Nature* 383:50–55.

Lee, J.E., Hollenberg, S.M., Snider, L., Turner, D.L., Lipnick, N., Weintraub, H. (1995).

Conversion of *Xenopus* extoderm into neurons by NeuroD, a basic helix-loop-helix protein. *Science* 268:836–844.

Li, S., Crenshaw, E.B., Rawson, E.J., Simmons, D.M., Swanson, L.W., Rosenfeld, M.G. (1990). Dwarf locus mutants lacking 3 pituitary cell-types result from mutations in the POU-domain gene Pit-1. *Nature* 347:528–533.

Lo, L., Johnson, J.E., Wuenschell, C.W., Saito, T., Anderson, D.J. (1991). Mammalian *achaete-scute* homolog 1 is transiently expressed by spatially-restricted subsets of early neuroepithelial and neural crest cells. *Genes Dev.* 5:1524–1537.

Ma, Q., Kintner, C., Anderson, D.J. (1996). *Neurogenin,* a vertebrate neuronal determination gene, acts upstream of *NeuroD* in a cascade. *Cell, 87:*43–52.

Mandel, G., McKinnon, D. (1993). Molecular basis of neural-specific gene expression. *Annu. Rev. Neurosci.* 16:323–345.

Martín-Bermundo, M., Martinez, C., Rodríguez, A., Jiménez, F. (1991). Distribution and function of the *lethal of scute* gene product during early neurogenesis in *Drosophila. Development* 113:445–454.

Michelson, A.M., Abmayr, S.M., Bate, M., Arias, A.M., Maniatis, T. (1990). Expression of a MyoD family member prefigures muscle pattern in *Drosophila* embryos. *Genes Dev.* 4:2086–2097.

Miller, D.M., Shen, M.M., Shamu, C.E., Burglin, T.R., Ruvkun, G., Dubois, M.L., Ghee, M., Wilson, L. (1992). *C. elegans unc-4* gene encodes a homeodomain protein that determines the pattern of synaptic input to specific motor neurons. *Nature* 355:841–845.

Mitani, S., Du, H., Hall, D.H., Driscoll, M., Chalfie, M. (1993). Combinatorial control of touch receptor neuron expression in *Caenorhabditis elegans. Development* 119:773–783.

Mitchell, P., Tjian, R. (1989). Transcriptional regulation in mammalian cells by sequence-specific DNA binding proteins. *Science* 245:371–378.

Mori, N., Schoenherr, C., Vandenbergh, D.J., Anderson, D.J. (1992). A common silencer element in the SCG10 and type II Na^+ channel genes binds a factor present in nonneuronal cells but not in neuronal cells. *Neuron* 9:1–10.

Moses, C., Ellis, M.C., Rubin, G.M. (1989). The *glass* gene encodes a zinc-finger protein required by *Drosophila* photoreceptor cells. *Nature* 340:531–536.

Murre, C., McCaw, P.S., Baltimore, D. (1989a). A new DNA binding and dimerization motif in immunoglobin enhancer binding, *daughterless, MyoD* and *myc* proteins. *Cell* 56:777–783.

Murre, C., McCaw, P.S., Vaessin, H., Caudy, M., Jan, L.Y., Jan, Y.N., Cabrera, C.V., Buskin, J.N., Hauschka, S.D., Lassar, A.B., Weintraub, H., Baltimore, D. (1989b). Interactions between heterologous helix-loop-helix proteins generate complexes that bind specifically to a common DNA sequence. *Cell* 58:537–544.

Nakao, K., Campos-Ortega, J.A. (1996). Persistent expression of genes of the *Enhancer of Split* complex suppresses neural development in *Drosophila. Neuron* 16:275–286.

Nambu, J.R., Lewis, J.O., Wharton, K.A., Crews, S.T. (1991). The *Drosophila single minded* gene encodes a helix-loop-helix protein that acts as a master regulator of CNS midline development. *Cell* 67:1157–1167.

O'Kane, C., Gehring, W. (1987). Detection in situ of genomic regulatory elements in *Drosophila. Proc. Natl. Acad. Sci. U.S.A.* 84:9123–9127.

Ohsako, S., Hyer, J., Panganiban, G., Oliver, I., Caudy, M. (1994). *hairy* function as a DNA-binding helix-loop-helix repressor of *Drosophila* sensory organ formation. *Genes Dev.* 8:2743–2755.

Parks, A.L., Muskavitch, M.A.T. (1993). *Delta* function is required for bristle organ determination and morphogenesis in *Drosophila. Dev. Biol.* 157:484–496.

Patterson, P.H. (1978). Environmental determination of autonomic neurotransmitter functions. *Annu. Rev. Neurosci.* 1:1–17.

Posakony, J.A. (1994). Nature versus nurture: asymmetric cell divisions in *Drosophila* bristle development. *Cell* 76:415–418.

Rao, M.S., Patterson, P.H., Landis, S.C. (1992). Multiple cholinergic differentiation factors are present in footpad extracts: comparison with known cholinergic factors. *Development* 116:731–744.

Read, D., Levine, M., Manley, J. (1992). Ectopic expression of the *Drosophila tramtrack* gene results in multiple embryonic defects, including repression of *even-skipped* and *fushi tarazu*. *Mech. Dev.* 38:183–196.

Rhyu, M., Jan, L.Y., Jan, Y.N. (1994). Asymmetric distribution of numb protein during division of the sensory organ precursor cell confers distinct fates to daughter cells. *Cell* 76:477–491.

Roark, M., Sturtevant, M.A., Emery, J., Vaessin, H., Grell, E., Bier, E. (1995). *scratch*, a panneural gene encoding a zinc finger protein related to *snail*, promotes neuronal development. *Genes Dev.* 9:2384–2398.

Ruiz-Gómez, M., Ghysen, A. (1993). The expression and role of a proneural gene, *achaete*, in the development of the larval nervous system of *Drosophila*. *EMBO J.* 12:1121–1130.

Sasai, Y., Kageyama, R., Tagawa, Y., Shigemoto, R., Nakanishi, S. (1992). Two mammalian helix-loop-helix factors structurally related to *Drosophila hairy* and *enhancer of split*. *Genes Dev.* 6:2620–2634.

Sasai, Y., Lu, B., Steinbeisser, H., DeRobertis, E.M. (1995). Regulation of neural induction by the Chd and Bmp-4 antagonistic patterning signals in *Xenopus*. *Nature* 376:333–336.

Schoenherr, C.J., Anderson, D.J. (1995a). The neuron-restrictive silencer factor (NRSF): a coordinate repressor of multiple neuron-specific genes. *Science* 267:1360–1363.

Schoenherr, C.J., Anderson, D.J. (1995b). Silencing is golden: negative regulation in the control of neuronal gene transcription. *Curr. Opin. Neurobiol.* 5:566–571.

Schoenherr, C.J., Paquette, A.J., Anderson, D.J. (1996). Identification of potential target genes for the neuron-restrictive silencer factor. *Proc. Natl. Acad. Sci. U.S.A.* 93:9881–9886.

Schotzinger, R., Landis, S.C. (1988). Cholinergic phenotype developed by noradrenergic sympathetic neurons after innervation of a novel cholinergic target in vivo. *Nature* 335:637–639.

Schotzinger, R.J., Landis, S.C. (1990). Acquisition of cholinergic and peptidergic properties by sympathetic innervation of rat sweat glands requires interaction with normal target. *Neuron* 5:91–100.

Schweisguth, F. (1995). *Suppressor of Hairless* is required for signal reception during lateral inhibition in the *Drosophila* pupal notum. *Development* 121:1875–1884.

Schweisguth, F., Posakony, J.W. (1994). Antagonistic activities of *Suppressor of Hairless* and *Hairless* control alternative cell fates in *Drosophila* adult epidermis. *Development* 120: 1433–1441.

Shah, N., Groves, A., Anderson, D.J. (1996). Alternative neural crest cell fates are instructively promoted by TGFβ superfamily members. *Cell*, 85:331–343.

Shimizu, C., Akazawa, C., Nakanishi, S., Kageyama, R. (1995). MATH-2, a mammalian helix-loop-helix factor structurally related to the product of *Drosophila* proneural gene *atonal*, is specifically expressed in the nervous system. *Eur. J. Biochem.* 229:239–248.

Skeath, J.B., Panganiban, G., Selegue, J., Carroll, S.B. (1992). Gene regulation in two dimensions: the proneural *achaete* and *scute* genes are controlled by combinations of axis-patterning genes through a common intergenic control region. *Genes Dev.* 6:2606–2619.

Sommer, L., Shah, N., Rao, M., Anderson, D.J. (1995). The cellular function of MASH1 in autonomic neurogenesis. *Neuron* 15:1245–1258.

Spana, E.P., Doe, C.Q. (1995). The *prospero* transcription factor is asymmetrically localized to the cell cortex during neuroblast mitosis in *Drosophila*. *Development* 121:3187–3195.

Spana, E.P., Doe, C.Q. (1996). Numb antagonizers Notch signalling to specify sibling neuron cell fates. *Neuron* 17:21–26.

Spana, E.P., Kopczynski, C., Goodman, C.S., Doe, C.Q. (1995). Asymmetric localization of *numb* autonomously determines sibling neuron identity in the *Drosophila* CNS. *Development* 121:3489–3494.

Stern, C.D., Artinger, K.B., Bronner-Fraser, M. (1991). Tissue interactions affecting the migration and differentiation of neural crest cells in the chick embryo. *Development* 113: 207–216.

Tata, F., Hartley, D.A. (1995). Inhibition of cell fate in *Drosophila* by *Enhancer of split* genes. *Mech. Dev.* 51:305–315.

Tissier-Seta, J.-P., Hirsch, M.-R., Valarché, I., Brunet, J.-F., Goridis, C. (1993). A possible link between cell adhesion receptors, homeodomain proteins and neuronal identity. *C.R. Acad. Sci. Paris* 316:1306–1315.

Tsuchida, T., Ensini, M., Morton, S.B., Baldassare, M., Edlund, T., Jessell, T.M., Pfaff, S.L. (1994). Topographic organization of embryonic motor neurons defined by expression of LIM homeobox genes. *Cell* 79:957–970.

Turner, D.L., Weintraub, H. (1994). Expression of *achaete-scute* homolog 3 in *Xenopus* embryos converts ectodermal cells to a neural fate. *Genes Dev.* 8:1434–1447.

Uemura, T., Shepherd, S., Ackerman, L., Jan, L.Y., Jan, Y.N. (1989). *numb,* a gene required in determination of cell fate during sensory organ formation in *Drosophila* embryos. *Cell* 58:349–360.

Vaessin, H., Grell, E., Wolff, E., Bier, E., Jan, L.Y., Jan, Y.N. (1991). prospero is expressed in neuronal precursors and encodes a nuclear protein that is involved in the control of axonal outgrowth in *Drosophila. Cell* 67:941–953.

Vaessin, H., Brand, M., Jan, L.Y., Jan, Y.N. (1994). daughterless is essential for neuronal precursor differentiation but not for initiation of neuronal precursor formation in *Drosophila* embryo. *Development* 120:935–945.

Valarché, I., Tissier-Seta, J.-P., Hirsch, M.-R., Martinez, S., Goridis, C., Brunet, J.-F. (1993). The mouse homeodomain protein Phox2 regulates NCAM promoter activity in concert with Cux/CDP and is a putative determinant of neurotransmitter phenotype. *Development* 119:881–896.

VanDoren, M., Ellis, H.M., Posakony, J.W. (1991). The *Drosophila* extramacrochaetae protein antagonize sequence-specific DNA binding by the daughterless/achaete-scute protein complexes. *Development* 113:245–255.

VanDoren, M., Powell, P.A., Pasternak, D., Singson, A., Posakony, J.W. (1992). Spatial regulation of proneural gene activity: auto- and cross-activation of *achaete* is antagonized by *extramacrochaete. Genes Dev.* 6:2592–2605.

VanDoren, M., Bailey, A.M., Esnayra, J., Ede, K., Posakony, J.W. (1994). Negative regulation of proneural gene activity: hairy is a direct transcriptional repressor of *achaete. Genes Dev.* 8:2729–2742.

Vervoort, M., Zink, D., Pujol, N., Victoir, K., Dumont, N., Ghysen, A., Dambly-Chaudiere, C. (1995). Genetic determinants of sense organ identity in *Drosophila:* regulatory interactions between *cut* and *poxn. Development* 121:3111–3120.

Way, J.C., Chalfie, M. (1988). *mec-3,* a homeobox-containing gene that specifies differentiation of the touch receptor neurons in C. elegans. *Cell* 54:5–16.

Weintraub, H. (1993). The MyoD family and myogenesis: redundancy, networks and thresholds. *Cell* 75:1241–1244.

Wilson, P.A., Hemmati-Brivanlou, A. (1995). Induction of epidermis and inhibition of neural fate by Bmp-4. *Nature* 376:331–333.

Xue, D., Tu, Y., Chalfie, M. (1993). Cooperative interactions between the *Caenorhabditis elegans* homeoproteins UNC-86 and MEC-3. *Science* 261:1324–1328.

Younger-Shepherd, S., Vaessin, H., Bier, E., Jan, L.Y., Jan, Y.N. (1992). *deadpan,* an essential pan-neural gene encoding an HLH protein, acts as a denominator in *Drosophila* sex determination. *Cell* 70:911–922.

Zellmer, E., Zhang, Z., Greco, D., Rhodes, J., Cassel, S., Lewis, E.J. (1995). A homeodomain protein selectively expressed in noradrenergic tissue regulates transcription of neurotransmitter biosynthetic genes. *J. Neurosci.* 15:8109–8120.

Zhao, C., Emmons, S.W. (1995). A transcription factor controlling development of peripheral sense organs in *C. elegans*. *Nature* 373:74–78.

Zhong, W., Feder, J.N., Jiang, M., Jan, L.Y., Jan, Y.N. (1996). Asymmetric localization of a mammalian Numb homolog during mouse cortical neurogenesis. *Neuron* 17:43–53.

Zimmerman, K., Shih, J., Bars, J., Collazo, A., Anderson, D.J. (1993). *XASH-3,* a novel *Xenopus achaete-scute* homolog, provides an early marker of planar neural induction and position along the medio-lateral axis of the neural plate. *Development* 119:221–232.

3

Neuron–glial interactions

BARBARA A. BARRES

INTRODUCTION

A view held by many neurobiologists is that the complexity of our behavior will ultimately be accounted for by simple interactions between neurons and glia. Whereas neurons interconnect to form electrically active circuits, the role of glia is a mystery. Do glial cells just provide a passive framework that supports, nourishes, and insulates neurons, or, in addition, do glial cells play more active roles in signaling and plasticity? The potential importance of glia is suggested by their increase in number during evolution; glial cells constitutes 25%, 65%, and 90% of cells in *Drosophila*, rodent, and human brain, respectively.

In this review, I will focus on recent studies that suggest that it is no longer tenable to consider glial cells as passive support cells. These studies have shown that interactions between neurons and glial cells control neurogenesis, myelination, node of Ranvier formation, synapse formation, and probably even neuronal signaling. Thus I will argue that the development, structure, and function of the brain all depend on an intimate neuron–glia partnership and that full understanding of nearly all neurobiological processes must include consideration of the role of glial cells.

GENESIS OF NEURONS AND GLIA

Neuron–Glial Interactions Control the Matching of Neurons and Glia

During development the number of oligodendrocytes is matched to the number and length of axons requiring myelination. Oligodendrocytes are postmitotic, myelinating cells (Temple and Raff, 1986; Gard and Pfeiffer, 1990; Hardy and Reynolds, 1991) that develop from proliferating precursor cells, which migrate into developing white matter from germinal zones around the ventricles of the brain and the central canal of the spinal cord (Paterson et al., 1973; Raff et al., 1983; Small et al., 1987; Levine and Goldman, 1988; Hardy and Reynolds, 1991; Pringle et al., 1992; Grove et al., 1993).

One would expect that neurons play an important role in controlling oligodendrocyte development because the function of oligodendrocytes is to demyelinate axons, and most neurons are generated before most oligodendrocytes. This has been surprisingly difficult to demonstrate. For instance, oligodendrocyte precursor cells can divide in neuron-free cultures and differentiate into oligodendrocytes with the same timing as occurs in vivo (Raff et al., 1985), and newly formed oligodendrocytes go on to produce myelin in the absence of neurons (Mirsky et al., 1980; Dubois-Dalcz et al., 1986; Hudson et al., 1989). Thus developing oligodendrocytes in culture can survive, proliferate, and differentiate in the absence of neurons. Nonetheless, neurons can stimulate oligodendrocyte precursor cells to divide in vitro by both soluble and contact-mediated signals (Edgar and Pfeiffer, 1985; Wood and Bunge, 1986; Guilian and Young, 1986; Hunter and Bottenstein, 1989; Levine, 1989; Hardy and Reynolds, 1991, 1993). In addition, few oligodendrocytes develop in neonatal optic nerves that have been transected (David et al., 1984; Fulcrand and Private, 1977; Private et al., 1981; Valat et al., 1983), demonstrating that the migration, proliferation, differentiation, or survival of oligodendrocytes in vivo depends on axons.

Recent studies have focused on the role of axons in promoting oligodendrocyte development in the rat optic nerve, one of the simplest parts of the CNS. It contains two main cell types—astrocytes and oligodendrocytes—in addition to the axons of retinal ganglion cells. Cultures of optic nerve cells generally contain two types of astrocytes: type-1 astrocytes develop from optic stalk cells, while type-2 astrocytes develop from oligodendrocyte precursor cells, which are therefore called O-2A progenitor cells (Raff et al., 1983). There is still no convincing evidence that type-2 astrocytes develop in the normal CNS (Skoff, 1990; Fulton et al., 1992). Type-1 astrocytes first appear in the nerve about a week before birth, while oligodendrocytes first appear just after birth and increase in number for 6 weeks (Miller et al., 1985; Barres et al., 1992a,b). The adult rat optic nerve contains about 100,000 axons (Lam et al., 1982; Linden and Perry, 1983), 150,000 astrocytes, and 300,000 oligodendrocytes (Barres et al., 1992a).

The Proliferation of Oligodendrocyte Precursor Cells Depends on Axons

The proliferation of an oligodendrocyte precursor cell depends on signals from other cells: When a single precursor cell is cultured on its own in the absence of mitogens, it immediately stops dividing and differentiates prematurely into an oligodendrocyte (Temple and Raff, 1985). When precursor cells that have been purified to greater than 99.9% purity are cultured in the absence of mitogens or other cell types, no matter how high the cell density, they stop dividing and differentiate into oligodendrocytes (Barres et al., 1992a,b), indicating that the proliferation of a precursor cell depends on signals from other types of cells.

At least four peptide growth factors have been found to promote DNA synthesis by oligodendrocyte precursor cells in vitro—platelet-derived growth factor (PDGF; Noble et al., 1988; Raff et al., 1988; Richardson et al., 1988), basic fibroblast growth factor (bFGF; Bogler et al., 1990; McKinnon et al., 1990), neurotrophin-3 (NT-3; Barres et al., 1993a, 1994a), and insulin-like growth factor (IGF-1; McMorris and Dubois-Dalcz, 1988; Barres et al., 1994a), although NT-3 and IGF-1 do not do so on their own. All of these have been shown to be made by astrocytes in vitro and to be present in the developing optic nerve (Pringle et al., 1989; Mudhar et al., 1993; Hannson et al., 1989); PDGF is also made by retinal ganglion cells (Mudhar et al., 1993). Although no one of the factors on its own is able to stimulate the sustained proliferation of purified precursor cells in culture, when used in combination they do: In particular, when purified precursor cells are cultured in plateau concentrations of PDGF, NT-3, and IGF-1 their proliferation and differentiation behavior resembles that in the developing nerve (Barres et al., 1994a). PDGF and NT-3 both help to promote the proliferation of oligodendrocyte precursor cells in vivo. When neutralizing NT-3 antibodies are delivered into the developing nerve for 1 week, the number of oligodendrocyte precursor cells that are synthesizing DNA is decreased and the numbers of oligodendrocytes and their precursor cells in the nerve are decreased by 50%, indicating that NT-3 is required for optimal precursor cell proliferation and normal oligodendrocyte development (Barres et al., 1994a). In spinal cord white matter of transgenic mice deficient in PDGF-A, there are 100-fold less oligodendrocyte precursor cells, whereas transgenic mice that overexpress PDGF-A have a proportional increase in their number of oligodendrocyte precursor cells (Calver et al., 1995).

As oligodendrocyte cell number must be matched to the number (and length) of axons requiring myelination, one might expect that axons would promote the proliferation of oligodendrocyte precursor cells, and this is the case. When developing optic nerves are transected, the number of mitotic oligodendrocyte precursor cells falls by 90% in 4 days (Barres et al., 1993b). If the same experiment is performed in mutant mice whose axons do not degenerate after a transection (Perry et al., 1991), proliferation of oligodendrocyte precursor cells also decreases by 90%, raising the possibility that the proliferation depends on electrical activity in axons. Consistent with this possibility, intraocular injection of tetrodotoxin (TTX), which silences the electrical activity of retinal ganglion cells and their axons, decreases oligodendrocyte precursor cell proliferation by about 80%; the effect of TTX is prevented by experimentally increasing the concentration of PDGF in the developing optic nerve, suggesting that axonal electrical activity normally controls the production and/or release of the growth factors that are responsible for the proliferation of oligodendrocyte precursor cells (Barres and Raff, 1992). Activity could conceivably regulate either astrocyte or axonal mitogens. Recent experiments, however, have shown that the levels of PDGF-A protein in control optic nerves and optic nerves 4 days after a transection do not differ (S. Shen, B. Barres, unpublished observations), raising the possibility that activity regulates an axonal mitogen instead.

The Survival of Oligodendrocytes Depends on Axons

Oligodendrocyte numbers are not solely determined by controls on the proliferation of their precursor cells; the survival of oligodendrocyte lineage cells is also controlled. If a single oligodendrocyte or precursor cell is cultured on its own in the absence of exogenous signaling molecules, it rapidly dies with the morphological features of apoptosis, suggesting that these cells need signals from other cells to avoid programmed cell death (PCD; Barres et al., 1992a,b; Gard and Pfeiffer, 1993). Similarly, if purified oligodendrocytes or precursor cells are cultured in the absence of other cell types or exogenous signaling molecules, the majority undergo apoptosis within a day, even when they are cultured at high density, suggesting that the cells require signals from other types of cells to avoid PCD. These deaths are prevented by the addition of medium conditioned by cultures of optic nerve cells depleted of oligodendrocyte lineage cells, indicating that nonoligodendrocyte lineage cells, in vitro at least, secrete factors that promote the survival of oligodendrocytes and their precursors.

The three growth factors, PDGF, IGF-1, and NT-3, which together promote the proliferation of oligodendrocyte precursor cells, also promote the survival of these cells in culture (Barres et al., 1992a, 1993a). They, as well as ciliary neurotrophic factor (CNTF) and its relatives leukemia inhibitory factor (LIF) and interleukin-6 (IL-6), also promote the survival of purified oligodendrocytes in culture (Barres et al., 1993a; Louis et al., 1993); whereas PDGF supports the survival of newly formed oligodendrocytes, it does not support the survival of more mature oligodendrocytes, which no longer express PDGF receptors (Hart et al., 1989; McKinnon et al., 1990). Although any one of these growth factors or cytokines only supports the short-term survival of newly formed oligodendrocytes or their precursors, a combination of NT-3, IGF-1 (or IGF-2 or high concentrations of insulin), and CNTF (or LIF or IL-6) supports oligodendrocyte survival for weeks in culture (Barres et al., 1993a). Thus both cell proliferation and cell survival in this lineage depends on multiple signals acting together (Raff et al., 1993; Barres and Raff, 1994).

Apoptosis is also observed in developing oligodendrocytes in the rat optic nerve. When frozen sections of developing, postnatal rat optic nerves are stained with propidium iodide to label nuclear DNA, less than 0.3% of the nuclei are seen to be pyknotic (Barres et al., 1992a), a morphology that is characteristic of normal cell death (Wyllie et al., 1980). If the sections are stained at the same time with cell-type–specific antibodies, 90% of the dead cells are found to be oligodendrocytes and the remaining 10% are oligodendrocyte precursor cells, suggesting that normal cell death in the postnatal nerve is confined to the oligodendrocyte lineage (Barres et al., 1992a). The oligodendrocytes that die do so within 1 to 3 days of being produced from their dividing precursor cells, suggesting that this is a critical period of vulnerability for oligodendrocytes; most of those that survive this period probably live until the animal dies.

Why do newly formed oligodendrocytes die in the developing optic nerve? By analogy with developing neurons, it is possible that they require trophic factors to survive in vivo just as they do in vitro and that not all of them get enough. Consistent with this possibility, normal cell death in the developing nerve can be suppressed, at least temporarily, by treatment with a growth factor or cytokine that promotes the survival of newly formed oligodendrocytes in culture (Barres et al., 1992a). In these experiments the amount of the survival factor is increased by the transplanting of cells into the brain that have been genetically engineered to secrete large amounts of the factor. In experiments where transfected COS cells secreting PDGF are transplanted into the brain, for example, cell death in the optic nerve is decreased by about 80% without any effect on cell proliferation (Barres et al., 1992a). Over a 4-day period (between postnatal days 8 and 12), although there is no change in the number of astrocytes in the nerve, the number of oligodendrocytes is increased twofold, by about 40,000 cells, compared to the results of control animals. Each day, therefore, the excess PDGF saves about 10,000 newly formed oligodendrocytes that would normally have died if the PDGF concentration had not been artificially increased. Thus, at least 10,000 newly formed oligodendrocytes normally die in the optic nerve each day during this period, which is about half of the oligodendrocytes that are generated each day. Approximately the same proportion of newly formed oligodendrocytes seem to die in the nerve each day throughout the 6-week period of oligodendrocyte production (Barres et al., 1992a).

One reason that this massive oligodendrocyte death was initially missed is that, although 10,000 oligodendrocytes die each day in the optic nerve during the second postnatal week, one sees only about 400 dead cells in the nerve at any one time during this period (Barres et al., 1992a). This low number means that the time from which a cell dies to the time that it is phagocytized and degraded (so that it can no longer be recognized in a light microscope) is about 1 hour, which is the clearance time that has been directly observed for cells dying during normal development in the nematode *Caenorhabditis elegans* (Ellis et al., 1991). A high rate of apoptosis of newly formed oligodendrocytes has recently been confirmed by Bruce Trapp and his colleagues (personal communication), who have shown that a high percentage of newly formed oligodendrocytes in developing white matter, immunohistochemically identified by anti–DM-20 antibodies, have fragmented DNA when labeled with the terminal deoxytransferase end-labeling (TUNEL) method.

The findings that experimentally increasing the levels of PDGF, IGF-1, CNTF, or NT-3 in the developing optic nerve greatly decreases the death and increases the number of oligodendrocytes (Barres et al., 1992a, 1993b, 1994a) indicate that all of these signaling molecules are normally present in subsaturating amounts in the developing nerve and suggest that normally occurring oligodendrocyte death may reflect a competition for survival signals that are limited in amount or availability. Because PDGF, IGF-1, NT-3, and CNTF are all produced by optic nerve astrocytes (Stockli et al., 1991) and all promote the survival of oligodendrocyte lineage cells both in vitro and in vivo, it seems likely that astrocytes play a part in supporting the survival of this lineage in the nerve, at least during development.

As might be expected, however, the survival of oligodendrocytes also depends on axons (Barres et al., 1993b). If the postnatal optic nerve is transected behind the eye so that the axons degenerate, the oligodendrocytes die; the death is prevented if the levels of IGF-1 or CNTF are experimentally elevated. Oligodendrocytes do not die if the optic nerve is transected in mutant mice in which the axons do not degenerate, or if TTX is injected into the eye to electrically silence the retinal ganglion cells and thier axons. Thus, unlike the case for axon stimulation of oligodendrocyte precursor cell proliferation, the ability of axons to promote oligodendrocyte survival does not depend on electrical activity in the axons. Purified neurons, but not neuron-conditioned culture medium, promote the survival of purified oligodendrocytes in vitro, suggesting that oligodendrocytes might have to contact axons to survive. How does one reconcile the findings that both astrocyte-derived and axon-derived signals seem to promote oligodendrocyte survival? It is possible that signals from both sources collaborate to promote the survival of oligodendrocytes; alternatively, newly formed oligodendrocytes may depend on the astrocyte-derived signals while mature oligodendrocytes may lose their dependence on astrocytes and come to depend solely on axons for their survival.

Oligodendrocytes May Be Matched to Axons by a Competition for an Axon-Dependent Signal

A tentative model of how a competition for axon-dependent survival signals may help to match oligodendrocyte and axon numbers has been proposed (Barres et al., 1993b, 1994a). Once an oligodendrocyte precursor cell stops dividing and begins to differentiate into an oligodendrocyte, its specific requirements for survival signals change: it rapidly loses its PDGF receptors, for example, so that PDGF can no longer promote its survival (Hart et al., 1989; McKinnon et al., 1990). It now has only 2–3 days to contact a nonmyelinated region of axon that provides new signals that are required for its continued survival. A cell that fails to find an axon will kill itself (Fig. 3–1). Forcing newly formed oligodendrocytes to compete for axon-dependent survival signals that are limited in amount or availability would help to ensure that the final number of oligodendrocytes is precisely matched to the number (and length) of axons requiring myelination, just as a competition for target-cell–derived neurotrophic factors is thought to help ensure that the number of developing presynaptic neurons is matched to the number of postsynaptic cells requiring innervation (Cowan et al., 1984).

Two recent findings lend support to this hypothesis. First, it has recently been found that the number of oligodendrocytes in the optic nerve is precisely matched to the number of myelinated axons. The adult rat optic nerve contains about 300,000 ($\pm$30,000) oligodendrocytes (Barres et al., 1992a), each of which myelinates on average about 15 internodes, and each internode is about 200 μm in length (Butt and Ransom, 1993). As the rat optic nerve contains about 100,000 axons (Lam et al., 1982; Linden and Perry, 1983), each about 1 cm in length, almost all of which are myelinated, a minimum of 300,000 oligodendrocytes would be required to myelinate the axons. It appears that there are very few if any "extra" oligodendrocytes in the rat optic nerve. Second, another prediction of the target-matching hypothesis is that if the number of axons is increased, there will be less death of oligodendrocytes and an increase in the number of surviving oligodendrocytes. This prediction has been verified in recent studies of transgenic mice that overexpress bcl-2 in neurons. These mice have about an 80% increase in the number of retinal ganglion cells and axons, and remarkably,

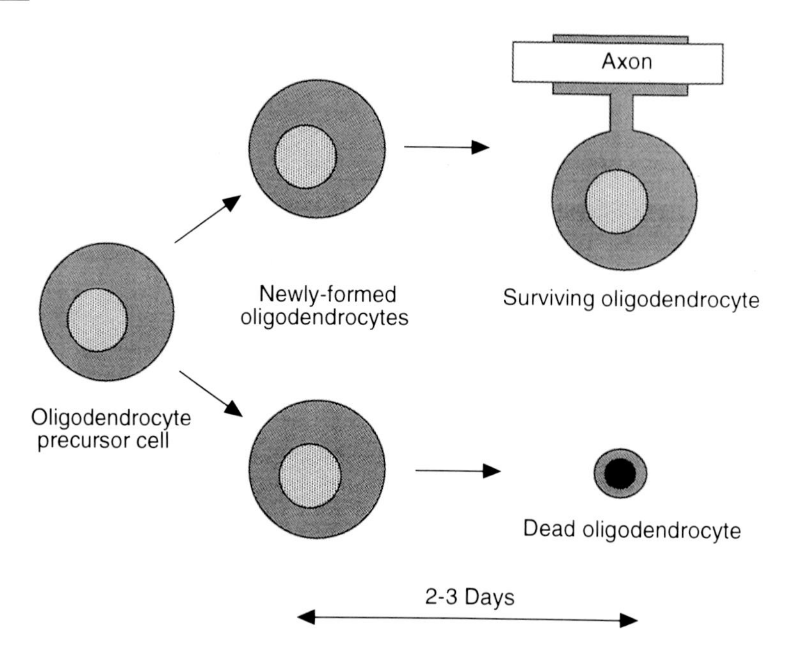

FIGURE 3–1 A tentative model for how the dependence of newly formed oligodendrocytes on an axon-derived survival signal may help to regulate oligodendrocyte cell numbers. Once an oligodendrocyte precursor cell stops dividing and begins to differentiate into an oligodendrocyte, it has 2–3 days to contact an unmyelinated region of axon, which provides a new signal that the cell requires for continued survival. About half of the oligodendrocytes normally fail to contact an axon and consequently die.

oligodendrocyte death is decreased and the number of oligodendrocytes that survive is increased proportionally (Burne et al., 1996). It remains to be shown directly, however, that the newly formed oligodendrocytes that die did in fact fail to establish a sufficient number of axonal connections.

In summary, the control of oligodendrocyte numbers in the rat optic nerve depends on complex cell–cell interactions mediated by multiple extracellular signaling molecules that collaborate to regulate cell proliferation, differentiation, and survival (Barres and Raff, 1994, 1995). An intrinsic cell suicide program, which is probably present in most cells (Raff, 1992; Raff et al., 1993), kills both oligodendrocytes and their precursors by default in the absence of sufficient survival signals produced by other cell types. Astrocytes increase oligodendrocyte numbers by secreting PDGF, IGF-1, NT-3, and possibly CNTF, which promote both the proliferation and survival of oligodendrocyte lineage cells. Axons control oligodendrocyte numbers by promoting both oligodendrocyte survival and precursor cell proliferation.

Neuronally Released Glial Growth Factor May Play a Pivotal Role in Matching the Number of Glial Cells to the Number of Neurons

Glial growth factor (GGF), also referred to as neuregulin, was first described because of its ability to promote the proliferation of Schwann cells in culture (Brockes et al., 1980); it is a peptide growth factor synthesized by many neurons in the CNS and PNS (Marchionni et al., 1993). Shah et al. (1994) showed that GGF suppresses the differentiation of rat neural crest stem cells in culture into neurons while simultaneously promoting glial differentiation. As neurons in peripheral ganglia are generated before glial cells, and each is surrounded by glial cells, these data suggest a lateral signaling model in which developing neurons actively secrete GGF, thus inhibiting neighboring uncommitted cells from adopting a neuronal fate and, in addition, committing them to a glial lineage (Shah et al., 1994). Consistent with a role for GGF in promoting the generation of Schwann cells, few Schwann cell precursors are generated in transgenic mice that lack GGF (Meyer and Birchmeier, 1995). Thus neuronally derived GGF helps to match glial cell number to neuronal number in developing peripheral ganglia.

GGF might also help to match the number of oligodendrocytes to axons in the CNS. It promotes the survival of oligodendrocytes and the proliferation of oligodendrocyte precursor cells purified from cortical cultures, which express neuregulin receptors, and the mitogenic effect of cortical neuron conditioned is inhibited by antibodies to GGF (Canoll et al., 1996). GGF may also help to promote oligodendrocyte differentiation (Vartanian et al., 1994). In the optic nerve, however, the situation is uncertain. GGF is produced by retinal ganglion cells and optic nerve oligodendrocyte lineage cells also express neuregulin receptors (P. Osterhoff, B. Barres, unpublished observations), but GGF does not promote the proliferation or survival of highly purified oligodendrocyte lineage cells in serum-free medium (J. Shi, B. Barres, unpublished observations). These results do not exclude a role for GGF in promoting oligodendrocyte development in the optic nerve, as the effect of GGF might be indirect or require a missing cofactor.

Lastly, GGF may mediate the matching of developing Schwann cells to their axons. The survival and proliferation of Schwann cell precursors (Dong et al., 1995; Mirsky and Jessen, 1996), as well as premyelinating Schwann cells (Brockes et al., 1980; Levi et al., 1995; Syroid et al., 1996; Grinspan et al., 1996; Cheng and Mudge, 1996), is promoted by GGF in culture as well as by dorsal root ganglia (DRG) neurons and their axons (Morrissey et al., 1995). GGF immunoreactivity is found along axons (Rosenbaum et al., 1996). The ability of DRG neurons in culture to promote Schwann cell proliferation and survival is significantly decreased by neutralizing GGF (Morrissey et al., 1995; Dong et al., 1995; Mirsky and Jessen, 1996). In vivo, the survival of premyelinating Schwann cells during development depends in part on axons, as many of these cells undergo apoptosis when axons are cut, and this death is prevented by delivery of exogenous GGF (Grinspan et al., 1996; Trachtenberg and Thompson, 1996). Together these observations suggest that axonally derived GGF may mediate axon–Schwann cell matching.

So far the mechanisms that control the matching of myelinating cells to axons appear quite similar in the PNS and CNS, as both cell types depend on axons for survival signals early in their development. After myelination has occurred, however, Schwann cells lose their dependence on the axon and do not die after transection, as do the oligodendrocytes (Grinspan et al., 1996). In contrast to oligodendrocytes, Schwann cells appear to gain the ability with maturation to produce their own survival factors (Ciutat et al., 1996; Mirsky and Jessen, 1996;

Chen and Mudge, 1996). This difference between CNS and PNS myelinating cells may be relevant to understanding the greater ability of the PNS to regenerate (see below).

Glia Guide Developing Axons and Promote Neuronal Differentiation and Survival

Glial cells are thought to guide developing axons and promote neuronal differentiation and survival. Recent studies have allowed these hypotheses to be tested in vivo (Pfrieger and Barres, 1995). Hosoya et al. (1995) and Jones et al. (1995) identified a *Drosophila* gene, *glial cells missing* (*gcm*), which encodes a novel nuclear protein that is transiently expressed early during the development of nearly all glia and is required for glial cell fate determination. In the absence of functional gcm protein, nearly all glial cells differentiate into neurons, whereas ectopic expression of *gcm* in immature neurons or their precursor cells transforms them into glia. Thus *gcm* functions as a binary genetic switch that determines whether developing neural cells will become neurons or glia. Although loss-of-function homozygous *gcm* mutations are ultimately lethal to the developing embryos, much of neural development is completed prior to their death, allowing a direct examination of the role of glia in neural development. In addition, mutation of another *Drosophila* glia-specific homeodomain protein, repo, has also been found to eliminate many glia (Campbell et al., 1994; Halter et al., 1995; Xiong and Montell, 1995).

In both insects and mammals, glial cells have been suggested to play a role in axon pathway formation by forming a scaffold that promotes axon extension and guides the growth cones of pioneer neurons (Dodd and Jessell, 1988; Jacobs and Goodman, 1989; Goodman and Doe, 1993). Consistent with this possibility, *gcm* was detected in a saturation screen for mutations that specifically affect the formation of axon pathways in the developing CNS of the *Drosophila* embryo. Although CNS pioneer neurons in the *gcm* mutant flies that lacked glia extended their axons appropriately in the correct longitudinal direction, serious axonal pathway defects appeared at later stages of development (Hosoya et al., 1995; Jones et al., 1995). Both CNS axonal tracts and PNS motor nerves showed defasciculation and misroutings. The axonal pathway defects most likely reflect the loss of two different kinds of glial signals. Whereas the misrouted and defasciculated axons probably result from the loss of glial signals that regulate axon pathway formation, some of the degenerated axon segments probably result from loss of glial signals that maintain neuronal survival (see below). It is unlikely that these glial signals will be unique to *Drosophila* since mammalian glial cells have recently been found to regulate axonal fasciculation in culture (Winslow et al., 1995).

In addition to a role in guiding developing axons, analysis of the mutant flies lacking glia suggests that glial also help to promote the survival and differentiation of developing neurons. The morphology of the accessible and easily identifiable peripheral sensory neuron known as the bipolar dendrite (BD) neuron was examined. Normally the BD neuron is closely associated with a glial support cell. In the mutant flies, this glial cell did not develop and the BD neurons lacked their normally distinctive bipolar dendrites, although their somata and axons were normal, suggesting that glial cells induce or maintain the dendrites of the BD neurons (Jones et al., 1995). Similarly, glial cells regulate the dendrite development of mammalian neurons in vivo. When sympathetic neurons are cultured in the absence of glial cells, they extend axons but not dendrites (Tropea et al., 1988). The addition of glial cells induces them to develop dendrites. A soluble signal released by glial cells, osteogenic protein–1 (OP-1, also referred to as bone morphogenetic protein–7), mimics this effect (Lein et al., 1995). OP-1 specifically promotes the differentiation but not the survival of sympathetic neurons in culture. Together, these findings indicate that glial cells are probably necessary for the induction of dendrites by at least some types of neurons.

During the development of the CNS of the late-stage *gcm* mutant embryos, the number of neurons decreases, the CNS begins to degenerate, and an unusually large number of macrophage-like cells appear, suggesting that neuronal cell death is increased (Jones et al., 1995). Glial cells fail to differentiate properly in the *Drosophila* mutant *drop-dead;* in these flies neurons die and the brain degenerates (Buchanan and Benzer, 1993). When repo, a glial-specific homeodomain protein necessary for glial differentiation and survival in *Drosophila,* is mutated, many, but not all, glial cells die (Campbell et al., 1994; Halter et al., 1995; Xiong and Montell, 1995). Subsequently, many neurons also undergo apoptosis. Together these studies show that neurons degenerate in the absence of glia, indicating that, at least in flies, glial cells provide signals necessary for neuronal survival in vivo.

Myelination and Node of Ranvier Formation

Myelinating Cells Control Axon Caliber by Regulating Neurofilament Number and Phosphorylation

A remarkable example of a reciprocal neuron–glial interaction is the relationship between axon diameter and myelination. Whether a Schwann cell makes myelin or not depends on a signal from the axon (Weinberg and Spencer, 1976); the nature of this signal is not known. The simplest idea is that axon caliber itself is the trigger. Developing axons become myelinated only after they reach a certain diameter, about 0.4 μm in the CNS and about 1 μm in the PNS (Samorajski and Friede, 1968). Although this idea remains controversial, enhancing axon caliber by increasing the target size is sufficient to induce myelination of some small axons that would otherwise not become myelinated (Voyvodic, 1989). There is also a constant relationship between the diameter of a myelinated axon and the thickness of its myelin sheath; the axon diameter is about 0.6 times the diameter of the myelinated fiber (Samorajski and Friede, 1968; Friede, 1972). These observations led Friede (1972) to propose that radial growth of the axon drives the number of turns of myelin lamellae that form.

However, myelination itself increases axon caliber as much as ten-fold. Speidel (1964) followed myelination in the translucent tadpole tail by time-lapse videomicroscopy. He showed that a segment of an axon being myelinated by a Schwann cell rapidly enlarged in diameter compared to the immediately adjacent unmyelinated segments of the same axon. This observation was paralleled by the tissue culture observations of Windebank et al. (1985), who showed that myelination is necessary for the development of axon diameters greater than 1.25 μm. Furthermore, when myelination of developing axons in the CNS is prevented by irradiation, growth of the axon diameter was substantially reduced (Colello et al., 1994). These observations point to a local regulation of axon diameter by Schwann cells and oligodendrocytes. The ability of a myelinating cell to reversibly control axonal caliber makes functional sense as it may help to ensure that axonal conduction is maintained regardless of the presence of myelin.

A molecular mechanism by which myelinating cells can induce radial growth of axons has recently been suggested. Axon caliber is directly correlated with the number of neurofilaments (NFs), which are the main structural component of axons (Hoffman et al., 1987; Wong et al., 1995; Zhao et al., 1995; Lee and Cleveland, 1996); as axon caliber increases, NF number increases proportionally. Neurofilaments are heteropolymers that consist of a complex of light, medium, and heavy neurofilament proteins (NF-L, NF-M, and NF-H). Mutant quails and transgenic mice that lack neurofilaments have severe inhibition of radial axon growth, developing small-caliber axons only (Eyer and Peterson, 1994; Zhao et al., 1995; Wong et al., 1995). Remarkably, axons in the PNS became strikingly hypermyelinated relative to their axon diameter, whereas axons in the CNS did not become hypermyelinated (Eyer and Peterson, 1994). Thus axon growth clearly does not drive the myelination process in the PNS, although this remains a provocative possibility in the CNS.

The mechanism by which neurofilaments mediate radial growth of axons is not yet clear, although phosphorylation of the NF-H has been suggested to play an important role by regulating the spacing and interactions between neurofilaments (Lee and Cleveland, 1996); consistent with this suggestion, myelination greatly increases the phosphorylation of the NF-H chain (de Waegh et al., 1992; Hsieh et al., 1994). Demyelination produced dephosphorylation of neurofilament protein, and this change was spatially restricted to the demyelinated region of the axon, indicating that local axon–glial cell contact modulates a kinase-phosphatase system that acts on neurofilaments and possibly other substrates (deWaegh et al., 1992; Cole et al., 1994). The local nature of this interaction is further demonstrated by the finding that neurofilaments are fewer and less phosphorylated at nodes of Ranvier (Hsieh et al., 1994). Similar changes in neurofilament phosphorylation correlating with myelination have been observed in the CNS (Sanchez et al., 1996). These results raise two interesting questions that remain unanswered: Do these kinases control the function or localization of axonal ion channels? Could the ability of an axon to grow and thus become myelinated depend on the presence of this kinase system?

These findings show that contact of axons by myelinating cells induces local cytoskeletal changes and axon growth, but is myelination itself necessary for the growth of axon caliber? This question has recently been addressed by examining CNS axon diameter in several strains of mutant mice where myelin formation is prevented, but limited wrapping by oligodendrocyte cytoplasmic processes can occur (Sanchez et al., 1996). In normal axons of the mouse optic nerve, axons expanded in radial area by about fourfold as they became myelinated, correlating with a fourfold increase in the number of neurofilaments and a 50-fold increase in the amount of phosphorylation of the NF-H subunit. In the mutant animals, axons that became ensheathed by oligodendrocyte cytoplasm, but were not myelinated, ultimately attained full axon caliber and neurofilament accumulation. Thus signals from oligodendrocytes, independent of myelin formation, are sufficient to induce full axon radial growth, and this growth appears to be mainly mediated by local accumulation of neurofilaments (Sanchez et al., 1996).

Taken together, it appears that, at least in the CNS, oligodendrocyte ensheathment is sufficient to initiate radial growth of axons, possibly by activating a membrane-bound kinase system that phosphorylates cytoskeletal molecules. Radial growth in turn may help to drive the myelination process in the CNS, although in the PNS, radial growth of axons is clearly not necessary to drive myelination. It remains unclear in the CNS and in the PNS, however, why certain axons become myelinated while others do not.

Myelination Is not a Default Process but Is Governed by Cell–Cell Interactions

For any given rate of conduction, a myelinated axon occupies only about one-hundredth the volume of an unmyelinated axon. Thus, the ability of increasingly complex nervous systems to maintain a small volume crucially depends on myelination. How is myelin formed? Despite the importance of myelin, little is known about the mechanism of myelination or the signals that regulate this complex process. Previous anatomical studies have demonstrated that myelination is a sequential multistep process in which a myelinating cell adheres to an axon, then ensheaths and wraps it, culminating with exclusion of the cytoplasm from the spiraling processes to form compact myelin. But how do myelinating cells know to wrap only axons and not dendrites or processes of astrocytes? How is the composition of myelin controlled and what is the nature of the protein and lipid interactions that allow compaction? Is the wrapping process passive or active? And how do myelinating cells know when to start laying down myelin and when to stop?

In lower organisms, genetics has been a powerful approach for elucidating the molecules that participate in such sequential processes, an approach that has not been possible for myelinating species. Despite this limitation, biochemists have found that myelin contains several main proteins: Protein zero (P0) in the PNS and proteolipid protein (PLP) in the CNS constitute 50% of this protein, while myelin–basic protein (MBP) constitutes 10 to 30% in the PNS and CNS, respectively (Martenson, 1992). In addition, peripheral myelin protein 22 (PMP22) has recently been discovered to constitute about 5% of the PNS myelin protein (Suter et al., 1993). Each of these proteins is distributed throughout the myelin sheath, and each protein when deficient, knocked out, or mutated results in the almost total failure of myelination (Snipes et al., 1993; Doyle and Colman, 1993). Myelin-associated glycoprotein (MAG) is found only in noncompacted myelin and constitutes only about 1% of myelin protein (Quarles and Trapp, 1992; Meyer-Franke and Barres, 1994); its axonal receptor has recently been identified (Yang et al., 1996; DeBellard et al., 1996). It is particularly concentrated at the periaxonal cytoplasmic collar, both in Schwann cells and oligodendrocytes. Because of its unique localization and properties, MAG was hypothesized to play an important role in initiating myelination, but it has recently been shown that transgenic mice lacking MAG undergo nearly normal myelination in the CNS and PNS although their myelin does degenerate in the adult (Montag et al., 1994; Meyer-Franke and Barres, 1994). While such experiments have confirmed the obvious importance of these major proteins in myelination, they have been disappointingly uninformative in teaching us much about the myelination process itself.

The process of myelination has been observed by time-lapse videomicroscopy in both PNS and CNS cultures, although this approach has so far also not been greatly informative. In the PNS it has been followed by time-lapse microscopy of cells in culture by Bunge and Wood, (1987), who observed back-and-forth rotatory movements of the Schwann cell nucleus during myelin wrapping of the axon, although the Schwann cell body did not revolve around the axon. Only recently has myelination by an oligodendrocyte process been visualized in culture (Asou et al., 1995). Three stages were observed. First, in the initial stage, a growth-cone–like structure forms at the end of the oligodendrocyte process and its lamellipodium repeatedly contacts and withdraws from the axon, as if searching for the appropriate region of the axon to myelinate. In the second stage, the lamellipodium appears to thicken and its filopodium anchors to the axon. Ruffling occurs during which the lamellipodium folds into layers. The angle between the anchoring filopodium and the axon abruptly changes and the lamellipodium wraps around the axon like a transverse wave in one motion, thus creating quickly up to five or six wraps of uncompacted myelin. During the completion of wrapping and compaction, the lamellipodium assumes a "bursting" form.

While the mechanism of the myelination process itself remains mysterious, some progress has been made in identifying signaling mechanisms that help to regulate the timing of myelination. Developing axons do not become myelinated until after they reach their targets; this time varies significantly among pathways. The proper timing of myelination is important because the development of oligodendrocytes or myelin before target innervation might inhibit growing axons. As oligodendrocytes secrete trophic signals (see below), their premature appearance might also interfere with the selection mechanism that helps to match presynaptic cells with their targets (Cowan et al., 1984).

Control of myelination in the CNS

The timing, rate, and amount of myelination are controlled by signals from neurons, astrocytes, and some endocrine cells. The recent finding that signals from the same cell types promote the survival of oligodendrocytes has raised the following question: Is myelination primarily regulated by permissive signals, which promote oligodendrocyte survival, or does myelination also require instructive signals, which regulate specific steps of the myelination program independently of survival? Because surviving oligodendrocytes synthesize myelin proteins in the absence of other cell types, a simple hypothesis is that myelination is a default program, which progresses inexorably forward in surviving oligodendrocytes.

The recent development of procedures to purify and culture defined types of postnatal CNS neurons and glial cells in serum-free medium (Barres et al., 1992a; Meyer-Franke et al., 1995) has allowed this hypothesis to be tested. Oligodendrocytes from the rat optic nerve were purified and cultured in the presence of saturating concentrations of oligodendrocyte survival signals in order to determine whether they would myelinate the axons of purified retinal ganglion cells in culture (Meyer-Franke and Barres, 1996). Although the axons freely grew over the oligodendrocytes in culture, the axons did not become ensheathed or myelinated. In contrast, when the cocultures were treated with conditioned medium from purified optic nerve astrocytes, the oligodendrocytes dramatically altered their morphology, aligning their processes parallel to the axons. Immunostaining with antibodies to myelin proteins and neurofilaments showed that the axons were ensheathed by the oligodendrocyte processes. Electron microscopic studies are needed to determine the extent, if any, to which wrapping has also occurred. These findings demonstrate that surviving oligodendrocytes do not myelinate axons by default, but that other cell–cell interactions are required.

Several other recent observations suggest that the initiation of myelination depends on local cell–cell interactions. First, several groups have recently shown that in some strains of rodents, although oligodendrocyte differentiation occurs in a chiasm-to-eye gradient along the rodent optic nerve, myelination proceeds in the opposite direction (Colello et al., 1995; Foran and Peterson, 1992). Astrocyte differentiate in a retinal-to-chiasm direction, consistent with a possible role of astrocytes in locally triggering ensheathment or myelination. In addition to signaling mechanisms that control ensheathment, the wrapping process may also be regulated. Inhibition of electrical activity with tetrodotoxin, either in vitro or in vivo, decreases the amount of myelination, whereas enhancement of electrical activity with α-scorpion toxin enhances myelination (Demerens et al., 1996). After TTX treatment, MBP-expressing oligodendrocytes had aligned along and ensheathed the axons, but their myelin sheaths were reduced in size and were poorly compacted.

The timing of differentiation of oligodendrocyte precursor cells into oligodendrocytes may be important in the control of timing of myelination. Thyroid hormone regulates a timing mechanism in oligodendrocyte precursor cells that limits the maximum number of times they may divide; in rodents and humans, the onset of thyroid function correlates closely with the appearance of the first oligodendrocytes (Barres et al., 1994b). In addition, the timing of myelination for any given fiber tract correlates closely with the time it innervates its target. A striking example is found in the corticospinal tract, which in the rat only innervates its target cells about 10 days after birth. Prior to this time, oligodendrocytes are not found in this tract, even though many nearby tracts contain oligodendrocytes and are actively myelinating (Schwab and Schnell, 1989). In the adult central nervous system, pathways that are not myelinated contain oligodendrocyte precursor cells but no oligodendrocytes. These observations point to as yet unidentified axon-dependent signals that inhibit oligodendrocyte precursor cells from differentiating into oligodendrocytes.

Control of myelination in the PNS

Myelin protein zero (P0), the major protein in peripheral myelin, is not expressed by Schwann cells until myelination begins (Brockes et al., 1980). Based on culture studies, it has generally been believed that Schwann cells require axonal signals to synthesize P0, because in neuron-free culture, purified Schwann cells do not express P0 (Brockes et al., 1980; Mirsky et al., 1980). Since pharmacological agents that elevate intracellular levels of cAMP in Schwann cells in culture induce them to synthesize P0, it was hypothesized that the axonal signal triggers myelination by elevating Schwann cell cAMP (Jessen et al., 1991). More recent studies, however, have come to a surprisingly opposite conclusion: when Schwann cells from postnatal day 1 rats, prior to the time of myelination, are acutely isolated by an immunopanning procedure and placed into serum-free culture they express P0 mRNA and protein in the absence of neurons and forskolin (Cheng and Mudge, 1996). Exposure of these cultures, even transiently, to fetal calf serum forever turns off P0 expression, unless the cells are treated with pharmacological agents that elevate their cAMP levels. This observation neatly explains the difference between the experiments of Cheng and Mudge (1996) and the previous studies, which were conducted with Schwann exposed to serum. As the Schwann cells at the time of isolation did not express P0, the studies of Cheng and Mudge (1996) indicate that Schwann cells normally express P0 by default unless they are prevented from doing so—for instance, before myelination, by inhibitory signals within the nerve. One inhibitory signal may be transforming growth factor–β, which suppresses myelination by promoting a nonmyelinating Schwann cell phenotype (Einheber et al., 1995; Guenard et al., 1995). It is likely that, at the time of myelination, axons provide other signals that enhance the amount of myelin production. For instance, the expression of myelin proteins by Schwann cells is controlled by the transcription factor Krox20 (Topilko et al., 1994), which is regulated by axons (Herdegen et al., 1993; Murphy et al., 1996). GGF/neuregulin is sufficient to upregulate Krox20 expression by Schwann cells and the neuron-derived Krox20-inducing activity is blocked by the GGF-binding domain of the erbB4 receptor, strongly implicating GGF as the axonal signal (Murphy et al., 1996). Thus axonal signals, both inhibitory and stimulatory, may help to regulate the timing, location and amount of myelination.

Node of Ranvier Formation

Whereas nonmyelinated and premyelinated axons have a relatively uniform structure, in myelinated axons sodium and potassium channels are segregated into complementary membrane domains. Sodium channels are about 25 times more concentrated in the nodal membrane (Shrager, 1987, 1989); potassium channels are located in the paranodal and internodal membrane (Chiu and Ritchie, 1981; Waxman and Black, 1995). Axon–glial interactions define and maintain nodal and internodal domains.

How are sodium channels clustered at nodes of Ranvier?

Saltatory conduction depends on the formation of regularly spaced clusters of sodium channels at gaps in myelin called nodes of Ranvier (Huxley and Stampfli, 1949; Stampfli, 1954). Nodes of Ranvier are spaced at axonal intervals of about 100 times the axon diameter (Hess and Young, 1952). It is not known whether the regular spacing of nodes results from regularly spaced glial contacts or is instead intrinsically specified by the axonal cytoskeleton. In the PNS, however, Schwann cell contact induces sodium channel clustering along the axons of DRG neurons in vitro (Joe and Angelides, 1992). Similarly it has been suggested that astrocyte contact induces sodium channel clustering along CNS axons (Black et al., 1989, 1995; Waxman and Black, 1995; Waxman, 1995).

The recent development of highly specific anti–sodium-channel antibodies that work well for immunostaining at the light and electron microscopic (EM) level has made it possible to directly analyze the process of sodium channel clustering in the PNS during both remyelination and normal development (England et al., 1990, 1991; Dugandzija-Novakovic et al., 1995; Vabnick et al., 1996). When focal demyelination of rat sciatic axons was induced by intraneural injection of lysolecithin, nodal sodium channel clusters disappeared (Dugandzija-Novakovic et al., 1995). Over the next few days as proliferating Schwann cells appeared and became adherent to axons, new clusters of sodium channels formed at the edges of the Schwann cells. As the Schwann cells elongated, the clusters appeared to move with them. Remarkably as the Schwann cells approached each other, the sodium channel clusters at the Schwann cell edges approached each other and ultimately fused to form a new nodal cluster in a region that previously was internodal. When Schwann cell proliferation was blocked by mitomycin, no new sodium channel clusters appeared within internodes. These results indicate that sodium channel aggregation and mobility in demyelinated nerve fibers is controlled by adhering Schwann cells, resulting in the formation of new nodes of Ranvier during remyelination.

Schwann cells also control sodium channel clustering during development (Vabnick et al., 1996). In rodents, few sodium channel clusters are observed along the axon at birth (P0), but over the next few days sodium channel clusters begin to appear, and these clusters are invariably associated with Schwann cell processes. By P4, small clusters at the edges of Schwann cells can be observed to fuse, forming nodal aggregates. The clustering correlates closely with the onset of Schwann cell ensheathment and the expression of myelin-associated glycoprotein. In order to test directly the role of Schwann cells in inducing the sodium channel clustering, transgenic animals that express diphtheria toxin in P0-expressing Schwann cells were studied. These animals failed to develop axonal sodium channel clusters, confirming the crucial role of Schwann cells in inducing sodium channel clustering in vivo.

How do Schwann cells control the localization of sodium channels? Previous studies have shown that sodium channels bind to a cytoskeletal protein, ankyrin$_G$, which is initially localized along premyelinated axons but ultimately becomes localized to nodes (Srinivasan et al., 1988; Lambert and Bennett, 1993; Bennett and Gilligan, 1993; Kordeli et al., 1995). In coculture studies and studies of developing peripheral nerve, Lambert et al. (1996) have recently demonstrated that the pattern of ankyrin$_G$ immunoreactivity parallels that of sodium channel immunoreactivity with ankyrin$_G$ clustering occurring at sites of axon–Schwann cell contact. Thus Schwann cell–axon contact may directly regulate the interactions of ankyrin with the cytoskeleton. Interestingly, the isoforms of ankyrin found in the axon are regulated by myelination. The 440-kD form of ankyrinB is found in unmyelinated and premyelinated axons and disappears from the axon consonant with myelination (Chan et al., 1993). The myelin-deficient mutant mouse *shiverer* displays greatly upregulated levels of ankyrinB axonal immunoreactivity, indicating that myelination is likely to directly regulate the axonal ankyrin cytoskeleton.

What is the origin of the sodium channels that aggregate in response to Schwann cell adhesion? Three possibilities are de novo synthesis by the neuron with axonal transport and insertion, transfer from the Schwann cell to the axolemma, and transfer from the low-density internodal pool. Synthesis of new channels seems not to be required as, at least in the frog, clustering can occur after axonal transection (Rubinstein and Shrager, 1990), although these studies do not exclude recruitment from an axoplasmic pool. Transfer of sodium channels from the Schwann cell to the axon has also been suggested (Ritchie et al., 1990), but seems unlikely because sodium channels appear absent from paranodal Schwann cell membrane (Wilson and Chiu, 1990). Thus the most likely possibility, based on recent EM studies, is that sodium channels present in the internodal region at low density become expelled by Schwann cell contact and accumulate at the Schwann cell tip (Vabnick et al., 1996). This possibility is consistent with the previous observations that demonstrated an elevated level of sodium channels along hypomyelinated axons in *shiverer* mutant mice (Westenbroek et al., 1992). This ensheathment of axons by Schwann cells probably excludes sodium channel from internodal regions, thereby forcing the channels into nodal regions.

Are the mechanisms that induce clustering of sodium channels along CNS axons similar to those occurring in the PNS? The ability to purify and culture a defined type of CNS neuron, retinal ganglion cells, in serum-free medium in the presence or absence of glial cells (Kaplan et al., 1996), together with the development of a highly specific sodium channel antibody (Dugandzija-Novakovic et al., 1995), has recently allowed investigation of the cell–cell interactions that induce sodium channel clustering along CNS axons. Retinal ganglion cells are intrinsically unable to form sodium channel clusters in the absence of glial cells, even though they express sodium channels, are excitable, form dendrites and axons, and target sodium channels appropriately to their axons (Meyer-Franke et al., 1995; Kaplan et al., 1996). Sodium channel clustering along RGC axons is induced by a soluble signal produced by optic nerve glial cells (Kaplan et al.,

sary for the correct localization of the paranodal potassium channels. Consistent with this hypothesis, myelination regulates the localization of axonal Kv1.1 and Kv1.2 channels (Wang et al., 1995). In dysmyelinating mutant mice that lack myelin, K channels became diffusely distributed along the axon membrane and were not localized to the paranode.

Differential localization of ion channels in Schwann cells at the node of Ranvier

Interaction of axons and glial cells helps to define the localization of glial, as well as axonal, ion channels. Using specific anti–potassium-channel antibodies, the localizations of three classes of potassium channels in Schwann cells were examined: voltage-gated potassium channels of the Kv family (Kv1.1, 1.2, and 1.5), inwardly rectifying potassium channels of the IRK family, and calcium-activated potassium channels of the *Slo* family (Mi, 1995). Kv1.1 protein was localized mainly intracellularly to the perinuclear region, whereas Kv1.5 protein was found in the Schwann cell abaxonal membrane near nodes of Ranvier and in the canaliculi (Mi et al., 1995). IRK1 and IRK3 immunoreactivity was highly concentrated in the Schwann cell microvilli (also known as fingers) that protrude from the Schwann cell membrane near the node and fill much of the nodal cleft (Mi et al., 1996). Lastly Slo potassium channel immunoreactivity was similar to the localization of Kv1.5 (Mi, 1995). The functional role of potassium channels in Schwann cells is not clear, although a role in potassium ion buffering has been suggested (Mi, 1995). The mechanisms by which K channels are localized to different domains of Schwann cells are unclear, but it is likely that axon–glial interactions will play an important role in defining cytoskeletal domains with which channels can interact (Kim et al., 1995; Shi et al., 1996).

Regeneration

Axon–glial interactions are currently thought to play a pivotal role in regulating axonal growth during regeneration. A central issue is understanding why the PNS has a greater ability to repair itself after injury than does the CNS. When PNS axons are cut, they regenerate and the cell body does not die; in contrast, after axotomy of CNS neurons, there is a failure of the axon to regenerate and the cell body dies. Although it was once believed that CNS neurons intrinsically lacked the ability to regenerate, this view was elegantly disproved by the experiments of Albert Aguayo and his colleagues, who showed that at least some retinal ganglion cells could regenerate their axons through fragments of peripheral nerve grafted into central pathways (David and Agauyo, 1981; Bray et al., 1991). Despite the initial success of this type of experiment, it remains unclear why only a small percentage of CNS neurons can regenerate successfully through peripheral nerve grafts. There are currently two general classes of explanation that have been put forth to help account for the difference between the ability of CNS and PNS axons to regenerate. There may be fundamental differences between CNS and PNS glia or there may be fundamental differences between CNS and PNS. Recent work suggests that both explanations may be correct.

Do Differences Between CNS and PNS Glia Affect the Success of Axonal Regeneration?

Glial inhibition of axon growth

The failure of CNS axons to regenerate could be due to a nonpermissive glial environment in the adult CNS. Oligodendrocytes but not Schwann cells express membrane proteins of 35 kD and 250 kD in CNS myelin that inhibit the growth of axons of sympathetic and DRG neurons (Schwab et al., 1993). Neutralizing monoclonal antibodies to these molecules allowed axons to grow over the oligodendrocytes in vitro and, to a much lesser degree, in vivo in a cut spinal cord regeneration model (Schnell and Schwab, 1990; Schwab, 1995). Presumably these signals would not be inhibitory to growing axons during development either because embryonic neurons lack receptors for these inhibitor molecules or because normally most oligodendrocytes do not develop until after target innervation. It is not yet clear what the identity or normal function of the inhibitory molecules on oligodendrocytes are. They might block axon sprouting in the CNS after myelination (Colello and Schwab, 1994) or keep axons in a pathway that has not yet become myelinated from inappropriately entering a nearby, already myelinated pathway (Schwab, 1995). However, although axons from PNS neurons such as DRG neurons and sympathetic neurons are inhibited by oligodendrocytes, the axons of retinal ganglion cells are not inhibited by oligodendrocytes in vitro (Kobayashi et al., 1995; Meyer-Franke and Barres, 1996). Thus the responsiveness of growth cones to the inhibitory signal of oligodendrocytes may strongly depend on neuron type. Attention should now be paid to determining whether CNS axons in general are much less inhibited than are PNS neurons. If so, perhaps these inhibitory proteins help to prevent the axons of PNS neurons from inappropriately entering into the CNS.

Unlike mammals, birds, and reptiles, the CNS axons of fish and amphibia can regenerate. Do fish oligodendrocytes express inhibitory proteins? The axons of fish retinal ganglion cells do grow over fish oligodendrocytes in culture, but they do not grow over mammalian oligodendrocytes unless they are treated with the IN-1 antibody (Bastmeyer et al., 1994; Stuermer, 1995), indicating that fish oligodendrocytes do not express axon-inhibitory molecules, consistent with the greater regenerative ability of the fish CNS. In addition, fish oligodendrocytes also actively promote axon regrowth (see below). Thus fish oligodendrocytes share important properties with Schwann cells that may be crucial for successful regeneration (Bastmeyer et al., 1994; Bunge, 1994).

In addition to the 35-kD and 250-kD proteins of unknown identity, another oligodendrocyte and Schwann cell protein, myelin-associated glycoprotein (MAG), is inhibitory to growing axons (Mukhopadhyay et al., 1994; McKerracher et al., 1994). MAG is a member of the immunoglobulin gene superfamily that promotes the growth of developing axons but is inhibitory to adult axons. For instance, the axons of adult DRG neurons are inhibited by MAG-expressing cells and this inhibition can be overcome by anti-MAG neutralizing antibodies or by small sialic-acid–bearing sugars (DeBellard et al., 1996). The effect of MAG on the growth of CNS axons has received little study so far except for its effects on postnatal cerebellar granule neurons, which do not normally interact with oligodendrocytes. This is an important issue because, at least in culture, the axons of retinal ganglion cells are not significantly inhibited by MAG (Kobayashi et al., 1995; Meyer-Franke and Barres, 1996). In addition, the extent of axon regrowth in lesioned optic nerve and corticospinal tract in transgenic MAG-deficient and wild-type mice was similarly poor (Bartsch et al., 1995). Thus it is not clear whether MAG is inhibitory to axons of specific types of CNS neurons that are normally myelinated. If not, MAG inhibition is not likely to help explain the failure of CNS axons to regenerate. In addition, in the developing, as well as to some extent also in the adult CNS, oligodendrocytes in a transected nerve rapidly undergo apoptosis after their axons degenerate (Barres et al., 1993b). Nonetheless, retinal ganglion cells fail to regenerate in postnatal nerves even though nearly all oligodendrocytes have undergone apoptosis within a week after transection. Clearly inhibition by oligodendrocytes cannot be the sole reason that CNS axons do not regenerate.

In addition to differences in the properties of myelinating cells in the CNS and PNS, another class of glial cells, astrocytes, is present in the CNS but not the PNS, and could be inhibitory to regrowing axons. Astrocytes form thick cellular scars at sites of traumatic damage in the CNS that potentially could impede regrowing axons (Bignami and Dahl, 1995). A number of molecules expressed on the surfaces of astrocytes including tenascin, keratan, and chondroitin sulfate proteoglycans inhibit growing axons (Schwab et al., 1993; Shewan et al., 1995; Smith-Thomas et al., 1995; Zeng et al., 1995).

Glial promotion of axon growth

Schwann cells actively guide and promote the growth of regenerating axons (Son and Thompson, 1995a,b). When muscle cells are denervated, the terminal Schwann cells overlying neuromuscular junctions sprout elaborate processes. These Schwann cell processes bridge to other nearby neuromuscular junctions. The motor axons use these processes as guides to grow along during regeneration, allowing them to grow between end plates to generate polyneuronal innervation. Implantation of Schwann cells into an innervated muscle also induces sprouting of the nerve terminal upon contact by the Schwann cell processes, suggesting that Schwann cells are actively inducing growth of the axons and not simply passively guiding them. Such axon–glial interactions could well be crucial in allowing successful regeneration. Consistent with this view, when terminal Schwann cells undergo apoptosis after denervation in neonatal animals, there is a deficiency of axonal sprouting and less successful reinnervation (Trachtenberg and Thompson, 1996).

How do Schwann cells promote axon growth? The ability of axons to grow depends on local trophic signaling (Campenot, 1994). Schwann cells secrete neurotrophic factors such as neurotrophins and cytokines that promote axon growth (Bunge, 1994). Stage-specific differences in the ability of glial cells to produce neurotrophic activities have been demonstrated in both the CNS and the PNS. Schwann cell precursors produce neurotrophins, whereas premyelinating and myelinating Schwann cells do not (Schechterson and Bothwell, 1992), and immature astrocytes appear more effective than mature astrocytes in promoting optic nerve axon regrowth when grafted in vivo (Sievers et al., 1995). Thus Schwann cells, even in the absence of axons, secrete a variety of peptide neurotrophic factors that promote axon growth.

The situation is less certain for CNS glial cells, including astrocytes and oligodendrocytes. Astrocytes in culture secrete neurotrophins, cytokines, and other peptide trophic factors, but it is not clear to what extent they do so in the absence of serum and whether axons or other signals normally regulate production of these factors in vivo. Several recent studies underscore the potential importance of such regulation. Astrocytes in culture can be induced to produce BDNF by elevation of their cAMP levels (Zafra et al., 1992). Neuronally derived glial growth factor can induce the production of NT-3 by non-neural cells, which in turn supports the survival and growth of nearby neurons (Verdi et al., 1996). Within 2 hours after axotomy, LIF mRNA increases in non-neuronal cells in sympathetic ganglia and peripheral nerve; this elevation is attributed to a LIF-inducing protein that is not yet identified that is released from the injured neurons within 1 hour after axotomy (Sun et al., 1996). Taken together these findings suggest that an important area of future research will be understanding how the production of trophic factors by glial cells is regulated and whether the nature of this regulation differs for CNS and PNS glia.

Although the ability of astrocytes and Schwann cells to produce neurotrophic peptides has been well documented, only in the past year has it become clear that oligodendrocytes also do this. Oligodendrocytes purified from rat optic nerve secrete a peptide trophic factor, not mimicked by known neurotrophins or cytokines, that is essential for the long-term survival and growth of the majority of purified retinal ganglion cells in vitro (Meyer-Franke et al., 1995). Similarly, Schwalb et al. (1995) have identified two novel trophic peptides, AF-1 and AF-2, produced by fish optic nerve oligodendrocytes that promote axonal regeneration in culture. Such oligodendrocyte-derived neurotrophic activities are of great interest as they could help to explain the differential ability of fish and mammalian CNS to regenerate: Whereas many oligodendrocyte die in injured mammalian CNS (Barres et al., 1993b), they do not die in appreciable number in the injured fish CNS (Nona et al., 1994; Nona and Stafford, 1995). Thus, like Schwann cells, after a transection, surviving oligodendrocytes in the fish could release neurotrophic factors that actively promote axon growth.

Do Differences Between CNS and PNS Neurons Affect the Success of Axonal Regeneration?

There are still advocates for the possibility that injured adult CNS axons do not regrow as do embryonic axons because of an age-dependent loss of ability to regenerate their axons (Feng Chen et al., 1995). This argument, however, has become untenable with the recent demonstration that purified rat retinal ganglion cells isolated from animals at ages at which regeneration does not occur can survive and rapidly regenerate their axons in serum-free medium containing neurotrophic peptides normally produced along the developing visual pathway (Meyer-Franke et al., 1995).

A simple potential explanation for why CNS neurons die after axotomy, whereas PNS neurons do not, might be that different signaling mechanisms control their survival. Purified, defined types of PNS neurons survive in serum-free culture in the presence of single peptide trophic factors: NGF, for instance, is sufficient to promote the survival of sympathetic neurons and DRG neurons, whereas CNTF is sufficient to promote the survival of ciliary neurons. They also survive in the absence of added growth factors when their intracellular levels of cAMP are elevated or they are depolarized. Less is known about the signals that are sufficient to promote survival of CNS neurons. The greater complexity of the CNS has made it harder to purify defined types of CNS neurons, and, in the few cases where this has been possible—for instance, in the case of purified cerebellar granule and Purkinje neurons—the cells rapidly undergo apoptosis in serum-free medium (Baptista et al., 1994). Culture of embryonic hippocampal neurons is possible in serum-free medium, but only in the presence of a monolayer of astrocytes to continuously condition the medium (Craig and Banker, 1994).

In order to begin to understand the nature of the extracellular signals that promote the survival of a defined type of CNS neuron, we have focused on the survival requirements of retinal ganglion cells in culture (Barres et al., 1988; Meyer and Franke et al., 1995). When highly purified postnatal rat retinal ganglion cells are cultured in serum-free medium they rapidly die in conditions that promote the survival of PNS neurons, including peptide trophic factor stimulation, cAMP elevation, or depolarization. Their survival is promoted by peptide trophic factors, however, if their intracellular cAMP is simultaneously increased, either by membrane-permeable cAMP analogues or by depolarization with K^+or glutamate agonists. Astrocyte and tectal-derived factors, such as BDNF, CNTF, IGF-1, and an additional protein made by oligodendrocytes, together with cAMP elevation, collaborate to promote the long-term survival of the majority of purified retinal ganglion cells in vitro. Both the cAMP- and depolarization-induced responsiveness to survival factors are blocked by protein kinase A inhibition. In preliminary experiments, we have found that purified cortical neurons behave identically to the RGCs, whereas the survival of purified DRG neurons can be promoted solely by NGF in the same serum-free medium. These results suggest that cAMP elevation may be necessary for optimal survival and growth of retinal ganglion cells and other CNS neurons. They also suggest that the mechanisms that control the survival of CNS and PNS neurons may be fundamentally different: Whereas the responsiveness of PNS neurons to their trophic factors may not depend strongly or at all on depolarization or cAMP elevation, the responsiveness of CNS neurons to their peptide trophic factors requires elevation of cAMP or depolarization.

While this hypothesis requires much further work, it immediately suggests a reason why CNS and PNS neurons may differ in their response to injury. Injured neurons might have lower levels of cAMP, either because they receive less synaptic input or become less active (Blinzinger and Kreutzberg, 1968; Northmore, 1987). Thus, an injured CNS neuron might undergo apoptosis and fail to regrow its axon not only because retrograde flow of peptide trophic signals from axonal glia and its target has been interrupted by axotomy but because it becomes less sensitive to what remaining peptide trophic signals it does receive. One prediction of this hypothesis is that delivery of exogenous trophic peptides to the injured neurons should be insufficient to promote their survival, which in most experiments has been found to be the case (Mey and Thanos, 1993; Mansour-Robaey et al., 1994; Weibel et al., 1995; Shen and Barres, 1996). A second prediction is that simultaneous delivery of exogenous trophic factors along with pharmacological agents to elevate cAMP levels would significantly promote survival of injured neurons. This second prediction has also been borne out (Shen and Barres, 1996), although it is not clear yet how long their survival can be promoted and whether the axons of the surviving neurons will regrow. At least in a dish, however, the trophic signals that control the survival of a neuron and the survival and growth of its processes are the same (Campenot, 1994; Meyer-Franke et al., 1995), providing grounds for optimism.

The apparent differences in the survival and growth requirements of CNS and PNS neurons could reflect the greater need of the CNS for activity-dependent plasticity. Obviously the simple cAMP and activity-dependent mechanism described above potentially constitutes a simple learning mechanism, whereby neuronal processes that are more active will survive and grow preferentially because they are more responsive to trophic factors. However, if growth depends on activity, and if damaged cells are less active, their ability to regenerate would become impaired. If so, perhaps the loss of regenerative ability during evolution was a trade-off for acquisition of a greater degree of activity-dependent plasticity.

In summary, the failure of the CNS to regenerate its axons after injury is likely to be multifactorial. On the one hand, injured CNS glia may inhibit regrowing axons, may fail to actively provide trophic support to the regrowing axons; and on the other, injured CNS neurons may fail to receive and respond to their trophic signals. In the future, simultaneous attention to each of these problems could result in effective promotion of CNS regeneration.

SYNAPSE FORMATION AND FUNCTION

Glial Cells Promote Synapse Formation

Although synapses throughout the brain are ensheathed by glial cells (Peters et al., 1991), the possible role of glia in the development and function of synapses has received little attention. There are two reasons why these questions have previously been hard to address. First, traditional methods for preparation of enriched neuronal cultures result in cultures that still contain some glial cells. Second, as the percentage of contaminating glial cells in neuronal cultures is decreased, neuronal survival decreases. Using recently identified neurotrophic factors that are normally expressed along the visual pathway, it has been possible to use defined serum-free conditions in order to promote the long-term survival of highly purified retinal ganglion cell (RGC) neurons from postnatal rats in culture. We have used these conditions to ask: Will synapses form in culture in the absence of glial cells (Pfrieger and Barres, 1996)?

To determine whether pure neuronal cultures will form functional synapses in the absence of glial cells in serum-free cultures, we monitored their synaptic activity using whole-cell patch clamp recording. In the absence of glial cells there was little synaptic activity. When the retinal ganglion cells were cocultured with purified glial cells from their target region, the superior colliculus, nearly all neurons exhibited synaptic activity. The frequency of spontaneous excitatory postsynaptic currents (EPSCs) was increased by about 100-fold after 20 days of coculture. The mean frequency of miniature EPSCs recorded in the presence of tetrodotoxin was about tenfold higher in the presence of glial cells, showing that the enhanced synaptic activity is not attributed to enhanced neuronal excitability. Similarly, little synaptic activity was observed when retinal ganglion cells were cocultured with purified tectal neurons in the absence of glial cells, showing that the failure of synapse formation in the absence of glia could not be simply attributed to the lack of their normal target. Similar effects on enhancing synaptic activity were seen with glial feeding layers of either highly purified astrocytes or oligodendrocytes, indicating that both types of glial cells can enhance synaptic activity by secretion of as yet unidentified soluble signals.

To determine whether glia increase the number of synapses that form, we counted the number of synaptic terminals per RGC after 2 weeks of culture by labeling presynaptic terminals in living neurons by synaptic vesicle immunostaining; dendrites were visualized by microtube associated protein 2 (MAP-2) immunostaining. In cultures lacking glial cells, few bouton-like terminals formed. Addition of glial cells enhanced by about tenfold the number of axodendritic bouton-like terminals contacting each neuron (Fig. 3–3). As the glial-induced increase in the number of terminals was approximately equivalent to the glial-induced increase in the frequency of mini-EPSCs, the effect of glial cells is in large part to induce an increase in the number of synapses that form. In addition, a glial enhancement of synaptic efficacy is also possible (see below).

These results suggest that purified neurons in serum-free culture have little ability to form functional synapses in the absence of glial cells. Glial cells greatly enhance the number of synapses that form. It is possible that synapse formation in vivo is also enhanced by glia as the vast majority of CNS synapses form only after the generation of astrocytes (Aghajanian and Bloom, 1967).

Do Glial Cells Regulate Synaptic Function?

Glial cells release substances that are expected to regulate the efficacy of transmission, but this has not yet been directly demonstrated (Barres, 1991; Bruner et al., 1993; Roberts et al., 1995; Schell et al., 1995). I will focus on several more fundamental glial–synapse interactions.

- Glia

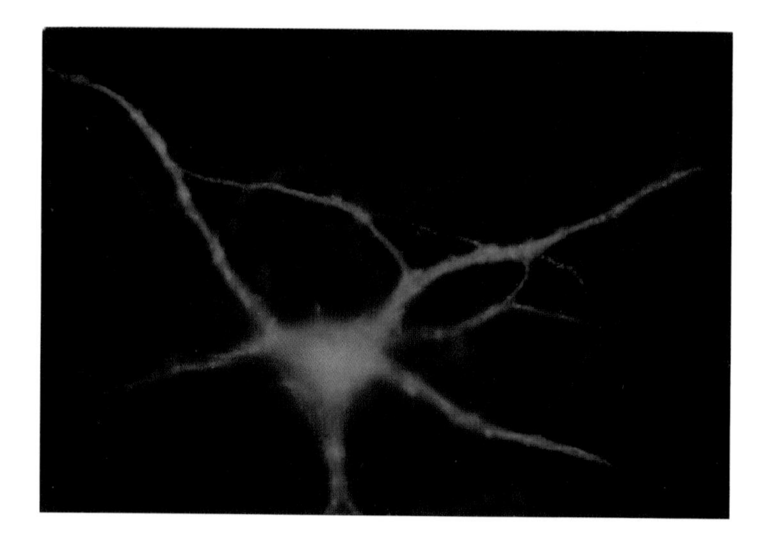

+ Glia

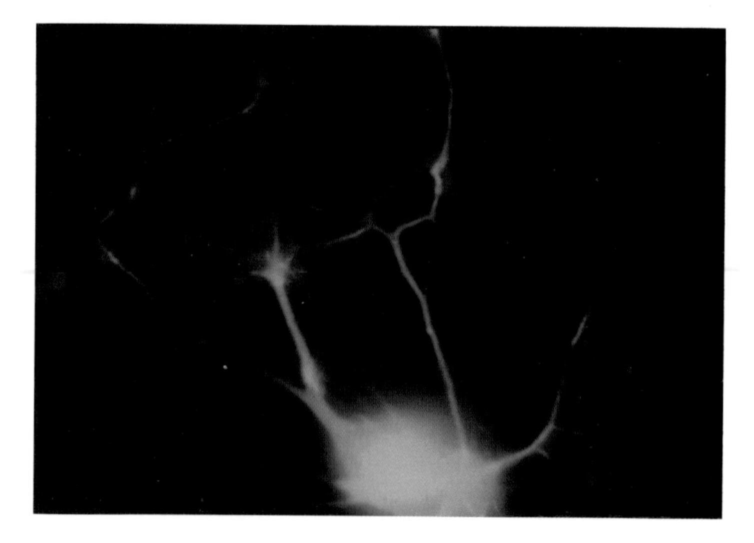

FIGURE 3–3 Soluble signals from glial cells enhance the formation of synapses in culture. The micrographs show purified retinal ganglion cells in culture. Their cell bodies and dendrites have been stained with anti–MAP-2 antibodies (green). Synaptotagmin immunoreactivity is shown in red. Retinal ganglion cells were cultured either in the absence (top) or presence (bottom) of a conditioning layer of glial cells (not shown). Scale bar, 20 μm.

The glutamine–glutamate shuttle and glutamate transport

Glutamate is the predominant excitatory transmitter in the mammalian CNS. Glutamate released from neurons is recycled; it is quickly taken up by glial cells and amidated to form the non-neuroactive compound glutamine. This amidation is catalyzed by glutamine synthetase, an enzyme found only in glial cells. The glutamine is released by the glial cells, taken up by the neurons, and converted back to glutamate by glutaminase, an enzyme found only in neurons and their terminals (Siegel et al., 1995). Three glutamate transporters have been identified including two glial transporters, GLAST and GLT-1, and a neuronal transmitter, EAAC1 (Arriza et al., 1994; Danbolt, 1994). The neuronal transporters are largely localized postsynaptically, whereas the glial GLT-1 transporter is localized to synaptic glia (Rothstein et al., 1994). Astrocyte membranes facing capillaries and pia have lower densities of glutamate transporters than do astrocyte membranes facing nerve terminals, axons, and spines (Chaudry et al., 1995). The spatial localization of the highest levels of GLT-1 immunoreactivity corresponds nearly exactly with that of glutamine synthetase, as would be expected for best operation of the glutamate–glutamine recycling mechanism. That these glial glutamate transporters become activated by synaptic release of glutamate has been vividly demonstrated by the recent experiments of Mennerick and Zorumski (1994), who demonstrated that excitatory synaptic events rapidly activate electrogenic glial glutamate transporter currents that produce large glial depolarizations. The crucial role of glial cells in glutamate uptake has been demonstrated recently. When the synthesis of GLT-1 was inhibited by chronic antisense oligonucleotide administration in vitro and in vivo, elevated extracellular glutamate levels ensued that were sufficient to produce excitotoxicity (Rothstein et al., 1996).

Who makes the neurotransmitters: glia or neurons?

Glutamate does not readily cross the blood–brain barrier and must be synthesized within the CNS. It is widely assumed that neurons synthesize glutamate themselves using aspartate or precursors derived from the Krebs cycle. Recent experiments, however, have suggested that glial cells may be important—possibly primary—synthesizers of glutamate or its immediate precursors. In provocative immunocytochemical studies in the retina, Pow and Robinson (1994) postulated that if neurons synthesize significant quantities of glutamate for themselves then inhibition of glutamine synthetase should have little effect on the neuronal content. In control retinas, immunoreactivity for glutamate was primarily localized to neurons and immunoreactivity for glutamine to the glial cells. When glutamine synthetase was inhibited pharmacologically, immunoreactivity for glutamate in the neurons was reduced below the level of detection but was now detectable in the glial cells. Furthermore, during development, neuronal glutamate immunoreactivity was detectable only after the appearance of glutamine synthetase immunoreactivity in the glial cells (Pow et al., 1994). Nearly identical findings have recently been reported for neurons and glial cells in hippocampal slices (Laake et al., 1995), arguing that glutamine is necessary for the maintenance of a normal level of glutamate in nerve terminals and that de novo synthesis of glutamate by other pathways other than glutaminase is insufficient.

While these findings are initially surprising, the brain exhibits a striking metabolic compartmentation with differential expression of fundamental metabolic and biosynthetic enzymes in neurons and glia (Berl, 1973; Schousboe et al., 1992, 1993; Pow and Robinson, 1994; Hamprecht and Dringen, 1995; Weisenger, 1995; Tsacopoulos and Magistretti, 1996). These observations have previously led to the proposal that there are essentially two separate Krebs cycle compartments in the brain: one in neurons that functions primarily for oxidative metabolism and one in glia that functions primarily anabolically to synthesize energy metabolites and transmitter precursors (Berl, 1973; Hamprecht and Dringen, 1995). For instance, glial cells are the main site of glycogen storage, of glycogen synthetic and degradative enzymes, of hexokinase necessary to retain and metabolize glucose after uptake, as well as of pyruvate carboxylase, which is the main CO_2-fixing enzyme in the brain. This latter pathway, in particular, implicates glia in one of the steps leading to the de novo synthesis of glutamate (Pow and Robinson, 1994). Moreover, several other essential enzymes that lead to the formation of glutamate are preferentially localized in glial cells including aconitase, which converts citrate to isocitrate, isocitrate dehydrogenase, which converts isocitrate to α-ketoglutarate, and glutamate dehydrogenase, which converts α-ketoglutarate to glutamate. In fact, treatment of hippocampal slices with an inhibitor of aconitase, fluoroacetate, which is thought to selectively block the Krebs cycle in glial cells, leads to a rundown of synaptic transmission within minutes (Keyser and Pellmar, 1994). Together, these findings argue that much of the de novo synthesis of glutamate (and probably aspartate and γ-aminobutyric acid—GABA) occurs principally in glia. If so, glial cells may play a crucial role in maintaining synaptic transmission as well as in modulating synaptic efficacy. The ability to purify and culture neurons with or without glia should allow direct testing of these hypotheses.

Energy metabolism

One of the most important advances in understanding the function of glial cells has come from recent studies that have established an activity-dependent metabolic coupling between neurons and glial cells. Glucose is nearly the exclusive energy source of the brain; as astrocytes processes ensheath and help to form the blood–brain barrier, nearly all glucose that enters the brain from the blood must first enter glial processes. Until recently, it was not clear to what extent this glucose was simply transferred to neurons or metabolically processed in the astrocytes into metabolic intermediates for neuronal transfer. Neurons are rich in mitochondria and devoid of glycogen, whereas the reverse is true for glia. Energy metabolism in neurons is thought to be largely aerobic. Uptake and phosphorylation of glucose can be followed by monitoring the accumulation of 2-deoxyglucose (2DG), an analog of glucose. Once phosphorylated by hexokinase it cannot be metabolized further and accumulates in the site of uptake. Consistent with previous studies that have localized hexokinase to glial cells, during neuronal activity, 2DG accumulation also occurs largely in glia, although oxygen is consumed largely by the neurons (Tsacopoulos and Magistretti, 1996). These findings suggest that neurons release a signal during activity that stimulates production of metabolic substrates by glial cells. In fact, glutamate has recently been shown to stimulate glycolysis in glial cells (Pellerin and Magistretti, 1994; Takahashi et al., 1995). Glutamate uptake by glia activates the Na-K ATPase and the elevated intracellular sodium concentration stimulates glycolysis, possibly by activating phosphoglycerate kinase in the plasma membrane (Pellerin and Magistretti, 1994). Thus neuronal activity induces release of glutamate, glutamate uptake stimulates glial aerobic glycolysis, which produces adenosine triphosphate (ATP) for use by the glial cells and lactate for release and uptake by the neurons, which convert the lactate to pyruvate for the Krebs cycle (Poitry et al., 1995). A wide variety of biochemical and electrophysiological observations are consistent with this model (Tsacopoulos and Magistretti, 1996). Thus glial cells play a pivotal role in coupling neuronal activity to energy metabolism.

Do Glial Calcium Waves Regulate Neuronal Excitability?

Neuronal excitation triggers waves of elevated intracellular calcium that propagate along a glial syncytium, both in vitro and in acutely isolated hippocampal slices (Smith, 1994; Dani and Smith, 1995). As these glial waves can be triggered by physiological levels of neuronal activity, glial calcium waves probably also occur in vivo. In the past few years, several groups have argued that the function of the glial calcium waves is to excite neurons. Parpura et al. (1994) and Hassinger et al. (1995) observed that glutamate is released by astrocytes during the calcium waves, thereby exciting overlying neurons. Nedergaard (1994) also found a rise in neuronal calcium in response to elevated astrocyte calcium, although in this case it appeared to be mediated by direct gap-junction coupling between the neurons and the glia. These latter culture conditions, however, were not optimal for the long-term culture of neurons, and the evidence that the excited cells were neurons was not convincing as they did not stain with neurofilament antibodies and could have been glial precursor cells.

Nonetheless, together the results of these groups convincingly show that elevation of intracellular calcium in astrocytes in culture leads to glutamate release from these astrocytes, which in turn stimulates overlying neurons. There are at least three reasons, however, why astrocytes in culture would be expected to artifactually accumulate and release glutamate. First, glial glutamine synthetase is downregulated in standard astrocyte cultures (Fages et al., 1988; Jackson et al., 1995). Second, culture medium contains high levels of glutamine (2 mM) which would be expected to inactivate glutamine synthetase by feedback inhibition (Waniewski, 1992). Lastly, glutaminase becomes aberrantly elevated in astrocytes in culture, while in vivo it has been found to be exclusively localized to neurons. Thus, intracellular glutamate levels appear to be elevated in astrocytes in culture, although there is much evidence that this does not normally occur in vivo. Activation of large conductance anion channels, permeable to glutamate, is a likely mechanism for the calcium-dependent release of glutamate from astrocytes in culture. In contrast, astrocytes in vivo express glutamine synthetase, but not glutaminase, and thus do not accumulate glutamate.

For these reasons, it is unlikely that glial calcium waves normally induce glutamate release and neuronal activation in vivo. There are pathological conditions, however, such as gliosis after neural injuries, in which astrocytes might downregulate glutamine synthetase or begin to express glutaminase. It would be interesting to determine whether reactive glial cells in vivo stain with glutamate or glutaminase antibodies; if so, perhaps glutamate release by reactive glial cells could help to explain the epileptic foci that are so often associated with injured brain.

What then is the function of syncytial astrocyte calcium waves? In view of the recent findings showing that glial cells play a pivotal role in coupling energy metabolism, and possibly neurotransmitter synthesis, to neuronal activity, a simple possibility is that elevation of astrocyte calcium might trigger glycogenolysis, thus enhancing the rate of generation of energy-generating precursors such as lactate and neurotransmitter precursors for the neurons (Magistretti et al., 1993). Increased release of substances such as lactate might also produce vasodilation of arterioles contacted by the astrocytes, thus enhancing local blood flow in response to neuronal activity (Bratitikos et al., 1993).

CONCLUSIONS

I have focused on the fundamental neuron–glial interactions that control the development and function of our brains. Taken together, recent studies indicate that a dynamic interplay between neurons and glial cells undoubtedly helps to shape developing neural circuits by controlling the survival and morphology of neurons, the growth of their axons, the number of synapses they form, the clustering and localization of their ion channels, and probably even the efficacy of their synapses (Fig. 3–4). While we have learned much, our knowledge of the development and function of glia is still rudimentary. In recent years we have learned a great deal about oligodendrocytes, but we still know remarkably little about the development and function of astrocytes. We have not yet identified astrocyte precursors or any of the signals that regulate their development in vivo. We cannot yet purify and culture astrocytes under serum-free conditions in which they survive and mimic their in vivo phenotype. This will certainly be crucial for future studies of their function and interactions with neurons. Particularly fundamental are a need to better understand the significance and nature of the metabolic compartmentation between neurons and glial cells, and the nature of the functional interactions of glial cells with synapses and blood vessels. While we have begun to identify a variety of important neuron–glial interactions, the molecules that mediate these processes in nearly all cases remain unidentified.

In addition to new in vitro approaches, new in vivo approaches will be important. Until recently there was no way to selectively eliminate glial cells in

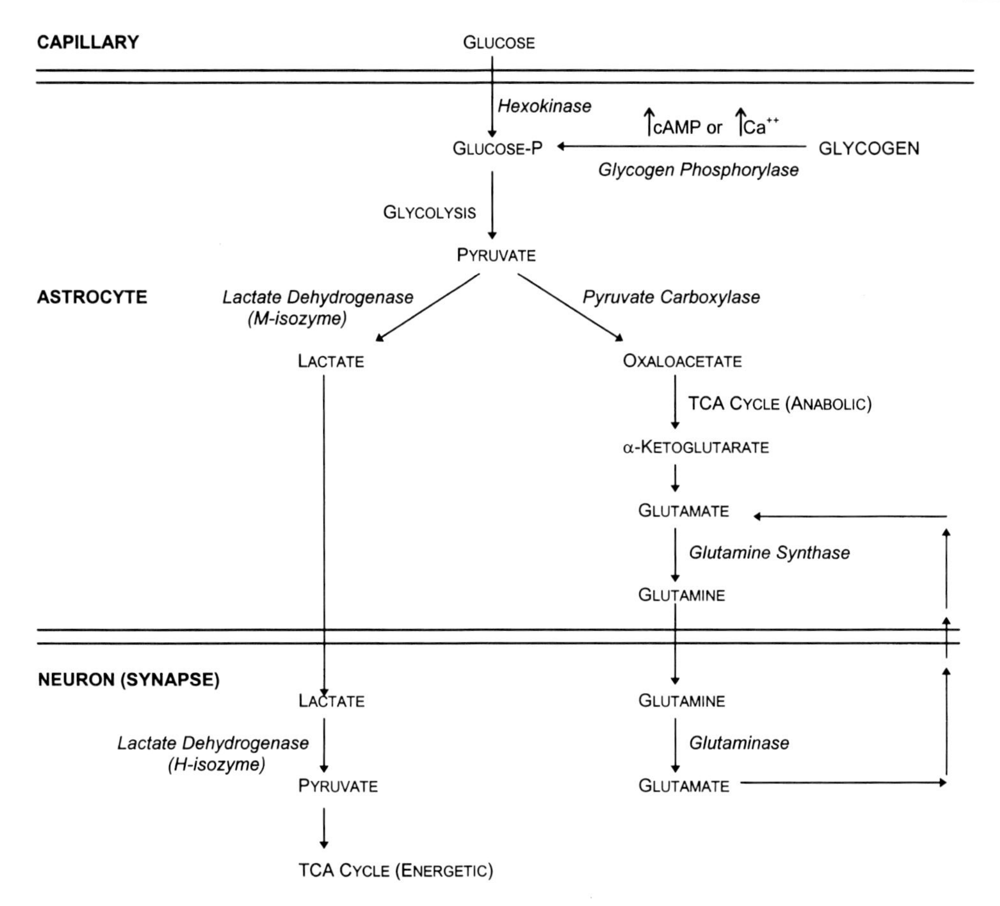

FIGURE 3–4 Metabolic compartmentation between neurons and glia. This figure summarizes current views on a metabolic division of labor between neurons and glial cells at the synapse (Pow and Robinson, 1994; Hamprecht and Dringen, 1995; Tsacopoulos and Magistretti, 1996). Enzymes that are specifically expressed by either neurons or glia are shown in italics. Neuronal activity is thought to elevate either Ca^{2+} or cAMP levels in glial cell. This triggers an increase in glial glycogen breakdown and aerobic glycolysis, generating ATP to meet the energy requirements of the glial cells, and pyruvate. The pyruvate serves as a precursor both for lactate and oxaloacetate. The lactate is released by the glia and taken up by neurons, which metabolize it via the TCA cycle to drive oxidative phosphorylation in synaptic mitochondria for neuronal energy production. In contrast, the oxaloacetate is metabolized in the glial cells via the TCA cycle to generate glutamine, which is released and taken up by the neurons to generate glutamate. Glutamate released by the neurons is also taken up by the glial cells and converted to glutamine (the glutamate–glutamine shuttle). Thus glucose that crosses the blood–brain barrier is metabolized by glial cells to generate energy precursors for both neurons and glia, and neurotransmitter precursors for the neurons.

vivo in order to explore how the brain develops and functions without them; it is likely that our ability to do this for *Drosophila* embryos will soon be extended to developing and adult mammals, particularly if a mammalian homologue of the *glial cells missing* gene exists. So there is much to do, but at least we can proceed knowing that it really does matter.

REFERENCES

Aghajanian, G.K., Bloom, F.E. (1967). The formation of synaptic junctions in developing rat brain. *Brain Res.* 6:716–727.

Arriza, J.L., Fairman, W.A., Wadiche, J.I., Murdoch, G.H., Kavanaugh, M.P., Amara, A.G. (1994). Functional comparisons of three glutamate transporter subtypes cloned from human motor cortex. *J. Neurosci.* 14:5559–5569.

Asou, H., Hamada, K., Sakota, T. (1995). Visualization of a single myelination process of an oligodendrocyte in culture by video microscopy. *Cell Struct. Func.* 20:59–70.

Baptista, C.A., Hatten, M.E., Blazeski, R., Mason, C.A. (1994). Cell-cell interactions influence survival and differentiation of purified Purkinje cells in vitro. *Neuron* 12:243–260.

Barres, B.A. (1991). New roles for glia. *J. Neurosci.* 11:3685–3694.

Barres, B.A., Raff, M.C. (1992). Proliferation of oligodendrocyte precursors depends on electrical activity in axons. *Nature* 361:258–260.

Barres, B.A., Raff, M.C. (1994). Control of oligodendrocyte number in the developing rat optic nerve. *Neuron* 12:935–942.

Barres, B.A., Raff, M.C. (1995). The axonal control of oligodendrocyte development. In *Glial Cell Development,* Jessen, K.R., Richardson, W.D., eds. Oxford: Bios Scientific.

Barres, B.A., Silverstein, B.E., Corey, D.P., Chun, L.L.Y. (1988). Morphological, immunological and electrophysiological characterization of rat retinal ganglion cells purified by panning. *Neuron* 1:791–803.

Barres, B.A., Hart, I.K., Coles, H.S.R., Burne, J.F., Voyvodic, J.T., Richardson, W.D., Raff, M.C. (1992a). Cell death and control of cell survival in the oligodendrocyte lineage. *Cell* 70:31–46.

Barres, B.A., Hart, I.K., Coles, H.S.R., Burne, J.F., Voyvodic, J.T., Richardson, W.D., Raff, M.C. (1992b). Cell death in the oligodendrocyte lineage. *J. Neurobiol.* 23:1221–1230.

Barres, B.A., Schmid, R., Sendtner, M., Raff, M.C. (1993a). Multiple extracellular signals are required for long-term oligodendrocyte survival. *Development* 118:283–295.

Barres, B.A., Jacobson, M.D., Schmid, R., Sendtner, M., Raff, M.C. (1993b). Does oligodendrocyte survival depend on axons. *Curr. Biol.* 3:489–497.

Barres, B.A., Raff, M.C., Gaese, F., Bartke, I., Dechant, G., Barde, Y.A. (1994a). A crucial role for neurotrophin-3 in oligodendrocyte development. *Nature* 367:371–375.

Barres, B.A., Lazar, M., Raff, M.C. (1994b). A novel role for thyroid hormone, glucocorticoids, and retinoic acid in timing oligodendrocyte differentiation. *Development* 120: 1097–1108.

Bartsch, U., Bandtlow C.E., Schnell, L., Bartsch, S., Spillman, A., Rubin, B.P., Hillenbrand, R., Montag, D., Schwab, M.E., Schachner, M. (1995). Lack of evidence that myelin-associated glycoprotein is a major inhibitor of axonal regeneration in the CNS. *Neuron* 15:1375–1381.

Bastmeyer, M., Jeserich, G., Stuermer C. (1994). Similarities and differences between fish oligodendrocytes and Schwann cells in vitro. *Glia* 11:300–314.

Bennett, V., Gilligan, D.M. (1993). The spectrin-based membrane skeleton and micron-scale organization of the plasma membrane. *Annu. Rev. Cell Biol.* 9:27–66.

Berl, S. (1973). Biochemical consequences of compartmentation of glutamate and associated metabolites. In *Metabolic Compartmentation in the Brain,* Balazs, R., Cremer, J.E., eds. London: MacMillan Press, pp. 3–20.

Bignami, A., Dahl, D. (1995). Gliosis. In *Neuroglia,* Kettenmann, H., Ransom, B., eds. New York: Oxford University Press, pp. 843–858.

Black, J.A., Friedman, B., Waxman, S.G., Elmer, L.W., Angelides, K.J. (1989). Immuno-ultrastructural localization of sodium channels at nodes of Ranvier and perinodal astrocytes in rat optic nerve. *Proc. R. Soc. Lond. Biol.* 238:38–57.

Black, J.A., Sontheimer, H., Oh, Y., Waxman, S.G. (1995). The oligodendrocyte, the peri-

nodal astrocyte, and the central node of Ranvier. In *The Axon*, Waxman, S.G., Kocsis, J.D., Stys, P.K., eds. New York: Oxford University Press.
Blinzinger, K.H., Kreutzberg, G.W. (1968). Displacement of synaptic terminals from regenerating motoneurons by microglial cells. *Z. Zellforsch. Mikrosk. Anat.* 85:145–147.
Bogler, O., Wren, D., Barnett, S., Land, H., Noble, M. (1990). Cooperation between two growth factors promotes extended self renewal and inhibits differentiation of oligodendrocyte-type-2 astrocyte (O-2A) progenitor cells. *Proc. Natl. Acad. Sci. U.S.A.* 87: 6368–6372.
Bray, G.M., Villegas-Perez, M.P., Vidal-Sanz, M., Carter, D.A., Aguayo, A.J. (1991). Neuronal and nonneuronal influences on retinal ganglion cell survival, axonal regrowth, and connectivity after axotomy. *Ann. N.Y. Acad. Sci.* 294:214–228.
Bratitikos, P.D., Pournaras, C.J., Munoz, J.L., Tsacopoulos, M. (1993). Microinjection of l-lactate in the preretinal vitreous induces segmental vasodilation in the inner retina of miniature pigs. *Invest. Ophthalmol. Vis. Sci.* 34:1744–1752.
Brockes, J.P., Lemke, G., Balzer, J. (1980). Purification and preliminary characterization of a glial growth factor from the bovine pituitary. *J. Biol. Chem.* 255:8374–8377.
Bruner, G., Simmons, M., Murphy, S. (1993). Astrocytes: targets and sources for purines, eicosanoids, and nitrosyl compounds. In *Astrocytes: Pharmacology and Function,* San Diego, Academic Press, pp. 89–108.
Buchanan, R.L., Benzer, S. (1993). Defective glia in the Drosophila brain degeneration mutant drop-dead. *Neuron* 10:839–850.
Bunge, R.P. (1994). The role of the Schwann cell in trophic support and regeneration. *J. Neurol.* 242 (1/ Suppl. 1):S19–21.
Bunge, R.P., Wood, P.M. (1987). Tissue culture studies of interactions between axons and myelinating cells of the central and peripheral nervous system. *Prog. Brain Res.* 71: 143–152.
Burne, J., Staple, J., Raff, M.C. (1996). Glial cells are increased and proportionally in transgenic optic nerves with increased numbers of axons. *J. Neurosci.* 16:2064–2073.
Butt, A.M., Ransom, B.R. (1993). Morphology of astrocytes and oligodendrocytes during development in the intact rat optic nerve. *J. Comp. Neurol.* 338:141–258.
Calver, A.R., Betsholtz, C., Heath, J.K., Duncan, I.D., Richardson, W.D. (1995). Control of oligodendrocyte development by PDGF-A in transgenic mice. *Soc. Neurosci. Abstr.* 228.3:552.
Campbell, G., Goring, H., Lin, T., Spana E, Andersson, S., Doe, C.Q., Tomlinson, A. (1994). RK2, a glial-specific homeodomain protein required for embryonic nerve cord condensation and viability in Drosophila. *Development* 120:2957–2966.
Campenot, R.B. (1994). NGF and the local control of nerve terminal growth. *J. Neurobiol.* 25:599–611.
Canoll, P.D., Musacchio, J.M., Hardy, R., Reynolds, R., Marchionni, M.A., Salzer, J. (1996). GGF/Neuregulin promotes the proliferation and survival and inhibits the differentiation of cells in the oligodendrocyte lineage. *Neuron* 17:229–243.
Chan, W., Kordeli, E., Bennett, V. (1993). 440kD Ankyrin-B, structure of the major developmentally regulated domain and selective localization in unmyelinated axons. *J. Cell Biol.* 123:1463–1473.
Charles, A.C. (1994). Glia-neuron intercellular calcium signaling. *Dev. Neurosci.* 16:196–206.
Chaudry, F.A., Lehre, K.P., van Lookeren Campagen, M., Ottersen, O.P., Danbolt, N.C., Storm-Mathisen, J. (1995). Glutamate transporters in glial plasma membranes: highly differentiated localizations revealed by quantitative ultrastructural immunocytochemistry. *Neuron* 15:711–720.
Chen, D.F., Jhaveri, S., Schneider, G.E. (1995). Intrinsic changes in developing retinal neurons result in regenerative failure of their axons. *Proc. Natl. Acad. Sci. U.S.A.* 92:7287–7291.

Cheng, L., Mudge, A.W. (1996). Cultured Schwann cells constitutively express the myelin protein P0. *Neuron* 16:309–319.

Chiu, S.Y., Ritchie, J.M. (1981). Evidence for the presence of potassium channels in the paranodal region of acutely demyelinated mammalian nerve fibers. *J. Physiol.* 313: 415–437.

Ciutat, D., Calderó, J., Oppenheim, R.W., Esquerda, J.E. (1996). Schwann cell apoptosis during normal development and after axonal degeneration induced by neurotoxins in the chick embryo. *J. Neurosci.* 16:3979–3990.

Cole, J.S., Messing, A., Trojanowski, J.Q., Lee, V. (1994). Modulation of axon diameter and neurofilaments by hypomyelinating Schwann cells in transgenic mice. *J. Neurosci.* 14: 6956–6966.

Colello, R., Schwab, M.E. (1994). A role for oligodendroglia in the stabilization of optic nerve axons. *J. Neurosci.* 14:6446–6452.

Colello, R., Pott, U., Schwab, M. (1994). The role of oligodendrocytes and myelin on axon maturation in developing rat retinofugal pathway. *J. Neurosci.* 14:2594–2605.

Colello, R.J., Devey, L.R., Imperato, E., Pott, U. (1995). The chronology of oligodendrocyte differentiation in the rat optic nerve: evidence for a signaling step initiating myelination in the CNS. *J. Neurosci.* 15:7665–7672.

Cowan, W.M., Fawcett, J.W., O'Leary, D.D.M., Stanfield, B.B. (1984). Regressive events in neurogenesis. *Science* 225:1258–1265.

Craig, A.M., Banker, G. (1994). Neuronal polarity. *Annu. Rev. Neurosci.* 17:267–310.

Danbolt, N.C. (1994). The high affinity uptake system for excitatory amino acids in the brain. *Prog. Neurobiol.* 44:377–396.

Dani, J.W., Smith, S.J. (1995). The triggering of astrocytic calcium waves by NMDA-induced neuronal activation. *Ciba Found. Symp.* 188:195–209.

David, S., Aguayo, A. (1981). Axonal elongation into peripheral nervous system bridges after central nervous system injury in adult rats. *Science* 214:931–933.

David, S., Miller, R.H., Patel, R., Raff, M.C. (1984). Effects of neonatal transection in the rat optic nerve: evidence that the oligodendrocyte-type-2 astrocyte cell lineage depends on axons for its survival. *J. Neurocytol.* 13:961–974.

DeBellard, M., Tang, S., Mukhopadhyay, G., Shen, Y., Filbin, M. (1996). Myelin-associated glycoprotein inhibits axonal regeneration from a variety of neurons via interaction with a sialoglycoprotein. *Mol. Cell. Neurosci.* 7:89–101.

Demerens, C., Stankoff, B., Logak, M., Anglade, P., Allinquant, B., Couraud, F., Zalc, B., Lubetzki, C. (1996). Induction of myelination in the central nervous system by electrical activity. *Proc. Natl. Acad. Sci. U.S.A.* 93:9887–9892.

deWaegh, S., Lee, V., Brady, S. (1992). Local modulation of neurofilament phosphorylation, axonal caliber, and slow axonal transport by myelinating Schwann cells. *Cell* 68: 451–463.

Dodd, J., Jessell, T.M. (1988). Axon guidance and the patterning of neuronal projections in vertebrates. *Science* 242:692–699.

Dong, Z., Brennan, A., Liu, N., Yarden, Y., Lefkowitz, G., Mirsky, R., Jessen, K.R. (1995). Neu differentiation factor is a neuron-glia signal and regulates survival, proliferation, and maturation of rat Schwann cell precursors. *Neuron* 15:585–596.

Doyle, J.P., Colman, D.R. (1993). Glial-neuron interactions and the regulation of myelin formation. *Curr. Opin. Cell. Biol.* 5:779–785.

Dubois-Dalcq, M., Behar, T., Hudson, L., Lazzarini, R.A. (1986). Emergence of three myelin proteins in oligodendrocytes cultured without neurons. *J. Cell Biol.* 102:384–392.

Dugandzija-Novakovic, S., Koszowski, A.G., Levinson, S.R., Shrager, P. (1995). Clustering of sodium channels and node of Ranvier formation in remyelinating axons. *J. Neurosci.* 15:492–503.

Edgar, A.D., Pfeiffer, S.E. (1985). Extracts from neuron-enriched cultures of chick telencephalon stimulate the proliferation of rat oligodendrocytes. *Dev. Neurosci.* 7:206–215.

Einheber, S., Hannocks, M., Metz, C., Rifkin, D., Salzer, J. (1995). Transforming growth factor-beta regulates axon-Schwann cell interactions. *J. Cell Biol.* 129:443–458.

Ellis, R.E., Yuan, J., Horvitz, H.R. (1991). Mechanisms and functions of cell death. *Annu. Rev. Cell Biol.* 7:663–698.

England, J.D., Gamboni, F., Levinson, S.R., Finger, T.E. (1990). Formation of new distribution of sodium channels along demyelinated axons. *Proc. Natl. Acad. Sci. U.S.A.* 87: 6777–6780.

England, J.D., Gamboni, F., Levinson, S.R. (1991). Increased numbers of sodium channels form along demyelinated axons. *Brain Res.* 548:334–337.

Eyer, J., Peterson, A. (1994). Neurofilament deficient axons and perikaryal aggregates in viable transgenic mice expressing a neurofilament beta galactosidase fusion protein. *Neuron* 12:389–405.

Fages, C., Khelil, M., Rolland, B., Bridoux, A.M., Tardy, M. (1988). Glutamine synthetase: a marker of an astroglial subpopulation in primary cultures of defined brain areas. *Dev. Neurosci.* 10:47–56.

Feng Chen, D., Jhaveris, S., Schneider, G.E. (1995). Intrinsic changes in developing retinal neurons result in regenerative failure of their axons. *Proc. Natl. Acad. Sci. U.S.A.* 92: 7287–7291.

Foran, D.R., Peterson, A.C. (1992). Myelin acquisition in the CNS of the mouse revealed by an MBP-LacZ transgene. *J. Neurosci.* 12:4890–4897.

Friede, R.L. (1972). Control of myelin formation by axon caliber (with a model of the control mechanism). *J. Comp. Neurol.* 144:233–252.

Fulcrand, J., Privat, A. (1977). Neuroglia reactions secondary to Wallerian degeneration in the optic nerve of the postnatal rat: ultrastructural and quantitative study. *J. Comp. Neurol.* 176:189–224.

Fulton, B.P., Burne, J.F., Raff, M.C. (1992). Visualization of O-2A progenitor cells in developing and adult rat optic nerve by quisqualate-stimulated cobalt uptake. *J. Neurosci.* 12:4816–4833.

Gard, A.L., Pfeiffer, S.E. (1990). Two proliferative stages of the oligodendrocyte lineage under different mitogenic control. *Neuron* 5:615–625.

Gard, A.L., Pfeiffer, S.E. (1993). Glial cell mitogens bFGF and PDGF differentially regulate development of O4+GalC+oligodendrocyte progenitors. *Dev. Biol.* 159:618–630.

Giuilian, D., Young, D.G. (1986). Brain peptides and glial growth. II. Identification of cells that secret glia-promoting factors. *J. Cell Biol.* 102:812–820.

Goodman, C.S., Doe, C.Q. (1993). Embryonic development of the Drosophila central nervous system. In *The Development of Drosophila Melanogaster,* Bate, M., Martinez Arias, A., eds. Cold Spring Harbor, NY: CSH Laboratory Press, pp. 1131–1206.

Grinspan, J., Marchionni, M., Reeves, M., Coulaloglou, M., Scherer, S. (1996). Axonal interactions regulate Schwann cell apoptosis in developing peripheral nerve: neuregulin receptors and the role of neuregulins. *J. Neurosci.* 16:6107–6118.

Grove, E.A., Williams, B.P., Li, D., Hajihosseini, M., Friedrich, A., Price, J. (1993). Multiple restricted lineages in the embryonic rat cerebral cortex. *Development* 117:5535–5561.

Guenard, V., Gwynn, L.A., Wood, P.M. (1995). Transforming growth factor beta blocks myelination but not ensheathment of axons by Schwann cells in vitro. *J. Neurosci.* 15: 419–428.

Halter, D.A., Urban, J., Rickert, C., Ner, S.S., Ito, K., Travers, A.A., Technau, G.M. (1995). The homeobox gene repo is required for the differentiation and maintenance of glia function in the embryonic nervous system of Drosophila melanogaster. *Development* 121:317–332.

Hamprecht, B., Dringen, R. (1995). Energy metabolism. In *Neuroglia,* Kettenmann, H., Ransom, B., eds. New York: Oxford University Press, pp. 473–487.

Hannson, H.A., Holmgren, A., Norstedt, G., Rosell, B. (1989). Changes in the distribution of insulin-like growth factor I, thioredoxin, thioredoxin reductase and ribonucleotide reductase during the development of the retina. *Exp. Eye Res.* 48:411–420.

Hardy, R., Reynolds, R. (1991). Proliferation and differentiation potential of rat forebrain oligodendrocyte progenitors both in vitro and in vivo. *Development* 111:1061–1080.

Hardy, R., Reynolds, R. (1993). Neuron-oligodendroglial interactions during central nervous system development. *J. Neurosci. Res.* 36:121–126.

Hart, I.K., Richardson, W.D., Heldin, C.H., Westermark, B., Raff, M.C. (1989). PDGF receptors on cells of the oligodendrocyte-type-2 astrocyte (O-2A) cell lineage. *Development* 105:595–603.

Hassinger, T.D., Atkinson, P., Strecker, G.J., Whalen, L.R., Dudek, F.E., Kossel, A.H., Kater, S.B. (1995). Evidence for glutamate-mediated activation of hippocampal neurons by glial calcium waves. *J. Neurobiol.* 28:159–170.

Herdegen, T., Kiessling, M., Bele, S., Bravo, R., Zimmermann, M., Gass, P. (1993). The KROX-20 transcription factor in the rat central and peripheral nervous systems: novel expression pattern of an immediate early gene-encoded protein. *Neuroscience* 57: 41–52.

Hess, A., Young, J.Z. (1952). The nodes of Ranvier. *Proc. R. Soc. Lond. Biol.* 140:301–319.

Hochstein, S., Shapley, R.M. (1976). Quantitative analysis of retinal ganglion cell classification. *J. Physiol.* 262:237–264.

Hoffman, P., Cleveland, D., Griffin, J.W., Landes, P.W., Cowan, N.J., Price, D.L. (1987). Neurofilament gene expression: a major determinant of axonal caliber. *Proc. Natl. Acad. Sci. U.S.A.* 84:3472–3476.

Hosli, E., Hosli, L. (1993). *Prog. Neurobiol.* 40:477–506.

Hosoya, T., Takizawa, K., Nitta, K., Hotta, Y. (1995). Glial cells missing: a binary switch between neuronal and glial determination in Drosophila. *Cell* 82:1025–1036.

Hsieh, S., Kidd, G., Crawford, T., Xu, Z., Lin, W., Trapp, B., Cleveland, D., Griffin, J.W. (1994). Regional modulation of neurofilament organization by myelination in normal axons. *J. Neurosci.* 14:6392–6401.

Hudson, L.D., Friedrich, V.L., Behar, T., Dubois-Dalcq, M., Lazzarini, R.A. (1989). The initial events in myelin synthesis: orientation of proteolipid protein in the plasma membrane of cultured oligodendrocytes. *J. Cell Biol.* 109:717–727.

Hunter, S.F., Bottenstein, J.E. (1989). Bipotential glial progenitors are targets of neuronal cell line-derived growth factors. *J. Neurosci. Res.* 28:574–582.

Huxley, A.F., Stampfli, R. (1949). Evidence for saltatory conduction in peripheral myelinated fibers. *J. Physiol.* 108:315–339.

Jacobs, J.R., Goodman, C.S. (1989). Embryonic development of axon pathways in the Drosophila CNS. *J. Neurosci.* 9:2402–2422.

Jackson, M.J., Zielke, H.R., Max, S.R. (1995). Effect of dbcAMP and dexamethasone on glutamine synthetase gene expression in rat astrocytes in culture. *Neurochem. Res.* 20: 201–207.

Jessen, K.R., Mirsky, R., Morgan, L. (1991). Role of cyclic AMP and proliferation controls in Schwann cell differentiation. *Ann. N.Y. Acad. Sci.* 633:78–89.

Joe, E., Angelides, K. (1992). Clustering of the voltage-dependent sodium channels on axons depends on Schwann cell contact. *Nature* 356:333–335.

Jones, B.W., Fetter, R.D., Tear, G., Goodman, C.S. (1995). Glial cells missing: a genetic switch that controls glial versus neuronal fate. *Cell* 82:1013–1024.

Kaplan, M., Meyer-Franke A., Lambert, S., Bennett, V., Levinson, S.R., Barres, B.A. (1996). Soluble glial signals induce regularly-spaced sodium channel clusters along CNS axons. *Soc. Neurosci. Abstr.* 22:32 (23.2).

Keyser, D.O., Pellmar, T.C. (1994). Synaptic transmission is the hippocampus: critical role for glial cells. *Glia* 10:237–243.

Kim, E., Niethammer, M., Rothschild, A., Jan, Y.N., Sheng, M. (1995). Clustering of Shaker type K channels by interaction with a family of membrane associated guanylate kinases. *Nature* 378:85–88.

Kobayashi, H., Watanabe, E., Murakami, F. (1995). Growth cones of dorsal root ganglion but not retina collapse and avoid oligodendrocytes in culture. *Dev. Biol.* 168:383–394.

Kordeli, E., Lambert, S., Bennett, V. (1995). Ankyrin-G: a new ankyrin gene with neural-specific isoforms localized at the axonal initial segment and node of Ranvier. *J. Biol. Chem.* 270:2352–2359.

Laake, J.H., Slyngstad, T.A., Haug, F., Ottersen, O.P. (1995). Glutamine from glial cells is essential for maintenance of the nerve terminal pool of glutamate. *J. Neurochem.* 65: 871–881.

Lam, K., Sefton, A.J., Bennet, M.R. (1982). Loss of axons from the optic nerve of the rat during early postnatal development. *Dev. Brain Res.* 3:487–491.

Lambert, S., Bennett, V. (1993). From anemia to cerebellar dysfunction. *Eur. J. Biochem.* 211: 1–6.

Lambert, S., Michaely, P., Davis, J.Q., Bennett, V. (1996). Ankyrin clustering in the coordinate recruitment of ion channels and adhesion molecules during morphogenesis of the node of Ranvier. *Am. Soc. Cell Biol. Abstr.* 568:98a.

Lee, M.K., Cleveland, D.W. (1996). Neuronal intermediate filaments. *Annu. Rev. Neurosci.* 19:187–212.

Lein, P., Johnson, M., Guo, X., Reuger, D., Higgins, D. (1995). Osteogenic protein-1 induced dendritic growth in rat sympathetic neurons. *Neuron* 15:597–605.

Levi, A.D., Bunge, R.P., Lofgren, J.A., Meima, L., Hefti, F., Nikolics, K., Sliwkowski, M.X. (1995). The influence of heregulins on human Schwann cell proliferation. *J. Neurosci.* 15:1329–1340.

Levine, J.M. (1989). Neuronal influences on glial progenitor cell development. *Neuron* 3:103–113.

Levine, S.M., Goldman, J.E. (1988). Embryonic divergence of oligodendrocyte and astrocyte lineages in developing rat cerebrum. *J. Neurosci.* 8:3992–4006.

Linden, R., Perry, V.H. (1983). Massive retinotectal projection in rats. *Brain Res.* 272:145–149.

Louis, J.C., Magal, E., Takayama, S., Varon, S. (1993). CNTF protection of oligodendrocytes against natural and tumor necrosis factor-induced death. *Science* 259:689–692.

Magistretti, P.J., Sorg, O., Martin, J.L. (1993). Regulation of glycogen metabolism in astrocytes. In *Astrocytes: Pharmacology and Function,* San Diego: Academic Press, pp. 243–265.

Mansour-Robaey, D.B., Clarke, Y.C., Clarke, Y.C., Bray, G.M., Aguayo, A. (1994). Effects of ocular injury and administration of brain-derived neurotrophic factor on survival and regrowth of axotomized retinal ganglion cells. *Proc. Natl. Acad. Sci. U.S.A.* 91:1632–1636.

Marchionni, M.A., Goodearl, A.D.J., Chen, M.S., Bermingham-McDonough, O., Kirk, C., Hendricks, M., Danehy, F., Misumi, D., Sudhalter, J., Kobyashi, K. (1993). Glial growth factors are alternatively spliced erbB2 ligands expressed in the nervous system. *Nature* 362:312–318.

Martenson, R.E. (1992). *Myelin: Biology and Chemistry.* Boca Raton, FL: CRC Press.

McKerracher, L., David, S., Jackson, D.L., Kottis, V., Dunn, R.J., Braun, P.E. (1994). Identification of myelin-associated glycoprotein as a major myelin-derived inhibitor of neurite outgrowth. *Neuron* 13:805–811.

McKinnon, R.D., Matsui, T., Dubois-Dalcq, M., Aaronson, S.A. (1990). FGF modulates the PDGF-driven pathway of oligodendrocyte development. *Neuron* 5:603–614.

McKinnon, R.D., Piras, G., Ida, J.A., Dubois-Dalcq, M. (1993). A role for TGF-B in oligodendrocyte differentiation. *J. Cell Biol.* 121:1397–1407.

McMorris, F.A., Dubois-Dalcq, M. (1988). Insulin-like growth factor I promotes cell proliferation and oligodendroglial commitment in rat glial progenitor cells developing in vitro. *J. Neurosci. Res.* 21:199–209.

Mennerick, S., Zorumski, C.F. (1994). Glial contributions to excitatory neurotransmission in cultured hippocampal cells. *Nature* 368:59–62.

Mey, J., Thanos, S. (1993). Intravitreal injections of neurotrophic factors support the survival of axotomized retinal ganglion cells in adult rats in vivo. *Brain Res.* 602:304–317.

Meyer, D., Birchmeier, C. (1995). Multiple essential functions of neuregulin in development. *Nature* 378:386–390.

Meyer-Franke, A., Barres, B.A. (1996). Oligodendrocytes do not myelinate axons by default. *Soc. Neurosci. Abstr. In press.*

Meyer-Franke, A., Barres, B.A. (1994). Myelination without myelin-associated glycoprotein. *Curr. Biol.* 4:847–850.

Meyer-Franke, A., Kaplan, M., Pfrieger, F., Barres, B.A. (1995). Characterization of the signaling interactions that promote the survival and growth of developing retinal ganglion cells in culture. *Neuron* 15:805–819.

Mi, H. (1995). Identification and localization of potassium channels in adult rat Schwann cells. Ph.D. thesis, Stanford University.

Mi, H., Deerinck, T.J., Ellisman, M.H., Schwarz, T.L. (1995). Differential distribution of closely related potassium channels in rat Schwann cells. *J. Neurosci.* 15:3761–3774.

Mi, H., Deerinck, T.J., Jones, M., Ellisman, M.H., Schwarz, T.L. (1996). Inwardly rectifying potassium channels that may participate in potassium buffering are localized in microvilli of Schwann cells. *J. Neurosci.* 16:2421–2429.

Miller, R.H., David, S., Patel, R., Abney, E.R., Raff, M.C. (1985). A quantitative immunohistochemical study of macroglial cell development in the rat optic nerve: in vivo evidence for two distinct astrocyte lineages. *Dev. Biol.* 111:35–41.

Mirsky, R., Jessen, K.R. (1996). Schwann cell development, differentiation and myelination. *Curr. Opin. Neurobiol.* 6:89–96.

Mirsky, R., Winter, J., Abney, E.R., Pruss, R.M., Gavrilovic, J., Raff, M.C. (1980). Myelin-specific proteins and glycolipids in rat Schwann cells and oligodendrocytes in culture. *J. Cell Biol.* 84:483–494.

Montag, D., Giese, K.P., Bartsch, U., Martini, R., Lang, Y., Bluthmann, H., Karthigasan, J., Kirschner, D.A., Wintergerst, E.S., Nave, K.A. (1994). Mice deficient for the myelin-associated glycoprotein show subtle abnormalities in myelin. *Neuron* 13:229–246.

Morrissey, T.K., Levi, A.D., Nuijens, A., Sliwkowski, M.X., Bunge, R.P. (1995). Axon-induced mitogenesis of human Schwann cells involves heregulin and p185erbB2. *Proc. Natl. Acad. Sci. U.S.A.* 92:1431–1435.

Mudhar, H.S., Pollock, R.A., Wang, C., Stiles, C.D., Richardson, W. (1993). PDGF and its receptors in the developing rodent retina and optic nerve. *Development* 118:539–552.

Mukhopadhyay, G., Doherty, P., Walsh, F.S., Crocker, P., Filbin, M. (1994). A novel role for myelin-associated glycoprotein as an inhibitor of axonal regeneration. *Neuron* 13:757–767.

Murphy, P., Topilko P., Schneider-Maunoury, S., Seitanidou, T., Baron-Van Evercooren, A., and Charnay, P. (1996). The regulation of Krox-20 expression reveals important steps in the control of peripheral glia development. *Development* 122:2847–2857.

Nedergaard, M. (1994). Direct signaling from astrocytes to neurons in cultures of mammalian brain cells. *Science* 263:1768–1771.

Noble, M., Murray, K. (1984). Purified astrocytes promote the in vitro divisions of a bipotential glial progenitor cell. *EMBO J.* 3:2243–2247.

Noble, M., Murray, K., Stroobant, P., Waterfield, M.D., Riddle, P. (1988). PDGF promotes division and motility and inhibits premature differentiation of the oligodendrocyte-type-2 astrocyte progenitor cell. *Nature* 333:560–562.

Nona, S., Stafford, C.A. (1995). Glial repair at the lesion site of regenerating goldfish spinal cord. *J. Neurosci. Res.* 42:350–356.

Nona, S.N., Stafford, C.A., Duncan, A., Cronly-Dillon, J.R., Scholes, J. (1994). Myelin repair in the regenerating goldfish visual pathway. *J. Neurocytol.* 23:400–409.

Northmore, D. (1987). Neural activity in the regenerating optic nerve of the goldfish. *J. Physiol.* 391:299–312.

Parpura, V., Basarsky, T.A., Liu, F., Jeftinija, K., Jeftinija, S., Haydon, P.D. (1994). Glutamate-mediated astrocyte-neuron signaling. *Nature* 369:744–747.

Paterson, J.A., Privat, A., Ling, A., Leblond, C.P. (1973). Investigation of glial cells in semithin sections. III. Transformation of subependymal cells into glial cells, as shown by radioautography after 3H-thymidine injection into the lateral ventricle of the brain of young rats. *J. Comp. Neurol.* 149:83–102.

Pellerin, L., Magistretti, P.J. (1994). Glutamate uptake into astrocytes stimulates aerobic glycolysis: a mechanism coupling neuronal activity to glucose utilization. *Proc. Natl. Acad. Sci. U.S.A.* 91:10625–10629.

Perry, V.H., Brown, M.C., Lunn, E.R. (1991). Very slow retrograde and Wallerian degeneration in the central nervous system of C57Bl/Ola mice. *Eur. J. Neurosci.* 3:102–105.

Peters, A., Palay, S.L., Webster, H.F. (1991). *The Fine Structure of the Nervous System.* New York: Oxford University Press.

Pfrieger, F., Barres, B.A. (1995). What the fly's glia tell the fly's brain. *Cell* 82:671–674.

Pfrieger, F., Barres, B.A. (1996). A role for astrocytes in synapse formation in vitro. *Soc. Neurosci. Abstr.* 22:1949 (766.5).

Poitry, C.L., Potry, S., Tsacopoulos, M. (1995). Lactate released by Muller glial cells is metabolized by photoreceptors from mammalian retina. *J. Neurosci.* 15:5179–5191.

Pow, D.V., Robinson, S.R. (1994). Glutamate in some retinal neurons is derived solely from glia. *Neuroscience* 60:355–366.

Pow, D.V., Crook, D.K., Wong, R.O.L. (1994). Early appearance and transient expression of putative amino acid neurotransmitters and related molecules in the developing rabbit retina. *Vis. Neurosci.* 11:1115–1134.

Pringle, N., Collarini, E.J., Mosley, M.J., Heldin, C.-H., Westermark, B., Richardson, W.D. (1989). PDGF A chain homodimers drive proliferation of bipotential (O-2A) glial progenitor cells in the developing rat optic nerve. *EMBO J.* 8:1049–1056.

Pringle, N.P., Mudhar, H.S., Collarini, E.J., Richardson, W.D. (1992). PDGF receptors in the rat CNS: during late neurogenesis, PDGF alpha-receptor expression appears to be restricted to glial cells of the oligodendrocyte lineage. *Development* 115:535–551.

Privat, A., Valat, J., Fulcrand, J. (1981). Proliferation of neuroglial cells in the degenerating optic nerve of young rat: an autoradiographic study. *J. Neuropathol. Exp. Neurol.* 40: 46–60.

Quarles, R.H., Colman, D.R., Salzer, J.L., Trapp, B.D. (1992). Myelin-associated glycoprotein: structure-function relationships and involvement in neurological diseases. In *Myelin: Biology and Chemistry,* Martenson, R.E., ed. Ann Arbor: CRC Press, pp. 413–448.

Raff, M.C. (1992). Social controls on cell survival and death: an extreme view. *Nature* 356: 397–400.

Raff, M.C., Abney, E.R., Fok-Seang, J. (1985). Reconstitution of a developmental clock in vitro: a critical role for astrocytes in the timing of oligodendrocyte differentiation. *Cell* 42:61–69.

Raff, M.C., Miller, R.H., Noble, M. (1983). A glial progenitor cell that develops in vitro into an astrocyte or an oligodendrocyte depending on culture medium. *Nature* 303:390–396.

Raff, M.C., Lillien, L.E., Richardson, W.D., Burne, J.F., Noble, M.D. (1988). Platelet-derived growth factor from astrocytes drives the clock that times oligodendrocyte development in culture. *Nature* 333:562–565.

Raff, M.C., Barres, B.A., Burne, J.F., Coles, H.S. Ishizaki, Y., Jacobsen, M.D. (1993). Programmed cell death and the control of cell survival: lessons from the nervous system. *Science* 262:695–700.

Richardson, W.D., Pringle, N., Mosley, M.J., Westermark, B., Dubois-Dalcq, M. (1988). A role for platelet-derived growth factor in normal gliogenesis in the central nervous system. *Cell* 53:309–319.

Ritchie, J.M., Black, J.A., Waxman, S.G., Angelides, K.J. (1990). Sodium channels in the cytoplasm of Schwann cells. *Proc. Natl. Acad. Sci. U.S.A.* 87:9290–9294.

Roberts, R.C., McCarthy, K.E., Du, F., Ottersen, O.P., Okuno, E., Schwarcz, R. (1995). 3-Hydroxyanthranilic acid oxygenase-containing astrocytic processes surround glutamate-containing axon terminals in the rat striatum. *J. Neurosci.* 15:1150–1161.

Rosenbaum, C., Krasnoelski, A.L., Marchionni, M.A., Brackenbury, R.W., Ratner, N. (1996). Neuregulins and Neu/erbB2 are required for response of Schwann cells to multiple growth factors. *Trans. Am. Soc. Neurochem.* 27.

Rothstein, J.D., Dykes-Hoberg, M., Pardo, C.A., Bristol, L.A., Jin, L., Kuncl, R.W., Kanai, Y., Hediger, M.A., Wang, Y., Schielke, J.P., Welty, D.F. (1996). *Neuron* 16:675–686.

Rothstein, J.D., Martin, L., Levey, A.I., Dykes-Hoberg, M., Jin, L., Wu, D., Nash, N., Kuncl, R.W. (1994). Localization of neuronal and glial glutamate transporters. *Neuron* 13:713–725.

Rubinstein, C.T., Shrager, P. (1990). Remyelination of nerve fibers in the transected frog sciatic nerve. *Brain Res.* 524:303–312.

Samorajski, T., Friede, R.L. (1968). A quantitative EM study of myelination in the pyramidal tract. *J. Comp. Neurol.* 134:323–338.

Sanchez, I., Hassinger, L., Paskevich, P., Shine, D., and Nixon, R. (1996). Oligodendroglia regulate the regional expansion of axon caliber and local accumulation of neurofilaments during development independently of myelin formation. *J. Neurosci.* In press.

Schecterson, L.C., Bothwell, M. (1992). Novel roles for neurotrophins are suggested by BDNF and NT-3 mRNA expression in developing neurons. *Neuron* 9:449–463.

Schell, M.J., Molliver, M.E., Snyder, S.H. (1995). D-serine, an endogenous synaptic modulator: localization to astrocytes and glutamate-stimulated release. *Proc. Natl. Acad. Sci. U.S.A.* 92:3948–3952.

Schnell, L., Schwab, M.E. (1990). Axonal regeneration in the rat spinal cord produced by an antibody against myelin-associated neurite growth inhibitors. *Nature* 343:269–272.

Schousboe, A., Westergaard, N., Sonnewald, U., Peterson, S., Huang, R., Peng, L., Hertz, L. (1993). Glutamate and glutamine metabolism and compartmentation if astrocytes. *Dev. Neurosci.* 15:359–366.

Schousboe, A., Westergaard, N., Sonnewald, U., Petersen, S.B., Yu, A., Hertz, L. (1992). Regulatory role of astrocytes for neuronal biosynthesis and homeostasis of glutamate and GABA. *Prog. Brain Res.* 94:199–211.

Schwab, M.E. (1995). Oligodendrocyte inhibition of nerve fiber growth and regeneration in the mammalian central nervous system. In *Neuroglia*, Kettenmann, H., Ransom, B., eds. New York: Oxford University Press, pp. 859–868.

Schwab, M.E., Schnell, L. (1989). Region-specific appearance of myelin constituents in the developing rat spinal cord. *J. Neurocytol.* 18:161–169.

Schwab, M.E., Kapfhammer, J.P., Bandtlow, C.E. (1993). Inhibitors of neurite growth. *Annu. Rev. Neurosci.* 16:565–595.

Schwalb, J.M., Boulis, N.M., Gu, M., Winickoff, J., Jackson, P.S., Irwin, N., Benowitz, L. (1995). Two factors secreted by the goldfish optic nerve induce retinal ganglion cells to regenerate axons in culture. *J. Neurosci.* 15:5514–5525.

Shah, N., Marchionni, M., Isaacs, I., Stroobant, P., Anderson, D.J. (1994). Glial growth factor restricts mammalian neural crest stem cells to a glial fate. *Cell* 77:349–360.

Shen, S., Barres, B.A. (1996). Promotion of the survival of developing retinal ganglion cells after axotomy. *Soc. Neurosci. Abstr.* 22:764 (305.11).

Shewan, D., Berry, M., Cohen, J. (1995). Extensive regeneration in vitro by early embryonic neurons on immature and adult CNS tissue. *J. Neurosci.* 15:2057–2062.

Shi, G., Nakahira, K., Hammond, S., Rohdes, K.J., Schechter, L.E., Trimmer, J.S. (1996). Beta subunits promote K channel surface expression through effects early in biosynthesis. *Neuron* 16:843–852.

Shrager, P. (1987). The distribution of Na and K channels in single demyelinated axons of the frog. *J. Physiol.* 392:587–602.

Shrager, P. (1989). Sodium channels in single demyelinated mammalian axons. *Brain Res.* 483:149–154.

Siegel, G.J., Agranoff, B.W., Albers, R.W., Molinoff, P.B. (1995). *Basic Neurochemistry. 5th ed.* New York: Raven Press.
Sievers, J., Bamberger, C., Debus, O.M., Lucius, R. (1995). Regeneration in the optic nerve of adult rats: influences of cultured astrocytes and optic nerve grafts of different ontogenetic stages. *J. Neurocytol.* 24:783–793.
Skoff, R.P. (1990). Gliogenesis in rat optic nerve: astrocytes are generated in a single wave before oligodendrocytes. *Dev. Biol.* 139:149–168.
Small, R.K., Riddle, P., Noble, M. (1987). Evidence for migration of oligodendrocyte-type-2 astrocyte progenitor cells into the developing rat optic nerve. *Nature* 328:155–157.
Smith, S.J. (1994). Neuromodulatory astrocytes. *Curr. Biol.* 4:807–810.
Smith-Thomas, L., Stevens, J., Fok-Seang, J., Faissner, A., Rogers, J., Fawcett, J.W. (1995). Increased axonal regeneration in astrocytes grown in the presence of proteoglycan synthesis inhibitors. *J. Cell Sci.* 108:1307–1315.
Snipes, G.J., Suter, U., Shooter, E.M. (1993). The genetics of myelin. *Curr. Opin. Neurobiol.* 3:694–702.
Son, Y.J., Thompson, W.J. (1995a). Nerve sprouting in muscle is induced and guided by processes extended by Schwann cells. *Neuron* 14:133–141.
Son, Y.J., Thompson, W.J. (1995b). Schwann cell processes guide regeneration of peripheral axons. *Neuron* 14:125–132.
Speidel, C.C. (1964). In vivo studies of myelinated fibers. *Int. Rev. Cytol.* 16:173–231.
Srinivasan, Y., Elmer, L., Davis, J., Bennett, V., Angelides, K. (1988). Ankyrin and spectrin associate with voltage dependent sodium channels in brain. *Nature* 333:177–180.
Stampfli, R. (1954). Saltatory conduction in nerve. *J. Physiol.* 34:101–112.
Stockli, K.A., Lillien, L.E., Noher-Noe, M., Breitfeld, G., Hughes, R.A., Raff, M.C., Thoenen, H., Sendnter, M. (1991). Regional distribution, developmental changes, and cellular localization of CNTF-mRNA and protein in the rat brain. *J. Cell Biol.* 115:447–459.
Stuermer, C. (1995). Glial cells and axonal regeneration in the central nervous system. In *Neuroglia,* Kettenmann, H., Ransom, B., eds. New York: Oxford University Press, pp. 905–918.
Sun, Y., Landis, S.C., Zigmond, R.E. (1996). Signals triggering the induction of leukemia inhibitory factor in sympathetic superior cervical ganglia and their nerve trunks after axonal injury. *Mol. Cell. Neurosci.* 7:152–163.
Suter, U., Welcher, A.A., Snipes, G.J. (1993). Progress in the molecular understanding of hereditary peripheral neuropathies reveals new insights into the biology of the peripheral nervous system. *Trends Neurosci.* 16:50–56.
Syroid, D., Maycox, P., Burrola, P., Liu, N., Wen, D., Lee, K., Lemke, G., Kilpatrick, T. (1996). Cell death in the Schwann cell lineage and its regulation by beta-neuregulin. *Proc. Natl. Acad. Sci. U.S.A.* 93:9229–9234.
Takahashi, S., Driscoll, B.F., Law, M.J., Sokoloff, L. (1995). Role of sodium and potassium ions in regulation of glucose metabolism in cultured astroglia. *Proc. Natl. Acad. Sci. U.S.A.* 92:4616–4620.
Temple, S., Raff, M.C. (1985). Differentiation of a bipotential glial progenitor cell in single cell microculture. *Nature* 313:223–225.
Temple, S., Raff, M.C. (1986). Clonal analysis of oligodendrocyte development in culture: evidence for a developmental clock that counts cell divisions. *Cell* 44:773–779.
Topilko, P., Schneider-Maunoury, S., Levi, G., Baron-Van Evercooren, A., Chennoufi, A.B., Seitanidou, T., Babinet, C., Charney, P. (1994). Krox-20 controls myelination in the peripheral nervous system. *Nature* 371:796–799.
Trachtenberg, J.T., Thompson, W.J. (1996). Schwann cell apoptosis at developing neuromuscular junctions is regulated by glial growth factor. *Nature* 379:174–177.
Tropea, M., Johnson, M., Higgins, D. (1988). Glial cells promote dendrite development in rat sympathetic neurons in vitro. *Glia* 1:380–392.
Tsacopoulos, M., Magistretti, P.J. (1996). Metabolic coupling between glia and neurons. *J. Neurosci.* 16:877–885.

Valat, J., Privat, A., Fulcrand, J. (1983). Multiplication and differentiation of glial cells in the optic nerve of the postnatal rat. *Anat. Embryol.* 167:335–346.

Vabnick, I., Novakovi, S.D., Levinson, S.R., Schachner, M., Shrager, P. (1996). The clustering of axonal sodium channels during development of the peripheral nervous system. *J. Neurosci.* In press.

Vartanian, T., Corfas, G., Li, Y., Fischbach, G., Stefansson, K. (1994). A role for the acetylcholine receptor-inducing protein ARIA in oligodendrocyte development. *Proc. Natl. Acad. Sci. U.S.A.* 91:11626–11630.

Verdi, J.M., Groves, A., Farinas, I., Jones, K., Marchionni, M.A., Reichardt, L.F., Anderson, D.J. (1996). A reciprocal cell-cell interaction mediated by NT-3 and neuregulins controls the early survival and development of sympathetic neuroblasts. *Neuron* 16:515–527.

Voyvodic, J.T. (1989). Target size regulates calibre and myelination of sympathetic axons. *Nature* 342:430–433.

Wang, H., Kunkel, D., Martin, T., Schwartzkroin, P., Tempel, B. (1993). Heteromultimeric K channels in terminal and juxtaparanodal regions of neurons. *Nature* 365:75–79.

Wang, H., Kunkel, D., Schwartzkroin, P., Tempel, B. (1994). Localization of kv1.1 and kv1.2, two K channel proteins, to synaptic terminals, somata, and dendrites in the mouse brain. *J. Neurosci.* 14:4588–4599.

Wang, H., Allen, M.L., Grigg, J., Noebels, J.L., Tempel, B. (1995). Hypomyelination alters K channel expression in mouse mutants shiverer and trembler. *Neuron* 15:1337–1347.

Waniewski, R.A. (1992). Physiological levels of ammonia regulate glutamine synthesis from extracellular glutamate in astrocyte cultures. *J. Neurochem.* 58:167–174.

Waxman, S.G. (1995). Voltage-gated ion channels in axons: localization, function and development. In *The Axon,* Waxman, S.G., Kocsis, J.D., Stys, P.K., eds. New York: Oxford University Press.

Waxman, S.G., Black, J.A. (1995). Axoglial interactions at the cellular and molecular levels in CNS myelinated fibers. In *Neuroglia,* Kettenmann, H., Ransom, B.R., eds. New York: Oxford University Press.

Weibel, D., Kreutzberg, G., Schwab, M. (1995). Brain-derived neurotrophic factor (BDNF) prevents lesion-induced axonal die-back in young rat optic nerve. *Brain Res.* 679:249–254.

Weinberg, H.J., Spencer, P.S. (1976). Studies on the control of myelinogenesis: evidence for neuronal regulation of myelin production. *Brain Res.* 113:363–378.

Weisinger, H. (1995). Glia specific enzyme systems. In *Neuroglia,* Kettenmann, H., Ransom, B., eds. New York: Oxford University Press, pp. 488–499.

Westenbroek, R., Noebels, J.L., Catterall, W.A. (1992). Elevated expression of type II Na channels in hypomyelinated axons of shiverer mouse brain. *J. Neurosci.* 12:2259–2267.

Wiley-Livingston, C.A., Ellisman, M.H. (1980). Development of axonal membrane specializations defines nodes of Ranvier and precedes Schwann cell myelin elaboration. *Dev. Biol.* 79:334–355.

Wilson, G.F., Chiu, S.Y. (1990). Ion channels in axon and Schwann cell membranes at paranodes of mammalian myelinated fibers studied with patch clamp. *J. Neurosci.* 10:3623–3234.

Windebank, A.J., Wood, P., Bunge, R.P., Dyck, P.J. (1985). Myelination determines the caliber of dorsal root ganglion neurons in culture. *J. Neurosci.* 5:1563–1569.

Winslow, J.W., Moran, P., Valverde, J., Shih, A., Yuan, J.Q., Wong, S.C., Tsai, S.P., Goddard, A., Henzel, W.J., Hefti, F., Beck, K.D., Carass, I.W. (1995). Cloning of AL-1, a ligand for an Eph-related tyrosine kinase receptor involved in axon bundle formation. *Neuron* 14:973–981.

Wood, P., Bunge, R. (1986). Evidence that axons are mitogenic for oligodendrocytes isolated from adult animals. *Nature* 320:756–758.

Wong, P., Marszalek, J., Crawford, T., Hsieh, S., Griffin, J.W., Cleveland, D.W. (1995). Increasing neurofilament subunit NF-M expression reduces axonal NF-H, inhibits radial growth, and results in neurofilamentous accumulation in motor neurons. *J. Cell Biol.* 130:1413–1422.

Wyllie, A.H., Kerr, J.F.R., Currie, A.R. (1980). Cell death: the significance of apoptosis. *Int. Rev. Cytol.* 68:251–307.

Yang, L., Zeller, C., Shaper, N., Kiso, M., Hasegawa, A., Shapiro, R., Schnaar, R. (1996). Gangliosides are neuronal ligands for myelin-associated glycoprotein. *Proc. Natl. Acad. Sci. U.S.A.* 93:814–818.

Xiong, W.C., Montell, C. (1995). Defective glia induce neuronal apoptosis in the repo visual system of Drosophila. *Neuron* 14:581–590.

Zafra, F., Lindholm, D., Castren, E., Hartikka, J., Thoenen, H. (1992). Regulation of brain-derived neurotrophic factor and nerve growth factor mRNA in primary cultures of hippocampal neurons and astrocytes. *J. Neurosci.* 12:4793–4799.

Zeng, R., Anderson, P., Campbell, G., Lieberman, A.R. (1995). Regenerative and other responses to injury in the retinal stump of the optic nerve in adult albino rats: transection of the intracranial optic nerve. *J. Anat.* 186:495–508.

Zhao, J.X., Ohnishi, A., Itakura, C., Mizutani, M., Yamamoto, T., Hojo, T., Murai, Y. (1995). Smaller axon and unaltered numbers of microtubules per axon in relation to number of myelin lamellae of myelinated fibers in the mutant quail deficient in neurofilaments. *Acta Neuropathol.* 89:305–312.

4

Molecular mechanisms of axon guidance and target recognition

COREY S. GOODMAN AND MARC TESSIER-LAVIGNE

INTRODUCTION

Neuronal growth cones traverse long distances along appropriate pathways to find and recognize their appropriate targets. The mechanisms and molecules that help them do so are the topics of this chapter. What molecular signals in the environment allow growth cones to find their way? What receptors endow growth cones with their exquisite ability to decipher these signals? Compounding these questions are several problems. First is the problem of diversity. Different growth cones, confronted with the same choice point, make divergent decisions in a stereotyped fashion. Second is the problem of quantity. The human brain contains over 10^{11} neurons, each making a thousand or more synaptic connections. Finally comes the problem of reliability. All of this occurs in a remarkably unerring fashion despite the extreme complexity of the process. Can the genome really possess sufficient information to orchestrate the complete wiring of the brain?

The answer of course is that it must, and we are beginning to understand how this wiring is achieved. Several features simplify this process. First, the migration of growth cones to their targets occurs stepwise. The intricate trajectories of axons are built up of small segments, perhaps each a few hundred microns in length. Thus, the daunting task of reaching a distant target is reduced to the simpler task of navigating each of these individual segments in turn. Second, once axons reach their targets, the selection of appropriate target cells within the target field also occurs stepwise. Axons first find the approximate location of their appropriate target using molecular cues. Then patterns of neuronal activity drive the refinement of these initial connections into highly tuned circuits, a process that continues throughout life. These activity-dependent mechanisms are discussed in later chapters of this book.

In this chapter, we discuss the molecular mechanisms that direct the initial migration of axons and establish the initial stereotyped, albeit unrefined, patterns of projections and connections. First we briefly review the cellular mechanisms that direct growing axons. An overview of the families of guidance molecules and receptors involved in this process follows. Finally, we discuss specific guidance events in which the precise guidance roles of different molecules have been defined through experimental analysis.

Cellular Analysis of Growth Cone Guidance

It is easy to lose sight of the fact that the field of growth cone guidance is a relatively young one. As we will discuss, over a century ago Ramón y Cajal (1893) recognized many of the central problems of growth cone guidance, and several decades ago Sperry (1963) formulated many of the principles that guide our current understanding of axon guidance and target recognition in his "chemoaffinity hypothesis." It is sobering to remember, however, that many of Sperry's ideas of specificity remained controversial throughout the 1970s. This was due in part to the fact that the outcomes of his experiments, which he attributed entirely to the operation of molecular guidance cues, actually reflected the operation of activity-dependent events as well. With this caveat, however, his ideas concerning the role and nature of specific guidance and targeting mechanisms are now seen to be remarkably prescient. New students of the field often take for granted many of our current ideas concerning the specificity of pathfinding and targeting, but many of these notions only emerged or became strongly rooted in the common wisdom of our field in the 1980s. It wasn't that long ago that neuroscientists were not even sure whether axons grow in a directed or a random manner.

Pathfinding is a Highly Directed Process

Ramón y Cajal (1893) first discovered the motile, ameboid-shaped tips of growing axons, and he named them growth cones. He observed in fixed material that growth cones often take circuitous routes as they navigate toward their targets, which prompted him to suggest that growth cones play an active role in this pathfinding behavior. Harrison (1910) invented the technique of tissue culture to test Cajal's hypothesis, and in so doing confirmed that growth cones do indeed extend from cell bodies and leave behind axons. Speidel (1941) provided further confirmation of the model by observing extending growth cones in a living tadpole tail. However, these ideas concerning specific axon pathfinding fell out of fashion in the 1930s and 1940s. Based largely on the work of Weiss (1941), an alternative model (called the "resonance hypothesis") gained favor in which specificity of connections was thought to arise not from directed migration of axons but rather from selective retention of connections that initially formed at random.

The reemergence of the idea of specificity in axon pathfinding and targeting came largely during the 1950s from the work of Sperry, one of Weiss's former students. Sperry focused on the regeneration of axon projections from retinal ganglion cells to the optic tectum in amphibians after the optic nerve had been cut, and was struck by the apparently high degree of specificity in the reformation of synaptic connections in this system. Based on these results, Sperry (1963) evoked "differential chemical attraction" to explain the specific homing behavior of growth cones. He proposed the "chemoaffinity hypothesis," which suggests the existence of specific surface markers used by neuronal growth cones for both pathway and target recognition.

Toward the end of the 1970s, the use of new model systems in which growth cone guidance could be observed and manipulated in the living embryo made it possible to move away from the analysis of regeneration and to focus on the initial development of connections. One of the first and best systems for the analysis of growth cone guidance at the single-cell level is the grasshopper embryo, in which it was found that individual identified growth cones follow specific pathways (Bate, 1976; Goodman and Spitzer, 1979). These initial studies were followed by a detailed experimental analysis of growth cone guidance in the developing central and peripheral nervous systems (CNS and PNS) of grasshopper and *Drosophila* (reviewed by Goodman et al., 1984; Bentley and O'Connor, 1992; see below).

Directed axonal pathfinding was also found to be the norm in vertebrate embryos. In the PNS, motor axons in both chick (Landmesser, 1978, 1980; Lance-Jones and Landmesser, 1981a,b; Tosney and Landmesser, 1985a,b) and zebrafish (Eisen et al., 1986, 1989) take specific pathways to innervate their appropriate muscles. Similarly, in the vertebrate CNS, growth cones in the developing spinal cord and brain follow specific axon pathways in zebrafish (e.g., Kuwada, 1986), as well as in a variety of other species that have been studied (e.g., Easter et al., 1993, and references therein). In particular, coming full circle back to Sperry (who had studied specificity in regeneration), studies started in the 1980s on the development of the retinotectal projection in vivo and in vitro confirmed the notion of specificity of projections in this system (Bonhoeffer and Huf, 1980, 1982; Holt and Harris, 1983; Holt, 1984).

These and other cellular and molecular studies led to the conclusions that axon pathfinding is highly specific and that common mechanisms of guidance appear to be used in all organisms (e.g., Harrelson and Goodman, 1988; Dodd and Jessell, 1988). As a result, over the last decade attention has shifted toward understanding the cellular and molecular events that determine this specificity. Here we review the cellular events that direct pathfinding; molecular mechanisms are discussed later in this chapter.

Cellular Sources of Guidance Information

Cells along a growth cone's path are not all created equal: Some are more important for guidance than others. As mentioned at the beginning of this chapter, axonal trajectories appear to be broken up into short segments of perhaps a few hundred microns in length. The axon navigates each segment in turn, reaching in each case an intermediate target (often an important cellular landmark) that marks the end of the segment and the beginning of the next. It is thought that it is less complicated for the growth cone to navigate each segment than to make the transition from one segment to the next at specific intermediate targets, since growth cones slow down and take on a more complex morphology at these choice points. Many of our ideas regarding the cellular mechanisms of guidance have been based on the analysis of particularly noteworthy intermediate targets and choice points in insects and vertebrates, which we now discuss.

Guidepost cells and intermediate targets

An early indication that axons can be influenced by specific cells or cell groups located at particular locations along their trajectories came from studies by Bate (1976), who observed that pioneering growth cones in the grasshopper migrate along a pathway marked by the presence of cells of specialized morphology which he termed "stepping stones" and which were later called guidepost cells. A demonstration that some of these cells do carry important guidance information was provided by Bentley and colleagues (reviewed by O'Connor et al., 1990; Bentley and O'Connor, 1992), who studied a pair of sensory neurons—the Ti1 pioneers—that migrate from the distal tip of the developing limb bud to the CNS (for diagram, see fig. 4–5). After extending along the proximodistal axis for a time, the Ti1 axons turn abruptly and at right angles in a ventral direction, growing along a specialized group of epidermal cells that form an apparent boundary to migration. To resume their extension toward the CNS, they must turn again at right angles and cross over the boundary, a process that is directed by the Cx1 guidepost cells that are located just beyond the boundary. Time-lapse studies suggest that the reorientation is triggered by contact of the growth cone filopodia with Cx1, an interpretation supported by the finding that ablation of these neurons prevents the Ti1 axons from making the turn. Thus, the Cx1 cells present guidance information that is crucial for forward progression of the Ti1 axons.

It is perhaps worth commenting that such studies have inspired the idea—often presented in reviews and textbooks—that growth cones can progress in saltatory fashion from one stepping stone to the next that is within filopodial grasp, but such accounts represent an oversimplification. Specialized guidepost cells function in a context of other cues that are distributed over a large number of cells. In the example cited above, additional guidance cues are provided by other intermediate targets, such as those that cause the axons to recognize a boundary to forward progression and instruct it to migrate ventrally. No case of actual stepping from one guidance post to the next has been documented to date.

In vertebrates there are few if any examples of isolated cells that function like insect guidepost cells. However, there are many examples of groups of cells—often large groups—that form intermediate targets for migrating growth cones and that appear to be sources of important guidance information. An example discussed in detail later in this chapter is the floor plate, a group of cells at the ventral midline of the central nervous system that appears to present a variety of different guidance cues with profound influences on the direction of migration of different classes of axons that navigate the midline (reviewed in Colamarino and Tessier-Lavigne, 1995b). Other examples of midline cells that function as important intermediate targets have been described in the vertebrate forebrain (e.g., Silver, 1993; Godement and Mason, 1993; Sretavan, 1993). Such intermediate targets might be considered the vertebrate analogues of the insect guidepost cells, although insects also clearly have intermediate targets made up of groups of cells, like the epidermal boundary discussed above, or cells at the midline of the insect CNS (e.g., Klämbt et al., 1991; Tear et al., 1993). In any case, the important point is that some of the more intricate decisions made by growth cones appear to be directed by specialized cells or groups of cells that appear to harbor particularly important guidance information and thus function as intermediate targets.

Selective fasciculation

For many axons, certain segments of their trajectories are defined by preexisting axon tracts along which they migrate. The process of axons bundling together into tracts or fascicles is termed fasciculation. In 1910, Harrison coined the term "pioneer" based on his examination of the formation of axon pathways in the frog. "The fibers which develop later," he wrote, "follow, in the main, the paths laid down by the pioneers." Since then, studies in a wide variety of organisms from insects to mammals have confirmed Harrison's observation that initial axon pathways are established early in development when distances are short and the terrain relatively simple and that later growing axons extend for long portions of their trajectories by fasciculating with axons in this initial scaffold (reviewed in Lin et al., 1995). Two questions, however, were left unanswered by these initial studies. How selective is the normal pattern of fasciculation? And do later growing axons possess the same pathfinding abilities, or is the selective fasciculation of follower growth cones with pioneer axons absolutely necessary for correct pathfinding?

Studies in insects first demonstrated that the patterns of fasciculation of follower axons with pioneers can be highly selective and provided evidence of selective affinities between particular followers and specific pioneers (e.g., Raper et al., 1984; Bastiani et al., 1984; Goodman et al., 1984). For instance, in the grasshopper embryo, the growth cone from a particular identified cell—the G neuron—comes in contact with up to 20 different longitudinal axon bundles after crossing the midline, but it invariably grows along one of these: the A/P fascicle (Raper et al., 1983a,b). Within this bundle of axons, the G growth cone displays a selective affinity for the three P axons over the two A axons (Bastiani et al., 1984). Ablation of the A neurons has no effect on the G growth cone, whereas ablation of the three P neurons causes the G growth cone to stall at the choice point (Raper et al., 1983c, 1984). Several additional examples of selective affinities were demonstrated for other identified growth cones in the grasshopper CNS (e.g., Bastiani et al., 1986; du Lac et al., 1986). These experiments suggested that recognition of specific axon pathways involves qualitative differences—some sort of distinct labels—a notion that was formalized in the "labeled pathways" hypothesis (Raper et al., 1984; Goodman et al., 1984).

Selective fasciculation was later demonstrated in the developing vertebrate CNS, initially in the fish embryo. The growth cones of certain neurons pioneer specific axon pathways and in so doing establish a simple scaffold of axon tracts; later follower growth cones make specific choices of which of these axon tracts to follow (e.g., Kuwada, 1986; Wilson et al., 1990; reviewed by Kuwada, 1992). These observations were later extended to the mouse embryo (Easter et al., 1993).

Are the pioneers absolutely required for pathfinding by the followers? There are certainly cases where followers are profoundly misrouted in the absence of pioneers, as in the example of the G growth cone discussed above. In particular, axons that arrive much later in development often need these guide fibers to find their way through a more complex and changing environment. For example, transection of a peripheral nerve in the insect *Rhodnius* causes subsequent sensory axons that normally follow this nerve to form large whirls and never reach their targets in the CNS (Wigglesworth, 1953). Likewise, in the zebrafish CNS a particular dorsal spinal cord pathway in the fish is pioneered by the axons from a population of transient Rohon-Beard neurons. When these neurons are ablated, the follower growth cones do not properly extend (Kuwada, 1986). Pioneers can also be required for target selection. In cats and ferrets, the transient population of subplate neurons helps to pioneer the pathway between the thalamus and cortex (McConnell et al., 1989). When the subplate neurons are ablated, the axons from a particular thalamic nucleus extend in the white matter under the cortex but fail to enter their correct target region of cortex (Ghosh et al., 1990).

There are, however, many cases in which the pioneers are not absolutely required. An example showing both extremes is provided by the Ti1 growth cones in the limb bud of the grasshopper embryo (discussed above), which function as pioneers for at least two distinct classes of followers. In the absence of these pioneers, follower growth cones of one class are able to reach the CNS (Keshishian and Bentley, 1983), whereas those of another class are not (Klose and Bentley, 1989). In related experiments, elimination of a brain tract in the zebrafish embryo increases pathfinding errors by follower growth cones but does not absolutely prevent them from finding their correct pathway (Chitnis and Kuwada, 1991).

Experiments in *Drosophila* have provided the most detailed analysis yet of the partial requirement of pioneers for guidance of followers. In these studies, pioneers were ablated genetically, using a rapid transgenic targeting of diphtheria toxin (Lin et al., 1995). The intersegmental nerve (ISN) pathway is pioneered by the aCC axon; the three U growth cones fasciculate in this pathway as followers. When aCC is ablated, the three U growth cones are delayed and make frequent pathfinding errors (73% of segments) in the short term. In the long term, however, the ISN pathway eventually does form in most segments (88%). Thus, in this case, the followers, although not as effective as the pioneer, are nevertheless ultimately able to form the pathway. The same technique was used to ablate the axons that pioneer the first longitudinal pathways within the CNS, and similar results were obtained. In the short term, the formation of longitudinal pathways is delayed and disorganized in most (70%) segments; however, in the long term, the longitudinal tracts ultimately form in most (80%) segments.

The conclusion from this study is that the pioneers appear to play an important role in facilitating the rapid and robust development of PNS and CNS axon pathways. In their absence, the followers make frequent errors. In the long term, however, the pioneers are not absolutely required, as the followers display a remarkable ability ultimately to correct and to compensate for the loss of these pioneering axons.

Guidance Forces: Attraction and Repulsion

The appreciation that axonal trajectories are made of small segments pushes the question back one step: How do axons navigate each small segment and choice point of their trajectory? Studies in the past decade have led to the view that guidance involves the coordinate action of four types of guidance cues: short-range (or local) cues and long-range (diffusible) cues, each of which can be either positive (attractive) or negative (repellent) (Fig. 4–1).

Local attraction

We have already discussed the operation of local cues in the process of selective fasciculation, in which an axon tract provides a substrate that channels the growth of other axons. Channeling of axons through permissive corridors occurs

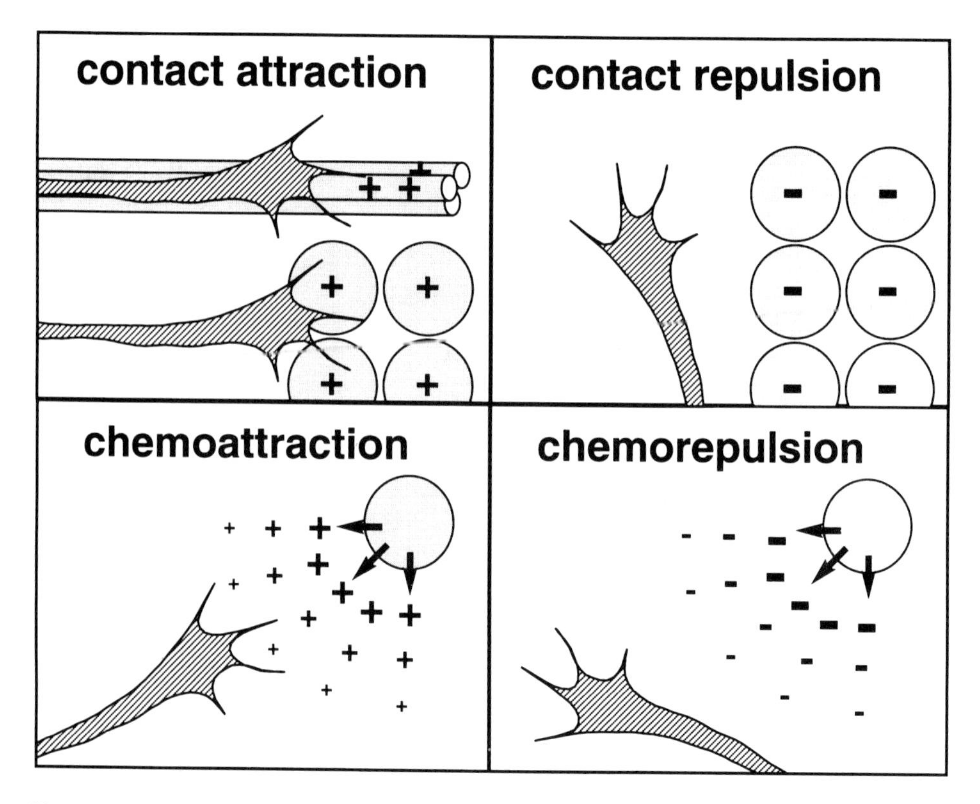

FIGURE 4–1 Guidance forces. Growth cone guidance mechanisms fall into four categories: contact attraction, chemoattraction, contact repulsion, and chemorepulsion, where the term "attraction" includes a range of permissive and attractive effects and the term "repulsion" a range of inhibitory and repulsive effects.

in many different regions of the nervous system. One well-characterized example in vertebrates is the anterior sclerotome of the somite, which provides a permissive environment for the growth of motor axons (Keynes and Stern, 1984).

Our thinking about how growth cones interact with permissive substrates has been conditioned by early in vitro experiments in the 1970s (Bray, 1979; Wessells and Nuttall, 1978; Letourneau, 1975). These experiments demonstrated that growth cone extension requires a physical substrate but that not any substrate will do. Mere binding to the substrate is not sufficient; rather, growth cone extension requires a *permissive* substrate. Moreover, growth cones are easily channeled in vitro within the boundaries of a permissive substrate surrounded by a nonpermissive substrate. They can also display a strong preference for one substrate over another when given a choice; a particular substrate that appears permissive in one pairwise test might be avoided when paired with a better alternative in another test. This phenomenon was first demonstrated by Letourneau (1975), who presented growth cones with boundaries of distinct substrates and showed that they displayed strong preferences that correlated with the differential adhesiveness of the two substrates. Thus was born the concept of differential adhesion as a mechanism to guide axons. Many of these early experiments, however, were performed with nonphysiological substrates. Subsequent experiments with molecules that are thought to provide more physiological substrates showed that substrate preference does not necessarily correlate with adhesiveness (Gundersen, 1987; Lemmon et al., 1992), suggesting that the selection of physiological substrates involves active recognition of the molecules by the growth cone.

Local repulsion

Although much of the early focus on axon guidance mechanisms was on positive influences on growth cone guidance, during the 1980s evidence began to accumulate suggesting the existence of negative influences as well (Kapfhammer et al., 1986; Walter et al., 1987b; Caroni and Schwab, 1988), similar to contact inhibition studied by cell biologists (e.g., Abercrombie, 1970).

An early and dramatic example of inhibitory influences on growth cones came from an extension of studies showing that axons from two different types of vertebrate neurons (sympathetic neurons and retinal ganglion cells) selectively fasciculate with their own kind in cell culture (Bray et al., 1980). When these experiments were repeated and examined with time-lapse photography, it was discovered that these two different types of axons fasciculate with their own kind because they selectively avoid the other kind of axon (Kapfhammer et al., 1986). Upon contact with the other (inhibitory) axons, the growth cones "collapse"; i.e., their filopodia and lamellipodia retract. The establishment of two in vitro assays—the growth cone collapse assay (Kapfhammer and Raper, 1987a) and the stripe assay (Walter et al., 1987a; see below)—led to numerous studies on the roles of inhibition and repulsion in growth cone guidance. Using these assays, as well as time-lapse analysis of interaction of growth cones with other cells, a wide variety of examples of contact-mediated repulsive and inhibitory cues in growth cone guidance have been discovered in the CNS and PNS of vertebrates (see reviews by Tessier-Lavigne, 1994; Luo and Raper, 1994; Keynes and Cook, 1995; Goodman, 1996; Kolodkin, 1996).

How is local repulsion (or inhibition) used as an axon guidance mechanism? First, it is used to channel axons down corridors. We discussed above how corridors of permissive molecules can channel axons. It appears that where axons are channeled in vivo, repulsive cues surrounding the corridor may serve to hem in the axons as well. Thus, in the example discussed above of motor axon growth through the anterior portion of the sclerotome, it is clear that this channeling is due in large part to the presence in posterior sclerotome of a repulsive cue that the axons avoid (Keynes and Stern, 1984; Oakley and Tosney, 1993). Second, local repulsive cues can be used as barriers to further growth. An example of this in vertebrates is provided by the optic chiasm, an intermediate target for retinal ganglion cell axons. Some axons cross the chiasm to project contralaterally, whereas others grow into the vicinity of the chiasm before turning away to remain ipsilaterally. It appears that those that remain ipsilaterally do so at least partly because of the presence of a local repellent at the chiasm that selectively repels these axons without effect on the contralaterally projecting axons (Godement et al., 1990; Sretavan, 1990; Sretavan et al., 1992; Wizenmann et al., 1993; Godement et al., 1994).

Finally, there are certain contexts in vivo in which growth cones come to a complete stop. For example, growth cones stop when they reach their targets (e.g., Baird et al., 1992) or when they reach transient targets or waiting zones (e.g., Allendoerfer and Shatz, 1994). Target-derived stop signals do not repel the growth cone but rather induce a shutdown in motility and a transformation of the growth cone into a presynaptic terminal arbor. It is not yet known whether target-derived stop signals use mechanisms that are more closely related to inhibitory guidance mechanisms or to attractive guidance mechanisms.

Long-range attraction

Thus far, we have largely considered mechanisms of short-range guidance in which the molecular cues are displayed on the surface of cells or the surrounding extracellular matrix and are detected by direct filopodial contact. There is, however, considerable evidence that guidance cues can also be diffusible.

At the turn of the century, Ramón y Cajal (1893) first proposed that growth cones might be guided by attractive gradients of diffusible factors emanating from distant targets. In the late 1970s, Gunderson and Barrett (1979) showed that the growth cones of regenerating sensory neurons can turn in vitro in response to a diffusible gradient of nerve growth factor (NGF). Although NGF is not likely to be a long-range chemoattractant in the developing organism, this study initiated a resurgence of interest in chemotropism. Direct evidence for the existence of target-derived attractants came from coculture studies in which neurons at appropriate stages of development were cultured together with their target cells placed at a distance, and the axons of the neurons were found to turn toward their targets. Many of these experiments involved culturing the tissues together in collagen gel matrices which allow for the establishment of stable gradients of target-derived factors (Ebendal and Jacobson, 1977). In this way, trigeminal sensory axons in mouse were shown to be attracted by a diffusible factor made by their final target, the whisker pad epithelium of the maxillary process (Lumsden and Davies, 1983, 1986). In the CNS, commissural axons were shown to be attracted by one of their intermediate targets, the floor plate, in both the spinal cord (Tessier-Lavigne et al., 1988; Placzek et al., 1990) and brainstem (Shirasaki et al., 1995; Tamada et al., 1995), establishing the principle that intermediate targets can be sources of chemoattractants as well. Another variation on the theme of target-derived attractants in vertebrates was provided by studies of the innervation of the basilar pons by cortical axons, which involves the collateral axon branches rather than the primary axon. In this case, layer 5 neurons of the cortex send axons to the spinal cord and later sprout a collateral branch that projects to innervate the pons. In vitro experiments showed that the pons secretes a factor(s) that induces collateral branches in these axons and that subsequently attracts the branch (Heffner et al., 1990; O'Leary et al., 1990; Sato et al., 1994). In addition to these cell culture experiments, in vivo perturbations in which target cells are ablated and axons get misrouted have provided evidence for target attraction, as in the case of innervation of somitic targets by motor axons (Tosney, 1987; reviewed in Tessier-Lavigne, 1992).

Long-range repulsion

The first examples of repulsion involved contact-mediated events. Subsequent experiments, however, revealed that growth cones can also be repelled by diffusible factors secreted from a distance, a process termed chemorepulsion (Pini, 1993). Three types of guidance function have been attributed to such chemorepulsive mechanisms. First, there are several examples where axons are repelled by cells that they grow away from in vivo. This is the case, for example, with olfactory tract axons that are repelled by cells in the septum (Pini, 1993) and with different classes of motor axons and other basal plate axons that are repelled by floor plate cells (Colamarino and Tessier-Lavigne, 1995a; Tamada et al., 1995; Guthrie and Pini, 1995; Shirasaki et al., 1996). In each case, the repellent is thought to provide a "push from behind" that helps set the axons off in the right direction as they initiate growth.

Second, two examples have been provided where axons initially project toward the source of repellent, only to be deflected away or to stall. Spinal sensory axons project from dorsal root ganglia into the spinal cord at dorsal levels and project ventrally into the spinal cord to specific termination sites along its dorsoventral axis. The ventral spinal cord secretes a diffusible repellent for sensory axons (Fitzgerald et al., 1993; Messersmith et al., 1995), whose function is thought to be to prevent subclasses of sensory axons that enter the spinal cord dorsally from projecting too far ventrally (Messersmith et al., 1995; Püschel et al., 1996). Likewise, alar plate neurons in the midbrain project ventrally toward the floor plate but are deflected along a longitudinal trajectory as they begin to approach the floor plate. One factor that is thought to contribute to this change of trajectory is a chemorepellent made by floor plate cells that can repel these axons at a distance in vitro (Tamada et al., 1995). In both of these examples, the axons appear to navigate head-on into the repellent, whose function appears to be to prevent their further progression.

Third, evidence has been obtained for a chemorepellent whose function appears more to be to provide a lateral deflection. This example was obtained in the case of a neuronal cell migration rather than an axon guidance event. In vertebrates, olfactory interneuron precursors migrate from near the septum to the olfactory bulb. The caudal septum was found to be the source of a chemorepellent that can repel these cells and redirect their migration in vitro (Hu and Rutishauser, 1996). The source of the repellent is located lateral to the path of migration of the cells and is presumed to prevent the cells from taking an inappropriate turn as they migrate toward the olfactory bulb.

Overview

In this subsection, we have discussed the four major forces that are thought to guide axons along individual segments of their trajectory. There is every reason to believe that axons are often guided by several of these forces acting in concert. For example, an axon might in principle receive a "push from behind" by a chemorepellent while simultaneously being attracted from afar by a chemoattractant, in what may be termed a "push–pull" mechanism. The same axon might also in principle be channeled in a corridor marked by an attractive factor and hemmed in by a local repulsive factor. Much of the present focus in cellular studies of axon guidance is directed at defining the complement of guidance forces acting on particular classes of axons. Later we discuss the identification of molecules that mediate these particular guidance forces.

Target Recognition

So far we have discussed the mechanisms that direct axons to their targets. Once at the target, growth cones appear to recognize their targets using two broad categories of information: topographic maps of graded cues and unique labels marking distinct targets. The operation of these cues is not mutually exclusive, and, in fact, these cues often operate simultaneously. For example, retinal axons that form topographic projections onto the tectum (as discussed below) also select discrete laminar termination sites within the tectum.

Topographic maps

Topographically organized patterns of neuronal connections, in which neighboring neurons project to neighboring sites in the target, occur throughout the nervous system. The best-studied example of the development of topographic projections is in the vertebrate visual system, where retinal ganglion cells make an orderly projection onto the optic tectum in fish, amphibians, and birds, and onto the homologous structure (the superior colliculus) in mammals. Neighboring retinal ganglion cells from a particular location in the retina connect to neighboring target neurons in a predictable location in the tectum, thus projecting the retina's map of visual space as a topographic map across the optic tectum. The map is inverted, so that nasal retina projects to posterior tectum, temporal retina to anterior tectum, dorsal retina to ventral tectum, and ventral retina to dorsal tectum.

Evidence that the establishment of this pattern of projections involves the recognition of positional information on the tectum was initially obtained in studies performed on the regeneration of connections by Sperry and others. A large number of different experiments were taken to provide support for the existence of positional cues, of which we mention only two. Sperry showed that when the optic nerve is cut and part of the retina is removed, the axons from the remaining portion of the retina regrow to their appropriate addresses of the tectum. Later, by reconstructing the trajectories of individual regenerating retinal axons, Fujisawa (1981) showed that axons that reentered the tectum from inappropriate directions reoriented their growth to project to their topographically appropriate termination site, in some cases making dramatic turns. These results were interpreted to indicate that the axons could detect positional information on the tectum that instructed them to project to their appropriate sites. These regeneration studies were later extended to the development of the retinotectal projection in ingenious experiments in the chicken. In a first experiment, retinal axons entering the tectum that were deflected laterally by a Teflon barrier were shown to correct their courses to project to their topographically appropriate sites (Thanos and Bonhoeffer, 1986). In a second experiment, small retinal explants from different dorsoventral and nasotemporal sites were grafted directly onto an uninnervated tectum of an enucleated host chick embryo. Axons emerging from these pieces of retina projected toward regions of the tectum that were appropriate for the site of origin of the retinal piece (Thanos and Dutting, 1987). Again, these experiments were taken to support the existence of cues on the tectum that can direct the growth of retinal axons.

How might these topographic projections be directed? Sperry first considered the possibility that each axon has a unique label that is complementary to another unique label on its appropriate target cell. There are two reasons why this model seemed inadequate. First was the implausibly large number of labels that was required to explain the precision of topographic mapping. A second equally vexing problem was the fact that this model did not provide a mechanism for each axon to find its target, except by wandering aimlessly around the tectum until it chanced upon its appropriate target cell. Thus, the directed growth of axons on the tectum would seem to rule out this model.

Instead, Sperry proposed that positional information might be encoded in the form of gradients of signaling molecules at the target that would "stamp each cell with its appropriate latitude and longitude" (Sperry, 1963). Detection of this information was postulated to involve complementary gradients of receptors on the axons. This model neatly solves the two problems facing the "unique labels" model: Positional information can be specified with a small number of molecules, and all axons are capable of reading the positional information at every point on the tectum. A more detailed analysis of gradient models of positional information has refined these models and suggested some important modifications (Gierer, 1987; Tessier-Lavigne, 1995; Nakamoto et al., 1996). In principle, topographic projections along each axis of the tectum could be directed by just one ligand gradient on the tectum and one receptor gradient on the retina, but in practice this would require that each axon seek out a specific concentration of ligand that is determined by its level of receptor expression (the "set point" for that axon). This would require that the axon tend to grow down-gradient at concentrations of ligand higher than the "set point" and that it tend to grow up-gradient at concentrations of ligand that are lower than the "set point." Although it is not impossible for this type of mechanism to operate, a simpler mechanism would involve the use of antagonistic effects of two gradients (along each axis). For example, if an axon is exposed to two gradients of repellent ligands with opposite slopes (along a single axis), it will tend to migrate to a point of minimum repulsion. Axons originating from different positions on the retina can be made to project to different locations along the axis by making their responses to one or both repellents dependent on their position of origin. The same result can be obtained using similar gradients of two ligands with opposite (attractive and repellent) actions (Gierer, 1987).

Can direct evidence be obtained for the existence of spatially graded positional cues that can influence axon growth on the tectum? In search for such gradients, Bonhoeffer and colleagues developed in vitro assays which demonstrated that some retinal ganglion cell axons are capable of discriminating between tectal cells from different regions (Bonhoeffer and Huf, 1980, 1982, 1985). They found that axons from the temporal retina could grow on cells from either the anterior or the posterior tectum, but when confronted with a choice of the two, they always opted to extend on the anterior tectal cells, their normal target. This preference was maintained when the axons were given a choice of alternating stripes of tectal membranes, rather than live tectal cells, in the so-called "stripe assay" (Walter et al., 1987a). The ability to use cell membranes in this assay made it possible to characterize the activity responsible for the preference, and it was shown that the preference of the temporal axons is based on avoidance of the posterior tectum, rather than attraction to the anterior tectum, since several nonspecific treatments of the posterior membranes (with heat, proteases, or phospholipase C) abolished the preference, whereas similar treatments of anterior tectal membranes did not (Walter et al., 1987b, 1990). The repulsive nature of the activity was further demonstrated by the finding that posterior membrane fragments caused collapse of temporal retinal axons, whereas anterior membrane fragments had much less collapse-inducing activity (Cox et al., 1990). Coming full circle, Baier and Bonhoeffer (1992) devised a clever method to generate smooth gradients of posterior tectal membranes and showed that temporal retinal axons growing up the gradient were deflected from their trajectory or could be made to stall in their progression, provided the slope of the gradient exceeded a certain value (about a 5% concentration change across the growth cone diameter).

The importance of these experiments was in demonstrating the existence of a graded activity that can guide axons and in showing directly that axons are capable of responding to smooth gradients of the activity. Of course, further characterization of the activity has required its molecular identification, and we therefore leave to later in this chapter a fuller discussion of the mechanisms operating in this system.

One additional aspect of the biology of this system deserves mention, however, which is that there is interspecies variability in the strategy used to generate the topographic map. In amphibians and fishes, retinal axons home in on appropriate regions of the target as soon as they invade it. In mammals (and to some extent in birds), many or even most axons tend instead to overshoot their targets, only later correcting their errors by sending out collateral branches in a highly directed manner toward topographically appropriate targets and by retracting inappropriate branches (Nakamura and O'Leary, 1989; Simon and O'Leary, 1992). Using a variant of the Bonhoeffer "stripe" assay, Roskies and O'Leary (1994) showed that posterior tectal membranes can not only direct a preference of temporal retinal axons through repulsion but can also inhibit branching of temporal axons, without effect on branching of nasal axons. It is tempting to speculate that the factor that repels the axons and the factor that inhibits their branching are one and the same (which is supported by the fact that both activities are lost by treatment of membranes with phospholipase C). This suggests that positional information on the tectum may be provided by molecules whose precise effects on axons (e.g., branching rather than repulsion) may be dictated by the response machinery of the axons and may vary from species to species.

Discrete targets

In some cases, growth cones also appear to be capable of recognizing discrete targets from within an array of possible targets. For example, in both vertebrates and invertebrates, motor axons have the ability to find and to innervate specific muscle targets in a highly stereotyped fashion. The cellular basis of this type of discrete target recognition has been studied by deleting or duplicating individual muscle targets in both grasshopper and *Drosophila*. The results of these experiments suggest that individual axons recognize specific labels expressed by their targets.

In the grasshopper embryo, motoneuron growth cones arrive at their target regions before the differentiation of mature muscle fibers, and they display affinities for specific multinucleate muscle pioneers (Ball et al., 1985). When a specific muscle pioneer (for limb muscle 133a) is ablated prior to innervation, the growth cone (Df) that would normally have left the main motor nerve and innervated that muscle instead continues to extend further distally along the main motor nerve, thus bypassing its normal target region.

In *Drosophila*, most studies of neuromuscular specificity have focused on the abdominal segments of the embryo (reviewed by Keshishian et al., 1996). The body wall musculature in each hemisegment is arranged in a stereotyped array of 30 individually identified muscles, each of which is a single, large multinucleate cell. Most of these muscles are innervated by one or a few of approximately 40 motoneurons, forming a stereotyped pattern of connections. To test the selective affinity of motoneuron growth cones for specific muscle targets, discrete sets of muscles were deleted either surgically or genetically. When muscles 6 and 7 were surgically ablated, the RP3 growth cone which normally innervates these muscles arborizes over nontarget muscles in a variable and abnormal fashion (Sink and Whitington, 1991a). Similarly, in mutant embryos in which muscles 12 and 13 are missing (*numb* mutants), the RP1 growth cone extends further distally in an abnormal fashion (Chiba et al., 1993). Although these motor axons initially behave abnormally in the absence of their correct muscle targets, there is evidence that target-deprived motoneurons do eventually form ectopic synapses on neighboring muscles (Cash et al., 1992). In complementary experiments, where muscle 13 was duplicated by heat shock, the RP1 motoneuron innervated both muscle 13 targets (Chiba et al., 1993). Taken together, these results from grasshopper and *Drosophila* argue for a high degree of specificity in the ability of motor axons to recognize discrete muscle targets.

MOLECULES THAT MODULATE AXON GROWTH

As recently as a few years ago, the cues in the extracellular environment that guide axons were thought to be divided into three nonoverlapping categories: diffusible factors that function in long-range chemotropic guidance, nondiffusible extracellular matrix (ECM) molecules that function in cell–ECM adhesion and as local guidance cues, and cell-surface molecules that function in cell–cell adhesion and as local guidance cues. Within those categories, it was thought that some molecules function as attractive cues while others function as repulsive cues.

As our knowledge has increased, however, the distinction between these categories has become blurred. Families like the semaphorins (described below) have been discovered that contain both cell-surface and diffusible members implicated in short- and long-range guidance, respectively. Likewise, a family of diffusible factors, the netrins, is closely related to the laminins, which are archetypical ECM molecules. At the same time, the distinction between attractive and repulsive factors has become blurred, as several diffusible, ECM, and cell-surface molecules have been shown to be bifunctional, displaying attractive effects on some axons and repulsive effects on others (see below). A further simplification in our understanding has come from the realization that the molecules involved in axon guidance are highly conserved across species, from worms and flies to birds and mammals, so that insights gained in one organism are relevant to other species (e.g., Goodman, 1994). In what follows, we therefore consider different species together.

In this section we briefly introduce several of the major classes of molecules that have been implicated as guidance cues in the extracellular environment or as receptors for guidance cues. We focus principally on the structural and biochemical features of these molecules; their roles in guidance and targeting are discussed later in this chapter.

Cell Adhesion Molecules as Ligands and Receptors

The search for cell-surface proteins that can mediate differential cell adhesion led in the 1980s to the identification of two major families of cell adhesion molecules (CAMs): the immunoglobulin (Ig) gene superfamily (Fig. 4–2) (reviewed

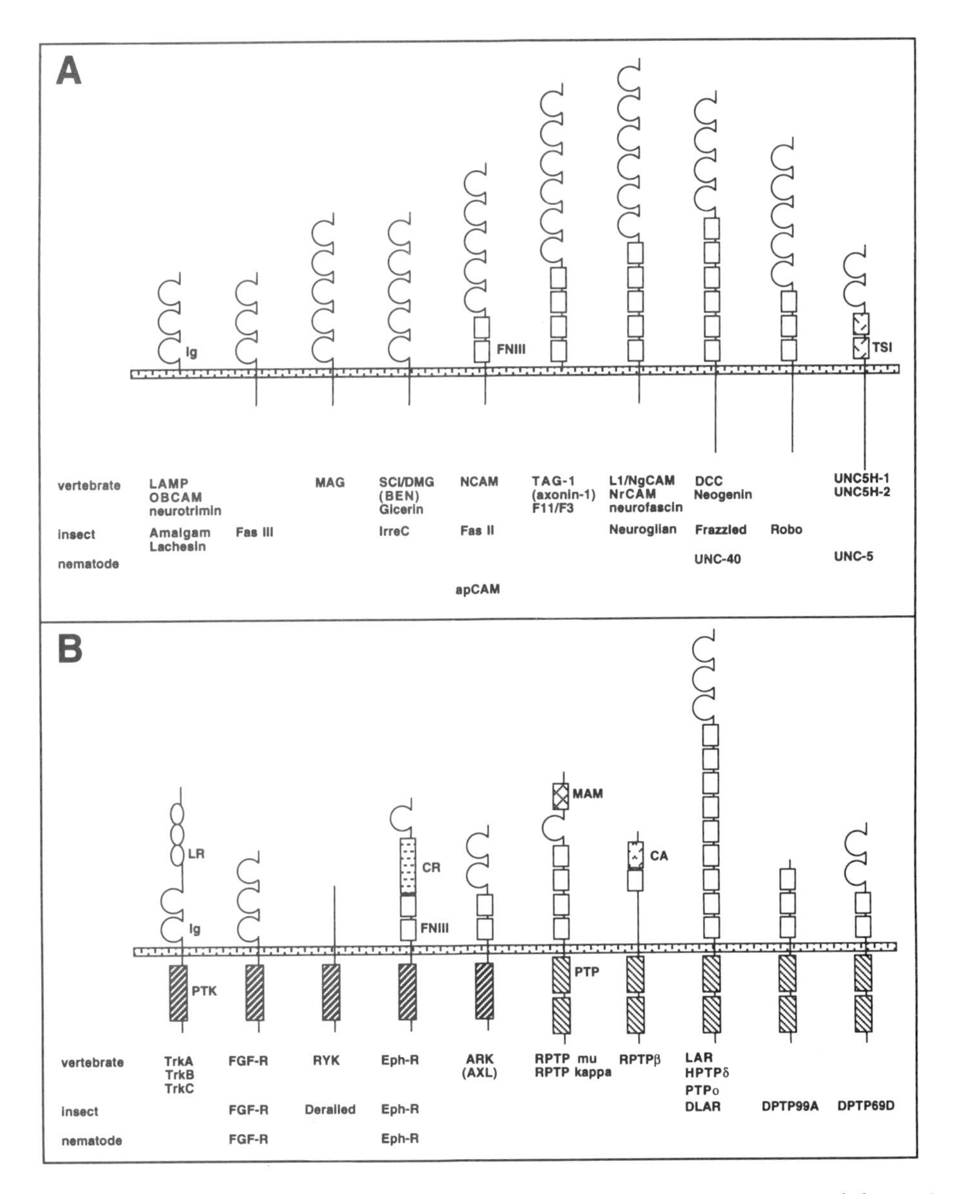

FIGURE 4–2 **Molecules that modulate axon growth.** (A) Representatives of the various subfamilies of the immunoglobulin (Ig) superfamily that modulate axon growth (not including Ig superfamily receptor tyrosine kinases and phosphatases, which are shown in panel B). Although members of some of these subfamilies have extracellular domains possessing only tandem Ig domains, many have both tandem Ig and fibronectin type III (FNIII) domains. For certain subfamilies, the first members were identified as proteins expressed on subsets of axons in the developing nervous system. For other subfamilies, the first members were identified in functional screens for adhesion molecules (CAMs). Yet other members (e.g., UNC-40, Robo, UNC-5) were identified as putative guidance receptors (the latter have longer cytoplasmic domains than adhesion molecules). Some Ig superfamily members are linked to the membrane by a GPI anchor. (B) Representatives of the several subfamilies of receptor protein tyrosine kinases (PTK) and tyrosine phosphatases (PTP) that are expressed on subsets of axons in the developing nervous system. Many of these receptors have extracellular domains comprising tandem Ig domains and/or FNIII domains. Note the conservation of subfamilies among vertebrates, insects, and nematodes. See text for details and references. Ig, immunoglobulin domain; FNIII, fibronectin type III domain; TSI, thrombospondin type I domain; LR, leucine-rich repeat; CR, cysteine-rich region; CA, carbonic anhydrase domain.

by Rutishauser, 1993; Edelman, 1993) and the cadherin superfamily (reviewed by Takeichi, 1995), which comprise both transmembrane and lipid-anchored proteins. The first members of these two families identified in the nervous system, the neural cell adhesion molecule NCAM (Cunningham et al., 1987) and N-cadherin (Hatta et al., 1988), were initially shown to function as homophilic adhesion molecules (e.g., NCAM binds to NCAM in a transassociation from one cell surface to another). Subsequent studies showed that several members of these families can engage in heterophilic interactions as well (see below).

These two CAM families appear to have independent origins, based on low similarity between the families in intron patterns and in amino acid sequence. The extracellular portions of cadherins comprise tandem arrays of so-called cadherin domains, whereas Ig superfamily members comprise tandem arrays of Ig domains and fibronectin type III (FNIII) domains. [The Ig superfamily is further divided into subfamilies defined by the number of Ig and FNIII domains they possess (Fig. 4–2; reviewed by Rathjen and Jessell, 1991; Brummendorf and Rathjen, 1995).] Furthermore, adhesion mediated by cadherins is calcium-dependent, whereas adhesion mediated by the Ig CAM family is calcium-independent. While these families are evolutionarily and biochemically distinct, a structure-based sequence analysis of cadherin domains, Ig domains, and FNIII domains suggests that the folding topology of these domains is quite similar, suggesting the selective convergence of a favorable folding topology for these two families of CAMs (Shapiro et al., 1995). Furthermore, these domains appear to be the determinants of adhesion. For example, all of the Ig domains of NCAM appear to participate in a pairwise fashion in homophilic binding (Ranheim et al., 1996).

Other apparently unrelated families of CAMs, with members expressed in the nervous system (neural CAMs) and capable of mediating homophilic adhesion, have been discovered more recently, including the leucine-rich repeat family (e.g., Chaoptin and Connectin; see Krantz and Zipursky, 1990; Nose et al., 1992) and the fasciclin I family (Zinn et al., 1988; Elkins et al., 1990a; Takeshita et al., 1993). The different families of neural CAMs appear to be highly conserved among invertebrates and vertebrates (e.g., Grenningloh et al., 1990; Oda et al., 1994). How many neural CAMs are encoded in any one genome is still unknown, although there are at least ten in *Drosophila* and more than 50 in mammals. In some cases, whereas *Drosophila* appears to have only one member in a particular Ig subfamily (e.g., neuroglian), gene duplications in the vertebrate lineage have led to the diversification of multiple genes in mammals encoding related subfamily proteins (e.g., L1, NgCAM, NrCAM, and neurofascin) (Hortsch, 1996) (Fig. 4–2).

The ability of neural CAMs to mediate homophilic adhesion made them good candidates for mediators of selective axon fasciculation and of adhesion of axons to pathway or target cells. Evidence described later in this chapter provides support for this idea. In addition, recent studies, primarily of Ig CAMs, have identified other important properties and potential functions of neural CAMs which are summarized in point form here.

Some CAMs can mediate adhesion by heterophilic binding to other CAMs
As mentioned above, heterophilic interactions have been described for a number of Ig CAMs (Fig. 4–3). For example, TAG-1 in mammals (and its homologue axonin-1 in chick) binds homophilically to itself and heterophilically to L1 subfamily members (L1, NgCAM, NrCAM) and to β1 integrins (e.g., Kuhn et al., 1991; Rader et al., 1993; Felsenfeld et al., 1994; Suter et al., 1995). Similarly, L1 can bind both to itself and to integrins (Montgomery et al., 1996). Other Ig CAMs also can bind in a heterophilic fashion (e.g., DeBernardo and Chang, 1996). Likewise, some cadherin superfamily members can engage in both homophilic and het-

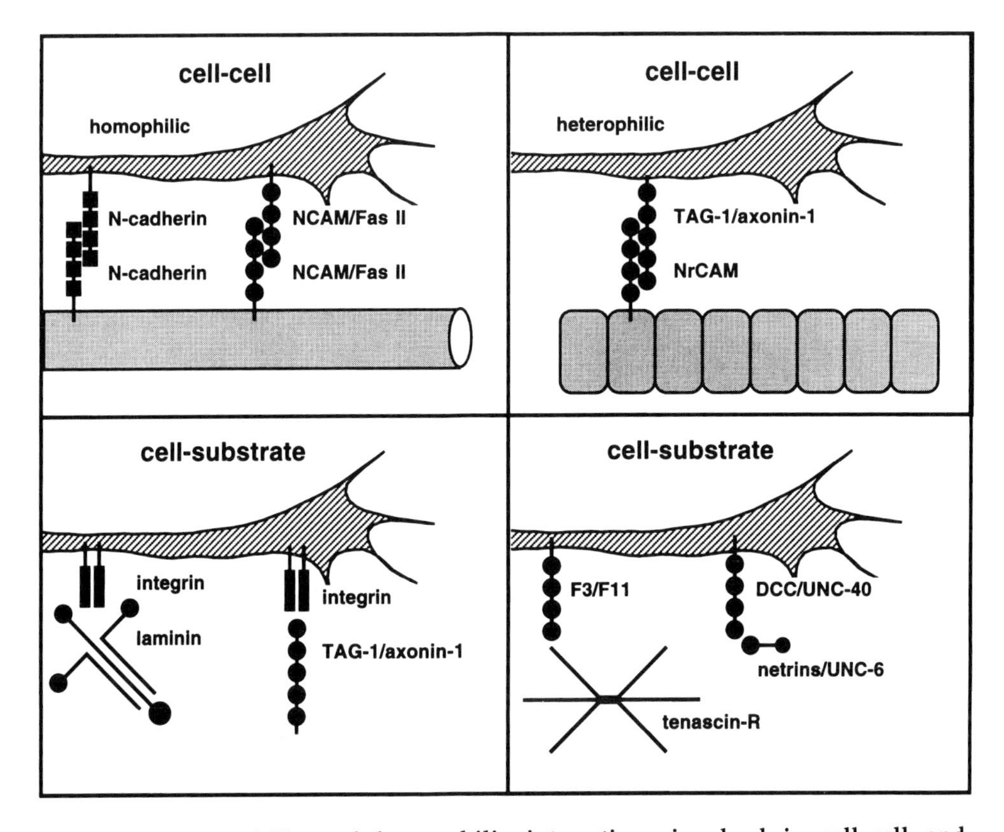

FIGURE 4–3 **Homophilic and heterophilic interactions involved in cell–cell and cell–substrate contacts during axon growth.** Top left: Examples of cell–cell adhesion of axons mediated by homophilic binding, in these examples showing representatives of the Ig superfamily (NCAM/Fas II) or cadherin superfamily (N-cadherin) binding to themselves in *trans*. Top right: Examples of cell–cell adhesion of axons mediated by heterophilic binding, in this case showing TAG-1/axonin-1 on the axon binding to NrCAM on intermediate target cells. Bottom left: Examples of heterophilic binding involved in cell–substrate adhesion of axons to extracellular matrix (ECM) molecules, in this case showing integrins on axons binding to laminin in the ECM, and integrins binding to secreted TAG-1/axonin-1. Bottom right: Further examples of heterophilic binding involved in cell–substrate adhesion of axons to ECM molecules, in this case showing F3/F11 binding to tenascin-R and DCC/UNC-40 binding to netrins/UNC-6. See text for details and references.

erophilic interactions with other cadherin family members (Murphy-Erdosh et al., 1995).

Some Ig CAMs can bind ECM molecules

F11 subfamily members bind members of the tenascin family of ECM molecules (Morales et al., 1993; Brümmendorf et al., 1993), an SC1 subfamily member (gicerin) binds a laminin-related molecule (Taira et al., 1994), an L1 subfamily member (NgCAM) binds laminin (Grumet et al., 1993), and DCC subfamily members (DCC and neogenin) can bind netrins (Fig. 4–3).

Ig CAMs can function as both ligands and receptors involved in stimulation or inhibition of axon extension

The Ig CAMs NCAM and L1, as well as N-cadherin, function as both ligands and receptors to stimulate extension of axons in vitro via homophilic interactions (e.g., Doherty and Walsh, 1994). In contrast, although TAG-1/axonin-1 can bind homophilically, it appears to function in vitro as a ligand to stimulate axon extension primarily via heterophilic interactions with L1 and β1 integrins (Felsenfeld et al., 1994). Furthermore, axonin-1 can function in vivo as a receptor to control axon guidance via a heterophilic interaction with NrCAM (Stoeckli and Landmesser, 1995). The Ig CAM F3 appears to function as part of a receptor that mediates an inhibitory effect of a tenascin family member (Pesheva et al., 1993). Another Ig CAM, MAG, can function as a ligand to stimulate growth of some axons and to inhibit growth of others (Mukhopadhyay et al., 1994; McKerracher et al., 1994), presumably by activating different receptors in the different axons.

Alternative splicing of Ig CAMs can give rise to isoforms with different functional properties

Members of the NCAM subfamily (NCAM, Fas II, apCAM) have different isoforms with alternatively spliced cytoplasmic domains, different membrane linkages, and alternatively spliced extracellular exons. Within the extracellular domain, different NCAM isoforms have different functional properties. For example, one isoform includes the 30 "VASE" microexon located in the fourth Ig domain. Neurite outgrowth in vitro is high on a substrate of the NCAM isoform lacking the VASE exon. In contrast, neurite outgrowth is much lower on a substrate of the VASE-containing isoform, suggesting that VASE functions to inhibit neurite outgrowth (Doherty et al., 1994; Saffell et al., 1994).

Different functional properties have also been reported for alternative CAM isoforms that differ in their cytoplasmic domains. For example, the cytoplasmic domain of fasciclin II is alternatively spliced and comes with or without a conspicuous PEST degradation sequence that appears to control protein stability in vivo (Helt et al., in preparation). Likewise, most vertebrate members of the L1 subfamily (L1, NgCAM, NrCAM, neurofascin) come in alternative isoforms with or without a short peptide sequence (-RSLE-) in the middle of their cytoplasmic domain. Although L1 isoforms with or without the RSLE insert support neurite outgrowth equally well, the form lacking this insert supports cell migration much better than the form containing the RSLE insert (Asou et al., 1996).

Polysialylation of NCAM modulates the function of NCAM and other CAMs

In mammals, NCAM can contain a very large carbohydrate moiety of polysialic acid (PSA) on its fifth Ig domain made up of as many as 100 sialic acid residues per chain (e.g., Rutishauser et al., 1988; Rutishauser, 1991; Nelson et al., 1995). NCAM's invertebrate counterparts (*Drosophila* fasciclin II and *Aplysia* apCAM) apparently do not possess this unusual carbohydrate. The polysialylation of NCAM is regulated independently of its core protein isoforms and appears to depend upon the activity of a particular sialyltransferase (reviewed by Fryer and Hockfield, 1996). During embryonic development, NCAM in general shifts from a PSA+ form to a PSA− form. PSA on NCAM has been proposed to modulate cell–cell adhesion by decreasing the adhesivity of NCAM itself and by decreasing the adhesivity of other neighboring CAMs (Yang et al., 1994; Rutishauser and Landmesser, 1996). The model proposed for PSA action is one of a cloud of hydration that interferes with other cell–cell interactions. Thus, there are two models for how NCAM might function during growth cone guidance. First, NCAM might function as a contact-mediated permissive or attractive signal. Second, in the PSA model, NCAM-PSA might function as a modulator of and in certain contexts as an inhibitor of other contact-mediated guidance molecules. These possibilities are discussed later in this chapter.

Some Ig CAMs have cytoplasmic domains that can function as tyrosine kinases and phosphatases

Considerable evidence has indicated that Ig CAMs function not just as adhesion proteins but also as signaling receptors. Certain members of the Ig superfamily that can function to mediate homophilic adhesion also contain cytoplasmic regions with protein tyrosine kinase or protein tyrosine phosphatase domains (Fig. 4–2). For example, the receptor protein tyrosine kinase (RPTK) ARK (or AXL) is expressed at high levels in the mouse brain and can bind in a homophilic fashion (Bellosta et al., 1995). Similarly, the receptor protein tyrosine phosphatases RPTP mu and RPTP kappa are expressed in the nervous system and can bind in a homophilic fashion (Zondag et al., 1995).

Most Ig CAMs lack obvious signal transduction sequences in their cytoplasmic domains, but nevertheless may function as signaling receptors

In contrast to the Ig CAMs that are RPTKs or RPTPs, most Ig CAMs have cytoplasmic domains that do not possess any obvious signaling motifs. Nevertheless, evidence now suggests that these Ig CAMs can participate in signal transduction (e.g., reviewed by Doherty and Walsh, 1994; Bixby et al., 1994). Nonreceptor protein tyrosine kinases (PTKs) such as Abl, Src, Fyn, and Yes are localized at high levels in growth cones (e.g., Maness et al., 1990; Bixby and Jhabvala, 1993) and have been proposed to function in CAM-mediated signaling. In *Drosophila,* embryos carrying mutations in the genes encoding either the Abl tyrosine kinase or the Fas I neural CAM have a relatively normal CNS, but embryos doubly mutant for both genes have major abnormalities in the development of their axon commissures (Elkins et al., 1990). More recent genetic analysis suggests that Abl may also function in a Fas II signaling pathway (García-Alonso et al., 1995). In mammals, NCAM-dependent neurite outgrowth is impaired in vitro in neurons from mice with a targeted mutation in the gene encoding Fyn (Beggs et al., 1994), whereas L1-dependent neurite outgrowth is impaired in vitro in neurons from mice defective in Src function (Ignelzi et al., 1994). Taken together, these studies suggest that different tyrosine kinases function downstream of specific CAMs. Cross-linking antibodies against L1 and NCAM (which are thought to function as activating reagents) also trigger phosphatase activity, suggesting that phosphatases may also be downstream of CAMs (Klinz et al., 1995).

Recent studies have also shown that neurite outgrowth stimulated by NCAM, L1, and N-cadherin in vitro is blocked by manipulations that block the fibroblast growth factor receptor (FGFR, a receptor protein tyrosine kinase) (Williams et al., 1994a). One possibility is that activation of these CAMs might involve activation of FGFR. The idea that CAMs might have a *cis* association (within the plane of the same membrane) with other transmembrane proteins like FGFR is not without precedent, as there is evidence of a *cis* association between NCAM and L1 (Kadmon et al., 1990). Other potential downstream components of CAM signaling have been identified through pharmacological manipulations (Williams et al., 1994b, 1995).

Receptor Protein Tyrosine Kinases and Phosphatases and Their Ligands

A variety of receptor protein tyrosine kinases and phosphatases (RPTKs and RPTPs), in addition to those IgCAMs mentioned above, can function either to modulate or to guide axon growth (Fig. 4–2). Here we describe these receptors and their ligands. Their functions are discussed later.

FGFs and their receptors

Above we discussed a requirement for FGF receptor function in axon outgrowth stimulated by CAMs (Williams et al., 1994a). At least four structurally related FGF receptors have been identified, characterized by the presence of three Ig domains in their extracellular region and by an intracellular tyrosine kinase domain (Fig. 4–2). FGFRs appear to have a variety of soluble ligands, including FGFs, which form a large family of structurally related ligands (reviewed by Basilico and Moscatelli, 1992) and other unrelated proteins (Kinoshita et al., 1995). Various FGFs have been shown to stimulate axon extension in vitro (Walicke et al., 1986). Recent studies have suggested that FGF exerts its growth-modulatory effect by binding to axonal FGFRs and is used in vivo to stimulate the growth of axons and to regulate target invasion in the retinotectal system (McFarlane et al., 1995, 1996).

Neurotrophins and their receptors

In addition to their survival-promoting effects, the neurotrophins NGF, BDNF, NT3, and NT4/5 have stimulatory effects on axon extension in vitro. NGF has also been shown to be capable of functioning as a chemoattractant for regenerating axons in vitro (Gundersen and Barrett, 1979). These factors produce their biological effects through receptor tyrosine kinases of the Trk family (Fig. 4–2). Despite their stimulatory effects, the evidence to date does not support a role for neurotrophins in directing axons to their targets (reviewed in Kennedy and Tessier-Lavigne, 1995). Much evidence indicates, however, that the neurotrophins play important roles in allowing axons to invade their target fields and in directing the elaboration of axon terminal arbors at the target (Hoyle et al., 1993; ElShamy et al., 1996).

Derailed

In *Drosophila,* the Derailed RPTK (Fig. 4–2; related to vertebrate Ryk) is expressed by a subset of neurons whose axons selectively fasciculate into two distinct CNS axon pathways (Callahan et al., 1995). Analysis of *derailed* mutant embryos has suggested a role for Derailed in axon guidance or fasciculation. The ligands for this subfamily of RTK are unknown.

Eph receptors and their ligands

In vertebrates, the largest subfamily of RPTKs is the Eph family, which comprises over a dozen members (Fig. 4–2); the ligands for these RPTKs are all membrane anchored via either a phospholipid (PI) anchor or a transmembrane domain (reviewed by van der Geer et al., 1994; Pandey et al., 1995; Tessier-Lavigne, 1995; Gale et al., 1996). There is considerable promiscuity in binding interactions: The receptors fall into two groups, one of which interacts largely indiscriminately with all transmembrane ligands and the other of which interacts largely indiscriminately with all GPI-anchored ligands (Brambilla et al., 1995; Gale et al., 1996). Both the Eph RPTKs and their ligands are expressed in the developing nervous system. Several of the ligands and their receptors have been implicated in axon growth, axon fasciculation, and topographic guidance, in the last case by functioning as contact inhibitors or repellents (e.g., Winslow et al., 1995; Drescher et al., 1995; Cheng et al., 1995; Nakamoto et al., 1996; Gao et al., 1996; Zhang et al., 1996; Monschau et al., in press).

Receptor tyrosine phosphatases

There is also growing evidence that receptor protein tyrosine phosphatases (RPTPs) play important roles in the control of growth cone guidance. A growing number of RPTPs have been discovered and many are expressed in the developing nervous system (Fig. 4–2) (reviewed by Krueger et al., 1990; Zinn, 1993). For example, the RPTP CRYPα is expressed on embryonic axons, growth cones, and filopodia (Stoker et al., 1995). Many RPTPs have extracellular domains consisting of tandem Ig domains followed by tandem FN type III domains (or tandem FN type III domains alone), similar to many CAMs of the Ig super family (see above), suggesting that they might transduce cell–cell interactions (mediated by either homophilic or heterophilic binding) into cytoplasmic phosphatase function (reviewed by Brady-Kalnay and Tonks, 1995). Little is known about the ligands for RPTPs or their mode of activation. Peles et al. (1995) have shown that a vertebrate RPTP, RPTPβ, binds the neural CAM contactin (F11), suggesting a possible link between CAMs and RPTPs. Five RPTPs have been identified in

Drosophila, four of which are expressed predominantly on axons and growth cones in the developing nervous system (Streuli et al., 1989; Yang et al., 1991; Tian et al., 1991; Hariharan et al., 1991; reviewed by Zinn, 1993). Genetic analysis has implicated several of these in the control of axon fasciculation and defasciculation (Desai et al., 1996; Krueger et al., 1996).

Extracellular Matrix Molecules and Their Receptors

Many extracellular matrix (ECM) molecules, including the laminin, tenascin, collagen, and thrombospondin families, as well as fibronectin, vitronectin, and a variety of proteoglycans, can act as either promoters or inhibitors of neurite outgrowth and extension. For example, laminin 1 and fibronectin have been shown to promote extension of different axonal classes (e.g., Lander, 1987), whereas tenascin-C and tenascin-R have in certain assays been shown to function as inhibitors or repellents (Faissner and Kruse, 1990; Schachner et al., 1994). Major receptor classes for ECM molecules are integrins (discussed below) and Ig superfamily members (discussed above). Proteoglycans are also receptors for some ECM molecules, though they may function primarily as binding or presenting molecules, rather than being involved in signaling by ECM molecules. In addition, some proteoglycans also appear to be able to function as ligands to inhibit extension of different axonal classes (Snow et al., 1990). The complexity and diversity of different ECM molecules and their growth cone receptors are not discussed in detail here (e.g., Bixby and Harris, 1991; Hynes and Lander, 1992). Rather, we focus on one of the best-known examples: laminins and their receptors.

Members of the laminin family are heterotrimeric, cruciform glycoproteins with constituent chains called α, β, and γ (Fig. 4–4). In vertebrates there are at least five α, three β, and two γ chains (see Burgeson et al., 1994, for a revised vertebrate nomenclature), which can assemble in different combinations to give rise to at least ten different laminin molecules. Laminin 1, for example, is a heterotrimer of subunits $\alpha 1$ (400 kG; formerly A), $\beta 1$ (220 kD; B1) and $\gamma 1$ (200 kD; B2). Other laminin complexes use different subunits (e.g., Engvall et al., 1990) including, for example, $\beta 2$ (S-laminin; Hunter et al., 1989), a subunit enriched at synaptic junctions, and $\alpha 2$ (merosin; Ehrig et al., 1990), a subunit enriched in striated muscle and peripheral nerve. The different forms of laminin have distinct adhesive or antiadhesive effects on different cell types (e.g., Calof and Lander, 1991). *Drosophila* contains at least one laminin (e.g., Montell and Goodman, 1988, 1989).

The axonal receptors implicated in mediating the outgrowth-stimulating effects of laminins on axons are members of the integrin family of α/β heterodimers. Several different $\beta 1$-containing heterodimers appear to function as laminin receptors (Fig. 4–3) (e.g., Reichardt and Tomaselli, 1991; Hynes and Lander, 1992). Other heterodimeric combinations of α and β subunits define integrin-binding specificities for other ECM molecules. Integrins appear to function quite generally to link ECM signals to the cytoskeleton and to initiate other sig-

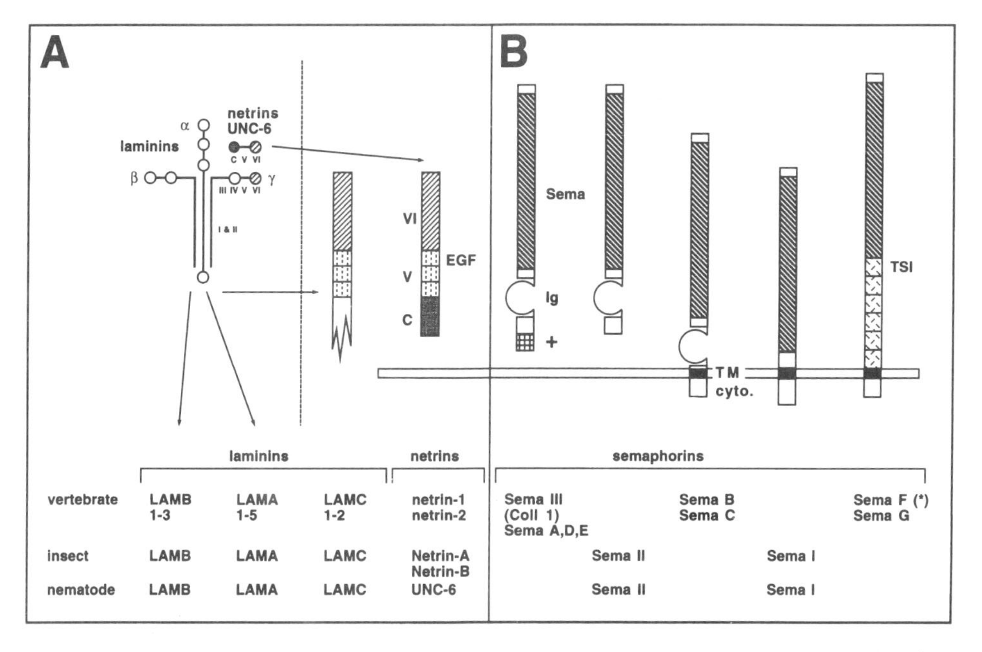

FIGURE 4–4 **Laminins, netrins, and semaphorins: three families of guidance molecules.** All three families of guidance molecules are conserved in structure and apparently in function from nematodes and insects to vertebrates. (A) The laminins are heterotrimeric, cruciform glycoprotein complexes with chains called α, β, and γ. In vertebrates, there are at least five α (LAMA), three β (LAMB), and two γ (LAMC) chains. The netrins, a family of ~600 amino-acid--secreted proteins, are related to the N-terminal domains of laminin chains (domains VI and V), although they then diverge from laminin sequences and are much shorter. (B) The semaphorins are a large family of cell-surface and secreted proteins. Most semaphorins are ~750 amino acids in length, and share a common ~500-amino-acid semaphorin domain; in several of these subfamilies, the semaphorin domain is followed by an Ig domain. One subfamily, however, contains members that are over 1,000 amino acids long; in these proteins, the semaphorin domain is followed by a set of tandem thrombospondin type I domains. For details and references, see text.

nal transduction pathways to control cell migration and proliferation (e.g., Yamada and Miyamoto, 1995; Clark and Brugge, 1995). A variety of other potential receptors or binding proteins for laminin have been identified, including the L1-related Ig superfamily member NgCAM (Grumet et al., 1993), but their functions in modulating neurite growth are poorly understood.

Based on their potent in vitro effects on axons and their expression in the developing nervous system, various ECM molecules including laminins and their integrin receptors are expected to function in axon guidance. For the most part, however, the precise roles of different ECM molecules in vivo are still unknown. Later we discuss the results of genetic studies that have provided some insights into the role of laminins in axon guidance in vivo. We now turn to two families of molecules, the netrins and semaphorins, which have members that may be thought of as ECM molecules, but whose axon guidance functions are better defined.

Netrins and Their Receptors

The netrins are a family of secreted proteins that are related to the laminins and are conserved from nematodes to vertebrates (Fig. 4–4) (reviewed in Culotti and Kolodkin, 1996). Netrins appear to be bifunctional guidance cues, attracting some axons toward the source of netrin while repelling others away (Colamarino and Tessier-Lavigne, 1995a; Wadsworth et al., 1996).

Netrins were discovered through two distinct approaches in nematode and chick. The first member of the family, UNC-6, was identified in *Caenorhabditis elegans* on the basis of the *unc-6* mutant phenotype, which involves defects in both cell migration and axon guidance (Hedgecock et al., 1990; Ishii et al., 1992). In a very different approach, an in vitro assay was established based on the ability of the floor plate (cells at the ventral midline of the spinal cord) to secrete a factor which induces both outgrowth and steering of dorsal spinal cord axons (Tessier-Lavigne et al., 1988; Placzek et al., 1990). The axon outgrowth assay was used to purify two proteins (netrin-1 and netrin-2) from embryonic chick brain, both of which have appropriate in vitro axon outgrowth and steering activity and in vivo spinal cord expression (Serafini et al., 1994; Kennedy et al., 1994). Based on the nematode and chick sequences, PCR was used to identify netrin family members in a variety of organisms, including *Drosophila* (Mitchell et al., 1996; Harris et al., 1996) and other vertebrates (Serafini et al., 1996).

Netrins are secreted proteins of ~600 amino acids. The N-termini of the netrins (~450 amino acids) are related to the N-termini of laminin γ (Fig. 4–4) subunits, including laminin domain VI (in netrins called domain VI) and the three epidermal growth factor (EGF)-like repeats in laminin domain V (in netrins called domains V-1, V-2, and V-3). The C-terminal ~150-amino-acid domain (domain C) is highly basic and diverges from laminins but is conserved amongst the netrins. A variety of experiments suggest that netrins can be both diffusible and cell associated and that the C-terminal domain regulates diffusibility of the netrins by mediating binding to cell surfaces or the extracellular matrix (C. Mirzayan and M. Tessier-Lavigne, unpublished observations).

In any one organism, the netrin family appears to be encoded by only one or a few genes. Thus far, there is one *netrin* gene in nematode, two in *Drosophila*, and two in vertebrates (for references, see above). The two tandem *Netrin* genes in *Drosophila* appear to represent a gene duplication that is independent of that which occurred in the vertebrate lineage (Mitchell et al., 1996; Harris et al., 1996).

Netrin receptors have been identified through genetic and biochemical analysis. There is evidence that members of the DCC subfamily of the Ig superfamily (DCC in vertebrates, UNC-40 in *C. elegans*, and Frazzled in *Drosophila;* Fig. 4–2) are components of receptors that mediate attractive effects of netrins (Chan et al., 1996; Keino-Masu et al., 1996; Kolodziej et al., 1996). Evidence from *C. elegans* has implicated the UNC-5 gene product (Leung-Hagesteijn et al., 1992; Hamelin et al., 1993), a transmembrane protein that possesses two Ig domains and two thrombospondin type I domains in its extracellular portion (Fig. 4–2), in mediating repulsive actions of the netrin UNC-6.

Semaphorins

The semaphorins are a family of cell-surface and secreted proteins that are conserved from insects to humans (Fig. 4–4) (Kolodkin et al., 1993). Many different members of the semaphorin family appear to function as chemorepellents or inhibitors, being capable of causing growth cone collapse (Coll 1/Sema III), inhibiting branching (Sema I), influencing steering decisions (Sema I and Sema III), preventing axons from entering certain target regions (Sema II and perhaps Sema III), or inhibiting the formation of synaptic terminal arborizations (Sema II) (Kolodkin et al., 1992; Luo et al., 1993; Matthers et al., 1995; Messersmith et al., 1995). It is possible that some semaphorins might also function as chemoattractants or contact attractants.

Like netrins, semaphorins were identified using two distinct approaches in invertebrates and vertebrates. A monoclonal antibody screen for surface antigens on subsets of fasciculating axons in the grasshopper embryo led to the identification and functional analysis of fasciclin IV (Kolodkin et al., 1992), a transmembrane protein that was subsequently renamed semaphorin I when it was discovered to be the first member of a new gene family. A PCR-based approach based on the Sema I sequence was used to identify two members of the family in *Drosophila* (Sema I and the secreted Sema II) and one in human (the secreted Sema III) (Kolodkin et al., 1993). In a very different approach, an in vitro growth cone collapse assay was established (Raper and Kapfhammer, 1990) and used to purify an activity from chick brain that causes the collapse of growth cones of sensory neurons in culture. This purification led to the identification of chick collapsin-1 (Luo et al., 1993), a secreted protein that is a member of the semaphorin family. Chick collapsin-1 and human Sema III are species homologues. Based on the insect, human, and chick sequences, additional semaphorins were identified in humans (Messersmith et al., 1995), mouse (Püschel et al., 1995), and chick (Luo et al., 1995).

Most semaphorins are ~750 amino acids in length (including signal sequence), although a few are over 1,000 amino acids in length (Fig. 4–4). The family is defined by a conserved ~500-amino-acid extracellular domain, called the semaphorin domain (from about amino acid 50 to 550), that contains 14–16 cysteines, many blocks of conserved residues, and no obvious repeats (Kolodkin et al., 1992, 1993). Semaphorins come in at least five different subtypes (Fig. 4–4B): (1) secreted with Ig domain and C-terminal basic domain (e.g., vertebrate Sema III/Collapsin), (2) secreted with Ig domain but without C-terminal basic domain (e.g., *Drosophila* Sema II), (3) transmembrane with Ig domain (e.g., vertebrate Sema B and Sema C), (4) transmembrane without Ig domain (e.g., insect Sema I), and (5) transmembrane without Ig domain but with six thrombospondin type I domains (e.g., mouse Sema F and Sema G). The diversification of semaphorins into secreted and transmembrane proteins appears to have occurred prior to the divergence of nematodes, insects, and vertebrates.

In contrast to the netrins, the semaphorins appear to be a large family. There are at least three in *Drosophila,* two in nematode, five in chick, seven in mouse, and ten in human (Kolodkin et al., 1993; P. Roy and J. Culotti, personal communication; Luo et al., 1993, 1995; Püschel et al., 1995; Messersmith et al., 1995; Sekido et al., 1996; Kolodkin et al., unpublished results; Püschel et al., unpublished results; and unpublished analysis of the dbest database). In addition, three semaphorins are encoded in viral genomes (Kolodkin et al., 1993; NIH database).

Nothing is yet known about the identity of the receptors for semaphorin family members. However, a probable downstream component of a semaphorin receptor, collapsin response mediator protein (CRMP)-62, a cytoplasmic protein, was identified using an oocyte expression screen (Goshima et al., 1995). Expression of CRMP-62 in oocytes makes the oocytes respond to collapsin-1 by activating an inward conductance (it is presumed that the oocytes must endogenously possess a semaphorin receptor), and antibodies against CRMP-62, when introduced into sensory neurons in vitro, prevent collapsin from inducing growth cone collapse (Goshima et al., 1995). CRMP-62 shows sequence similarity to the product of the *unc-33* gene of *C. elegans* (Li et al., 1992), mutations in which cause defects in the elongation, fasciculation, and branching of a wide variety of different axonal classes (e.g., Hedgecock et al., 1985; McIntire et al., 1992). Moreover, CRMP-62 expression appears to be restricted to the nervous system. Thus, CRMP-62 and UNC-33 represent potential downstream components in a semaphorin signaling pathway.

In Vivo Roles of Axon Growth Modulators

The previous sections have introduced the cellular interactions known to guide axons, as well as some of the best-characterized ligands and receptors thought to mediate guidance events. In many cases, these molecules have been implicated in guidance only indirectly, based, for example, on their patterns of expression in vivo and their ability to promote or inhibit neurite outgrowth in vitro. Insights into the precise functions of several of these molecules in guidance have started to be obtained through functional perturbations in vivo, including antibody or enzymatic perturbations, or genetic manipulations.

In this section, we focus on molecules whose guidance functions have been illuminated by in vivo perturbations, or, at the very least, by experimental paradigms in vitro that mimic relevant in vivo contexts. To discuss molecular function it is necessary in each case to describe a particular pathfinding decision in which the molecule is implicated. To this end, we have restricted our discussion to cases where some definitive insight into a pathfinding function has been obtained. In many cases, functional analysis of particular molecules has also provided greater insights into the other types of cellular and molecular interactions that must be involved in those decisions. We attempt to summarize both sets of insights.

We have excluded from our discussion cases where functional perturbations have so far failed to illuminate specific guidance functions. Thus, we do not discuss cases where loss-of-function mutations produce severe behavioral or anatomical defects but where specific pathfinding errors have not been pinpointed. This is the case, for example, for mutations in the *Drosophila Semall* gene (Kolodkin et al., 1993) or the human *L1* gene (Jouet et al., 1993, 1994; Vits et al., 1994), both of which have severe but poorly understood phenotypes. Likewise, we do not discuss loss-of-function mutations in putative guidance cues or receptors that have yielded more subtle phenotypes than those predicted based on the patterns of expression and in vitro activities of the molecules. Of course, where a phenotype is not observed it can be tempting to invoke the existence of functionally redundant cues, and there is certainly reason to believe that considerable redundancy does exist (see e.g., Elkins et al., 1990; Desai et al., 1996; Mitchell et al., 1996; discussed below). The extent to which functional redundancy operates in pathfinding will be understood fully only as we progress from the analysis of single mutations to the analysis of compound mutations.

We discuss local guidance, long-range guidance, axon fasciculation, and target recognition in turn.

Local Attraction and Repulsion

Local modulators of axon growth and guidance can be either on the cell surface or in the extracellular matrix (ECM) and can provide a variety of positive and negative signals. At one extreme, local signals are often thought of as permissive, providing an appropriate growth-promoting substrate to encourage axon extension, or inhibitory, leading to the cessation of axon extension. At the other extreme, local signals can be instructive, providing directional or steering information to guide growth cones.

But many axon growth modulators do not appear to function exclusively at one extreme or the other. Rather, many molecules function between these two extremes in a range of capacities to modulate fasciculation, axon sorting, axon branching, and growth cone responses to other guidance signals. Local signals can be presented in discrete step functions that form boundaries, or alternatively, in positional gradients where they can function to regulate topographic projections by modulating the speed of extension and stopping, the rate of axon branching, and steering decisions.

In some cases, the differences between a "permissive" and an "instructive" signal depend not so much on the molecule itself but rather upon the context in which the signal is presented to the growth cone. For example, we think of growth-promoting substrates as being "permissive," but in certain contexts the precise location of an appropriate growth-promoting substrate can influence the specific path taken by a growth cone, so that it becomes "instructive." Likewise, molecules that mediate axon adhesion or other kinds of cell interactions may either directly alter steering, or indirectly alter it by modulating the function of other effectors, or have little effect on directional decisions, depending on the context.

Below we focus on several examples in which the molecular events of growth cone guidance have been studied in some detail, and in which it can be shown that individual molecules play specific roles as local signals to encourage or inhibit axon extension, and to control steering. Two special cases of local guidance events, selective axon fasciculation and target recognition, require more detailed consideration and are discussed in separate sections below, after the section on long-range guidance.

Role of Permissive Substrates: Laminin as a Permissive Substrate for Sensory Axons in Drosophila

Laminin has been shown to be a potent promoter and substrate for neurite outgrowth in vitro (reviewed by Timpl and Brown, 1994; Gomez and Letourneau, 1994; Kuhn et al., 1995). These in vitro results, coupled with the abundance of laminin in regions of axon pathway formation (e.g., Rogers et al., 1986; McLoon et al., 1988; Letourneau et al., 1988), have led to the hypothesis that axons in vivo might use laminin as a growth-promoting substrate (e.g., reviews by Sanes, 1989; Bixby and Harris, 1991; Reichardt and Tomaselli, 1991; Hynes and Lander, 1992). Genetic analysis of laminin function in *Drosophila* has provided support for this hypothesis (García-Alonso et al., 1996).

In the eye-antenna imaginal disc in *Drosophila* pupae, the nerve from the three simple eyes (ocelli) to the brain is pioneered by a population of transient neurons (the ocellar pioneers) whose axons project toward the brain along a dorsal route by extending over a laminin-rich ECM covering the disc epithelium and ultimately take a ventral route to the brain.

In flies mutant in the *Laminin A* (*Lam A*) gene, the ECM over the disc is reduced or less stable (as assessed by its appearance in electron micrographs), and the ocellar pioneer axons, which normally extend on the ECM, display striking pathfinding defects, while their neighboring bristle axons appear normal. The ocellar pioneers now adhere to and extend along the epithelium, sometimes fasciculating with bristle axons, and thus begin to take the ventral route. Invariably, however, they fail to reach the brain (García-Alonso et al., 1996).

Evidently, the disc epidermis and bristle neuron axons provide suitable adhesive surfaces for the ocellar pioneers, but they do not provide the necessary growth-promoting signals for their complete extension into the brain (or alternatively, they might provide specific growth-inhibiting signals). These results lead to four conclusions.

First, for certain classes of sensory axons (ocellar pioneers), an ECM containing laminin A is a necessary growth-promoting substrate. This suggests that laminin can function as a local modulator of axon growth, although it is not yet established whether laminin acts directly or indirectly on these axons. The fact that laminins are potent stimulators of axon growth in cell culture argues for a direct effect, but demonstrating this conclusively will require identifying the laminin receptors expressed by these axons and perturbing their function in vivo.

Second, these results reveal a fundamental distinction between a substrate's adhesiveness and its ability to promote axonal extension. Adhesiveness on its own is not sufficient, but rather the substrate must also provide the appropriate growth-promoting signals. In vitro studies using various isoforms of NCAM have led to a similar conclusion (Doherty and Walsh, 1992).

Third, even in the absence of laminin A, although the ocellar pioneer axons stall and fail to reach the CNS, they nevertheless extend in the appropriate direction, suggesting that some other signal (perhaps a brain-derived chemoattractant?) provides the directional guidance cue toward the brain.

Fourth, even though some unknown cue makes the axons grow in the general direction of the brain, the two different substrates followed by the ocellar and the bristle axons lead these axons to enter the brain in two different locations. Thus, there appears to be a hierarchy of cues in this case: Some cue(s) signals the general direction of growth, while others subdivide possible pathways heading in that direction.

Integrins and FGF receptors influence extension but not steering of retinal ganglion cell axons

As just discussed, a full understanding of laminin function in axon guidance requires perturbation of the function of laminin receptors, i.e., of specific integrins. This has been achieved in the case of retinal ganglion axon guidance in *Xenopus* (Lilienbaum et al., 1995).

The axons of *Xenopus* retinal ganglion cells (RGCs) leave the retina, extend down the optic nerve, across the optic chiasm, through the optic tract, and to their target, the optic tectum. During their growth the axons express the α6β1 integrin, a laminin receptor. The function of integrins was perturbed in vivo by transfection of genes encoding various dominant forms of the β1 subunit into intact eye primordia of *Xenopus* embryos. Cells expressing these receptors showed a marked reduction of process outgrowth, with more than two-thirds failing to extend an axon. For those that did extend an axon, average axonal length was considerably shorter than that of control axons. Strikingly, however, there was no misrouting of the axons that did extend, implicating β1 integrins in axon extension but not pathfinding. In these experiments, the authors did not identify the ligands for β1 integrins whose functions were interfered with, but laminin is one factor that is likely to contribute since it is expressed along the axon pathway and is a potent promoter of RGC axon extension in vitro (cited in Lilienbaum et al., 1995). These results are therefore complementary to the results observed in *LamA* mutants discussed above in that in both cases the perturbation primarily affects axon extension rather than pathfinding.

Another factor that appears to contribute to stimulating extension of RGC axons, but not to pathfinding, is bFGF, which is abundantly expressed in a corridor that marks the optic tract of *Xenopus* embryos (McFarlane et al., 1995, 1996). RGCs express an FGF receptor, and bFGF stimulates the extension of these axons in vitro. The function of the FGF receptor was tested by transfection of genes encoding dominant forms of the FGFR into RGCs. Cells expressing these receptors showed a marked reduction in axon extension within the tract without apparent effect on pathfinding. Likewise, bFGF applied exogenously to an "exposed-brain" preparation to perturb the spatial distribution of bFGF did not cause misrouting of axons in the tract (although it did cause misrouting of axons at the target; see below). Thus, the role of both FGF and of as-yet-unidentified ligands for β1 integrins in this system appears to be to stimulate extension of axons as they grow down their path, rather than directing pathfinding. The next sections describe molecules more directly implicated in pathfinding per se.

Boundaries and sharp turns:
guidance of grasshopper sensory axons by semaphorin I

As discussed earlier, in the grasshopper embryo, the Ti1 pioneer neurons differentiate at the distal tip of the developing limb bud and extend axons that grow toward and into the CNS. On their way, these growth cones make a series of sharp turns that guide them along a specific pathway (Fig. 4–5). They first stall at a particular location and then make an abrupt turn from proximal to ventral ex-

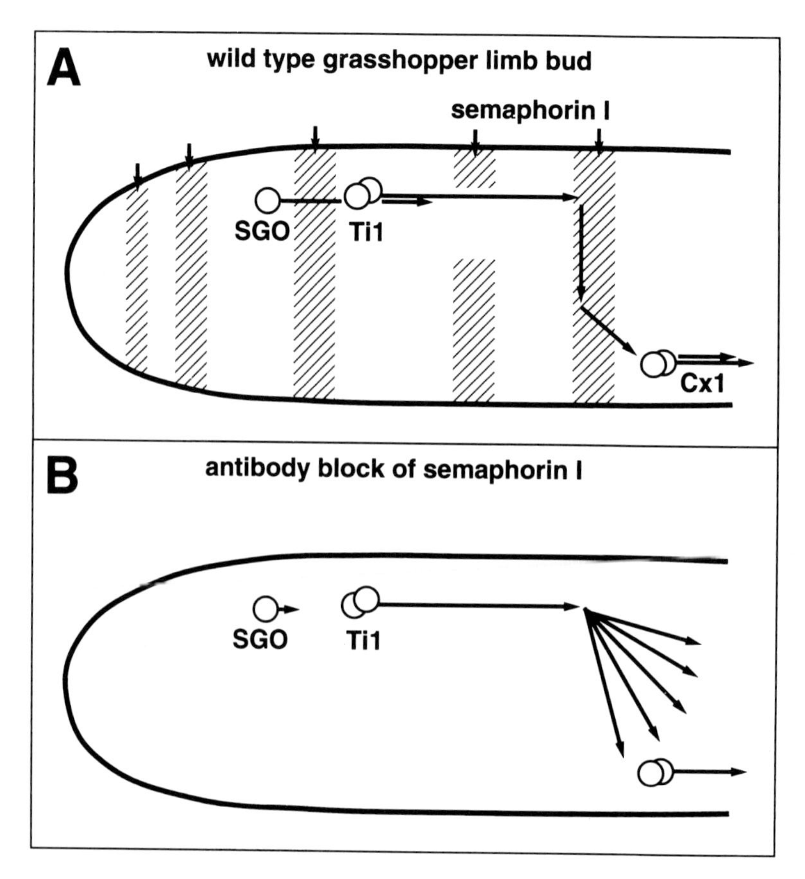

FIGURE 4–5 **Function of semaphorin I in axon guidance.** Schematic diagrams showing the functions of semaphorin I (Sema I) in the guidance of sensory axons in the grasshopper embryo. (**A,B**) Sema I, a transmembrane protein, is expressed on stripes of epithelial cells in the limb bud. Antibody-blocking experiments show that Sema I functions to stall and then steer the pair of Ti1 growth cones as they encounter epithelial cells expressing it; Sema I also prevents these axons from defasciculating and branching (Kolodkin et al., 1992). A recent study provides evidence that Sema I might also function as a positive factor for the SGO axons which encounter a different more distal stripe of Sema I in the lumb bud (J. Wong, W. Yu, and T. O'Connor, in preparation). When Sema I is blocked, these axons fail to grow onto the stripe of Sema I–expressing cells. See text for further discussion.

tension; they then make a second turn and continue their proximal extension toward the CNS. What causes them to stop and make their first sharp turn?

Semaphorin I (Sema I) was first identified based on its expression on a subset of fasciculating axons in the CNS, but Sema I is also expressed on stripes of epithelial cells in the limb bud. The distal boundary of one of these Sema I stripes corresponds to the location where the Ti1 growth cones stall and then make their first abrupt turn (Kolodkin et al., 1992). Antibody-blocking experiments show that Sema I indeed forms a molecular boundary recognized by the Ti1 growth cones. When Sema I function is impaired, these growth cones extend across the boundary, the two axons defasciculate and become highly branched on the epithelial cells, and they make a sweeping rather than a sharp turn in the ventral direction. Thus, Sema I appears to function to stall and then steer this pair of growth cones as they encounter the Sema I–expressing epithelial cells. In addition, it also prevents the axons that encounter it from defasciculating and branching. Furthermore, the fact that the axons do project ventrally indicates that some other guidance cue (either local or long range) must help these growth cones distinguish dorsal from ventral, even in the absence of Sema I function.

These in vivo experiments demonstrate an important role for Sema I in guiding the Ti1 axons but also raise an apparent paradox. Sema I appears to have a negative effect on the axons, in the sense that it prevents defasciculation and branching. At the same time, since the growth cones actually appear normally to extend along the stripe of Sema I–expressing cells but cross over the stripe in the presence of the antibodies, one might think that Sema I provides a favorable substrate for the growth of the axons. Is Sema I functioning here as both a positive and a negative factor? This question remains unresolved.

A recent study has, however, provided evidence that Sema I might indeed function as a positive factor for a distinct set of axons, the SGO axons, which encounter a different (more distal) stripe of Sema I in the limb bud (Fig. 4–5) (J. Wong, W. Yu, and T. O'Connor, in preparation). Normally the SGO axons grow over that stripe and then fasciculate with the Ti1 axons, but when the Ti1 neurons are ablated, the SGO axons grow onto the Sema I stripe but stop at the end of the stripe. Is Sema I a positive factor that the axons prefer to the rest of the environment, or is it a negative factor that inhibits motility of these axons? The growth cones never appear to collapse or to retract in any way when encountering the stripe. Moreover, in the presence of antibodies to Sema I, the axons fail to grow onto the stripe of Sema I–expressing cells. Together, these results suggest that Sema I is a positive factor for the SGO axons that makes the stripe a more attractive location for them to reside, and which they can leave only by fasciculating onto the Ti1 axons. Thus, these results argue that Sema I might have both attractive and repulsive effects on different classes of sensory axons and perhaps even on the same class.

Eph ligands as guidance receptors in vivo?

As mentioned earlier and discussed below, there has been considerable interest in the possibility that ligands for Eph-class receptor tyrosine kinases function as repulsive axon guidance cues to direct the formation of topographic projections and to regulate axon fasciculation. The distribution of these ligands and their cognate receptors has also suggested roles in pathfinding to the target (e.g., Gale et al., 1996). Some evidence for a role in pathfinding has been obtained through genetic manipulation of the function of an Eph receptor, the nuk receptor, in mice (Henkemeyer et al., 1996). The results, however, are surprising in that they suggest that nuk, rather than functioning as a receptor, actually functions as a guidance cue for a distinct receptor.

Mice carrying a null allele of *nuk* show a characteristic projection defect in the forebrain: Axons of the posterior portion of the posterior commissure fail to cross the midline and wander aberrantly ipsilaterally. The nuk receptor, however, is expressed not by the axons but rather by a group of cells at the midline that the axons grow on to cross the midline, whereas the axons express a transmembrane ligand for nuk called HTK-L. Even more remarkable, selective deletion of the cytoplasmic domain of nuk by the in-frame "knock-in" of sequences encoding β-galactosidase creates a *nuk* allele that does not have the guidance phenotype observed in the null. Thus, correct guidance of the axons only requires expression of the nuk extracellular domain on the midline cells. One interpretation of this result is that nuk functions as a simple adhesion molecule or signal for HTK-L on the axons. Subsequent in vitro experiments showed that binding of transfected Cos cells expressing HTK-L to cells expressing nuk results not just in the phosphorylation of nuk (the expected result, since nuk is a receptor tyrosine kinase) but also in the phosphorylation of the cytoplasmic domain of HTK-L (by an unknown cellular kinase endogenous to the cells) (Holland et al., 1996; K. Brueckner, E.B. Pasquale, R. Klein, personal communication). Taken together, these results suggest that HTK-L might function in vivo as a receptor used by the axons to detect the extracellular domain of nuk on pathway cells, which functions as a guidance cue for the axons.

While it remains to be shown that the misrouting of axons in the *nuk* knockout does occur at the midline and that the axons can use HTK-L as a receptor, it is nonetheless worth mentioning these results, both because of the great interest in the possible involvement of Eph receptors in axon guidance and because of the heretical nature of the model for the flow of information in this system. More recent evidence also suggests that the GPI-linked ligands might function as actual guidance cues in vivo as well (S. Park, J. Frisen, and M. Barbacid, personal communication).

To cross or not to cross: molecules that control steering at the midline of the CNS

Ventral midline structures of nervous systems as diverse as those of nematodes, fruitflies, and vertebrates provide a variety of different guidance signals. In principle, these might include signals that (1) attract some growth cones to extend toward the midline, (2) direct some growth cones to cross the midline, (3) prevent rostrocaudally (longitudinally) projecting growth cones from crossing the midline (either preventing the crossing of those axons that normally never cross the midline or preventing repeated crossing for the large number of axons that normally cross only once), and (4) repel some growth cones to extend away from the midline. Signals 1 and 4 appear to be long range and to be mediated in part by netrins and their receptors (see next section on long-range guidance). However, signals 2 and 3 appear to operate locally and to be mediated by a variety of molecules including Ig CAMs.

Ig CAMs direct crossing of the chick midline. The ventral midline of much of the CNS of vertebrates is populated by a specialized group of cells, floor plate cells, that influence the growth of axons that navigate the midline (reviewed by Colamarino and Tessier-Lavigne, 1995b). So-called commissural axons project to the midline, cross the midline, then turn to project longitudinally in close contact with the contralateral floor plate border (Bovolenta and Dodd, 1990). Although a long-range chemoattractant (netrin-1) is required to guide commissural axons toward the floor plate (see section below), local cues direct their growth cones at the floor plate.

A recent study in the chick has implicated two Ig CAMs in enabling axons to cross the floor plate: axonin-1 and NrCAM (Fig. 4–2). Commissural axons and growth cones express axonin-1, whereas floor plate cells express high levels of NrCAM (Fig. 4–6) (which is also expressed at low levels by many cells in the spinal cord). These two Ig CAMs can bind heterophilically. Blocking this axonin-1/NrCAM interaction with either antibodies against axonin-1, antibodies against NrCAM, or soluble axonin-1, led to dramatic pathfinding errors of the commissural growth cones in which up to 50% of the axons failed to cross the floor plate and instead turned along the ipsilateral border of the floor plate (Stoeckli and Landmesser, 1995). In addition, whereas in control embryos, commissural axons turn rostrally after crossing the midline, in the experimental embryos some axons turn caudally either before or after crossing the floor plate.

Further insight into the function of this axonin-1/NrCAM interaction has come from cell culture experiments in which commissural neurons were cultured with floor plate cells. Commissural axons normally grow onto the floor plate cells, but stall or collapse on contact with the floor plate cells in the presence of reagents that block this interaction (Stoeckli et al., 1997). Together, these in vitro experiments suggest that floor plate cells normally express an (unidentified) inhibitory factor on their surface whose function is normally masked by NrCAM to facilitate floor plate crossing.

What might the function of this inhibitory factor be? One possibility is that this factor serves to prevent commissural axons from recrossing the floor plate after they have crossed the midline once; there is in fact considerable indirect evidence to suggest that the floor plate does indeed provide a barrier to recrossing the midline (reviewed in Colamarino and Tessier-Lavigne, 1995b). A corollary of this model is that the axons must lose responsiveness to NrCAM after they have crossed the midline. One way in which this could be achieved would be to down-regulate axonin-1 function. In rat, expression of the axonin-1 homologue TAG-1 by commissural axons is in fact downregulated as the axons cross (Dodd et al., 1988), which would be consistent with a modulation of responsiveness to a positive factor on floor plate cells. In chick, however, axonin-1 expression is actually maintained after the axons cross. It might be predicted, then, that axonin-1

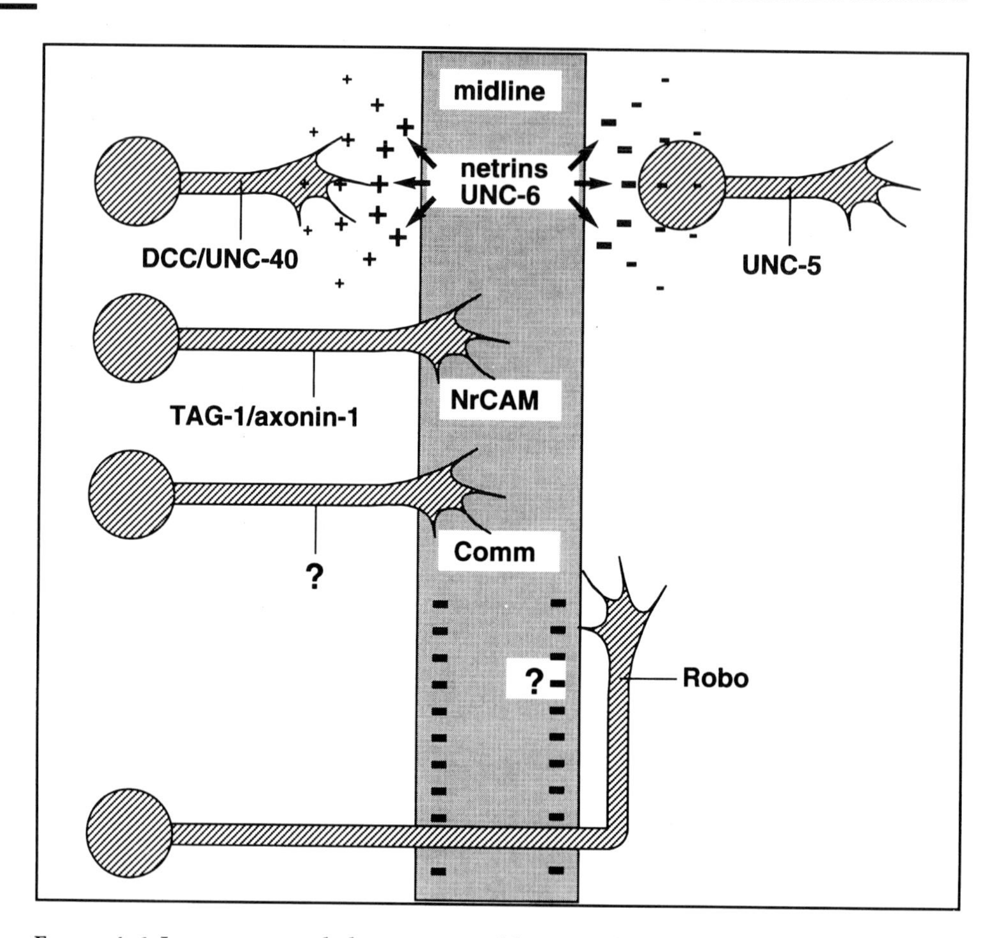

FIGURE 4–6 **Long-range and short-range guidance at the ventral midline.** A composite picture of guidance at the midline drawing on mechanisms identified in nematodes, fruitflies, and vertebrates, at least some of which (and possibly all of which) are conserved among these organisms. The netrins appear to function as both long-range chemoattractants and chemorepellents for different classes of axons. Attraction of growth cones by netrins appears to involve the DCC/UNC-40/Frazzled receptor (as shown in all three phyla), whereas repulsion of growth cones by netrins involves the UNC-5 receptor (as shown in *C. elegans*). In chick, crossing of the midline requires interaction of TAG-1/axonin-1 on commissural axons with NrCAM on the surface of midline cells. In *Drosophila*, it also requires the midline expression of commissureless (whose growth cone receptor is at present unknown). Finally, many commissural growth cones turn longitudinally along the midline after crossing. In *Drosophila*, they express the Robo receptor, which appears to function as a repulsive receptor for an unknown contact-mediated repellent at the midline, thus preventing these growth cones from recrossing the midline.

function is regulated in some other way. One way would be to increase the function of a receptor for the negative factor. Studies in *Drosophila*, which we now discuss, have provided strong evidence for this type of mechanism.

Robo and Comm are required to navigate the Drosophila midline. Netrins also function to attract commissural axons toward the ventral midline in *Drosophila* (see section below). In addition, local cues expressed by distinctive cells at the midline of the CNS are also required to guide commissural growth cones across the midline and to prevent them from recrossing after they have crossed (e.g., Thomas et al., 1988; Crews et al., 1988; Jacobs and Goodman, 1989; Klämbt et al., 1991). A large-scale genetic screen yielded numerous mutants that disrupt commissural axon pathways, but only a few of these mutants display highly penetrant phenotypes (Seeger et al., 1993). The phenotypes of two of these genes discussed below, *roundabout (robo)* and *commissureless (comm)*, suggest that they encode components of repulsive and attractive signaling systems at the midline (Fig. 4–6).

In *roundabout* mutants, many growth cones that normally extend only on their own side instead now project across the midline. Moreover, in *robo* mutants, axons that normally cross the midline only once are instead capable of crossing and recrossing the midline multiple times. *robo* encodes a novel transmembrane Ig superfamily member characterized by the presence of a long cytoplasmic domain (Fig. 4–2) (T. Kidd, K.J. Mitchell, C.S. Goodman, and G. Tear, in preparation). Expression of Robo protein is low or nearly absent on the commissural portions of axons, but it is detected at high levels on the longitudinal portions of axons. Robo is expressed from the onset of axon outgrowth on those axons that never cross the midline (~10% of CNS axons). However, on axons that normally cross the midline just once (~90% of axons), Robo expression on their growth cones increases dramatically after they cross the midline, and is then maintained at high levels on the postcommissural (longitudinal) portions of the axons (T. Kidd et al., in preparation).

Since expression of Robo correlates with the inability of axons to cross the midline and since its absence results in aberrant crossing, the simplest model is that Robo is a receptor or a necessary component of a receptor for an inhibitory factor at the midline that prevents axon crossing. Axons can cross only in the absence of Robo function, as occurs for commissural axons which begin to express detectable levels of Robo only after they cross the midline.

A possible insight into the signal transduction mechanisms downstream of Robo has been obtained from a study in which inhibitors of calmodulin were specifically expressed in subsets of growth cones. These mutant axons project inappropriately across the midline in a manner reminiscent of the behavior of axons in *robo* mutants, suggesting that Ca^{2+} and calmodulin may function downstream of Robo to control midline guidance (Van Berkum and Goodman, 1995).

Mutations in the *commissureless* gene have the opposite phenotype: They lead to a complete absence of all embryonic CNS axon commissures (with the exception of one anterior brain commissure). Although the midline cells are present and appear to differentiate normally, growth cones that would normally project across the midline instead now extend only on their own side. In *comm* mutant embryos, commissural growth cones initially orient toward the midline, but none actually cross is. Rather, any short medially oriented processes are retracted and the axons instead extend exclusively on their own side. The *comm* gene encodes a novel protein which lacks a signal sequence, has a transmembrane domain, and biochemically copurifies with membranes (Tear et al., 1996). The Comm protein is dynamically expressed during embryogenesis in a variety of locations, including the CNS midline cells during the formation of axon commissures. Anti-Comm antibodies reveal strong staining of organelles likely to include the Golgi complex and endosomes and weaker staining of the cell surface. As commissural growth cones contact and traverse the CNS midline, Comm protein is apparently transferred from midline glia to commissural axons.

What is the function of Comm? One clue is derived from the finding that double mutants of *comm* and *robo* display a *robo*-like phenotype. Thus, although Comm is normally essential for axons to cross the midline, in the absence of Robo it is not at all required for crossing. One interpretation is that Comm normally functions to antagonize a putative midline repulsive cue that requires Robo for its effects. In this model, the function of Comm is not unlike that postulated for NrCAM at the midline of the vertebrate CNS. What this explanation requires, however, is that in the absence of Comm the axons are sensitive to the repulsive cue even before they have crossed the midline. Perhaps Comm normally functions to downregulate what little Robo is expressed on growth cones before they cross the midline, or alternatively, to delay Robo expression by these axons until they have crossed the midline.

Finally, other mutations that alter midline guidance have also been identified. For instance, fasciclin I is a lipid-anchored cell adhesion molecule expressed on subsets of axons including some commissural pathways. Abl is a cytoplasmic tyrosine kinase expressed in embryonic axons. As discussed earlier, neither of these genes alone gives a strong mutant phenotype, but the double mutant has a striking, almost completely commissureless, phenotype (Elkins et al., 1990b). This suggests that Fas I and Abl function in redundant signaling pathways that are required for midline crossing, but their precise functions and possible interactions with Comm and Robo remain to be elucidated.

Long-Range Attraction and Repulsion

Although the existence of several diffusible attractants and repellents for developing axons has been demonstrated through tissue culture experiments in several different parts of the vertebrate nervous system, until recently the biochemical nature of these factors was unknown. The first identified diffusible attractants and repellents for developing axons are relatively large (~100 kD) molecules that are members of the netrin and semaphorin families. Interestingly, these molecules are closely related to other nondiffusible molecules involved in axon guidance (laminins in the case of netrins and transmembrane semaphorins like Sema I in the case of diffusible semaphorins). It is of course dangerous to generalize on the basis of just a few examples, and it is possible that other families of chemoattractants and repellents exist that are less related than are netrins and semaphorins to other guidance cues. Nonetheless, the impression derived from this initial identification is that during evolution long-range guidance may have evolved from short-range guidance through the emergence of diffusible relatives of short-range guidance cues. Reinforcing this impression of evolutionary parsimony is the finding that, at least in the case of the netrins, the same molecules have been co-opted to function simultaneously as attractants and repellents for different classes of axons, as we now describe.

Guidance toward and away from the ventral midline: attraction and repulsion by netrins

The netrins, described earlier (Fig. 4–4), have an evolutionary conserved role in guiding developing axons to the ventral midline of the nervous system in *C. elegans, Drosophila,* and vertebrates. In each organism, cells at the ventral midline of the developing nervous system express a netrin family member (Fig. 4–6) (UNC-6 in *C. elegans,* Netrin-A and -B in *Drosophila,* and netrin-1 in birds, mammals and amphibia) (Wadsworth et al., 1996; Mitchell et al., 1996; Harris et al., 1996; Kennedy et al., 1994; Serafini et al., 1996; J. de la Torre, M. Tessier-Lavigne, and A. Hemmati-Brivanlou, in preparation). In each organism, a genetic perturbation has provided direct evidence for an involvement of netrins in guiding axons to the ventral midline. In *C. elegans,* mutations in *unc-6* result in a disruption of virtually all classes of axons that project along a circumferential, ventrally directed trajectory, such that the axons project longitudinally rather than ventrally (Hedgecock et al., 1990). The effect is only partially penetrant, however, as many axons project normally (and different classes of neurons are affected to different degrees). Similarly, in *Drosophila,* a genomic deficiency that deletes the two *Netrin* genes results in a thinning, though not a complete loss, of commissures; the posterior commissure in every segment is more severely affected than is the anterior commissure. (This defect appears to be due to loss of function of *Netrin* genes rather than some other genes, as it can be rescued by reintroduction of either *NetrinA* or *NetrinB* function at the midline; Mitchell et al., 1996; Harris et al., 1996.) Similarly, loss of *netrin-1* function in the mouse has a severe effect on axon guidance, as very few commissural axons eventually reach the midline. These axons start out along their normal trajectory in the dorsal spinal cord, but when they reach the ventral spinal cord, about ~150 μm from the floor plate, many fail to progress further and appear to stall out, while others continue to extend but appear to get misrouted, with some projecting medially and others laterally. Only a few are directed toward the ventral midline or eventually reach the floor plate (Serafini et al., 1996).

What is the evidence that the defects are direct, and that the netrins are functioning in each case as guidance cues? The first piece of evidence comes from tissue culture experiments in which it was found that purified netrin proteins or cells secreting netrins can promote the outgrowth of axons into collagen matrices and can reorient the growth of these axons (Serafini et al., 1994; Kennedy et al., 1994). Thus, netrins appear to have direct effects on the axons. A second piece of evidence comes from the identification of a family of netrin receptors, the DCC subfamily of the Ig superfamily (Fig. 4–1). DCC is expressed on commissural axons and can bind netrin-1. Furthermore, an antibody to DCC blocks the outgrowth-promoting effect of netrin-1 on commissural axons (Keino-Masu et al., 1996). Moreover, mice carrying a null allele of *DCC* generated by targeted mutation show defects in commissural axon growth that are similar to (although more severe than) those observed in the *netrin-1*–deficient mice (Fazeli et al., 1997). Together, these results argue that netrin-1 is a guidance cue that acts directly on commissural axons and that the defects observed in *netrin-1*–deficient mice are direct guidance defects.

Similar results have been obtained in invertebrates. In *C. elegans*, the DCC homologue UNC-40 is expressed by ventrally directed axons, and null mutations in *unc-40* result in misrouting of those axons that is similar to that observed in *unc-6* mutants; moreover, there is evidence that *unc-40* acts cell autonomously, consistent with a receptor function (Chan et al., 1996). In *Drosophila*, the DCC homologue Frazzled is likewise expressed by commissural axons, and null mutations in *Frazzled* phenocopy loss of *Netrin* function (Kolodziej et al., 1996). Together, these studies imply that in each organism the different netrin family members act directly to guide the various axons that are expressing DCC or its homologues.

Netrins have also been shown to act as bifunctional guidance cues, acting as either attractants or repellents for different classes of axons. This was first implied by studies in *C. elegans*, where mutations in *unc-6* impair not just ventrally directed migrations but also dorsally directed migrations (Hedgecock et al., 1990; McIntire et al., 1992) away from the source of UNC-6 protein (Wadsworth et al., 1996). This suggests that UNC-6 functions to guide these axons by repelling them away from the ventral midline, i.e., providing a "push from behind" as described earlier. Like the defects in ventral guidance, the defects in dorsal migrations in *unc-6* mutants are only partially penetrant. In vertebrates, netrin-1 has been shown to be capable of repelling in culture a set of axons, trochlear motor axons, that grow dorsally away from the floor plate (Colamarino and Tessier-Lavigne, 1995a). However, no major deficits are observed in trochlear motor axon guidance in *netrin-1*–deficient animals. This may be explained by the expression of another diffusible repellent made by floor plate cells, since floor plate cells dissected from *netrin-1*–deficient animals still possess repellent activity in vitro (Serafini et al., 1996). Thus, the importance of the repulsive activity of netrin-1 in vivo has not yet been determined in vertebrates. Likewise, it is not yet known whether netrins guide axons through repulsion in vivo in *Drosophila*. Other evidence for a floor-plate–derived chemorepellent(s) distinct from netrin-1 has been provided (Hu and Rutishauser, 1996; Shirasaki et al., 1996; Tucker et al., 1997).

How do netrins repel axons? Three lines of evidence from genetic studies in *C. elegans* have implicated the *unc-5* gene product in mediating the (presumed) repulsive actions of UNC-6 on dorsally directed axons: (1) Null mutations in *unc-5* impair dorsal migrations to the same extent as null mutations in *unc-6*, without effect on ventral migrations (Hedgecock et al., 1990); (2) *unc-5* encodes a transmembrane protein (Leung-Hagesteijn et al., 1992) that is a member of the Ig superfamily (Fig. 4–2); (3) ectopic expression of *unc-5* in touch neurons which normally extend axons longitudinally or ventrally causes their axons to project dorsally in an *unc-6*–dependent fashion (Hamelin et al., 1993). Together, these results strongly imply that UNC-5 is a receptor or a component of a receptor that mediates presumed repulsive actions of UNC-6 on dorsally migrating cells and axons. Two vertebrate homologues of UNC-5 have been identified, and one of these (UNC5H-2) is a netrin-1–binding protein, providing support for the idea that UNC-5 in *C. elegans* is a netrin receptor (E.D. Leonardo, L. Hinck, M. Masu, K. Keino-Masu, and M. Tessier-Lavigne, unpublished results). Interestingly, genetic evidence in *C. elegans* has also implicated the DCC homologue UNC-40 in mediating dorsal migrations, since UNC-40 is expressed by those axons (Chan et al., 1996) and mutations in *unc-40* also impair dorsal migrations, although to a much more limited extent than in *unc-6* and *unc-5* mutants (Hedgecock et al., 1990). It remains to be determined whether UNC-5 and UNC-40 form a receptor complex involved in dorsal migrations, and, indeed, how signaling by netrin proteins occurs through DCC and UNC-5 family members.

Halting axonal progression: sensory axon patterning by semaphorin III
Whereas netrins appear to provide a push from behind for some axons, a semaphorin family member, Sema III, has been implicated in a different kind of chemorepulsion discussed earlier: the halting of axonal growth from in front.

Messersmith et al. (1995) examined the effects of semaphorin III on different classes of rat sensory axons. Distinct classes of primary sensory neurons in dorsal root ganglia (DRG) subserve different sensory modalities, terminate in different dorsoventral locations in the spinal cord, and display different neurotrophin response profiles. Both types of sensory axons described below enter the developing spinal cord in the dorsal root. The small-diameter afferents subserving pain and temperature terminate in the dorsal spinal cord, and these axons are NGF-responsive. The large-diameter muscle afferents extend ventrally and terminate in the ventral spinal cord, and these axons are NT3-responsive.

As described earlier, previous in vitro studies showed that the developing ventral spinal cord secretes a long-range diffusible factor that inhibits the growth of sensory axons (Fitzgerald et al., 1993). It was found that this diffusible factor repels NGF-responsive axons but has little effect on NT3-responsive axons (Messersmith et al., 1995). The in vivo pattern of expression of semaphorin III made it a good candidate to be this long-range repellent, as Sema III is expressed by ventral spinal cord cells (but not the floor plate) at the appropriate stages of development. Consistent with this possibility, COS cells secreting Sema III were found to mimic the repellent effect of the ventral spinal cord on axons extending in response to NGF, with little effect on NT3-responsive axons. NT3-responsive axons can in fact respond to Sema III at earlier stages (Püschel et al., 1996), but at the relevant stages described here, they require much higher concentrations (K.H. Wang and M.T.-L, unpublished observations).

These results suggest that Sema III functions to pattern sensory projections by selectively repelling axons that normally terminate dorsally. This model is supported by the finding that a subset of sensory axons that normally terminate dorsally instead project further ventrally in *Sema III* knockout mice (Behar et al., 1996). Thus, Sema III is likely to function in the developing spinal cord as a long-range "stop" signal, preventing certain sensory axons from progressing further ventrally.

Axon Fasciculation and Defasciculation

As discussed earlier, growth cones can be guided by other axons. Several aspects of this process of selective fasciculation must be highly regulated.

First, the decision by an axon to initiate fasciculation with another axon must be highly regulated. For example, as described above, in the developing nerve cord of insects and the spinal cord of vertebrates, the growth cones of many neurons first project across the midline before turning either rostrally or caudally in one of the longitudinal axon tracts. The growth cones of these neurons typically contact, but show no affinity for, the homologous longitudinal axon pathway on their own side. The simplest hypothesis to explain this dramatic change in growth cone behavior is to postulate that the expression and/or function of surface receptors can be controlled in a dynamic and regional fashion, as discussed above for molecules implicated in midline guidance.

Second, the process of selective fasciculation must be mediated by a sufficient diversity of axonal glycoproteins to account for a remarkable degree of specificity. (A dramatic example of specificity in this process by the G growth cone is described earlier in this chapter.) Whether these axon pathway labels are all attractive or a combination of attractive and repulsive is not known. The later seems more likely, however, given both the complexity of the process and the expression on subsets of axon pathways of proteins that are thought to function in either an attractive (fasciclin II; Harrelson and Goodman, 1988; Lin et al., 1994) or a bifunctional (Semaphorin I; Kolodkin et al., 1992; Wong et al., in preparation) fashion.

Finally, the process of defasciculation must be as highly regulated as the process of fasciculation. In some contexts, such as motor axons entering the plexus at the base of the limb bud in the chick, all of the fasciculated axons de-adhere at the same time, suggesting a global process. In other cases, such as motor axons at particular choice points along the common motor pathway in *Drosophila*, one specific subset of axons de-adheres from the other axons in a bundle, and yet within these two groups, the axons remain fasciculated with their own kind. In cases such as these, the process is not global, but rather highly specific, and, not surprisingly, its regulation appears to be very complex, as described below.

Mediating selective fasciculation in insects and vertebrates: affinities mediated by Ig CAMs

Members of the Ig superfamily have been implicated in mediating axon fasciculation, primarily through homophilic adhesion. So far the most detailed understanding has been obtained in the analysis of fasciclin II (Fas II), an NCAM-like molecule in insects (Fig. 4–2) that was initially identified on the basis of its dynamic pattern of expression on a subset of fasciculating axons in the grasshopper embryo (Bastiani et al., 1987; Harrelson and Goodman, 1988). In the *Drosophila* CNS, Fas II is expressed on a subset of embryonic axons, many of which selectively fasciculate in three distinct longitudinal axon pathways (Grenningloh et al., 1991; Lin et al., 1994).

Lin et al. (1994) examined the in vivo growth cone guidance function of Fas II using genetic analysis. When the levels of Fas II are decreased in *FasII* loss-of-function mutants, the axons that normally fasciculate together in the three Fas II-expressing CNS pathways instead fail to do so and these axon fascicles do not form. Nevertheless, these growth cones extend in the normal direction at a normal rate, indicating that other guidance cues control directional steering.

In complementary gain-of-function experiments, two types of phenotypes are observed. First, transgenic constructs that specifically drive Fas II expression on the axons in these same three pathways can rescue the defasciculation phenotype in a *FasII* loss-of-function background, thus creating a refasciculation of these three fascicles. Second, in both wild-type and *FasII* mutant backgrounds, these transgenic constructs can lead to a gain-of-function phenotype in which these axons fasciculate incorrectly. Pairs of pathways that should remain separate instead become abnormally joined together, and a pair of pathways that normally start together and then defasciculate into two separate pathways instead remain fasciculated as one.

The complementary phenotypes produced by the loss-of-function and gain-of-function conditions define an in vivo function for Fas II as a molecule that controls specific patterns of selective fasciculation and axon sorting in the CNS. They also demonstrate that several aspects of growth cone steering do not require Fas II function, since at least for these early CNS pathways, the follower growth cones can find their way independently without absolutely requiring fasciculation with the pioneers.

Members of the Ig superfamily have been implicated in axon fasciculation in vertebrates as well. Blocking the function of various Ig superfamily members can interfere with the fasciculation of axons in vivo (e.g., Brittis et al., 1995; Honig and Rutishauser, 1996). A particularly complete study has been performed on the L1 subfamily member Ng-CAM (Fig. 4–2) by Stoeckli and Landmesser (1995). Commissural neurons express NgCAM on their surface as they extend toward and across the midline. Injection of antibodies against NgCAM resulted in the defasciculation of the commissural axons approaching the floor plate. As a result, these defasciculated axons often approached the floor plate from a much more lateral position than in control embryos and without the guidance provided by fasciculation with earlier fibers. Although defasciculated, all of the commissural axons nevertheless appeared to be attracted individually toward the floor plate, which they crossed normally. The antibodies also caused the defasciculation of the axons that had crossed the floor plate as they turned and projected anteriorly but, again, did not cause major pathfinding errors (under conditions where interfering with other Ig CAMs did cause pathfinding errors—see above). Thus, in these two well-described examples in chick and *Drosophila,* axonal fasciculation appears to be dissociable from many other aspects of pathfinding.

Mediating selective fasciculation in insects and vertebrates: affinities modulated by inhibitory surroundings

Although Ig CAMs can clearly drive fasciculation, it is becoming equally clear that the function of these or other fasciculation molecules may not be absolute and can be modulated by the environment through which the axons migrate: If that environment provides a highly favorable substrate for growth, the axons may prefer to grow on that substrate rather than on one another; conversely, if the environment is nonpermissive, axons that might otherwise ignore one another might prefer to grow on each other. Two sets of experiments support the idea that the extent of fasciculation reflects the balance between the attraction of the axons to one another and the attraction of the axons to the environment.

The first experiment has already been described above in the discussion of Sema I function in the grasshopper limb bud. The Ti1 axons that grow on a stripe of Sema I are more highly fasciculated when they are on the stripe than during earlier portions of their trajectory, but in the presence of antibodies to Sema I they defasciculate and branch. Given the inhibitory effects of various semaphorin proteins (discussed above for Sema III and below for Sema II), the simplest interpretation of this result is that Sema I is a negative factor that makes the environment less favorable and therefore drives the axons to fasciculate.

The second experiment that adds credence to this balance model involves analysis of a lipid-anchored ligand of the Eph family called AL-1. Winslow et al. (1995) studied AL-1 function in cultures of cortical neurons on monolayers of astrocytes in experimental conditions in which axons grow out in thick fascicles. The cortical neurons and their axons express a receptor for AL-1, whereas the substrate astrocytes express the ligand AL-1. When soluble AL-1 or a soluble form of the receptor was used as the antagonist of AL-1 function in these cultures, the axons defasciculated (Winslow et al., 1995). Since AL-1 has collapse-inducing activity for cortical axons (Meima et al., in press), the simplest interpretation is that AL-1 functions as a repellent for cortical axons that makes the astrocytes a less attractive substrate than axonal surfaces and therefore encourages the axons to grow on one another. Whether AL-1 can also function to drive fasciculation in vivo remains to be determined.

Although still preliminary, these results suggest that the expression of molecules that create an inhibitory environment can drive axons to fasciculate and that, by inference, the expression of molecules that create a favorable environment might drive axons to defasciculate. Thus, fasciculation should be thought of as any other type of guidance decision, the result of which will be determined by the balance of attraction and inhibition operating at the decision point.

Initiating selective fasciculation in *Drosophila*: the role of the Derailed receptor protein tyrosine kinase

As discussed above, the initiation of selective fasciculation is a tightly regulated process. The behavior of growth cones that project across the midline provides a particularly good example of the switching on and off of the expression of specific axonal glycoproteins that control fasciculation. The mechanisms that control this regulation are poorly understood, but an entry point into their analysis may have been obtained with the *derailed* (*drl*) mutant in *Drosophila* (Callahan et al., 1995).

As described earlier, *drl* encodes a receptor protein tyrosine kinase that is related to vertebrate RYK (Fig. 4–2) and is normally expressed by a subset of CNS interneurons (~20 per hemisegment) whose axons project across the anterior commissure and then turn anteriorly and fasciculate into one of two longitudinal axon pathways (called the DD and DV pathways). Derailed (Drl) protein is expressed on the growth cones and axons of the *drl* neurons as they cross the anterior commissure; after the neurons have turned anteriorly, Drl cannot be detected on the portions of the axons within the longitudinal axon pathways, and the protein remains restricted to the commissural segments of these axons.

In *drl* mutant embryos, the neurons that normally express Drl behave normally in a number of respects: They cross the midline in the anterior commissure and then turn anteriorly in the longitudinal connectives. But they behave abnormally in that they fail to properly fasciculate together into the DD and DV longitudinal axon pathways. In contrast, the axons in three other longitudinal axon pathways defined by Fas II expression fasciculate normally in the *drl* mutant embryos.

The Drl protein appears to participate indirectly in the fasciculation of the *drl* neurons within the DD and DV pathways since it is not expressed on the fasciculating portions of the axons. Apparently, Drl functions on the commissural growth cones to help control the selection of the appropriate longitudinal axon pathway as they turn anteriorly. One simple model, based on admittedly limited evidence, is that signaling via the Drl RPTK control the expression or function of as yet unidentified CAMs that control the selective fasciculation of these axons.

Modulating the global defasciculation of motor axons in the chick: the role of PSA

The regulation of defasciculation can also have a profound impact on axon pathfinding. We have already seen above how forced expression of Fas II can cause misrouting of interneurons in the *Drosophila* CNS by preventing axons from defasciculating at the appropriate juncture. In recent years considerable advances have been made in our understanding of how this process is regulated (Fig. 4–7). These studies have implicated polysialic acid (PSA) as a key regulator of this process (along with receptor protein tyrosine phosphatases and the secreted factor Beat: see next section).

Earlier we discussed the fact that NCAM can be highly sialylated, and that addition of PSA chains to NCAM is believed to decrease the ability of NCAM and of other CAMs like L1 to mediate adhesion. Support for this hypothesis has been obtained in studies of motor axon pathfinding in chick embryos, where PSA appears to help motor axons defasciculate, enabling different axons to select divergent pathways.

During the development of the chick embryo, motoneurons from the lumbosacral region of the spinal cord exit the CNS, and axons destined to innervate different muscles are intermingled within eight spinal nerves. At the base of the limb bud these nerves converge into two groups to form the crural and sciatic plexuses. To reach their appropriate muscle targets, motor axons that leave the plexus region must make a number of appropriate pathway choices, starting with a decision to grow in either the dorsal or the ventral nerve trunk. One important local regulator of this sorting out process is changes in axon–axon adhesion (Tang et al., 1992, 1994). Upon reaching the plexus, the tightly fasciculated and intermingled motor axons begin to defasciculate. This occurs concomitant with a dramatic increase in the levels of PSA on motoneuron growth cones and axons (Fig. 4–7). When PSA was enzymatically removed during the period of axonal sorting out in the plexus region, an increase in the number of projection errors was observed, which was associated with altered growth cone behavior within the plexus region (Tang et al., 1992). Rather than making dramatic, divergent changes in direction, motor axons instead maintained relatively straight paths. Thus, the removal of PSA appeared to prevent growth cones from responding appropriately to guidance cues at the base of the limb bud. Interestingly, when PSA was removed at later stages, during the formation of specific muscle nerves, projection errors were not observed.

Tang et al. (1994) went on to show that the defects in pathfinding were likely due to an increase in fasciculation caused by removal of PSA. In the case of com-

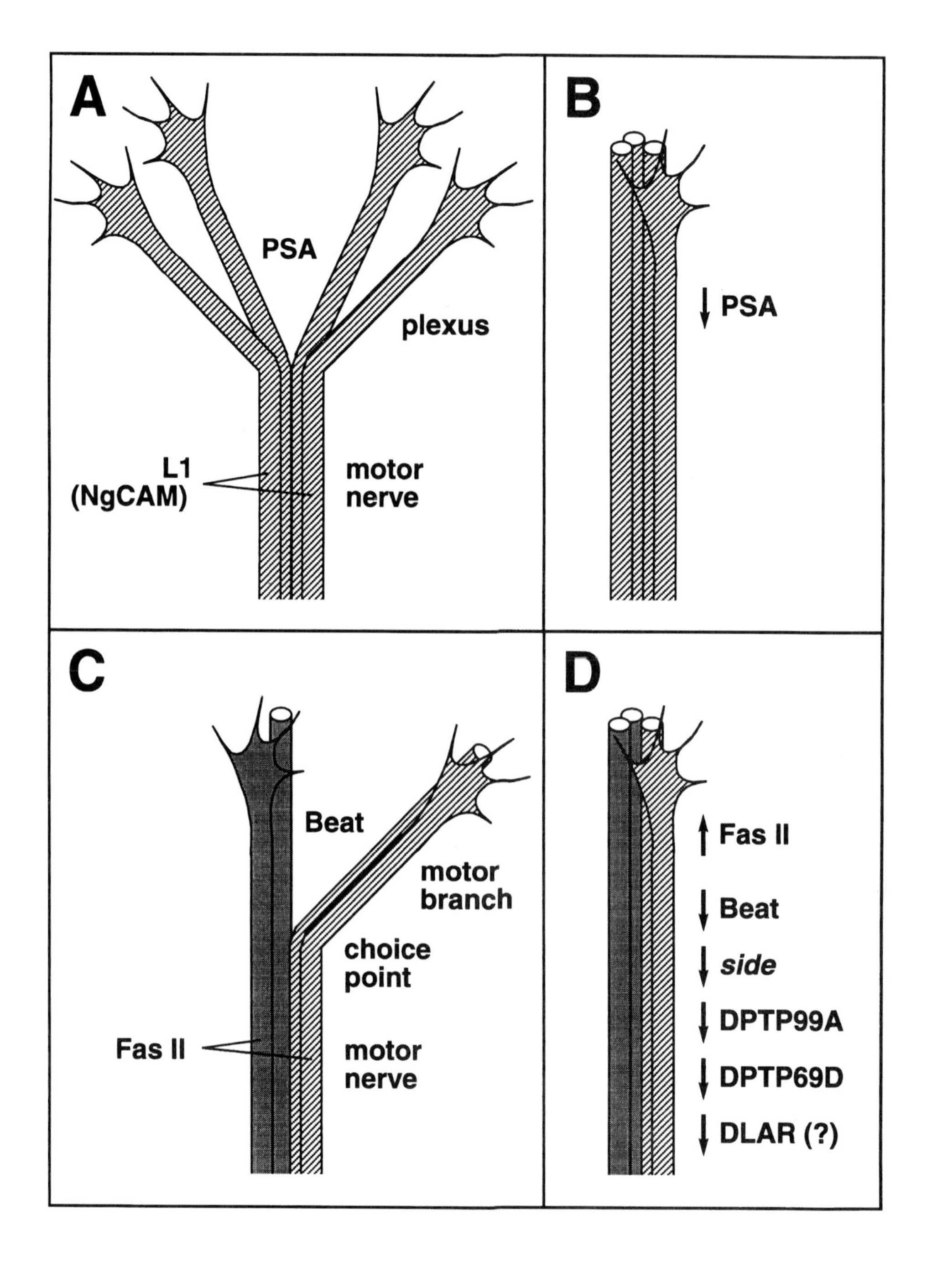

missural axons discussed above, the Ig CAM NgCAM/L1 was found to be a major mediator of axon–axon adhesion in this system. When antibodies to NgCAM were injected in embryos in which PSA had also been removed, the effects of PSA removal on both growth cone trajectories and projection errors were reversed. The simplest interpretation is that PSA plays a permissive role in this system, causing a decrease in NgCAM-mediated axon fasciculation that increases the ability of motor growth cones to respond to specific (unidentified) guidance cues in the plexus and to sort out into their appropriate nerves. In a similar fashion, PSA-mediated modulation of axon adhesion is implicated in the patterning of intramuscular motor branches (Landmesser et al., 1990), as removal of PSA leads to increased fasciculation and reduced nerve branching. A PSA-mediated decrease in adhesion is also apparently required for certain cell migrations (Ono et al., 1994).

Modulating the specific defasciculation of motor axons in *Drosophila:* the roles of receptor protein tyrosine phosphatases and the secreted factor Beat

Further insights into the regulation of defasciculation have come from genetic studies of motor axons in the *Drosophila* embryo. As described in an earlier section, in each abdominal hemisegment, approximately 40 motor axons grow into the periphery to specifically innervate 30 unique muscle targets. Motor axons exit the CNS in two distinct motor nerves—the intersegmental nerve (ISN) and the segmental nerve (SN)—which extend into the periphery. Subsets of motor axons leave the two main motor nerves at particular choice points to form specific motor branches. In particular, the motor axons of the SNb initially follow

FIGURE 4–7 **Molecules that mediate defasciculation.** (A) Motor axons in the chick embryo enter the plexus at the base of the limb bud in tightly fasciculated bundles where they defasciculate, sort, and exit in specific motor nerves. Polysialic acid (PSA) is required for this global defasciculation and appears to function by modulating the axon–axon adhesion mediated by L1/NgCAM. (B) When PSA is enzymatically removed, these axons fail to defasciculate (Tang et al., *1992, 1994*). (C) One group of motor axons (the Snb) in the *Drosophila* embryo selectively defasciculates from the major motor nerve (the ISN) at a specific choice point as they steer into their appropriate muscle target region. (D) Six genes have been identified thus far that modulate SNb guidance at this choice point. The neural cell adhesion molecule fasciclin II is normally expressed at high levels on motor axons; when the Fas II levels on motor axons are increased, the SNb axons fail to defasciculate at the choice point, and instead continue to extend along the ISN, suggesting that defasciculation at the choice point requires modulation of Fas II function. The other five genes encode candidate regulators of Fas II function at this choice point. Three receptor protein tyrosine phosphatases—Dlar, DPTP69D, and DPTP99A—appear to control defasciculation at this choice point, although Dlar may also control steering. Mutations in two other genes—*beaten path (beat)* and *sidestep (side)*—result in a SNb defasciculation phenotype as well. The beat gene encodes a novel secreted protein which is expressed by motoneurons. Genetic interactions between *beat* and *FasII* suggest that secretion of Beat functions to decrease axon–axon adhesion that is mediated in large part by Fas II. See text for further discussion.

and adhere to ISN axons, but at their choice point in the ventral muscle region they defasciculate from the ISN axons and form a separate bundle which steers into the ventral muscle region. The SNb choice point has proved to be an excellent model system for the in vivo genetic dissection of guidance and steering decisions (e.g., Vactor et al., 1993; Lin and Goodman, 1994; Nose et al., 1994; Krueger et al., 1996; Desai et al., 1996; Fambrough and Goodman, 1996). Six genes have been identified so far that can affect SNb guidance at this choice point; five appear primarily to regulate defasciculation, and one primarily steering (Fig. 4–7). We discuss these genes in turn.

The first gene implicated in this process is *FasII,* introduced above in our discussion of selective fasciculation in the CNS. In the PNS, Fas II is normally expressed at high levels on all motor axons (Vactor et al., 1993). Loss of *FasII* function causes defasciculation of axons in Fas II–expressing pathways without strongly perturbing pathfinding (Lin et al., 1994). In contrast, using transgenic methods to increase the levels of Fas II on motor axons prevents SNb axons from defasciculating at the choice point, so that they now continue to extend dorsally along the ISN and fail to invade the ventral muscle region (Lin and Goodman, 1994). Remarkably, the levels of Fas II (as detected by immunocytochemistry) normally appear unchanged on SNb and ISN motor axons at this choice point. Taken together, these studies suggest that defasciculation at this choice point requires the modulation of Fas II function, rather than the removal of Fas II from the surface of these axons.

Five genes have been identified that encode candidate regulators of Fas II function at this choice point, as loss-of-function mutations in these genes give similar SNb defasciculation phenotypes to those observed in the increased (gain-of-function) Fas II experiments. The first two genes encode the receptor protein tyrosine phosphatases (RPTPs) DPTP69D and DPTP99A, which are expressed on motor axons. Single mutations in the genes encoding these receptors have no or little effect on the axons, but embryos mutant in both *Dptp69D* and *Dptp99A* show the same phenotype as that produced by increased expression of Fas II on these same axons: The SNb fails to defasciculate from the ISN and thus fails to invade the ventral target region (Desai et al., 1996).

Single mutations in either of two other genes—*beaten path* (*beat,* Vactor et al., 1993; Fambrough and Goodman, 1996) and *sidestep* (*side;* Sink and Goodman, 1994)—result in the same phenotype: a failure of the SNb to defasciculate from the ISN. *beat* encodes a novel secreted protein (Beat) that is expressed by motoneurons (Fambrough and Goodman, 1996). The *beat* mutant phenotype can be rescued by expressing *beat* in motoneurons. Moreover, the SNb defasciculation phenotype in *beat* mutant embryos can be largely rescued by genetically removing 90% of Fas II protein. Together, these results suggest that Beat, secreted by motoneurons, functions as an antiadhesive factor to facilitate the ability of specific subsets of axons to selectively defasciculate from other axons. Other ectopic expression and biochemical experiments (Fambrough and Goodman, 1996) suggest a model in which Beat does not directly bind to Fas II, but rather controls selective defasciculation by binding to some other unknown receptor.

Another RPTP, Dlar, has also been implicated in the process of SNb axon defasciculation and steering at the choice point, since loss-of-function mutations in *Dlar* result in pathfinding errors at the choice point such that SNb axons fail to invade the ventral target region. Interestingly, however, the phenotype is subtly different from that described above for the other five genes in that the SNb axons often (but not always) appear to defasciculate from the ISN and form a separate bundle, but this bundle then extends parallel to the ISN and past the ventral muscles (Krueger et al., 1996).

These differences in phenotypes raise the possibility that *Dlar* might control steering into the target region, while the other five genes (*beat, side, Dptp69D, Dptp99A,* and *FasII*) might control defasciculation, a necessary step that precedes the change in steering. However, it cannot be excluded that some of the other genes (*Fas II* excepted) control steering in addition to defasciculation, since a defasciculation phenotype will obscure any steering phenotype.

Target Recognition: Invading the Target Field

Earlier, we discussed the issue of target recognition mainly from the point of view of the selection of synaptic partners within a target field. That issue will be discussed below. However, even prior to the selection of synaptic partners, some cues appear to be involved specifically in the initial recognition of the target field and invasion of the target field by the axons, as we now discuss.

Invading the target field: a role for neurotrophins and FGF

As discussed earlier, the molecular era in the study of axon guidance started with the demonstration that the neurotrophin NGF can act as a chemoattractant for regenerating axons in culture (Gundersen and Barrett, 1979). Subsequent experiments failed to support a guidance role for NGF during development, as NGF appears to be expressed in vivo only after the axons reach their targets, and the axons are not NGF responsive as they grow to their targets; likewise, there is as yet no evidence implicating other neurotrophins in getting axons to their targets (reviewed in Kennedy and Tessier-Lavigne, 1995). Of course, at the targets neurotrophins are thought to play important roles in controlling the process of naturally occurring cell death and in controlling the extent of collateralization of axons in the target region; these processes are the subject of other chapters in this volume.

Whereas a role for neurotropins in guidance has not been substantiated, there is mounting evidence that neurotrophins might play a role in the process of target invasion, a process properly included under the rubric of axon guidance and target recognition. A clear demonstration of this has been obtained by ElShamy et al. (1996), through analysis of target invasion by sympathetic axons in mice carrying mutations in the *neurotrophin-3* gene. In *NT3* −/− mice, sympathetic fibers of the carotid nerve failed to invade the pineal gland and external ear, whereas other targets of this nerve, such as the iris, were normally innervated. Interestingly, in heterozygous animals (carrying just one functional copy of

NT3), sympathetic fibers did invade the pineal but failed to branch and form a plexus. Both defects in pineal innervation were rescued by exogenous administration of NT3 to the mutants. These results suggested a model in which the target is normally inhibitory for sympathetic fibers and NT3 overrides this inhibition to allow invasion of the target field. The elaboration of axon collaterals is also controlled by NT3 but appears to require higher concentrations of the factor (since two copies of the *NT3* gene are required).

Further support for the notion that neurotrophins can control target invasion was obtained in studies of sympathetic innervation of various peripheral targets, including the pancreas, for which NGF rather than NT3 appears to be the important target-derived neurotrophin. Hoyle et al. (1993) generated transgenic mice in which NGF expression was driven in the sympathetic neurons themselves under the control of the dopamine β-hydroxylase promoter and found that this caused a reduction in invasion of the target tissue. The defect in the pancreas could, however, be rescued by crossing these animals to other transgenic mice in which NGF was expressed at high levels in the pancreas. This suggests that NGF is required for target invasion and, more specifically, that an upward gradient of NGF is required for the axons to invade the target.

Another factor that acts via a receptor tyrosine kinase and that has been implicated in controlling target invasion is bFGF. In the case of bFGF, however, it is a *downward* gradient that appears to be required for invasion. We discussed earlier the evidence that bFGF plays a role in stimulating axon extension by retinal ganglion cells (RGCs) in the *Xenopus* optic tract, which is marked by a corridor of bFGF that ends at the junction between the optic tract and the target tectum (McFarlane et al., 1995). Addition of exogenous bFGF to an "exposed brain" preparation had the surprising effect of causing the axons to skirt around the target. Remarkably, the same result was obtained when the opposite manipulation was performed, i.e., when FGF receptor function was impaired by transfecting constructs encoding dominant negative forms of the FGFR into the RGCs (McFarlane et al., 1996). As discussed above, the axons of transfected cells were on average shorter than wild-type axons, but many of those that did make it all the way to the tectum showed the same skirting behavior at the target that was observed in the presence of exogenously added FGF. Thus, opposite manipulations of FGF function appear to have the same result, causing the axons to avoid the target. How might this be explained? The authors suggest that these paradoxical observations can be reconciled if the axons must detect a *decrease* in bFGF function in order to invade the target, the opposite of what Hoyle et al. (1993) postulate for NGF's role in target invasion. Obviously further studies are required to determine whether this unusual mechanism really does operate, especially since the misrouting phenotype of the transfected axons was not fully penetrant (which the authors attribute to variability in expression levels of the dominant negative construct).

While preliminary, these studies together support the idea that invasion of a target field by a cohort of axons is regulated by a combination of target- and pathway-derived signals.

Recognition of limbic system targets in mammals:
a role for the Ig CAM LAMP

Levitt and colleagues (Pimenta et al., 1995) have identified an Ig CAM that appears to function in the recognition of a discrete region of the forebrain, the limbic system. The limbic-system–associated membrane protein (LAMP) has three Ig domains and a GPI anchor (Fig. 4–2). In cell culture, LAMP can mediate homophilic binding and can induce neurite outgrowth of appropriate limbic neurons (Zhukareva and Levitt, 1995). In the developing forebrain, LAMP is expressed mostly by neurons in limbic-associated cortical and subcortical regions. Administration of anti-LAMP antibodies in vivo result in abnormal growth of the mossy fiber projection from developing granule neurons in the dentate gyrus (which express LAMP) to pyramidal neurons in the hippocampus (which also express LAMP), suggesting that LAMP is involved in proper targeting of this pathway. The nature of the misrouting, in which the axons overshoot the target, suggests that LAMP serves as a target recognition molecule for the formation of limbic connections. However, it appears to serve the role of labeling a large region, with other molecular guidance cues and target labels helping to determine the specificity of projections and connections within the region.

Target Recognition: Selecting Addresses Within the Target Field

Once axons have started to invade a target field, they must then select specific partner with which to form synaptic connections. Earlier, we presented evidence that growth cones appear to recognize their targets using two broad categories of signals: topographic maps of graded cues and discrete targets with distinct labels. Here we discuss molecules implicated in these two types of processes.

Eph ligands and topographic projections in the retinotectal system

Earlier we described the widespread formation of topographic projections, in which axons of neighboring neurons project to neighboring areas in the target region, thus maintaining their spatial order. We also described, in the case of the retinotectal projection, the arguments supporting Sperry's hypothesis that the initial formation of a crude topographic map involves the recognition of positional information in the form of gradients of signaling molecules at the target and a corresponding complementary gradient of responsivity of the axons.

Recent studies have implicated ligands for Eph-family-receptor tyrosine kinases as strong candidates for the positional labels that Sperry postulated. A strong link came initially from studies by Bonhoeffer and colleagues, following their demonstration in chicken that posterior tectum is enriched in a repellent activity for temporal retinal axons and that this activity appears to reside in a lipid-anchored protein. Drescher et al. (1995) sought to identify the repellent by isolating GPI-linked proteins that are enriched in posterior tectum during the period of retinal axon invasion. Using two-dimensional gels to compare the profiles of GPI-linked proteins in posterior and anterior tectum, they identified a 25-

kD protein enriched in posterior tectum which they termed RAGS. Purification, cloning, and expression studies demonstrated that the protein is the chick homologue of AL-1 (an Eph ligand discussed above in the context of axon fasciculation: Winslow et al., 1995) and that, as expected, RAGS mRNA is enriched in posterior tectum (Fig. 4–8). Importantly, membrane preparations of COS cells expressing recombinant RAGS were found to cause collapse of retinal axons and to repel the axons in the "stripe" assay (in which the cells were given a choice of transfected and untransfected membranes). Interestingly, RAGS does not have the specificity of effects observed for posterior tectal membranes, which only repel temporal retinal axons. Instead, RAGS was found to repel and cause the collapse of all retinal axons (Drescher et al., 1995), but there is a smooth gradient of sensitivity of retinal axons across the A/P axis, with temporal axons more sensitive than nasal axons (Monschau et al., in press), as postulated by antagonistic gradient models.

Flanagan and colleagues had independently focused on the possible involvement of Eph family members in retinal mapping because of their discovery of complementary position-specific gradients in expression for an Eph receptor–ligand pair. They had identified ELF-1 as a ligand for the related Mek4 and Sek receptors (Cheng and Flanagan, 1994). Expression studies in chickens demonstrated that ELF-1 mRNA is also expressed in a gradient in the tectum with highest expression in posterior tectum, in the retinorecipient layers (Cheng et al., 1995). ELF-1, like RAGS, was also found to function as a repellent for retinal axons in vitro (in both the stripe and the collapse assays), but in contrast to RAGS, this repulsion has the specificity observed for posterior tectal membranes: Tem-

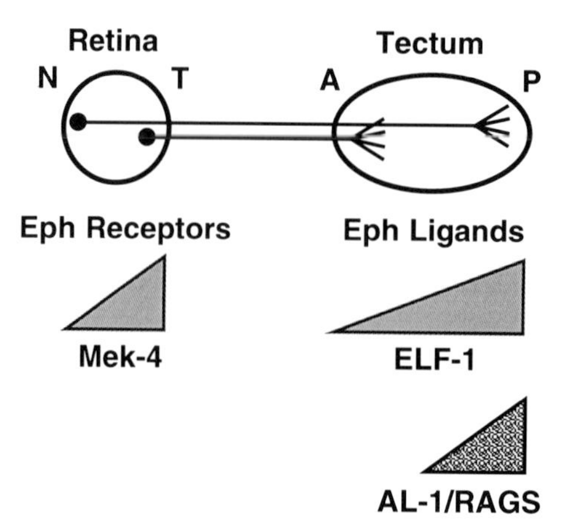

FIGURE 4–8 **Eph ligands and receptors in the retinotectal system.** Several Eph ligands are expressed in topographic gradients on the tectum, and appropriate Eph receptors are expressed in matching gradients on retinal neurons. ELF-1 is expressed in a gradient across the entire tectum, with highest expression in posterior (P) tectum. The ELF-1 receptor, Mek-4, is expressed in a gradient across the retina with highest expression in the temporal (T) retina. RAGS/AL-1 is expressed in a gradient across the posterior half of the tectum, with highest expression at the posterior end. N, nasal retina; A, anterior tectum. See text for further details.

poral retinal axons are repelled but nasal ones are not (Nakamoto et al., 1996; Monschau et al., in press). This conclusion is supported by ectopic expression experiments in vivo in which ELF-1 was expressed by retroviral-mediated gene transfer in patches in the tectum. Mirroring the in vitro results, temporal retinal axons avoided the patches of ectopic ELF-1, whereas nasal axons were unaffected (Nakamoto et al., 1996).

Do these studies solve the problem of topographic projections in the retinotectal system? Not yet, as there are several outstanding questions. First, the precise functions of ELF-1 and RAGS in retinal axon guidance need to be defined. ELF-1 is a strong candidate for a causal factor in the visual system that contributes to confining temporal axons to the anterior tectum (Nakamura and O'Leary, 1989), without effect on nasal axons. RAGS expression is actually confined to more caudal regions of the tectum than is ELF-1 expression (Monschau et al., in press; Fig. 4–5), and since retinal axons show graded responses to RAGS it is possible that it functions to help axons in the posteriormost portions of the tectum find their precise targets. Loss-of-function studies should help resolve these issues. It is possible that the gradient of responsiveness of retinal axons to RAGS results from a decreasing temporal-to-nasal gradient of expression of the receptor Mek4 in the retina (Cheng et al., 1995), although it is unclear at a mechanistic level how these different responses to RAGS and ELF-1 are elicited in the axons.

A second outstanding question is whether the decreasing posterior to anterior gradients of repellents that have been discovered are sufficient to explain how the axons terminate at their correct addresses along the A/P axis. As discussed earlier, the simplest gradient models require that axons be exposed to antagonistic gradients along the A/P axis: a first gradient that tends to make axons grow from A to P and which could affect all axons equally, and a second antagonistic gradient that tends to make the axons grow from P to A, but where axons show a gradient of sensitivity such that temporal axons are more sensitive than are nasal axons. A RAGS-like molecule could provide the second gradient, but there is at present no good candidate for the first gradient, involving a molecule that tends to make all axons project from A to P.

Finally, as discussed earlier, how positional information on the tectum is read by axons is likely to differ from species to species, with effects on growth cone migration dominating in some and effects on axonal branching being more important in others. Whether Eph ligands can account for all of these effects remains to be determined.

Molecular diversity controlling neuromuscular specificity in *Drosophila*

As discussed earlier, studies in a variety of organisms have shown that motor axons can recognize their specific muscle targets with a high degree of precision. The molecular mechanisms underlying neuromuscular target recognition have been studied in most detail in *Drosophila* (e.g., Vactor et al., 1993; Sink and Goodman, 1994; Nose et al., 1992a,b; Chiba et al., 1995; Matthes et al., 1995; Mitchell et al., 1996), where 30 individual muscle targets are specifically innervated by ~40 motor axons in each abdominal hemisegment.

The strongest candidates for bona fide targeting molecules in this system are the two *Drosophila* Netrin proteins (Mitchell et al., 1996). The two *Netrin* genes are expressed by different subsets of muscles, *NetA* strongly by dorsal muscle 1 and weakly by dorsal muscle 2, and *NetB* by dorsal muscle 2 (but not muscle 1) and strongly by ventral muscles 6 and 7. Embryos carrying a small deficiency that deletes both genes show defects in the projections of motor axons that normally innervate muscles 1 and 2 and by those innervating muscles 6 and 7 (Mitchell et al., 1996); these targeting phenotypes are only partially penetrant (~40% of segments). The same loss-of-function phenotypes are observed in embryos mutant in the *frazzled* gene which encodes a putative Netrin receptor (Kolodziej et al., 1996; see above). Ectopic expression of either *Netrin* gene in all muscles results in aberrant motor projections, particularly of the motor axons that normally innervate dorsal muscles 1 and 2. Thus, the two Netrins appear to function as part of the normal targeting system for the motor axons that innervate muscles 1 and 2, and probably for the axons innervating muscles 6 and 7 as well.

Three other genes—*Connectin, FasIII,* and *Semall*—are expressed by subsets of muscles and have been implicated in targeting by gain-of-function (ectopic expression) phenotypes, although they show no easily observable loss-of-function defects (Nose et al., 1994; Chiba et al., 1995; Matthes et al., 1995).

Fasciclin III. A subset of muscles and motoneurons expresses the Ig CAM *Fas III* (Patel et al., 1987; Jacobs and Goodman, 1989; Halpern et al., 1991). In FasIII loss-of-function mutations, motor axons find and innervate their appropriate targets. However, when Fas III levels are increased on all muscles, one motoneuron that normally expresses Fas III makes stable contacts with inappropriate muscle targets (Chiba et al, 1995), suggesting that Fas III might function as an attractive targeting molecule.

Semaphorin II. A single muscle normally expresses high levels of the secreted semaphorin Sema II (Kolodkin et al., 1993). In *Sema II* loss-of-function mutations, motor axons find and innervate their appropriate targets. However, when Sema II levels are increased on many ventral muscles, certain growth cones extend close to their muscle targets but are inhibited from forming their synaptic terminal arborizations on these ectopic Sema II–expressing muscles (Matthes et al., 1995), suggesting that Sema II might function as a repulsive targeting molecule.

Connectin. A distinct subset of muscles and motoneurons expresses the leucine-rich repeat molecule Connectin, which in vitro can function as a homophilic CAM (Nose et al., 1992). In *conn* loss-of-function mutations, motor axons find and innervate their appropriate targets. However, when Connectin levels are increased on some or all muscles during the stage of growth cone exploration, two results are observed. A growth cone that does not normally express Connectin now avoids ectopic Connectin-expressing muscles, and certain Connectin-expressing axons inappropriately innervate a muscle that now expresses ectopic Connectin (Nose et al., 1994; A. Nose, unpublished results).

These results suggest that Connectin might function as both an attractive and a repulsive targeting molecule.

Thus, only for the two *Netrin* genes are there complementary loss-of-function and gain-of-function phenotypes to confirm definitively their role in target recognition. Interestingly, only two *Netrin* genes are known in *Drosophila,* and they are expressed by only four of the 30 muscles, suggesting that other types of molecules must control targeting as well. For the other three candidate genes (*FasIII, SemaII,* and *Connectin*) to be involved in targeting, one must invoke the explanation that they function in redundant systems, since they display gain-of-function phenotypes but not loss-of-function phenotypes. Two observations support the possibility of redundancy. First, even the *Netrin-A* and *Netrin-B* double mutant phenotype is only partially penetrant, indicating the existence of other cues that collaborate with the Netrins. Second, large-scale genetic screens thus far have failed to uncover other genes that encode targeting ligands or receptors (Vactor et al., 1993; Sink and Goodman, 1994). The only two mutants that were isolated that have strong and specific targeting phenotypes, when cloned, have been shown to encode a transcription factor (*clueless/abrupt;* Hu et al., 1995) and a target-region specific chaperonin (*walkabout;* D. Van Vactor and C.S. Goodman, unpublished results). The failure to isolate mutations in targeting molecules in these screens is at least consistent with the possibility of redundancy.

Thus, although available evidence suggests that specificity in this system may rely on the balance of overlapping patterns of expression of attractive, repulsive, and bifunctional targeting signals from a wide variety of gene families, we are still a long way from understanding the molecular code of target labels used in this or, indeed, in any other system.

Conclusions

We have reviewed the explosion of new information regarding the guidance roles of specific ligands and receptors (Fig. 4–9). Although much of the analysis must still be considered incomplete, several generalizations—some of which have been alluded to earlier—seem warranted at this time.

Four guidance mechanisms

First, axons appear to be guided through the combined operation of four guidance mechanisms (short- and long-range attraction and short- and long-range repulsion), and the outcome of any particular guidance decision reflects the balance of attraction and repulsion operating at the decision point. In one sense, this conclusion might be considered a truism, since these four mechanisms together appear to cover all possibilities. However, it is a truism that is well worth reiterating, since until 1993 there was no mention anywhere in the literature that axons might be guided by long-range repulsive mechanisms. This possibility even eluded Ramón y Cajal, who seems otherwise to have thought of almost every possible mechanism. In addition, what is not a truism is that, based on in vivo analysis, all four types of mechanisms appear to operate in concert to guide many different classes of axons throughout the nervous system. More informa-

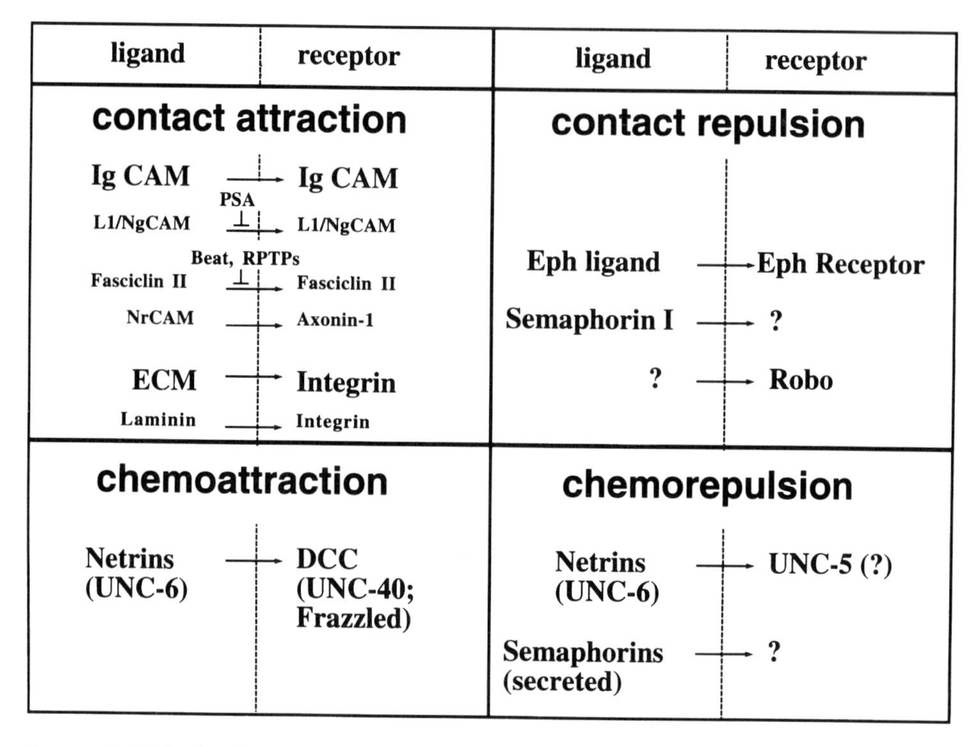

FIGURE 4–9 **Mechanisms and molecules that mediate growth cone guidance.** Diagram categorizing many of the guidance molecules and receptors discussed in this chapter. There is not a one-to-one match between molecules and mechanisms since some guidance molecules are not exclusively attractive or repulsive, but rather bifunctional, and some families of guidance cues possess both secreted and transmembrane members. Under "contact attraction," molecules (PSA, Beat, and receptor protein tyrosine phosphatases) that function as negative regulators of primary effectors (Ig CAMs) are also indicated.

tion is needed at this point to define how widespread long-range attraction and repulsion are: There are still just a few examples of the operation of these mechanisms and only two classes of ligands that are known to mediate such effects.

In addition, the conclusion that guidance decisions are influenced by the balance of attraction and repulsion contains within it a hidden agenda: the need to determine the hierarchy of guidance effects of different molecules that are presented simultaneously to a growth cone. We are still far from knowing how any one growth cone in the developing organism integrates the array of guidance cues at each of its choice points and makes a series of decisions of which way to go. For instance, in the case of commissural axon growth to the floor plate, when does long-range chemoattraction end and short-range guidance take over? Also, floor plate cells appear to possess an inhibitory molecule whose effects are normally masked by NrCAM. How does NrCAM dominate this interaction? Is it a quantitative effect (more NrCAM) or a qualitative one (does NrCAM "gate" the inhibitory effects)? How is the integration performed at the growth cone, and what are the rules for computing the effects of multiple guidance cues?

Evolutionary and mechanistic conservation

Second, these four guidance mechanisms appear to be mechanistically related and phylogenetically conserved. Neither of these conclusions could have been considered obvious as little as 3 years ago. Only recently has it been shown that molecules that function as long-range attractants or repellents (netrins and semaphorins) are structurally related to molecules that function as short-range attractants and repellents (laminins and other semaphorins). This conclusion is further reinforced by the discovery that receptors implicated in mediating relevant attractive and repulsive actions of the netrins in vivo are members of the Ig superfamily and are therefore close relatives of Ig superfamily members that are receptors (and ligands) implicated in several short-range guidance events, as well as axon fasciculation. In addition, there are now several striking parallels between pathfinding events in nematodes, insects, and vertebrates that illustrate vividly the high degree of evolutionary conservation in guidance mechanisms.

It is of course much too early to assume that all guidance mechanisms will be related to those currently implicated in guidance in vivo (or that all guidance mechanisms are necessarily phylogenetically conserved). In particular, it will be of interest to determine whether guidance molecules in non-neural systems that are unrelated to those we have discussed will be found to function in the nervous system as well. For instance, are any guidance events in the nervous system mediated by molecules implicated in immune cell guidance, like the chemokines (which act via serpentine receptors) and the selectins?

Fasciculation as a guidance mechanism

Third, axonal fasciculation, although superficially distinct from other guidance events, might be conceptually and mechanistically related to those events. Thus, the analysis of axon fasciculation and defasciculation presented here implicates Ig CAMS, as well as inhibitory influences mediated by Eph family receptors and semaphorins, in directing the extent of fasciculation of axons. These same molecules have been implicated in short- and long-range guidance and even in some aspects of target recognition. The importance of this observation is that, if true, the analysis of fasciculation is likely to provide complementary insights to those obtained from analysis of other guidance events. Several molecules implicated in fasciculation like PSA and Beat are best considered modulators of the function of effectors like Fas II and NgCAM, rather than effectors themselves. It will therefore be interesting to determine to what extent they function as modulators of the guidance functions of effectors implicated in other local and long-range guidance events. In addition, since fasciculation and defasciculation phenotypes are particularly well suited to genetic studies, their analysis might serve as one of the best approaches to the identification of modulators that have more subtle influences on guidance.

Redundancy

Fourth, there is a large amount of functional redundancy in guidance mechanisms. Some of the redundancy, including the simultaneous operation of several of the four guidance forces described above, is presumably present to ensure a high degree of fidelity in axonal projections. There are, however, also examples of what might be termed gratuitous redundancy, arising from gene duplications. Thus, the two *Drosophila Netrin* genes appear to play functionally redundant roles at the midline of the CNS. Although redundancy is clearly present, it is worth pointing out that some of the worst fears about redundancy have not been borne out. Historically, studies of axon guidance progressed in the 1980s from an initial identification of candidate guidance molecules (often based on distribution and in vitro activities) to functional perturbations of these candidates. In many cases, strong phenotypes that have been expected were not observed. This raised the fear that the guidance mechanisms might be massively overspecified and that it might be all but impossible to pinpoint the guidance functions of any particular molecules. More recent analyses indicate that this is not generally true. Many guidance molecules have now been identified, mutations in which result in a range of pathfinding and targeting phenotypes from dramatic to only partially penetrant. These studies have given us hope that an understanding of guidance mechanisms might be within reach.

In closing, we have in a very short time come a long way in our understanding of guidance mechanisms. It is, however, also important to reiterate that many of the conclusions we have attempted to draw are based largely on what are still very preliminary insights. A major lesson of the studies of recent years is that elucidating the function of a candidate guidance cue requires identification not just of the cue but also of its receptor and analysis of both, using loss-of-function and gain-of-function experiments in vivo and in vitro. A glance at the list of receptor–ligand pairs implicated in guidance (Fig. 4–9) illustrates that this standard of analysis has yet to be obtained for any guidance system. We can, however, expect this standard to be applied quite widely in the coming years, justifying a guarded optimism that the logic and mechanisms of axonal pathfinding might soon be elucidated.

ACKNOWLEDGMENT

Special thanks to Katja Brose, who spent considerable time discussing the manuscript with us. We also thank Lindsay Hinck, Kuan-Hong Wang, and Thorsten Erpel for thoughtful comments on the manuscript; Laura Bauer for help with manuscript preparation; and many colleagues both within our laboratories and around the world who have discussed many of the ideas presented here or who have allowed us to refer to their unpublished results. Supported by NIH grants to M.T.-L. and C.S.G. M.T.-L. is an Assistant Investigator and C.S.G. is an Investigator with the Howard Hughes Medical Institute.

REFERENCES

Abercrombie, M. (1970). Contact inhibition in tissue culture. *In Vitro* 6:128–142.

Allendoerfer, K.L., Shatz, C.J. (1994). The subplate, a transient neocortical structure: its role in the development of connections between thalamus and cortex. *Annu. Rev. Neurosci.* 17:185–218.

Asou, H., Takeda, Y., Sugawa, M., Kobayashi, M., Miura, M., Uyemura, K. (1996). The RSLE sequence in the cytoplasmic domain of cell adhesion molecule L1 is involved in cell migration on L1 substrates. *J. Cell. Physiol., in press.*

Baier, H., Bonhoeffer, F. (1992). Axon guidance by gradients of a target-derived component. *Science* 255:472–475.

Baird, D.H., Baptista, C.A., Wang, L.C., Mason, C.A. (1992). Specificity of a target cell-derived stop signal for afferent axonal growth. *J. Neurobiol.* 23:579–591.

Ball, E.E., Ho, R.K., Goodman, C.S. (1985). Development of neuromuscular specificity in the grasshopper embryo: guidance of motoneuron growth cones by muscle pioneers. *J. Neurosci.* 5:1808–1819.

Basilico, C., Moscatelli, D. (1992). The FGF family of growth factors and oncogenes. *Adv. Cancer Res.* 59:115–165.

Bastiani, M.J., Raper, J.A., Goodman, C.S. (1984). Pathfinding by neuronal growth cones in grasshopper embryos. III. Selective affinity of the G growth cone for the P cells within the A/P fascicle. *J. Neurosci.* 4:2311–2328.

Bastiani, M.J., du Lac, S., Goodman, C.S. (1986). Guidance of neuronal growth cones in the grasshopper embryo. I. Recognition of a specific axonal pathway by the pCC neuron. *J. Neurosci.* 6:3518–3531.

Bastiani, M.J., Harrelson, A.L., Snow, P.M., Goodman, C.S. (1987). Expression of fasciclin I and II glycoproteins on subsets of axon pathways during neuronal development in the grasshopper. *Cell* 48:745–755.

Bate, C.M. (1976). Embryogenesis of an insect nervous system. I. A map of the thoracic and abdominal neuroblasts in Locusta migratoria. *J. Embryol. Exp. Morphol.* 35:107–123.

Beggs, H.E., Soriano, P., Maness, P.F. (1994). NCAM-dependent neurite outgrowth is inhibited in neurons from Fyn-minus mice. *J. Cell Biol.* 127:825–833.

Behar, O., Golden, J.A., Mashimo, H., Schoen, F.J., Fishman, M.C. (1996). Semaphorin III is needed for normal patterning and growth of nerves, bones, and heart. *Nature* 383:525–528.

Bellosta, P., Costa, M., Lin, D.A., Basilico, C. (1995). The receptor tyrosine kinase ARK mediates cell aggregation by homophilic binding. *Mol. Cell. Biol.* 15:614–625.

Bentley, D., O'Connor, T.P. (1992). Guidance and steering of peripheral pioneer growth cones in grasshopper embryos. In *The Nerve Growth Cone,* Letourneau, P.C., Kater, S.B., Macagno, E.R., eds. New York: Raven Press, pp. 265–282.

Bentley, D., O'Connor, T.P. (1994). Cytoskeletal events in growth cone steering. *Curr. Opin. Neurobiol.* 4:43–48.

Bixby, J.L., Harris, W.A. (1991). Molecular mechanisms of axon growth and guidance. *Annu. Rev. Cell Biol.* 7:117–159.

Bixby, J.L., Jhabvala, P. (1993). Tyrosine phosphorylation in early embryonic growth cones. *J. Neurosci.* 13:3421–3432.

Bixby, J.L., Grunwald, G.B., Bookman, R.J. (1994). Ca2+influx and neurite growth in response to purified N-cadherin and laminin. *J. Cell Biol.* 127:1461–1475.

Bonhoeffer, F., Huf, J. (1980). Recognition of cell types by axonal growth cones in vitro. *Nature* 288:162–164.

Bonhoeffer, F., Huf, J. (1982). In vitro experiments on axon guidance demonstrating an anterior-posterior gradient on the tectum. *EMBO J.* 1:427–431.

Bonhoeffer, F., Huf, J. (1985). Position-dependent properties of retinal axons and their growth cones. *Nature* 315:409–410.

Bovolenta, P., Dodd, J. (1990). Guidance of commissural growth cones at the floor plate in embryonic rat spinal cord. *Development* 109:435–447.

Bovolenta, P., Dodd, J. (1991). Perturbation of neuronal differentiation and axon guidance in the spinal cord of mouse embryos lacking a floor plate: analysis of Danforth's short-tail mutation. *Development* 113:625–639.

Brady-Kalnay, S.M., Tonks, N.K. (1995). Protein tyrosine phosphatases as adhesion receptors. *Curr. Opin. Cell Biol.* 7:650–657.

Brambilla, R., Schnapp, A., Casagranda, F., Labrador, J.P., Bergemann, A.D., Flanagan, J.G., Pasquale, E.B., Klein, R. (1995). Membrane-bound LERK2 ligand can signal through three different Eph-related receptor tyrosine kinases. *EMBO J.* 14:3116–3126.

Bray, D. (1979). Mechanical tension produced by nerve cells in tissue culture. *J. Cell Sci.* 37: 391–410.

Bray, D., Wood, P., Bunge, R.P. (1980). Selective fasciculation of nerve fibers in culture. *Exp. Cell Res.* 130:241–250.

Brittis, P.A., Lemmon, V., Rutishauser, U., Silver, J. (1995). Unique changes of ganglion cell growth cone behavior following cell adhesion molecule perturbations: a time-lapse study of the living retina. *Mol. Cell. Neurosci.* 6:433–449.

Brummendorf, T., Rathjen, F.G. (1995). Cell adhesion molecules. 1: immunoglobulin superfamily. *Protein Profile* 2:963–1108.

Brummendorf, T., Hubert, M., Treubert, U., Leuschner, R., Tarnok, A., Rathjen, F.G. (1993). The axonal recognition molecule F11 is a multifunctional protein: specific domains mediate interactions with Ng-CAM and restrictin. *Neuron* 10:711–727.

Burgeson, R.E., Chiquet, M., Deutzmann, R., Ekblom, P., Engel, J., Kleinman, H., Martin, G.R., Meneguzzi, G., Paulsson, M., Sanes, J., et al. (1994). A new nomenclature for the laminins. *Matrix Biol.* 14:209–211.

Callahan, C.A., Muralidhar, M.G., Lundgren, S.E., Scully, A.L., Thomas, J.B. (1995). Control of neuronal pathway selection by a Drosophila receptor protein-tyrosine kinase family member. *Nature* 376:171–174.

Calof, A.L., Lander, A.D. (1991). Relationship between neuronal migration and cell-substratum adhesion: laminin and merosin promote olfactory neuronal migration but are anti-adhesive. *J. Cell Biol.* 115:779–794.

Caroni, P., Schwab, M.E. (1988). Antibody against myelin-associated inhibitor of neurite growth neutralizes nonpermissive substrate properties of CNS white matter. *Neuron* 1:85–96.

Cash, S., Chiba, A., Keshishian, H. (1992). Alternate neuromuscular target selection following the loss of single muscle fibers in Drosophila. *J. Neurosci.* 12:2051–2064.

Chan, S.S.-Y., Zheng, H., Su, M.-W., Wilk, R., Killeen, M.T., Hedgecock, E.M., Culotti, J.G. (1996). UNC-40, a C. elegans homolog of DCC (Deleted in Colorectal Cancer), is required in motile cells responding to UNC-6 netrin cues. *Cell* 18:187–195.

Cheng, H.J., Flanagan, J.G. (1994). Identification and cloning of ELF-1, a developmentally expressed ligand for the Mek4 and Sek receptor tyrosine kinases. *Cell* 79:157–168.

Cheng, H.J., Nakamoto, M., Bergemann, A.D., Flanagan, J.G. (1995). Complementary gradients in expression and binding of ELF-1 and Mek4 in development of the topographic retinotectal projection map. *Cell* 82:371–381.

Chiba, A., Hing, H., Cash, S., Keshishian, H. (1993). Growth cone choices of Drosophila motoneurons in response to muscle fiber mismatch. *J. Neurosci.* 13:714–732.

Chiba, A., Snow, P., Keshishian, H., Hotta, Y. (1995). Fasciclin III as a synaptic target recognition molecule in Drosophila. *Nature* 374:166–168.

Chitnis, A.B., Kuwada, J.Y. (1991). Elimination of a brain tract increases errors in pathfinding by follower growth cones in the zebrafish embryo. *Neuron* 7:277–285.

Clark, E.A., Brugge, J.S. (1995). Integrins and signal transduction pathways: the road taken. *Science* 268:233–239.

Colamarino, S.A., Tessier-Lavigne, M. (1995a). The axonal chemoattractant netrin-1 is also a chemorepellent for trochlear motor axons. *Cell* 81:621–629.

Colamarino, S.A., Tessier-Lavigne, M. (1995b). The role of the floor plate in axon guidance. *Annu. Rev. Neurosci.* 18:497–529.

Cox, E.C., Muller, B., Bonhoeffer, F. (1990). Axonal guidance in the chick visual system: posterior tectal membranes induce collapse of growth cones from the temporal retina. *Neuron* 4:31–37.

Crews, S.T., Thomas, J.B., Goodman, C.S. (1988). The Drosophila single-minded gene encodes a nuclear protein with sequence similarity to the per gene product. *Cell* 52:143–151.

Culotti, J.G., Kolodkin, A.L. (1996). Functions of netrins and semaphorins in axon guidance. *Curr. Opin. Neurobiol.* 6:81–88.

Cunningham, B.A., Hemperly, J.J., Murray, B.A., Prediger, E.A., Brackenbury, R., Edelman, G.M. (1987). Neural cell adhesion molecule: structure, immunoglobulin-like domains, cell surface modulation, and alternative RNA splicing. *Science* 236:799–806.

DeBernardo, A.P., Chang, S. (1996). Heterophilic interactions of DM-GRASP: GRASP-NgCAM interactions involved in neurite extension. *J. Cell Biol.* 133:657–666.

Desai, C.J., Gindhart, J.G., Jr., Goldstein, L.S., Zinn, K. (1996). Receptor tyrosine phosphatases are required for motor axon guidance in the Drosophila embryo. *Cell* 84:599–609.

Dodd, J., Jessell, T.M. (1988). Axon guidance and the patterning of neuronal projections in vertebrates. *Science* 242:692–699.

Dodd, J., Morton, S.B., Karagogeos, D., Yamamoto, M., Jessell, T.M. (1988). Spatial regulation of axonal glycoprotein expression on subsets of embryonic spinal neurons. *Neuron* 1:105–116.

Doherty, P., Walsh, F.S. (1992). Cell adhesion molecules, second messengers and axonal growth. *Curr. Opin. Neurobiol.* 2:595–601.

Doherty, P., Walsh, F.S. (1994). Signal transduction events underlying neurite outgrowth stimulated by cell adhesion molecules. *Curr. Opin. Neurobiol.* 4:49–55.

Doherty, P., Furness, J., Williams, E.J., Walsh, F.S. (1994). Neurite outgrowth stimulated by the tyrosine kinase inhibitor herbimycin A requires activation of tyrosine kinases and protein kinase C. *J. Neurochem.* 62:2124–2131.

Drescher, U., Kremoser, C., Handwerker, C., Loschinger, J., Noda, M., Bonhoeffer, F. (1995). In vitro guidance of retinal ganglion cell axons by RAGS, a 25 kDa tectal protein related to ligands for Eph receptor tyrosine kinases. *Cell* 82:359–370.

du Lac, S., Bastiani, M.J., Goodman, C.S. (1986). Guidance of neuronal growth cones in the grasshopper embryo. II. Recognition of a specific axonal pathway by the aCC neuron. *J. Neurosci.* 6:3532–3541.

Easter, S.S., Jr., Ross, L.S., Frankfurter, A. (1993). Initial tract formation in the mouse brain. *J. Neurosci.* 13:285–299.

Easter, S., Jr., Burrill, J., Marcus, R.C., Ross, L.S., Taylor, J.S., Wilson, S.W. (1994). Initial tract formation in the vertebrate brain. *Prog. Brain Res.* 102:79–93.

Ebendal, T., Jacobson, C.O. (1977). Tissue explants affecting extension and orientation of axons in cultured chick embryo ganglia. *Exp. Cell Res.* 105:379–387.

Edelman, G.M. (1993). A golden age for adhesion. *Cell Adhes. Commun.* 1:1–7.

Ehrig, K., Leivo, I., Argraves, W.S., Ruoslahti, E., Engvall, E. (1990). Merosin, a tissue-specific basement membrane protein, is a laminin-like protein. *Proc. Natl. Acad. Sci. U.S.A.* 87:3264–3268.

Eisen, J.S., Myers, P.Z., Westerfield, M. (1986). Pathway selection by growth cones of identified motoneurones in live zebra fish embryos. *Nature* 320:269–271.

Eisen, J.S., Pike, S.H., Debu, B. (1989). The growth cones of identified motoneurons in embryonic zebrafish select appropriate pathways in the absence of specific cellular interactions. *Neuron* 2:1097–1104.

Elkins, T., Zinn, K., McAllister, L., Hoffmann, F.M., Goodman, C.S. (1990). Genetic analysis of a Drosophila neural cell adhesion molecule: interaction of fasciclin I and Abelson tyrosine kinase mutations. *Cell* 60:565–575.

ElShamy, W.M., Linnarsson, S., Lee, K.F., Jaenisch, R., Ernfors, P. (1996). Prenatal and postnatal requirements of NT-3 for sympathetic neuroblast survival and innervation of specific targets. *Development* 122:491–500.

Engvall, E., Earwicker, D., Haaparanta, T., Ruoslahti, E., Sanes, J.R. (1990). Distribution and isolation of four laminin variants; tissue restricted distribution of heterotrimers assembled from five different subunits. *Cell Regul.* 1:731–740.

Faissner, A., Kruse, J. (1990). J1/tenascin is a repulsive substrate for central nervous system neurons. *Neuron* 5:627–637.

Fambrough, D., Goodman, C.S. (1996). The Drosophila *beaten path* gene encodes a novel secreted protein that regulates defasciculation at motor axon choice points. *Cell,* 87: 1049–1058.

Fazeli, A., Dickinson, S.L., Hermiston, M., Tighe, R., Steen, R., Small, C., Stoeckli, E., Keino-Masu, K., Masu, M., Rayburn, H., Simons, J., Bronson, R., Gordon, J., Tessier-Lavigne, M., Weinberg, R.A. (1997). Phenotype of mice lacking functional *Deleted in Colorectal Cancer (Dcc)* gene. *Nature,* in press.

Felsenfeld, D.P., Hynes, M.A., Skoler, K.M., Furley, A.J., Jessell, T.M. (1994). TAG-1 can mediate homophilic binding, but neurite outgrowth on TAG-1 requires an L1-like molecule and beta 1 integrins. *Neuron* 12:675–690.

Fitzgerald, M., Kwiat, G.C., Middleton, J., Pini, A. (1993). Ventral spinal cord inhibition of neurite outgrowth from embryonic rat dorsal root ganglia. *Development* 117:1377–1384.

Fryer, H.J.L., Hockfield, S. (1996). Protein determinants for specific polysialylation of the neural cell adhesion molecule. *Curr. Opin. Neurobiol.* 6:113–118.

Fujisawa, H. (1981). Retinotopic analysis of fiber pathways in the regenerating retinotectal system of the adult newt cynops Pyrrhogaster. *Brain Res.* 206:27–37.

Gale, N.W., Holland, S.J., Valenzeula, D.M., Flenniken, A., et al. (1996). EPH receptors and ligands comprise two major specificity subclasses and are reciprocally compartmentalized during embryogenesis. *Neuron* 17:9–19.

Gao, P.P., Zhang, J.H., Yokoyama, M., Racey, B., Dreyfus, C.F., Black, I.B., Zou, R. (1996). Regulation of topographic projection in the brain: Elf-1 in the hippocamposeptal system. *Proc. Natl. Acad. Sci. U.S.A.,* 93:11161–11166.

Garcia-Alonso, L., VanBerkum, M.F., Grenningloh, G., Schuster, C., Goodman, C.S. (1995). Fasciclin II controls proneural gene expression in Drosophila. *Proc. Natl. Acad. Sci. U.S.A.* 92:10501–10505.

Garcia-Alonso, L., Fetter, R.D., Goodman, C.S. (1996). Genetic analysis of *Laminin A* in Drosophila: extracellular matrix containing Laminin A is required for ocellar axon pathfinding. *Development* 122:2611–2621.

Ghosh, A., Antonini, A., McConnell, S.K., Shatz, C.J. (1990). Requirement for subplate neurons in the formation of thalamocortical connections. *Nature* 347:179–181.

Gierer, A. (1987). Directional cues for growing axons forming the retinotectal projection. *Development* 101:479–489.

Godement, P., Mason, C.A. (1993). Guidance of retinal fibers in the optic chiasm. *Perspect. Dev. Neurobiol.* 1:217–225.

Godement, P., Salaun, J., Mason, C.A. (1990). Retinal axon pathfinding in the optic chiasm: divergence of crossed and uncrossed fibers. *Neuron* 5:173–186.

Godement, P., Wang, L.C., Mason, C.A. (1994). Retinal axon divergence in the optic chiasm: dynamics of growth cone behavior at the midline . *J. Neurosci.* 14:7024–7039.

Gomez, T.M., Letourneau, P.C. (1994). Filopodia initiate choices made by sensory neuron growth cones at laminin/fibronectin borders in vitro. *J. Neurosci.* 14:5959–5972.

Goodman, C.S. (1994). The likeness of being: phylogenetically conserved molecular mechanisms of growth cone guidance. *Cell* 78:353–356.

Goodman, C.S. (1996). Mechanisms and molecules that control growth cone guidance. *Annu. Rev. Neurosci.* 19:341–377.

Goodman, C.S., Spitzer, N.C. (1979). Embryonic development of identified neurones: differentiation from neuroblast to neurone. *Nature* 280:208–214.

Goodman, C.S., Bastiani, M.J., Doe, C.Q., du Lac, S., Helfand, S.L., Kuwada, J.Y., Thomas, J.B. (1984). Cell recognition during neuronal development. *Science* 225:1271–1279.

Goshima, Y., Nakamura, F., Strittmatter, P., Strittmatter, S.M. (1995). Collapsin-induced growth cone collapse mediated by an intracellular protein related to UNC-33. *Nature* 376:509–514.

Grenningloh, G., Bieber, A.J., Rehm, E.J., Snow, P.M., Traquina, Z.R., Hortsch, M., Patel, N.H., Goodman, C.S. (1990). Molecular genetics of neuronal recognition in Drosophila: evolution and function of immunoglobulin superfamily cell adhesion molecules. *Cold Spring Harb. Symp. Quant. Biol.* 55:327–340.

Grenningloh, G., Rehm, E.J., Goodman, C.S. (1991). Genetic analysis of growth cone guidance in Drosophila: fasciclin II functions as a neuronal recognition molecule. *Cell* 67: 45–57.

Grumet, M., Friedlander, D.R., Edelman, G.M. (1993). Evidence for the binding of Ng-CAM to laminin. *Cell Adhes. Commun.* 1:177–190.

Gundersen, R.W. (1987). Response of sensory neurites and growth cones to patterned substrata of laminin and fibronectin in vitro. *Dev. Biol.* 121:423–431.

Gundersen, R.W., Barrett, J.N. (1979). Neuronal chemotaxis: chick dorsal-root axons turn toward high concentrations of nerve growth factor. *Science* 206:1079–1080.

Guthrie, S., Pini, A. (1995). Chemorepulsion of developing motor axons by the floor plate. *Neuron* 14:1117–1130.

Halpern, M.E., Chiba, A., Johansen, J., Keshishian, H. (1991). Growth cone behavior underlying the development of stereotypic synaptic connections in Drosophila embryos. *J. Neurosci.* 11:3227–3238.

Hamelin, M., Zhou, Y., Su, M.W., Scott, I.M., Culotti, J.G. (1993). Expression of the UNC-5 guidance receptor in the touch neurons of C. elegans steers their axons dorsally. *Nature* 364:327–330.

Hariharan, I.K., Chuang, P.T., Rubin, G.M. (1991). Cloning and characterization of a receptor-class phosphotyrosine phosphatase gene expressed on central nervous system axons in Drosophila melanogaster. *Proc. Natl. Acad. Sci. U.S.A.* 88:11266–11270.

Harrelson, A.L., Goodman, C.S. (1988). Growth cone guidance in insects: fasciclin II is a member of the immunoglobulin superfamily. *Science* 242:700–708.

Harris, R., Sabatelli, L.M., Seeger, M.A. (1996). Guidance cues at the Drosophila CNS midline—identification and characterization of two Drosophila Netrin/UNC-6 homologs. *Neuron* 17:217–228.

Harrison, R.G. (1910). The outgrowth of the nerve fiber as a mode of protoplasmic movement. *J. Exp. Zool.* 9:787–846.

Hatta, K., Nose, A., Nagafuchi, A., Takeichi, M. (1988). Cloning and expression of cDNA encoding a neural calcium-dependent cell adhesion molecule: its identity in the cadherin gene family. *J. Cell Biol.* 106:873–881.

Hedgecock, E.M., Culotti, J.G., Thomson, J.N., Perkins, L.A. (1985). Axonal guidance mutants of Caenorhabditis elegans identified by filling sensory neurons with fluorescein dyes. *Dev. Biol.* 111:158–170.

Hedgecock, E.M., Culotti, J.G., Hall, D.H. (1990). The unc-5, unc-6, and unc-40 genes guide circumferential migrations of pioneer axons and mesodermal cells on the epidermis in C. elegans. *Neuron* 4:61–85.

Heffner, C.D., Lumsden, A.G., O'Leary, D.D. (1990). Target control of collateral extension and directional axon growth in the mammalian brain. *Science* 247:217–220.

Henkemeyer, M., Orioli, D., Henderson, J.T., Saxton, T.M., et al. (1996). NUK controls pathfinding of commissural axons in the mammalian central nervous system. *Cell* 86: 35–46.

Holland, S.J., Gale, N.W., Mbamalu, G., Yancopoulos, G.D., Henkemeyer, M., Pawson, T. (1996). Bi-directional signalling through the Eph family receptor Nuk and its transmembrane ligands. *Nature* 383:722–725.

Holt, C.E. (1984). Does timing of axon outgrowth influence initial retinotectal topography in Xenopus? *J. Neurosci.* 4:1130–1152.

Holt, C.E., Harris, W.A. (1983). Order in the initial retinotectal map in Xenopus: a new technique for labelling growing nerve fibers. *Nature* 301:150–152.

Honig, M.G., Rutishauser, U.S. (1996). Changes in the segmental pattern of sensory neuron projections in the chick hindlimb under conditions of altered cell adhesion molecule function. *Dev. Biol.* 175:325–337.

Hortsch, M. (1996). The L1 family of neural cell adhesion molecules: old proteins performing new tricks. *Neuron* 17:587–593.

Hoyle, G.W., Mercer, E.H., Palmiter, R.D., Brinster, R.L. (1993). Expression of NGF in sympathetic neurons leads to excessive axon outgrowth from ganglia but decreased terminal innervation within tissues. *Neuron* 10:1019–1034.

Hu, H., Rutishauser, U. (1996). A septum-derived chemorepulsive factor for migrating olfactory interneuron precursors. *Neuron* 16:933–940.

Hu, S., Fambrough, D., Atashi, J.R., Goodman, C.S., Crews, S.T. (1995). The Drosophila abrupt gene encodes a BTB-zinc finger regulatory protein that controls the specificity of neuromuscular connections. *Genes and Development* 9:2936–48.

Hunter, D.D., Shah, V., Merlie, J.P., Sanes, J.R. (1989). A laminin-like adhesive protein concentrated in the synaptic cleft of the neuromuscular junction. *Nature* 338:229–234.

Hynes, R.O., Lander, A.D. (1992). Contact and adhesive specificities in the associations, migrations, and targeting of cells and axons. *Cell* 68:303–322.

Ignelzi, M., Jr., Miller, D.R., Soriano, P., Maness, P.F. (1994). Impaired neurite outgrowth of src-minus cerebellar neurons on the cell adhesion molecule L1. *Neuron* 12:873–884.

Ishii, N., Wadsworth, W.G., Stern, B.D., Culotti, J.G., Hedgecock, E.M. (1992). UNC-6, a laminin-related protein, guides cell and pioneer axon migrations in C. elegans. *Neuron* 9:873–881.

Jacobs, J.R., Goodman, C.S. (1989). Embryonic development of axon pathways in the Drosophila CNS. I. A glial scaffold appears before the first growth cones. *J. Neurosci.* 9:2402–2411.

Jouet, M., Rosenthal, A., MacFarlane, J., Kenwrick, S., Donnai, D. (1993). A missense mutation confirms the L1 defect in X-linked hydrocephalus (HSAS). *Nat. Genet.* 4:331.

Jouet, M., Rosenthal, A., Armstrong, G., MacFarlane, J., Stevenson, R., Paterson, J., Metzenberg, A., Ionasescu, V., Temple, K., Kenwrick, S. (1994). X-linked spastic paraplegia (SPG1), MASA syndrome and X-linked hydrocephalus result from mutations in the L1 gene. *Nat. Genet.* 7:402–407.

Kadmon, G., Kowitz, A., Altevogt, P., Schachner, M. (1990). The neural cell adhesion molecule N-CAM enhances L1-dependent cell-cell interactions. *J. Cell Biol.* 110:193–208.

Kapfhammer, J.P., Raper, J.A. (1987a). Collapse of growth cone structure on contact with specific neurites in culture. *J. Neurosci.* 7:201–212.

Kapfhammer, J.P., Raper, J.A. (1987b). Interactions between growth cones and neurites growing from different neural tissues in culture. *J. Neurosci.* 7:1595–1600.

Kapfhammer, J.P., Grunewald, B.E., Raper, J.A. (1986). The selective inhibition of growth cone extension by specific neurites in culture. *J. Neurosci.* 6:2527–2534.

Keino-Masu, K., Masu, M.L.H., Leonardo, E.D., Chan, S.S.-Y., Culotti, J.G., Tessier-Lavigne, M. (1996). *Deleted in colorectal carcinomas (DCC)* encodes a netrin receptor. *Cell* 87:175–185.

Kennedy, T.E., Tessier-Lavigne, M. (1995). Guidance and induction of branch formation in developing axons by target-derived diffusible factors. *Curr. Opin. Neurobiol.* 5:83–90.

Kennedy, T.E., Serafini, T., de la Torre, J.R., Tessier-Lavigne, M. (1994). Netrins are diffusible chemotrophic factors for commissural axons in the embryonic spinal cord. *Cell* 78:425–435.

Keshishian, H., Bentley, D. (1983). Embryogenesis of peripheral nerve pathways in grasshopper legs. III. Development without pioneer neurons. *Dev. Biol.* 96:116–124.

Keshishian, H., Broadie, K., Chiba, A., Bate, M. (1996). The drosophila neuromuscular junction—a model system for studying synaptic development and function. *Annu. Rev. Neurosci.* 19:545–575.

Keynes, R., Cook, G.M. (1995). Axon guidance molecules. *Cell* 83:161–169.

Keynes, R.J., Stern, C.D. (1984). Segmentation in the vertebrate nervous system. *Nature* 310:786–789.

Kinoshita, N., Minshull, J., Kirschner, M.W. (1995). The identification of two novel ligands of the FGF receptor by a yeast screening method and their activity in Xenopus development. *Cell* 83:621–630.

Klambt, C., Jacobs, J.R., Goodman, C.S. (1991). The midline of the Drosophila central nervous system: a model for the genetic analysis of cell fate, cell migration, and growth cone guidance. *Cell* 64:801–815.

Klinz, S.G., Schachner, M., Maness, P.F. (1995). L1 and N-CAM antibodies trigger protein phosphatase activity in growth cone-enriched membranes. *J. Neurochem.* 65:84–95.

Klose, M., Bentley, D. (1989). Transient pioneer neurons are essential for formation of an embryonic peripheral nerve. *Science* 245:982–984.

Kolodkin, A.L. (1996). Semaphorins—mediators of repulsive growth cone guidance. *Trends Cell Biol.* 6:15–22.

Kolodkin, A.L., Matthes, D.J., O'Connor, T.P., Patel, N.H., Admon, A., Bentley, D., Goodman, C.S. (1992). Fasciclin IV: sequence, expression, and function during growth cone guidance in the grasshopper embryo. *Neuron* 9:831–845.

Kolodkin, A.L., Matthes, D.J., Goodman, C.S. (1993). The semaphorin genes encode a family of transmembrane and secreted growth cone guidance molecules. *Cell* 75:1389–1399.

Kolodziej, P.A., Timpe, L.C., Mitchell, K.J., Fried, S.A., Goodman, C.S., Jan, L.Y., Jan, Y.N. (1996). *frazzled* encodes a Drosophila member of the DCC immunoglobulin subfamily and is required for CNS and motor axon guidance. *Cell* 87:197–204.

Krantz, D.E., Zipursky, S.L. (1990). Drosophila chaoptin, a member of the leucine-rich repeat family, is a photoreceptor cell-specific adhesion molecule. *Embo. J.* 9:1969–1977.

Krueger, N.X., Streuli, M., Saito, H. (1990). Structural diversity and evolution of human receptor-like protein tyrosine phosphatases. *EMBO J.* 9:3241–3252.

Krueger, N.X., Van Vactor, D., Wan, H.I., Gelbart, W.M., Goodman, C.S., Saito, H. (1996). The transmembrane tyrosine phosphatase DLAR controls motor axon guidance in Drosophila. *Cell* 84:611–622.

Kuhn, T.B., Stoeckli, E.T., Condrau, M.A., Rathjen, F.G., Sonderegger, P. (1991). Neurite outgrowth on immobilized axonin-1 is mediated by a heterophilic interaction with L1(G4). *J. Cell Biol.* 115:1113–1126.

Kuhn, T.B., Schmidt, M.F., Kater, S.B. (1995). Laminin and fibronectin guideposts signal sustained but opposite effects to passing growth cones. *Neuron* 14:275–285.

Kuwada, J.Y. (1986). Cell recognition by neuronal growth cones in a simple vertebrate embryo. *Science* 233:740–746.

Kuwada, J.Y. (1992). Growth cone guidance in the zebrafish central nervous system. *Curr. Opin. Neurobiol.* 2:31–35.

Lance-Jones, C., Landmesser, L. (1981a). Pathway selection by chick lumbosacral motoneurons during normal development. *Proc. Royal Soc. Lond.* B214:1–18.

Lance-Jones, C., Landmesser, L. (1981b). Pathway selection by embryonic chick motoneurons in an experimentally altered environment. *Proc. Royal Soc. Lond.* B214:19–52.

Lander, A.D. (1987). Molecules that make axons grow. *Mol. Neurobiol.* 1:213–245.

Landmesser, L. (1978). The development of motor projection patterns in the chick hind limb. *J. Physiol.* 284:391–414.

Landmesser, L.T. (1980). The generation of neuromuscular specificity. *Annu. Rev. Neurosci.* 3:279–302.

Landmesser, L., Dahm, L., Tang, J.C., Rutishauser, U. (1990). Polysialic acid as a regulator of intramuscular nerve branching during embryonic development. *Neuron* 4:655–667.

Lemmon, V., Burden, S.M., Payne, H.R., Elmslie, G.J., Hlavin, M.L. (1992). Neurite growth on different substrates: permissive versus instructive influences and the role of adhesive strength. *J. Neurosci.* 12:818–826.

Letourneau, P.C. (1975). Cell-to-substratum adhesion and guidance of axonal elongation. *Dev. Biol.* 44:92–101.

Letourneau, P.C., Madsen, A.M., Palm, S.L., Furcht, L.T. (1988). Immunoreactivity for laminin in the developing ventral longitudinal pathway of the brain. *Dev. Biol.* 125: 135–144.

Leung-Hagesteijn, C., Spence, A.M., Stern, B.D., Zhou, Y., Su, M.W., Hedgecock, E.M., Culotti, J.G. (1992). UNC-5, a transmembrane protein with immunoglobulin and thrombospondin type 1 domains, guides cell and pioneer axon migrations in C. elegans. *Cell* 71:289–299.

Li, W., Herman, R.K., Shaw, J.E. (1992). Analysis of the Caenorhabditis elegans axonal guidance and outgrowth gene unc-33. *Genetics* 132:675–689.

Lilienbaum, A., Reszka, A.A., Horwitz, A.F., Holt, C.E. (1995). Chimeric integrins expressed in retinal ganglion cells impair process outgrowth in vivo. *Mol. Cell. Neurosci.* 6:139–152.

Lin, C.H., Forscher, P. (1995). Growth cone advance is inversely proportional to retrograde F-actin flow. *Neuron* 14:763–771.

Lin, D.M., Goodman, C.S. (1994). Ectopic and increased expression of Fasciclin II alters motoneuron growth cone guidance. *Neuron* 13:507–523.

Lin, D.M., Fetter, R.D., Kopczynski, C., Grenningloh, G., Goodman, C.S. (1994). Genetic analysis of Fasciclin II in Drosophila: defasciculation, refasciculation, and altered fasciculation. *Neuron* 13:1055–1069.

Lin, D.M., Auld, V.J., Goodman, C.S. (1995). Targeted neuronal cell ablation in the Drosophila embryo: pathfinding by follower growth cones in the absence of pioneers. *Neuron* 14:707–715.

Lumsden, A.G., Davies, A.M. (1983). Earliest sensory nerve fibres are guided to peripheral targets by attractants other than nerve growth factor. *Nature* 306:786–788.

Lumsden, A.G., Davies, A.M. (1986). Chemotropic effect of specific target epithelium in the developing mammalian nervous system. *Nature* 323:538–539.

Luo, Y., Raper, J.A. (1994). Inhibitory factors controlling growth cone motility and guidance. *Curr. Opin. Neurobiol.* 4:648–654.

Luo, Y., Raible, D., Raper, J.A. (1993). Collapsin: a protein in brain that induces the collapse and paralysis of neuronal growth cones. *Cell* 75:217–227.

Luo, Y., Shepherd, I., Li, J., Renzi, M.J., Chang, S., Raper, J.A. (1995). A family of molecules related to collapsin in the embryonic chick nervous system. *Neuron* 14:1131–1140.

Maness, P.F., Shores, C.G., Ignelzi, M. (1990). Localization of the normal cellular src protein to the growth cone of differentiating neurons in brain and retina. *Adv. Exp. Med. Biol.* 265:117–125.

Matthes, D.J., Sink, H., Kolodkin, A.L., Goodman, C.S. (1995). Semaphorin II can function as a selective inhibitor of specific synaptic arborizations. *Cell* 81:631–639.

McConnell, S.K., Ghosh, A., Shatz, C.J. (1989). Subplate neurons pioneer the first axon pathway from the cerebral cortex. *Science* 245:978–982.

McFarlane, S., McNeill, L., Holt, C.E. (1995). FGF signaling and target recognition in the developing Xenopus visual system. *Neuron* 15:1017–1028.

McFarlane, S., Cornel, E., Amaya, E., Holt, C.E. (1996). Inhibition of FGF receptor activity in retinal cell axons causes errors in target recognition. *Neuron* 17:245–254.

McIntire, S.L., Garriga, G., White, J., Jacobson, D., Horvitz, H.R. (1992). Genes necessary for directed axonal elongation or fasciculation in C. elegans. *Neuron* 8:307–322.

McKerracher, L., David, S., Jackson, D.L., Kottis, V., Dunn, R.J., Braun, P.E. (1994). Identification of myelin-associated glycoprotein as a major myelin-derived inhibitor of neurite growth. *Neuron* 13:805–811.

McLoon, S.C., McLoon, L.K., Palm, S.L., Furcht, L.T. (1988). Transient expression of laminin in the optic nerve of the developing rat. *J. Neurosci.* 8:1981–1990.

Meima, L., Kljavin, I.J., Moran, P., Shih, A., Winslow, J.W., Caras, I.C. (1996). AL-1 induced growth cone collapse of rat cortical neurons is correlated with REK7 expression and rearrangement of the actin cytoskeleton. *Eur. J. Neurosci., in press.*

Messersmith, E.K., Leonardo, E.D., Shatz, C.J., Tessier-Lavigne, M., Goodman, C.S., Kolodkin, A.L. (1995). Semaphorin III can function as a selective chemorepellent to pattern sensory projections in the spinal cord. *Neuron* 14:949–959.

Mitchell, K.J., Doyle, J.L., Serafini, T., Kennedy, T., Tessier-Lavigne, M., Goodman, C.S., Dickson, B.J. (1996). Genetic analysis of *Netrin* genes in Drosophila: Netrins guide CNS commissural axons and peripheral motor axons. *Neuron* 17:203–215.

Monschau, B., Kremoser, C., Ohta, K., Tanaka, H., Kaneko, T., Yamada, T., Handwerker, C., Hornberger, M.H., Löschinger, J., Pasquale, E.B., Siever, D.A., Verdame, M.F.,

Müller, B., Bonhoeffer, F., Drescher, U. (1996). Shared and distinct functions of RAGS and ELF-1 in guiding retinal axons. *EMBO J., in press.*

Montell, D.J., Goodman, C.S. (1988). Drosophila substrate adhesion molecule: sequence of laminin B1 chain reveals domains of homology with mouse. *Cell* 53:463–473.

Montell, D.J., Goodman, C.S. (1989). Drosophila laminin: sequence of B2 subunit and expression of all three subunits during embryogenesis. *J. Cell Biol.* 109:2441–2453.

Montgomery, A.M.P., Becker, J.C., Siu, C.-H., Lemmon, V.P., Cheresh, D.A., Pancook, J.D., Zhao, X., Reisfeld, R.A. (1996). Human neural cell adhesion molecule L1 and rat homolog NILE are ligands for integrin alpha-v beta-3. *J. Cell Biol.* 132:475–485.

Morales, G., Hubert, M., Brummendorf, T., Treubert, U., Tarnok, A., Schwarz, U., Rathjen, F.G. (1993). Induction of axonal growth by heterophilic interactions between the cell surface recognition proteins F11 and Nr-CAM/Bravo. *Neuron* 11:1113–1122.

Mukhopadhyay, G., Doherty, P., Walsh, F.S., Crocker, P.R., Filbin, M.T. (1994). A novel role for myelin-associated glycoprotein as an inhibitor of axonal regeneration. *Neuron* 13: 757–767.

Murphy-Erdosh, C., Yoshida, C.K., Paradies, N., Reichardt, L.F. (1995). The cadherin-binding specificities of B-cadherin and LCAM. *J. Cell Biol.* 129:1379–1390.

Nakamoto, M., Cheng, H.-J., Friedman, G.C., McLaughlin, T., Hansen, M.J., Yoon, C.H., O'Leary, D.D.M., Flanagan, J.G. (1996). Topographically specific effects of ELF-1 on retinal axon guidance in vitro and retinal axon mapping in vivo. *Cell* 86:755–766.

Nakamura, H., O'Leary, D.D. (1989). Inaccuracies in initial growth and arborization of chick retinotectal axons followed by course corrections and axon remodeling to develop topographic order. *J. Neurosci.* 9:3776–3795.

Nelson, R.W., Bates, P.A., Rutishauser, U. (1995). Protein determinants for specific polysialylation of the neural cell adhesion molecule. *J. Biol. Chem.* 270:17171–17179.

Nose, A., Mahajan, V.B., Goodman, C.S. (1992). Connectin: a homophilic cell adhesion molecule expressed on a subset of muscles and the motoneurons that innervate them in Drosophila. *Cell* 70:553–567.

Nose, A., Takeichi, M., Goodman, C.S. (1994). Ectopic expression of connectin reveals a repulsive function during growth cone guidance and synapse formation. *Neuron* 13:525–539.

O'Connor, T.P., Duerr, J.S., Bentley, D. (1990). Pioneer growth cone steering decisions mediated by single filopodial contacts in situ. *J. Neurosci.* 10:3935–3946.

O'Leary, D.D., Bicknese, A.R., De Carlos, J.A., Heffner, C.D., Koester, S.E., Kutka, L.J., Terashima, T. (1990). Target selection by cortical axons: alternative mechanisms to establish axonal connections in the developing brain. *Cold Spring Harb. Symp. Quant. Biol.* 55:453–468.

Oakley, R.A., Tosney, K.W. (1993). Contact-mediated mechanisms of motor axon segmentation. *J. Neurosci.* 13:3773–3792.

Oda, H., Uemura, T., Harada, Y., Iwai, Y., Takeichi, M. (1994). A Drosophila homolog of cadherin associated with armadillo and essential for embryonic cell-cell adhesion. *Dev. Biol.* 165:716–726.

Ono, K., Tomasiewicz, H., Magnuson, T., Rutishauser, U. (1994). N-CAM mutation inhibits tangential neuronal migration and is phenocopied by enzymatic removal of polysialic acid. *Neuron* 13:595–609.

Pandey, A., Lindberg, R.A., Dixit, V.M. (1995). Cell signalling—receptor orphans find a family. *Curr. Biol.* 5:986–989.

Park, S., Frisén, J., Barbacid, M. (1996). Aberrant axonal pathfinding in mice lacking the PTK-4/EEK receptor, a member of the Eph family of tyrosine protein kinases. *Submitted.*

Patel, N.H., Snow, P.M., Goodman, C.S. (1987). Characterization and cloning of fasciclin III: a glycoprotein expressed on a subset of neurons and axon pathways in Drosophila. *Cell* 48:975–988.

Peles, E., Nativ, M., Campbell, P.L., Sakurai, T., Martinez, R., Lev, S., Clary, D.O., Schilling, J., Barnea, G., Plowman, G.D., et al. (1995). The carbonic anhydrase domain of receptor tyrosine phosphatase beta is a functional ligand for the axonal cell recognition molecule contactin. *Cell* 82:251–260.

Pesheva, P., Gennarini, G., Goridis, C., Schachner, M. (1993). The F3/11 cell adhesion molecule mediates the repulsion of neurons by the extracellular matrix glycoprotein J1-160/180. *Neuron* 10:69–82.

Pimenta, A.F., Zhukareva, V., Barbe, M.F., Reinoso, B.S., Grimley, C., Henzel, W., Fischer, I., Levitt, P. (1995). The limbic system-associated membrane protein is an Ig superfamily member that mediates selective neuronal growth and axon targeting. *Neuron* 15: 287–97.

Pini, A. (1993). Chemorepulsion of axons in the developing mammalian central nervous system. *Science* 261:95–98.

Placzek, M., Tessier-Lavigne, M., Jessell, T., Dodd, J. (1990). Orientation of commissural axons in vitro in response to a floor plate-derived chemoattractant. *Development* 110: 19–30.

Puschel, A.W., Adams, R.H., Betz, H. (1995). Murine semaphorin D/collapsin is a member of a diverse gene family and creates domains inhibitory for axonal extension. *Neuron* 14:941–948.

Puschel, A.W., Adams, R.H., Betz, H. (1996). The sensory innervation of the mouse spinal cord may be patterned by differential expression of and differential responsiveness to semaphorins. *Mol. Cell. Neurosci.* 7:419–431.

Rader, C., Stoeckli, E.T., Ziegler, U., Osterwalder, T., Kunz, B., Sonderegger, P. (1993). Cell-cell adhesion by homophilic interaction of the neuronal recognition molecule axonin-1. *Eur. J. Biochem.* 215:133–141.

Ramòn y Cajal, S. (1892). La rétine des vertébrés. *La Cellule* 9:119–258.

Ranheim, T.S., Edelman, G.M., Cunningham, B.A. (1996). Homophilic adhesion mediated by the neural cell adhesion molecule involves multiple immunoglobulin domains. *Proc. Natl. Acad. Sci. U.S.A.* 93:4071–4075.

Raper, J.A., Kapfhammer, J.P. (1990). The enrichment of a neuronal growth cone collapsing activity from embryonic chick brain. *Neuron* 4:21–29.

Raper, J.A., Bastiani, M., Goodman, C.S. (1983a). Pathfinding by neuronal growth cones in grasshopper embryos. I. Divergent choices made by the growth cones of sibling neurons. *J. Neurosci.* 3:20–30.

Raper, J.A., Bastiani, M., Goodman, C.S. (1983b). Pathfinding by neuronal growth cones in grasshopper embryos. II. Selective fasciculation onto specific axonal pathways. *J. Neurosci.* 3:31–41.

Raper, J.A., Bastiani, M.J., Goodman, C.S. (1983c). Guidance of neuronal growth cones: selective fasciculation in the grasshopper embryo. *Cold Spring Harb. Symp. Quant. Biol.* 2:587–598.

Raper, J.A., Bastiani, M.J., Goodman, C.S. (1984). Pathfinding by neuronal growth cones in grasshopper embryos. IV. The effects of ablating the A and P axons upon the behavior of the G growth cone. *J. Neurosci.* 4:2329–2345.

Rathjen, F.G., Jessell, T.M. (1991). Glycoproteins that regulate the growth and guidance of vertebrate axons: domains and dynamics of the immunoglobulin/fibronectin type III subfamily. *Semin. Neurosci.* 3:297–308.

Reichardt, L.F., Tomaselli, K.J. (1991). Extracellular matrix molecules and their receptors: functions in neural development. *Annu. Rev. Neurosci.* 14:531–570.

Reinke, R., Krantz, D.E., Yen, D., Zipursky, S.L. (1988). Chaoptin, a cell surface glycoprotein required for Drosophila photoreceptor cell morphogenesis, contains a repeat motif found in yeast and human. *Cell* 52:291–301.

Rogers, S.L., Edson, K.J., Letourneau, P.C., McLoon, S.C. (1986). Distribution of laminin in the developing peripheral nervous system of the chick. *Dev. Biol.* 113:429–435.

Roskies, A.L., O'Leary, D.D. (1994). Control of topographic retinal axon branching by inhibitory membrane-bound molecules. *Science* 265:799–803.

Rutishauser, U. (1991). Pleiotropic biological effects of the neural cell adhesion molecule (NCAM). *Semin. Neurosci.* 3:265–270.

Rutishauser, U. (1993). Adhesion molecules of the nervous system. *Curr. Opin. Neurobiol.* 3:709–715.

Rutishauser, U., Landmesser, L. (1996). Polysialic acid in the vertebrate nervous system: a promoter of plasticity in cell-cell interactions. *Trends Neurosci.* 19:422–427.

Rutishauser, U., Acheson, A., Hall, A.K., Mann, D.M., Sunshine, J. (1988). The neural cell adhesion molecule (NCAM) as a regulator of cell-cell interactions. *Science* 240:53–57.

Saffell, J.L., Walsh, F.S., Doherty, P. (1994). Expression of NCAM containing VASE in neurons can account for a developmental loss in their neurite outgrowth response to NCAM in a cellular substratum. *J. Cell Biol.* 125:427–436.

Sanes, J.R. (1989). Extracellular matrix molecules that influence neural development. *Annu. Rev. Neurosci.* 12:491–516.

Sato, M., Lopez-Mascaraque, L., Heffner, C.D., O'Leary, D.D. (1994). Action of a diffusible target-derived chemoattractant on cortical axon branch induction and directed growth. *Neuron* 13:791–803.

Schachner, M., Taylor, J., Bartsch, U., Pesheva, P. (1994). The perplexing multifunctionality of janusin, a tenascin-related molecule. *Perspect. Dev. Neurobiol.* 2:33–41.

Seeger, M., Tear, G., Ferres-Marco, D., Goodman, C.S. (1993). Mutations affecting growth cone guidance in Drosophila: genes necessary for guidance toward or away from the midline. *Neuron* 10:409–426.

Sekido, Y., Bader, S., Latif, F., Chen, J.Y., Duh, F.M., Wei, M.H., Albanesi, J.P., Lee, C.C., Lerman, M.I., Minna, J.D. (1996). Human semaphorins A(V) and IV reside in the 3p21.3 small cell lung cancer deletion region and demonstrate distinct expression patterns. *Proc. Natl. Acad. Sci. U.S.A.* 93:4120–4125.

Serafini, T., Kennedy, T.E., Galko, M.J., Mirzayan, C., Jessell, T.M., Tessier-Lavigne, M. (1994). The netrins define a family of axon outgrowth-promoting proteins homologous to C. elegans UNC-6. *Cell* 78:409–424.

Serafini, T., Colamarino, S.A., Leonardo, E.D., Wang, H., Beddington, R., Skarnes, W.H., Tessier-Lavigne, M. (1996). Netrin-1 is required for commissural axon guidance in the developing vertebrate nervous system. *Cell* 87:1001–1014.

Shapiro, L., Kwong, P.D., Fannon, A.M., Colman, D.R., Hendrickson, W.A. (1995). Considerations on the folding topology and evolutionary origin of cadherin domains. *Proc. Natl. Acad. Sci. U.S.A.* 92:6793–6797.

Shirasaki, R., Tamada, A., Katsumata, R., Murakami, F. (1995). Guidance of cerebellofugal axons in the rat embryo: directed growth toward the floor plate and subsequent elongation along the longitudinal axis. *Neuron* 14:961–972.

Shirasaki, R., Mirzayan, C., Tessier-Lavigne, M., Murakami, F. (1996). Guidance of circumferentially growing axons by netrin-dependent and -independent floor plate chemotropism in the vertebrate brain. *Neuron* 17:1079–1088.

Silver, J. (1993). Glia-neuron interactions at the midline of the developing mammalian brain and spinal cord. *Perspect. Dev. Neurobiol.* 1:227–236.

Simon, D.K., O'Leary, D.D. (1992). Responses of retinal axons in vivo and in vitro to position-encoding molecules in the embryonic superior colliculus. *Neuron* 9:977–989.

Sink, H., Goodman, C.S. (1994). Mutations in *sidestep* lead to defects in pathfinding and synaptic specificity during the development of neuromuscular connectivity in Drosophila. *Soc. Neurosci. Abstr.* 20:1283.

Sink, H., Whitington, P.M. (1991a). Early ablation of target muscles modulates the arborisation pattern of an identified embryonic Drosophila motor axon. *Development* 113: 701–707.

Sink, H., Whitington, P.M. (1991b). Pathfinding in the central nervous system and periphery by identified embryonic Drosophila motor axons. *Development* 112:307–316.

Snow, D.M., Lemmon, V., Carrino, D.A., Caplan, A.I., Silver, J. (1990). Sulfated proteoglycans in astroglial barriers inhibit neurite outgrowth in vitro. *Exp. Neurol.* 109:111–130.

Speidel, C.C. (1941). Adjustments of nerve endings. *Harvey Lect.* 36:126–158.

Sperry, R.W. (1963). Chemoaffinity in the orderly growth of nerve fiber patterns and connections. *Proc. Natl. Acad. Sci. U.S.A.* 50:703–710.

Sretavan, D.W. (1990). Specific routing of retinal ganglion cell axons at the mammalian optic chiasm during embryonic development. *J. Neurosci.* 10:1995–2007.

Sretavan, D.W. (1993). Pathfinding at the mammalian optic chiasm. *Curr. Opin. Neurobiol.* 3:45–52.

Sretavan, D., Siegel, M., Reichardt, L. (1992). Retinal ganglion cell axons fail to form an optic chiasm following embryonic development. *Soc. Neurosci. Abstr.* 18:1274.

Stoeckli, E.T., Landmesser, L.T. (1995). Axonin-1, Nr-CAM, and Ng-CAM play different roles in the in vivo guidance of chick commissural neurons. *Neuron* 14:1165–1179.

Stoeckli, E.T., Sonderegger, P., Pollerberg, G.E., Landmesser, L.T. (1997). Interference with axonin-1 and NrCAM interactions unmasks a floor plate activity inhibitory for commissural axons. *Neuron*, in press.

Stoker, A.W., Gehrig, B., Haj, F., Bay, B.H. (1995). Axonal localisation of the CAM-like tyrosine phosphatase CRYP alpha: a signalling molecule of embryonic growth cones. *Development* 121:1833–1844.

Streuli, M., Krueger, N.X., Tsai, A.Y., Saito, H. (1989). A family of receptor-linked protein tyrosine phosphatases in humans and Drosophila. *Proc. Natl. Acad. Sci. U.S.A.* 86: 8698–8702.

Suter, D.M., Pollerberg, G.E., Buchstaller, A., Giger, R.J., Dreyer, W.J., Sonderegger, P. (1995). Binding between the neural cell adhesion molecules axonin-1 and Nr-CAM/ Bravo is involved in neuron-glia interaction. *J. Cell Biol.* 131:1067–1081.

Taira, E., Takaha, N., Taniura, H., Kim, C.H., Miki, N. (1994). Molecular cloning and functional expression of gicerin, a novel cell adhesion molecule that binds to neurite outgrowth factor. *Neuron* 12:861–872.

Takeichi, M. (1995). Morphogenetic roles of classic cadherins. *Curr. Opin. Cell Biol.* 7:619–627.

Takeshita, S., Kikuno, R., Tezuka, K., Amann, E. (1993). Osteoblast-specific factor 2: cloning of a putative bone adhesion protein with homology with the insect protein fasciclin I. *Biochem. J.* 294:271–278.

Tamada, A., Shirasaki, R., Murakami, F. (1995). Floor plate chemoattracts crossed axons and chemorepels uncrossed axons in the vertebrate brain. *Neuron* 14:1083–1093.

Tang, J., Landmesser, L., Rutishauser, U. (1992). Polysialic acid influences specific pathfinding by avian motoneurons. *Neuron* 8:1031–1044.

Tang, J., Rutishauser, U., Landmesser, L. (1994). Polysialic acid regulates growth cone behavior during sorting of motor axons in the plexus region. *Neuron* 13:405–414.

Tear, G., Seeger, M., Goodman, C.S. (1993). To cross or not to cross: a genetic analysis of guidance at the midline. *Perspect. Dev. Neurobiol.* 1:183–194.

Tear, G., Harris, R., Sutaria, S., Kilomanski, K., Goodman, C.S., Seeger, M.A. (1996). Commissureless controls growth cone guidance across the CNS midline in Drosophila and encodes a novel membrane protein. *Neuron* 16:501–514.

Tessier-Lavigne, M. (1992). Axon guidance by molecular gradients. *Curr. Opin. Neurobiol.* 2:60–65.

Tessier-Lavigne, M. (1994). Axon guidance by diffusible repellants and attractants. *Curr. Opin. Genet. Dev.* 4:596–601.

Tessier-Lavigne, M. (1995). Eph receptor tyrosine kinases, axon repulsion, and the development of topographic maps. *Cell* 82:345–348.

Tessier-Lavigne, M., Placzek, M., Lumsden, A.G., Dodd, J., Jessell, T.M. (1988). Chemotropic guidance of developing axons in the mammalian central nervous system. *Nature* 336:775–778.

Thanos, S., Bonhoeffer, F. (1986). Course corrections of deflected retinal axons on the tectum of the chick embryo. *Neurosci. Lett.* 72:31–36.

Thanos, S., Dutting, D. (1987). Outgrowth and directional specificity of fibers from embryonic retinal transplants in the chick optic tectum. *Brain Res.* 429:161–179.

Thomas, J.B., Crews, S.T., Goodman, C.S. (1988). Molecular genetics of the single-minded locus: a gene involved in the development of the Drosophila nervous system. *Cell* 52: 133–141.

Tian, S.S., Tsoulfas, P., Zinn, K. (1991). Three receptor-linked protein-tyrosine phosphatases are selectively expressed on central nervous system axons in the Drosophila embryo. *Cell* 67:675–680.

Timpl, R., Brown, J.C. (1994). The laminins. *Matrix Biol.* 14:275–281.

Tosney, K.W., Landmesser, L.T. (1985a). Specificity of early motoneuron growth cone outgrowth in the chick embryo. *J. Neuroscience* 5:2336–44.

Tosney, K.W., Landmesser, L.T. (1985b). Growth cone morphology and trajectory in the lumbosacral region of the chick embryo. *J. Neuroscience* 5:2345–58.

Tosney, K.W. (1987). Proximal tissues and patterned neurite outgrowth at the lumbosacral level of the chick embryo: deletion of the dermamyotome. *Dev. Biol.* 122:540–558.

Tosney, K.W. (1988). Proximal tissues and patterned neurite outgrowth at the lumbosacral level of the chick embryo: partial and complete deletion of the somite. *Dev. Biol.* 127: 266–286.

Tucker, A., Varela-Echevarria, A., Puschel, A.W., Guthrie, S. (1997). Motor axon subpopulations respond differentially to the chemorepellents netrin-1 and semaphorin D. *Neuron, in press.*

Vactor, D.V., Sink, H., Fambrough, D., Tsoo, R., Goodman, C.S. (1993). Genes that control neuromuscular specificity in Drosophila. *Cell* 73:1137–1153.

VanBerkum, M.F., Goodman, C.S. (1995). Targeted disruption of Ca(2+)-calmodulin signaling in Drosophila growth cones leads to stalls in axon extension and errors in axon guidance. *Neuron* 14:43–56.

Van der Geer, P., Hunter, T., Lindberg, R.A. (1994). Receptor protein-tyrosine kinases and their signal transduction pathways. *Annu. Rev. Cell Biol.* 10:251–337.

Vits, L., Van Camp, G., Coucke, P., Fransen, E., De Boulle, K., Reyniers, E., Korn, B., Poustka, A., Wilson, G., Schrander-Stumpel, C., et al. (1994). MASA syndrome is due to mutations in the neural cell adhesion gene L1CAM. *Nat. Genet.* 7:408–413.

Wadsworth, W.G., Bhatt, H., Hedgecock, E.M. (1996). Neuroglia and pioneer neurons express UNC-6 to provide global and local netrin cues for guiding migrations in C. elegans. *Neuron* 16:35–46.

Walicke, P., Cowan, W.M., Ueno, N., Baird, A., Guillemin, R. (1986). Fibroblast growth factor promotes survival of dissociated hippocampal neurons and enhances neurite extension. *Proc. Natl. Acad. Sci. U.S.A.* 83:3012–3016.

Walter, J., Henke-Fahle, S., Bonhoeffer, F. (1987a). Avoidance of posterior tectal membranes by temporal retinal axons. *Development* 101:909–913.

Walter, J., Kern-Veits, B., Huf, J., Stolze, B., Bonhoeffer, F. (1987b). Recognition of position-specific properties of tectal cell membranes by retinal axons in vitro. *Development* 101: 685–696.

Walter, J., Muller, B., Bonhoeffer, F. (1990). Axonal guidance by an avoidance mechanism. *J. Physiol.* 84:104–110.

Weiss, P. (1941). Nerve patterns: the mechanics of nerve growth. *Growth* 5(Suppl.):163–203.

Wessells, N.K., Nuttall, R.P. (1978). Normal branching, induced branching and steering of cultured parasympathetic motor neurons. *Exp. Cell Res.* 115:111–122.

Wigglesworth, V.B. (1953). The origin of sensory neurones in an insect *Rhodnius prolixis* (Hemiptera). *Q. J. Microscop. Sci.* 94:93–112.

Williams, E.J., Furness, J., Walsh, F.S., Doherty, P. (1994a). Activation of the FGF receptor underlies neurite outgrowth stimulated by L1, N-CAM, and N-cadherin. *Neuron* 13: 583–594.

Williams, E.J., Furness, J., Walsh, F.S., Doherty, P. (1994b). Characterisation of the second messenger pathway underlying neurite outgrowth stimulated by FGF. *Development* 120:1685–1693.

Williams, E.J., Mittal, B., Walsh, F.S., Doherty, P. (1995). A Ca2+/calmodulin kinase inhibitor, KN-62, inhibits neurite outgrowth stimulated by CAMs and FGF. *Mol. Cell. Neurosci.* 6:69–79.

Wilson, S.W., Ross, L.S., Parrett, T., Easter, S., Jr. (1990). The development of a simple scaffold of axon tracts in the brain of the embryonic zebrafish, Brachydanio rerio. *Development* 108:121–145.

Winslow, J.W., Moran, P., Valverde, J., Shih, A., Yuan, J.Q., Wong, S.C., Tsai, S.P., Goddard, A., Henzel, W.J., Hefti, F., et al. (1995). Cloning of AL-1, a ligand for an Eph-related tyrosine kinase receptor involved in axon bundle formation. *Neuron* 14:973–981.

Wizenmann, A., Thanos, S., von Boxberg, Y., Bonhoeffer, F. (1993). Differential reaction of crossing and non-crossing rat retinal axons on cell membrane preparations from the chiasm midline: an in vitro study. *Development* 117:725–735.

Yamada, K.M., Miyamoto, S. (1995). Integrin transmembrane signaling and cytoskeletal control. *Curr. Opin. Cell Biol.* 7:681–689.

Yang, X.H., Seow, K.T., Bahri, S.M., Oon, S.H., Chia, W. (1991). Two Drosophila receptor-like tyrosine phosphatase genes are expressed in a subset of developing axons and pioneer neurons in the embryonic CNS. *Cell* 67:661–673.

Yang, P., Major, D., Rutishauser, U. (1994). Role of charge and hydration in effects of polysialic acid on molecular interactions on and between cell membranes. *J. Biol. Chem.* 269:23039–23044.

Zhang, J.-H., Cerretti, D.P., Yu, T., Flanagan, J.G., Zhou, R. (1996). Detection of ligands in regions anatomically connected to neurons expressing the Eph receptor Bsk: potential roles in neuron-target interaction. *J. Neurosci.* 16:7182–7190.

Zinn, K. (1993). Drosophila protein tyrosine phosphatases. *Semin. Cell Biol.* 4:397–401.

Zinn, K., McAllister, L., Goodman, C.S. (1988). Sequence analysis and neuronal expression of fasciclin I in grasshopper and Drosophila. *Cell* 53:577–587.

Zondag, G.C., Koningstein, G.M., Jiang, Y.P., Sap, J., Moolenaar, W.H., Gebbink, M.F. (1995). Homophilic interactions mediated by receptor tyrosine phosphatases mu and kappa. A critical role for the novel extracellular MAM domain. *J. Biol. Chem.* 270: 14247–14250.

5

Synapse formation: a molecular perspective

JOSHUA R. SANES AND RICHARD H. SCHELLER

The most striking physical feature of the nervous system is the precision of its synaptic connections. At a cellular level, axons choose appropriate postsynaptic partners from a myriad of candidate targets, thereby forming the circuits from which behavior is generated. At a subcellular level, individual synapses are formed by and on stereotyped portions of the cell surface—some, for example, on dendritic spines and others on dendritic shafts—as required for integration of synaptic signals within individual neurons. At each individual synaptic contact, specializations of the pre- and postsynaptic elements form and mature in precise apposition, as required to sustain high-speed, high-efficiency transmission of information.

How are these precise intercellular relationships generated and maintained? In vertebrates, it is clear that autonomous behaviors of pre- and postsynaptic cells play subsidiary roles. Instead, a large amount of morphogenetically important information passes between the pre- and postsynaptic elements as the synapse forms in the embryo, matures in early postnatal life, persists throughout the life of the animal, and rearranges in response to experience. It is reasonable to suppose that this information is conveyed in the form of molecules that are externalized by one of the synaptic partners and perceived by the other. In principle, one might imagine that these molecules are of at least two types: synaptic recognition molecules that mediate choices of appropriate synaptic partners, and synaptic organizing molecules that regulate differentiation and maturation of the synapse. At present, relatively little is known about the recognition processes that account for selective synapse formation. However, the past several years have witnessed impressive progress in the identification and characterization of candidate organizing molecules at one particular synapse, the skeletal neuromuscular junction. Studies on the structure, function, and development of this relatively simple synapse have now enabled several groups to ask, and begin to answer, mechanistic questions about how the developing synaptic partners signal to each other and how those signals act.

Accordingly, our main aim in this chapter is to review recent molecular analyses of neuromuscular synaptogenesis. To a limited extent, we have also considered the extent to which lessons derived from the neuromuscular junction can be applied to interneuronal synapses in the brain. Because of space limitations, however, we have had to focus on vertebrate synapses and to omit discussion of synapse elimination, a critical late step in the sequence of processes that leads to the adult pattern of connectivity. In addition, we summarize all too briefly the extensive histological and electrophysiological studies that have provided the foundation and context for the molecular studies. For a fuller account of work in these areas, the reader is referred to reviews by Hall and Sanes (1993), Grinnell (1995), Keshishian et al. (1996), and Nguyen and Lichtman (1996). In addition, synapse elimination and reorganization are discussed in the context of the visual system in Chapter 15 of this volume.

MOLECULAR ARCHITECTURE OF THE NERVE TERMINAL

The presynaptic nerve terminal is highly specialized to meet the needs of chemical synaptic transmission (Fig. 5–1). Major morphological features of the terminal include mitochondria, which are necessary for the energy requirements of this metabolically active part of the nerve cell. Most nerve terminals also contain a membrane organelle which is thought to function as a presynaptic endosome in support of the dynamic membrane cycling that accompanies synaptic transmission. The most prominent features of the presynaptic terminal, though, are the synaptic vesicles that contain chemical messengers secreted by neurons. These are of two types: small, clear, and large, dense-core vesicles. In general, the small vesicles contain fast-acting transmitters, such as γ-aminobutyric acid (GABA), glutamate, or acetylcholine, whereas the large vesicles contain peptides or amines which act on metabotropic receptors to elicit activities with slower onsets and longer durations.

The small, clear synaptic vesicles have emerged as the best-characterized organelles at the molecular level in all of biology (Scheller, 1995; Südhof, 1995). Their homogeneous size and density, along with their abundance in brain tissue and the electric organ from marine rays, have made it possible to purify these relatively simple organelles in quantities sufficient for biochemical analyses. Typical mammalian synaptic vesicles have a protein-to-lipid ratio of about 1 : 1 (Floor et al., 1988). There are approximately 100 protein molecules per synaptic vesicle, most of which belong to one of about ten gene families (Fig. 5–2). These synaptic vesicle proteins can be divided into two very general classes: those involved in regulating transmitter content and others involved in the membrane trafficking of organelles within the nerve terminal. A brief description of the molecules in each of these classes follows.

Neurotransmitter Transport

There are four distinct vesicular transport systems; one each for acetylcholine, biogenic amines, and glutamate and a fourth that packages both GABA and glycine (McMahon and Nicholls, 1991). All of these systems require the electro-

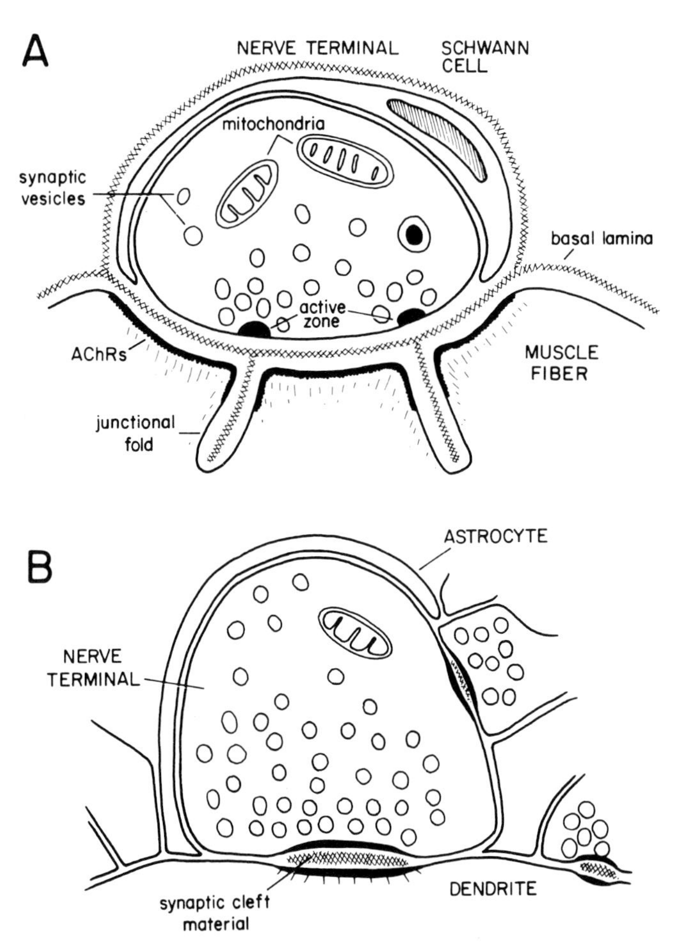

FIGURE 5–1 The structure of the synapse, as sketched from electron micrographs. (A) The skeletal neuromuscular junction. (B) An interneuronal synapse in the brain. At both neuromuscular and interneuronal synapses, a vesicle-laden nerve terminal is capped by processes of a glial cell and faces a specialized postsynaptic membrane across a synaptic cleft. Both pre- and postsynaptic membranes are thickened, reflecting an abundance of specialized membrane-bound and cytoskeletal elements. The glial cell is an astrocyte at interneuronal synapses and a Schwann cell at the neuromuscular junction. The synaptic cleft material is amorphous at interneuronal synapses but associated with a discrete basal lamina at the neuromuscular junction. Section A is modified from Hall and Sanes (1993).

chemical gradient generated by the vesicular proton pump. The proton pump is an approximately 500-kD complex comprised of at least nine integral and peripheral membrane subunits (Nelson, 1992). Approximately six proton-pump complexes are present per synaptic vesicle (Floor et al., 1990). Hydrolysis of adenosine triphosphate (ATP) is coupled to translocation of a proton from the cytoplasm into the lumen of the vesicle. Acidification of the vesicle interior generates both pH and potential gradients which are differentially utilized by the four transmitter systems. Uptake of glutamate, which is negatively charged, is driven by the potential gradient; transport of positively charged amines and

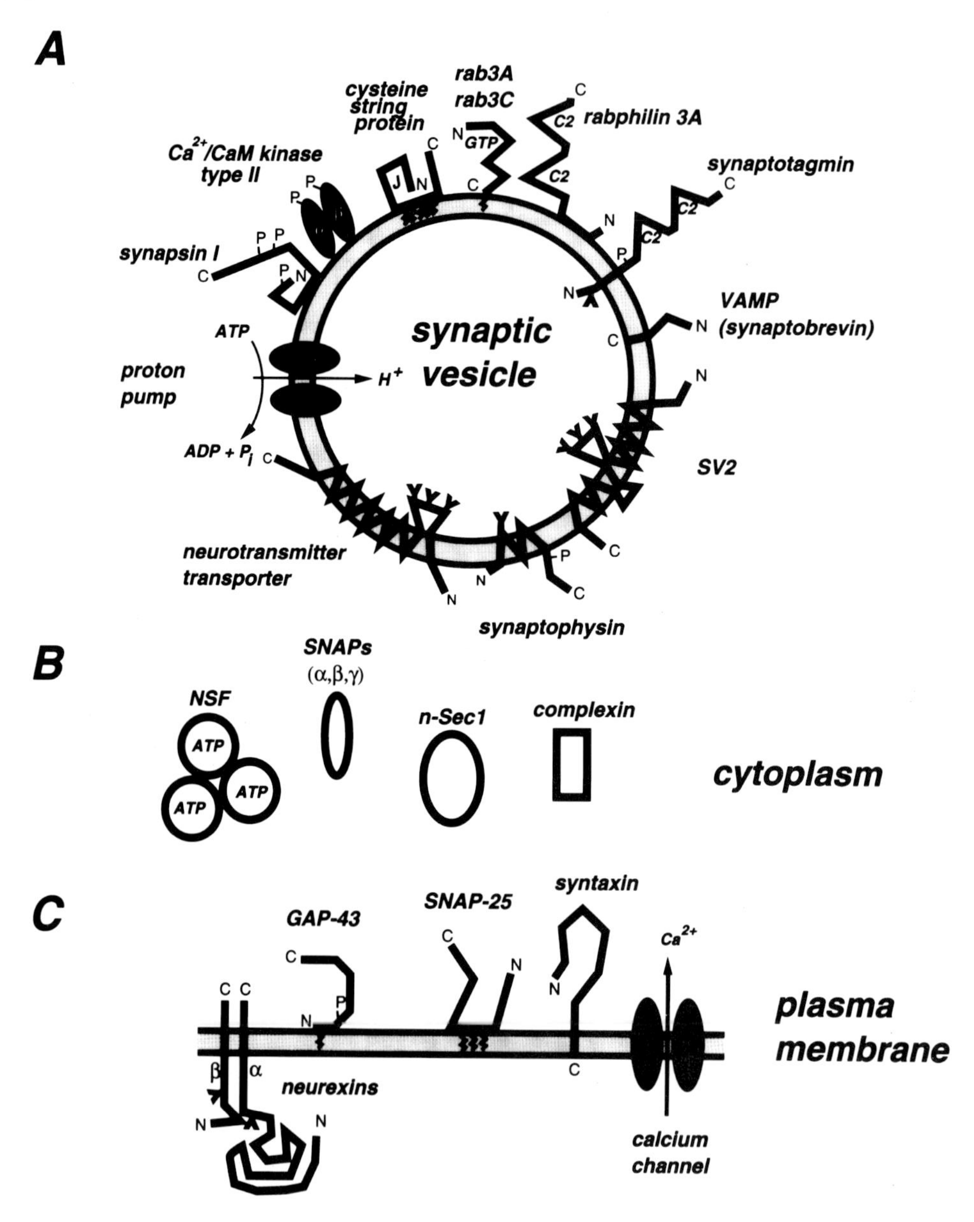

FIGURE 5–2 Components of the secretory apparatus of nerve terminals. (A) Vesicle-associated proteins include integral membrane, lipid-modified proteins, and membrane-associated proteins. (B) Cytoplasmic proteins which associate with vesicle and plasma membrane proteins. (C) Proteins concentrated in the plasma membrane that are critical for membrane fusion or the regulation of neurotransmitter release. See text for detailed dis-

acetylcholine uses energy derived from the pH gradient; and GABA and glycine transport requires both gradients to function optimally.

Genes encoding amine and acetylcholine transporters have been cloned. The amine transporter was initially isolated based on its ability to sequester the toxic compound N-methyl-4-phenylpyridinium (MPP-) into vesicles (Liu et al.,

1992a,b). A putative ACh transporter was isolated using genetic strategies in *Caenorhabditis elegans* (Alfonso et al., 1993). Both transporter proteins are predicted to span the lipid bilayer 12 times and share a low level of amino acid sequence identity with bacterial drug-resistance transporters. This sequence similarity may reflect the fact that bacteria, like vesicles, utilize a proton gradient to drive the movement of compounds across their plasma membranes.

Another family of vesicle proteins, the SV2s, are also predicted to be involved in transport based on their 12-membrane-spanning domain topology and their homology to bacterial drug-resistance and sugar transporters (Bajjalieh et al., 1992, 1993; Feany et al., 1992). The distribution of neither of the SV2s matches that of known transmitters, suggesting that if there is a transported substrate, it is not a known neurotransmitter (Bajjalieh et al., 1994). Ions or molecules that might be transported by the SV2s include Ca^{2+} and ATP. In addition, it remains a distinct possibility that SV2s are not transporters and that their topological and sequence similarity to this class of proteins does not reflect a functional similarity.

Synaptic Vesicle Trafficking

Many of the remaining proteins associated with synaptic vesicles mediate neurotransmitter release by regulating vesicle trafficking and membrane fusion in the presynaptic nerve terminal. These proteins, along with a series of soluble and plasma membrane factors (Fig. 5–2), will first be discussed in the context of current hypotheses regarding their functions in transmitter release. Synaptic vesicles in the nerve terminal occur in two morphologically distinct states. Some appear docked, that is, directly opposed to the plasma membrane, while others are clustered in the vicinity of the active zone (Kelly, 1993). This organization provides a means of rapidly replenishing the releasable pool when the neuron fires at high frequency.

How are synaptic vesicles localized to these specific regions of the cytoplasm, and what regulates their movement to the plasma membrane where they enter the releasable pool? While microtubules are rich in the axon, they terminate at its distalmost preterminal aspect, penetrating only slightly into the terminal. Morphological studies reveal that the plasma membrane of the nerve terminal is surrounded by a filamentous actin network which extends into the synaptoplasm (Gotow et al., 1991). The vesicles may be tethered to the actin filaments by the synapsins, a family of neuron-specific synaptic-vesicle–associated proteins which have been shown to bind actin and synaptic vesicles in a phosphorylation-dependent fashion (Greengard et al., 1993). Physiological (Pieribone et al., 1995), genetic (Rosahl et al., 1995; Li et al., 1995), and biochemical (Benfenati et al., 1993) studies support the hypothesis that one of the functions of the synapsins is to regulate vesicle recruitment and docking at the active zone. This regulation is proposed to occur through phosphorylation of synapsin by calmodulin-dependent protein kinase II (CaM kinase II), which decreases the affinity of synapsin for synaptic vesicles. In contrast to the clear hypothesis as to how cytoplasmic vesicle clusters form, little is known about how the clusters are linked to the active zone.

The biochemical machinery responsible for docking and fusion of synaptic vesicles is now being characterized (Scheller, 1995; Südhof, 1995). A fundamental observation is that the biochemical machinery which mediates neurotransmitter secretion in the mammalian nervous system is similar to the set of proteins that regulates vesicle trafficking from the Golgi complex to the plasma membrane in all organisms from yeast to man (Bennett and Scheller, 1993). Thus, synaptic transmission is a specially regulated form of the vesicle trafficking found in all cells.

The protein syntaxin (t-SNARE), a target-soluble N-ethyl maleinide stimulatory factor (NSF) attachment protein receptor, is a receptor component for vesicles on the plasma membrane (Bennett et al., 1992). Prior to vesicle docking, syntaxin is proposed to be associated with a soluble factor called nsec1 (Pevsner et al., 1994a,b) or munc-18 (Hata et al., 1993). As the vesicle docks, a heterotetrameric complex that migrates at 7S on glycerol gradients is formed. This 7S complex is comprised of two vesicle proteins—VAMP (a vesicle-soluble NSF attachment protein receptor) and v-SNARE (also called synaptobrevin)—and synaptotagmin, plus two plasma membrane proteins, syntaxin and SNAP-25 (Söllner et al., 1993a; McMahon and Südhof, 1995). As this complex forms, nsec1 likely dissociates from syntaxin and may play a chaperone or regulatory role in the docking process (Pevsner et al., 1994b). After the 7S particle forms, a soluble factor, αSNAP, binds to it, concurrent with the dissociation of synaptotagmin (Söllner et al., 1993b). Following ATP hydrolysis, the complex dissociates, leading to a series of intermediates which are hypothesized to precede the fusion of vesicle and plasmalemma membranes. While components of the 7S particle are thought to be specific for fusion events with the plasma membrane, NSF and αSNAP function at many stages within the secretory pathway. Thus, the function of NSF may be to rearrange the VAMP, syntaxin, and SNAP-25 molecules to favor fusion of the opposing membranes.

An additional set of vesicle-associated proteins likely to play key roles in vesicle trafficking are the low-molecular-weight guanosine triphosphatases (GTPases) or Rab proteins. Over 30 members of the Rab family have been characterized in mammalian species and many of these are specifically associated with discrete compartments in the secretory pathway (Pfeffer, 1992). In yeast, mutations in a low-molecular-weight GTPase, sec4p, block the fusion of Golgi-derived vesicles with the plasma membrane (Ferro-Novick and Novick, 1993). Similarly, overexpression or microinjection of mutant Rabs into PC12 (Johannes et al., 1994) or chromaffin (Holz et al., 1994) cells severely perturbs Ca^{2+}-dependent secretion. Given these results, it was somewhat surprising to find that the only observable phenotype of the Rab3a knockout mouse is that its synapses fatigue more rapidly than normal, suggesting that the protein may modulate the rate at which vesicles become available for release (Geppert et al., 1994a).

The mechanism of Rab function is not clearly understood. GTP-bound Rab3A binds rabphilin-3A, and the level of rabphilin-3A in nerve terminals is reduced in the Rab3A knockout mice. Rabphilin-3A, like synaptotagmin, contains two C2 domains, suggesting a link to Ca^{2+}-dependent phospholipid binding (Shirataki et al., 1994). Further work is necessary to clarify the biochemistry and physiology of the Rab system in the nerve terminal. In particular, the link between the GTPase cycle and the NSF ATPase cycle remains to be clearly defined.

Synaptic transmission has a series of regulatory requirements that distinguish it from constitutive secretion. Most notably, transmitter release is coupled to the action potential through voltage-dependent ion channels which allow the influx of extracellular Ca^{2+}. Freeze-fracture studies reveal a series of particles at the neuromuscular junction which are organized into parallel linear arrays along the active zone (Pumplin et al., 1981). These particles are positioned very close to the exocytotic pits and are thought to be the Ca^{2+} channels. Pharmacological and molecular biological studies have defined a family of structurally related Ca^{2+} channels that are expressed in diverse tissues (Birnbaumer et al., 1994). The channels most likely to be responsible for transmitter release are the N and Q types, which contain the α_{1B} and α_{1A} subunits, respectively, as their major ion-conducting subunits (Wheeler et al., 1994). The distance from the vesicle to the channel source of Ca^{2+} influx is very small, leading to the proposal that components of the vesicle secretion machinery may be physically associated with the channel. In support of this hypothesis, immunoprecipitation and in vitro binding studies suggest that a fraction of the syntaxin on the plasma membrane is bound to Ca^{2+} channels (Bennett et al., 1992). In addition, expression studies in frog oocytes suggest that syntaxin binds and stabilizes the inactivated state of the N- and Q- but not of the L-type channels (Bezprozvanny et al., 1995). Furthermore, a synaptic vesicle protein, the cysteine string protein (CSP), modulates Ca^{2+} currents. The CSP is characterized by domains similar to molecular chaperones involved in the folding of proteins or the regulation of macromolecular complexes (Gundersen and Umbach, 1992). Thus, the secretory machinery of synaptic vesicles may be in communication with the plasma membrane channels and a docking event may signal the Ca^{2+} channels that the vesicle is ready for release.

Definition of the precise mechanism of Ca^{2+}-triggered exocytosis awaits a better understanding of the molecular mechanism of membrane fusion. The best candidate Ca^{2+} sensors are a subset of the family of synaptotagmins, vesicle integral membrane proteins that contain two C2 domains (Geppert et al., 1994b). C2 domains were first defined in isoforms of protein kinase C which are regulated by Ca^{2+} and phospholipid. Genetic mutant studies of synaptotagmin in *C. elegans* (Nonet et al., 1993), *Drosophila* (DiAntonio et al., 1993b), and mice (Geppert et al., 1994b) all show a reduction, but not elimination, of the Ca^{2+}-triggered secretion of neurotransmitter. In mice, the analysis was sufficiently precise to demonstrate that a fast component of secretion is dramatically reduced in magnitude, while a slow component is essentially unchanged. These studies establish synaptotagmin as a critical component of the transmitter release process and support its role as a Ca^{2+} sensor.

A single round of exo/endocytosis can occur in less than 1 minute, so all of the molecular events of vesicle cycling must be very efficient (Ryan et al., 1993). While endocytosis is less well understood than exocytosis, specific molecular models have recently been proposed (David et al., 1996). The proteins required to assemble the clathrin coat on the endocytic vesicle are components of the AP2 or plasma membrane clathrin adaptor complex. The AP2 complex has been proposed to bind the cytoplasmic tail of synaptotagmin and may interact with other vesicle proteins as well. The bound AP2 then acts as a template for the formation of the clathrin lattice. AP2 may also recruit the GTPase dynamin, which forms a ring around the stalk of the budding vesicle. Dynamin may actually physically constrict or choke the neck of the endocytic membrane, resulting in the pinching off of a vesicle. Proteins such as amphiphysin and synaptojanin are also likely to be involved in the endocytosis process (McPherson et al., 1996), but their precise roles are yet not well understood.

ASSEMBLY OF THE NERVE TERMINAL

Once a growing axon contacts its target, a dramatic reorganization of the morphology and behavior of the growth cone ensues, resulting in differentiation of the presynaptic nerve terminal. At the neuromuscular junction, this process takes up to 3 weeks (in rodents) and involves the bidirectional exchange of signals between the pre- and postsynaptic elements. The initial contact is a close and unspecialized apposition with no intervening extracellular matrix, making possible the direct interaction of plasma membrane proteins. As the terminal matures, vesicles accumulate and an electron-dense membrane at the region of the active zone appears, presumably due to an increase in the concentration of membrane proteins at this site. Additionally, the synaptic cleft widens towards its mature, 60-nm, dimension and basal lamina is deposited within it. As differentiation proceeds, the cytoskeletal organization characteristic of growth cones evolves into that of the mature terminal, and more vesicles accumulate. Physiological signs of maturation include increases in the frequency of spontaneous end-plate potentials and the size of evoked responses. Multiple growth cones follow the lead of the pioneering axon, eventually resulting in polyinnervation. Supernumerary axons are eliminated prenatally, leaving each muscle fiber with a single input (reviewed in Hall and Sanes, 1993).

Relatively little is known about the molecular mechanisms that underlie differentiation of the nerve terminal. Certainly regulation of the expression patterns of many synaptic genes is critical. An interesting regulatory mechanism is found for the proteins that control ACh synthesis and packaging into vesicles. A portion of the enzyme choline acetyltransferase (ChAT), which synthesizes acetylcholine, is associated with the vesicle membrane. This raises the question of how the expression and activity of ChAT and the vesicular acetylcholine transporter (VAChT), which are both localized to vesicles and required for cholinergic function, are regulated. Insight into this question is found in the linkage of the ChAT and VAChT genes. Interestingly, the VChAT gene is within an intron of the ChAT gene. Moreover, both mRNAs make use of a single 5′ untranslated sequence which is spliced onto the exons encoding either VAChT or ChAT (Erickson et al., 1994; Benjamin et al., 1994). This genomic organization is conserved from *C. elegans* to mammalian species and implies a mechanism for coordinating expression of these two genes at the levels of transcription and

RNA splicing. Other mechanisms for more rapid regulation of transport activity are likely to involve direct phosphorylation of the transporters, but these processes are as yet poorly understood.

Of particular interest are the developmental changes that switch the membrane-vesicle–trafficking pathways from those of the growth cone, where new membrane is added to the axon, to those of the mature nerve terminal, where vesicles recycle locally as neurotransmitter is released. It is likely that some membrane-trafficking proteins are specific for either axon elongation or synaptic transmission, while others function in both processes. Because functional transmission can occur soon after contact of the growth cone with the muscle, expression of synaptic vesicle proteins is likely to occur prior to synaptogenesis. In fact, membrane patches or myocytes on patch electrodes have been used to detect low levels of transmitter secretion from growth cones even prior to contact with their targets (Hume et al., 1983; Young and Poo, 1983a). In vivo, the interval between nerve–muscle contact and transmission is 1–2 hours (Chow and Cohen, 1983), while in vitro transmission can occur a few minutes after contact of the nerve and muscle (Kidokoro and Yeh, 1982). Thus, while we do not yet understand how similar the vesicles which release transmitter from growth cones are to those of mature synaptic vesicles, it is likely that these two organelles partially overlap in their molecular composition and membrane-trafficking mechanisms.

The mechanisms which underlie the biogenesis of transport vesicles during axon outgrowth and the relationships of these vesicles to those found in mature synapses are beginning to be investigated using specific molecular markers. Numerous studies indicate that synaptic vesicle proteins are not simply assembled into mature and functional vesicles in the cell body of the neuron as the proteins exit the trans-Golgi network. In PC12 cells, biogenesis of the synaptic-like vesicles has been extensively studied using synaptophysin as a marker (Régnier-Vigouroux et al., 1991). Synaptophysin appears to undergo multiple rounds of exocytic and endocytic cycling, traveling through an early endosome compartment prior to its appearance in the synaptic vesicle fraction (Clift-O'Grady et al., 1990; Johnston et al., 1989). A similar process is likely to occur in neurons as exo/endocytic cycling of vesicle proteins occurs in hippocampal neurons in culture, both along the axon and at the growth cone (Matteoli et al., 1992). These data suggest that as particular synaptic vesicle proteins are being transported to the nerve terminal, perhaps in many different types of vesicles, they are fusing with the plasma membrane and recycling. Perhaps the vesicle proteins are only assembled into mature organelles upon arrival at their final destination in the synapse. Thus, the release of transmitter from the growing axons may be from immature vesicles, each carrying components which will eventually reside together in the mature synaptic vesicles.

Blocking studies have provided evidence that SNAP-25 and synapsin, which are components of the exocytotic machinery, also play roles in the formation of the nerve terminal. Antisense oligonucleotide inhibition studies suggest that SNAP-25 is required for axon outgrowth in rat cortical neurons and PC-12 cells in vitro and in amacrine cells of the retina in vivo (Osend-Sand et al., 1993). Preliminary studies show that syntaxin is expressed early in development as well, suggesting that the receptor for vesicles on the growing axon may be the same or similar to that of the active zone in the adult. In a separate series of functional studies, antisense oligonucleotides to the synapsins inhibited synapse formation in hippocampal cultures (Ferreira et al., 1995). In addition, overexpression of synapsin 1b leads to an increase in the number of varicosities, as well as in the number of vesicles within each varicosity (Han et al., 1991). Synaptic currents at contacts between neurons overexpressing synapsin 1b are larger than currents at control sites (Lu et al., 1992). While these data suggest a role for the synapsins in synaptic development, more recent investigations of knockout mice have led to conflicting conclusions regarding this hypothesis (Rosahl et al., 1995; Chin et al., 1995).

It is important to document the expression patterns of the synaptic components discussed above as their differential regulation may influence membrane-trafficking pathways in growth cones and mature synaptic terminals. Several investigations of the developmental expression patterns of the genes encoding presynaptic proteins have been conducted in a variety of organisms, most notably *Drosophila* (DiAntonio et al., 1993a) and rat (Lou and Bixby, 1993). In *Drosophila*, expression of synaptotagmin, Rab 3 and VAMP begins between 9 and 12 hours of development (DiAntonio et al., 1993a). The initial expression of these genes coincides approximately with the maturation of the nervous system and the onset of synaptogenesis. In mammalian systems, the developmental expression patterns of SV2 (Bajjalieh et al., 1994), synaptotagmin (Bixby and Reichardt, 1985), synapsin (Haas and DeGennaro, 1988), synaptophysin (Knaus et al., 1986), nsec1 (Pevsner et al., 1994a), and SNAP-25 (Oyler et al., 1991) have been studied using Northern blotting, Western blotting, and/or in situ hybridization. These and other studies of the genes encoding the synaptic membrane-trafficking proteins demonstrate that the expression levels parallel the appearance of mature synapses, reaching adult levels between 7 and 30 days after birth in the rat.

In contrast, the onset of expression of some vesicle proteins occurs prior to synapse formation, and some vesicle protein mRNAs are present even prior to neurite extension (Marazzi and Buckley, 1993). For example, SV2B is expressed in all layers of the external germinal layer of the cerebellum at postnatal day 15 in rats (Bajjalieh et al., 1994). Cells in this region have not yet formed synapses; many are extending parallel fibers and beginning to migrate to their final position in the granule cell layer, while others are still proliferating.

One way of regulating the membrane-trafficking patterns of neurons during development might be to change the isoforms of the synaptic-membrane–trafficking proteins that are expressed. As noted above, many vesicle proteins are encoded by gene families or are alternatively spliced, and recent studies have provided evidence for developmentally regulated isoform switching. For example, SV2A and SV2B are expressed in the dentate gyrus of the hippocampus in postnatal day 16 rats, while only SV2A is expressed in these cells in the adult (Bajjalieh et al., 1994). Additionally synaptotagmin and synaptophysin isoforms switch in relative abundance during late stages of embryogenesis, coincident with the maturation of synapses in the ciliary ganglion (Lou and Bixby, 1995). Finally, two alternatively spliced isoforms of SNAP-25 have different patterns of palmitylated cysteine residues and switch in their relative abundance during embryonic development (Oyler et al., 1991).

While further studies are required to understand the developmental significance of the synaptic vesicle protein isoform expression patterns, several functions are possible. First, vesicle cycling in the growth cone appears to be relatively independent of Ca^{2+}, suggesting that the switch between constitutive secretion and transmitter release in the mature nerve terminal is likely to involve modification of Ca^{2+} regulatory processes (Taylor and Gordon-Weeks, 1989). Second, secretion from the mature nerve terminal is largely restricted to active zones and, therefore, the requirements for spatial localization of membrane fusion are much greater in the mature synapse. Third, the secretion process itself may need to be more robust in adult brain, requiring more efficient isoforms of the docking and fusion machinery. Finally, isoform switchings could contribute to a general transformation in the vesicle-trafficking pattern, from the addition of new membrane to a growth cone to local recycling in the adult synapse.

Signals that Regulate Presynaptic Differentiation

Several signaling mechanisms have been suggested to mediate presynaptic nerve terminal differentiation. Candidate signaling molecules include integral membrane proteins, secreted diffusible factors, and components of the extracellular matrix. An attractive possibility is that these signaling molecules act sequentially with the growth cone being influenced first by diffusible factors, then by surface interactions with membrane proteins, and finally by matrix components. Alternatively, particular aspects of synapse development might be influenced by individual signaling components which sum to produce the final functioning terminal.

As mentioned above, secreted factors may play an important role in synapse formation. While the specific roles of particular factors are not clearly defined, insulin-like growth factor–2 (IGF-2) (Caroni and Grandes, 1990; Caroni et al., 1994), basic fibroblast growth factor (FGF) (Gurney et al., 1992), FGF5 (Hughes et al, 1993), leukemia inhibitory factor (LIF) (Martinou et al., 1992; Kwon et al., 1995), ciliary neurotrophic factor (CNTF) (Oppenheim et al., 1991), and glia-derived neurotrophic factor (GDNF) (Oppenheim et al, 1995) exert effects on motor neurons in a variety of assays (Houenou et al., 1994). Several of these factors induce motor neuron sprouting and at least one, basic FGF, can induce both nerve terminal and postsynaptic differentiation Peng et al., 1995; Dai and Peng, 1995). Thus, these or related secreted factors may play roles in regulating the sprouting of nerve terminals that occurs during growth or following denervation. However, mutant mice lacking these genes are generally viable (e.g., LIF, CNTF, FGF5), have normal numbers of motoneurons (Ip and Yancopoulos, 1996), and undergo synapse elimination roughly on schedule (Kwon et al., 1995). It remains to be determined whether double mutants will have more severe phenotypes and/or whether detailed analyses will reveal subtle defects in neuromuscular development.

Upon initial contact of the nerve and muscle, direct protein–protein interactions may mediate signaling. At this stage of development the membranes are directly opposed; extracellular matrix accumulates in the synaptic cleft only later. Neural cell adhesion molecules, such as N-CAM or N-cadherin, are candidates for mediating these early adhesion events (Bixby et al., 1987). Again, however, the knockout phenotype does not support a critical role for at least one of these molecules, N-CAM, in synaptic development (Moscoso, 1996).

An additional series of candidate adhesion proteins are the neurexins (Ushkaryov et al., 1992). The first neurexin was isolated as a cell-surface protein which binds to α-latrotoxin, a component of black widow spider toxin that promotes massive secretion from nerve terminals. Molecular genetic studies then revealed a family of neurexin proteins which have a small cytoplasmic domain, a membrane-spanning region, and a large extracellular domain which contains several laminin-like domains and an O-linked sugar domain. Interestingly, the three known neurexin genes are transcribed into RNAs which undergo extensive alternative splicing. The splicing has been predicted to give rise to over 1,000 distinct proteins, suggesting a large potential for the generation of specificity of synaptic contacts (Ullrich et al., 1995). While the neurexins are expressed differentially in the developing nervous system, critical tests of their functions in synaptic transmission and development remain to be performed.

Recently, genetic studies of *Drosophila* have revealed a gene, late bloomer (lbl) (Kopczynski et al., 1996), which is a member of the "tetraspanin" family of surface proteins, so named because they are predicted to bear four membrane-spanning segments. LBL protein is expressed transiently on motoneuron axons, growth cones, and terminal arbors. In mutant embryos, synapse formation is delayed in particular cells while nearby axons show an increased level of sprouting. LBL is most homologous to the tetraspanins, suggesting it may be a component of an adhesion complex which regulates synapse formation upon contact with target tissues. Several members of the tetraspanin family are subunits of receptor complexes in the immune system, where they may function to deliver coactivation signals (Imai et al., 1995). Tetraspanins may also be important in mammalian neuronal development where CD9 is dynamically expressed in both neurons and glia (Kaprielian et al., 1995). Additionally the four-membrane-spanning-domain structure of LBL is reminiscent of the gap-junction subunits, connexins, suggesting the protein may form a pore for signaling ions or molecules such as Ca^{2+}. In fact, gap junctions have been observed between nerve and muscle in culture (Allen and Warner, 1991), and it is possible that passage of Ca^{2+} or electrical signals through such channels regulates early steps in synapse formation.

Finally, it is clear that many proteins in the extracellular matrix are recognized by the presynaptic nerve terminal. The first experimental evidence in support of this idea came from analysis of the century-old observation that regenerating axons innervate denervated muscle fibers at original synaptic sites, even though such sites occupy only a minute fraction of the total muscle area. Sanes et al. (1978) showed that motor axons preferentially synapsed at original sites on basal lamina ghosts that persist following damage of nerve and muscle; moreover, the axons differentiated into electrically responsive; functional nerve terminals at such sites (Glicksman and Sanes, 1983). Thus, matrix appears to contain sufficient information to direct the regeneration of presynaptic active zones, including the organization of synaptic vesicle clusters.

Few components of the matrix have been analyzed in sufficient detail to allow comment on their ability to interact with motoneuron growth cones. An exception is laminin β2 (originally called s-laminin), a subunit of laminin related to the β1 chain (Hunter et al., 1989a). While conventional laminin (now called laminin-1) is adhesive for most neuronal types, laminin β2 is selectively adhesive for motoneurons. The motoneuron adhesive site has been mapped to a tripeptide sequence, leucine-arginine-glutamate (LRE) (Hunter et al., 1989b). Subsequent studies showed that recombinant laminin β2 fragments function to stop laminin-promoted outgrowth of motor axons (Porter et al., 1995). The stopping activity, like the adhesive activity, is attributable to the LRE sequence. Moreover, laminin β2 can initiate the differentiation of the arrested axonal segments into nerve terminals (Patton et al., 1995). These studies suggest that laminin β2 may be an organizer of presynaptic differentiation. Consistent with this notion, deletion of the laminin β2 gene in mice resulted in aberrant differentiation of nerve terminals at the neuromuscular junction (Noakes et al., 1995). In addition, Schwann cells extended into the synaptic cleft in mutants, raising the possibility that a further role of laminin β2 may be to prevent glial cell processes from invading the region of the synaptic cleft and limiting the domain available for differentiation of the active zone (Patton and Sanes, 1992).

A recent study has suggested that agrin, which is known to affect postsynaptic differentiation (see below), may also have presynaptic effects (Campagna et al., 1995). Ciliary ganglion neurons, but not sensory neurons, selectively adhere to fibroblast cells that had been transfected with an agrin expression vector. Growth of neurites was inhibited upon contact with the agrin-expressing cells and a synaptic vesicle protein marker became concentrated at contact sites, suggesting the formation of vesicle clusters. Interestingly, avian, mammalian and *Torpedo* agrins all contain the LRE sequences initially characterized in laminin β2; it remains to be determined whether this sequence mediates the effects of agrin on neurons.

Molecular Architecture of the Postsynaptic Apparatus

Most chemical synapses have a specialized postsynaptic membrane that is rich in neurotransmitter receptors, associated with a cytoplasmic "postsynaptic density," and surmounted by extracellular synaptic cleft material. The molecular architecture and development of this postsynaptic apparatus has been best studied at the vertebrate neuromuscular junction. We will, therefore, focus the following sections on this "model" peripheral synapse, returning at the end to consider central synapses in light of what we have learned at the neuromuscular junction.

The most prominent molecular component of the postsynaptic membrane at the neuromuscular junction is the nicotinic acetylcholine receptor (AChR). The AChR is a ligand-gated ion channel which transduces the chemical signal provided by the neurotransmitter (acetylcholine) into an electrical signal (depolarization) that in turn triggers the muscle action potential. In adults, the AChR is a pentamer of subunits with the composition $\alpha_2\beta\delta\varepsilon$ (reviewed in Duclert and Changeux, 1995). All subunits are homologous and form a symmetrical array around a central pore that opens when acetylcholine binds to the α subunit. AChRs are packed at a density of 10,000–20,000/μm^2 in the postsynaptic membrane, but there are fewer than 10 AChRs/μm^2 in extrasynaptic membrane. More remarkable is that the density falls more than tenfold within a few microns of the terminal's edge (Salpeter et al., 1988). The massive synaptic concentration of AChRs implies that the muscle has remarkable mechanisms for anchoring receptors in the postsynaptic membrane. Consistent with this notion, the cytoskeletal framework that underlies this membrane is highly specialized and differs from that beneath extrasynaptic membrane (Hall and Sanes, 1993). The precise apposition of AChR-rich membrane with nerve terminals implies that the nerve plays a critical role in organizing the postsynaptic specialization. As discussed below, the last few years have brought great progress in understanding the molecular basis of the neural influence and the muscle's response to it.

Initial calculations based on the dense packing of AChRs led to the speculation that there would be room for little else in the postsynaptic membrane. Over the past several years, however, at least ten other proteins have been identified that are present, and concentrated to varying degrees, in the postsynaptic membrane (Fig. 5–3B). These include voltage-sensitive sodium channels, the neural cell adhesion molecule (N-CAM), two integrin heterodimers (α7Aβ1 and α7Bβ1), three EGF-receptor–like tyrosine kinases (erbB2–4), a muscle-specific tyrosine kinase (MuSK), and a set of dystrophin-associated glycoproteins (α- and β-dystroglycan and α-, β-, and γ-sarcoglycan) (Covault et al., 1986; Valenzuela et al., 1995; Martin et al., 1996; Moscoso et al., 1995b; Zhu et al., 1995; Altiok et al., 1995; Ohlendieck et al., 1991). The sodium channels are likely present at high concentration in the synaptic region to facilitate initiation of the action potential following a synaptic potential. The other proteins, in contrast, are not obviously involved in neurotransmission per se. Instead, they presumably play roles in the formation and maintenance of the synapse. For example, N-CAM has been implicated in formation of junctional folds (Moscoso, 1996), while dystroglycan and the tyrosine kinases are putative receptors for nerve-derived signals described below.

The synaptic cleft that separates the pre- and postsynaptic membranes at the neuromuscular junction is traversed by the basal lamina that coats the entire muscle fiber. Synaptic and extrasynaptic portions of the basal lamina are ultrastructurally similar but molecularly distinct (Sanes, 1995; Fig. 5–3C). For example, the enzyme acetylcholinesterase, which hydrolyzes the neurotransmitter to terminate its action, is selectively associated with synaptic basal lamina. The association is likely mediated by a collagenous subunit which associates with the

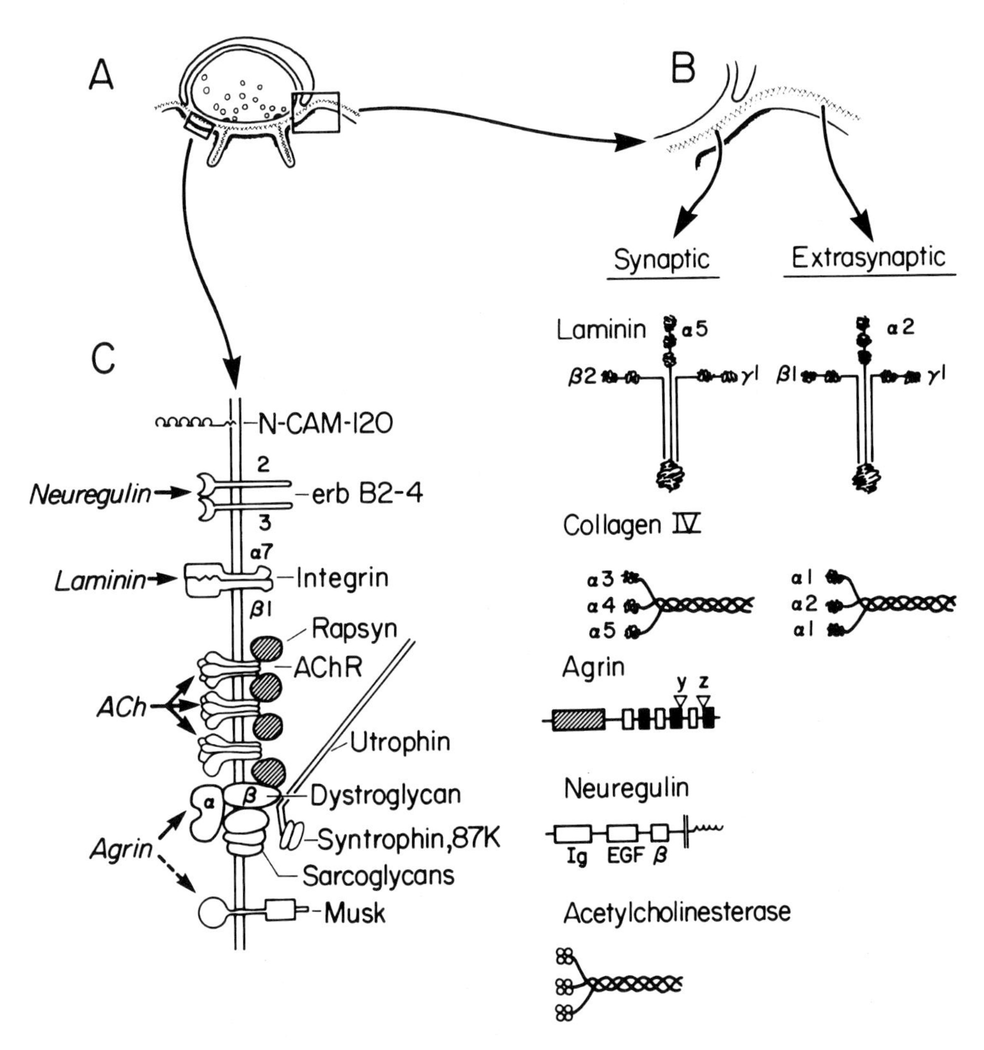

FIGURE 5–3 Components of the postsynaptic apparatus at the neuromuscular junction. (A) Sketch of the neuromuscular junction, redrawn from Figure 5–1, indicating positions of higher-magnification views in parts B and C. (B) Both synaptic and extrasynaptic basal lamina contain laminin and collagen IV heterotrimers, but their isoform composition differs. Synaptic basal lamina also contains collagen-tailed acetylcholinesterase, agrin, and ARIA/neuregulin. The A_{12} form of acetylcholinesterase is shown, in which 12 catalytic subunits are associated with a short triple-helical collagen. For agrin, positions of alternative splicing that affect bioactivity are indicated as y and z; they correspond to the A and B positions in chick agrin. Follistatin-like (hatched), EGF-like (open boxes), and laminin G–like (black boxes) domains are also shown. For neuregulin, a bioactive β isoform that contains an EGF-like and an Ig-like (Ig) domain is illustrated. Synaptic basal lamina also contains an abundance of GalNAc-terminated glycoconjugates. (C) Proteins of the postsynaptic membrane include AChRs, N-CAM, integrin α7β1, α and β dystroglycan, the sarcoglycans (α, β, and γ), and the tyrosine kinases MuSK and erb B2–4. Known ligands are ACh for AChRs, agrin for dystroglycan, and ARIA neuregulin for erbB kinases. The cytoskeletal apparatus concentrated beneath the postsynaptic membrane includes the AChR-associated protein rapsyn and three members of the dystrophin associated protein complex: utrophin, syntrophin, and an 87-kD protein, recently renamed "dystrobrevin."

catalytic subunit of the enzyme in muscle but not in brain (Massoulié et al., 1993). In addition, synaptic basal lamina contains proteins and proteoglycans that are contributed by the nerve, including agrin and acetylcholine-receptor–inducing factor (ARIA)/heregulin (see below).

Interestingly, even the major structural components of basal laminae, laminin and collagen IV, differ between synaptic and extrasynaptic regions. Both of these glycoproteins are heterotrimers of related chains that have recently been shown to be products of multigene families. Laminins are cruciform molecules that contain one α, one β, and one γ subunit, drawn from a set of at least five α, three β, and two γ subunits (Burgeson et al., 1994; Miner et al., 1995). In muscle, the predominant laminin of synaptic basal lamina is α5/β2/γ1 whereas the predominant extrasynaptic laminin is α2/β1/γ1 (Hunter et al., 1989a; Sanes et al., 1990; Patton, Miner and Sares, unpublished). We now suspect that the initial identification of α1 in synaptic basal lamina (Sanes et al., 1990) may reflect cross-reaction of the anti–α1 monoclonal antibodies with a newly identified *Drosophila* A–like chain, α5 (Miner et al., 1995; Miner and Sanes, unpublished). Likewise, collagens IV are triple helices formed from products of the α1–6(IV) genes. In muscle, synaptic basal lamina is rich in the α3–5(IV) chains, whereas extrasynaptic basal lamina contains predominantly α1 and α2(IV) chains, probably assembled as $(\alpha 1)_2(\alpha 2)_1$IV heterotrimers (Sanes et al., 1990; Miner and Sanes, 1994). These isoform variations appear to provide an elegant way of achieving functional specialization within a common structural framework.

ASSEMBLY OF THE POSTSYNAPTIC APPARATUS

How do synaptic components become concentrated in the tiny fraction of the myotube surface that becomes the postsynaptic apparatus? This question has been addressed in greatest detail for the AChRs. At least five processes have been found to contribute to establishment of the adult distribution of AChRs.

Myogenesis

First, AChR genes are activated as part of the program that controls the transformation of the myoblast into a myotube. The subunits are then translated, assembled into pentamers, glycolysated, and inserted in the myotube membrane. Important mediators of this process are the myogenic factors, a family of four homologous transcription factors of the basic helix-loop-helix family (Rudnicki and Jaenisch, 1995). These proteins bind to consensus motifs called E boxes in the promoters of numerous muscle-specific genes. All of the AChR-subunit gene promoters contain E boxes and, for most of the subunits, mutation of the E box has been shown to block expression in muscle (Duclert and Changeux, 1995). Of the myogenic factors, myogenin appears to be most important for differentiation; the others (myoD, myf5, and mrf4) play larger roles in myoblast determination. As expected from this scheme, AChR-subunit genes are particularly sensitive to myogenin (Huang et al., 1994), and most of the subunit genes remain inactive in mutant mice that lack myogenin (Hasty et al., 1993; Nabeshima et al., 1993).

Aggregation

Following their insertion into the plasma membrane, some of the AChRs are redistributed in the plane of the membrane to form high-density clusters. Such clusters, called "hot spots," occur in cultured myotubes that never encounter neurons, an observation that led to the speculation that nerves might preferentially innervate preformed synaptic specializations. Direct tests of this idea showed, however, that axons contact myotubes more or less at random, and that high-density AChR aggregates form subsynaptically only following contact (Anderson and Cohen, 1977; Frank and Fischbach, 1979). Thus, the nerve induces AChR clustering by a short-range signal. In addition, the nerve apparently exerts a longer-range effect, which leads to dispersal of nonsynaptic hot spots once the synapse has been established (Kuromi and Kidokoro, 1984). Such "ectopic" hot spots are not generally seen in developing muscle in vivo. Their absence might reflect an additional inhibitory constraint that is lacking in cultures, but more likely it results from the fact that myotubes in vivo are innervated nearly as soon as they form, whereas standard culture protocols leave a considerable interval between the formation and innervation of myotubes.

A key player in the AChR clustering mechanism is a 43-kD protein called rapsyn. Initially isolated as a protein that copurified with AChRs (Sobel et al., 1978), rapsyn was subsequently shown to be a peripheral membrane protein of the cytoplasmic face. Rapsyn and AChRs are precisely colocalized in synaptic membranes from the earliest stages of synaptogenesis (Noakes et al., 1993). Moreover, they are present in equimolar amounts, and may interact directly. Removal of rapsyn from synaptic membranes increases the lateral and rotational mobilities of AChRs (Rousselet et al., 1982), suggesting that it serves as part of an anchoring mechanism that could be involved in clustering. Direct support for this idea came from studies in nonmuscle cells. Following transfection of such cells with AChR expression vectors, AChRs appear diffusely on the cell surface. In contrast, coexpression of AChRs with rapsyn results in the formation of high-density AChR–rapsyn coclusters (Froehner et al., 1990; Phillips et al., 1991). Confirmation that rapsyn is crucial for clustering in vivo came from characterization of a rapsyn–null mutant mouse: No AChR clusters form in muscle fibers from such mice, or in myotubes cultured from them (Gautam et al., 1995). Moreover, other components of the postsynaptic membrane and cytoskeleton, such as dystroglycan, erbB3, utrophin, and syntrophin, are dispersed in the absence of rapsyn, suggesting that it does not merely anchor AChRs to a preformed cytoskeleton but is crucial to the organization of the entire postsynaptic apparatus (Gautam et al., 1995; Moscoso et al., 1995b). As expected, rapsyn-deficient animals die within hours of birth.

The functions of the other components of the postsynaptic cytoskeleton remain to be determined. Recently, attention has focused on utrophin, which is a homologue of dystrophin, the gene mutated in human Duchenne muscular dystrophy and *mdx* mutant mice. Both dystrophin and utrophin bind to dystroglycan, which in turn binds to components of the basal lamina (Campbell, 1995). Like rapsyn, utrophin is precisely colocalized with AChRs from the onset of synaptogenesis, whereas dystrophin is more diffusely distributed through the myotube cytoskeleton. In vitro, utrophin is present at large but not small hot spots, suggesting that it is involved in the growth of AChR "microclusters" into large aggregates found subsynaptically (Phillips et al., 1993). In addition, the AChR clustering factor agrin binds tightly to dystroglycan (see below), suggesting that utrophin might be part of a signal transduction pathway that leads to clustering. Inconsistent with this hypothesis, however, is the recent observation that utrophin-deficient mutant mice are viable and have nearly normal densities of AChRs in their postsynaptic membranes (Grady et al., in press; Decanunch et al., in press).

Local Synthesis

Although the myogenic program initially leads to activation of AChR synthesis throughout the myotube, ultimately AChRs are selectively inserted into the membrane in synaptic areas of the cell (Role et al., 1985). One major contributor to local insertion of AChRs, in turn, is their selective synthesis in subsynaptic areas. Evidence for local synthesis in adult muscle came initially from the demonstration that AChR-subunit mRNAs are more abundant in synaptic than in extrasynaptic regions of the muscle fiber (Merlie and Sanes, 1985). Subsequent studies using in situ hybridization showed that the mRNAs are highly concentrated near the few nuclei (of the several hundred per muscle fiber) that directly underlie the postsynaptic membrane (e.g., Fontaine et al., 1988), and that this synaptic localization is established soon after synapses form (Piette et al., 1993). These results were consistent with the idea that innervation modifies the transcriptional potential of the subsynaptic nuclei, but it also remained possible that AChR mRNAs were being transported to or stabilized near synaptic areas. To distinguish these alternatives, transgenic mice were generated in which regulatory sequences from AChR-subunit genes direct expression of reporter. For the α, γ, δ, and ε subunits, fragments have now been identified that lead to selective accumulation of reporter in (for nuclear localized β-galactosidase) or near (for growth hormone) synaptic nuclei (Klarsfeld et al., 1991; Sanes et al., 1991; Simon et al., 1992; Missias et al., 1996). Because other promoters direct widespread expression of the same reporters in myotubes (Sanes et al., 1991; Tang et al., 1994), these studies generated direct evidence for the unique transcriptional potential of synaptic nuclei. In addition, these constructs provided starting points from which to search for *cis*-acting sequences responsible for synaptic expression. This enterprise is now well underway (e.g., Koike et al., 1995) and hopefully will lead, in turn, to identification of the "synaptic" transcription factors.

Extrasynaptic Repression

The selective transcription of AChR-subunit genes by synaptic nuclei results from both positive and negative regulation: Soon after synaptic nuclei begin transcribing AChR genes at higher levels than extrasynaptic nuclei, the latter begin to shut off AChR expression altogether, which soon leads to the loss of AChRs from extrasynaptic membrane. The loss is neurally mediated, as shown originally by the reappearance of extrasynaptic AChRs following denervation (denervation supersensitivity). The nature of the neural signal was controversial for decades, but the main regulator is now accepted to be electrical activity (Fambrough, 1979; Goldman et al., 1988). That is, the action potentials triggered by synaptic potentials initiate an intracellular signaling cascade that results in repression of AChR transcription by extrasynaptic nuclei. The initial signal in this pathway is likely to be calcium entry into the cytoplasm, either from extracellular space or from intracellular stores. The calcium then activates protein kinase C– and cAMP/protein kinase A–dependent pathways (e.g., Chahine et al., 1993; Walke et al., 1994), leading to changes in levels and activities (via phosphorylation) of the myogenic factors. Finally, these factors act on E boxes in AChR-subunit genes to affect their transcription (reviewed in Duclert and Changeux, 1995). Thus, the initial myogenic activation and the subsequent activity-dependent suppression of AChR expression appear to converge on the same genomic elements.

An apparent paradox is that AChR transcription is neurally activated in synaptic nuclei but neurally repressed in extrasynaptic nuclei within the same muscle fiber cytoplasm. In other words, synaptic nuclei appear to be immune from the suppressive effects of activity. Two general models could account for this phenomenon. In one, expression levels would be computed by the AChR-subunit gene promoters from the relative strengths of positive and negative signals; activity would ensure that the negative influence is uniformly distributed, whereas the positive signal would be concentrated at the synapse. Alternatively, synaptic nuclei might be stably modified in a way that renders them insensitive to activity. They might lack, for example, some component of the activity-triggered signaling pathway. Synaptic nuclei appear to interact with the AChR-associated cytoskeleton (Englander and Rubin, 1987) and remain specialized after denervation (Gundersen et al., 1993). No experiment reported to date, however, has provided a clear distinction between these alternatives.

Subunit Switching

The AChRs of newly formed neuromuscular junctions differ from those of the adult in several respects. First, the density of AChRs in synaptic membrane is ~2,000–5,000/μm^2 at newly formed synapses but then increases to its adult value of >10,000/μm^2. This increase occurs prenatally in rodents. Second, the half-life of synaptic and extrasynaptic receptors is initially equal (~1 day), but synaptic receptors become stabilized perinatally and acquire a half-life of ~10 days. Third, at about the same time they become metabolically stable, synaptic AChRs become increasingly resistant to dispersion by collagenase or calcium chelators. Finally, postnatally in mammals, the mean channel open time of ACh-activated synaptic channels decreases severalfold, and the mean channel conductance increases (Hall and Sanes, 1993).

The fact that these changes occur at distinct times suggests they reflect different alterations in AChRs or AChR-associated components. To date, however, a molecular basis has been established only for the change in channel properties: the AChR γ-subunit gene is shut off and a new subunit gene, ε, is activated, leading to replacement of α2βγδ AChRs by α2βεδ AChRs (Mishina et al., 1986; Gu and Hall, 1988). The shut-off of the γ-subunit gene appears to reflect its special sensitivity to activity, and the induction of ε expression appears to require both activity-dependent and activity-independent neural influences (Duclert and Changeux, 1995). In addition, the subunit switch may be part of the general maturational program of the muscle, which also results in replacement of fetal by adult myosin isoforms.

Assembly of the Postsynaptic Membrane

No other components of the postsynaptic membrane have been studied in as much detail as the AChRs. There is, however, reason to believe that processes originally studied as contributors to the synaptic concentration of AChRs also affect the distribution of other synaptic membrane proteins. Relevant observations include the following: (1) N-CAM, like AChRs, is initially diffusely distributed in the myotube membrane, then lost from extrasynaptic regions by an activity-dependent mechanism (Covault and Sanes, 1985). (2) The N-CAM and MuSK genes are selectively transcribed by subsynaptic nuclei (Valenzuela et al., 1995; Moscoso et al., 1995a). (3) Dystroglycan and erbB3 are aggregated in the postsynaptic membrane by a rapsyn-dependent mechanism (Apel et al., 1995; Gautam et al., 1995; Moscoso et al., 1995b). Thus, as hoped, detailed studies of AChR seem likely to provide general insight into assembly of the postsynaptic membrane.

Assembly of the Synaptic Cleft

Relatively little is known about how the synaptic basal lamina is assembled. Clearly, both the nerve and the muscle contribute to it: Laminin, for example, is derived largely if not exclusively from the muscle, whereas agrin and the acetylcholine-receptor–inducing activity (ARIA) are synthesized by, and probably contributed by, both synaptic partners (Lieth and Fallon, 1993; Moscoso et al., 1995b). Once secreted, the synaptic components are stably retained in the synaptic cleft. Interactions with other components of the basal lamina are likely to play important roles in this process. Predominant among these are likely to be the major structural components of the basal lamina—laminin, collagen IV, and heparan sulfate proteoglycans. Indeed, there is evidence that agrin binds to laminin (Denzer et al., 1995), and that ARIA, agrin, and acetylcholinesterase all bind to heparan moieties (Loeb and Fischbach, 1995; Campanelli et al., 1996; Rossi and Rotundo, 1996). In addition, dystroglycan and integrins present in the postsynaptic membrane are clearly competent to bind components of synaptic basal lamina, including agrin and laminins (Martin et al., 1996; Campanelli et al., 1994; Gee et al., 1994; Sugiyama et al., 1994).

For the laminins and collagens IV, the question arises as to how some isoforms become restricted to the synaptic portion of the basal lamina while other, similar isoforms are present extrasynaptically and/or excluded from synaptic sites. Martin et al. (1995) addressed this question for laminin β1 and β2 by assessing the localizing of chimeric proteins following transfection of expression vectors into cultured muscle cells. In culture, as in vivo, laminin β2 (presumably β2-containing trimers) is selectively associated with AChR-rich domains, whereas β1-containing trimers are broadly distributed. Use of chimeras led to the identification of a small "synaptic localization" domain in laminin β2 that appears to be necessary and sufficient to localize β chains to synaptic sites. Remarkably, the localization domain contains the tripeptide site, LRE, which had previously been implicated in the interactions of laminin β2 with neurites (see above). This conjoining of function and localization raises the possibility that axons and myotubes may use similar receptor mechanisms to interact with laminin β2. Moreover, in that other synaptic basal lamina components, including agrin, contain the LRE sequence, it is possible that results obtained with laminin β2 may reflect the presence of a general mechanism for localization or assembly of synaptic cleft components.

Signals That Regulate Postsynaptic Differentiation

The nerve clearly provides numerous signals to the muscle as synaptogenesis proceeds. Myogenesis itself, long thought to be nerve independent, now appears to be stimulated or regulated by signals that arise from the neural tube (Pourquié et al., 1996) and from motor axons (e.g., Duxson and Sheard, 1995). Several nerve-derived factors have been shown to be capable of inducing aggregation of AChRs, including FGF (Peng et al., 1991), heparin binding growth factor (HB-GF)/pleiotrophin (Peng et al., 1995), and agrin (Bowe and Fallon, 1995). Separate factors induce AChR synthesis at the transcriptional level, including the neuropeptide CGRP (Changeux et al., 1992) and the AChR-inducing activity, ARIA (Falls et al., 1993). Acetylcholine itself affects postsynaptic differentiation by eliciting electrical activity, as discussed above. Finally, some neural influences are likely to trigger late events in synaptic maturation, such as the subunit switch in AChRs, and formation of junctional folds. Two of these factors, agrin and ARIA, have been studied with special intensity over the past few years, so the following discussion will focus on them.

Agrin

In studies of damaged muscle, Burden et al. (1979) found that contact with cell-free patches of synaptic basal lamina triggered regenerating myotubes to form a specialized postsynaptic apparatus. This result indicated that components of synaptic basal lamina can regulate post- as well as presynaptic (see above) differentiation. Based on this study, McMahan and colleagues set out to identify the AChR clustering agent(s) in synaptic basal lamina. Using the formation of "hot spots" in cultured embryonic myotubes as an assay, they isolated an active protein, which they named agrin, from basal lamina extracts (Godfrey et al., 1984; Nitkin et al., 1987). Pure agrin was capable of triggering formation of specialized domains on the myotube surface in which high densities of AChRs were associated with cytoskeletal proteins (such as rapsyn) and basal lamina components (such as acetylcholinesterase) that were associated with the postsynaptic density and synaptic cleft, respectively, in vivo (Wallace, 1989). Immunohistochemical studies showed that agrin was present in motoneurons, transported down motor axons, and stably retained in synaptic basal lamina (Magill-Solc and McMahan, 1988). Moreover, antibodies to agrin inhibited nerve-induced AChR clustering in vitro (Reist et al., 1992). Thus, agrin became an attractive candidate for a nerve-derived factor that organizes postsynaptic differentiation in developing as well as regenerating muscle.

One difficulty with this "agrin hypothesis" (McMahan, 1990) was that muscles, as well as neurons, make agrin. Although agrin is selectively associated with synaptic basal lamina in adult muscle, it is present at substantial levels throughout the basal lamina in embryos (Ferns et al., 1993). Moreover, muscles as well as nerves contribute to the deposits of agrin associated with synapses in vivo and in vitro (Lieth and Fallon, 1993). These results raised the possibility that agrin might function as a second messenger activated by a distinct nerve-derived signal, or as a cofactor that promotes stability of the basal lamina (Anderson et al., 1995).

Molecular cloning defined the primary sequence of agrin, provided a ready source of recombinant material, and suggested a satisfying means of reconciling the abundance of "muscle agrin" with the "agrin hypothesis." The agrin polypeptide is about 2,000 amino acids long and contains numerous domains homologous to sequences found in other proteins (Rupp et al., 1991; Tsim et al., 1992; Smith et al., 1992). These include nine repeats that resemble domains of kazal type protease inhibitors or follistatin; a domain that resembles a cysteine-rich region of laminin; four EGF-like repeats of a type found in numerous membrane and matrix proteins; and three repeats homologous to the subunits of the "G" domain found in all laminin αsubunits. The role of the follistatin-like repeats is unknown, as recombinant protein lacking all of these repeats is highly active in aggregating assays in vivo. In contrast, the aggregating and heparin-binding activities of agrin have been localized to the C-terminal of the molecule, which is composed largely of two "G" and two EGF repeats (Hoch et al., 1994; Gesemann et al., 1995).

Perhaps the most important result from analysis of cDNAs was the finding that agrin is alternatively spliced at several sites. Most striking is a site called "B" in chick and "z" in rodents, which borders the C-terminal–most G domain (Rüegg et al., 1992; Ferns et al., 1992). This site can be occupied by neither, either, or both of two inserts (exons) which encode eight and 11 amino acids. Agrin containing the eight-amino-acid exon and the B/z site, here termed B/z+, is 1,000-fold more active than agrin with no B/z insert (B/z−) (Rüegg et al., 1992; Ferns et al., 1993; Hoch et al., 1994; Gesemann et al., 1995). Moreover, B/z+ agrin is found only in the nervous system. Although some non-neural tissues, including muscle, express readily detectable levels of agrin, it is all of the B/z− form (Rüegg et al., 1992; Hoch et al., 1993). Indeed, even within the nervous system, it is likely that B/z+ agrin is confined to neurons, with glia expressing only B/z− forms (Tsim et al., 1992; Smith and O'Dowd, 1994). Thus, alternative splicing provides a means of reconciling the idea that agrin is a nerve-derived signal with the presence of significant depots of muscle-derived agrin: Only the former is highly active as a synaptic organizer. Interestingly, a newly discovered site of alternative splicing near the N-terminus of agrin may regulate its binding to synaptic basal lamina (Tsen et al., 1995b; Denzer et al., 1995).

Generation of an agrin-deficient mutant mouse has now provided strong evidence that agrin is essential for synapse formation in vivo (Gautam et al., 1996). An allele was generated in which the z exons were deleted, in order to ensure that no "highly active" z+ agrin could be expressed. In fact, the mutant allele is either inefficiently transcribed or generates an unstable transcript, because levels of agrin mRNA and protein are greatly reduced in the mutant. Thus, the mouse not only lacks all z+ agrin but is a severe hypomorph for agrin in general. Homozygous agrin mutants die during the last fetal day, have never been seen to move, and have severely disorganized neuromuscular relations. Most striking is that the number, size, and density of AChR aggregates are greatly reduced on mutant myotubes. Moreover, most motor nerve terminals end blindly on myotubes, having induced no detectable postsynaptic specialization. Thus, agrin apparently serves as an inducer of postsynaptic differentiation in vivo. There is, however, a minority of nerve terminals that are opposed to a specialized postsynaptic apparatus that is rich in AChRs and accompanied by synaptic cytoskeletal and basal lamina components. Thus, either z− agrin or a second agent is capable of inducing postsynaptic differentiation to a limited degree.

Three additional results of initial analysis of this mutant are also worth noting. First, there are numerous hot spots on uninnervated portions of myotubes, raising the possibility that agrin is not only a neural signal that induces AChR clustering but also the signal that causes dispersal of nonsynaptic specializations. Second, the innervation of mutant muscles is aberrant in pattern, suggesting either that agrin affects pre- as well as postsynaptic differentiation (as suggested by Campagna et al., 1995) or that the absence of postsynaptic specializations leads secondarily to excessive axonal growth. Finally, AChR-subunit genes are transcribed throughout the mutant myotubes rather than (nearly) only subsynaptically. Thus, even though agrin has no detectable effects on AChR gene expression in vitro, it may be necessary for this step in postsynaptic differentiation in vivo.

The mechanism by which agrin acts remains to be determined. A large body of correlative evidence favors the possibility that agrin signals through a tyrosine kinase, and that phosphorylation of the AChR β subunit is important for clustering (e.g., Wallace, 1994, 1995). At least in nonmuscle cells, however, mutant AChRs that lack the phosphorylatable residue can form coclusters with rapsyn (Yu and Hall, 1994). It is likely that a satisfying understanding of how agrin works will require identification of its receptors. To date, there are several candidate receptors but no consensus on what roles any of them play.

First is the dystrophin-associated glycoprotein, α-dystroglycan, which has been shown to be the major agrin-binding protein on the myotube surface (Gee et al., 1994; Campanelli et al., 1994; Sugiyama et al., 1994; Bowe et al., 1994). The importance of the dystroglycan complex for muscle function is indisputable, as mutations in any of several of its components lead to muscular dystrophy in humans and mice (Campbell, 1995). Moreover, dystroglycan is well suited to serve as a receptor, in that it links the matrix to the cytoskeleton and can be shown to cocluster with rapsyn in nonmuscle cells (Apel et al., 1995). However, attempts to block the effects of agrin with antibodies to α-dystroglycan have yielded conflicting results (Gee et al., 1994; Campanelli et al., 1994; Sugiyama et al., 1994). Moreover, while the binding of agrin to α-dystroglycan is dependent on alternatively spliced inserts at the z site, it is the less active forms that bind with highest affinity (Campanelli et al., 1996; Gesemann et al., 1996). In addition, while the less active forms of agrin block binding of the highly active forms to α-dystroglycan, they do not block the activity of the highly active forms in clustering AChR (Hoch et al., 1994). This result suggests that some agrin-initiated signals are transduced by a pathway that does not require α-dystroglycan.

Second, there is evidence that heparan sulfate proteoglycans may serve as agrin receptors: Heparin blocks the ability of agrin or neurons to cluster AChRs, and muscle cell lines selected for a deficiency in proteoglycans are deficient in their ability to respond to agrin (Ferns et al., 1992, 1993). One complication is that agrin itself has recently been shown to be a heparan sulfate proteoglycan (Tsen et al., 1995a; Denzer et al., 1995). Moreover, even though heparin blocks binding of agrin to α-dystroglycan (Campanelli et al., 1994; Sugiyama et al., 1994) it has recently been shown that heparin and dystroglycan bind to distinct sites on agrin (Campanelli et al., 1996). However, recombinant agrin fragments that bear no glycosaminoglycan are nonetheless heparin sensitive. Thus, the inhibitory effects of heparin cannot be explained by agrin's own heparan chains or its ability to bind α-dystroglycan. One attractive possibility is that proteoglycans serve as agrin coreceptors, as has been shown to be the case for growth factors such as FGF.

Third, glycoconjugates that terminate in an N-acetylgalactosaminyl (GalNAc) residue have been implicated in agrin signaling. Initially, lectin-binding studies revealed that GalNAc-bearing carbohydrates are highly concentrated at the neuromuscular junction and are present on multiple species in the pre- and postsynaptic membranes, as well as in the synaptic basal lamina (Scott et al., 1988). In seeking the function of these sugars, Martin and Sanes (1995) found that GalNAc-specific lectins cause AChR clustering, whereas either treatment of live cells with hexosaminidase or incubation with GalNAc-decorated proteins inhibits the effects of agrin. These results suggest that GalNAc itself, rather than merely the moiety to which it is attached, interacts with agrin. Neither muscle dystroglycan nor heparan sulfate glycosaminoglycans interact appreciably with the GalNAc-specific lectins, and the identity of the function of lectin-binding species remains unknown.

Finally, a muscle-specific tyrosine kinase, MuSK, has recently been identified that appears to be part of the agrin signal transduction pathway. Cloned in a search for tyrosine kinases regulated by innervation, MuSK is concentrated in the postsynaptic membrane of adult muscle and selectively transcribed by synaptic nuclei (Valenzuela et al., 1995). Most striking is the finding that mutant mice lacking MuSK exhibit few signs of postsynaptic differentiation. Moreover, MuSK-deficient mice resemble the agrin-deficient mutants described above in having aberrant nerve-branching patterns (DeChiara et al., 1996). Importantly, myotubes from agrin mutant mice are responsive to exogenous agrin, whereas MuSK mutant mice are not. These results place MuSK "downstream" of agrin in the signaling pathway. However, numerous methods have so far failed to demonstrate binding of agrin to MuSK, and agrin stimulates phosphorylation of recombinant MuSK in muscle but not in nonmuscle cells (G. Yancopoulos, personal communication). Thus, although MuSK is probably associated with the agrin receptor, there must be an additional component or posttranslational modification needed to render it functional.

In considering roles that these putative receptors may play, it is important to note that agrin may exert several discrete effects on synapse formation. First and most thoroughly analyzed is its ability to cluster AChRs. Second, studies in vitro have shown that agrin-induced AChR clusters are accompanied by concentrations of cytoskeletal and basal lamina components that are synapse-specific in vivo. It is simplest to imagine that a single mechanism accounts for all of agrin's clustering activity. In rapsyn-deficient mutant mice, however, laminin β2 and acetylcholinerase are synaptically localized whereas AChRs do not form clusters (Gautam et al., 1995). It is, therefore, possible that agrin causes AChR clustering by a rapsyn-dependent mechanism and aggregation of matrix components by a distinct, rapsyn-independent mechanism. Third, even though agrin has no detectable effect on AChR gene expression in vitro, synaptic nuclei fail to become transcriptionally specialized in agrin-deficient mice. Because transcriptional specialization is unaffected in the rapsyn-deficient mice, it is possible that agrin affects transcription by a rapsyn-independent interaction with the ARIA–erbB kinase pathway (discussed below). Fourth, studies in vitro have suggested that agrin affects axonal behavior (Campagna et al., 1995) and, consistent with these data, intramuscular axons grow abnormally in agrin-deficient mice. Thus, it is not unreasonable to imagine that several distinct receptors mediate agrin's effects.

Finally, agrin may trigger even the apparently unitary phenomenon of AChR clustering in at least two ways. The aggregating activity of soluble, recombinant agrin is highly dependent on the presence of the alternatively spliced B/z inserts, as detailed above. In contrast, both z-containing and z-minus forms are highly active when presented in a cell-bound form (Campanelli et al., 1991; Ferns et al., 1992, 1993). Because binding of agrin to dystroglycan does not require—and indeed is impeded by—the z inserts (Campanelli et al., 1996; Gesemann et al., 1996), it seems unlikely that dystroglycan is the sole receptor for z-dependent clustering. However, dystroglycan might well be responsible for z-independent clustering, perhaps by a mechanism that involves the formation of specialized, utrophin-rich cytoskeletal domains. In vivo, clustering by neural, z-containing agrin is likely to be of primary importance, but z-independent clustering could play an ancillary or stabilizing role.

In summary, agrin is a multidomain protein that has numerous effects and is likely to interact with several receptors. Determination of which receptors are biologically relevant is perhaps the most important next step in unraveling the roles that agrin plays in synaptogenesis.

ARIA/neuregulin

The starting point for isolation of ARIA was the observation that innervation of cultured myotubes not only causes clustering of AChRs but also increases the total number of AChRs on the cell surface. Using the number (rather than the distribution) of AChRs as an assay, Fischbach and colleagues purified receptor-inducing activities from brain and spinal cord. The purified material was a glycoprotein of ~42 kD that increased AChR-subunit gene expression but had little effect on AChR distribution or on the levels of other synaptic markers (Usdin and Fischbach, 1986). Molecular cloning (Falls et al., 1993) revealed that ARIA is encoded by the same gene that encodes several other independently isolated proteins: neu differentiation factor, heregulin, and glial growth factor (Holmes et al., 1992; Marchionni et al., 1993). Neu differentiation factor and heregulin were isolated as ligands of the *neu* proto-oncogene, and glial growth factor stimulates proliferation of Schwann cells. This gene is alternatively spliced at numerous positions, and the precise relationships of isoforms to biological functions remain to be determined. It has recently been proposed that all products of this gene be referred to as neuregulins (Carraway and Burden, 1995).

ARIA mRNA has been demonstrated in embryonic motoneurons and ARIA protein in adult motor nerve terminals by in situ hybridization and immunohistochemistry, respectively (Falls et al., 1993; Chu et al., 1995; Jo et al., 1995; Sandrock et al., 1995; Goodearl et al., 1995). In embryonic and neonatal muscle, however, ARIA is synthesized by muscle cells (Moscoso et al., 1995b). This situation is reminiscent of that described above for agrin and suggests that ARIA could act in an autocrine or paracrine fashion. However, it is possible for ARIA, as for agrin, that the muscle- and nerve-derived species differ in potency: At least in terms of binding to *neu*, some alternatively spliced forms of ARIA (called β) are far more potent than others (α forms; Marikovsky et al., 1995). The composition of muscle ARIA remains to be defined, but at least some are of the α variety (Moscoso et al., 1995b).

The identity of ARIA with ligands of *neu* suggested strongly that ARIA receptors are related to *neu*. *Neu* is one of a family of three receptor tyrosine kinases that are all related to the EGF receptor and all bind ARIA: *neu* is now called erbB2, and erbB3 and 4 were isolated subsequently. The EGF receptor itself is sometimes called erbB1. The erbB peptides form homo- and heterodimers that bind neuregulins and transduce signals to the cell's interior (Carraway and Burden, 1995). Application of ARIA to muscle cells stimulates phosphorylation of the erbB kinases, and all three kinases have been reported to be concentrated in the postsynaptic membrane (Moscoso et al., 1995b; Altiok et al., 1995; Zhu et al., 1995). Thus, although proof is lacking, it is quite likely that ARIA's effects are mediated by erbB kinases. Downstream of the erbB kinases lies a complex pathway involving the ras, raf, mek, and erk kinases, as well as Pl-3-kinase (Tansey et al., 1996). Ultimately, the pathway leads to activation of still-unknown regulators that affect gene expression. Recombinant ARIA has now been shown to stimulate expression of reporter genes linked to promoter fragments from several AChR-subunit genes in muscle cell lines and primary cultures (Jo et al., 1995; Chu et al., 1995; Missias et al., 1996). These culture systems are well suited for defining minimal ARIA-responsive *cis*-acting sequences, and ultimately for isolating the transcriptional regulators that bind to them.

Studies on the roles of ARIA in vivo have not yet been reported. A complication is that null mutations of the neuregulin, erbB2, and erbB4 genes all lead to embryonic lethality at a stage prior to myogenesis (Meyer and Birchmeier, 1995; Lee et al., 1995; Gassmann et al., 1995). Conditional or chimeric mutagenesis strategies may therefore be necessary to study ARIA's role directly. Interestingly, however, injection of a neuregulin into neonatal muscle has profound effects on the survival of Schwann cells (Trachtenburg and Thompson, 1996). In this instance, the protein is acting as would be expected for glial growth factor. Effects of AChR expression were not assayed in these experiments, but could be in the future.

The Postsynaptic Apparatus of Interneuronal Synapses

The postsynaptic apparatus of interneuronal synapses resembles that of the neuromuscular junction in several reports. Moreover, interneuronal synapses develop in a series of steps that imply the passage of signals between the synaptic partners (Rees, 1978; Vaughn, 1989). Until recently, however, it remained unclear how helpful lessons from neuromuscular synaptogenesis would be in understanding synapse formation in the brain. Fortunately, the last few years have revealed numerous similarities. Most striking is mounting evidence that neurotransmitter receptors in the brain are regulated by processes similar to those documented for the neuromuscular junction.

First, GABA, and glutamate receptors, like AChRs, are aggregated in the subsynaptic membrane of dendrites by an influence of the nerve (Craig et al., 1994), whereupon they become anchored by interactions with the cytoskeleton. Moreover, dystrophin, dystroglycan, and utrophin are associated with central postsynaptic densities (Montanaro et al., 1995; Lidov et al., 1990; Kamakura et al., 1994). Although rapsyn has not been found in the brain, two proteins have been implicated as playing a similar role. One is gephyrin, originally isolated as a protein associated with the glycine receptor, and later shown to be highly concentrated in postsynaptic densities at glycinergic and possibly GABAergic synapses (Kirsch and Betz, 1993). Gephyrin is not homologous to rapsyn (Prior et al., 1992), but its overexpression in non-neural cells leads to clustering of glycine receptors (Kirsch et al., 1995), and its removal from cultured neurons (by use of antisense oligonucleotides) leads to loss of receptor clusters (Kirsch et al., 1993). Gephyrin apparently binds directly to GABA receptor and to tubulin, and this anchors receptors to the cytoskeleton as rapsyn does for AChRs (Kirsch et al., 1991; Kirsch and Betz, 1995; Meyer et al., 1995). Similarly, a protein called PSD-95 or SAP-90 has now been suggested to cluster glutamate receptors (Kornau et al., 1995; Niethammer et al., 1996).

Second, there is evidence for local synthesis of some components of the postsynaptic apparatus at central synapses. In that few neurons are multinucleated, the mechanism is clearly different from that seen at the neuromuscular junction, but the effect may be similar. Some mRNAs are transported into dendrites and become associated with polyribosomes at the base of dendritic spines, whereupon they are translated. The mechanism of RNA localization is poorly understood, but at least one of the dendritic RNAs encodes CaM kinase, which is a major component of the postsynaptic density (Steward and Banker, 1992).

Third, innervation and electrical activity regulate levels of neurotransmitter receptor expression in both peripheral ganglia (reviewed in Froehner, 1993; Sargent, 1993) and central neurons (reviewed in Armstrong and Montminy, 1993), as they do at the neuromuscular junction. In at least some cases, activity (DeKoninck and Cooper, 1995) or innervation increases levels of receptors, as seen in nerve–muscle cultures (Role, 1988; Levey et al., 1995). In other cases, axotomy leads to "supersensitivity" of the denervated neurons. In the brain, though, supersensitivity often reflects loss of transmitter inactivation mechanisms, and the extent to which it involves inactivity-triggered receptor upregulation remains unclear.

Fourth, the subunit composition of both glycine and glutamate receptors changes during development, reminiscent of the γ-to-ε subunit switch documented for AChRs at the neuromuscular junction (Béchade et al., 1994; Sheng et al., 1994).

Finally, some of the anterograde signals that operate at the neuromuscular junction may also function in the brain. Both agrin and ARIA are widely distributed in the brain, and neither is confined to cholinergic neurons (Chen et al., 1994; Corfas et al., 1995; Ma et al., 1994; O'Connor et al., 1994; Pinkas-Kramarski et al., 1994). In addition, agrin is present in some synaptic clefts in retina (Kröger et al., 1996). To date, no clear effects of agrin or ARIA on neurons or neuronal receptors have been documented, but it seems likely that such demonstrations are not far off.

Conclusions

During the last 15 years, many components of one synapse, the neuromuscular junction, have been defined at the molecular level. We are beginning to achieve an understanding of the signals that regulate the synthesis and organization of these molecules both during development and regeneration. An appreciation of the molecular architecture of brain synapses is also becoming available, and studies of central synapse assembly are underway in many laboratories. It is already clear that the presynaptic elements of central and peripheral synapses are similar in organization, but the postsynaptic apparatus at the neuromuscular junction differs in several respects from that at interneuronal synapses. Thus, it remains to be determined whether the lessons learned from the neuromuscular junction will be restricted to general principles or whether some of the same regulatory proteins will function both centrally and peripherally.

References

Alfonso, A., Grundahl, K., Duerr, J.S., Han, H.P., Rand, J.B. (1993). The *Caenorhabditis elegans unc-17* gene: a putative vesicular acetylcholine transporter. *Science* 261:617–619.

Allen, F., Warner, A. (1991). Gap junctional communication during neuromuscular junction formation. *Neuron* 6:101–111.

Altiok, N., Bessereau, J.L., Changeux, J.P. (1995). ErbB3 and ErbB2/neu mediate the effect of heregulin on acetylcholine receptor gene expression in muscle: differential expression at the endplate. *EMBO J.* 14:4258–4266.

Anderson, M.J., Cohen, M.W. (1977). Nerve-induced and spontaneous redistribution of acetylcholine receptors on cultured muscle cells. *J. Physiol.* 268:757–773.

Anderson, M.J., Shi, Z.Q., Grawel, R., Zackson, S.L. (1995). Erratic deposition of agrin during the formation of *Xenopus* neuromuscular junctions in culture. *Dev. Biol.* 170:1–20.

Apel, E.D., Roberds, S.L., Campbell, K.P., Merlie, J.P. (1995). Rapsyn may function as a link between the acetylcholine receptor and the agrin-binding dystrophin-associated glycoprotein complex. *Neuron* 15:115–126.

Armstrong, R.C., Montminy, M.R. (1993). Transsynaptic control of gene expression. *Annu. Rev. Neurosci.* 16:17–29.

Bajjalieh, S.M., Peterson, K., Shinghal, R., Scheller, R.H. (1992). SV2, a brain synaptic vesicle protein homologous to bacterial transporters. *Science* 257:1271–1273.

Bajjalieh, S.M., Peterson, K., Linial, M., Scheller, R.H. (1993). Brain contains two forms of synaptic vesicle protein 2. *Proc. Natl. Acad. Sci. U.S.A.* 90:2150–2154.

Bajjalieh, S.M., Frantz, G.D., Weimann, J.M., McConnell, S.K., Scheller, R.H. (1994). Differential expression of synaptic vesicle protein 2 (SV2) isoforms. *J Neurosci.* 14:5223–5235.

Béchade, C., Sur, C., Triller, A. (1994). The inhibitory neuronal glycine receptor. *Bioessays* 16:735–744.

Benjamin, S., Cervini, R., Mallet, J., Berrard, S. (1994). A unique gene organization for two cholinergic markers, choline acetyltransferase and a putative vesicular transporter of acetylcholine. *J. Biol. Chem.* 269:21944–21947.

Benfenati, F., Valtora, F., Rossi, M.C., Onofri, F., Sihra, T., Greengard, P. (1993). Interactions of synapsin I with phospholipids: possible role in synaptic vesicle clustering and in the maintenance of bilayer structures. *J. Cell Biol.* 123:1845–1855.

Bennett, M.K., Scheller, R.H. (1993). The molecular machinery for secretion is conserved from yeast to neurons. *Proc. Natl. Acad. Sci. U.S.A.* 90:2559–2563.

Bennett, M.K., Calakos, N., Scheller, R.H. (1992). Syntaxin: a synaptic protein implicated in docking of synaptic vesicles at presynaptic active zones. *Science* 257:255–259.

Bezprozvanny, I., Scheller, R.H., Tsien, R.W. (1995). Functional impact of syntaxin on gating of N-type and Q-type calcium channels. *Nature* 378:623–626.

Birnbaumer, L., Campbell, K.P., Catterall, W.A., Harpold, M.M., Hofmann, F., Horne, W.A., Mori, Y., Schwartz, A., Snutch, T.P., Tanabe, T., Tsien, R.W. (1994). The naming of voltage-gated calcium channels. *Neuron* 13:505–506.

Bixby, J.L., Reichardt, L.F. (1985). The expression and localization of synaptic vesicle antigens at neuromuscular junctions *in vitro*. *J. Neurosci.* 5:3070–3080.

Bixby, J.L., Pratt, R.S., Lilien, J., Reichardt, L.F. (1987). Neurite outgrowth on muscle cell surfaces involves extracellular matrix receptors as well as Ca^{2+}-dependent and -independent cell adhesion molecules. *Proc. Natl. Acad. Sci. U.S.A.* 84:2555–2559.

Bowe, M.A., Fallon, J.R. (1995). The role of agrin in synapse formation. *Annu. Rev. Neurosci.* 18:443–462.

Bowe, M.A., Deyst, K.A., Leszyk, J.D., Fallon, J.R. (1994). Identification and purification of an agrin receptor from *Torpedo* postsynaptic membranes: a heteromeric complex related to the dystroglycans. *Neuron* 12:1173–1180.

Burden, S.J., Sargent, P.B., McMahan, U.J. (1979). Acetylcholine receptors in regenerating muscle accumulate at original synaptic sites in the absence of the nerve. *J. Cell Biol.* 82: 412–425.

Burgeson, R.E., Chiquet, M., Deutzmann, R., Ekblom, P., Engel, J., Kleinman, H., Martin, G.R., Meneguzzi, G., Paulsson, M., Sanes, J., Timpl, R., Tryggvason, K., Yamada, Y., Yurchenco, P.D. (1994). A new nomenclature for the laminins. *Matrix Biol.* 14:209–211.

Campagna, J.A., Rüegg, M.A., Bixby, J.L. (1995). Agrin is a differentiation-inducing "stop signal" for motoneurons *in vitro*. *Neuron* 15:1365–1374.

Campanelli, J.T., Gayer, G.G., Scheller, R.H. (1996). Alternative RNA splicing that determines agrin activity regulates binding to heparin and α-dystroglycan. *Development* 122:1663–1672.

Campanelli, J.T., Hoch, W., Rupp, F., Kreiner, T., Scheller, R.H. (1991). Agrin mediates cell contact-induced acetylcholine receptor clustering. *Cell* 67:909–916.

Campanelli, J.T., Roberds, S.T., Campbell, K.P., Scheller, R.H. (1994). A role for dystrophin-associated glycoproteins and utrophin in agrin-induced AChR clustering. *Cell* 77:663–674.

Campbell, K.P. (1995). Three muscular dystrophies: loss of cytoskeleton-extracellular matrix linkage. *Cell* 80:675–679.

Caroni, P., Grandes, P. (1990). Nerve sprouting in innervated adult skeletal muscle induced by exposure to elevated levels of insulin-like growth factors. *J. Cell Biol.* 110: 1307–1317.

Caroni, P., Schneider, C., Kiefer, M.C., Zapf, J. (1994). Role of muscle insulin-like growth factors in nerve sprouting: suppression of terminal sprouting in paralyzed muscle by IGF-binding Protein 4. *J. Cell Biol.* 4:893–902.

Carraway, K.L., III, Burden, S.H. (1995). Neuregulins and their receptors. *Curr. Opin. Neurobiol.* 5:606–612.

Chahine, K.G., Baracchini, E., Goldman, D. (1993). Coupling muscle electrical activity to gene expression via a cAMP-dependent second messenger system. *J. Biol. Chem.* 268: 2893–2898.

Changeux, J.-P., Duclert, A., Sekine, S. (1992). Calcitonin gene-related peptides and neuromuscular interactions. *Ann. N.Y. Acad. Sci.* 657:361–378.

Chen, M.S., Bermingham-McDonogh, O., Danehy, F.T., Nolan, C., Scherer, S.S., Lucas, J., Gwyne, D., Marchionni, M.A. (1994). Expression of multiple NRG transcripts in postnatal rat brains. *J. Comp. Neurol.* 349:389–400.

Chin, L.S., Li, L., Ferreira, A., Kosik, K.S., Greengard, P. (1995). Impairment of axonal development and of synaptogenesis in hippocampal neurons of synapsin I-deficient mice. *Proc. Natl. Acad. Sci. U.S.A.* 92:9230–9234.

Chow, I., Cohen, M.W. (1983). Developmental changes in the distribution of acetylcholine receptors in the myotomes of *Xenopus laevis. J. Physiol.* 339:553–571.

Chu, G.C., Moscoso, L.M., Sliwkowski, M.X., Merlie, J.P. (1995). Regulation of the acetylcholine receptor subunit gene by recombinant ARIA: an *in vitro* model for transsynaptic gene regulation. *Neuron* 14:329–339.

Clift-O'Grady, L., Linstedt, A.D., Lowe, A.W., Grote, E., Kelly, R.B. (1990). Biogenesis of synaptic vesicle-like structures in a pheochromocytoma cell line PC-12. *J. Cell Biol.* 110: 1693–1703.

Corfas, G., Rosen, K.M., Aratake, H., Frauss, R., Fischbach, G.D. (1995). Differential expression of ARIA isoforms in the rat brain. *Neuron* 14:103–115.

Covault, J., Sanes, J.R. (1985). Neural cell adhesion molecule (N-CAM) accumulates in denervated and paralyzed skeletal muscles. *Proc. Natl. Acad. Sci.* 82:4544–4548.

Covault, J., Merlie, J.P., Goridis, C., Sanes, J.R. (1986). Molecular forms of N-CAM and its RNA in developing and denervated skeletal muscle. *J. Cell Biol.* 102:731–739.

Craig, A.M., Blackstone, C.D., Huganir, R.L., Banker, G. (1994). Selective clustering of glutamate and γ-aminobutyric acid receptors opposite terminals releasing the corresponding neurotransmitters. *Proc. Natl. Acad. Sci. U.S.A.* 91:12373–12377.

Dai, Z., Peng, H.B. (1995). Presynaptic differentiation induced in cultured neurons by local application of basic fibroblast growth factor. *J. Neurosci.* 15:5466–5475.

David, C., McPherson, P.S., Mundigl, O., De Camili, P. (1996). A role of amphiphysin in synaptic vesicle endocytosis suggested by its binding to dynamin in nerve terminals. *Proc. Natl. Acad. Sci. U.S.A.* 93:331–335.

DeChiara, T.M., Bowen, D.C., Valenzuela, D.M., Simmons, M.V., Poueymirou, W.T., Thomas, S., Kinetz, E., Smith, C., Compton, D.L., Park, J.S., DiStefano, P.S., Glass, D.J., Burden, S.J., Yancopoulos, G.D. (1996). The receptor tyrosine kinase, MuSK, is required for all aspects of neuromuscular junction formation *in vivo. Cell* 85:501–512.

Deconinck, A.E., Potter, A.C., Tinsley, J.M., Wood, S.J., Vater, R., Young C., Metzinger, L., Vincent A., Slater, C.R., Davies, K.E. Postsynaptic abnormalities at the neuromuscular junctions of utrophin deficient mice. *J. Cell. Biol.* in press (1997).

DeKoninck, P., Cooper, E. (1995). Differential regulation of neuronal nicotinic acetylcholine receptor subunit genes in cultured neonatal rat sympathetic neurons: specific induction of alpha 7 by membrane depolarization through a Ca^{2+}/calmodulin-dependent kinase pathway. *J. Neurosci.* 15:7966–7978.

Denzer, A.J., Gesemann, M., Schumacher, B., Rüegg, M.A. (1995). An amino-terminal extension is required for the secretion of chick agrin and its binding to extracellular matrix. *J. Cell Biol.* 131:1547–1560.

DiAntonio, A., Burgess, R.W., Chin, A.C., Deitcher, D.L., Scheller, R.H., Schwarz, T.L. (1993a). Identification and characterization of *Drosophila* genes for synaptic vesicle proteins. *J. Neurosci.* 13:4924–4935.

DiAntonio, A., Parfitt, K.D., Schwarz, T.L. (1993b). Synaptic transmission persists in synaptotagmin mutants of *Drosophila. Cell* 73:1281–1290.

Duclert, A., Changeux, J.-P. (1995). Acetylcholine receptor gene expression at the developing neuromuscular junction. *Physiol. Rev.* 75:339–368.

Duxson, M.J., Sheard, P.W. (1995). Formation of new myotubes occurs exclusively at the multiple innervation zones of an embryonic large muscle. *Dev. Dyn.* 204:391–405.

Englander, L.L., Rubin, L.L. (1987). Acetylcholine receptor clustering and nuclear movement in muscle fibers in culture. *J. Cell Biol.* 104:87–95.

Erickson, J., Varoqui, H., Schafer, M., Modi, W., Diebler, M., Weihe, E., Rand, J., Eiden, L., Bonner, T., Usdin, T. (1994). Functional identification of a vesicular acetylcholine transporter and its expression from a "cholinergic" gene locus. *J. Biol. Chem.* 269: 21929–21932.

Falls, D.L., Rosen, K.M., Corfas, G., Lane, W.S., Fischbach, G.D. (1993). ARIA, a protein that stimulates acetylcholine receptor synthesis, is a member of the neu ligand family. *Cell* 72:801–815.

Fambrough, D.M. (1979). Control of acetylcholine receptors in skeletal muscle. *Physiol. Rev.* 59:165–227.

Feany, M.B., Lee, S., Edwards, R.H., Buckley, K.M. (1992). The synaptic vesicle protein SV2 is a novel type of transmembrane transporter. *Cell* 70:861–867.

Ferns, M., Hoch, W., Campanelli, J.T., Rupp, F., Hall, Z.W., Scheller, R.H. (1992). RNA splicing regulates agrin-mediated acetylcholine receptor clustering activity in cultured myotubes. *Neuron* 8:1079–1086.

Ferns, M., Campanelli, J.T., Hoch, W., Scheller, R.H., Hall, Z.W. (1993). The ability of agrin to cluster AChRs depends on alternative splicing and on cell surface proteoglycans. *Neuron* 11:491–502.

Ferreira, A., Han, H.Q., Greengard, P., Kosik, K.S. (1995). Suppression of synapsin II inhibits the formation and maintenance of synapses in hippocampal culture. *Proc. Natl. Acad. Sci. U.S.A.* 92:9225–9229.

Ferro-Novick, S., Novick, P. (1993). The role of GTP-binding proteins in transport along the exocytic pathway. *Annu. Rev. Cell Biol.* 9:575–599.

Floor, E., Schaeffer, S.F., Feist, B.E., Leeman, S.E. (1988). Synaptic vesicles from mammalian brain: large-scale purification and physical and immunochemical characterization. *J. Neurochem.* 50:1588–1596.

Floor, E., Leventhal, P.S., Schaeffer, S.F. (1990). Partial purification and characterization of the vacuolar H^+-ATPase of mammalian synaptic vesicles. *J. Neurochem.* 55:1663–1670.

Fontaine, B., Sasson, D., Buckingham, M., Changeux, J.P. (1988). Detection of the nicotinic acetylcholine receptor a-subunit mRNA by *in situ* hybridization at neuromuscular junctions of 15-day-old chick striated muscles. *EMBO J.* 7:603–609.

Frank, E., Fischbach, G.D. (1979). Early events in neuromuscular junction formation *in vitro.* Induction of acetylcholine receptor clusters in the postsynaptic membrane and morphology of newly formed nerve-muscle synapses. *J. Cell Biol.* 83:143–158.

Froehner, S.C. (1993). Regulation of ion channel distribution at synapses. *Annu. Rev. Neurosci.* 16:347–368.

Froehner, S.C., Luetje, C.W., Scotland, P.B., Patrick, J. (1990). The postsynaptic 43K protein clusters muscle nicotinic acetylcholine receptors in *Xenopus oocytes. Neuron* 5:403–410.

Gassmann, M., Casagranda, F., Orioli, D., Simon, H., Lai, C., Klein, R., Lemke, G. (1995). Aberrant neural and cardiac development in mice lacking the ErbB4 neuregulin receptor. *Nature* 378:390–394.

Gautam, M., Noakes, P.G., Moscoso, L., Rupp, F., Scheller, R.H., Merlie, J.P., Sanes, J.R. (1996). Defective neuromuscular synaptogenesis in agrin-deficient mutant mice. *Cell* 85:525–535.

Gautam, M., Noakes, P.G., Mudd, J., Nichol, M., Chu, G.C., Sanes, J.R., Merlie, J.P. (1995). Failure of postsynaptic specialization to develop at neuromuscular junctions of rapsyn-deficient mice. *Nature* 377:232–236.

Gee, S.H., Montanaro, F., Lindenbaum, M.H., Carbonetto, S. (1994). Dystroglycan-alpha, a dystrophin-associated glycoprotein, is a functional agrin receptor. *Cell* 77:675–686.

Geppert, M., Bolshakov, V.Y., Siegelbaum, S.A., Takei, K., De Camilli, P., Hammer, R.E., Südhof, T.C. (1994a). The role of Rab3A in neurotransmitter release. *Nature* 369:493–497.

Geppert, M., Goda, Y., Hammer, R.E., Li, C., Rosahl, T.W., Stevens, C.F., Südhof, T.C. (1994b). Synaptotagmin I: a major calcium sensor for transmitter release at a central synapse. *Cell* 79:717–727.

Gesemann, M., Denzer, A.J., Rüegg, M.A. (1995). Acetylcholine receptor aggregating activity of agrin isoforms and mapping of the active site. *J. Cell Biol.* 128:625–636.

Gesemann, M., Cavalli, V., Denzer, A.J., Brancaccio, A., Schumacher, B., Ruegg, M.A.

(1996). Alternative splicing of agrin alters its binding to heparin, dystroglycan, and the putative agrin receptor. *Neuron* 16:755–767.
Glicksman, M., Sanes, J.R. (1983). Differentiation of motor nerve terminals formed in the absence of muscle fibers. *J. Neurocytol.* 12:661–671.
Godfrey, E.W., Nitkin, R.M., Wallace, B.G., Rubin, L.L., McMahan, U.J. (1984). Components of *Torpedo* electric organ and muscle that cause aggregation of acetylcholine receptors on cultured muscle cells. *J. Cell Biol.* 99:615–627.
Goldman, D., Brenner, H.R., Heinemann, S. (1988). Acetylcholine receptor α, β, γ, σ subunit mRNA levels are regulated by muscle activity. *Neuron* 1:329–333.
Goodearl, A.D.J., Yee, A.G., Sandrock, A.W., Jr., Corfas, G., Fischback, G.D. (1995). ARIA is concentrated in the synaptic basal lamina of the developing chick neuromuscular junction. *J. Cell Biol.* 130:1423–1434.
Gotow, T., Miyaguchi, K., Hashimoto, P.H. (1991). Cytoplasmic architecture of the axon terminal: filamentous strands specifically associated with synaptic vesicles. *Neuroscience* 40:587–598.
Grady, R.M., Merlie, J.P., Sanes, J.R. Subtle neuromuscular defects in utrophin-deficient mice. *J. Cell Biol.* in press (1997).
Greengard, P., Valtorta, F., Czernik, A.J., Benfenati, F. (1993). Synaptic vesicle phosphoproteins and regulation of synaptic function. *Science* 259:780–785.
Grinnell, A.D. (1995). Dynamics of nerve-muscle interaction in developing and mature neuromuscular junctions. *Physiol. Rev.* 75:789–834.
Gu, Y, Hall, Z.W. (1988). Immunological evidence for a change in subunits of the acetylcholine receptor in developing and denervated rat muscle. *Neuron* 1:117–125.
Gundersen, C.B., Umbach, J.A. (1992). Suppression cloning of the cDNA for a candidate subunit of a presynaptic calcium channel. *Neuron* 9:527–537.
Gundersen, K., Sanes, J.R., Merlie, J.P. (1993). Neural regulation of muscle acetylcholine receptor ε- and α-subunit gene promoters in transgenic mice. *J. Cell Biol.* 123:1535– 1544.
Gurney, M.E., Yamamoto, H., Kwon, Y. (1992). Induction of motor neuron sprouting *in vivo* by ciliary neurotrophic factor and basic fibroblast growth factor. *J. Neurosci.* 12: 3241–3247.
Haas, C.A., DeGennaro, L.J. (1988). Multiple synapsin I messenger RNAs are differentially regulated during neuronal development. *J. Cell Biol.* 106:195–203.
Hall, Z.W., Sanes, J.R. (1993). Synaptic structure and development: the neuromuscular junction. *Cell* 72:*Neuron* 10(Suppl):99–121.
Han, H.-Q., Nichols, R.A., Rubin, M.R., Bähler, M., Greengard, P. (1991). Induction of formation of presynaptic terminals in neuroblastoma cells by synapsin IIb. *Nature* 349: 697–700.
Hasty, P., Bradley, A., Morris, J.H., Venuti, J.M., Olson, E.N., Klein, W.H. (1993). Muscle deficiency and neonatal death in mice with a targeted mutation in the *myogenin* gene. *Nature* 364:501–506.
Hata, Y., Slaughter, C.A., Südhof, T.C. (1993). Synaptic vesicle fusion complex contains unc-18 homolog bound to syntaxin. *Nature* 366:347–351.
Hoch, W., Ferns, M., Campanelli, J.T., Hall, Z.W., Scheller, R.H. (1993). Developmental regulation of highly active alternatively spliced forms of agrin. *Neuron* 11:479–490.
Hoch, W., Campanelli, J.T., Harrison, S., Scheller, R.H. (1994). Structural domains of agrin required for clustering of nicotinic acetylcholine receptors. *EMBO J.* 13:2814–2821.
Holmes, W.E., Sliwkowski, M.X., Akita, R.W., Henzel, W.J., Lee, J., Park, J.W., Yansura, D., Abadi, N., Raab, H., Lewis, G.D., Shepard, H.M., Kuang, W.-J., Wood, W.I., Goeddel, D.V., Vandlen, RL. (1992). Identification of heregulin, a specific activator of $p185^{erbB2}$. *Science* 256:1205–1210.
Holz, R.W., Brondyk, W.H., Senter, R.A., Kuizon, L., Macara, I.G. (1994). Evidence for the involvement of Rab3A in Ca(2+)-dependent exocytosis from adrenal chromaffin cells. *J. Biol. Chem.* 269:10229–10234.

Houenou, L.J., Li, L., Lo, A.C., Yan, Q., Oppenheim, R.W. (1994). Naturally occurring and axotomy-induced motoneuron death and its prevention by neurotrophic agents: a comparison between chick and mouse. *Prog. Brain Res.* 102:217–226.

Huang, C.F., Lee, Y.S., Schmidt, M.M., Schmidt, J. (1994). Rapid inhibition of myogenin-driven acetylcholine receptor subunit gene transcription. *EMBO J.* 13:634–640.

Hughes, R.A., Sendtner, M., Goldfarb, M., Lindholm, D., Thoenen, H. (1993). Evidence that fibroblast growth factor 5 is a major muscle-derived survival factor for cultured spinal motoneurons. *Neuron* 10:369–377.

Hume, R.I., Role, L.W., Fischbach, G.D. (1983). Acetylcholine release from growth cones detected with patches of acetylcholine receptor-rich membranes. *Nature* 305:632–634.

Hunter, D.D., Shah, V., Merlie, J.P., Sanes, J.R. (1989a). A laminin-like adhesive protein concentrated in the synaptic cleft of the neuromuscular junction. *Nature* 338:229–234.

Hunter, D.D., Porter, B.E., Bulock, J.W., Adams, S.P., Merlie, J.P., Sanes, J.R. (1989b). Primary sequence of a motor neuron-selective adhesive site in the synaptic basal lamina protein s-laminin. *Cell* 59:905–913.

Imai, T., Kakizaki, M., Nishimura, M., Yoshi, O. (1995). Molecular analyses of the association of CD4 with two members of the transmembrane 4 superfamily, CD81 and CD82. *J. Immunol.* 155:1229–1239.

Ip, N.Y., Yancopoulos, G.D. (1996). The neutrotrophins and CNTF: two families of collaborative neurotrophic factors. *Annu. Rev. Neurosci.* 19:491–515.

Jo, S.A., Zhu, X., Marchionni, M.A., Burden, S.J. (1995). Neuregulins are concentrated at nerve-muscle synapses and activate ACh-receptor gene expression. *Nature* 373:158–161.

Johannes, L., Lledo, P.-M., Roa, M., Vincent, J.-D., Henry, J.-P., Darchen, H., Darchen, F. (1994). The GTPase Rab3a negatively controls calcium-dependent exocytosis in neuroendocrine cells. *EMBO J.* 13:2029–2037.

Johnston, P.A., Cameron, P.L., Stukenbrok, H., Jahn, R., De Camilli, P., Südhof, T.C. (1989). Synaptophysin is targeted to similar microvesicles in CHO and PC 12 cells. *EMBO J.* 8: 2863–2872.

Kamakura, K., Tadano, Y., Kawai, M., Ishiura, S., Nakamura, R., Miyamoto, K., Nagata, N., Sugita, H. (1994). Dystrophin-related protein is found in the central nervous system of mice at various developmental stages, especially at the postsynaptic membrane. *J. Neurosci. Res.* 37:728–734.

Kaprielian, Z., Cho, K.O., Hadjiargyrou, M., Patterson, P.H. (1995). CD9, a major platelet cell surface glycoprotein, is a ROCA antigen and is expressed in the nervous system. *J. Neurosci.* 15:562–573.

Kelly, R.B. (1993). Storage and release of neurotransmitters. *Cell* 72:43–53.

Keshishian, H., Broadie, K., Chiba, A., Bate, M. (1996). The *Drosophila* neuromuscular junction: a model system for studying synaptic development and function. *Annu. Rev. Neurosci.* 19:545–575.

Kidokoro, Y., Yeh, E. (1982). Initial synaptic transmission at the growth cone in *Xenopus* nerve-muscle cultures. *Proc. Natl. Acad. Sci. U.S.A.* 79:6727–6731.

Kirsch, J., Betz, H. (1993). Widespread expression of gephyrin, a putative glycine receptor-tubulin linker protein, in rat brain. *Brain Res.* 621:301–310.

Kirsch, J., Betz, H. (1995). The postsynaptic localization of the glycine receptor-associated protein gephyrin is regulated by the cytoskeleton. *J. Neurosci.* 15:4148–4156.

Kirsch, J., Langosch, D., Prior, P., Littauer, U.Z., Schmitt, B., Betz, H. (1991). The 93-kDa glycine receptor-associated protein binds to tubulin. *J. Biol. Chem.* 266:22242–22245.

Kirsch, J., Wolters, I., Triller, A., Betz, H. (1993). Gephyrin antisense oligonucleotides prevent glycine receptor clustering in spinal neurons. *Nature* 366:745–748.

Kirsch, J., Kuhse, J., Betz, H. (1995). Targeting of glycine receptor subunits to gephyrin-rich domains in transfected human embryonic kidney cells. *Mol. Cell. Neurosci.* 6:450–461.

Klarsfeld, A., Bessereau, J.-L., Salmon, A.-M., Triller, A., Babinet, C., Changeux, J.-P. (1991). An acetylcholine receptor α-subunit promoter conferring preferential synaptic expression in muscle of transgenic mice. *EMBO J.* 10:625–632.

Knaus, P., Betz, Y.H., Rehm, H. (1986). Expression of synaptophysin during postnatal development of the mouse brain. *J. Neurochem.* 47:1302–1304.

Koike, S., Schaeffer, L., Changeux, J.-P. (1995). Identification of a DNA element determining synaptic expression of the mouse acetylcholine receptor σ-subunit gene. *Proc. Natl. Acad. Sci. U.S.A.* 92:10624–10628.

Kopczynski, C.C., Davis, G.W., Goodman, C.S. (1996). A neural tetraspanin encoded by late bloomer, that facilitates synapse formation. *Science* 271:1867–1870.

Kornau, H.-C., Schenker, L.T., Kennedy, M.B., Seeburg, P.H. (1995). Domain interaction between NMDA receptor subunits and the postsynaptic density protein PSD-95. *Science* 269:1737–1740.

Kröger, S., Horton, S.E., Honig, L.S. (1996). The developing avian retina expresses agrin isoforms during synaptogenesis. *J. Neurobiol.* 29:165–182.

Kuromi, H., Kidokoro, Y. (1984). Nerve disperses pre-existing acetylcholine receptor clusters prior to induction of receptor accumulation in *Xenopus* muscle cultures. *Dev. Biol.* 103:53–61.

Kwon, Y.W., Abbondanzo, J., Stewart, C.L., Gurney, M.E. (1995). Leukemia inhibitory factor influences the timing of programmed synapse withdrawal from neonatal muscles. *J. Neurobiol.* 28:35–50.

Lee, K.F., Simon, H., Chen, H., Bates, B., Hung, M.C., Hauser, C. (1995). Requirement for neuregulin receptor erbB2 in neural and cardiac development. *Nature* 378:394–398.

Levey, M.S., Brumwell, C.L., Dryer, S.E., Jacob, M.H. (1995). Innervation and target tissue interactions differentially regulate acetylcholine receptor subunit mRNA levels in developing neurons *in situ*. *Neuron* 14:153–162.

Li, L., Chin, L.S., Shupliakov, O., Brodin, L., Sihra, T.S., Hvalby, O., Jensen, V., Zheng, D., McNamara, J.O., Greengard, P. (1995). Impairment of synaptic vesicle clustering and of synaptic transmission, and increased seizure propensity, in synapsin I-deficient mice. *Proc. Natl. Acad. Sci. U.S.A.* 92:9235–9239.

Lidov, H.G.W., Byers, T.J., Watkins, S.C., Kunkel, L.M. (1990). Localization of dystrophin to postsynaptic regions of central nervous system cortical neurons. *Nature* 348:725–728.

Lieth, E., Fallon, J.R. (1993). Muscle agrin: neural regulation and localization at nerve-induced acetylcholine receptor clusters. *J. Neurosci.* 13:2509–2514.

Liu, Y., Peter, D., Roghani, A., Schuldiner, S., Privé, G.G., Eisenberg, D., Brecha, N., Edwards, R.H. (1992a). A cDNA that suppresses MPP^+ toxicity encodes a vesicular amine transporter. *Cell* 70:539–551.

Liu, Y., Roghani, A., Edwards, R.H. (1992b). Gene transfer of a reserpine-sensitive mechanism of resistance to N-methyl-4-phenylpyridinium. *Proc. Natl. Acad. Sci. U.S.A.* 89: 9074–9078.

Loeb, J.A., Fischbach, G.D. (1995). ARIA can be released from extracellular matrix through cleavage of a heparin-binding domain. *J. Cell Biol.* 130:127–135.

Lou, X.J., Bixby, J.L. (1993). Coordinate and noncoordinate regulation of synaptic vesicle protein genes during embryonic development. *Dev. Biol.* 159:327–337.

Lou, X.J., Bixby, J.L. (1995). Patterns of presynaptic gene expression define two stages of synaptic differentiation. *Mol. Cell Neurosci.* 6:252–262.

Lu, B., Greengard, P., Poo, M.-M. (1992). Exogenous synapsin I promotes functional maturation of developing neuromuscular synapses. *Neuron* 8:521–529.

Ma, E., Morgan, R., Godfrey, E.W. (1994). Distribution of agrin mRNAs in the chick nervous system. *J. Neurosci.* 14:2943–2952.

Magill-Solc, C., McMahan, U.J. (1988). Motor neurons contain agrin-like molecules. *J. Cell Biol.* 107:1825–1833.

Marazzi, G., Buckley, K.M. (1993). Accumulation of mRNAs encoding synaptic vesicle-specific proteins precedes neurite extension during early neuronal development. *Dev. Dyn.* 197:115–124.

Marchionni, M.A., Goodearl, A.D.J., Chen, M.S., Bermingham-McDonogh, O., Kirk, C., Hendricks, M., Danehy, F., Misumi, D., Sudhalter, J., Kobayashi, K., Wroblewski, D., Lynch, C., Baldassare, M., Hiles, I., Davis, J.B., Hsuan, J.J., Totty, N.F., Otsu, M., McBurney, R.N., Waterfield, M.D., Stroobant, P., Gwynne, D. (1993). Glial growth factors are alternatively spliced erbB2 ligands expressed in the nervous system. *Nature* 362:312–318.

Marikovsky, M., Lavi, S., Pinkas-Kramarski, R., Karunagaran, D., Liu, N., Wen, D., Yarden, Y. (1995). ErB-3 mediates differential mitogenic effects of NDF/heregulin isoforms on mouse keratinocytes. *Oncogene* 10:1403–1411.

Martin, P.T., Sanes, J.R. (1995). Role for a synapse-specific carbohydrate in agrin-induced clustering of acetylcholine receptors. *Neuron* 14:743–754.

Martin, P.T., Ettinger, A.M., Sanes, J.R. (1995). A synaptic localization domain in the synaptic cleft protein, s-laminin/laminin 2. *Science* 269:413–416.

Martin, P.T., Kaufman, S.J., Kramer, R.H., Sanes, J.R. (1996). Synaptic integrins in developing, adult, and mutant muscle: selective association of α1, α7A, and α7B integrins with the neuromuscular junction. *Dev. Biol.* 174:125–139.

Martinou, J.-C., Martinou, I., Kato, A.C. (1992). Cholinergic differentiation factor (CDF/LIF) promotes survival of isolated rat embryonic motoneurons *in vitro*. *Neuron* 8:737–744.

Massoulié, J., Pezzementi, L., Bon, S., Krejci, E., Vallette, F.-M. (1993). Molecular and cellular biology of cholinesterases. *Prog. Neurobiol.* 41:31–91.

Matteoli, M., Takei, K., Perin, M.S., Südhof, T.C., De Camilli, P. (1992). Exo-endocytotic recycling of synaptic vesicles in developing processes of cultured hippocampal neurons. *J. Cell Biol.* 117:849–861.

McMahan, U.J. (1990). The agrin hypothesis. Cold Spring Harb. Symp. *Quant. Biol.* 55: 407–418.

McMahon, H.T., Nicholls, D.G. (1991). The bioenergetics of neurotransmitter release. *Biochem. Biophys. Acta* 1059:243–264.

McMahon, H.T., Südhof, T.C. (1995). Synaptic core complex of synaptobrevin, syntaxin and SNAP25 forms high affinity α-SNAP binding site. *J. Biol. Chem.* 270:2213–2217.

McPherson, P.S., Garcia, E.P., Slepnev, V.I., David, C., Zhang, X., Grabs, D., Sossin, W.S., Bauerfeind, R., Nemoto, Y., De Camilli, P. (1996). A presynaptic inositol-5-phosphatase. *Nature* 379:353–357.

Merlie, J.P., Sanes, J.R. (1985). Concentration of acetylcholine receptor mRNA in synaptic regions of adult muscle fibers. *Nature* 317:66–68.

Meyer, D., Birchmeier, C. (1995). Multiple essential functions of neuregulin in development. *Nature* 378:386–390.

Meyer, G., Kirsch, J., Betz, H., Langosch, D. (1995). Identification of a gephyrin binding motif on the glycine receptor subunit. *Neuron* 15:563–572.

Miner, J.H., Sanes, J.R. (1994). Collagen IV α3, α4 and α5 chains in rodent basal laminae: sequence, distribution, association with laminins, and developmental switches. *J. Cell Biol.* 127:879–891.

Miner, J.H., Lewis, R.M., Sanes, J.R. (1995). Molecular cloning of a novel laminin chain, α5, and widespread expression in adult mouse tissues. *J. Biol. Chem.* 270:28523–28526.

Mishina, M., Takai, T., Imoto, K., Noda, M., Takahashi, T., Numa, S., Methfessel, C., Sakmann, B. (1986). Molecular distinction between fetal and adult forms of muscle acetylcholine receptor. *Nature* 321:406–411.

Missias, A.C., Chu, G.C., Klocke, B., Sanes, J.R., Merlie, J.P. (1996). Maturation of the acetylcholine receptor in developing skeletal muscle: regulation of the AChR γ-to-ε switch. *Dev. Biol.* 179:223–238.

Montanaro, F., Carbonetto, S., Campbell, K.P., Lindenbaum, M.H. (1995). Dystroglycan expression in the wild type and MDX mouse neural retina: synaptic colocalization with dystrophin, dystrophin-related protein but not laminin. *J. Neurosci. Res.* (in press).

Moscoso, M., Merlie, J.P., Sanes, J.R. (1995a). N-CAM, 43K-rapsyn, and s-laminin mRNAs are concentrated at synaptic sites in muscle fibers. *Mol. Cell. Neurosci.* 6:80–89.

Moscoso, L.M., Chu, G.C., Gautam, M., Noakes, P.G., Merlie, J.P., Sanes, J.R. (1995b). Synapse-associated expression of an acetylcholine receptor-inducing protein, ARIA/heregulin, and its putative receptors, ErbB2 and ErbB3, in developing mammalian muscle. *Dev. Biol.* 172:158–169.

Moscoso, L.M. (1996). Molecules that influence neuromuscular interaction in vitro: molecular studies of their regulation and roles *in vivo*. PhD Thesis, Washngton University, St. Louis, MO.

Nabeshima, Y.K., Hanaoka, K., Hayasaka, M., Esumi, S., Li, S., Nonake, I. (1993). *Myogenin* gene disruption results in perinatal lethality because of severe muscle defect. *Nature* 364:532–535.

Nelson, N. (1992). The vacuolar H^+-ATPase—one of the most fundamental ion pumps in nature. *J. Exp. Biol.* 172:19–27.

Nguyen, Q.T., Lichtman, J.W. (1996). Mechanism of synapse disassembly at the developing neuromuscular junction. *Curr. Opin. Neurobiol.* 6:104–112.

Niethammer, M., Kim, E., Sheng, M. (1996). Interaction between the C terminus of NMDA receptor subunits and multiple members of the PSD-95 family of membrane-associated guanylate kinases. *J. Neurosci.* 16:2157–2163.

Nitkin, R.M., Smith, M.A., Magill, C., Fallon, J.R., Yao, MY.-M., Wallace, B.G., McMahan, U.J. (1987). Identification of agrin, a synaptic organizing protein from *Torpedo* electric organ. *J. Cell Biol.* 105:2471–2478.

Noakes, P.G., Phillips, W.D., Hanley, T.A., Sanes, J.R., Merlie, J.P. (1993). 43k protein and acetylcholine receptors co-localize during the initial stages of neuromuscular synapse formation *in vivo*. *Dev. Biol.* 155:275–280.

Noakes, P.G., Gautam, M., Mudd, J., Sanes, J.R., Merlie, J.P. (1995). Aberrant differentiation of neuromuscular junctions in mice lacking s-laminin/lamininβ2. *Nature* 374:258–262.

Nonet, M.L., Grundahl, K., Meyer, B.J., Rand, J.B. (1993). Synaptic function is impaired but not eliminated in *C. elegans* mutants lacking synaptotagmin. *Cell* 73:1291–1305.

O'Connor, L.T., Lauterborn, J.C., Gall, C.M., Smith, M.A. (1994). Localization and alternative splicing of agrin mRNA in adult rat brain: transcripts encoding isoforms that aggregate acetylcholine receptors are not restricted to cholinergic regions. *J. Neurosci.* 14: 1141–1152.

Ohlendieck, K., Ervasti, J.M., Matsumura, K., Kahl, S.D., Leveille, C.J., Campbell, K.P. (1991). Dystrophin-related protein is localized to neuromuscular junctions of adult skeletal muscle. *Neuron* 7:499–508.

Oppenheim, R.W., Prevette, D., Quin-Wei, Y., Collins, F., MacDonald, J. (1991). Control of embryonic motoneuron survival *in vivo* by ciliary neurotrophic factor. *Science* 251: 1616–1618.

Oppenheim, R.W., Houenou, L.J., Johnson, J.E., Lin, L.-F.H., Li, L., Lo, A.C., Newsome, A.L., Prevette, D.M., Wang, S. (1995). Developing motor neurons rescued from programmed and axotomy-induced cell death by neurotrophic factor derived from a glial cell line. *Nature* 373:344–346.

Osend-Sand, A., Catsicas, M., Staple, J.K., Jones, K.A., Ayala, G., Knowles, J., Grenningloh, G., Catsicas, S. (1993). Inhibition of axonal growth by SNAP-25 antisense oligonucleotides *in vitro* and *in vivo*. *Nature* 364:445–448.

Oyler, G.A., Polli, J.W., Wilson, M.C., Billingsley, M.L. (1991). Developmental expression of the 25-kDa synaptosomal-associated protein (SNAP-25) in rat brain. *Proc. Natl. Acad. Sci. U.S.A.* 88:5247–5251.

Patton, B.L., Liu, J., Sanes, J.R. (1995). The synaptic cleft protein s-laminin/laminin β2 promotes synaptic responses by motoneurons, muscle cells, and Schwann cells *in vitro*. *Neuroscience*. (*Abstract*) 20:799.

Patton, B.L., Sanes, J.R. (1996). Evidence that s-laminin/laminin-β2 has a direct role in synaptic regulation of schwann cells *in vivo* at the neuromuscular junction. *Neurosci. Abst.* 21:1474.

Peng, H.B., Baker, L.P., Chen, Q. (1991). Induction of synaptic development in cultured muscle cells by basic fibroblast growth factor. *Neuron* 6:237–246.

Peng, H.B., Afshan Ali, A., Dai, Z., Daggett, D.F., Raulo, E., Rauvala, H. (1995). The role of heparin-binding growth-associated molecule (HB-GAM) in the postsynaptic induction in cultured muscle cells. *J. Neurosci.* 15:3027–3038.

Pevsner, J., Hsu, S.-C., Scheller, R.H. (1994a). N-Sec1: a neural-specific syntaxin-binding protein. *Proc. Natl. Acad. Sci. U.S.A.* 91:1445–1449.

Pevsner, J., Hsu, S.-C., Braun, J.E.A., Calakos, N., Ting, A.E., Bennett, M.K., Scheller, R.H. (1994b). Specificity and regulation of a synaptic vesicle docking complex. *Neuron* 13: 353–361.

Pfeffer, S.R. (1992). GTP binding proteins in intracellular transport. *Trends Cell Biol.* 2:41–46.

Phillips, W.D., Kopta, C., Blount, P., Gardner, P.D., Steinbach, J.H., Merlie, J.P. (1991). ACh receptor-rich domains organized in fibroblasts by recombinant 43-kilodalton protein. *Science* 251:568–570.

Phillips, W.D., Noakes, P.G., Roberds, S.L., Campbell, K.P., Merlie, J.P. (1993). Clustering and immobilization of acetylcholine receptors by the 43-kD protein: a possible role for dystrophin-related protein. *J. Cell Biol.* 123:729–740.

Pieribone, V.A., Shupllakov, O., Brodin, L., Hilfiker-Rothenfluh, S., Czernik, A.J., Greengard, P. (1995). Distinct pools of synaptic vesicles in neurotransmitter release. *Nature* 375:493–497.

Piette, J., Huchet, M., Houzelstein, D., Changeux, J.P. (1993). Compartmentalized expression of the α and δ-subunits of the acetylcholine receptor in recently fused myotubes. *Dev. Biol.* 157:205–213.

Pinkas-Kramarski, R., Eilam, R., Spiegler, O., Lavi, S., Liu, N., Chang, D., Wen, D., Schwartz, M., Yarden, Y. (1994). Brain neurons and glial cells express neu differentiation factor/heregulin: a survival factor for astrocytes. *Proc. Natl. Acad. Sci. U.S.A.* 91: 9387–9391.

Porter, B.E., Weis, J., Sanes, J.R. (1995). A motoneuron-selective stop signal in the synaptic protein S-laminin. *Neuron* 14:549–559.

Pourquié, O., Fan, C.-M., Coltey, M., Hirsinger, E., Watanabe, Y., Bréant, C., Francis-West, P., Brickell, P., Tessier-Lavigne, M., Le Douarin, L.M. (1996). Lateral and axial signals involved in avian somite patterning: a role for BMP4. *Cell* 84:461–471.

Prior, P., Schmitt, B., Grenningloh, G., Pribilla, I., Multhaup, G., Beyreuther, K., Maulet, Y., Werner, P., Langosch, D., Kirsch, J., Betz, H. (1992). Primary structure and alternative splice variants of gephyrin, a putative glycine receptor-tubulin linger protein. *Neuron* 8:1161–1170.

Pumplin, D.W., Reese, T.S., Llinás, R. (1981). Are the presynaptic membrane particles the calcium channels? *Proc. Natl. Acad. Sci. U.S.A.* 78:7210–7213.

Rees, R.P. (1978). The morphology of interneuronal synaptogenesis: a review. *FASEB J.* 37: 2000–2009.

Régnier-Vigouroux, A., Tooze, S.A., Huttner, W.B. (1991). Newly synthesized synaptophysin is transported to synaptic-like microvesicles via constitutive secretory vesicles and the plasma membrane. *EMBO J.* 10:3589–3601.

Reist, N.E., Werle, M.J., McMahan, U.J. (1992). Agrin released by motor neurons induces the aggregation of acetylcholine receptors at neuromuscular junctions. *Neuron* 8:865–868.

Role, L.W. (1988). Neural regulation of acetylcholine sensitivity in embryonic sympathetic neurons. *Proc. Natl. Acad. Sci. U.S.A.* 85:2825–2829.

Role, L.W., Matossian, R.R., O'Brien, R.J., Fischbach, G.D. (1985). On the mechanism of acetylcholine receptor accumulation at newly formed synapses of chick myotubes. *J. Neurosci.* 5:2197–2204.

Rousselet, A., Cartaud, J., Devaux, P.F., Changeux, J.-P. (1982). The rotational diffusion of the acetylcholine receptor in *Torpedo marmorata* membrane fragments studied with a spin-labelled alpha-toxin: importance of the 43,000 protein(s). *EMBO J.* 1:439–445.

Rosahl, T.W., Spillane, D., Missler, M., Herz, J., Selig, D.K., Wolff, J.R., Hammer, R.E., Malenka, R.C., Südhof, T.C. (1995). Essential functions of synapsins I and II in synaptic vesicle regulation. *Nature* 375:488–493.

Rossi, S.G., Rotundo, R.L. (1996). Transient interactions between collagen-tailed acetylcholinesterase and sulfated proteoglycans prior to immobilization on the extracellular matrix. *J. Biol. Chem.* 271:1979–1987.

Rudnicki, M.A., Jaenisch, R. (1995). The MyoD family of transcription factors and skeletal myogenesis. *Bioessays* 17:203–209.

Rüegg, M.A., Tsim, K.W., Horton, S.E., Kröger, S., Escher, G., Gensch, E.M., McMahan, U.J. (1992). The agrin gene codes for a family of basal lamina proteins that differ in function and distribution. *Neuron* 8:691–699.

Rupp, F., Payan, D.G., Magill-Solc, C., Cowan, D.M., Scheller, R.H. (1991). Structure and expression of a rat agrin. *Neuron* 6:811–823.

Ryan, T.A., Reuter, H., Wendland, B., Schweitzer, F.E., Tsien, R.W., Smith, S.J. (1993). The kinetics of synaptic vesicle recycling measured at single presynaptic boutons. *Neuron* 11:713–724.

Salpeter, M.M., Marchaterre, M., Harris, R. (1988). Distribution of extrajunctional acetylcholine receptors on a vertebrate muscle: evaluated by using a scanning electron microscope autoradiographic procedure. *J. Cell Biol.* 106:2087–2093.

Sandrock, A.W., Jr., Goodearl, A.D., Yin, Q.W., Chang, D., Fischbach, G.D. (1995). ARIA is concentrated in nerve terminals at neuromuscular junctions and at other synapses. *J. Neurosci.* 15:6124–6136.

Sanes, J.R. (1995). The synaptic cleft of the neuromuscular junction. *Semin. Dev. Biol.* 6:163–173.

Sanes, J.R., Marshall, L.M., McMahan, U.J. (1978). Reinnervation of muscle fiber basal lamina after removal of myofibers. Differentiation of regenerating axons at original synaptic sites. *J. Cell Biol.* 78:176–198.

Sanes, J.R., Engvall, E., Butkowski, R., Hunter, D.D. (1990). Molecular heterogeneity of basal laminae: isoforms of laminin and collagen IV at the neuromuscular junction and elsewhere. *J. Cell Biol.* 111:1685–1699.

Sanes, J.R., Kotzbauer, P.T., Mudd, J., Hanley, T., Martinou, J.-C., Merlie, J.P. (1991). Selective expression of an acetylcholine receptor-lacZ transgene in synaptic nuclei of adult muscle fibers. *Development* 113:1181–1191.

Sargent, P.B. (1993). The diversity of neuronal nicotinic acetylcholine receptors. *Annu. Res. Neurosci.* 16:403–443.

Scheller, R.H. (1995). Membrane trafficking in the presynaptic nerve terminal. *Neuron* 14: 893–897.

Scott, L.J.C., Bacou, F., Sanes, J.R. (1988). A synapse-specific carbohydrate at the neuromuscular junction: association with both acetylcholinesterase and a glycolipid. *J. Neurosci.* 8:932–944.

Sheng, M., Cummings, J., Roldan, L.A., Jan, Y.N., Jan, L.Y. (1994). Changing subunit composition of heteromeric NMDA receptors during development of rat cortex. *Nature* 368:144–147.

Shirataki, H., Yamamoto, T., Hagi, S., Miura, H., Oishi, H., Jin-no, Y., Senbonmatsu, T., Takai, Y. (1994). Rabphilin-3A is associated with synaptic vesicles through a vesicle protein in a manner independent of Rab 3A. *J. Biol. Chem.* 269:32717–32720.

Simon, A.M., Hoppe, P., Burden, S.J. (1992). Spatial restriction of AChR gene expression to subsynaptic nuclei. *Development* 114:545–553.

Smith, M.A., O'Dowd, D.K. (1994). Cell specific regulation of agrin mRNA splicing in the chick ciliary ganglion. *Neuron* 12:795–804.

Smith, M.A., Magill-Solc, C., Rupp, F., Yao, Y.M.-M., Schilling, J.W., Snow, P., McMahan, U.J. (1992). Isolation and characterization of a cDNA that encodes an agrin homolog in the marine ray. *Mol. Cell. Neurosci.* 3:406–417.

Sobel, A., Heidmann, T., Hofler, J., Changeux, J.P. (1978). Distinct protein components from *Torpedo marmorata* membranes carry the acetylcholine receptor site and the binding site for local anesthetics and histrionicotoxin. *Proc. Natl. Acad. Sci. U.S.A.* 75:510–514.

Söllner, T., Bennett, M.K., Whiteheart, S.W., Scheller, R.H., Rothman, J.E. (1993a). A protein assembly-disassembly pathway *in vitro* that may correspond to sequential steps of vesicle docking, activation, and fusion. *Cell* 75:409–418.

Söllner, T., Whiteheart, S.W., Brunner, M., Erdjument-Bromage, H., Geromanos, S., Tempst, P., Rothman, J.E. (1993b). SNAP receptors implicated in vesicle targeting and fusion. *Nature* 362:318–324.

Steward, O., Banker, G. (1992). Getting the message from the gene to the synapse: sorting and intracellular transport of RNA in neurons and other spatially complex cells. *Trends Neurosci.* 15:180–186.

Südhof, T.C. (1995). The synaptic vesicle cycle: a cascade of protein-protein interactions. *Nature* 375:645–653.

Sugiyama, J., Bowen, D.C., Hall, Z.W. (1994). Dystroglycan binds nerve and muscle agrin. *Neuron* 13:103–115.

Tang, J., Jo, S.A., Burden, S.J. (1994). Separate pathways for synapse-specific and electrical activity-dependent gene regulation. *Development* 120:1799–1804.

Tansey, M.G., Chu, G.C., Merlie, J.P. (1996). ARIA/HRG regulates AChR ϵ subunit gene expression at the neuromuscular synapse via activation of phosphatidylinositol 3-kinase and Ras/MAPK pathway. *J. Cell Biol.* 134:465–476.

Taylor, J., Gordon-Weeks, P.R. (1989). Developmental changes in the calcium dependency of γ-aminobutyric acid release from isolated growth cones: correlation with growth cone morphology. *J. Neurochem.* 53:834–843.

Trachtenberg, J.T., Thompson, W.J. (1996). Schwann cell apoptosis at developing neuromuscular junctions is regulated by glial growth factor. *Nature* 379:174–177.

Tsen, G., Halfter, W., Kröger, S., Cole, G.J. (1995a). Agrin is a heparan sulfate proteoglycan. *J. Biol. Chem.* 270:3392–3399.

Tsen, G., Napier, A., Halfters, W., Cole, G.J. (1995b). Identification of a novel alternatively spliced agrin mRNA that is preferentially expressed in non-neuronal cells. *J. Biol. Chem.* 270:15934–15937.

Tsim, K.W.K., Rüegg, M.A., Escher, G., Kröger, S., McMahan, U.J. (1992). cDNA that encodes active agrin. *Neuron* 8:677–689.

Ullrich, B., Ushkaryov, Y.A., Südhof, T.C. (1995). Cartography of neurexins: more than 1000 isoforms generated by alternative splicing and expressed in distinct subsets of neurons. *Neuron* 14:497–507.

Usdin, T.B., Fischbach, G.D. (1986). Purification and characterization of a polypeptide from chick brain that promotes the accumulation of acetylcholine receptors in chick mykotubes. *J. Cell Biol.* 103:493–507.

Ushkaryov, Y.A., Petrenko, A.G., Geppert, M., Südhof, T.C. (1992). Neurexins: synaptic cell surface proteins related to the a-Latrotoxin receptor and laminin. *Science* 257:50–56.

Valenzuela, D.M., Stitt, T.N., DiStefano, P.S., Rojas, E., Mattsson, K., Compton, D.L., Nuñez, L., Park, J.S., Stark, J.L., Gies, D.R., Thomas, S., Le Beau M.M., Fernald, A.A., Copeland, N.G., Jenkins, N.A., Burden, S.J., Glass, D.J., Yancopoulos, G.D. (1995). Receptor tyrosine kinase specific for the skeletal muscle lineage: expression in embryonic muscle, at the neuromuscular junction, and after injury. *Neuron* 15:573–584.

Vaughn, J.E. (1989). Review: fine structure of synaptogenesis in the vertebrate central nervous system. *Synapse* 3:255–285.
Walke, W., Staple, J., Adams, L., Gnegy, M., Chahine, K., Goldman, D. (1994). Calcium-dependent regulation of rat and chick muscle nicotinic acetylcholine receptor (nAChR) gene expression. *J. Biol. Chem.* 269:19447–19456.
Wallace, B.G. (1989). Agrin-induced specializations contain cytoplasmic, membrane, and extracellular matrix-associated components of the postsynaptic apparatus. *J. Neurosci.* 9:1294–1302.
Wallace, B.G. (1994). Staurosporine inhibits agrin-induced acetylcholine receptor phosphorylation and aggregation. *J. Cell Biol.* 125:661–668.
Wallace, B.G. (1995). Regulation of the interaction of nicotinic acetylcholine receptors with the cytoskeleton by agrin-activated protein tyrosine kinase. *J. Cell Biol.* 128:1121–1129.
Wheeler, D.B., Randall, A., Tsien, R.W. (1994). Roles of N-type and Q-type $CA2^{+}$ channels in supporting hippocampal synaptic transmission. *Science* 264:107–111.
Young, S.H., Poo, M.-M. (1983a). Spontaneous release of transmitter from growth cones of embryonic neurons. *Nature* 305:634–637.
Yu, X.-M., Hall, Z.W. (1994). The role of the cytoplasmic domains of individual subunits of the acetylcholine receptor in 43 kDa protein-induced clustering in COS cells. *J. Neurosci.* 14:785–795.
Zhu, X., Lai, C., Thomas, S., Burden, S.J. (1995). Neuregulin receptors, erbB3 and erbB4, are localized at neuromuscular synapses. *EMBO J.* 14:5842–5848.

6

Neurotrophic factors and their receptors

Roles in neuronal development and function

LOUIS F. REICHARDT AND ISABEL FARIÑAS

Neurotrophic factors are important regulators of the development and maintenance of vertebrate nervous systems (reviewed in Korsching, 1993; Eide et al., 1993); Fariñas and Reichardt, 1996. During the development of the nervous system, neuronal populations undergo a process of naturally occurring cell death at a time when their axons are innervating target areas. It is believed that this mechanism ensures a balance between the size of an innervating population and the size of its target territory. As illustrated in Figure 6–1, the production of limited amounts of survival factors by target organs is thought to regulate this matching process and constitutes the central concept in the Neurotrophic Factor Hypothesis which was first formulated by Hamburger, Levi-Montalcini, and their colleagues (reviewed in Levi-Montalcini, 1987; Purves, 1988). Nerve growth factor (NGF), the first neurotrophic factor to be discovered, was identified during a search for such survival factors. NGF and its family members have been named neurotrophins. More recent work has shown that neurotrophins are not the only molecules likely to be secreted by target organs that regulate neuronal survival and differentiation as described above (Table 6–1). To cite only a few examples, GDNF, and cytokines that activate LIF-receptor-β/gp130–mediated signaling appear to have similar critical roles in regulating neuronal survival and development (e.g., Moore et al., 1996; Durbec et al., 1996; DeChiara et al., 1995). This chapter focuses on the neurotrophins with brief descriptions of the actions of these other molecules.

Seminal experiments of Levi-Montalcini and colleagues in which antibodies to NGF were injected into neonatal rodents demonstrated that this factor is essential for the survival of virtually all sympathetic neurons in vivo (reviewed in Levi-Montalcini, 1987; Purves, 1988). Subsequent experiments demonstrated that prenatal exposure to anti-NGF also resulted in loss of nociceptive sensory neurons. Definitive experiments have since confirmed each of the major predictions of the Trophic Factor Hypothesis for NGF. First target ablation, axotomy, or inhibition of retrograde axonal transport has been shown to result in death of neonatal sympathetic and sensory neurons. Second, administration of NGF has

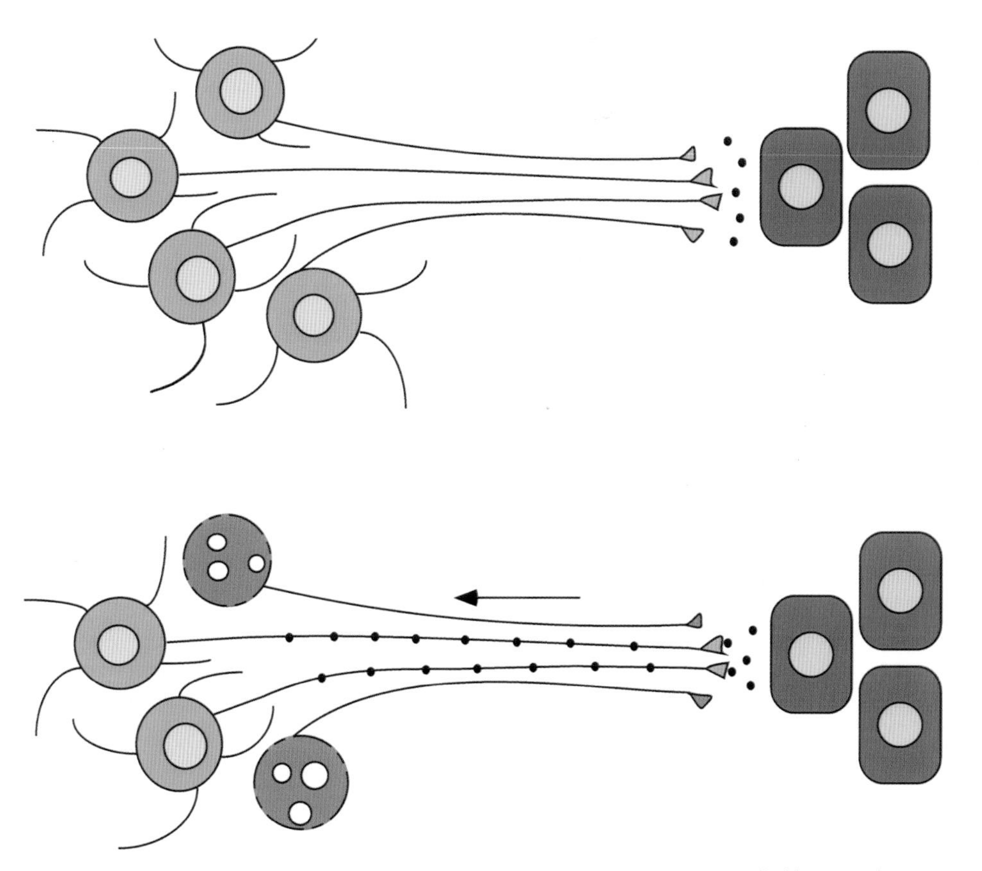

FIGURE 6–1 The Neurotrophic Factor Hypothesis. (Top) Neurons (left) extend axons to target cells (right) which secrete limiting amounts of neurotrophic factors (solid circles). (Bottom) These factors bind specific receptors and are internalized and transported retrogradely to the neuronal cell somata (direction indicated by solid arrow), where they promote neuronal survival. Neurons without access to adequate amounts of these factors die by apoptotic mechanisms (depicted as dark cells with membrane blebs and fragmenting nuclei). Apart from the regulation of the process of target-dependent neuronal survival, several additional functions have been proposed for these molecules, which are described in the text.

been shown to prevent both experimentally induced and naturally occurring cell death. NGF applied to sympathetic or sensory nerve terminals has been shown to be bound to specific receptors, internalized, and transported retrogradely to the cell body, resulting in a variety of changes including induction of transmitter enzymes. Finally, a sensitive two-site enzyme-linked immunosorbent assay (ELISA), RNA blot assays, and in situ analyses have demonstrated that minute, but biologically relevant quantities of NGF protein and mRNA are present in targets of sympathetic and sensory neurons and that NGF protein, but not mRNA, is present in the cell bodies of these same neurons (see Korsching, 1993). Maintenance of NGF protein levels in these neurons is dependent on retrograde axonal transport.

TABLE 6–1. Neurons Responsive to Different Neurotrophic Factors

Neurotrophic Factor	*Peripheral Nervous System*	*Central Nervous System*
NGF	Sympathetic Sensory (neural crest derived)	Basal forebrain cholinergic Striatal cholinergic Cerebellar Purkinje
BDNF	Sensory neurons (neural crest and placode derived)	Basal forebrain cholinergic Trigeminal mesencephalic Substantia nigra dopaminergic Cortical and hippocampal Cerebellar granule Retinal ganglion Motor
NT-3	Sympathetic Sensory Enteric	Basal forebrain cholinergic Locus coeruleus adrenergic Motor
CNTF	Parasympathetic (ciliary) Sympathetic Sensory	Striatal cholinergic Motor
GNDF	Sympathetic Sensory Enteric	Substantia nigra dopaminergic Locus coeruleus adrenergic Motor

After injury, however, neurons innervating peripheral tissues utilize an important alternative source of several neurotrophic molecules, including NGF, whose synthesis in Schwann and other cells in peripheral nerve is strongly induced as part of an inflammatory response mediated by macrophage infiltration, activation, and cytokine release (reviewed by Korsching, 1993). This alternate source of trophic factor support is believed to be essential for survival of many injured neurons and for successful regeneration of their axons.

Neurotrophic factors have been implicated in the regulation of developmental processes other than target-tissue–regulated neuronal survival, as shown by in vitro and in vivo studies. First, these factors have been shown to regulate survival of neuroblasts, O2A cells, and immature neurons before innervation of their final targets (e.g., Birren et al., 1993; DiCicco-Bloom et al., 1993; Barres, 1994; Buchman and Davies, 1993). Thus they act as survival factors for a wider spectrum of cells than envisioned in the initial formulation of the Neurotrophic Factor Hypothesis. In addition, like other factors which activate tyrosine kinases, neurotrophic factors have been shown to regulate proliferation of cells such as neural crest cells and O2A oligodendrocyte precursors (Kalcheim et al., 1992; Barres et al., 1994). They have also been shown to regulate the pathways of differentiation selected by certain precursors (e.g., Sieber-Blum et al., 1993; Anderson, 1993; Patterson and Nawa, 1993) and to regulate the differentiation process, helping determine the levels of expression of many proteins expressed in differentiated neurons that are important for their physiological function, such as neurotransmitters and calcium-binding proteins (e.g., Ip et al., 1993a; Nawa et al., 1993). Neurotrophins, but not other neurotrophic factors, also have been shown to regulate axon growth in vitro and innervation of target tissues in vivo (e.g., Campenot, 1977; Gundersen and Barrett, 1979; Edwards et al., 1989; Hoyle et al., 1993). They also regulate collateral branching and sprouting after nerve or spinal cord transection (Diamond et al., 1992; Schnell et al., 1994). In the CNS, neurotrophins have been shown to regulate formation of ocular dominance columns (Cabelli et al., 1995).

Since many neurotrophic factors continue to be expressed in adult animals, they could potentially modulate the function and plasticity of the mature nervous system. Particularly intriguing are recent results indicating that neurotrophins and CNTF can regulate synaptic communication. Neurotrophins and CNTF regulate exocytosis by motor neurons and hippocampal neurons (Lohof et al., 1993; Kang and Schuman, 1995, 1996). Recent work also implicates these molecules in long-term potentiation (LTP) in the CA1 region of the hippocampus (Korte et al., 1995; Patterson et al., 1996). Neurotrophins have also been shown to mediate sensitivity of adult sensory neurons to nociceptive stimuli (e.g., Lewin and Mendell, 1994). Of particular importance for studies on aging and neurodegenerative diseases are results showing that the neurotrophins, GDNF, and LIF/CNTF cytokines protect, at least partially, responsive neurons from axotomy and exposure to neurotoxins (reviewed in Eide et al., 1993).

Factors, Receptors, and Signaling Mechanisms

At present, five neurotrophins have been isolated: nerve growth factor (NGF); brain-derived neurotrophic factor (BDNF), neurotrophin-3 (NT-3), NT-4/5, and NT-6. Each appears to function as a noncovalently associated homodimer. The tertiary structures of NGF and NT-4/5 have been solved and novel features of their structures—a tertiary fold and cystine knot—have been identified in structures of several other growth factors, including PDGF and TGF-β(reviewed in McDonald and Chao, 1995). At the present time NT-6 has only been identified in fish and NT-4/5 has not yet been detected in avian species. Evidence indicates that NGF and NT-6 act on similar, perhaps identical populations of neurons (e.g., Gotz et al., 1994). Similarly, BDNF and NT-4/5 appear to have almost identical targets (e.g., Ip et al., 1993b). Thus the specificities of individual neurotrophins divide them into three general classes and all vertebrate species examined to date have at least one neurotrophin in each class. GDNF is a distant relative of TGF-β and therefore shares some structural features with the neurotrophins (Lin et al., 1993). While originally believed to be a specific trophic factor for midbrain dopaminergic neurons, it has since been shown to act on many distinct neuronal populations, many of which are also responsive to a neurotrophin (e.g., Henderson et al., 1994; Buj-Bello et al., 1995; Yan et al., 1995). CNTF and LIF are structurally related to several hematopoietic cytokines including interleukin-6 and granulocyte-colony–stimulating factor (Bazan, 1991). A broad spectrum of neurons are responsive to these factors in vitro, including many also responsive to GDNF or neurotrophins (e.g., Ernsberger et al., 1989; Murphy et al., 1993; Barbin et al., 1994). Thus the in vitro data suggest the potential for substantial functional overlap and redundancy in vivo.

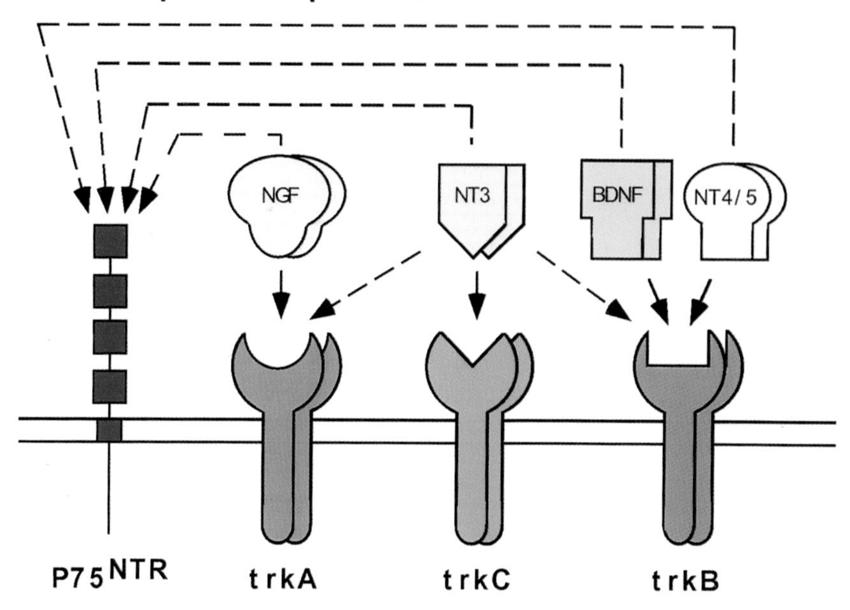

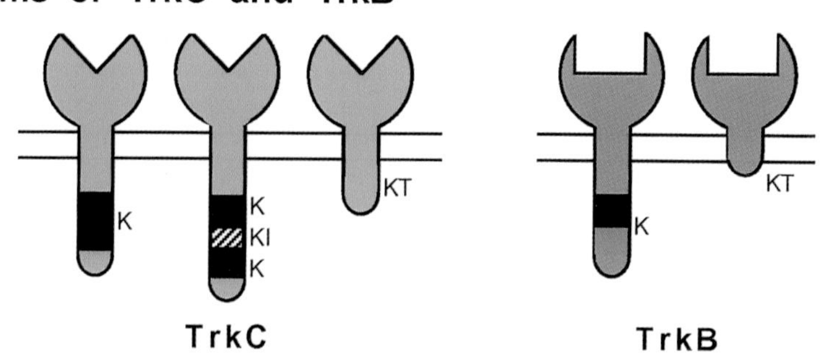

FIGURE 6–2 Neurotrophic factors and their receptors. (A) Interactions of neurotrophins with trk receptors. This figure illustrates the interactions of each member of the neurotrophin family with the protein products of trk receptor genes. Strong interactions are depicted with solid arrows; weaker interactions with broken arrows. The interactions of NT-6 are not included in this figure because they have not been characterized in ligand-receptor–binding assays. The observed specificity of NT-6 in neuronal survival and axon outgrowth assays indicates that it interacts with $gp140^{trkA}$ but not other trk receptors. In addition, all neurotrophins can bind to the low-affinity NGF receptor $p75^{NTR}$. (B) Isoforms of trkC and trkB. Differential splicing of the trkB and trkC genes generates truncated (KT) receptor isoforms, lacking kinase domains (K), or receptors with kinase domain inserts (KI).

After the discovery of the neurotrophins, their apparently unique actions on neurons made it seem likely that their receptors and signal transduction pathways would prove to be completely different from those of mitogenic growth factors, such as PDGF or EGF, whose receptors are protein tyrosine kinases. Seminal discoveries during the past decade, however, have shown that the neurotrophins and many other neurotrophic factors bind, dimerize, and thereby ac-

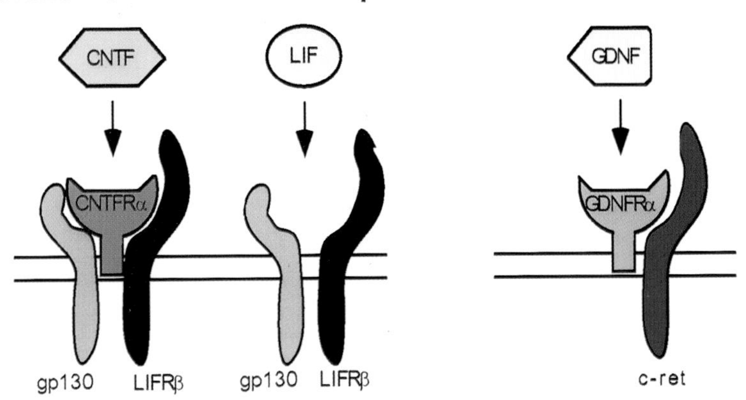

FIGURE 6–2 (*Continued*) (C) Cytokine and GDNF–receptor interactions. (Left) CNTF and LIF induce association of multisubunit heteromeric receptor complexes consisting of an LIF-receptor b and gp130 heterodimer. An additional receptor subunit, CNTFR-a, is required for CNTF to induce association of the complex. (Right) GDNFRα binds GDNF specifically and mediates activation of the protein tyrosine kinase c-ref

tivate protein tyrosine kinases, which act on substrates and signaling pathways shared, at least in part, with other growth and differentiation factors. As depicted in Figure 6–2, the major signal-transducing receptors for the neurotrophins are a family of three homologous receptor tyrosine kinases named trkA, trkB, and trkC (reviewed by Bothwell, 1995; Greene and Kaplan, 1995). Despite its homology to TGF-β, which activates a receptor with serine-threonine kinase activity, very recent work has shown that GDNF, in a complex with a GPI-linked binding protein named GDNF receptor-α, activates the receptor tyrosine kinase c-ret (Durbec et al., 1996b; Trupp et al., 1996; Jing et al., 1996; Treanor et al., 1996). CNTF, LIF, and other members of the IL-6 family of cytokines interact with a completely distinct family of receptor subunits (reviewed in Schindler and Darnell, 1995). In brief, LIF associates with LIF receptor-β and this complex associates with a subunit named gp130 to form a transmembrane receptor heterodimer. The dimerized receptor complexes activate associated cytoplasmic JAK protein tyrosine kinases which phosphorylate many substrates including the receptor complex and STAT transcription factors. CNTF acts through the same receptor heterodimer, but CNTF interactions with LIF receptor-β and gp130 depend upon prior association of CNTF with an additional protein, the CNTF receptor-α subunit. Thus GDNF, the cytokines and neurotrophins all act by promoting dimerization of receptors, which results in activation of protein tyrosine kinases. The trk receptors have been shown to activate many of the same signaling pathways as the EGF and PDGF receptors, including phosphatidylinositol-3-kinase (PI-3-kinase), phospholipase-C-γ1 (PLC-γ1), ras, and kinases activated by ras (illustrated in Fig. 6–3). After phosphorylation, specific tyrosines in the cytoplasmic domains of trk receptors have been shown to recruit specific signaling molecules. Mutagenesis of these tyrosines has been used to inhibit specific signaling pathways and determine the roles of each pathway in mediating cell sur-

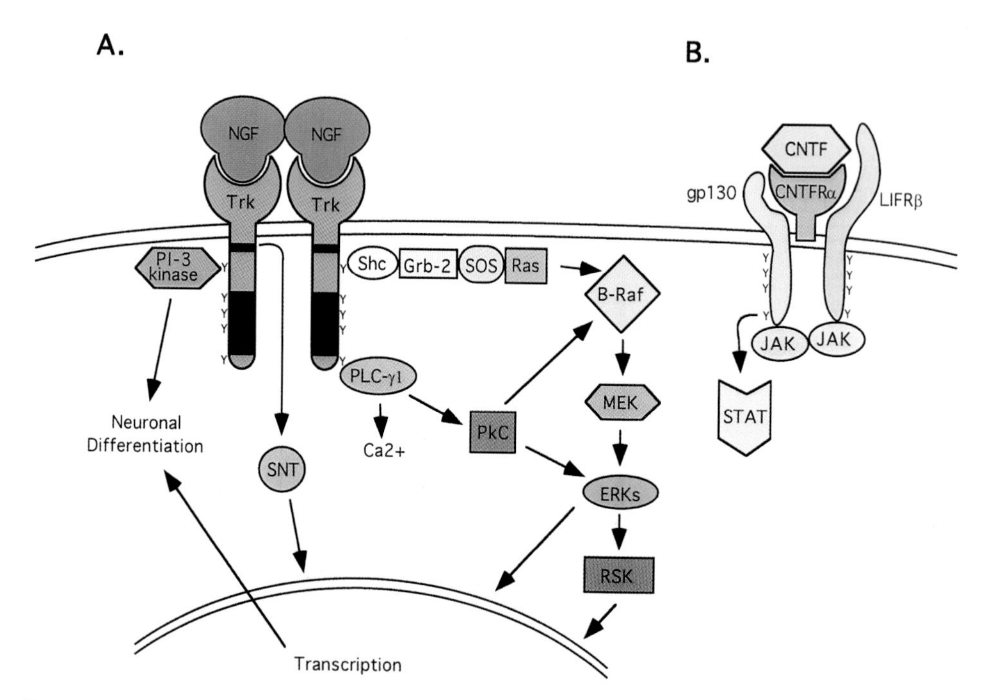

FIGURE 6–3 Signaling pathways related to the activation of neurotrophic factor receptors. (A) Neurotrophin signal transduction by trk receptors. Upon ligand binding, trk dimerization and subsequent transphosphorylation lead to kinase activation and receptor association with the cytoplasmic proteins SHC, PI-3-kinase, and PLC-g. These proteins activate several signaling pathways including ras. Trk activation also induces the phosphorylation of SNT, a target whose phosphorylation is associated with neuronal differentiation. (B) Signal transduction by cytokine receptors. Receptor dimerization activates associated cytoplasmic JAK tyrosine kinases which phosphorylate, among other substrates, receptor complexes and STAT transcription factors.

vival, neurite outgrowth, gene induction, and other functions of neurotrophins (reviewed in Greene and Kaplan, 1995). Phosphorylation of specific tyrosines on cytokine receptors can similarly provide docking sites for proteins, such as PI-3-kinase (e.g., Damen et al., 1995), but it is not certain which, if any, of these are recruited directly by the gp130/LIF-receptor-β complex.

While each of these neurotrophic factors acts, at least in part, through activation of a tyrosine kinase signaling cascade, there are significant differences in the consequences of kinase activation. For example, while NGF was originally isolated as a survival factor for sympathetic neurons, LIF was isolated as a factor that promotes cholinergic differentiation, but not survival of these same neurons (reviewed in Patterson and Nawa, 1993). During the past year, the range of differences in signals transmitted to neurons or their targets has expanded still further with the demonstration that activation of eph-family tyrosine kinases can inhibit neurite outgrowth (Drescher et al., 1995) and demonstration of an essential role for a receptor-like tyrosine kinase in synapse differentiation at the neuromuscular junction (DeChiara et al., 1996). These quite different actions are likely to depend not only upon specific signals conveyed by each kinase, but also upon properties of each cell in which a kinase is expressed.

In general, the expression patterns and ligand-binding properties of the receptors described above are sufficient to explain the responsiveness of different neuronal populations to individual neurotrophic factors. Consistent with the patterns of biological responsiveness described above, each neurotrophin binds strongly to and activates one of three homologous trk receptors. As summarized in Figure 6–2, NGF interacts specifically with trkA; since NT-6 (not depicted) appears to act only on NGF-responsive neurons, it is also likely to interact specifically with trkA; BDNF and NT-4/5 activate primarily trkB; NT-3 specifically activates trkC but can also activate less efficiently trkA and trkB, at least in some cellular contexts (see Ip et al., 1993b; Benedetti et al., 1993; Clary and Reichardt, 1994). Neurons responsive to NGF invariably express trkA; those responsive to BDNF and NT-4 express trkB; those responsive to CNTF must express the CNTF receptor-α subunit; and so on.

This overview, however, is not sufficient to explain all aspects of neurotrophin responsiveness. In addition to the receptors described above, the neurotrophins interact with two additional classes of cell-surface receptors depicted in Figure 6–2—isoforms of trk receptors lacking functional kinase domains and the pan-neurotrophin receptor $p75^{NTR}$ (reviewed by Bothwell, 1995; McDonald and Chao, 1995). Differential splicing of the trkB and trkC genes has been shown to generate receptor isoforms lacking functional kinase domains or with kinase domain inserts which modify their specificity. Receptor isoforms lacking functional kinase domains are expressed in many non-neuronal cells which do not exhibit detectable responsiveness to the neurotrophins. In such cells, the functions of these isoforms are poorly understood but could include neurotrophin degradation and/or presentation to adjacent neurons. In neurons, these isoforms almost certainly inhibit neurotrophin signaling by forming nonproductive heterodimers with trk receptor monomers containing a kinase domain (e.g., Eide et al., 1996). Most characterizations of trk receptor expression patterns have not used reagents capable of distinguishing between these isoforms. Consequently, it seems likely that there are discrepancies between trk expression on neurons and their responsiveness to neurotrophins which reflects the presence of these isoforms. As an extreme example of possible complexity, some neurons may express a functional trk receptor but lack neurotrophin responsiveness because they also express high amounts of a trk isoform with inhibitory activity.

Each neurotrophin binds with similar affinity to the low-affinity neurotrophin receptor $p75^{NTR}$ whose extracellular domain is related to that of the TNF receptor but is not structurally related to the trk family members. In vitro studies on $p75^{NTR}$ have led to many models, but comparatively little definitive evidence on its functions. Its best-documented role is presentation of NGF to trkA receptors (see Bothwell, 1995; McDonald and Chao, 1995). Kinetic analysis indicates that $p75^{NTR}$ exhibits fast association and dissociation rates with NGF, while trkA exhibits association and dissociation rates about 100-fold slower. Coexpression of the two receptors generates a high-affinity binding site detected in Scatchard analyses which may reflect presentation by $p75^{NTR}$ of NGF to trkA receptors. Consistent with these measurements, $p75^{NTR}$ appears to be important in potentiating trkA activation by low, but not saturating, concentrations of NGF (e.g., Barker and Shooter, 1994). Surprisingly, $p75^{NTR}$ does not appear to be important in potentiating interactions of other neurotrophins with trkB or trkC. Analysis of sensory neuron cultures from $p75^{NTR}$-deficient homozygous and heterozygous mice has provided strong evidence for the importance of $p75^{NTR}$ in potentiating neuronal responses to NGF in vitro (Davies et al., 1993; Lee et al., 1994a). In culture, reduction or elimination of $p75^{NTR}$ shifted the dose response of sensory and sympathetic neurons to NGF but not to other neurotrophins. The major phenotype of $p75^{NTR}$-deficient homozygotes is loss of NGF-dependent sensory neurons (Lee et al., 1992), which is consistent with $p75^{NTR}$ enhancing NGF responsiveness in vivo.

Studies in cell culture have also indicated that the presence of $p75^{NTR}$ refines the specificity of neurotrophin action by reducing activation of trkA by the nonpreferred ligand NT-3 (e.g., Benedetti et al., 1993; Clary and Reichardt, 1994). Analysis of neuronal cultures from the $p75^{NTR}$-deficient mice also indicates that sympathetic neurons are more responsive to NT-3 than similar neurons from control animals. The importance of this function in vivo has not been critically analyzed using the $p75^{NTR}$-deficient animals. In addition, $p75^{NTR}$ may transmit signals to cells directly. In one set of experiments, NGF binding to $p75^{NTR}$ has been reported to activate the sphingomyelin cycle (Dobrowsky et al., 1994). Perhaps through this signaling pathway, NGF activation of $p75^{NTR}$ also promotes migration by Schwann cells which do not express trkA (Anton et al., 1994). In addition, it has been proposed that $p75^{NTR}$ in the absence of NGF can promote apoptosis (Rabizadeh et al., 1993). In recent work, NGF, but not other neurotrophins, has been shown to promote apoptosis of adult rat oligodendrocytes in vitro and embryonic chick retinal ganglion cells in vivo by activation of p75NTR (Casaccia-Bonnefil et al., 1996; Frade et al., 1996). In addition, a population of basal forebrain cholinergic neurons has recently been identified that expresses p75NTR, but not trkA, is normally lost during postnatal development, but is spared in mutants lacking p75NTR (Van der Zee et al., 1996). Finally, $p75^{NTR}$ has been shown to mediate retrograde transport of neurotrophins and appears particularly important in facilitating retrograde transport of NT-4/5 (Curtis et al., 1995). Analysis, however, of the $p75^{NTR}$-deficient mice has not revealed an obvious deficiency in any population of neurons dependent on NT-4/5, as might be predicted by these results (compare Lee et al., 1992 and 1994a to Conover et al., 1995; Liu et al., 1995).

ESSENTIAL EFFECTS ON NEURONAL SURVIVAL

The landmark experiments (described earlier) of Hamburger, Levi-Montalcini, and later workers demonstrated conclusively that NGF has essential roles in vivo in maintaining the viability of nociceptive sensory and sympathetic neurons (see Levi-Montalcini, 1987; Purves, 1988). The development by Thoenen and colleagues of a sensitive two-site ELISA capable of detecting picomolar concentrations of NGF made it possible to test critically the major predictions of the Neurotrophic Factor Hypothesis (see Korsching, 1993). Generalized, this hypothesis predicts that many or all neurons will depend for survival on trophic factors provided through afferent or efferent contacts. Difficulties in characterizing and purifying additional factors, though, delayed experiments designed to examine the generality of this model for almost three decades. Because of technological advances, the major work in the 1990s has utilized gene targeting rather than specific antibodies that are still difficult to prepare for many factors. Gene targeting has been used to generate mouse strains carrying mutations in the genes encoding the neurotrophins, GDNF, CNTF, LIF, and their receptors. Examination of these mutants has provided many examples where neurons appear to require trophic support from their final targets but has also demonstrated requirements for trophic support by neural precursors, and young neurons whose axons have not yet contacted their final targets.

Sensory Neurons

Spinal sensory neurons, grouped in dorsal root ganglia (DRG) along both sides of the spinal cord, include different subpopulations within each ganglion specialized for transfer of different modalities of sensory information (reviewed in Scott, 1992). Different classes of neurons innervate separate types of peripheral sensory organs and distinct laminae in the spinal cord. Previous work had indicated that these different neuronal populations exhibit different patterns of trk receptor expression and different neurotrophin dependencies. Examination of the mutant mice, summarized in Table 6–2 and illustrated in Figure 6–4, has confirmed and extended these findings.

In the DRGs of normal postnatal mice, approximately 50% of the neurons express trkA (Mu et al., 1993; McMahon et al., 1994). However, in the DRGs in NGF- or trkA-deficient mice, approximately 70% of the normal complement of neurons is missing (Crowley et al., 1994; Smeyne et al., 1994). These include essentially all the neurons which express trkA postnatally, which are small-diameter neurons with unmyelinated axons which mediate pain perception. As expected, the mutant mice have extremely reduced sensitivity to painful stimuli. Consistent with this behavioral deficit, there is a remarkable reduction in the levels of calcitonin-gene–related peptide (CGRP) and substance P (SP) (Crowley et al., 1994; Silos-Santiago et al., 1995; Minichiello et al., 1995) and complete absence of axonal projections to layers I and II of the spinal cord as shown by axonal tracing (Silos-Santiago et al., 1995). In addition to the nociceptive CGRP- and SP-positive neurons, almost all of which express trkA postnatally (McMahon et al., 1994), the missing neurons in the trkA mutant include a small-diameter neuronal population with C-fibers, which mediate non-nociceptive thermal and low-threshold mechanoreceptive stimuli (Silos-Santiago et al., 1995). Although these cells do not express trkA postnatally, they can be identified because they bind certain lectins. Two explanations have been proposed to explain this unexpected deficit. One possibility is that these neurons require a

TABLE 6–2. Neuronal Losses in Ligand- and Receptor-Deficient Mice

	trkA[1]	*NGF*[2]	*trkB*[3]	*BDNF*[4]	*NT-4/5*[5]	*trkC*[6]	*NT-3*[7]	*p75*^{NTR}[8]	*GDNF*[9]	*CNTFRα*[10]	*LIFR*[11]
Sensory											
Trigeminal	75%	75%	60%	30%	n.s.	ND	60%	ND	n.s.	n.s.	ND
Nodose	ND	n.s.	90%	45%	40%	ND	30%	ND	40%	ND	ND
Vestibular	ND	ND	60%	85%	n.s.	15%	20%	ND	n.s.	ND	ND
Cochlear	ND	ND	15%	7%	ND	50%	85%	ND	ND	ND	ND
DRG	70%	70%	30%	35%	n.s.	20%	60%	smaller	20%	n.s.	ND
	Small CGRP (nociceptive)- and BSI (thermoceptive)-positive neurons missing		ND	Myelinated and non-myelinated axons lost; Ia afferents present		Proprio-ceptive neurons missing	Proprio-ceptive and cutaneous mechano-receptors	Nociceptive neurons missing	ND		
Symp.											
SCG	>95%	>95%	ND	n.s.	n.s.	n.s.	50%	n.s.	35%	n.s.	ND
Motor											
Facial	ND	ND	n.s.	n.s.	n.s.	ND	n.s.	ND	n.s.	40%	35%
Spinal cord	ND	ND	n.s.	n.s.	n.s.	ND	ND	ND	20%	35%	40%
CNS	Cholinergic basal fore-brain neurons present.		ND	Deficits in NPY, calbindin, parvalbumin	ND	ND	No clear deficits	Increase in the number of forebrain cholinergic neurons	No deficit in TH^{+} neurons	ND	ND
	Reduced hippocampal innervation										
Viability											
	poor	poor	very poor	poor to moderate	good	moderate	very poor	good	very poor	very poor	very poor

Neuronal losses are expressed as the percentage of neurons lost in the mutant compared to wild type controls. "Nodose" is the complex formed by the petrosal and nodose ganglia; DRG, dorsal root ganglia; Symp., sympathetics; SCG, superior cervical ganglion; NPY, neuropeptide Y. ND, not determined; n.s. not significantly different. Original references from which information was extracted: [1]Smeyne et al., 1994; Silos-Santiago et al., 1995; Minichiello et al., 1995; Fagan et al., 1996. [2]Crowley et al., 1994. [3]Klein et al., 1993; Minichiello et al., 1995; Schimmang et al., 1995; Erickson et al., 1996. [4]Jones et al., 1994; Ernfors et al., 1994a; Conover et al., 1995; Liu et al., 1995; Bianchi et al., 1996; Erickson et al., 1996. [5]Conover et al., 1995; Liu et al., 1995; Erickson et al., 1996. [6]Klein et al., 1994; Minichiello et al., 1995; Schimmang et al., 1995; Fagan et al., 1996. [7]Fariñas et al., 1994; Ernfors et al., 1994b; Tessarollo et al., 1994; Tojo et al., 1995; Airaksinen et al., 1996. [8]Lee et al., 1992; Ven der Zee et al., 1996. [9]Moore et al., 1996; Sanchez et al., 1996. [10]DeChiara et al., 1995. [11]Li et al., 1995.

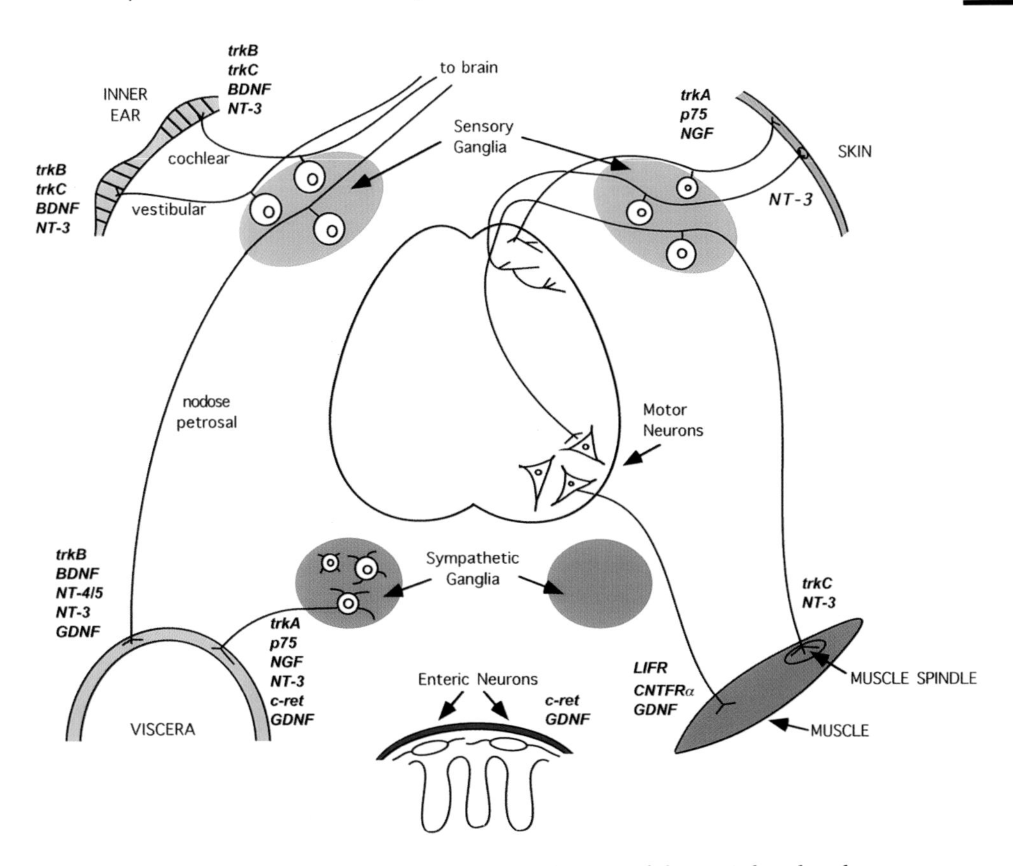

FIGURE 6–4 Diagram showing the neuronal populations of the peripheral and motor systems that are reduced in number in the targeted mutants for different neurotrophic factors and their receptors. Indicated at the target sites are the genes whose inactivation led to loss of neurons of that particular projection. Sensory ganglia include spinal (DRG) (right) and cranial (left) ganglia. See text for details and original references.

trophic factor supplied by the trkA population. Alternatively, these neurons might express trkA and require NGF transiently during embryogenesis. The latter possibility seems plausible because highly specific anti-trkA antibodies (Clary et al., 1994) label more than 80% of the neurons present in embryonic DRGs (Fariñas and Reichardt, unpublished). Considering the nociceptive neurons, the results are consistent with previous evidence that NGF serves as a target-derived trophic factor for these cells (reviewed in Scott, 1992). NGF mRNA and protein have been detected in their differentiated targets, and trkA-expressing sensory neurons have been shown to retrogradely transport NGF (reviewed in Korsching, 1993). However, we will argue below that a significant fraction of the total neuronal deficit in the trkA-deficient animals is likely to reflect a requirement for trophic support by neuroblasts and neurons before they make contact with their final targets. NT-3, not NGF, appears to provide this support.

The generation of mutations in the trkC and NT-3 genes has shown the absolute requirement of this signaling pathway for the generation and survival of proprioceptive neurons. In NT-3–deficient mice, virtually all group Ia afferents are absent, as assessed by the absence of muscle spindles and the Ia projection to the ventral spinal cord, and by the presence of abnormal movements and postures (Fariñas et al., 1994; Ernfors et al., 1994b; Tessarollo et al., 1994; Tojo et al., 1995a). Golgi tendon organs are also absent in these animals (Ernfors et al., 1994b). The Ia afferent population is also missing in a trkC kinase mutant mouse lacking the exon encoding the kinase domain of trkC (Klein et al., 1994). While not examined, tendon organs seem likely to be missing also in these animals. Since NT-3 expression is observed in both muscle spindles and the ventral spinal cord (illustrated in Figs. 6–8 and 6–9; see also Schecterson and Bothwell, 1992; Ernfors et al., 1992; Copray and Brouwer, 1994), the loss of the Ia afferent population would be predicted if NT-3 functioned as a target-derived trophic factor for these neurons. While this is probably true, the Ia projection never appears to develop in mutant animals (Tessarollo et al., 1994; Kucera et al., 1995b), so the loss of these neurons almost certainly reflects an earlier requirement for trophic support which will be discussed in more depth below. Interestingly, D-hair cutaneous afferents have been recently shown to be deficient in heterozygous NT-3 mutant mice (Airaksinen et al., 1996). In addition, the NT-3–deficient homozygous and heterozygous animals develop postnatally a severe deficit in Merkel cells, which are induced and maintained by slowly adapting (SA) afferent sensory endings. Thus, NT-3 deficiency results in deficits in specific cutaneous (D-hair and SA) and proprioceptive deficits, some of which are only manifested postnatally. NT-3 is expressed in peripheral targets of these neurons. The time course of development of these deficits indicates that NT-3 functions as a target-derived trophic factor for these cells.

Comparison of the phenotypes of the NT-3 and trkC deficiencies has revealed that many more sensory neurons are lost in the DRGs of NT-3-deficient mice (ca. 60%) (e.g., Fariñas et al., 1994; Fig. 5C,D) compared to trkC-deficient animals, in which only the population of proprioceptive neurons (approximately 20%) seems to be missing. The difference between the phenotype of the neurotrophin and the receptor knockout suggests physiological importance in vivo for the comparatively weak interactions of NT-3 with other trk receptors observed in vitro. As will be discussed below, these deficiencies appear to reflect requirements for NT-3 at earlier stages in development than would be expected if it only functioned as a factor present in final targets of these cells.

The identification of populations of DRG sensory neurons lost in animals with impaired trkB signaling has been more elusive. In lumbar DRGs of 2-week-old BDNF- or trkB-deficient mice, approximately 35% of the neurons are missing (Klein et al., 1993; Ernfors et al., 1994a; Jones et al., 1994). In contrast, NT-4/5–deficient animals do not appear to have a DRG neuron deficit (Liu et al., 1995; Conover et al., 1995). In addition to the deficits in cell number, approximately 35% of the myelinated axons are absent from lumbar dorsal roots of 2-week-old BDNF-deficient mice (Jones et al., 1994). As assessed by the number of muscle spindles, the proprioceptive neurons which induce and innervate these spindles appear normal (Jones et al., 1994). Thus, other neurons with myelinated axons, such as mechanoreceptors, must be affected. In addition, since only about 50% of the axons are myelinated, the similarities in loss of neurons and myelinated ax-

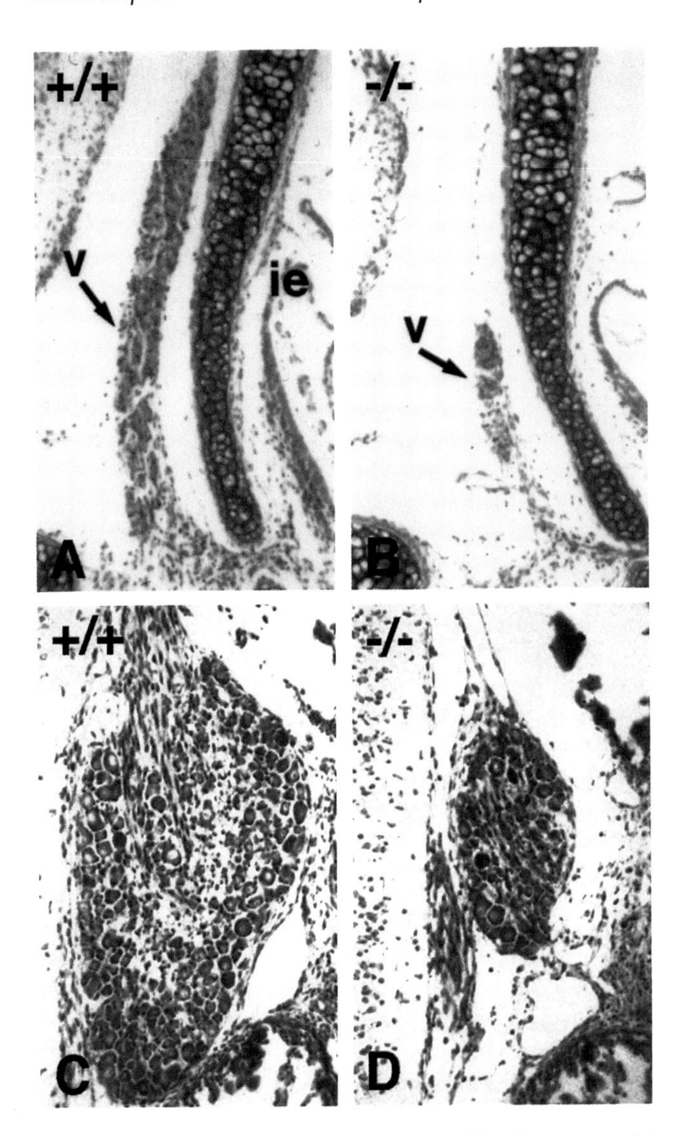

FIGURE 6–5 Deficiencies in sensory ganglia of neurotrophin mutant mice. (A, B) Vestibular ganglia (v) in wild-type (+/+) and BDNF-deficient (−/−) 2-week-old mice. Note that the mutant ganglion is virtually absent. ie, inner ear. (C, D) Lumbar 5 DRG of wild-type (+/+) and of NT-3 mutant (−/−) neonatal mice. There is a loss of approximately 60% of the neuronts in the mutant ganglion. Reproduced by permission from Jones et al. (1994) and Fariñas et al. (1994).

ons indicate that a population of small neurons with nonmyelinated axons is also absent in this mutant. In recent work, double trkB/trkA homozygous mutants have been reported to have essentially the same number of neurons as trkA homozygotes (Minichiello et al., 1995). This indicates that a substantial fraction of the BDNF-dependent neurons also require NGF at some period in development. Consistent with this, trkB and trkA have been shown to be coexpressed in DRG neurons of adult rats (McMahon et al., 1994). At birth, the deficits observed in BDNF or trkB animals are marginal (Minichiello et al., 1995; A, Silos-Santiago and M. Barbacid, personal communication; Fariñas and Reichardt, unpublished), suggesting the deficits occur postnatally. Available data are also consistent with the possibility that the BDNF deficiency does not reflect a target neurotrophin dependence but may instead reflect the importance in vivo of an autocrine BDNF-trkB loop that DRG neurons have been shown to develop postnatally (Acheson et al., 1995).

In contrast to DRG neurons, cranial sensory neurons which transmit different modalities of sensory information tend to be segregated into different ganglia and have been shown to have different requirements for neurotrophins in vitro. As expected, individual cranial ganglia are differentially affected by deficiencies in the different neurotrophins and trk receptors. Neural-crest–derived cells of the superior and jugular ganglia respond only to NGF in vitro (Davies and Lindsay, 1985) and, consistent with this, are present in normal numbers in mice lacking BDNF or NT-3 (Jones et al., 1994; Fariñas et al., 1994). The placode-derived visceral neurons of the petrosal and nodose ganglia, however, respond to BDNF, NT-4/5, and, to a lesser extent, NT-3 in vitro, and, consistent with this, their survival is only reduced in the absence of BDNF, NT-3, or NT-4/5 but not NGF in vivo (Ernfors et al., 1994a,b; Jones et al., 1994; Fariñas et al., 1994; Crowley et al., 1994; Conover et al., 1995; Liu et al., 1995; Erickson et al., 1996).

A striking example of trophic selectivity is seen in the vestibular and auditory sensory systems located in the inner ear. BDNF and NT-3 have distinct complementary and nonoverlapping roles in the survival of neurons in the vestibular and spiral ganglia. Neurons in these ganglia express both trkB and trkC, and both neurotrophins are expressed by hair cells in the cochlea and vestibular organs. Accordingly, in BDNF/NT-3 double homozygous mice, all neurons die in these ganglia (Ernfors et al., 1995). However, each neurotrophin appears to promote the survival of an individual nonoverlapping population of neurons. Lack of BDNF is reflected in the loss of most vestibular neurons (Fig. 6–5A,B; and Ernfors et al., 1994a; Jones et al., 1994; Ernfors et al., 1995; Bianchi et al., 1996), and lack of NT-3 causes the loss of the remaining small subpopulation (Ernfors et al., 1995). Similarly, most spiral ganglion neurons are lost in the NT-3 mutant mice (Fariñas et al., 1994; Ernfors et al., 1995) but only a few such cells are lost in the BDNF mutant mice (Ernfors et al., 1995).

In summarizing dependencies of sensory neurons on neurotrophins, it is worth noting that the neurotrophin and trk receptor mutant phenotypes are generally consistent with in vitro observations on ligand and receptor specificities. NGF and trkA mutants appear to lack the same sensory neuron populations, consistent with the specificity of NGF for trkA. The phenotype of the trkB mutant is more severe than that of the BDNF or NT-4/5 mutant and is consistent with observed ligand redundancy. This is clearly seen in the trigeminal and petrosal-nodose ganglia where larger deficits in neuron numbers are found in the receptor-deficient than in either of the neurotrophin-deficient mice. As expected, the deficiency in the petrosal-nodose ganglion in a BDNF/NT-4/5 double homozygous mouse appears similar or identical to that in trkB mutant animals (Liu et al., 1995; Conover et al., 1995; Erickson et at., 1996). Although both NT-3–

and trkC-deficient animals lack Ia afferent sensory neurons, NT-3 mutants lack a large additional population of spinal sensory neurons. Since NT-3 is also able to activate trkA and trkB (e.g., Ip et al., 1993b; Clary and Reichardt, 1994), it is perhaps not surprising that this is the one instance where ligand deficiency has more severe consequences than receptor loss. Recently, Davies et al. (1995) have shown that nodose and trigeminal neurons isolated from trkC-deficient mice can survive in the presence of NT-3 at developmental times when they are responsive to BDNF and NGF, respectively. In addition, responses of nodose neurons to NT-3 are lost in trkB/trkC double mutant neurons, while responses of trigeminal neurons to NT-3 are lost in trkA/trkC mutants. The concentration of NT-3 needed to achieve these responses is higher than that of the other ligands, but the experiments elegantly demonstrate that NT-3 can mediate survival responses through trkA and trkB, supporting the probable importance of these interactions in vivo.

As summarized in Table 6–1, neurotrophins are not the only trophic factors which promote sensory neuron survival and differentiation in vitro. CNTF, LIF, and GDNF (e.g., Murphy et al., 1993; Barbin et al., 1984; Buj-Bello et al., 1995) have each been shown to be active in these assays. Analyses of mice deficient in these factors or in the CNTF receptor-α, LIF receptor-β, or gp130 receptor subunit has indicated that some, but not all, of these factors are important in vivo. In GDNF-deficient mice, significant deficits have been observed in the petrosal-nodose ganglion (40%) and spinal lumbar DRGs (ca. 20%), but it has not been determined whether these partial deficiencies reflect reductions in trophic factors for neurons in contact with targets or deficits arising at earlier developmental stages (e.g., Moore et al., 1996). Statistically significant deficits have not been reported in sensory ganglia in mice deficient in c-ret or in mice with deficiencies in cytokines of the LIF-CNTF family or their receptors (e.g., Durbec et al., 1996; DeChiara et al., 1995). It is not certain whether partial deficiencies could have been missed in these analyses.

Sympathetic Neurons

In cell culture assays of neuronal survival, sympathetic neurons require NGF during late gestation and early postnatal life and early postnatal application of inhibitory antibodies results in virtually complete loss of these neurons in vivo (e.g., Levi-Montalcini, 1987). As expected, the vast majority of sympathetic neurons express trkA at these stages (Ernfors et al., 1992; Birren et al., 1993; DiCicco-Bloom et al., 1993). This period of NGF dependence, as assessed in vitro, correlates with the period during which these neurons innervate their final targets and are reduced by cell death. Considering these results, it is not surprising that sympathetic ganglia are virtually absent in 2-week-old mice carrying mutations in the NGF or the trkA genes (Crowley et al., 1994; Smeyne et al., 1994). In each mutant, extensive cell death occurs perinatally, indicating that differentiated sympathetic neurons are dying in these animals. In the trkA mutants a significant deficit is already found at embryonic day 17.5 (E17.5) and develops progressively after birth (Fagan et al., 1996). Interestingly, significant numbers of sympathetic neurons do not appear to be lost in mice deficient in p75 receptors (Lee et al., 1992).

Earlier in development, however, NT-3, but not NGF or BDNF, promotes the survival of sympathetic neuroblasts and neurons (Birren et al., 1993; DiCicco-Bloom et al., 1993; Verdi and Anderson, 1994). Consistent with this, trkC is expressed in sympathetic ganglia prior to trkA (e.g., Birren et al., 1993). Consistent with the observations in vitro, approximately 50% of the normal complement of sympathetic neurons is missing from NT-3–deficient animals (Ernfors et al., 1994b; Fariñas et al., 1994). The time course of the deficiency is temporally more restricted than that caused by the lack of NGF because the same percentage of neurons is lost in the superior cervical ganglion if animals are analyzed at birth (Fariñas et al., 1994) or at 2 weeks of age (Ernfors et al., 1994b). It is not clear whether the partial cell loss reflects a stochastic process or unidentified differences between the surviving and dying cells. As neural cells appear to be lost before contact with final targets, these observations will be discussed in more detail in a later section.

Again, development of sympathetic neurons has been shown to depend upon trophic factors in addition to the neurotrophins. Of particular interest, GDNF-deficient mice have a partial deficit and c-ret–deficient neurons a virtually complete deficit in neurons in the superior cervical ganglion (Moore et al., 1996; Durbec et al., 1996). Since these deficits have been shown to occur before neurogenesis, they will not be further discussed in this section.

Other Neural Crest Derivatives

Interestingly, other trkC-expressing cell populations derived from the neural crest are not lost in the NT-3 mutant (Fariñas et al., 1994). For example, adrenal chromaffin cell precursors have been shown to express trkC (Tessarollo et al., 1993), but chromaffin cells are clearly present and express tyrosine hydroxylase (Fariñas et al., 1994). TrkC is the only trk receptor detected in enteric neurons (Tessarollo et al., 1993). However, as visualized in whole mounts of intestines using neuronal markers, the myenteric plexuses of neonatal NT-3 mutant homozygotes appear normal (Fariñas et al., 1994). The GDNF-deficient and c-ret– deficient homozygotes, however, do lack intestinal enteric neurons (e.g., Moore et al., 1996; Sanchez et al, 1996; Pichel et al., 1996; Durbec et al., 1996). Since these deficits occur before neurogenesis, they will be discussed in a later section.

Motoneurons

Each of the neurotrophins, except NGF, has been shown to promote the survival of embryonic motoneurons when cultured in vitro (Henderson et al., 1993). Moreover, the same neurotrophins reduce cell death in motor populations deprived of target contact by axotomy (e.g., Yan et al., 1993). Consistent with these results, motoneurons express both trkB and trkC (e.g., Henderson et al., 1993; Yan et al., 1993) and would thus be expected to respond to all neurotrophins except NGF. Despite this, cell counts in motoneuron populations of mice lacking BDNF, NT-3, or NT-4/5 (Ernfors et al., 1994a,b; Jones et al., 1994; Fariñas et al., 1994; Conover et al., 1995; Liu et al., 1995) revealed no detectable differences. Also, a normal complement of motor neurons was found in double BDNF/NT-4/5 homozygous mice (Conover et al., 1995; Liu et al., 1995). In addition, it now appears that no deficits in motor neuron number are found in the trkB mutant mice (A. Silos-Santiago, M. Barbacid, personal communication). In BDNF mutant homozygotes, normal numbers of motor neurons survive, develop a cholinergic phenotype, and innervate skeletal muscle (Jones et al., 1994). Postnatal polyneuronal synapse elimination in skeletal muscle, which is dependent on normal activity and synaptic transmission by the innervating motoneurons, also takes place normally.

In spite of the lack of effects of neurotrophin deficiencies on large α-motoneurons, in the NT-3 mutants there appear to be reduced numbers of γ-motoneurons which innervate muscle spindles (Kucera et al., 1995a). It is not clear whether this deficit is a direct consequence of NT-3 deficiency or is caused secondarily by the absence of muscle spindles in these animals. A roughly 30% deficit in axon numbers was reported in the ventral roots of trkC mutant animals, which may also reflect reductions in γ-motoneuron axons (Klein et al., 1994).

Receptor–ligand interactions extending beyond the neurotrophin and trk families seem very likely to reconcile the strong in vivo evidence that motoneurons depend on target-derived trophic support with their survival in these mutants. In vitro, survival of motoneurons is promoted not only by neurotrophins but by other factors including CNTF, LIF, GDNF, and FGF-5 (e.g., Henderson et al., 1993, 1994). Recently, characterization of CNTF-receptor-α-subunit– and LIF-receptor-β-subunit–deficient mice has revealed significant, but incomplete, loss of motoneurons (DeChiara et al., 1995; Li et al., 1996). Increased numbers of apoptotic figures are observed in motor populations of perinatal animals deficient in the LIF receptor-β subunit, which strongly suggests that this receptor mediates survival of differentiated motor neurons (Li et al., 1996). Mice with a targeted mutation in gp130 have been isolated and appear to have more severe deficits than LIF-receptor-β–deficient animals (Yoshida et al., 1996). While not yet characterized, the neural deficits seem almost certain to be at least as severe as those in animals lacking other components of this receptor complex. The similarities in motoneurons deficiencies between the LIF-receptor-β– and CNTF-receptor-α–deficient animals indicate that a ligand that interacts with both of these subunits has an important role in vivo. Since CNTF is the only identified member of this cytokine family that requires the CNTF receptor-αsubunit for signaling and a perinatal motoneuron deficiency is not seen in animals lacking CNTF (Masu et al., 1993; DeChiara et al., 1995), this ligand cannot be CNTF and must be a novel member of this family.

Mild deficiencies have also been seen in some, but not all, cranial motor nuclei and in spinal motoneurons in GDNF-deficient animals (e.g., Moore et al., 1996; Sanchez et al., 1996). Even though c-ret mediates signaling by GDNF (Durbec et al., 1996; Trupp et al., 1996), similar deficiencies have not been reported in c-ret–deficient mice. However, the publications to date on the c-ret mutants have focused on the dramatic deficits in morphogenesis of the kidney, enteric nervous system, and superior cervical ganglion (e.g., Durbec et al., 1996a). It is not certain whether the cell counts needed to reveal partial motoneuron deficits have been performed. Since the evidence that motoneurons do depend on target-derived trophic support is so strong, it is important to determine whether the etiology of the motoneuron deficiency in the GDNF-deficient homozygotes is compatible with predictions of the neurotrophic factor model. The incompleteness of the deficiencies observed in the GDNF-, c-ret–, CNTF-receptor-α– and LIF-receptor-β–deficient mice indicates that there must be considerable redundancy in motoneuron survival factors present in vitro.

Other CNS Neurons

The survival of many classes of embryonic or neonatal central neurons has been shown to be promoted by neurotrophic factors in vitro (reviewed in Korsching, 1993; Eide et al., 1993). NGF-responsive neurons include most prominently basal forebrain and striatal cholinergic neurons; BDNF- and NT-4/5–responsive neurons include basal forebrain cholinergic neurons, cerebellar granule cells, mesencephalic dopaminergic neurons, and retinal ganglion cells. Similarly, cerebellar granule cells, hippocampal neurons, adrenergic neurons in the locus coeruleus, and other populations have been shown to be responsive to NT-3. In many instances neurotrophins have been shown to be effective on these same cell populations in vivo, reducing cell death after axon lesion or neurotoxin application. Strikingly, though, no deficits in number of any cell population have been detected in neurotrophin or trk mutants at birth in vivo. Recently, however, unusually large numbers of neurons have been shown to undergo apoptosis during postnatal development of trkB and trkB-trkC double mutants (Minichiello and Klein, 1996; I. Silos-Santiago, personal communication).

GDNF was originally isolated as a potent survival factor for midbrain dopaminergic neurons (Lin et al., 1993) and has since been shown to protect these neurons partially from the dopaminergic neurotoxin 1-methyl-4-phenyl-1,2,3,6-tetrahydropyridine (MPTP) (e.g., Tomac et al., 1995). Despite this, these neurons are present in normal numbers in neonatal GDNF-deficient mice (Moore et al., 1996; Sanchez et al. 1996). CNTF and LIF have also been reported to promote survival of many populations of CNS neurons (e.g., Hagg et al., 1992). While it is not certain how carefully the central nervous systems of the mutant animals have been examined, no deficits have been reported in mice deficient in any of the LIF/CNTF cytokine family or their receptors.

While the postnatal lethality of many neurotrophic factor and receptor mutants has made it difficult to examine functions of these molecules through the whole extent of postnatal development, the most surprising single result of the mutant analyses must be the apparent survival during embryogenesis of all of these populations in vivo in the absence of individual factors or receptors, despite their responsiveness in vitro, suggesting that deficiencies in individual trophic factors in the CNS in vivo are easily compensated.

Essential Roles in Neuronal Differentiation

Neurotrophic factors have been shown to regulate the pathways of differentiation selected by neural precursors and to regulate the differentiation process, helping to determine the levels of expression of many proteins important for the physiological function of differentiated neurons such as neurotransmitters and proteins affecting second messengers. In vitro and in vivo, application of NGF has been shown to promote the differentiation of sympathoadrenal precursors into sympathetic neurons as opposed to adrenal chromaffin cells (reviewed in Levi-Montalcini, 1987; Anderson, 1993). In contrast, glucocorticoids suppress responsiveness to NGF, inhibit differentiation of these precursors into neurons, and promote their differentiation into mature chromaffin cells. Similarly, application of NGF or anti-NGF during early postnatal rodent development has a profound influence on the fate of nociceptive sensory neurons (Lewin et al., 1992, 1994a). Anti-NGF treatment results in loss of A-δ high-threshold mechanoreceptors which appear to be converted in D-hair afferents. C-mechanoheat fibers are converted into unusual pressure-sensitive fibers. These effects on sensory neuron phenotype persist for many weeks and appear to be permanent. While its relevance in vivo is less certain, BDNF has been shown to enhance differentiation of sensory neurons from sensory precursors in vitro (Sieber-Blum, 1991). Many similar effects of neurotrophins and other neurotrophic factors, too numerous to cite here, have been observed in cultures of embryonic CNS neurons or neuronal precursors.

While neurotrophins and GDNF were first identified as factors that promoted neuronal survival, appreciation of the important role of cytokines in the nervous system derives from work of Patterson and colleagues demonstrating that LIF promotes the permanent conversion of adrenergic sympathetic neurons into neurons which no longer express catecholamines but instead synthesize and secrete acetylcholine (reviewed in Patterson and Nawa, 1993). CNTF was shown to have the same cholinergic-inducing activity on these neurons. Expression of several peptide neurotransmitters is also affected by these and related cytokines. In normal animals, sympathetic neurons innervating sweat glands undergo a similar conversion which appears to be mediated by a factor related to, but not identical to, LIF and CNTF, since conversion occurs normally in both CNTF and LIF mutant homozygotes (reviewed in Patterson and Nawa, 1993). When LIF is targeted to islet cells in the pancreas using the insulin promoter, it induces transmitter switching of these same neurons in vivo (Bamber et al., 1994). Thus the effects of this family of cytokines on sympathetic neurons in vitro and in vivo appear to be fundamentally different from those of NGF, which promotes increased expression in these cells of proteins involved in generating catecholamines or acetylcholine but does not regulate the cholinergic switch.

Neurotrophins regulate expression of neurotransmitters, neuropeptides, cell adhesion molecules, transcription factors, and numerous other proteins in peripheral sensory and sympathetic neurons, but effects of neurotrophin deficiency on expression of these molecules have not been pursued using the mutant mice because of the difficulty of separating direct effects on differentiation from secondary responses in dying neurons. Since all major populations of neurons in the CNS examined appear to be present in approximately normal numbers in these mutant animals, however, it has been possible to determine whether neurotrophins are essential for their normal differentiation. In the mutant animals, qualitative deficits in differentiation in the CNS appear rare. For example, both NGF and BDNF promote differentiation of basal forebrain cholinergic neurons in vitro and in vivo (reviewed in Korsching, 1993), but NGF-, trkA-, BDNF-, and trkB-deficient mice appear to have normal complements of these cells expressing proteins necessary for cholinergic function. Similarly, despite effects of GDNF, BDNF, and NT-4/5 on mesencephalic dopaminergic neurons and of NT-3 on adrenergic neurons in the locus coeruleus, observed in vitro and in vivo (e.g., Arenas and Persson, 1994), absence of GDNF or any of these neurotrophins does not prevent differentiation of normal numbers of neurons expressing tyrosine hydroxylase in either region (e.g., Jones et al., 1994; Fariñas et al., 1994; Moore et al., 1996; Liu et al., 1995). One limitation of the analyses described above is that the vast majority of work to date has examined brains of early postnatal mice sacrificed before completion of brain development because, except for NT-4/5, neurotrophin deficiency strongly compromises the health of postnatal animals and even their survival. Thus, many deficits that arise later in development or in adulthood may have been missed. With ingenuity, though, it has proven possible to maintain a proportion of some neuro-trophin-deficient animals for extended periods (Tojo et al., 1995a,b). These animals also do not appear to have obvious deficits in cortical or cerebellar development (e.g., Tojo et al., 1995b).

In contrast to analyses of other animals, examination of BDNF-deficient mice has shown that this neurotrophin is required for normal differentiation of several populations of CNS neurons in vivo. In particular, reductions in immunoreactivity for neuropeptide Y and for the calcium-binding proteins calbindin and parvalbumin (Fig. 6–6), whose expression rises dramatically during postnatal brain development, are seen in several discrete populations of neurons (Jones et al., 1994). Expression of neuropeptide Y appears normal in the striatum but significantly altered in the cerebral cortex and hippocampus. This reduced expression with respect to the wild-type does not change between postnatal days 15 and 21, indicating that maturation of neuropeptide expression is not simply delayed but is permanently affected. Arguing further that the mutation specifically affects differentiation in the cerebral cortex, expression of neuropeptide Y in all layers and regions of the cerebral cortex appears to be reduced in the BDNF mutant, and layer VI is almost devoid of immunoreactive cells. During development of normal mice, neuropeptide Y immunoreactivity first appears in layer VI of the cerebral cortex and subsequently in the upper layers (Parnavelas et al., 1988). Thus, an indirect effect of the mutation, delaying brain maturation, would be expected to result in more pronounced deficits in upper layers, a result that is opposite to that actually observed. In most instances, it seems likely that the neurons are present but fail to differentiate normally because other markers characteristic of these same cells, such as GABA, are expressed. When examined, BDNF has proven to increase expression of these same differentiation markers in

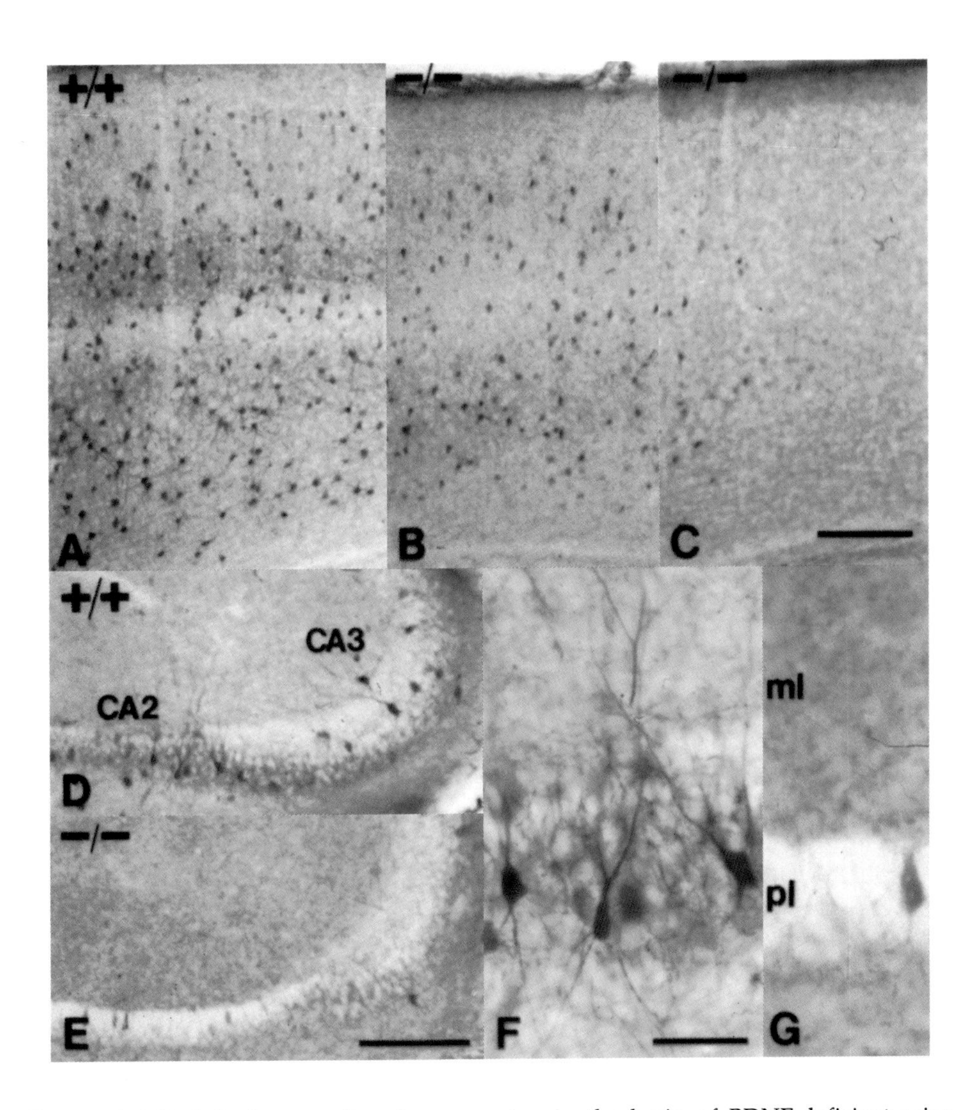

FIGURE 6–6 Deficits in parvalbumin expression in the brain of BDNF-deficient mice. (A–C) Immunoreactivity for parvalbumin in the cerebral cortex of wild type (+/+) and BDNF mutant (−/−) 2-week-old mice. A and B are equivalent sections through the somatosensory cortex. Note the reduction in staining in the mutant. This reduction is even more accentuated in areas like the visual cortex, as shown in C. (D–G) Staining for parvalbumin in the hippocampus of wild-type (D, F) and BDNF mutant (E, G) mice. Note the striking reduction in the level of expression in the mutant. Bars: A–E, 100 mm; F–G, 50 mm. Reproduced by permission from Jones et al. (1994).

cultured neurons (e.g., Nawa et al., 1993; Ip et al., 1993a). These reductions in neuropeptides and calcium-binding proteins almost certainly have physiological consequences in vivo, since they modulate neuronal physiology and synaptic function (e.g., Baimbridge et al., 1992). Thus, there is strong evidence, though limited in scope, that at least one neurotrophin is essential for normal CNS neuron differentiation in vivo.

EFFECTS ON AXONOGENESIS AND TARGET INNERVATION

The neurotrophins, GDNF, and the LIF/CNTF family of cytokines all promote neurite outgrowth by responsive neurons in vitro. Elegant experiments have demonstrated that local NGF regulates the advance of sympathetic neuron growth cones (Campenot, 1977). In addition, neurotrophin gradients appear able to steer growth cones in vitro (Gundersen and Barrett, 1979). In some cases, however, increased NGF has also been shown to stimulate transient growth cone retraction (Griffith and Letourneau, 1980). Thus activation of trkA appears to have complex actions on cell motility. Similarly, the PDGF receptor has been shown to activate pathways that stimulate and inhibit both cell motility and chemotaxis in non-neural cells (e.g., Kundra et al., 1994). In chemotactic assays, activation of gp 120rasGAP has been shown to be inhibitory, while activation of ras through either PLC-γ1 or SHC-1/grb-2/SOS is stimulatory (e.g., Kundra et al., 1994). These studies on PDGF signaling provide a conceptual framework for understanding mechanisms by which neurotrophic factors may regulate axon and dendrite growth and guidance. They may also provide insight into the inhibitory actions of other tyrosine kinases, such as the eph receptors (e.g., Drescher et al., 1995).

Over the past few years, considerable progress has been made in characterizing the signaling pathways activated by trk receptors which mediate neurite outgrowth, and these appear similar, but not identical to, those activated by the PDGF receptor in PC12 cells (e.g., Greene and Kaplan, 1995). Activation of phospholipase-C-γ1 or the SHC-1/grb-2/SOS complex results in activation of ras and the MAP kinase pathway, which is essential for neurite outgrowth. At least one site on trkA not containing a tyrosine has also been shown to be crucial (Peng et al., 1995). Comparable studies to characterize pathways involved in promotion of neurite outgrowth by CNTF, LIF, or GDNF have not been published. As for more distal events, neurotrophic factor stimulation of growth cone motility almost certainly involves the same molecules utilized by non-neural cells, where motility appears to be regulated by controlled polymerization and depolymerization of the action cytoskeleton through actions of small G proteins of the CDC-42/rac/rho family, targets of these G proteins, actin-binding proteins, and phosphoinositides (reviewed in Hall, 1994; Mitchison and Cramer, 1996; Lauffenburger and Horwitz, 1996). Except for the phosphoinositides, whose metabolism is regulated directly by phospholipase-C-γ1, pathways leading from trkA to these effectors are not well delineated.

Early work showed that systemically applied neurotrophins affect innervation patterns in vivo (see Levi-Montalcini, 1987). NGF was shown to enhance innervation of tissues that receive sympathetic or sensory innervation normally and to induce aberrant patterns of innervation involving tissues that normally receive very little innervation, such as the exocrine pancreas (reviewed in Levi-Montalcini, 1987). Analyses of transgenic animals expressing neurotrophins ectopically have confirmed and extended this early work. As one example, using the insulin promoter to increase NGF in pancreatic islets has been shown to induce dense sympathetic innervation (e.g., Edwards et al., 1989). In contrast, in transgenic mice in which the dopamine-β-hydroxylase promoter targets NGF expression to sympathetic neurons, creating a potential autocrine loop, these neurons project to their target areas but fail to innervate them (Hoyle et al., 1993). Target innervation can be restored by overexpression of NGF within a target, suggesting that a gradient of NGF may be important for development of the final innervation pattern (Hoyle et al., 1993). The observations suggest that NGF is not required for sympathetic axons to reach the vicinity of their targets but is important for invasion and development of the final innervation pattern. Similar observations have not been made with CNTF, LIF, or GDNF.

Analyses of neurotrophin-deficient and trk-deficient mice has confirmed the important, but limited role of neurotrophins in regulating axon growth and target innervation. In BDNF mutant mice, almost all vestibular neurons are lost, but a small number of NT-3-dependent neurons survive (Ernfors et al., 1995). These neurons, however, fail to innervate the epithelial hair cells and terminate instead in the connective tissue adjacent to the sensory epithelium (Ernfors et al., 1994a, 1995). The lack of BDNF does not affect the outgrowth of these fibers or the target organ encounter but alters the terminal innervation pattern. A developmental analysis indicates that this is likely to reflect initial failure, not subsequent retraction, of the axons (Ernfors et al., 1995). In trkA mutants, deficits in sympathetic innervation of distal target tissues are observed during embryogenesis, but it is not certain whether the deficit is direct or is instead an indirect consequence of the initiation of apoptosis in these neurons. In more puzzling results, $p75^{NTR}$ deficiency does not appear to reduce survival of sympathetic neurons and sympathetic innervation of most target tissues appears normal (Lee et al., 1992). However, pineal glands lack sympathetic innervation and innervation of some, but not all, foot-pad sweat glands is reduced or absent (Lee et al., 1994b). While these results suggest a function for this receptor in supporting axon growth, it also seems possible that limited, target-regulated apoptosis could account for the absence of sympathetic innervation at these sites. Direct effects, not complicated by apoptosis, can be observed, however, in development of innervation patterns by basal forebrain cholinergic neurons which remain viable in trkA-deficient and NGF-deficient animals. In each animal, cholinergic fibers reach the hippocampus but fail to establish normal innervation within it (Smeyne et al., 1994; Crowley et al., 1994; H. Phillips, personal communication). Similarly, the cholinergic innervation of the cerebral cortex is also strongly reduced in each of these animals. In NT-3–deficient mice, deficits have also been reported in innervation of some, but not all, sympathetic targets in perinatal animals (Fariñas et al., 1994; ElShamy et al., 1996). NT-3 has also been shown to affect collateral branching by the axons of corticospinal neurons and has been proposed to function as a chemotropic factor for these branches (D. D. O'Leary, personal communication).

In summary, observations using a variety of experimental approaches have revealed several instances where a neurotrophin plays a crucial role in establishing a final innervation pattern in vivo, consistent with in vitro observations.

Other complex aspects of the morphological differentiation of neurons, such as dendritic growth and differentiation, have also been shown to be regulated by neurotrophins (e.g., McAllister et al., 1995, 1996).

EFFECTS ON NEURONAL PRECURSOR POPULATIONS

Growing evidence suggests that neurotrophic factors can regulate the survival, proliferation, and differentiation of neuronal and glial progenitor cells. Initially, the very early expression of some neurotrophic factors, such as GDNF, BDNF, and NT-3 and their receptors had suggested that they might play a role in the development of precursor populations in both central and peripheral nervous systems. More recently, NT-3, but not other neurotrophins, has been shown to promote neuronal differentiation of cortical precursors from rodents (Ghosh and Greenberg, 1995) and to increase the numbers of motoneurons that differentiate from quail neural-tube progenitor cells (Pinco et al., 1993; Averbuch-Heller et al., 1994). In addition, nestin-positive precursor cells from the rat striatum can be induced to proliferate with NGF (Cattaneo and McKay, 1990), while precursors from the hippocampus are induced to differentiate with NT-3, BDNF, or NT-4/5 (Vicario-Abejon et al., 1995). These observations suggest that neurotrophins regulate neurogenesis in the CNS. Other cell lineages are also affected by neurotrophic factors. The proliferation and survival of oligodendrocyte precursors (O-2A progenitor cells) in the optic nerve have been shown to be regulated by CNTF in vitro and by NT-3, both in vitro and in vivo (Barres et al., 1994). So far, though, no obvious abnormalities have been detected in the brains of the various homozygous mutant mice.

In studies on peripheral sensory or sympathetic neuron precursors, every neurotrophic factor discussed in this review has been shown to promote survival, proliferation, or differentiation in vitro (Sieber-Blum, 1991; Kalcheim et al., 1992; Wright et al., 1992; Birren et al., 1993; DiCicco-Bloom et al., 1993; Verdi and Anderson, 1994; Karavanov et al., 1995; Memberg and Hall, 1995). In vivo, the most striking deficiencies are observed in c-ret– and GDNF-deficient animals which lack all intestinal enteric neurons (e.g., Moore et al., 1996; Pichel et al., 1996; Sanchez et al., 1996; Durbec et al., 1996a). The GDNF-deficient mice have a partial deficit and c-ret–deficient animals a virtually complete deficit in neurons in the superior cervical ganglion. The sympathetic deficit observed in c-ret homozygotes is specific for the superior cervical ganglion. No deficiency is seen in other neurons in the sympathetic chain, reflecting differences in dependence on this signaling pathway between the pools of neural crest cells that give rise to superior cervical and other sympathetic chain ganglia. In both mutants, the deficits in the enteric nervous system have been shown to be fully developed by E12.5, i.e., neural precursors never enter the intestine. In the c-ret homozygotes, the deficiencies in the intestinal enteric nervous system and in the superior cervical ganglion have been shown to arise in a pool of hindbrain-derived neural crest cells before migration, neurogenesis, or axonogenesis. Activation by GDNF of the c-ret signaling pathway appears essential for the commitment, survival, or migration of these precursors. The presence of a small superior cervical ganglion in GDNF, but not in c-ret, mutants suggests that more than one ligand can activate this signaling pathway. Recently, a second member of the GDNF family named neurturin has been isolated that is a strong candidate to activate this signaling pathway (Kotzbauer et al., 1996).

Effects of NT-3 on sympathetic precursors in vitro provide the best-documented example of neurotrophin effects on peripheral neuron precursors (Fig. 6–7). In the sympathetic nervous system, trkC is the only trk receptor expressed during the early stages of ganglion development (Birren et al., 1993; DiCicco-Bloom et al., 1993). It is expressed before neurons acquire dependence on NGF, much earlier than trkA can be detected. The trkC receptor is expressed by sympathoblasts, cells that express characteristic traits of catecholaminergic neurons but still retain their proliferative capacity (Roehrer and Thoenon, 1987). Consistent with this, the survival in vitro of these precursor cells is dependent on NT-3 (Birren et al., 1993; DiCicco-Bloom et al., 1993; Verdi and Anderson, 1994). In addition, if high concentrations of NT-3 are used, these cells withdraw from the cell cycle and differentiate, thereby inducing the expression of trkA (Verdi and Anderson, 1994). This antimitotic effect is also obtained by application of CNTF (Ernsberger et al., 1989; Verdi and Anderson, 1994). NT-3 may be an essential mediator of some, but not all, of these effects on sympathetic neuroblasts and neurons in vivo. Consistent with the observations in vitro, around 50% of the normal complement of sympathetic neurons is missing from NT-3–deficient animals (Ernfors et al., 1994b; Fariñas et al., 1994). The observations in vitro predict that sympathetic neurons will be lost before E17 in both NT-3– and trkC-deficient mice. Consistent with this prediction, it has been reported that there is excessive apoptosis of sympathetic neuroblasts from E11.5 to E14.5 in NT-3 mutants resulting in loss of 50% of the normal complement of neurons before E17.5 (ElShamy et al., 1996). However, in analyses of NT-3 mutant homozygotes, two other groups have not observed deficiencies at E15.5 although they do observe deficits at later times (reviewed in Davies, 1997). Thus, the initial report of early effects of NT-3 on sympathetic neuroblasts appears problematic. Moreover, it appears certain that NT-3 activation of trk C is not required to form a normal sympathetic ganglion because mice lacking trkC do not develop detectable sympathetic deficiencies (Fagan et al., 1996). Potentially, NT-3 signaling through trkA, the only other trk receptor expressed in developing sympathetic neurons, could be the crucial signaling pathway required for normal development. The number of neurons in the SCG of trkA mutants remains normal to E15.5 and only decreases at later stages (Fagan et al., 1996). Thus a deficiency in this signaling pathway seems unlikely to induce excessive apoptosis of neuroblasts before this time, but could be crucial for maintaining neurons at later developmental stages.

Whatever the resolution of this discrepancy, NT-3 and trkC are not essential in vivo for expression of trkA or withdrawal of cells from the mitotic cycle. Absence of NT-3 does not delay expression of trkA in surviving sympathetic neurons in the NT-3 mutant mice (Fariñas and Reichardt, unpublished), perhaps because other factors control exit of these neurons from the cell cycle in vivo. Activation of trkC by NT-3 is not essential for acquisition of NGF responsiveness since NGF has been shown recently to support survival of perinatal sympathetic neurons isolated from trkC-deficient mice (Davies et al., 1995).

Studies on neural crest cell cultures in vitro and analysis of trkC expression in vivo, using in situ hybridization with probes that would detect all isoforms of trkC have suggested that NT-3 has essential early actions on neural crest cells in vivo. In the chick, different isoforms of the trkC receptor are found in neural crest cells, particularly in those with neurogenic potential (Henion et al., 1995). When avian neural crest cells are cultured in vitro, effects of NT-3 promoting survival and/or proliferation have been reported (Kalcheim et al., 1992). The trkC receptor is also expressed in migrating neural crest cells in the mouse (Fig. 6–7) (Tessarollo et al., 1993). However, we have analyzed but failed to find deficiencies in trunk neural crest cells in NT-3–deficient mice. (Fariñas et al., 1996). As visualized with antibodies to $p75^{NTR}$, neural crest cell migration seems qualitatively normal in E9 mutant embryos. Moreover, newly formed DRGs have the same number of cells as their wild-type counterparts, indicating that migration, proliferation, and survival of neural crest cells prior to ganglion formation do not require this neurotrophin.

Sensory ganglia (Fig. 6-7) deficiencies are observed, however, soon after completion of ganglion condensation in NT-3 deficient mice (Fariñas et al., 1996). These deficiencies are the result of two different effects caused by the mutation. First, the lack of NT-3 causes elevated neuronal apoptosis that significantly reduces the numbers of already generated neurons. This loss occurs very early in development and it is likely to include at least the population of proprioceptive neurons, which are the earliest born neurons in the ganglion. However, an additional unexpected effect of the lack of NT3 is detected that specifically affects the precursor population. In NT-3 deficient embryos precursors abandon the cell cycle and differentiate into neurons prematurely. This depletes the precursor pool before the normal round of cell divisions occurs leading to a final diminished number of neurons. This result indicates that, in normal development, NT-3 keeps the sensory precursors cycling either as a mitogen or as a repressor of their differentiation program. At times when precusor cells are actively dividing in normal embryos, high expression of NT-3 is found in areas immediately surrounding the ganglion (Fig. 6-8). Interestingly, at the time of maximum neurogenesis, when most precursors actually differentiate into neurons, NT-3 expression in mesenchymal tissues adjacent to the ganglia is sharply reduced (Fariñas et al., 1996), suggesting NT-3 regulates the number of cell divisions that occur during normal development of these ganglia. Thus, the effect on precursor differentiation in the NT-3 mutant mice indicates that much of the final deficit in neuron number found in these mutants occurs because some of the neurons are never generated. In the trigeminal ganglion of NT-3 deficient embryos, however, neuronal apoptosis, but not effects on precursors, accounts for the entire deficit in the number of neurons (Wilkinson et al., 1996). Other reports have suggested that NT-3 is essential in vivo for the survival of precursor cells in trigeminal and dorsal root sensory ganglia (ElShamy and Ernfors, 1996a, b) because cells that incorporated the nucleotide analog bromo-deoxyuridine in mutant animals were observed to be also positive for a marker of apoptotic cells. However, when lower, non-toxic concentrations of the analog were used, proliferating cells did

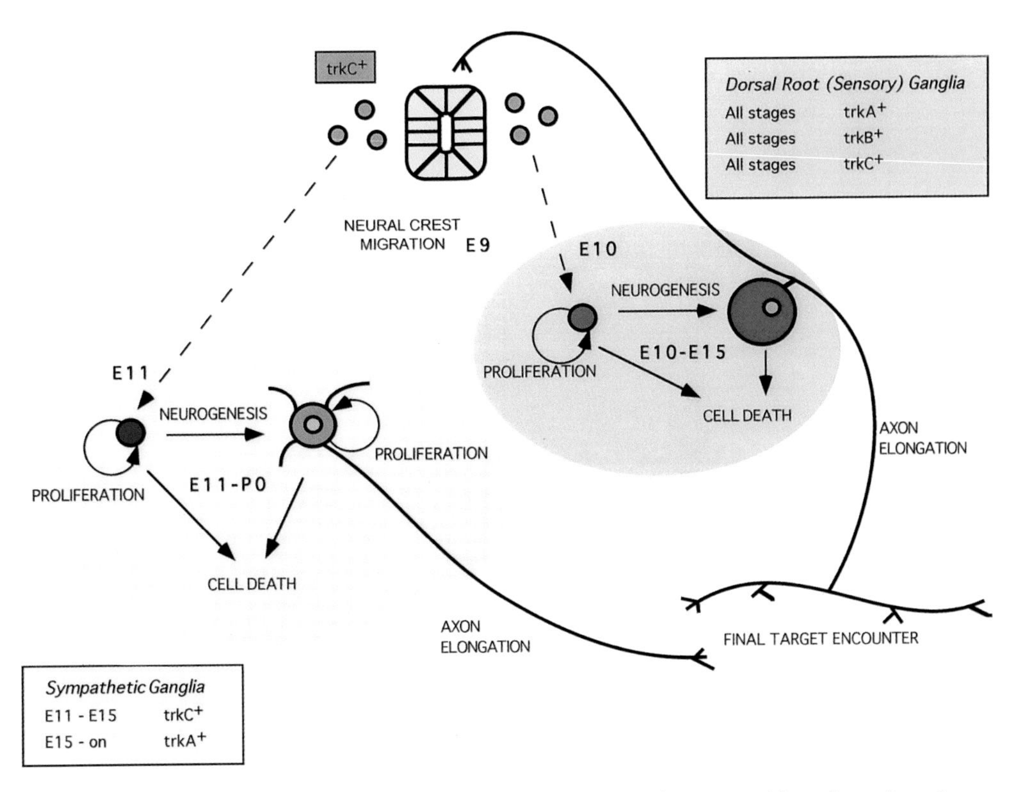

FIGURE 6–7 Schematic representation of the development of sensory (dorsal root) and sympathetic ganglia in the mouse. The ganglia form after coalescence of migrating neural crest cells which then give rise to both neuronal and glial precursor populations. These progenitors proliferate to increase the precursor pool and give rise to neurons. In the sympathetic system, newly formed neurons, or sympathoblasts, acquire some characteristics of differentiated neurons but still have proliferative capacity. Expression of trk receptors is well segregated in the sympathetic system. Early sympathoblasts express trkC initially and later switch to express trkA. In the DRGs, all trk receptors are expressed at all times. However, the specific pattern of expression for each one changes over time. E, embryonic stage.

not colabel with the apoptotic marker (Fariñas et al., 1996). Therefore, survival of precursors is not compromised in these animals. In vitro studies have also shown that BDNF and LIF can promote or accelerate the differentiation of sensory neuron progenitor cells (Sieber-Blum, 1991; Murphy et al., 1991). At present there is no evidence these proteins play essential roles in this process in vivo. As illustrated above, these potential functions are difficult to study in vivo, because so many other developmental pro-cesses occur simultaneously.

Neuronal Dependence on Trophic Factors Before Target Encounter

The neurotrophic hypothesis postulates that neurons become dependent on a particular neurotrophin when their axons come into contact with their final target, as initially shown for the trigeminal ganglion neurons and their projection to the whisker pad (Davies and Lumsden, 1984). More recently, studies in vitro have indicated that these same neurons may be dependent on neurotrophins other than NGF before their axons have established contact with their final targets (see Davies, 1994, for a review). In particular, they suggest that BDNF and NT-3 play a critical early roles in regulating neuronal survival that reflects a unique regulatory circuit in which embryonic neuroblasts and neurons induce NT-3 expression in surrounding mesenchyme and cells contacted by axons growing toward their final targets. While some actions of NT-3 may be mediated through trkC, analyses of mutants indicate that some are mediated through interactions with other trk receptors. The results, therefore, indicate that the unique ability of NT-3 to interact with all trk receptors is essential for understanding its functions in vivo.

This model predicts that NT-3 will be expressed in close proximity to neuroblasts and growing axons at early times in development and that NT-3 deficiency will result in loss of neurons before their axons contact their final targets. Our analysis of the development of sensory neuron deficiencies in the NT-3 knockout mouse indicates that the final deficit in cell number is indeed achieved at very early times, before their axons reach their targets. Replacement of the NT-3 coding exon with lacZ has made it possible to analyze the expression of the endogenous NT-3 gene at times when the neuronal deficits are being generated (Fariñas et al., 1996). A close correlation is seen between NT-3 expression and axon extension during development of the peripheral innervation pattern in these animals. Figure 6–8 shows different aspects of the expression between E10 and E13. In whole-mount preparations, lacZ-monitored NT-3 expression in E10 heads is present in the mesencephalon, branchial arches, and otic vesicle. In the trunk, expression is seen in the most rostral part of the embryo in dorsal areas, close to where DRGs develop. The level of expression increases in these areas between E10 and E11, at the same time as the deficits are beginning to be generated, and new expression appears in the snout and the proximal third of the limbs. In addition, expression is initiated dorsally in more caudal regions of the embryo. The initial staining along the dorsal axis of the body corresponds to the mesenchyme surrounding the dorsal root ganglia. Later, as the axons grow peripherally, the staining becomes more widespread, and it is found in mesenchymal/muscle precursor cells surrounding the growth cones. Finally, NT-3 is found in the target areas at the time of innervation (Fig. 6–9).

Our developmental analysis shows that the cellular deficits in the NT-3–deficient mice are already completed at the time of innervation of final targets (Fariñas et al., 1996; see also Kucera et al., 1995b). Thus, the expression of NT-3 during axonal growth appears to be essential to keep neurons alive before they reach their final targets. Similar early deficits have also been seen in chick embryos injected with antibodies to NT-3 before the period of naturally occurring cell death (Gaese et al., 1994). Some of the neurons lost early in NT-3–deficient mice almost certainly continue to require this factor at later times, when it is probably derived from targets, such as muscle spindles. Direct evidence for this continued requirement has been obtained in chick embryos where injections of antibodies to NT-3 after target innervation have been shown to cause cell death of trkC-expressing neurons (Oakley et al., 1995).

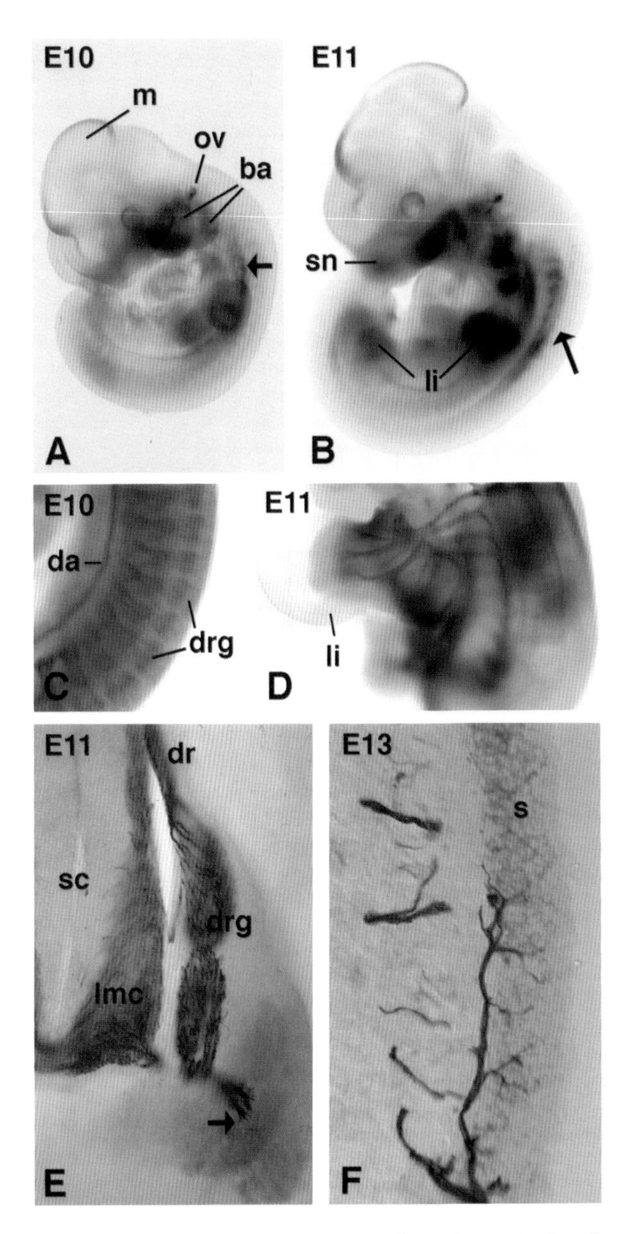

FIGURE 6–8 NT-3 expression in early embryonic development. (A–D) Whole-mount stain for b-gal of NT-3lacZneo heterozygous embryos at different embryonic (E) stages. (A) At E10, the staining can be seen in the mesencephalon (m), otic vesicle (ov), branchial arches (ba), and next to the most rostral dorsal root ganglia (arrow). (B) At E11, there is increased staining in the same areas and new staining is found in the snout and the most proximal half of the limbs (li). (C, D) Combination of the lacZ staining (blue) with antibodies to neuronal specific tubulin (brown) at the same stages. At E10 the expression of NT-3 is found adjacent to the developing dorsal root ganglia (drg) as visualized with the neuronal marker. Another site of expression is the dorsal aorta (da). At E11, NT-3 expression visualized as the blue precipitate of the X-gal histochemical reaction is present in the area where sensory and motor axons are growing in the limb (li). Reproduced by permission from Fariñas et al., 1994. (E, F) Transverse sections through the thoracic region stained with the lacZ histochemical reaction and with antibodies to neurofilaments. In E, it can be seen how the growth cones are exposed to NT-3 during their trajectory. Another site of strong expression is the motoneurons themselves in the lateral motor column (lmc) of the spinal cord (sc). drg, dorsal root ganglion; dr, dorsal root. In F, expression is found for the first time in the skin (s) at E13, when the cutaneous axons arrive.

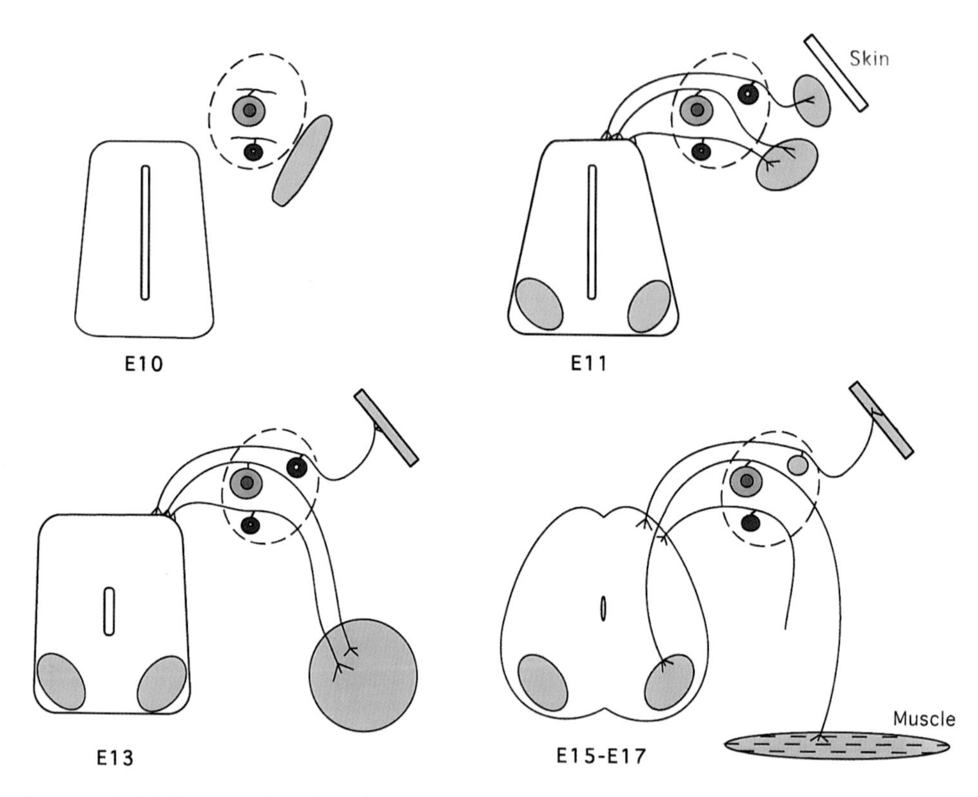

FIGURE 6–9 Diagram showing the relationship between the growth of sensory neuron projections and expression of NT-3. Sensory neurons of the dorsal root ganglia (dotted lines) are represented adjacent to cross sections through the spinal cord (yellow) at different embryonic (E) stages. The growth of the axons over time is indicated and the areas of NT-3 expression are represented in blue. NT-3 is found very close to the ganglia since the time of initial formation. Throughout development expression of NT-3 is tightly associated with the growing axons. It is present also in the final targets at the time of innervation. Cellular deficits in the DRGs of NT-3 mutant mice are completed by E13, suggesting that neurons die before final target innervation. A few neurons in the ganglion also express NT-3 at late stages.

The tight association between growing axons and areas of NT-3 expression suggests that there might be an inductive interaction between the growth cones and the mesenchymal/muscle precursor cells that produce NT-3. Recent experiments in vitro have provided compelling evidence for an inductive interaction between sympathoblasts and their surrounding mesenchyme (Verdi et al., 1996). These neuroblasts have been shown to secrete neuregulins that promote the production by mesenchymal cells of NT-3 which, in turn, promotes sympathoblast survival. Since many neurons have been shown to express neuregulins, this inductive interaction provides a mechanism which is likely to be important in regulating survival transiently during embryogenesis of many populations of neurons, including those whose survival is regulated at later stages by neurotrophic factors expressed in their final targets.

Multiple Neurotrophin Dependency

One of the most intriguing observations to emerge from the comparisons of the various neurotrophin mutants is strong evidence that a large fraction of sympathetic and sensory neurons require more than one neurotrophin for survival during development (Table 6–3). As one example, essentially all sympathetic neurons require NGF and are absent in the NGF and trkA mutant homozygotes. Approximately half of these are also absent in the NT-3 mutant. Thus, a large fraction of these neurons require two neurotrophins. For sympathetic neurons, observed early expression of trkC provides a mechanism by which survival of neuroblasts and neurons is supported by this neurotrophin. As discussed above, though, the presence of normal numbers of sympathetic neurons in trkC-deficient, but not NT-3–deficient, animals argues that essential survival-promoting effects of NT-3 must be mediated through another receptor, almost certainly trkA.

A large fraction of cranial and spinal sensory neurons also exhibit multiple dependencies: for example, approximately 75% of trigeminal sensory neurons require the NGF signaling pathway, while around 30% and 60% are lost in the BDNF and NT-3 mutants, respectively. Similarly, approximately 70%, 30%, and 60% of lumbar sensory neurons are lost in the NGF, BDNF, and NT-3 mutants, respectively. In each example, a much larger number of neurons are lost than are dependent on NT-3 at birth (e.g., Davies, 1994). This loss must include NGF- and/or BDNF-dependent neurons. Since temporal shifts in dependence have been seen in vitro (Buchman and Davies, 1993; Buj-Bello et al., 1994), sequential, not simultaneous, dependence provides the most likely explanation for these observations.

Since NT-3 appears to play crucial roles in development not mediated by interactions with trkC, it is clearly important to characterize the factors that facilitate its interactions with other trk receptors. Studies in cell culture indicate that the presence of a small, differentially spliced extracellular exon in trkA strongly increases its responsiveness to NT-3 without reducing its responsiveness to NGF (Clary and Reichardt, 1994). The presence of high levels of $p75^{NTR}$ appears to inhibit trkA activation in response to NT-3 even though this protein enhances re-

Table 6–3. Neuronal Losses in Neurotrophin-Deficient Mice

	NGF	*BDNF*	*NT-3*	*NT-4/5*
Trigeminal	75%	30%	60%	0%
Nodose	0%	45%	30%	40%
DRG	70%	35%	60%	0%
SCG	95%	0%	50%	0%

Neuronal deficiencies are expressed as the percentage of neurons lost in mutant ganglia compared to wild-type controls. Nodose is the petrosal-nodose complex; DRG, dorsal root ganglia; SCG, superior cervical ganglia. See Table 6–2 for original references.

sponsiveness to NGF. Since non–kinase-containing isoforms of trkC must be coexpressed with trkA in embryonic DRG neurons, it will be important to determine whether they inhibit or potentiate trkA-mediated responses to NT-3.

Analyses have not yet critically addressed the possibility that there may also be neurons with a simultaneous dependence on more than one factor. The concept that a single neuron may require more than one neurotrophin is conceptually appealing because it provides an attractive means of refining neuronal specificity. If, for example, a sensory neuron receives separate trophic support from central and peripheral targets, a neuron failing to establish either connection would be eliminated. A fraction of neurons in the nodose region of the petrosal-nodose complex has been shown to require both BDNF and NT-4/5 for survival during development (Conover et al., 1995; Liu et al., 1995). While dual dependence may be observed because each factor is present in limiting amounts or has a nonredundant signaling function, sequential expression of the two survival factors seems more likely to explain the results. Since trkB is the receptor for each of these neurotrophins, changes in receptor expression are not likely to explain these observations.

AUTOCRINE/PARACRINE ACTIONS OF NEUROTROPHINS

BDNF and NT-3 expression has been detected within cells in DRGs, raising the possibility that these factors have autocrine and/or paracrine actions (Ernfors et al., 1992; Schecterson and Bothwell, 1992). Possibly death of a neurotrophin-dependent population may cause also loss of neighboring neurons dependent on a different factor expressed in the neurotrophin-dependent cells, resulting in a larger-than-expected deficit. Our observations on lacZ expression in NT-^{3lacZneo} mice (Fariñas et al., 1996) indicate, however, that paracrine or autocrine actions of NT-3 are not important in regulating the extent of cell death during the major period of cell death in these ganglia. Confirming earlier in situ analysis (Schecterson and Bothwell, 1992), NT-3 (as assessed by lacZ expression) is detected in a few DRG neurons at E15 and later. However, sensory neuron loss in normal or mutant mice is complete before NT-3 expression is detectable in these neurons. Thus, paracrine actions of NT-3 cannot be important in regulating neuronal survival during the major period of naturally occurring cell death. It remains possible that such mechanisms may be important, however, in other ganglia.

As mentioned earlier, strong evidence has been obtained in vitro for an autocrine action of BDNF on a subpopulation of postnatal sensory neurons (Acheson et al., 1995). The dependence of these neurons on BDNF corresponds to the time they are probably lost in the BDNF mutant homozygotes in vivo.

EFFECTS ON ADULT NERVOUS SYSTEM FUNCTION

Many neurons continue to express receptors for neurotrophic factors into adulthood and all of the neurotrophic factors discussed in this chapter have been detected in adult animals. Regulation of expression of several neurotrophins in the central nervous system has been shown to be regulated by neuronal activity, neurotransmitters, hormones, and other neurotrophins (reviewed in Lindholm et al., 1994). Perhaps the most definitive example of the continued role for neurotrophic factors in mediating adult nervous system function is provided by work of Mandel and colleagues who have shown that NGF mediates hyperalgesic responses induced by inflammation (e.g., Lewin et al., 1992, 1994). Single injections of NGF have been shown to induce hyperalgesia, while injections of anti-NGF suppress hyperalgesic responses to inflammatory stimuli. NGF induces both thermal and mechanical hyperalgesia. The former response has been shown to involve degranulation of mast cells which have been shown to be associated with nociceptive nerve endings. Both survival and degranulation of mast cells is promoted by NGF activation of trkA which is expressed on the surface of these cells (e.g., Horigome et al., 1993, 1994).

Continued expression of neurotrophic factors, particularly of BDNF, in adult brain has suggested that neurotrophic factors may modulate adult brain plasticity and function in addition to brain development. The selective reduction of BDNF in several brain regions of Alzheimer's disease victims has increased interest in this possibility (Phillips et al., 1991). Recent results indicating that neurotrophins and CNTF can regulate the efficiency of exocytosis by motor and hippocampal neurons suggest one cellular role for neurotrophic factors in the adult CNS (e.g., Lohof et al., 1993; Kang and Schuman, 1995, 1996). Regulation of expression levels of neurotransmitters, calcium-binding proteins, and other proteins shown to be dependent on neurotrophins during development is another probable mechanism (e.g., Jones et al., 1994). Of particular interest, recent papers report that long-term potentiation (LTP) is impaired in the CA1 region of the hippocampus of heterozygous BDNF mutant mice (Korte et al., 1995; Patterson et al., 1996). Numerous parameters of hippocampal anatomy and pharmacology appeared normal in these healthy animals. While the observations do not fit simply into current models of CA1 LTP (reviewed in Nicoll and Malenka, 1995), the results suggest a functional role for BDNF in hippocampal LTP. To date, similar studies have not been performed using other heterozygous mutant animals and the deficit can be rescued by exogenous BDNF (Patterson et al., 1996). While results in this paper are exciting and encouraging, use of genetics to reach absolutely definitive conclusions on roles of neurotrophic factors in adult brain function will require temporal and spatial control of gene manipulation.

Summation

The Neurotrophic Factor Hypothesis was proposed to explain the loss of many populations of embryonic neurons in response to target ablation (see Levi-Montalcini, 1987; Purves, 1988). Later it was generalized to explain observed dependencies of neurons on both afferent and efferent connections. The concept that limiting amounts of survival factors could prune neurons to match the demands for their functions was attractive conceptually and was believed to be a novel mechanism utilized by vertebrates that helped make possible development of their large and complex nervous systems. NGF, the first of the neurotrophic factors to be isolated, has proven to match surprisingly well the predictions of this model. NGF is synthesized in targets of sympathetic and sensory neurons. Exogenous NGF can prevent the apoptosis of these neurons which would otherwise be induced by target ablation or axotomy. Moreover, these neurons express specific receptors for NGF which mediate NGF signaling and, in addition, mediate NGF internalization and retrograde transport to the cell somata. Interference with NGF action, using antibodies or targeted mutations, results in death of differentiated sympathetic and sensory neurons, fulfilling strong predictions of this model.

The discovery of NGF predated discoveries of other growth and differentiation factors by many years and in many instances decades. The experiments performed to demonstrate that it fulfilled the stringent predictions of a target-derived trophic factor resulted in discoveries of fundamental importance to cell and developmental biology. During the past few years, for example, it has become appreciated that apoptotic cell death shapes development of most tissues in an organism (e.g., Raff et al., 1993). Its importance is not restricted to the nervous system. Receptor-mediated internalization of a growth factor was discovered in studies attempting to identify the mechanisms of communication between sympathetic targets and neurons (e.g., Korsching, 1993). We now appreciate that the same mechanisms are used to regulate surface expression of receptors for almost all polypeptide growth factors. Similarly, retrograde transport from the nerve terminal to the cell somata was first observed using radiolabeled NGF, but, after a delay of more than a decade, resulted in discovery that the same pathway is present in every eukaryotic cell, where it is powered by the motor protein dynein.

With the appreciation that neurons share far more similarities than differences with other cells has come the similar appreciation that NGF and other trophic factors do not act by unique mechanisms. Perhaps the most important recent discovery in neurotrophic factor research has been the identification of trkA, a ligand-activated receptor tyrosine kinase as the major functional NGF receptor (see Greene and Kaplan, 1995; Bothwell, 1995). After this discovery it was quickly shown that NGF activates many of the same intracellular signaling pathways utilized by mitogenic growth factors to regulate proliferation of nonneural cells. In addition, some of these factors, most notably the LIF-related cytokines, PDGF, and members of the fibroblast growth factor family have been shown to have survival- and differentiation-promoting activities on neurons similar to those of NGF. Thus, while NGF has provided a powerful focus for research into molecular mechanisms directing neuronal survival and differentiation, it has proven to share far more similarities with other factors than was believed possible at the time of its discovery.

Finally, while Hamburger, Levi-Montalcini, and their colleagues appreciated that they were initiating studies of fundamental importance, it could not have been appreciated that NGF and other neurotrophic factors regulated so many facets of neural development: survival and proliferation of precursors, cell-fate decisions, the differentiation process, axon growth, exocytosis, and synaptic plasticity. Understanding mechanisms by which they influence each of these important processes is a major challenge for the future. In addition, our understanding of their roles in development presently exceeds by far our understanding of their roles in adult nervous system function. To address critically the functions of these molecules in adult animals, it will be necessary to ensure that development occurs normally. For such studies, genetics will undoubtedly make major contributions since it appears increasingly possible to generate animals where gene loss can be regulated temporally (e.g., Gossen and Bujard, 1992; Spencer et al., 1993) or can be limited to subpopulations of neurons (e.g., Gu et al., 1994).

References

Acheson, A, Conover, J.C., Fandl, J.P., DeChiara, T.M., Russell, M., Thadani, A., Squinto, S.P., Yancopoulos, G.D., Lindsay, R.M. (1995). A BDNF autocrine loop in adult sensory neurons prevents cell death. *Nature* 374:450–453.

Airaksinen, M.S., Koltzenburg, M., Lewin, G.R., Masu, Y., Helbig, C., Wolf, E., Brem, G., Toyka, K.V., Thoenen, H., Meyer, M. (1996). Specific subtypes of cutaneous mechanoreceptors require neurotrophin-3 following peripheral target innervation. *Neuron* 16:287–295.

Anderson, D.J. (1993). Cell fate determination in the peripheral nervous system: the sympathoadrenal progenitor. *J. Neurobiol.* 24:185–198.

Anton, E.S., Weskamp, G., Reichardt, L.F., Matthew, W.D. (1994). Nerve growth factor receptor promote Schwann cell migration. *Proc. Natl. Acad. Sci. U.S.A.* 91:2795–2799.

Arenas, E., Persson, H. (1994). Neurotrophin-3 prevents the death of adult central noradrenergic neurons in vivo. *Nature* 367:368–371.

Averbuch-Heller, L., Pruginin, M., Kahane, N., Tsoulfas, P., Parada, L., Rosenthal, A., Kalcheim, C. (1994). Neurotrophin 3 stimulates the differentiation of motoneurons from avian neural tube progenitor cells. *Proc. Natl. Acad. Sci. U.S.A.* 91:3247–3251.

Baimbridge, K.G., Celio, M.R., Rogers, J.H. (1992). Calcium-binding proteins in the nervous system. *Trends Neurosci.* 15:303–308.

Bamber, B.A., Masters, B.A., Hoyle, G.W., Brinster, R.L., *Palmiter P.D.* (1994). Leukemia inhibitory factor induces neurotransmitter switching in transgenic mice. *Proc. Natl. Acad. Sci. U.S.A.* 91:7839–7843.

Barbacid, M. (1995). Neurotrophic factors and their receptors. *Curr. Opin. Cell Biol.* 7:148–155.

Barbin, G., Manthorpe, M., Varon, S. (1984). Purification of chick eye ciliary neuronotrophic factor. *J. Neurochem.* 43:1468–1478.

Barker, P.A., Shooter, E.M. (1994). Disruption of NGF-binding to the low affinity neurotrophin receptor p75LNTR reduces NGF binding to TrkA on PC12 cells. *Neuron* 13:203–215.

Barres, B.A., Raff, M.C., Gaese, F., Bartke, I., Dechant, G., Barde, Y.A. (1994). A crucial role for neurotrophin-3 in oligodendrocyte development. *Nature* 367:371–375.

Bazan, J.F. (1991). Neuropoietic cytokines in the hematopoietic fold. *Neuron* 7:197–208.

Bendetti, M., Levi, A., Chao, M.V. (1993). Differential expression of nerve growth factor receptors leads to altered binding affinity and neurotrophin responsiveness. *Proc. Natl. Acad. Sci. U.S.A.* 90:7859–7863.

Bianchi, L.M., Conover, J.C., Fritzsch, B., DeChiara, T., Lindsay, R.M., Yancopoulos, G.D. (1996). Degeneration of vestibular neurons in late embryogenesis of both heterozygous and homozygous BDNF null mutant mice. *Development* 122:1965–1973.

Birren, S.J., Lo, L., Anderson, D.J. (1993). Sympathetic neuroblasts undergo a developmental switch in trophic dependence. *Development* 119:597–610.

Blochl, A., Thoenen, H. (1996). Localization of cellular storage compartments and sites of constitutive and activity-dependent release of nerve growth factor (NGF) in primary cultures of hippocampal neurons. *Mol. Cell. Neurosci.* 7:173–190.

Bothwell, M. (1995). Functional interactions of neurotrophins and neurotrophin receptors. *Annu. Rev. Neurosci.* 18:223–253.

Buchman, V.L., Davies, A.M. (1993). Different neurotrophins are expressed and act in a developmental sequence to promote the survival of embryonic sensory neurons. *Development* 118:989–1001.

Buj-Bello, A., Pinon, L.G., Davies, A.M. (1994). The survival of NGF-dependent but not BDNF-dependent cranial sensory neurons is promoted by several different neurotrophins early in their development. *Development* 120:1573–1580.

Buj-Bello, A., Buchman, V.L., Horton, A., Rosenthal, A., Davies, A.M. (1995). GDNF is an age-specific survival factor for sensory and autonomic neurons. *Neuron* 15:821–882.

Cabelli, R.J., Hohn, A., Shatz, C.J. (1995). Inhibition of ocular dominance column formation by infusion of NT-4/5 or BDNF. *Science* 267:1662–1666.

Campenot, R.B. (1977). Local control of neurite development by nerve growth factor. *Proc. Natl. Acad. Sci. U.S.A.* 74:4516–4519.

Casaccia-Bonnefil, P., Carter, B.D., Dobrowsky, R.T., Chao, M.V. (1996). Death of oligodendrocytes mediated by the interaction of nerve growth factor with its receptor p75. *Nature* 383:716–719.

Cattaneo, E., McKay, R. (1990). Proliferation and differentiation of neuronal stem cells regulated by nerve growth factor. *Nature* 347:762–765.

Clary, D.O., Reichardt, L.F. (1994). An alternatively spliced form of the nerve growth factor receptor TrkA confers an enhanced response to neurotrophin 3. *Proc. Natl. Acad. Sci. U.S.A.* 91:11133–11137.

Clary, D.O., Weskamp, G., Austin, L.R., Reichardt, L.F. (1994). TrkA cross-linking mimics neuronal responses to nerve growth factor. *Mol. Biol. Cell* 5:549–563.

Conover, J.C., Erickson, J.T., Katz, D.M., Bianchi, L.M., Poueymirou, W.T., McClain, J., Pan, L., Helgren, M., Ip, N.Y., Boland, P., Friedman, B., Wiegand, S., Vejsada, R., Kato, A.C., DeChiara, T.M., Yancopoulos, G.D. (1995). Neuronal deficits, not involving motor neurons, in mice lacking BDNF and/or NT4. *Nature* 375:235–238.

Copray, J.C., Brouwer, N. (1994). Selective expression of neurotrophin-3 messenger RNA in muscle spindles of the rat. *Neuroscience* 63:1125–1135.

Crowley, C., Spencer, S.D., Nishimura, M.C., Chen, K.S., Pitts-Meek, S., Armanini, M.P., Ling, G.H., McMahon, S.B., Shelton, D.L., Levinson, A.D., Phillips, H.S. (1994). Mice lacking nerve growth factor display perinatal loss of sensory and sympathetic neurons yet develop basal forebrain cholinergic neurons. *Cell* 76:1001–1011.

Curtis, R., Adryan, K.M., Stark, J.L., Park, J.S., Compton, D.L., Weskamp, G., Huber, L.J., Chao, M.V., Jaenisch, R., Lee, K.-L., Lindsay, R.M., DiStephano, P.S. (1995). Differential role of the low affinity neurotrophin receptor (p75) in retrograde axonal transport of the neurotrophins. *Neuron* 14:1201–1211.

Damen, J.E., Cutler, R.L., Jiao, H., Yi, T., Krystal, G. (1995). Phosphorylation of tyrosine 503 in the erythropoietin receptor (EpR) is essential for binding the P85 subunit of phosphatidylinositol (Pl) 3-kinase and for EpR-associated Pl 3-kinase activity. *J. Biol. Chem.* 270:23402–23408.

Davies, A.M. (1997). Neurotrophin Switching: where does it stand? *Curr. Op. Neurobiol.* in press.

Davies, A.M. (1994). The role of neurotrophins in the developing nervous system. *J. Neurobiol.* 25:1334–1348.

Davies, A.M., Lumsden, A. (1984). Relation of target encounter and neuronal death to nerve growth factor responsiveness in the developing mouse trigeminal ganglion. *J. Comp. Neurol.* 223:124–137.

Davies, A.M., Lindsay, R.M. (1985). The cranial sensory ganglia in culture: differences in the response of placode-derived and neural crest-derived neurons to nerve growth factor. *Dev. Biol.* 111:62–72.

Davies, A.M., Lee, K.F., Jaenisch, R. (1993). p75-deficient trigeminal sensory neurons have an altered response to NGF but not to other neurotrophins. *Neuron* 11:565–574.

Davies, A.M., Minichiello, L., Klein, R. (1995). Developmental changes in NT3 signalling via TrkA and TrkB in embryonic neurons. *EMBO J.* 14:4482–4489.

DeChiara, T.M., Vejsada, R., Poueymirou, W.T., Acheson, A., Suri, C., Conover, J.C., McClain, J., Pan, L., Stahl, N., Ip, N.Y., Kato, A., Yancopoulos, G.D. (1995). Mice lacking the CNTF receptor, unlike mice lacking CNTF, exhibit profound motor neuron deficits at birth. *Cell* 83:313–322.

DeChiara, T.M., Bowen, D., Valenzuela, D.M., Simmons, M.V., Poueymirou, W.T., Thomas, S., Kinetz, E., Compton, D.L., Park, J.S., Smith, C., DiStephano, P.S., Glass, D.J., Burden, S.J., Yancopoulos, G.D. (1996). The receptor tyrosine kinase, MuSK, is required for all aspects of neuromuscular junction formation in vivo. *Cell* 85:501–513.

Diamond, J., Holmes, M., Coughlin, M. (1992). Endogenous NGF and nerve impulses regulate the collateral sprouting of sensory axons in the skin of the adult rat. *J. Neurosci.* 12:1454–1466.

DiCicco-Bloom, E., Friedman, W.J., Black, I.B. (1993). NT-3 stimulates sympathetic neuroblast proliferation by promoting precursor survival. *Neuron* 11:1101–1111.

Drescher, U., Kremoser, C., Handwerker, C., Loschinger, J., Noda, M., Bonhoeffer, F. (1995). In vitro guidance of retinal ganglion cell axons by RAGS, a 25 kDa tectal protein related to ligands for Eph receptor tyrosine kinases. *Cell* 82:359–703.

Dobrowsky, R.T., Werner, M.H., Castellino, A.M., Chao, M.V., Hannun, Y.A. (1994). Activation of the sphingomyelin cycle through the low-affinity neurotrophin receptor. *Science* 265:1596–1599.

Durbec, P.L., Larsson-Blomberg, L.B., Schuchardt, A., Costantini, F., Pachnis, V. (1996a). Common origin and developmental dependence on c-ret of subsets of enteric and sympathetic neuroblasts. *Development* 122:349–358.

Durbec, P., Marcos-Gutierrez, C.V., Kilkenny, C., Grigoriou, M., Wartiowaara, K., Suvanto, P., Smith, D., Ponder, B., Costantini, F., Saarma, M., Sariola, H., Pachnis, V. (1996b). GDNF signalling through the Ret receptor tyrosine kinase. *Nature* 381:789– 793.

Edwards, R.H., Rutter, W.J., Hanahan, D. (1989). Directed expression of NGF to pancreatic beta cells in transgenic mice to selective hyperinnervation of the islets. *Cell* 58:161–170.

Eide, F.F., Lowenstein, D.H., Reichardt, L.F. (1993). Neurotrophins and their receptors—current concepts and implications for neurologic disease. *Exp. Neurol.* 121:200–214.

Eide, F.F., Vining, E., Eide, B.L., Zang, K., Wang, X.-Y., Reichardt, L.F. (1996). Naturally occurring truncated trkB receptors have dominant inhibitory effects on brain-derived neurotrophic factor signaling. *J. Neurosci.* 16:3123–3129.

ElShamy, W. M., Ernfors, P. (1996a). A local action of neurotrophin-3 prevents the death of proliferating sensory neuron precursor cells. *Neuron* 16:963–972.

ElShamy, W.M. and Ernfors, P. (1996b). Requirement of neurotrophin-3 for the survival of proliferating trigeminal ganglion progenitor cells. *Development* 122:2405–2414.

ElShamy, W.M., Linnarsson, S., Lee, K.-F., Jaenisch, R., Ernfors, P. (1996). Prenatal and postnatal requirements of NT-3 for sympathetic neuroblast survival and innervation of specific targets. *Development* 122:491–500.

Erickson, J.T., Conover, J.C., Borday, V., Champagnat, J., Barbacid, M., Yancopoulos, G., Katz, D.M. (1996). Mice lacking brain-derived neurotrophic factor exhibit visceral sensory neuron losses distinct from mice lacking NT4 and display a severe developmental deficit in control of breathing. *J. Neurosci.* 16:5361–5371.

Ernfors, P., Merlio, J.-P., Persson, H. (1992). Cells expressing mRNA for neurotrophins and their receptors during embryonic rat development. *Eur. J. Neurosci.* 4:1140–1158.

Ernfors, P., Lee, K.F., Jaenisch, R. (1994a). Mice lacking brain-derived neurotrophic factor develop with sensory deficits. *Nature* 368:147–150.

Ernfors, P., Lee, K.F., Kucera, J., Jaenisch, R. (1994b). Lack of neurotrophin-3 leads to deficiencies in the peripheral nervous system and loss of limb proprioceptive afferents. *Cell* 77:503–512.

Ernfors, P., Van De Water, T., Loring, J., Jaenisch, R. (1995). Complementary roles of BDNF and NT-3 in vestibular and auditory development. *Neuron* 14:1153–1164.

Ernsberger, U., Sendtner, M., Rohrer, H. (1989). Proliferation and differentiation of embryonic chick sympathetic neurons: effects of ciliary neurotrophic factor. *Neuron* 2:1275–1284.

Fagan, A.M., Zhang, H., Landis, S., Smeyne, R.J., Silos-Santiago, I., Barbacid, M. (1996). TrkA,. but not trkC receptors are essential for survival of sympathetic neurons in vivo. *J. Neurosci.* 16:6208–6218.

Fariñas, I., Jones, K.R., Backus, C., Wang, X.Y., Reichardt, L.F. (1994). Severe sensory and sympathetic deficits in mice lacking neurotrophin-3. *Nature* 369:658–661.

Fariñas, I., Yoshida, C.K., Backus, C., Reichardt, L.F. (1996). Lack of neurotrophin-3 results in death of spinal sensory neurons and premature differentiation of their precursors. *Neuron* 17:1065–1078.

Frade, J.M., Rodriguez-Tebar, A., Barde, Y.-A. (1996). Induction of cell death by endogenous nerve growth factor through its p75 receptor. *Nature* 383:166–168.

Gaese, F., Kolbeck, R., Barde, Y.A. (1994). Sensory ganglia require neurotrophin-3 early in development. *Development* 120:1613–1619.

Ghosh, A., Greenberg, M.E. (1995). Distinct roles for bFGF and NT-3 in the regulation of cortical neurogenesis. *Neuron* 15:89–103.

Gossen, M., Bujard, H. (1992). Tight control of gene expression in mammalian cells by tetracycline-responsive promoters. *Proc. Natl. Acad. Sci. U.S.A.* 89:5547–5551.

Gotz, R., Koster, R., Winkler, C., Raulf, F., Lottspeich, F., Schartl, M., Thoenen, H. (1994). Neurotrophin-6 is a new member of the nerve growth factor family. *Nature* 372: 266–269.

Greene, L.A., Kaplan, D.R. (1995). Early events in neurotrophin signalling via Trk and p75 receptors. *Curr. Opin. Neurobiol.* 5:579–587.

Griffith, C.G., Letourneau, P.C. (1980). Rapid retraction of neurites by sensory neurons in response to increased concentrations of nerve growth factor. *J. Cell Biol.* 86:156–161.

Gu, H., Marth, J.D., Orban, P.C., Mossmann, H., Rajewsky, K. (1994). Deletion of a DNA polymerase beta gene segment in T cells using cell type-specific gene targeting. *Science* 265:103–106.

Gundersen, R.W., Barrett, J.N. (1979). Neuronal hemotaxis: chick dorsal root ganglion axons turn toward high concentrations of nerve growth factor. *Science* 206:1079–1080.

Hagg, T., Quon, D., Higaki, J., Varon, S. (1992). Ciliary neurotrophic factor prevents neuronal degeneration and promotes low affinity NGF receptor expression in the adult rat CNS. *Neuron* 8:145–158.

Hall, A. (1994). Small GTP-binding proteins and the regulation of the actin cytoskeleton. *Annu. Rev. Cell. Biol.* 10:31–54.

Henderson, C.E., Camu, W., Mettling, C., Gouin, A., Poulsen, K., Karihaloo, M., Rullamas, J., Evans, T., McMahon, S.B., Armanini, M.P., Berkemeier, L., Phillips, H.S., Rosenthal, A. (1993). Neurotrophins promote motor neuron survival and are present in embryonic limb bud. *Nature* 363:266–270.

Henderson, C.E., Phillips, H.S., Pollock, R.A., Davies, A.M., Lemeuelle, C., Armanini, M., Simpson, L.C., Moffet, B., Vandlen, R.A., Koliatsos, V.E., Rosenthal, A. (1994). GDNF: a potent survival factor for motoneurons present in peripheral nerve and muscle. *Science* 266:1062–1064.

Henion, P.D., Garner, A.S., Large, T.H., Weston, J.A. (1995). trkC-mediated NT-3 signaling is required for the early development of a subpopulation of neurogenic neural crest cells. *Dev. Biol.* 172:602–613.

Horigome, K., Pryor, J.C., Bullock, E.D., Johnson, E.M., Jr. (1993). Mediator release from mast cells by nerve growth factor. Neurotrophin specificity and receptor mediation. *J. Biol. Chem.* 268:14881–14887.

Horigome, K., Bullock, E.D., Johnson, E.M., Jr. (1994). Effects of nerve growth factor on rat peritoneal mast cells. Survival promotion and immediate-early gene induction. *J. Biol. Chem.* 269:2695–2702.

Hoyle, G.W., Mercer, E.H., Palmiter, R.D., Brinster, R.L. (1993). Expression of NGF in sympathetic neurons leads to excessive axon outgrowth from ganglia but decreased terminal innervation within tissues. *Neuron* 10:1019–1034.

Ip, N.Y., Li, Y., Yancopoulos, G.D., Lindsay, R.M. (1993a). Cultured hippocampal neurons show responses to BDNF, NT-3, and NT-4, but not NGF. *J. Neurosci.* 13:3394–3405.

Ip, N.Y., Stitt, T.N., Tapley, P., Klein, R., Glass, D.J., Fandl, J., Green, L.A., Barbacid, M., Yancopoulos, G.D. (1993b). Similarities and differences in the way neurotrophins interact with the Trk receptors in neuronal and nonneuronal cells. *Neuron* 10:137–149.

Jing, S., Wen, D., Yu, Y., Holst, P.L., Luo, Y., Fang, M., Tamir, R., Antonio, L., Hu, Z., Cupples, R., Louis, J.-C., Hu S., Altrock, B.W., Fox, G.M. (1996). GDNF-induced activation of the ret protein tyrosine kinase is mediated by GDNFR-α, a novel receptor for GDNF. *Cell* 85:1113–24.

Jones, K.R., Fariñas, I., Backus, C., Reichardt, L.F. (1994). Targeted disruption of the BDNF gene perturbs brain and sensory neuron development but not motor neuron development. *Cell* 76:989–999.

Kang, H., E. M. Schuman. (1996). A requirement for local protein synthesis in 140 neurotrophin-induced hippocampal synaptic plasticity. *Science* 273:1402–6.

Kang, H., Schuman, E.M. (1995). Long-lasting neurotrophin-induced enhancement of synaptic transmission in the adult hippocampus. *Science* 267:1658–1662.

Kalcheim, C., Carmeli, C., Rosenthal, A. (1992). Neurotrophin 3 is a mitogen for cultured neural crest cells. *Proc. Natl. Acad. Sci. U.S.A.* 89:1661–1665.

Karavanov, A., Sainio, K., Palgi, J., Saarma, M., Saxen, L., Sariola, H. (1995). Neurotrophin 3 rescues neuronal precursors from apoptosis and promotes neuronal differentiation in the embryonic metanephric kidney. *Proc. Natl. Acad. Sci. U.S.A.* 92:11279–11283.

Klein, R., Emeyne, R.J., Wurst, W., Long, L.K., Auerbach, B.A., Joyner, A.L., Barbacid, M. (1993). Targeted disruption of the trkB neurotrophin receptor gene results in nervous system lesions and neonatal death. *Cell* 75:113–122.

Klein, R., Silos-Santiago, I., Smeyne, R.J., Lira, S.A., Bambrilla, R., Bryant, S., Zhang, L., Snider, W.D. Barbacid, M. (1994). Disruption of the neurotrophin-3 receptor gene trkC eliminates Ia muscle afferents and results in abnormal movements. *Nature* 368: 249–251.

Korsching, S. (1993). The neurotrophic factor concept: a reexamination. *J. Neurosci.* 13: 2739–2748.

Korte, M., Carroll, P., Wolf, E., Brem, G., Thoenen, H., Bonhoeffer, T. (1995). Hippocampal long-term potentiation is impaired in mice lacking brain-derived neurotrophic factor. *Proc. Natl. Acad. Sci. U.S.A.* 92:8856–8860.

Kotzbauer, P.T., Lampe, P.A., Heuckeroth, R.O., Golden, J.P., Creedon, D.J., Johnson Jr., E.M., Milbrandt, J. (1996). Neurturin, a relative of glial-cell-derived neurotrophic factor. *Nature* 384:467–470.

Kucera, J., Ernfors, P., Walro, J., Jaenisch, R. (1995a). Reduction in the number of spinal motor neurons in neurotrophin-3-deficient mice. *Neuroscience* 69:321–330.

Kucera, J., Fan, G., Jaenisch, R., Linnarsson, S., Ernfors, P. (1995b). Dependence of developing group Ia afferents on neurotrophin-3. *J. Comp. Neurol.* 363:307–320.

Kundra, V., Escobedo, J.A., Kazlauskas, A., Kim, H.K., Rhee, S.G., Williams, L.T., Zetter, B.R. (1994). Regulation of chemotaxis by the platelet-derived growth factor receptor-beta. *Nature* 367:474–476.

Lamballe, F., Smeyne, R.J., Barbacid, M. (1994). Developmental expression of trkC, the neurotrophin-3 receptor, in the mammalian nervous system. *J. Neurosci.* 14:14–28.

Lauffenburger, D.A., Horwitz, A.F. (1996). Cell migration: a physically integrated molecular process. *Cell* 84:359–369.

Lee, K.-F., Li, E., Huber, J., Landis, S., Sharpe, A.H., Chao, M., Jaenisch, R. (1992). Targeted mutation of the gene encoding the low affinity NGF receptor p75 leads to deficits in the peripheral sensory nervous system. Cell 69:737–749.

Lee, K.F., Davies, A.M., Jaenisch, R. (1994a). p75-deficient embryonic dorsal root sensory and neonatal sympathetic neurons display a decreased sensitivity to NGF. *Development* 120:1027–1033.

Lee, K.F., Bachman, K., Landis, S., Jaenisch, R. (1994b). Dependence on p75 for innervation of some sympathetic targets. *Science* 263:1447–1449.

Levi-Montalcini, R. (1987). The nerve growth factor 35 years later. *Science* 237:1154–1162.

Lewin, G.R., Mendell, L.M. (1994). Regulation of cutaneous C-fiber heat nociceptors by nerve growth factor in the developing rat. *J. Neurophysiol.* 71:941–949.

Lewin, G.R., Ritter, A.M., Mendell, L.M. (1992). On the role of nerve growth factor in the development of myelinated nociceptors. *J. Neurosci.* 12:1896–1905.

Lewin, G.R., Rueff, A., Mendell, L.M. (1994). Peripheral and central mechanisms of NGF-induced hyperalgesia. *Eur. J. Neurosci.* 6:1903–1912.

Li, M., Sendtner, M., Smith, A. (1996). Essential function of LIF receptor in motor neurons. *Nature* 378:724–727.

Lin, L.F., Doherty, D.H., Lile, J.D. Bektesh, S., Collins, F. (1993). GDNF: a glial cell line-derived neurotrophic factor for midbrain dopaminergic neurons. *Science* 260:1130–1132.

Lindholm, D., Castrén, E., Berzaghi, M., Blöchl, A., Thoenen, H. (1994). Activity-dependent and hormonal regulation of neurotrophin mRNA levels in the brain—implications for neuronal plasticity. *J. Neurobiol.* 25:1362–1372.

Liu, X., Ernfors, P., Wu, H., Jaenisch, R. (1995). Sensory but not motor neuron deficits in mice lacking NT4 and BDNF. *Nature* 375:238–241.

Lohof, A.M., Ip, N.Y., Poo, M.M. (1993). Potentiation of developing neuromuscular synapses by the neurotrophins NT-3 and BDNF. *Nature* 363:350–353.

Masu, Y., Wolf, E., Holtmann, B., Sendtner, M., Brem, G., Thoenen, H. (1993). Disruption of the CNTF gene results in motor neuron degeneration. *Nature* 365:27–32.

McAllister, A.K., Katz, L.C., Lo, D.C. (1996). Neurotrophin regulation of cortical dendritic growth requires activity. *Neuron* 17:1057–1064.

McAllister, A.K., Lo, D.C., Katz, L.C. (1995). Neurotrophins regulate dendritic growth in developing visual cortex. *Neuron* 15:791–803.

McDonald, N.Q., Chao, M.V. (1995). Structural determinants of neurotrophin action. *J. Biol. Chem.* 270:19669–19672.

McMahon, S.B., Armanini, M.P., Ling, L.H., Phillips, H.S. (1994). Expression and coexpression of trk receptors in subpopulations of adult primary sensory neurons projecting to identified peripheral targets. *Neuron* 12:1161–1171.

Memberg, S.P., Hall. A. K. (1995). Proliferation, differentiation, and survival of rat sensory neuron precursors in vitro require specific trophic factors. *Mol. Cell. Neurosci.* 6:323–335.

Minichiello, L., Klein, R. (1996). TrkB and TrkC neurotrophin receptors cooperate in promoting survival of hippocampal and cerebellar granule neurons. *Genes & Development* 15:2849–2858.

Minichiello, L., Piehl, F., Vazquez, E., Schimmang, T., Hökfelt, T., Represa, J., Klein, R. (1995). Differential effects of combined trk receptor mutations on dorsal root ganglion and inner ear sensory neurons. *Development* 121:4067–4075.

Mitchison, T.J., Cramer, L.P. (1996). Actin-based cell motility and cell locomotion. *Cell* 84:371–379.

Moore, M.W., Klein, R.D., Fariñas, I., Sauer, H., Armanini, M., Phillips, H., Reichardt, L.F., Ryan, A.M., Carver-Moore, K.C., Rosenthal, A. (1996). Renal and neuronal abnormalities in mice lacking GDNF. Nature 382:76–79.

Mu, X., Silos, S.I., Carroll, S.L., Snider, W.D. (1993). Neurotrophin receptor genes are expressed in distinct patterns in developing dorsal root ganglia. *J. Neurosci.* 13: 4029–4041.

Murphy, M., Reid, K., Hilton, D., Bartlett, P.F. (1991). Generation of sensory neurons is stimulated by leukemia inhibitory factor. *Proc. Natl. Acad. Sci. U.S.A.* 88:3498–3501.

Murphy, M., Reid, K., Brown, M.A., Bartlett, P.F. (1993). Involvement of leukemia inhibitory factor and nerve growth factor in the development of dorsal root ganglion neurons. *Development* 117:1173–1182.

Nawa, H., Bessho, Y., Carnahan, J., Nakanishi, S., Mizuno, K. (1993). Regulation of neuropeptide expression in cultured cerebral cortical neurons by brain-derived neurotrophic factor. *J. Neurochem.* 60:772–775.

Nicoll, R.A., Malenka, R.C. (1995). Contrasting properties of two forms of long-term potentiation in the hippocampus. *Nature* 377:115–118.

Oakley, R.A., Garner, A.S., Large, T.H., Frank, E. (1995). Muscle neurons require neurotrophin-3 from peripheral tissues during the period of normal cell death. *Development* 121:1341–1350.

Ockel, M., Lewin, G.R., Barde, Y.-A. 1996. In vivo effects of neurotrophin-3 during sensory neurogenesis. *Development* 122:301–307.

Parnavelas, J.G., Papadopoulos, G.C., Cavanagh, M.E. (1988). Changes in neurotransmitters during development. In *Cerebral Cortex,* vol 7, Peters, A., Jones, E.G., eds. New York: Plenum Press.

Patterson, P.H., Nawa, H. (1993). Neuronal differentiation factors/cytokines and synaptic plasticity. *Cell/Neuron* 72/10 (*suppl.*):123–137.

Patterson, S.L., Abel, T., Deuel, T.A.S., Martin, K.C., Rose, J.C., Kandel, E.R. (1996). Recombinant BDNF rescues deficits in basal synaptic transmission and hippocampal LTP in BDNF knockout mice Neuron 16:1137–1145.

Peng, X., Greene, L.A., Kaplan, D.R., Stephens, R.M. (1995). Deletion of a conserved juxtamembrane sequence in trk abolishes NGF-promoted neuritogenesis. *Neuron* 15: 395–406.

Phillips, H.S., Hains, J.M., Armanini, M., Laramee, G.R., Johnson, S.A., Winslow, J.W. (1991). BDNF mRNA is decreased in the hippocampus of individuals with Alzheimer's disease. *Neuron* 7:695–702.

Pichel, J.G., Shen, L., Sheng, H.Z., Granholm, A.-C., Drago, J., Grinberg, A., Lee, E.J., Huang, S.P., Saarma, M., Hoffer, B.J., Sariola, H., Westphal, H. (1996). Defects in enteric innervation and kidney development in mice lacking GDNF. *Nature* 382:73–76.

Pinco, O., Carmeli, C., Rosenthal, A., Kalcheim, C. (1993). Neurotrophin-3 affects proliferation and differentiation of distinct neural crest cells and is present in the early neural tube of avian embryos. *J. Neurobiol.* 24:1626–1641.

Purves, D. (1988). *Body and Brain. A Trophic Theory of Neural Connections.* Cambridge: Harvard University Press.

Rabizadeh, S., Oh, J., Zhong, L., Yang, J., Bitler, C.M., Butcher, L.L., Bredesen, D.E. (1993). Induction of apoptosis by the low-affinity NGF receptor. *Science* 261:345–348.

Raff, M.C., Barres, B.A., Burne, J.F., Coles, H.S., Ishizaki, Y., Jacobson, M.D. (1993). Programmed cell death and the control of cell survival: lessons from the nervous system. *Science* 262:695–700.

Rohrer, H., Thoenen, H. (1987). Relationship between differentiation and terminal mitosis: chick sensory and ciliary neurons differentiate after terminal mitosis of precursor cells, whereas sympathetic neurons continue to divide after differentiation. *J. Neurosci.* 7:3739–3748.

Sanchez, M.P., Silos-Santiago, I., Frisen, J., He, B., Lira, S.A., Barbacid, M. 1996. Renal agenesis and the absence of enteric neurons in mice lacking GDNF. *Nature* 382:70–72.

Schecterson, L.C., Bothwell, M. (1992). Novel roles for neurotrophins are suggested by BDNF and NT-3 mRNA expression in developing neurons. *Neuron* 9:449–466.

Schimmang, T., Minichiello, L., Vazquez, E., San Jose, I., Giraldez, F., Klein, R., Represa, J. (1995). Developing inner ear sensory neurons require TrkB and TrkC receptors for innervation of their peripheral targets. *Development* 121:3381–3391.

Schindler, C., Darnell, J.E., Jr. (1995). Transcriptional responses to polypeptide ligands: the JAK-STAT pathway. *Annu. Rev. Biochem.* 64:621–651.

Schnell, L., Schneider, R., Kolbech, R., Barde, Y.-A., Schwab, M.E. (1994). Neurotrophin-3 enhances sprouting of corticospinal tract during development and after adult spinal cord lesion. *Nature* 367:170–173.

Scott, S.A. (1992). *Sensory Neurons. Diversity, Development, and Plasticity.* New York: Oxford University Press.

Sieber-Blum, M. (1991). Role of the neurotrophic factors BDNF and NGF in the commitment of pluripotent neural crest cells. *Neuron* 6:949–955.

Silos-Santiago, A., Molliver, D.C., Ozaki, S., Smeyne, R.J., Fagan, A.M., Barbacid, M., Snider, W.D. (1995). Non-trkA-expressing small DRG neurons are lost in trkA deficient mice. *J. Neurosci.* 15:5929–5942.

Smeyne, R.J., Klein, R., Schnapp, A., Long, L.K., Bryant, S., Lewin, S.A., Barbacid, M. (1994). Severe sensory and sympathetic neuropathies in mice carrying a disrupted Trk/NGF. *Nature* 368:246–248.

Spencer, D.M., Wandless, T.J., Schreiber, S.L., Crabtree, G.R. (1993). Controlling signal transduction with synthetic ligands. *Science* 262:1019–1024.

Tessarollo, L., Tsoulfas, P., Martin-Zanca, D., Gilbert, D.J., Jenkins, N.A., Copeland, N.G., Parada, L.F. (1993). trkC, a receptor for neurotrophin-3, is widely expressed in the developing nervous system and in non-neuronal tissues. *Development* 118:463–475.

Tessarollo, L., Vogel, K.S., Palko, M.E., Reid, S.W., Parada, L.F. (1994). Targeted mutation in the neurotrophin-3 gene results in loss of muscle sensory neurons. *Proc. Natl. Acad. Sci. U.S.A.* 91:11844–11848.

Tojo, H., Kaisho, Y., Nakata, M., Matsuoka, K., Kitagawa, M., Abe, T., Takami, K., Yamamoto, M., Shino, A., Igarashi, K., Aizawa, S., Shiho, O. (1995a). Targeted disruption of the neurotrophin-3 gene with lacZ induces loss of trkC-positive neurons in sensory ganglia but not in spinal cords. *Brain Res.* 669:163–175.

Tojo, H., Takami, K., Kaisho, Y., Nakata, M., Abe, T., Shiho, O., Igarashi, K. (1995b). Neurotrophin-3 is expressed in the posterior lobe of mouse cerebellum, but does not affect the cerebellar development. *Neurosci. Lett.* 192:169–172.

Tomac, A., Lindqvist, E., Lin, L.F., Ogren, S.O., Young, D., Hoffer, B.J., Olson, L. (1995). Protection and repair of the nigrostriatal dopaminergic system by GDNF in vivo. *Nature* 373:335–339.

Treanor, J.J.S., Goodman, L., deSauvage, F., Stone, D.M., Poulsen, K.T., Beck, C.D., Gray, C., Armanini, M.P., Pollock, R.A., Hefti, F., Phillips, H.S., Goddard, A., Moore, M.W., Buj-Bello, A., Davies, A.M., Asai, N., Takahashi, M., Vandlen, R., Henderson, C.E., Rosenthal, A. (1996). Characterization of a multicomponent receptor for GDNF. Nature 382:80–83.

Trupp, M., Arenas, E., Fainzilber, M., Nilsson, A.-S., Sieber, B.-A., Grigoriou, M., Kilkenny, C., Salazar-Grueso, E., Pachnis, V., Arumae, U., Sariola, H., Saarma, M., Ibañez, C.F. (1996). Functional receptor for GDNF encoded by the c-ret proto-oncogene. Nature 381:785–789.

Van der Zee, C.E.E.M., Ross, G.M., Riopelle, R.J., Hagg, T. (1996). Survival of cholinergic forebrain neurons in developing p75NGFR-deficienct mice. *Science* 274:1729–1732.

Verdi, J.M., Anderson, D.J. (1994). Neurotrophins regulate sequential changes in neurotrophin receptor expression by sympathetic neuroblasts. *Neuron* 13:1359–1372.

Verdi, J.M., Groves, A.K., Fariñas, I., Jones, K.R., Marchionni, M.A., Reichardt, L.F., Anderson, D.J. (1996). A reciprocal cell-cell interaction mediated by NT-3 and neuregulins controls the early survival and development of sympathetic neuroblasts. *Neuron* 16:515–527.

Vicario-Abejón, C., Johe, K.K., Hazel, T.G., Collazo, D., McKay, R.D.G. (1995). Functions of basic fibroblast growth factor and neurotrophins in the differentiation of hippocampal neurons. *Neuron* 15:105–114.

Ware, C.B., Horowitz, M.C., Renshaw, B.R., Hunt, J.S., Liggitt, D., Koblar, S.A., Gliniak, B.C., McKenna, H.J., Papayannopoulou, T., Thoma, B., Cheng, L., Donovan, P.J., Peschon, J.J., Bartlett, P.F., Willis, C.R., Wright, B.D., Carpenter, M.K., Davison, B.L., Gearing, D.P. (1995). Targeted disruption of the low-affinity leukemia inhibitory factor receptor gene causes placental, skeletal, neural and metabolic defects and results in perinatal death. *Development* 121:1283–1299.

Wilkinson, G.A., Fariñas, I., Backus, C., Yoshida, C.K., Reichardt, L.F. (1996). Neurotrophin-3 is a survival factor in vivo for early mouse trigeminal neurons. *J. Neurosci.* 16:7661–7669.

Wright, E.M., Vogel, K.S., Davies, A.M. (1992). Neurotrophic factors promote the maturation of developing sensory neurons before they become dependent on these factors for survival. *Neuron* 9:139–150.

Yan, Q., Elliott, J.L., Matheson, C., Sun, J., Zhang, L., Mu, X.M., Rex, K.L., Snider, W.D. (1993). Influences of neurotrophins on mammalian motoneurons in vivo. *J. Neurobiol.* 24:1555–1577.

Yan, Q., Matheson, C., Lopez, O.T. (1995). In vivo neurotrophic effects of GDNF on neonatal and adult facial motor neurons. *Nature* 373:341–344.

Yoshida, K., Taga, T., Saito, M., Suematsu, S., Kumanogoh, A., Tanaka, T., Fujiwara, H., Hirata, M., Yamagami, T., Nakahata, T., Hirabayashi, T., Yoneda, Y., Tanaka, K., Wang, W.-Z., Mori, C., Shiota, K., Yoshida, N., and Kishimoto, T. (1996). Targeted disruption of gp130, a common signal transducer for the interleukin 6 family of cytokines, leads to myocardial and hematological disorders. *Proc. Natl. Acad. Sci. U.S.A.* 93:407–411.

7

Neuronal cell death

JULIE AGAPITE AND HERMANN STELLER

Vast numbers of cells die during normal development of the nervous system in both vertebrate and invertebrate animals (Hamburger and Levi-Montalcini, 1949; Glucksmann, 1951; Cowan, 1970; Lockshin, 1991; Truman and Schwartz, 1984; Cowan et al., 1984; Stewart et al., 1987; Stuart et al., 1987; Clarke, 1990; Oppenheim, 1991; Raff et al., 1993; Truman et al., 1992; Steller and Grether, 1994; Blaschke et al., 1996; Yaginuma et al., 1996). In the last few years, considerable progress has been made in understanding the extrinsic factors that regulate neuronal cell death, and also the mechanism by which these deaths occur. Although some morphological variances have been reported (Clarke, 1990), most of the cells that die in the nervous system exhibit the characteristic morphology of apoptosis, such as cell shrinkage, chromatin condensation, cellular fragmentation, and phagocytosis of the apoptotic bodies (Kerr et al., 1972; Wyllie et al., 1980; Oppenheim, 1991; Abrams et al., 1993; Raff et al., 1993; Sloviter et al., 1993; Blaschke et al., 1996; Yaginuma et al., 1996). Neuronal apoptosis is thought to serve an important role in sculpting the developing nervous system and in regulating its number of cells. For example, during the development of the vertebrate peripheral nervous system an excess of peripheral neurons competes for limiting amounts of specific target-derived survival factors, the paradigm of which is nerve growth factor (NGF) (Purves, 1988; Barde, 1989, 1990; Oppenheim, 1991; Raff et al., 1993; Lindsay et al., 1994; Snider, 1994; Heumann, 1994). NGF binds to and activates a receptor tyrosine kinase, encoded by the *TrkA* gene, that is expressed on the cell surface of distinct populations of peripheral neurons. This activation of TrkA blocks the onset of apoptosis, which is otherwise the default state of the cell. Because of the minute quantities of NGF, or other neurotrophins, at the target site, many of the innervating neurons do not receive an adequate supply and die. As a result of this competition for limited amounts of a survival factor, the number of innervating neurons is adjusted to a size that properly matches the postsynaptic cell population. More recently, it has been recognized that most, if not all, cultured mammalian cells require extracellular survival factors to avoid apoptosis, and it has been suggested that trophic interactions may be a general mechanism that regulates the numbers of many different cell types in other tissues (Raff, 1992; Raff et al., 1993). This dependence on specific survival factors, the "social controls" on cell survival, is thought to ensure that cells only live when and where they are needed.

In addition to the withdrawal of trophic factors, there are numerous other circumstances that lead to neuronal death. Neuronal apoptosis can be induced by a variety of different agents and cellular insults that include steroid hormones, oxidative stress, excitotoxicity, ischemia, and ionizing radiation (Oppenheim, 1991; Truman et al., 1992; Linnick et al., 1993; MacManus et al., 1993, 1994; Macaya et al., 1994; Steller and Grether, 1994; Bonfoco et al., 1995; Ankarcrona et al., 1995). Remarkably, all these different death-inducing stimuli activate the same distinctive structural changes of apoptosis, suggesting that they induce a common death program. One of the major advances of the last few years has involved the gradual discovery—the emergence—of a basic biochemical pathway for apoptosis. This chapter focuses on recent progress that has been made in identifying genes that are directly involved in the control and execution of apoptosis. Although we emphasize death in the nervous system, the genes and mechanisms that we describe are not unique to neurons, and we refer extensively to data obtained from studying the death of non-neuronal cells. For all the variety and complexity in the control of cell death, it now appears that all apoptotic deaths occur by a common effector pathway that arose early during evolution (Ameisen, 1996) and that has been largely conserved between animals as different as nematodes, insects, and mammals. A central component of this pathway is the activation of CED-3/ICE-like cysteine proteases (ICE—interleukin-1β–converting enzyme), a conclusion that was initially derived from molecular genetic studies of programmed cell death in the nematode *Caenorhabditis elegans* (Ellis and Horvitz, 1986; Yuan et al., 1993). There is also growing evidence for the evolutionary conservation of proteins that control apoptosis, most notably the Bcl-2 family of anti-apoptotic regulators (Hengartner and Horvitz, 1994a; Vaux et al., 1994). Unfortunately, the precise mechanism by which Bcl-2 family members exert their protective function remains unknown. Also, it is not clear how CED-3/ICE-like cystine proteases are activated in cells that are doomed to die. One reason for the difficulties in resolving these questions could be that many elements of the apoptotic pathway still remain to be discovered. Genetic screens for mutations affecting cell death can represent a very powerful approach for identifying these missing components. Until lately, forward genetic approaches have been limited to the nematode *C. elegans* (Ellis and Horvitz, 1986; Ellis et al., 1991; Hengartner and Horvitz, 1994b,c; Hengartner, 1996). Even there, most of the mutations that cause abnormal cell deaths have not been molecularly characterized. More recently, genetic studies of apoptosis in *Drosophila* have led to the identification of several new death genes (White et al., 1994; Grether et al., 1995; Hay et al., 1995). In *Drosophila,* as in vertebrates, the regulation of apoptosis is highly plastic and involves a wide variety of stimuli originating both within a cell, as well as from its environment (Steller and Grether, 1994; White and Steller, 1995; Steller, 1995). Therefore, this organism is particularly well suited to the study of how many different death-inducing signals converge onto a common apoptotic pathway.

It is likely that a better understanding of the mechanism of apoptosis may ultimately be useful for therapeutic purposes. Much of the current interest in neuronal apoptosis stems from observations that a number of human diseases, including stroke, Alzheimer's disease, Parkinson's disease, retinitis pigmentosa, and possibly epilepsy, are associated with excessive apoptosis (reviewed by Thompson, 1995). With the realization that these deaths result from an active, gene-directed process, the loss of many of these neurons no longer has to be accepted as an unavoidable fate. The rapidly expanding list of cell-death regulator and effector proteins provides an increasing number of promising targets for enhancing neuronal survival in these diseases.

PROPERTIES OF THE APOPTOTIC PROGRAM

Most, perhaps all, neurons have the ability to undergo apoptosis. In some cases, the induction of cell death requires new protein synthesis since it can be blocked by inhibitors of RNA and protein synthesis (Fahrbach and Truman, 1987; Martin et al., 1988; Oppenheim et al., 1990; Schwartz et al., 1990; Svendsen et al., 1994). Cell death is an active, gene-directed process (Lockshin, 1969). But apoptosis can be induced in the absence of de novo protein synthesis upon exposure to a strong apoptotic stimulus, such as high concentrations of the kinase inhibitor staurosporine (Jacobson et al., 1993, 1994). Moreover, cells whose nuclei have been removed can still undergo all of the characteristic cytoplasmic changes of apoptosis if they are deprived of survival factors or treated with staurosporine (Jacobson et al., 1994). These findings strongly suggest that major aspects of apoptosis do not require the synthesis of new genes and indicate that all the proteins required to carry out apoptosis are constitutively expressed. How can this be reconciled with earlier work demonstrating a dependence of cell death on macromolecular synthesis? The evidence becomes less confusing if one distinguishes between apoptotic effectors, those components that are directly responsible for carrying out the program of apoptosis, and regulators that control the activity of this program. The simplest interpretation is then to assume that the inhibition of macromolecular synthesis blocks the production of cell-death regulators which are needed to activate or derepress the death program in response to certain apoptotic stimuli. Apparently, these activators are dispensable in response to strong, perhaps unphysiologically severe, insults.

CONTROL OF NEURONAL APOPTOSIS

The induction of apoptosis is influenced by numerous distinct signals that originate both within cells and from their environment. As discussed earlier, it appears that cell death is the default state of many, if not all, mammalian cells: Every mammalian cell type tested to date undergoes apoptosis in the absence of extracellular survival factors, such as neurotrophins (Raff, 1992; Raff et al., 1993, 1994). Despite impressive advances in understanding the mechanism of neurotrophin signaling, the precise mechanism by which survival factors inhibit apoptosis remains unknown. Recent advances in the analysis of cell death in *Drosophila* indicate that this may be a very promising system in which to explore the connection between extracellular survival signals and the basic cell-death machinery. Many cell types in *Drosophila,* including cultured cells, require extracellular survival factors to avoid apoptosis. Furthermore, the number of neurons in the *Drosophila* nervous system is not genetically predetermined and may vary widely depending on environmental circumstances (Power, 1943; Nordlander and Edwards, 1968; Fischbach and Technau, 1984; Steller et al., 1987; Wolff and Ready, 1991; Truman et al., 1992; Steller and Grether, 1994). Not surprisingly, apoptosis is a prominent mechanism in the control of neuronal cell number in the imaginal nervous system, and both anterograde and retrograde neurotrophic interactions have been described during visual system development (Fischbach, 1983; Fischbach and Technau, 1984; Steller et al., 1987; Campos et al., 1992). The molecular basis of the observed trophic interactions is not known, but tyrosine kinase receptor genes that are related to the Trk family of neurotrophin receptors have been isolated (Pulido et al., 1992; Wilson et al., 1993). As discussed below, the discovery of a genetic pathway for apoptosis in *Drosophila* provides a unique opportunity to explore how extracellular signals regulate the apoptotic program.

In addition to signals that promote survival, there are also numerous stimuli that induce apoptosis. Interestingly, recent results indicate that the 75-kD "low-affinity" neurotrophin receptor ($p75^{NTR}$) may function to promote neuronal death (reviewed in Bothwell, 1996). This receptor was initially studied as an NGF receptor, with the expectation that it may promote neuronal survival. More recently, it was recognized that $p75^{NTR}$ is structurally related to a family of lymphotoxin receptors, most notably Fas and tumor necrosis factor receptors (TNFR) (Smith et al., 1990; Tartaglia and Goeddel, 1992; Smith et al., 1994; Beutler and Van Huffel, 1994; Nagata and Golstein, 1995). Both Fas and TNFR1 induce apoptosis upon binding to oligomeric ligands that activate these receptors by cross-linking. The cytoplasmic domains of both receptors contain a short segment of approximately 80 amino acids, termed the "death domain," which is necessary and sufficient for both receptor aggregation and the induction of apoptosis (Itoh and Nagata, 1993; Tartaglia et al., 1993). Significantly, a region of $p75^{NTR}$ was found to be structurally homologous to this "death domain" (Chapman, 1995), suggesting the possibility that $p75^{NTR}$ may promote apoptosis. Support for this idea has come from expression of $p75^{NTR}$ in several neuronal cell lines and from examining its role during apoptosis in two populations of developing neurons (Rabizadeh et al., 1993a; Barrett and Bartlett, 1994; von-Bartheld et al., 1994). Furthermore, there is now good evidence (Carter et al., 1996) that $p75^{NTR}$ signaling involves some of the same components as the Fas and TNF receptors, notably the generation of ceramide and NF-κB activation (reviewed in Bothwell, 1996). Finally, it has been reported that the production of ceramide, a complex second-messenger lipid that is generated by hydrolysis of sphingomyelin (Mathias and Kolesnick, 1993), is correlated with the induction of apoptosis in several circumstances (Pronk et al., 1996; Bothwell, 1996). All these intriguing observations point to possible similarities in the activation of apoptosis by Fas, TNFR1, and $p75^{NTR}$. However, there are also very important differences in the characteristics of deaths triggered by these different receptors (Nagata and Golstein, 1995), and the major details of the relevant signaling mechanisms remain to be worked out.

Circulating hormones are another important group of extracellular cell-death regulators. Both peptide and steroid hormones have been shown to affect neuronal cell death in vertebrates as well as in invertebrates (Oppenheim, 1991; Truman et al., 1992). The steroid hormones that control metamorphosis—namely, the molting hormone ecdysone in insects, and thyroxine in vertebrates—have major effects on the onset of cell death (Kollros 1981; Truman and Schwartz, 1984; Truman et al., 1992). In holometabolous insects, such as *Drosophila* and *Manduca,* complete metamorphosis occurs in two major phases. In the initial transition from larva to pupa, a crude adult form is produced. This is then followed by maturation of the pupa into an adult. Both of these stages involve large-scale cell death that is controlled by the levels of circulating ecdysteroid hormone. The initial wave of cell death is triggered by the surge of ecdysone at pupation (Truman and Schwartz, 1984; Truman et al., 1992). Interestingly, however, ecdysteroid levels have the opposite effect on postmetamorphic cell death. Upon eclosion of the adult, ecdysone levels drop, and it is this decline that induces the late wave of cell death (Truman and Schwartz, 1984; Robinow et al., 1993). The injection of 20-hydroxyecdysone can block normally occurring postmetamorphic cell death (Robinow et al., 1993).

The effects of ecdysteroids are mediated through heterodimeric nuclear hormone receptors encoded by two genes that are members of the steroid receptor superfamily, the *Drosophila ecdysone receptor (EcR)* gene (Koelle et al., 1991) and the *ultraspiracle* gene (Yao et al., 1992; Thomas et al., 1993). The *EcR* gene encodes three different isoforms, EcR-A, EcR-B1, and EcR-B2 (Talbot et al., 1993). While the DNA- and hormone-binding domains are common, each of the three receptor isoforms has a distinct N-terminus. Using isoform-specific monoclonal antibodies, Talbot et al. (1993) showed that EcR-A and EcR-B1 are expressed with distinct spatial and temporal profiles during metamorphosis. Interestingly, almost all the cells in the central nervous system that will undergo postmetamorphic cell death express high levels of the EcR-A protein (Robinow et al., 1993). This selective hyperexpression of the EcR-A isoform in doomed cells may explain why only some neurons die in response to a decline in hormone titer. However, elevated levels of EcR-A expression cannot be sufficient for determining which neurons die, since two neurons, n6 and n7, survive despite expressing high levels of EcR-A. Furthermore, cell death can be delayed in the absence of ecdysone by enforcing sustained ecdysis behavior (Truman, 1983; Kimura and Truman, 1990). Finally, decapitation of adults soon after eclosion will prevent postmetamorphic death in the CNS, even though ecdysone titers will decline under these conditions. This has led to the postulation of a "head factor" that is required for death. The molecular nature of this factor remains unknown, but it is possible that neuronal activity is required for the death-inducing activity.

Truman et al. (1992) have proposed a two step model of how EcR-A acts to effect programmed cell death. In this model, the binding of ecdysteroid to EcR-A induces the transcription of one or several early genes which are transcriptional regulators of the genes needed for cell killing. The continued presence of an ecdysteroid/EcR-A receptor complex, however, also serves to suppress the expression of the death-related genes regulated by the hypothetical early gene(s). Only once ecdysteroid levels drop is this suppression relieved, and the early gene(s) can activate transcription of death genes. The observation that treatment with either actinomycin D or cycloheximide can reduce the amount of cell death in the abdomen (Fahrbach et al., 1993) is consistent with the model's requirement of de novo RNA and protein synthesis to carryout postmetamorphic cell death. The relevant downstream targets of the ecdysone receptor complex have yet to be identified, but the cell-death gene *reaper* (see below) is an excellent candidate for such a gene.

In vertebrates, steroids have been implicated in controlling sex-specific differences in the brain of rats and in the forebrain nuclei of songbirds (reviewed in Oppenheim, 1991). However, in these cases it is not clear that the hormonal control of cell death is due to a direct action on these neurons rather than an indirect effect mediated through their targets.

A variety of conditions involving stress or toxic insults to cells can induce neuronal apoptosis. Of particular interest is oxidative stress, since it is associated with a number of human degenerative disorders of the CNS, including Alzheimer's disease, Parkinson's disease, and amyotrophic lateral sclerosis, and also with ischemia (reviewed in Coyle and Puttfarcken, 1993; Thompson, 1995). During ischemic brain injury, the accumulation of glutamate leads to an overstimulation of postsynaptic glutamate receptors, resulting in the massive influx of extracellular Ca^{2+} that overloads a neuron and leads to its death (Choi, 1988 1995; Coyle and Puttfarcken, 1993; Lipton and Rosenberg, 1994). Immediately following the exposure to glutamate, a subpopulation of neurons die by necrosis. However, after a period of recovery, large numbers of neurons undergo apoptosis (Ankarcrona et al., 1995). Such mixed responses to the same type of insult have been observed in several other instances (Dypbukt et al., 1994; Bonfoco et al., 1995). Those cells that die instantly by necrosis are presumably so severely damaged that they lose their membrane integrity and undergo necrosis. However, if given a choice, cells appear to prefer to die by apoptosis, the "socially correct" way to end their lives. The exact mechanism of excitotoxic death is currently not known, but it has been shown that glutamate neurotoxicity involves the formation of reactive oxygen species (Patel et al., 1996).

There are numerous other conditions that induce apoptosis by inflicting stress or damage to cells. Among these, the damage to DNA by ionizing radiation has received particular attention. DNA damage triggers apoptosis via the p53 tumor-suppressor protein (reviewed in Bates and Vousden, 1996). Cells that are deficient for p53 are resistant to radiation-induced death but undergo apoptosis in response to various other stimuli, suggesting a specialized role of p53 in regulating apoptosis (Lowe et al., 1993; Clarke et al., 1993). Other genes that play a role in the regulation of cell growth and mitosis can affect apoptosis in specific situations. This has led to the suggestion that apoptosis may be mechanistically related to mitosis, and that both may share common components (Ucker, 1991; Rubin et al., 1993; Heintz, 1993; Shi et al., 1994; Freeman et al., 1994). However, a clear demonstration that growth control or cell-cycle genes normally function in apoptosis still remains to be performed. Rather, it appears that the aberrant expression or activity of such genes is detected at important checkpoints, which somehow leads to the generation of a signal for the activation of apoptosis. In general, cells appear to have a remarkable ability to detect problems or errors in their mitotic or developmental program. Presumably, this represents a powerful selection mechanism to remove any unfit, damaged, and potentially dangerous cells from the organism. At this point, little is known about the precise molecular nature of these checkpoints. However, given the possible involvement of these pathways in a variety of diseases, one should expect significant advances in this area over the course of the next few years.

GENETIC ANALYSIS OF PROGRAMMED CELL DEATH IN *C. ELEGANS*

The initial evidence for the existence of genes that function specifically in the regulation of cell death has come from genetic analyses of programmed cell death in the nematode *C. elegans*. Mutations in about a dozen different genes have been found to affect the normal pattern of developmental cell deaths in this organism (Ellis and Horvitz, 1986; Ellis et al., 1991; Hengartner and Horvitz, 1994b,c; Hengartner, 1996). Mutations in three of these genes, *ced-3, ced-4,* and *ced-9,* affect the onset of all somatic cell deaths, suggesting that these genes play a central role in the initiation of death. Mutations that inactivate either *ced-3* or *ced-4* result in the survival of cells that normally die, indicating that the activity of these two genes is required for cell killing. In contrast, loss-of-function mutations in *ced-9* cause widespread ectopic cell deaths (Hengartner et al., 1992). However, either a gain-of-function mutation in the *ced-9* gene or overexpression of the wild-type gene can block cell death (Hengartner et al., 1992, 1994a). Taken together, these results suggest that *ced-9* function is both necessary and sufficient to prevent cells from undergoing apoptosis.

These genes have been ordered into a genetic pathway by examining the phenotypes of double mutant combinations. If *ced-9* loss-of-function mutations are combined with mutations in either *ced-3* or *ced-4,* the ectopic cell deaths that normally result from the lack of *ced-9* function are prevented (Hengartner et al., 1992). Therefore, *ced-9* function is not required in the absence of either *ced-3* or *ced-4* activity. If these genes were to act in a linear pathway, then these results would suggest that *ced-9* acts upstream of *ced-3* and *ced-4* to inhibit their activities, thereby preventing the inappropriate deaths of cells that have been selected to live. Consistent with this model, all three genes appear to act cell autonomously to kill or protect cells (Yuan and Horvitz, 1990; Shaham and Horvitz, 1996).

MECHANISM OF CELL KILLING

Molecular Characterization of ced-3 *and* ced-4

The properties of the two *C. elegans* genes, *ced-3* and *ced-4,* indicate that they encode killer activities. Molecular analysis of *ced-4* indicates that this gene encodes a 63-kD protein that shows no striking similarities to other currently known proteins (Yuan and Horvitz, 1992). The predicted CED-4 sequence contains a nucleotide-binding and Ca^{2+}-binding domain, but it is currently unclear whether these domains are important for *ced-4* function. However, sequencing of *ced-3* indicated that this gene encodes a protein which is 29% identical to a mammalian cysteine protease, interleukin-1β (IL-1β)-converting enzyme (ICE) (Yuan et al., 1993). ICE was previously characterized for its role in cleaving pro IL-1β to produce the mature form during the inflammatory response (Cerretti et al., 1992; Thornberry et al., 1992). This sequence similarity between CED-3 and ICE has raised the exciting possibility that CED-3 functions as a cysteine protease in the induction of cell death. Indeed, several recent studies strongly support this idea. First, it has been demonstrated that CED-3 protein has protease activity in vitro (Xue et al., 1996). Second, the protease activity of mutant CED-3 proteins in vitro is correlated with their killing ability in vivo (Xue et al., 1996). Furthermore, inhibition of CED-3 protease activity by the baculovirus p35 protein (see below) inhibits programmed cell death in vivo (Sugimoto et al., 1994; Xue and Horvitz, 1995).

Mammalian Cysteine Proteases Function in Apoptosis

The discovery that *ced-3* encodes an ICE-like cysteine protease also suggested the possibility that ICE and/or related proteases function in mammalian apoptosis. Some support for this idea has come from the observation that overexpression of ICE in cultured mammalian cells induces apoptosis (Miura et al., 1993). The caveat with these experiments is that it is very difficult to deduce a physiological function in apoptosis based on effects that result from the unphysiologically high levels of expression obtained in transfected cultured cells. As discussed earlier, cells have the ability to activate apoptosis in response to a variety of insults or damage, including intracellular proteolysis (Williams and Henkart, 1994). Therefore, one has to consider the possibility that overexpression of ICE may simply represent an insult that only indirectly leads to the activation of the death program. However, compelling evidence for a role of ICE-like cysteine proteases in mammalian apoptosis has come from studies using specific inhibitors of these enzymes. For example, expression of the cowpox virus crmA protein, which inhibits ICE-like protease activity, blocks the death of chicken dorsal root ganglion neurons induced by trophic factor withdrawal (Gagliardini et al., 1994) and Fas-induced death of rat fibroblasts (Enari et al., 1995). Furthermore, small peptide inhibitors of ICE-like proteases block the death of chicken motoneurons deprived of trophic support in culture and also block naturally occurring motoneuron death in vivo (Milligan et al., 1995). In addition, the baculovirus p35 protein has been shown to function as an inhibitor of ICE family members (Bump et al., 1995; Xue and Horvitz, 1995). Significantly, p35 is capable of suppressing cell death in every paradigm tested so far, including different cell types of insects, nematodes, and mammals (Clem et al., 1991; Rabizadeh et al., 1993b; Hay et al., 1994; Sugimoto et al., 1994; Martinou et al., 1995; Beidler et al., 1995). This provides strong evidence for an evolutionarily conserved role of CED-3/ICE-like cysteine proteases in the activation of most, if not all, apoptotic cell deaths.

A question that has been more difficult to resolve is, Which mammalian ICE family member is the functional equivalent of *ced-3*? The discovery of homologies between CED-3 and ICE has spurred intensive efforts that have identified numerous additional family members, including Ich1/nedd-2 (Wang et al. 1994; Kumar et al., 1994), Ich2/TX/ICErelII (Kamens et al., 1995; Faucheu et al., 1995; Munday et al., 1995), ICErelIII (Munday et al., 1995), CPP32/yama/apopain (Fernandes-Alnemri et al., 1994; Tewari et al., 1995; Nicholson et al., 1995), Mch2 (Fernandes-Alnemri et al., 1995a), and ICE-LAP3/Mch3 (Duan et al., 1996; Fernandes-Alnemri et al., 1995b). One obvious approach to implicate a particular family member in apoptosis is to generate mutant ("knockout") mice by targeted gene replacement. So far, ICE-deficient mice have been generated but found to develop normally and, except for a specific defect in Fas-induced death (Kuida et al., 1995), have no gross abnormalities in apoptosis (Kuida et al., 1995; Li et al., 1995). This result indicates that ICE itself is not strictly required for most mammalian cell deaths. It may be that one of the other mammalian ICE family members is the functional mammalian equivalent of *ced-3*. For example, CPP-32, Mch-2, and Mch-3 share extensive homology among themselves and are more similar to CED-3 than ICE is (Nicholson et al., 1995; Fernandes-Alnemri et al., 1995a,b; Duan et al., 1996). Furthermore, a function of these proteins in apoptosis was also suggested based on their substrate specificity and inhibitor profile (Nicholson et al., 1995), but this point remains controversial (Takahashi and Earnshaw, 1996). Attempts to generate the relevant knockouts to test the functions of these family members are almost certainly in progress at the time of this writing. However, given the sheer numbers of mammalian ICE family members it seems likely that there is significant redundancy in the cell death pathway, and one may have to be prepared for disappointing phenotypes.

Substrates for Cysteine Proteases

How do CED-3/ICE-like cysteine proteases activate the apoptotic program? Several substrates of the ICE cysteine protease family have been identified including the DNA repair enzyme poly(ADP-ribose) polymerase (PARP) (Lazebnik et al., 1994), the 70-kD component of the U1 small nuclear ribonucleoprotein (Casciola-Rosen et al. 1994), and the nuclear lamins (Ucker et al., 1992; Oberhammer et al., 1994; Lazebnik et al., 1995) which are filament proteins that provide support to the nuclear envelope. Cleavage of these substrates is coincident with cell death (reveiwed in Takahashi and Earnshaw, 1996). Of particular interest is the cleavage of nuclear lamins because of the dramatic changes in nuclear morphology during apoptosis. Furthermore, nuclear fragmentation in vitro fails to occur when lamin cleavage is inhibited (Lazebnik et al., 1995). However, it remains to be determined whether cleavage of these substrates in vivo is required for apoptosis. An emerging view is that different ICE-like proteases have distinct substrate and inhibitor specificities (Nicholson et al., 1995; Takahashi and Earnshaw, 1996). These differences, together with possibly differential expression of some ICE family members, may explain to some degree the morphological variances that have been observed between different cell types undergoing programmed cell death.

Activation of Cysteine Proteases

How is the killing activity of cysteine proteases restricted to cells that have been selected to die? One possibility is that these genes are only expressed in doomed cells. However, most of the ICE-like cysteine protease genes are widely expressed and can be even detected in a single cell type (Fernandes-Alnemri et al., 1994; Munday et al., 1995; Faucheu et al., 1995; Takahashi and Earnshaw, 1996). Furthermore, even though the exact tissue distribution of *ced-3*, has not been reported, genetic evidence suggests that the expression of this gene is also not restricted to cells that will die (Shaham and Horvitz, 1996). However, ICE-like protease activity has been detected in apoptotic, but not live, cells (Lazebnik et al., 1994; Nicholson et al., 1995). These properties of CED-3/ICE-like proteases are consistent with the idea that cell-death effectors are constitutively expressed and activated upon receiving an apoptotic stimulus (Raff et al., 1993; Jacobson, et al., 1994).

One explanation for why the mere transcription/translation of physiological levels of CED-3/ICE-like proteases is insufficient to induce apoptosis comes from the observation that all ICE family members discovered to date are synthesized as inactive precursor proteins (Takahashi and Earnshaw, 1996). These proenzymes are then themselves cleaved by ICE-like proteases to form the active enzyme, which is a heterotetramer composed of two p10 and two p20 subunits (Walker et al., 1994; Wilson et al., 1994). In addition, differential splicing generates at least five different ICE mRNA forms, two of which encode polypeptides that are catalytically inactive but can assemble with an active p20 subunit, yielding an inactive protease (Alnemri et al., 1995). Therefore, much, if not all, of the regulation of ICE-like protease activity seems to occur at the post-translational level.

A central unresolved question is, Precisely how is the conversion of the inactive pro-form to the active enzyme regulated in response to death-inducing stimuli? Some family members like ICE itself are able to autocleave, whereas others like CPP32 rely on other ICE-like proteases for activation (Takahashi and Earnshaw, 1996). This requirement for activation appears to provide one level of regulation to ensure the appropriate induction of cell death. Indeed, both CPP32 and MCH3 are activated in response to apoptotic stimuli (Chinnaiyan et al., 1996). The number of ICE-like proteases and the inability of some of them to autoactivate raises the possibility that they may function in a cascade where one family member, once activated in response to a death signal, activates another. Evidence for such a cascade has recently been documented. Upon induction of apoptosis by Fas, ICE-like and CPP32-like proteases are activated sequentially. ICE-like activity peaks first and declines and is then followed by a rise in CPP32-like activity. Furthermore, treatment with ICE-specific inhibitors abolishes the appearance of both ICE- and CPP32-like activities, while treatment with CPP32 inhibitors blocks only the appearance of the CPP32-like activity, suggesting that a CPP32-like protease acts downstream of an ICE-like protease during Fas-mediated apoptosis (Enari et al., 1996). Interestingly, Bcl-2 and Bcl-x-L (see below) can prevent the activation of CPP32 and MCH3 in response to the death inducer staurosporine (Chinnaiyan et al., 1996). This result places Bcl-2-like molecules upstream of the activation of ICE-like proteases, which is consistent with the pathway order worked out genetically in the nematode (Hengartner et al., 1992).

Negative Regulators of Apoptosis: ced-9 *and Bcl-2*

The third gene implicated in the general control of *C. elegans* cell deaths, *ced-9*, encodes a homologue of the mammalian proto-oncogene *bcl-2*. The human *bcl-2* gene was initially identified at the t(14;18) translocation breakpoint common in many B-cell lymphomas. As a result of this translocation, *bcl-2* comes under the control of the immunoglobulin heavy-chain enhancer and is constitutively expressed in B cells (Bakhshi al., 1985; Tsujimoto et al., 1985; Cleary et al., 1986). Expression of *bcl-2* protects cells from death in a number of cells types and in response to a number of stimuli (reviewed in Davies, 1995). Thus *bcl-2* shares functional as well as sequence similarity with *ced-9*. In fact, *bcl-2* can partially substitute for *ced-9* in *ced-9* mutant animals (Vaux et al., 1992; Hengartner and Horvitz, 1994a). However, although *bcl-2* is effective in a number of systems, it fails to prevent apoptosis in some circumstances and is therefore not a universal protector (reviewed in Davies, 1995).

ced-9 and *bcl-2* are members of a large and growing gene family. Many new family members have been recently identified, including the vertebrate *bax* (Oltvai et al., 1993), *bcl-x* (Boise et al., 1993), *bak* (Chittenden et al., 1995a; Farrow et al., 1995; Kiefer et al., 1995), *bad* (Yang et al., 1995), *MCL1* (Kozopas et al., 1993), and *A1* (Lin et al., 1993), as well as the viral genes E1B19K from adenovirus (White et al., 1992), BHRF1 from Epstein-Barr virus (Henderson et al., 1993), and LMW5-HL from African swine fever virus (Neilan et al., 1993). The highest degree of conservation among different family members occurs in three domains, BH1, BH2, and BH3 (Oltvai et al., 1993; Chittenden et al., 1995b; Boyd et al., 1995). Bcl-2 family proteins have the ability to form homo- and heterodimers with each other, and these domains are important for these interactions (Oltvai et al., 1993; Chittenden et al., 1995b; Yin et al., 1994). The C-terminal hydrophobic domain of Bcl-2, which is required for its association with intracellular membranes, is less well conserved and absent from some family members.

Several of these family members, including *bcl-2, bcl-x-L,* and *bax* are expressed in the nervous system (Merry et al., 1994; Boise et al., 1993; Oltvai et al., 1993; Gonzalez-Garcia et al., 1994), suggesting that they may be involved in the regulation of neuronal apoptosis. The function of different Bcl-2 family members has been extensively studied in tissue-culture transfection experiments. Overexpression of many, but not all, of these genes inhibits apoptosis of cultured cells. For example, Bcl-2 protects cultured neurons from death induced by NGF, NT-3, or BDNF deprivation but not CNTF deprivation (Garcia et al., 1992; Allsopp et al., 1993; Farlie et al., 1995). Overexpression of *bcl-2* in the nervous system of transgenic animals results in hypertrophy of the brain and inhibits apoptosis in response to axotomy (Martinou et al., 1994; Dubois-Dauphin et al., 1994). Surprisingly, eliminating *bcl-2* function in knockout mice does not lead to a striking reduction of neuronal viability and produces viable mice (Veis et al., 1993; Nakayama et al., 1994). Perhaps other family members can compensate for the loss of *bcl-2* function. In fact, *bcl-x* appears to play a more prominent role for neuronal survival: *bcl-x*-deficient mice die during embryogenesis and exhibit massive cell death in the nervous system (Motoyama et al., 1995). As previously discussed for the mammalian ICE family, the presence of multiple Bcl-2 homologues has made it difficult to unequivocally assign in vivo functions to individual proteins.

Overexpression of *bax, bad,* and *bak* can increase the susceptibility of cultured cells to death. Based on these results it has been suggested that the proteins encoded by these genes may function as antagonists of protectors, such as Bcl-2 and Bcl-X, and promote apoptosis (Oltvai et al., 1993; Farrow et al., 1995; Chittenden et al., 1995a; Kiefer et al., 1995; Yang et al., 1995). However, some caution is warranted in interpreting these results. Upon transfection, Bax may form heterodimers and thereby inactivate a "resident" Bcl-2 family member that the recipient cell prefers for protection. It appears that most cells that normally express Bax are viable, and they may even be protected by this protein. The latter possibility is suggested by the recent finding that Bax can protect neurons against apoptosis from NGF withdrawal (Middleton et al., 1996). Furthermore, Bax-deficient mice were found to have excessive germ cell death (Knudson et al., 1995). These observations again reveal the difficulties of deducing the normal function of a gene from overexpression studies, in particular for families of proteins that can form heteromeric aggregates with each other.

The biochemical function of Bcl-2 family members has remained enigmatic. It has been suggested that Bcl-2 functions in an antioxidant pathway to protect against death (Hockenberry et al., 1993). However, cell death and protection by Bcl-2 can occur in the absence of mitochondrial function and also under anaerobic conditions, indicating that protection from death by bcl-2 does not require antioxidant function (Jacobson et al., 1993; Jacobson and Raff, 1995). Another model proposes that Bcl-2 protects against cell death by binding to and inactivating Bax, the presumed "death inducer." A mutation in the BH1 domain of Bcl-2 which disrupts heterodimerization with Bax, but not Bcl-2/Bcl-2 homodimerization, abolishes the protective activity of Bcl-2 (Oltvai et al., 1993). One interpretation of these results is that Bcl-2 homodimers are not important in preventing death but rather that Bax homodimers are the active death-inducing factors which are disrupted by heterodimerization with Bcl-2. However, it cannot be excluded that this *bcl-2* mutation disrupts the protective function of Bcl-2 by preventing the interaction with another protein that is required to inhibit apoptosis. While the proposed concept of competition between protectors and inducers is attractive, it will be important to identify interacting molecules outside the Bcl-2 family to understand the regulation and mode of action of these proteins.

Another potential negative regulator of cell death is *dad-1*. This gene was initially identified in a temperature-sensitive mutant hamster cell line that undergoes apoptosis at the restrictive temperature (Nakashima et al., 1993). Overexpression of *dad-1* was found to partially prevent cell death in *C. elegans* (Sugimoto et al., 1995). Homologues of *dad-1* have been identified in nematodes, plants, and even yeast. The yeast homologue, OST2, appears to be involved in the amino-linked glycosylation of proteins within the endoplasmic reticulum (Silberstein et al., 1995). The mechanism by which Dad-1 protein inhibits apoptosis is not clear, and it action may be rather indirect.

CELL DEATH INHIBITORS FROM BACULOVIRUS

Molecular genetic approaches in baculoviruses have yielded two types of genes, *p35* and *iap*, which are capable of inhibiting cell death (Clem et al., 1991; Crook et al., 1993; Birnbaum et al., 1994). *p35* was identified as an inhibitor of cell death through the characterization of a baculovirus mutant, named *Annihilator (Anh)*, with greatly diminished growth. Normally, insect cells infected with wild-type baculovirus lyse several days postinfection. Cells infected with the *Anh* mutant instead die prematurely by apoptosis. It was later shown that a lesion in the *p35* gene causes the mutant phenotype (Clem et al., 1991). The *p35* gene encodes a novel protein with no homology to other known genes and no readily identifiable sequence motifs.

Expression of *p35* blocks cell death in wide variety of situations, organisms, and cell types. It can block programmed cell death in nematodes and partially substitute for the loss of *ced-9* function (Sugimoto et al., 1994; Xue and Horvitz, 1995), and it prevents both developmental and X-ray–induced cell death (Hay et al., 1994) as well as apoptosis induced by overexpression of the *Drosophila* cell-death genes *hid* and *rpr* (Grether et al., 1995; Hay et al., 1995; White et al., 1996; Pronk et al., 1996). In mammals, p35 can protect a neuronal cell line (Rabidazeh et al., 1993b) and cultured sympathetic neurons (Martinou et al., 1995) against growth factor deprivation. Finally, p35 can block TNF- and Fas-mediated apoptosis (Beidler et al., 1995). As described earlier, p35 appears to exert these protective effects by inhibiting the activity of CED-3/ICE-like proteases (Bump et al., 1995; Xue and Horvitz, 1995).

Members of the *iap* family were cloned on the basis of their ability to functionally complement the *p35* mutant (Crook et al., 1993; Birnbaum et al., 1994). *iap* genes from different baculovirus species were cloned in this manner. The *iap* genes also encode novel proteins, but certain sequence motifs can be recognized. The C-terminal portion of these proteins contains a RING finger motif which is preceded by a unique repeated domain termed BIR (baculovirus iap repeat). Both of these sequence elements appear to be important for the protective activity of *iap* genes (Clem and Miller, 1994). While the function of the RING finger is unknown, BIRs appear to be involved in protein–protein interactions (see below).

IAP function, like that of p35, is conserved across species. The viral *iap* genes can block cell death in higher organisms. For example, the expression of a viral *iap* gene can block Sindbis virus–induced death of mammalian cells (Duckett et al., 1996). Moreover, cellular homologues of the *iap* genes have been found in both *Drosophila* and humans, and these homologues can have death protection activity (Roy et al., 1995; Rothe et al., 1995b; Hay et al., 1995; Liston et al., 1996; Duckett et al., 1996). Overexpression of the *Drosophila iap* homologues *diap1* and *diap2* can block both the naturally occurring cell deaths in the *Drosophila* compound eye and the deaths induced by the ectopic expression of the *Drosophila* cell-death genes *rpr* and *hid* (Hay et al., 1995). The human *iap* homologues *hiap-1/c-iap2, hiap-2/c-iap1,* and *NAIP* can protect mammalian cells from death induced by serum withdrawal and by menadione treatment (Liston et al., 1996).

While most of the *iap* homologues were identified based on sequence homology to the viral genes, the human *NAIP* gene was cloned based on its involvement in spinal muscular atrophy (Roy et al., 1995), a disease in which motoneurons degenerate. This raises the possibility that these neurons undergo apoptosis as a result of the loss of this protective factor. *c-iap1* and *c-iap2* were cloned based on an interaction between their gene products and the tumor necrosis factor type-2 receptor, TNFR2 (Rothe et al., 1995b). This interaction was shown to be indirect, as it required the presence of two other molecules, TRAF1 and TRAF2. Interestingly, the direct interaction between TRAF2 and the *c-iap*s require the *c-iap* BIR repeats, suggesting that these repeats are important for protein–protein interactions. The relevance of these interactions for the control of apoptosis, however, is not yet clear.

Although the IAPs can block cell death in a number of species and in response to diverse stimuli, their function is not as universal as that of p35. For example, expression of these genes does not inhibit death in the neuronal cell line CSM (Clem et al., 1996), nor do they protect against staurosporine-induced death (Liston et al., 1996). Perhaps the IAPs target regulatory proteins, and not central components of the apoptotic pathway.

Molecular Genetics of Apoptosis in *Drosophila*

Besides *C. elegans, Drosophila* is the only other metazoan animal in which systematic screens for mutations affecting programmed cell death are technically feasible at this point. As discussed earlier, cell death in *Drosophila* is plastic and under epigenetic control. This provides an opportunity to apply the power of genetics to identify both components of the apoptotic program, as well as the signaling pathways that regulate it.

A large fraction of the *Drosophila* genome has been screened for mutations that affect apoptosis (White et al., 1994). By analyzing the pattern of cell death in embryos homozygous for previously identified chromosomal deletions, one region—75C1,2—was found to be essential for virtually all cell deaths that normally occur during embryogenesis. In addition, these deletion mutant embryos were also extensively protected against the ectopic cell deaths that are normally induced upon radiation or in developmental mutants. Molecular analyses of the corresponding interval led to the isolation of two novel cell-death genes, *reaper (rpr)* and *head involution defective (hid)* (White et al., 1994; Grether et al., 1995). In the *Drosophila* embryo, *reaper* mRNA is specifically found in cells that are doomed to die, preceding the onset of death typically by 1–2 hours. The expression of this gene is also induced in response to a variety of other death-inducing stimuli, including X-irradiation, block of cellular differentiation, and steroid-hormone–induced deaths (J. M. Abrams, A. F. Lamblin, H. Steller, unpublished results; E. Baehrecke, C. Thummel, personal communication; S. Robinow, J. Truman, personal communication). This indicates that the integration of different death-inducing signals occurs, at least in part, by a transcriptional mechanism.

The *reaper* gene encodes a small peptide of 65 amino acids (White et al., 1994) that bears some similarity to the aforementioned "death domain" of Fas and the type-1 tumor necrosis factor receptor (TNFR1) (Golstein et al., 1995; Cleveland and Ihle, 1995). Interestingly, like the "death domain" recombinant REAPER protein has strong self-aggregating properties in vitro (C. Carboy-Newcomb, C.L. Wei, H. Steller, unpublished results). This suggests the possibility that the active form of REAPER, like TNFR1 and Fas, is a multimer. Furthermore, it is possible that the mechanism by which REAPER activates apoptosis is similar to that of TNFR1/Fas. Some support for this idea has come from the observation that *reaper*-induced apoptosis is associated with increased ceramide production (Pronk et al., 1996). However, it has not been ruled out that the generation of ceramide is simply a consequence, rather than a cause, of apoptosis.

The second gene in the 75C1,2 interval *hid*, encodes a 410-amino-acid protein with no significant homologies to other known proteins (Grether et al., 1995). There is a small region of similarity between the N-terminus of HID and REAPER, but the functional relevance of this is unclear. *hid* mutants show an overall decrease in the level of cell death, and that decrease is most dramatic in the head region (Grether et al., 1995). It is possible that this decrease of cell death causes the head involution defect that was originally described for this mutant (Abbott and Lengyel, 1991).

The ectopic expression of *reaper* induces apoptosis in cultured cells (Pronk et al., 1996) and in many different cell types in transgenic animals (White et al., 1996). Likewise, ectopic expression of *hid* is sufficent for the induction of apoptosis in *Drosophila* (Grether et al., 1995). Overexpression of either *rpr*, or *hid* from the *hsp70* heat-shock promoter results in extensive cell death in the embryo. Interestingly, expression of either gene causes widespread cell death in H99 mutant embryos as well as in wild-type embryos (Grether et al., 1995; White et al., 1996). Since the H99 deletion removes both *hid* and *rpr*, this result indicates that *hid* and *rpr* can act independently of each other and suggests that both genes act in parallel to activate apoptosis.

Expression of either *rpr* or *hid* under the control of an eye-specific promoter results in a dose-dependent eye ablation (Grether et al., 1995; White et al., 1996). At intermediate levels of expression, a reduced, rough eye phenotype is obtained. Because the overexpression of a gene can represent a cellular insult that may indirectly activate apoptosis, the specificity of *reaper*- and *hid*-mediated killing has been investigated. Coexpression of the baculovirus p35 protein completely suppressed both of these phenotypes and allowed for essentially normal ommatidial differentiation (Grether et al., 1995; White et al., 1996). This demonstrates that the ectopic expression of these genes does not have general adverse effects on eye development, differentiation, or pattern formation but rather acts specifically and directly in the cell-death pathway. Furthermore, these results indicate that REAPER and HID activate a CED-3/ICE-like protease activity. This notion is also supported by the observation that cell death induced by expression of *reaper* in *Drosophila* tissue culture cells is blocked by small peptide inhibitors of ICE as well as by p35 (Pronk et al., 1996). These results have prompted us to directly investigate the presence of CED-3/ICE-like protease in *Drosophila*, and we have recently identified a gene that is highly homologous to the CPP-32 subfamily of cysteine proteases (Z. Song, K. McCall, H. Steller, unpublished results).

How do *reaper* and *hid* activate an ICE-protease pathway? The eye phenotypes caused by the ectopic expression of *rpr* and *hid* provide a very sensitive and rapid assay for identifying additional cell-death genes that should facilitate answering this question. We have initiated a large-scale screen for mutations that either suppress or enhance these reduced, rough eye phenotypes (J. Agapite, K. McCall, C. Hynds, K. Mejia, H. Steller, unpublished results). Mutations that promote apoptosis can be identified as enhancers of the *hid-* and/or *reaper-*induced eye defects, while mutations that inhibit death suppress these phenotypes. A large number of mutations have already been isolated, and we are in the process of characterizing several of the corresponding genes. Meanwhile, mutations in the *thread* locus, which encodes a *Drosophila* homologue of the baculovirus *iap* gene, *diap1*, were found to enhance both the *rpr* and *hid* induced eye phenotypes (Hay et al., 1995). This indicates that mutant screens for genetic modifiers of *reaper* and *hid* hold great promise for the identification of novel genes involved in apoptosis. We expect that this type of genetic approach in *Drosophila* will make a fundamental contribution to understanding both the molecular nature of the cell death program and how that program is regulated by a variety of distinct signaling pathways.

References

Abbott, M.K., Lengyel, J.A. (1991). Embryonic head involution and rotation of male terminalia require the *Drosophila* locus *head involution defective. Genetics* 129:783–789.

Abrams, J.M., White, K., Fessler, L.I., Steller, H. (1993). Programmed cell death during *Drosophila* embryogenesis. *Development* 117:29–43.

Allsopp, T.E., Wyatt, S., Paterson, H.F., Davies, A.M. (1993). The proto-oncogene *bcl-2* can selectively rescue neurotrophic factor-dependent neurons from apoptosis. *Cell* 73:295–307.

Alnemri, E., Fernandes-Alnemri, S.T., Litwack, G. (1995). Cloning and expression of four novel isoforms of human interleukin-1b converting enzyme with different apoptotic activities. *J. Biol. Chem.* 270:4312–4317.

Ameisen, J.C. (1996). The origin of programmed cell death. *Science* 272:1278–1279.

Ankarcrona, M., Dypbukt, J.M., Bonfoco, E., Zhivotovsky, B., Orrenius, S., Lipton, S., Nicotera, P. (1995). Glutamate-induced neuronal death: a succession of necrosis or apoptosis depending on mitochondrial function. *Neuron* 15:961–973.

Bakhshi, A., Jensen, J.P., Goldman, P., Wright, J.J., McBride, O.W., Epstein, A.L., Korsmeyer, S.J. (1985). Cloning the chromosomal break-point of the t[14;18] human lymphomas: clustering around JH on chromosome 14 and near a transcriptional unit on 18. *Cell* 41:889–906.

Barde, Y.A. (1989). Trophic factors and neuronal survival. *Neuron* 2:1525–1534.

Barde, Y.A. (1990). The nerve growth factor family. *Prog Growth Factor Res* 2:237–248.

Barrett, G.L., Bartlett, P.F. (1994). The p75 nerve growth factor receptor mediates survival or death depending on the stage of sensory neuron development. *Proc. Natl. Acad. Sci. U.S.A.* 91:6501–6505.

Bates, S., Vousden, K. (1996). p53 signaling checkpoint arrest or apoptosis. *Curr. Op. in. Genet. Dev.* 6:12–18.

Beidler, D.R., Tewari, M., Friesen, P.D., Poirier, G., Dixit, V.M. (1995). The Baculovirus p35 protein inhibits fas and tumor necrosis factor-induced apoptosis. *J. Biol. Chem.* 270: 16526–16529.

Beutler, B., Van Huffel, C. (1994). Unraveling function in the TNF ligand and receptor families. *Science* 264:667–668.

Birnbaum, M.J., Clem, R.J., Miller, L.K. (1994). An apoptosis-inhibiting gene from a nuclear polyhedrosis virus encoding a peptide with cys/his sequence motifs. *J. Virol.* 68: 2521–2528.

Blaschke, A.J., Staley, K., Chun, J. (1996). Widespread programmed cell death in proliferative and postmitotic regions of the fetal cerebral cortex. *Development* 122:1165–1174.

Boise, L.H., Gonzalez-Garcia, M., Postema, C.E., Ding, L., Lindsten, T., Turka, L.A., Mao, X., Nunez, G., Thompson, C. (1993). *bcl-x,* a *bcl-2* related gene that functions as a dominant regulator of apoptotic death. *Cell* 74:597–608.

Bonfoco, E., Krainc, D., Ankarcrona, M., Nicotera, P., Lipton, S.A. (1995). Apoptosis and necrosis: two distinct events induced respectively by mild and intense insults with N-methyl-D-aspartate or nitric oxide/superoxide in cortical cultures. *Proc. Natl. Acad. Sci. U.S.A.* 92:7162–7166.

Bothwell, M. (1996). p75^{NTR}: a receptor after all. *Science* 272:506–507.

Boyd, J.M., Gallo, G.J., Elangovan, B., Houghton, A.B., Malstrom, S., Avery, B.J., Ebb, R.G., Subramanian, T., Chittenden, T., Lutz, R.J., Chinnadurai, G. (1995). Bik1, a novel death-inducing protein shares a distinct sequence motif with Bcl-2 family proteins and interacts with viral and cellular survival-promoting proteins. *Oncogene* 11:1921–1927.

Bump, N.J., Hackett, M., Huginin, M., Seshagiri S., Brady, K., Chen, P., Ferenz, C., Franklin, S., Ghayur, T., Li, P., Licari, P., Mankovich, J., Shi, L., Greenberg, A.H., Miller, L.K. and Wong, W.W. (1995). Inhibition of ICE family proteases by baculovirus antiapoptotic protein p35. *Science* 269:1885–1888.

Campos, A.R., Fischbach, K.-F., Steller, H. (1992). Survival of photoreceptor neurons in the compound eye of *Drosophila* depends on connections with the optic ganglia. *Development* 114:355–366.

Carter, B.D., Kaltschmidt, C., Kaltschmidt, B., Offenhäuser, N., Böhm-Matthaei, Baeuerle, P.A., Barde, Y.A. (1996). Selective activation of NF-κB by nerve growth factor through the neurotrophin receptor p75. *Science* 272:542–545.

Casciola-Rosen, L.A., Miller, D.K., Anhalt, G.J., Rosen, A. (1994). Specific cleavage of the 70-kDa protein component of the U1 small nuclear ribonucleoprotein is a characteristic biochemical feature of apoptotic cell death. *J. Biol. Chem.* 69:30,757–30,760.

Cerretti, D.P., Koziosky, C.J., Mosley, B., Neslon, N., Ness, K., Greenstreet, T.A., March, C.J., Kronheim, S.R., Druck, T., Cannizzaro, L.A., Huebner, K., Black, R.A. (1992). Molecular cloning of the interleukin-1b converting enzyme. *Science* 256:97–100.

Chapman, B.S. (1995). A region of the 75 kDa neurotrophin receptor homologous to the death domains of TNFR-I and Fas. *FEBS Lett.* 374:215–220.

Chinnaiyan, A.M., Orth, K., O'Rourke, K., Duan, H., Poirier, G.G., Dixit, V.M. (1996). Molecular ordering of the cell death pathway: bcl-2 and bcl-xL function upstream of the *ced-3*-like apoptotic proteases. *J. Biol. Chem.* 271:4573–4576.

Chittenden, T., Harrington, E.A., O'Connor, R., Flemington, C., Lutz, R.J., Evan, G.I., Guild, B.C. (1995a). Induction of apoptosis by the Bcl-2 homolog Bak. *Nature* 374:733–736.

Chittenden, T., Flemington, C., Houghton, A.B., Ebb, R.G., Gallo, G.J., Elangovan, B., Chinnadurai, G., Lutz, R.J. (1995b). A conserved domain in Bak, distinct from BH1 and BH2, mediates cell death and protein binding functions. *EMBO J.* 14:5589–5596.

Choi, D.W. (1988). Glutamate neurotoxicity and diseases of the nervous system. *Neuron* 1: 623–634.

Clarke, P.G.H. (1990). Developmental cell death: morphological diversity and multiple mechanisms. *Anat. Embryol.* 181:195–213.

Clarke, A.R., Purdie, C.A., Harrison, D.J., Morris, R.G., Bird, C.C., Hooper, M.L., Wyllie, A.H. (1993). Thymocyte apoptosis induced by p53-dependent and independent pathways. *Nature* 362:849–851.

Cleary, M.L., Smith, S.D., Sklar, J. (1986). Cloning and structural analysis of cDNAs for bcl-2 and a hybrid bcl-2/immunoglobulin transcript resulting from the t[14;18] translocation. *Cell* 47:19–28.

Clem, R.J. Miller, L.K. (1994). Control of programmed cell death by the baculovirus genes *p35* and *iap*. *Mol. Cell. Biol.* 14:5212–5222.

Clem, R.J., Fechheimer, M., Miller, L.K. (1991). Prevention of apoptosis by a baculovirus gene during infection of insect cells. *Science* 254:1388–1390.

Clem, R.J., Hardwick, J.M., Miller, L.K. (1996). Anti-apoptotic genes of baculoviruses. *Cell Death Differ.* 3:9–16.

Cleveland, J.L., Ihle, J.N. (1995). Contenders in FasL/TNF death signaling. *Cell* 81:479–482.

Cowan, W.M. (1970). Anterograde and retrograde transneuronal degeneration in the central and peripheral nervous system. In *Contemporary Research Methods in Neuroanatomy*, Naute, J.H., Ebbenson, S.O.E., eds. Berlin: Springer Verlag, pp. 215–251.

Cowan, W.M., Fawcett, J.W., O'Leary, D.M., Stanfield, B.B. (1984). Regressive events in neurogenesis. *Science* 225:1258–1265.

Coyle, J.T., Puttfarcken, P. (1993). Oxidative stress, glutamate, and neurodegenerative disorders. *Science* 262:689–695.

Crook, N.E., Clem, R.J., Miller, L.K. (1993). An apoptosis-inhibiting baculovirus gene with a zinc finger-like motif. *J. Virol.* 67:2168–2174.

Davies, A.M. (1995). The Bcl-2 family of proteins, and the regulation of neuronal survival. *Trends Neurosci.* 18:355–358.

Duan, H., Chinnaiyan, A.M., Hudson, P.L., Wing, J.P., He, W., Dixit, V.M. (1996). ICE-LAP3, a novel mammalian homolog of the *Caenorhabditis elegans* cell death protein CED-3 is activated during Fas- and Tumor Necrosis Factor-induced apoptosis. *J. Biol. Chem.* 271:35013–35035.

Dubois-Dauphin, M., Frankowski, H., Tsujimoto, Y., Huarte, J., Martinou, J.-C. (1994). Neonatal motoneurons overexpressing the *bcl-2* protooncogene in transgenic mice are protected from axotomy-induced cell death. *Proc. Natl. Acad. Sci. U.S.A.* 91:3309–3313.

Duckett, C.S., Nava, V.E., Gedrich, R.W., Clem, R.J., Van Dongen, J.L., Gilfillan, M.C., Shiels, H., Hardwick, J.M., Thompson, C.B. (1996). A conserved family of cellular genes related to the baculovirus IAP gene and encoding apoptosis inhibitors. *EMBO J.* 15:2685–2694.

Dypbukt, J.M., Ankarcrona, M., Burkitt, M., Sjöholm, A., Ström, K., Orrenius, S., Nicotera, P. (1994). Different prooxidant levels stimulate growth, trigger apoptosis, or produce necrosis of insulin-secreting RINm5F cells: the role of intracellular polyamines. *J. Biol. Chem.* 269:30553–30560.

Ellis, R.E., Horvitz, H.R. (1986). Genetic control of programmed cell death in the nematode *C. elegans*. *Cell* 44:817–829.

Ellis, R.E., Yuan, J., Horvitz, H.R. (1991). Mechanisms and functions of cell death. *Annu. Rev. Cell Biol.* 7:663–698.

Enari, M., Hug, H., Nagata, S. (1995). Involvement of an ICE-like protease in Fas-mediated apoptosis. *Nature* 375:78–81.

Enari, M., Talanian, R.V., Wong, W.W., Nagata, S. (1996). Sequential activation of ICE-like and CPP32-like proteases during Fas-mediated apoptosis. *Nature* 380:723–726.

Fahrbach, S.E., Choi, M.K., Truman, J.W. (1993). Inhibitory effects of actinomycin D and cycloheximide on neuronal death in adult *Manduca sexta*. *J. Neurobiol.* 25:59–69.

Fahrbach S.E., Truman, J.W. (1987). Mechanisms for programmed cell death in the nervous system of a moth. *Ciba Found. Symp.* 126:65–81.

Farlie, P.G., Dringen, R., Rees, S.M., Kannourakis, G., Bernard, O. (1995). *bcl-2* transgene expression can protect neurons against developmental and induced cell death. *Proc. Natl. Acad. Sci. U.S.A.* 92:4397–4401.

Farrow, S.N., White, J.H.M., Martinou, I., Raven, T., Pun, K.-T., Grinham, C.J., Martinou, J.-C., Brown, R. (1995). Cloning of a bcl-2 homolog by interaction with adenovirus E1B 19K. *Nature* 374:731–733.

Faucheu, C., Diu, A., Chan, A.W.E., Blanchet, A.-M., Miossec, C., Herve, F., Collard-Dutilleul, Gu, Y., Aldape, R.A., Lipke, J.A., Rocher, C., Su, M.S.-S., Livingston, D.J., Hercend, T., Lalanne, J.-L. (1995). A novel human protease similar to the interleukin-1β converting enzyme induces apoptosis in transfected cells. *EMBO J.* 14:1914–1922.

Fernandes-Alnemri, T., Litwack, G., Alnemri, E.S. (1994). CPP32, a novel human apoptotic protein with homology to *Caenorhabditis elegans* cell death protein Ced-3 and mammalian interleukin-1ς-converting enzyme. *J. Biol. Chem.* 269:30761–30764.

Fernandes-Alnemri, T., Litwack, G., Alnemri, E.S. (1995a). *Mch2*, a new member of the apoptotic *Ced-3/ICE* cysteine protease gene family. *Cancer Res.* 55:2737–2742.

Fernandes-Alnemri, T., Takahashi, A., Armstrong, R., Krebs, J., Fritz, L., Tomaselli, K.J., Wang, L., Yu, Z., Croce, C.M., Salveson, G., Earnshaw, W.C., Litwack, G., Alnemri, E.S. (1995b). Mch3, a novel human apoptotic cysteine protease highly related to CPP32. *Cancer Res.* 55:6045–6052.

Fischbach, K.F. (1983). Neural cell types surviving congenital sensory deprivation in the optic lobes of *Drosophila melanogaster*. *Dev. Biol.* 95:1–18.

Fischbach, K.-F., Technau, G. (1984). Cell degeneration in the developing optic lobes of the *sine oculis* and *small-optic-lobes* mutants of *Drosophila melanogaster*. *Dev. Biol.* 104: 219–239.

Freeman, R.S., Estus, S., Johnson, E.M. (1994). Analysis of cell cycle-related gene expression in postmitotic neurons: selective induction of cyclin D1 during programmed cell death. *Neuron* 12:343–355.

Gagliardini, V., Fernandez, P.A., Lee, R.K.K., Drexler, H.C.A., Rotello, R.J., Fishman, M.C., Yuan, J. (1994). Prevention of vertebrate neuronal death by the *crmA* gene. *Science* 263:826–828.

Garcia, I., Martinou, I., Tsujimoto, Y., Martinou, J.-C. (1992). Prevention of programmed cell death of sympathetic neurons by the *bcl-2* proto-oncogene. *Science* 258:302–304.

Glucksmann, A. (1951). Cell deaths in normal vertebrate ontogeny. *Biol. Rev.* 26:59–86.

Golstein, P., Marguet, D., Depraetere, V. (1995). Homology between Reaper and the cell death domains of Fas and TNFR1. *Cell* 81:185–186.

Gonzalez-Garcia, M., Perez-Ballestero, R., Ding, L., Duan, L., Boise, L.H., Thompson, C.B., Nunez, G. (1994). bcl-xL is the major bcl-x mRNA form expressed during murine development and its product localizes to mitochondria. *Development* 120:3033–3042.

Grether, M.E., Abrams, J.M., Agapite, J., White, K., Steller, H. (1995). The *head involution defective* gene of *Drosophila melanogaster* functions in programmed cell death. *Genes Dev.* 9:1694–1708.

Hamburger, V., Levi-Montalcini, R. (1949). Proliferation, differentiation and degeneration in the spinal ganglia of the chick embryo under normal and experimental conditions. *J. Exp. Zool.* 111:457–502.

Hay, B.A., Wolff, T., Rubin, G.M. (1994). Expression of baculovirus P35 prevents cell death in *Drosophila*. Development 120:2121–2129.

Hay, B.A., Wassarman, D.A., Rubin, G.M. (1995). Drosophila homologs of baculoviral inhibitor of apoptosis proteins function to block cell death. *Cell* 83:1253–1262.

Heintz, N. (1993). Cell death and the cell cycle: a relationship between transformation and neurodegeneration? *Trends Biochem. Sci.* 18:157–159.

Henderson, S., Huen, D., Rowe, M., Dawson, C., Johnson, G., Rickinson, A. (1993). Epstein-Barr virus-coded BHRF1 protein, a viral homolog of Bcl-2, protects human B cells from programmed cell death. *Proc. Natl. Acad. Sci. U.S.A.* 90:8479–8483.

Hengartner, M.O. (1996). Programmed cell death in invertebrates. *Curr. Opin. Genet. Dev.* 6:34–38.

Hengartner, M.O., Horvitz, H.R. (1994a). *C. elegans* cell survival gene *ced-9* encodes a functional homolog of the mammalian proto-oncogene *bcl-2*. *Cell* 76:665–674.

Hengartner, M.O., Horvitz, H.R. (1994b). Programmed cell death in *Caenorhabditis elegans*. *Curr. Opin. Genet. Dev.* 4:581–586.

Hengartner, M.O., Horvitz, H.R. (1994c). The ins and outs of programmed cell death during *C. elegans* development. *Philos. Trans. R. Soc. Lond. Biol* 345:243–246.

Hengartner, M.O., Ellis, R.E., Horvitz, H.R. (1992). *C. elegans* gene *ced-9* protects cells from programmed cell death. *Nature* 356:494–499.

Heumann, R. (1994). Neurotrophin signalling. *Curr. Opin. Neurobiol.* 4:668–679.

Hockenberry, D.M., Oltvai, Z.N., Yin, X.-M., Milliman, C.L., Korsmeyer, S.J. (1993). Bcl-2 functions in an antioxidant pathway to prevent apoptosis. *Cell* 75:241–251.

Itoh, N., Nagata, S. (1993). A novel protein domain required for apoptosis. Mutational analysis of human Fas antigen. *J. Biol. Chem.* 269:10932–10937.

Jacobson, M.D., Raff, M.C. (1995). Programmed cell death and Bcl-2 protection in very low oxygen. *Nature* 361:365–369.

Jacobson, M.D., Burne, J.F., King, M.P., Miyashita, T., Reed, J.C., Raff, M.C. (1993). Bcl-2 blocks apoptosis in cells lacking mitochondrial DNA. *Nature* 361:365–369.

Jacobson, M.D., Burne, J.F., Raff, M.C. (1994). Programmed cell death and Bcl-2 protection in the absence of a nucleus. *EMBO J.* 13:1899–1910.

Kamens, J., Paskind, M., Hugunin, M., Talanian, R.V., Allen, H., Banach, D., Bump, N., Hackett, M. Johnston, C.G., Li, P., Mankovitch, J.A., Terranova, M., Ghayur, T. (1995). Identification and characterization of ICH-2, a novel member of the interleukin-1β-converting enzyme family of cysteine proteases. *J. Biol. Chem.* 270:15250–15256.

Kerr, J.F.R., Wyllie, A.H., Currie, A.R. (1972). Apoptosis: a basic biological phenomenon with wide ranging implications in tissue kinetics. *Br. J. Cancer* 26:239–257.

Kiefer, M.C., Brauer, M.J., Powers, V.C., Wu, J.J., Umansky, S.R., Tomei, L.D., Barr, P.J. (1995). Modulation of apoptosis by the widely distributed Bcl-2 homolog Bak. *Nature* 374:736–739.

Kimura, K., Truman, J.W. (1990). Postmetamorphic cell death in the nervous and muscular systems of *Drosophila melanogaster*. *J. Neurosci.* 10:403–411.

Knudson, C.M., Tung, K.S.K., Tourtellotte, W.G., Brown, G.A.J., Korsmeyer, S.J. (1995). Bax deficient mice with lymphoid hyperplasia and male germ cell death. *Science* 270:96–99.

Koelle, M.R., Talbot, W.S., Segraves, W.A., Bender, M.T., Cherbas, P., Hogness, D.S. (1991). The *Drosophila EcR* gene encodes an ecdysone receptor, a new member of the steroid receptor superfamily. *Cell* 67:59–77.

Kollros, J.J. (1981). Transitions in the nervous system during amphibian metamorphosis. In *Metamorphosis: A problem in Developmental Biology*, Gilbert, L.I., Frieden, E., eds. New York: Plenum Press, pp. 445–459.

Kozopas, K.M., Yang, T., Buchan, H.L., Zhou, P., Craig, R.W. (1993). MCL1, a gene expressed in programmed myeloid cell differentiation, has sequence similarity to BCL-2. *Proc. Natl. Acad. Sci. U.S.A.* 90:3516–3520.

Kuida, K., Lippke, J.A., Ku, G., Harding, M.W., Livingston, D.J., Su, M.S.-S., Flavell, R.A. (1995). Altered cytokine export and apoptosis in mice deficient in ICE. *Science* 267:2000–2003.

Kumar, S., Kinoshita, M., Noda, M., Copeland, N.G., Jenkins, M.A. (1994). Induction of apoptosis by the mouse *Nedd2* gene, which encodes a protein similar to the product of the *Caenorhabditis elegans* cell death gene *ced-3* and the mammalian IL-1β converting enzyme. *Genes Dev.* 8:1613–1626.

Lazebnik, Y.A., Kaufmann, S.H., Desnoyers, S., Poirier, G.G., Earnshaw, W.C. (1994). Cleavage of poly(ADP-ribose) polymerase by a proteinase with properties like ICE. *Nature* 371:346–347.

Lazebnik, Y.A., Takahashi, A., Moir, R., Goldman, R., Poirier, G.G., Kaufmann, S.H., Earnshaw, W.C. (1995). Studies of the lamin proteinase reveal multiple parallel biochemical pathways during apoptotic execution. *Proc. Natl. Acad. Sci. U.S.A.* 92:9042–9046.

Li, P., Hamish, A., Banerjee, S., Franklin, S., Herzog, L., Johnston, C., McDowell, J., Paskind, M., Rodman, L., Salfeld, J., Towne, E., Tracey, D., Wardwell, S., Wei, F.-Y., Wong, W., Kamen, R., Seshardi, T. (1995). Mice deficient in IL-1B-converting enzyme are defective in production of mature IL-1B and resistant to endotoxic shock. *Cell* 80: 401–411.

Lin, E.Y., Orlofsky, A., Berger, M.S., Prystowsky, M.B. (1993). Characterization of A1, a novel hematopoietic-specific early response gene with sequence similarity to *bcl-2*. *J. Immunol.* 151:1979–1988.

Lindsay, R.M., Wiegand, S.J., Altar, C.A., DiStefano, P.S. (1994). Neurotrophic factors: from molecule to man. *Trends NeuroSci.* 17:182–190.

Linnik, M.D., Zobrist, R.H., Hatfield, M.D. (1993). Evidence supporting a role for programmed cell death in focal cerebral ischemia in rats. *Stroke* 24:2002–2009.

Lipton, S.A., Rosenberg, P.A. (1994). Mechanisms of disease: excitatory amino acids as a final common pathway for neurologic disorders. *N. Engl. J. Med.* 330:613–622.

Liston, P., Roy, N., Tamai, K., Lefebvre, C., Baird, S., Cherton-Horvat, G., Farahani, R., McLean, M., Ikeda, J.-E., MacKenzie, A., Korneluk, R.G. (1996). Suppression of apoptosis in mammalian cells by NAIP and a related family of IAP genes. *Nature* 379: 349–353.

Lockshin, R.A. (1969). Programmed cell death. Activation of lysis by a mechanism involving the synthesis of protein. *J. Insect Physiol.* 15:1505–1516.

Lockshin, R.A., Zakeri, Z. (1991). Programmed Cell Death and Apoptosis. In *Apoptosis The Molecular Basis of Cell Death* (ed. L.D. Tomei, F.O. Cope) pp. 47–60. Cold Spring Harbor Laboratory Press, Cold Spring Harbor, NY

Lowe, S.W., Schmitt, E.M., Smith, S.W., Osborne, B.A., Jacks, T. (1993). p53 is required for radiation-induced apoptosis in mouse thymocytes. *Nature* 362:847–849.

Macaya, A., Munell, F., Gubits, R.M., Burke, R.E. (1994). Apoptosis in substantia nigra following developmental striatal excitotoxic injury. *Proc. Natl. Acad. Sci. U. S. A.* 91:8117–8121.

MacManus, J.P., Buchan, A.M., Hill, I.E., Rashquinha, I., Preston, E. (1993). Global ischemia can cause DNA fragmentation indicative of apoptosis in rat brain. *Neurosci. Lett.* 164:89–92.

MacManus, J.P., Hill, I.E., Huang, Z.G., Rashquinha, I., Xue, D., Buchan., A.M. (1994). DNA damage consistent with apoptosis in transient focal ischemic neurocortex. *Neuroreport* 5:493–496.

Martin, D.P., Schmidt, R.E., DiStefano, P.S., Lowry, O.H., Carter, J.G., Johnson, E.M. (1988). Inhibitors of protein synthesis and RNA synthesis prevent neuronal death caused by nerve growth factor deprivation. *J. Cell Biol.* 106:829–844.

Martinou, J.-C., Dubois-Dauphin, M., Staple, J.K., Rodriguez, I., Frankowski, H., Missotten, M., Albertini, P., Talabot, D., Catsicas, S., Pietra, C., Huarte, J. (1994). Overexpression of BCL-2 in transgenic mice protects neurons from naturally occurring cell death and experimental ischemia. *Neuron* 13:1017–1030.

Martinou, I., Fernandex, P.A., Missotten, M., White, E., Allet, B., Sadoul, R., Martinou, J.C. (1995). Viral proteins E1B19K and p35 protect sympathetic neurons from cell death induced by NGF deprivation. *J. Biol. Chem.* 128:201–208.

Mathias, S., Kolesnick, R. (1993). Ceramide: a novel second messenger. *Adv. Lipid Res.* 25:65–90.

Merry, D.E., Veis, D.J., Hickey, W.F., Korsmeyer, S.J. (1994). bcl-2 protein expression is widespread in the developing nervous system and retained in the adult PNS. *Development* 120:301–311.

Middleton, G., Nunez, G., Davies, A.M. (1996). Bax promotes neuronal survival and antagonizes the survival effects of neurotrophic factors. *Development* 122:695–701.

Milligan, C.E., Prevette, D., Yaginuma, H., Homma, S., Cardwell, C., Fritz, L.C., Tomaselli, K.J., Oppenheim, R.W., Schwartz, L.M. (1995). Peptide inhibitors of the ICE protease family arrest programmed cell death of motoneurons in vivo and in vitro. *Neuron* 15:385–393.

Miura, M., Zhu, H., Rotello, R., Hartwieg, E.A., Yuan, J. (1993). Induction of apoptosis in fibroblasts by IL-1β converting enzyme, a mammalian homolog of the *C. elegans* cell death gene *ced-3*. *Cell* 75:653–660.

Munday, N.A., Vaillancourt, J.P., Ali, A., Casano, F.J., Miller, D.K., Molineaux, S.M., Yamin, T.-T., Yu, V.L., Nicholson, D.W. (1995). Molecular cloning and pro-apoptotic activity of ICErelII and ICErelIII, members of the ICE/*ced-3* family of cysteine proteases. *J. Biol. Chem.* 270:15870–15876.

Motoyama, N., Wang, F., Roth, K.A., Sawa, H., Nakayama, K., Nakayama, K., Negishi, I., Senju, S., Zhang, Q., Fujii, S., Loh, D.Y. (1995). Massive cell death of immature hematopoietic cells and neurons in Bcl-x-deficient mice. *Science* 267:1506–1510.

Nagata, S., Golstein, P. (1995). The Fas death factor. *Science* 267:1449–1456.

Nakashima, T., Sekiguchi, T., Kuraoka, A., Fukushima, K., Shibata, Y., Komiyama, S., Nishimoto, T. (1993). Molecular cloning of a human cDNA encoding a novel protein, DAD-1, whose effect causes apoptosis cell death in hamster BHK21 cells. *Mol. Cell. Biol.* 13:6367–6374.

Nakayama, K., Nakayama, K., Negishi, I., Kuida, K., Sawa, H., Loh, D.Y. (1994). Targeted disruption of bcl-2 alpha beta in mice—occurrence of grey hair, polycystic kidney disease, and lymphocytopenia. *Proc. Natl. Acad. Sci. U.S.A.* 91:3700–3704.

Neilan, J.G., Lu, Z., Afonzo, C.L., Kutish, G.F., Sussman, M.D., Rock, D.L. (1993). An African swine fever virus gene with similarity to the proto-oncogene *bcl-2* and the Epstein-Barr virus gene BHRF1. *J. Virol.* 67:4391–4394.

Nicholson, D.W., Ali, A., Thornberry, N.A., Vaillancourt, J.P., Ding, C.K., Gallant, M., Gareau, Y., Griffin, P.R., Labelle, M., Lazebnik, Y.A., Munday, N.A., Raju, S.M., Smulson, M.E., Yamin, T.-T., Yu, V.L., Miller, D.K. (1995). Identification and inhibition of the ICE/CED-3 protease necessary for mammalian apoptosis. *Nature*. 376:37–43.

Nordlander, R.H., Edwards, J.S. (1968). Morphological cell death in the post-embryonic development of the insect optic lobes. *Nature* 218:780–781.

Oberhammer, F.A., Hochegger, K., Froschl, G., Tiefenbacher, R., Pavelka, M. (1994). Chromatin condensation during apoptosis is accompanied by degradation of lamin A + B, without enhanced activation of cdc2 kinase. *J. Cell. Biol.* 126:827–837.

Oltvai, Z.N., Millman, C.L., Korsmeyer, S.J. (1993). Bcl-2 heterodimerizes in vivo with a conserved homolog, Bax, that accelerates programmed cell death. *Cell* 74:609–619.

Oppenheim, R.W. (1991). Cell death during development of the nervous system. *Annu. Rev. Neurosci.* 14:453–501.

Oppenheim, R.W., Prevette, D., Tytell, M., Homma, S. (1990). Naturally occurring and induced neuronal death in the chick embryo *in vivo* requires protein and RNA synthesis: evidence for the role of cell death genes. *Dev. Biol.* 138:104–113.

Patel, M., Day, B.J., Crapo, J.D., Fridovich, I., McNamara, J.O. (1996). Requirement for superoxide in excitotoxic cell death. *Neuron* 16:345–355.

Power, M.E. (1943). The effect of reduction in numbers of ommatidia upon the brain of *Drosophila* melanogaster. *J. Exp. Zool.* 94:33–72.

Pronk, G.J., Ramer, K., Amiri, P., Williams, L.T. (1996). Requirement of an ICE-like protease for induction of apoptosis and ceramide generation by REAPER. *Science* 271: 808–810.

Pulido, D., Campuzano, S., Koda, T., Modolell, J., Barbacid, M. (1992). *Dtrk*, a *Drosophila* gene related to the *trk* family of neurotrophin receptors, encodes a novel class of neural cell adhesion molecule. *EMBO J.* 11:391–404.

Purves, D. (1988). *Body and Brain: A Trophic Theory of Neural Connections*. Cambridge, MA: Harvard University Press.

Purves, D., Lichtman, J.W. (1985). Neuronal Death During Development In *Principles of Neural Development* pp. 131–154. Sunderland, MA: Sinauer.

Rabizadeh, S., Oh, J., Zhong, L., Yang, J., Bitler, C.M., Bucher, L.L., Bredesen, D.E. (1993a). Induction of apoptosis by the low-affinity NGF receptor. *Science* 261:345–348.

Rabizadeh, S., La Count, D.J., Friesen, P.D., Bredesen, D.E. (1993b). Expression of the baculovirus p35 gene inhibits mammalian cell death. *J. Neurochem.* 61:2318–2321.

Raff, M.C. (1992). Social controls on cell survival and cell death. *Nature* 356:397–400.

Raff, M.C., Barres, B.A., Burne, J.F., Coles, H.S.R., Ishizaki, Y., Jacobson, M.D. (1993). Programmed cell death and the control of cell survival: lessons from the nervous system. *Science* 262:695–700.

Raff, M.C., Barres, B.A., Burne, J.F., Coles, H.S.R., Ishizaki, Y., Jacobson, M.D. (1994). Programmed cell death and the control of cell survival. *Philos. Trans. R. Soc. Lond. Biol.* 345:265–268.

Robinow, S., Talbot, W.S., Hogness, D.S., Truman, J.W. (1993). Programmed cell death in the *Drosophila* CNS is ecdysone-regulated and coupled with a specific ecdysone receptor isoform. *Development* 119:1251–1259.

Rothe, M., Pan, M.-G., Henzel, W.J., Ayres, T.M., Goeddel, D.V. (1995b). The TNFR2-TRAF signaling complex contains two novel proteins related to baculoviral inhibitor of apoptosis proteins. *Cell* 83:1243–1252.

Roy, N., Mahadevan, M.S., McLean, M., Shutler, G., Yaraghi, Z., Farahani, R., Baird, S., Besner-Johnston, A., Lefebvre, C., Kang, X., Salih, M., Aubry, H., Tamai, K., Guan, X., Ioannou, P., Crawford, T.O., de Jong, P.J., Surh, L., Ikeda, J.-E., Korneluk, R.G., MacKenzie, A. (1995). The gene for neuronal apoptosis inhibitory protein is partially deleted in individuals with spinal muscular atrophy. *Cell* 80:167–178.

Rubin, L.L., Philpott, K.L., Brooks, S.F. (1993). The cell cycle and cell death. *Curr. Biol.* 3:391–394.

Saunders, J.W. (1966). Death in embryonic systems. *Science* 154:604–612.

Schwartz, L.M., Kosz, L., Kay, B.K. (1990). Gene activation is required for developmentally programmed cell death. *Proc. Natl. Acad. Sci. U.S.A.* 87:6594–6598.

Shaham, S., Horvitz, H.R. (1996). Developing *Caenorhabditis elegans* neurons may contain both cell-death protective and killer activities. *Genes Devel.* 10:578–591.

Shi, L., Nishioka, W.K., Th'ng, J., Bradbury, E.M., Litchfield, D.W., Greenberg, A.H. (1994). Premature p34^{cdc2} activation required for apoptosis. *Science* 263:1143–1145.

Silberstein, S., Collins, P.G., Kellecher, D.J., Gilmore, R. (1995). The essential OST2 gene encodes the 16kD subunit of the yeast oligosaccharyltransferase, a highly conserved protein expressed in diverse eukaryotic organisms. *J. Cell. Biol.* 131:371–383.

Sloviter, R.S., Dean, E., Neubort, S. (1993). Electron microscopic analysis of adrenalectomy-induced hippocampal granule cell degeneration in the rat: apoptosis in the adult central nervous system. *J. Comp. Neurol.* 330:337–351.

Smith, C.A., Davis, T., Anderson, D., Solam, E., Beckmann, M.P., Jerzy, R., Dower, S.K., Cosman, D., Goodwin, R.G. (1990). A receptor for tumor necrosis factor defines an unusual family of cellular and viral proteins. *Science* 248:1019–1023.

Smith, C.A., Farrah, T., Goodwin, R.G. (1994). The TNF receptor superfamily of cellular and viral proteins: activation, costimulation, and death. *Cell* 76:959–962.

Snider, W.D. (1994). Functions of the neurotrophins during nervous system development: what the knockouts are teaching us. *Cell* 77:627–638.

Steller, H. (1995). Mechanisms and genes of cellular suicide. *Science* 267:1445–1449.

Steller, H., Grether, M.E. (1994). Programmed cell death in *Drosophila*. *Neuron* 13:1269–1274.

Steller, H., Fischbach, K.-F., Rubin, G. M. (1987). *disconnected*: a locus required for neuronal pathway formation in the visual system of *Drosophila*. *Cell* 50:1139–1153.

Stewart, R.R., Gao, W.O., Peinado, A., Zipser, B., Macagno, E.R. (1987). Cell death during gangliogenesis in the leech: bipolar cells appear and then degenerate in all ganglia. *J. Neurosci.* 7:1919–1927.

Stuart, D.K., Blair, S.S., Weisblat, D.A. (1987). Cell lineage, cell death, and the developmental origin of identified serotonin- and dopamine-containing neurons in the leech. *J. Neurosci.* 7:1107–1122.

Sugimoto, A., Friesen, P.D., Rothmans, J.H. (1994). Baculovirus p35 prevents developmentally regulated cell death and rescues a ced-9 mutant in the nematode *Caenorhabditis elegans. EMBO J.* 13:2023–2028.

Sugimoto, A., Hozak, R.R., Nakashima, T., Nishimoto, R., Rothmans, J.H. (1995). *dad-1*, an endogenous programmed cell death suppressor in *Caenorhabditis elegans* and vertebrates. *EMBO J.* 14:2023–2028.

Svendsen, C.N., Kew, J.N.C., Staley, K., Sofroniew, M.V. (1994). Death of developing septal cholinergic neurons following NGF withdrawal: protection by protein synthesis inhibition. *J. Neurosci.* 14:75–87.

Talbot, W.S., Swyryd, E.A., Hogness, D.S. (1993). *Drosophila* tissues with different metamorphic responses to ecdysone express different ecdysone receptor isoforms. *Cell* 73:1323–1337.

Takahashi, A., Earnshaw, W.C. (1996). ICE-related proteases in apoptosis. *Curr. Opin. Gen. Dev.* 6:50–55.

Tartaglia, L.A., Goeddel, D.V. (1992). Two TNF receptors. *Immunol. Today* 13:151–153.

Tartaglia, L.A., Ayres, T.M., Wong, G.H.W., Goeddel, D.V. (1993). A novel domain within the 55-kD TNF receptor signals death. *Cell* 74:845–853.

Tewari, M., Quan, L.T., O'Rourke, K., Desnoyers, S., Zeng, Z., Beidler, D.R., Poirier, C.G., Salvesen, G.S., Dixit, V.M. (1995). Yama/CPP32β, a mammalian homolog of CED-3, is a CrmA-inhibitable protease that cleaves the death substrate poly(ADP-ribose) polymerase. *Cell* 81:801–809.

Thomas, H.E., Stunnenberg, H.G., Stewart, A.F. (1993). Heterodimerization of the *Drosphila* ecdysone receptor with retinoid X receptor and ultraspiracle. *Nature* 362: 471–475.

Thompson, C.B. (1995). Apoptosis in the pathogenesis and treatment of disease. *Science* 267:1456–1462.

Thornberry, N.A., Bull, H.G., Calaycay, J.R., Chapman, K.T., Howard, A.D., Kostura, M.J., Miller, D.K., Molineaux, S.M., Weidner, J.R., Aunins, J., Elliston, K.O., Ayala, J.M., Casano, F.J., Chin, J., Ding, J.F., Egger, L.A., Gaffney, E.P., Limjuco, G., Palyha, O.C., Raju, S.M., Rolando, A.M., Salley, J.P., Yamin, T., Lee, T.D., Shively, J.E., MacCross, M.M., Mumford, R.A., Schmidt, J.A., Tocci, M.J. (1992). A novel heterodimeric cystine protease is required for interleukin-1β processing in monocytes. *Nature* 356:768–774.

Truman, J.W. (1983). Programmed cell death in the nervous system of an adult insect. *J. Comp. Neurol.* 216:445–452.

Truman, J.W., Schwartz, L.M. (1984). Steroid regulation of neural death in the moth nervous system. *J. Neurosci.* 4:274–280.

Truman, J.W., Thorn, R.S., Robinow, S. (1992). Programmed neuronal death in insect development. *J. Neurobiol.* 23:1295–1311.

Tsujimoto, Y., Gorham, J., Cossman, J., Jaffe, E., Croce, C.M. (1985). The t[14;18] chromosome translocations involved in B cell neoplasms result from mistakes in VDJ joining. *Science* 229:1390–1393.

Ucker, D.S. (1991). Death by suicide: one way to go in mammalian cellular development? *New Biol.* 3:103–109.

Ucker, D.S., Meyers, J., Obermiller, P.S. (1992). Activation driven T cell death. II. Quantitative differences alone distinguish stimuli triggering nontransformed T cell proliferation or death. *J. Immunol.* 149:1583–1592.

Vaux, D.J., Weissman, I.L., Kim, S.K. (1992). Prevention of programmed cell death in *Caenorhabditis elegans* by human bcl-2. *Science* 258:1955–1957.

Vaux, D.L., Haecker, G., Strasser, A. (1994). An evolutionary perspective on apoptosis. *Cell* 76:777–779.

Veis, D.J., Sorenson, C.M., Shutter, J.R., Korsmeyer, S.J. (1993). Bcl-2-deficient mice demonstrate fulminant lymphoid apoptosis, polycystic kidneys, and hypopigmented hair. *Cell* 75:229–240.

von Bartheld C.S., Kinoshita, Y., Prevette, D., Yin, Q.W., Oppenheim, R.W., Bothwell, M. (1994). Positive and negative effects of neurotrophins of the isthmo-optic necleus in chick embryos. *Neuron* 12:639–654.

von-Bartheld et al., (1994). *Neuron* 12:639.

Walker, N.P.C., Talanian, R.V., Brady, K.D., Dang, L.C., Bump, N.J., Ferenz, C.R., Franklin, S., Ghayur, T., Hackett, M.C., Hammill, L.D., et al. (1994). Crystal structure of the cysteine protease interleukin-1L-1β converting enzyme: a $(p20/p10)_2$ homodimer. *Cell* 78:343–352.

Walker, N.P.C., Talanian, R.V., Brady, K.D., Dang, L.C., Bump, N.J., Ferenz, C.R., Franklin, S., Ghayur, T., Hackett, M.C., Hammill L.D., Herzog, L., Hugunin, M., Houy, W., Mankovich, J.A., McGuiness, L., Orlewicz, E., Paskind, M., Pratt, C.A., Reis, P., Summani, A., Terranova, M., Welch, J.P., Xiong, L., Möller, A., Tracey, D.E., Kamen, R., Wong, W., W. (1994). Crystal structure of the cysteine protease intrleukin 1β-converting enzyme: a $(p20/p10)_2$ homodimer. *Cell* 78:343–352.

Wang, L., Miura, M., Bergeron, L., Zhu, H., Yuan, J. (1994). *Ich-1*, an *Ice/ced-3-* related gene, encodes both positive and negative regulators of programmed cell death. *Cell* 78:739–750.

White, E., Sabbatini, P., Debbas, M., Wold, W.S.M., Kusher, D.I., Gooding, L. (1992). The 19-kilodalton adenovirus E1B transforming protein inhibits programmed cell death and prevents cytolysis by tumor necrosis factor α. *Mol. Cell. Biol.* 12:2570–2580.

White, K., Grether, M., Abrams, J., Young, L., Farrell, K., Steller, H. (1994). Genetic control of cell death in *Drosophila. Science* 264:677–683.

White, K., Steller, H. (1995). The control of apoptosis in *Drosophila. Trends Cell Biol.* 5: 74–77.

White, K., Tahaoglu, E., Steller, H. (1996). Cell killing by *Drosophila reaper. Science* 271:805–807.

Williams, M.S., Henkart, P.A. (1994). Apoptotic cell death induced by intracellular proteolysis. *J. Immunol.* 153:4247–4255.

Wilson, K.P., Black, J.A., Thomson, J.A., Kim, E.E., Griffith, J.P., Navia, M.A., Murcko, M.A., Chambers, S.P., Aldape, R.A., Raybuck, S.A., et al. (1994). Structure and mechanism of interleukin-1L-1β converting enzyme. *Nature* 370:270–275.

Wilson, K.P., Black J.F., Thomson, J.A., Kim, E.E., Griffith, J.P., Navia, M.A., Murcko, M.A., Chambers, S.P., Aldape, R.A., Raybuck, S.A., Livingston, D.J. (1994). Structure and mechanism of interleukin-1β converting enzyme. *Nature* 370:270–275.

Wilson, C., Goberdhan, D.C.I., Steller, H. (1993). *Dror,* a potential neurotrophic receptor gene, encodes a *Drosophila* of the vertebrate Ror family of Trk-related receptor tyrosine kinases. *Proc. Natl. Acad. Sci. U.S.A.* 90:7109–7113.

Wolff, T., Ready, D.F. (1991). Cell death in normal and rough eye mutants of *Drosophila. Development* 113:825–839.

Wyllie, A.H., Kerr, J.F.R., Currie, A.R. (1980). Cell death: the significance of apoptosis. *Int. Rev. Cytol.* 68:251–306.

Xue, D., Horvitz, H.R. (1995). Inhibition of the *Caenorhabditis elegans* cell-death protease CED-3 by a CED-3 cleavage site in baculovirus p35 protein. *Nature* 377:248–251.

Xue, D., Shaham, S., Horvitz, H.R. (1996). The *Caenorhabditis elegans* cell-death protein CED-3 is a cysteine protease with substrate specificities similar to those of the human CPP32 protease. *Genes Dev.* 10:1073–1083.

Yaginuma, H., Tomita, M., Takashita, N., McKay, S.E., Cardwell, C., Yin, Q.W., and Oppenheim, R.W. (1996). A novel type of programmed neuronal death in the cervical spinal cord of the chick embryo. *J. Neurosci.* 16:3685–3703.

Yao, T.P., Segraves, W.A., Oro, A.E., McKeown, M., Evans, R.M. (1992). *Drosophila* ultraspiracle modulates ecdysone receptor function via heterodimer formation. Cell 71: 63–72.

Yang, E., Zha, J., Jockel, J., Boise, L.H., Thompson, C.B., Korsmeyer, S.J. (1995). Bad, a heterodimeric partner for Bcl-xL and Bcl-2, displaces Bax and promotes death. *Cell* 80:285–291.

Yin, X.-M., Oltvai, Z., Korsmeyer, S.J. (1994). BH1 and BH2 domains of Bcl-2 are required for inhibition of apoptosis and heterodimerization with Bax. *Nature* 369:321–323.

Yuan, J., Horvitz, H.R. (1990). The *Caenorhabditis elegans* genes *ced-3* and *ced-4* Act cell autonomously to cause programmed cell death. *Dev. Biol.* 138:33–41.

Yuan, J., Horvitz, H.R. (1992). The *Caenorhabditis elegans* cell death gene *ced-4* encodes a novel protein and is expressed during the period of extensive programmed cell death. *Development* 116:309–320.

Yuan, J., Shaham, S., Ellis, H.M., Horvitz, H.R. (1993). The *C. elegans* cell death gene *ced-3* encodes a protein similar to mammalian interleukin-1beta converting enzyme. *Cell* 75:641–652.

8

Inductive signals and the assignment of cell fate in the spinal cord and hindbrain

An axial coordinate system for neural patterning

THOMAS M. JESSELL AND ANDREW LUMSDEN

The assembly of functional circuits during neural development depends critically on the generation of distinct classes of neurons at defined positions within the neural epithelium. The past decade has seen the emergence of an increasingly coherent view of the principles that control the identity and patterning of neural cells in vertebrate embryos (McConnell, 1995; Anderson and Jan, 1996). It is now apparent that the fate of neural progenitor cells is not rigidly preordained: Instead, distinct neural identities are acquired through a progressive restriction in the developmental potential of cells under the control of local environmental signals. Such signals are thought to specify cell fate by inducing the expression of cell-intrinsic factors, among which are transcriptional regulatory proteins. In turn, cell-specific combinations of transcription factors orchestrate the expression of effector proteins, notably surface receptors and components of signal transduction pathways that endow axons and growth cones with the ability to select specific pathways and form precise target connections (Tessier-Lavigne and Goodman, 1996).

Recent studies of the molecular mechanisms that control the identity and pattern of neural cell types generated in vertebrate embryos have focused on three main issues: (1) defining the source and identity of signals that control the fate of nearby cells, (2) defining the cell-intrinsic factors that commit neural cells to specific fates, and (3) determining how intrinsic regulatory proteins direct the expression of the sensory machinery required in and on the growth cone for the recognition of local guidance cues. These basic issues are under analysis at all axial levels of the neural tube. In this review we focus on the patterning mechanisms implicated in the generation of distinct cell types at caudal levels of the neural tube that give rise to the segmentally organized structures of the spinal cord and hindbrain. Many of the principles that have emerged from studies of the organization of cell types at caudal levels of the neuraxis, however, appear applicable to the patterning processes that operate at more rostral levels that give rise to the midbrain and forebrain (Alvarado-Mallart, 1993; Joyner, 1996; Rubenstein and Shimamura, 1996).

Induction and Axial Organization of the Neural Plate

The patterning of spinal cord and hindbrain is initiated by signaling events that occur at the time of neural plate formation (Schoenwolf and Smith, 1990; Doniach et al., 1992; Ruiz i Altaba, 1994), a process triggered by inductive signals from the axial mesoderm (Harland, 1996). Although the identity of relevant neural-inducing molecules remains uncertain, three proteins expressed by the axial mesoderm, follistatin, noggin, and chordin, each have the ability to induce neural plate cells of anterior character from naive ectodermal cells in the absence of accompanying mesoderm (Hemmati-Brivanlou et al., 1994; Sasai et al., 1995; Knecht et al., 1995). These three proteins exhibit no obvious common structural features but each can bind to and antagonize the actions of members of the TGF-β family of signaling proteins (Harland, 1996). These and other findings have led to a proposal that the induction of anterior neural plate differentiation involves the inhibition of TGF-β–like signals which function in a tonic manner to promote epidermal and repress neural differentiation. A fourth class of secreted factor, the FGFs, can induce neural cells of posterior character, apparently independently of any inhibition of TGF-β–like signals (Kengaku and Okamoto, 1995; Cox and Hemmati-Brivanlou, 1995; Lamb and Harland, 1995). The early regional identity of the neural plate along its anteroposterior (A/P) axis may therefore be established by the coordinate actions of these distinct inductive signals and as discussed below, by other patterning signals that act independently of their ability to induce neural tissue. A detailed account of the molecular steps of neural induction is given by Harland (1996).

Once induced, the neural plate spans the midline of the embryo and occupies almost its entire AP extent. Fate maps of the neural plate derived in many vertebrate species indicate that its most anterior region gives rise to forebrain structures, with progressively more posterior regions giving rise to the midbrain, hindbrain, and spinal cord (Schoenwolf et al., 1989; Eagleson and Harris, 1990). Analysis of neural cell fate after rotation of regions of the neural plate along the A/P and/or mediolateral (M/L), prospective dorsoventral (D/V), axes has provided evidence that restrictions in the regional fate of cells along the A/P axis occur before and independently of restrictions along the D/V axis (Roach, 1945; Jacobson, 1964; Simon et al., 1995).

These observations have suggested that the patterning of cell types in the neural tube at caudal levels of the neuraxis is organized on the basis of a Cartesian grid of positional information, the coordinates of which correspond to the A/P and D/V axes of the neural tube (Simon et al., 1995). The position that a cell occupies along these two orthogonal axes may therefore define its eventual fate by virtue of the early and positionally restricted inductive signals to which it is exposed. This scheme has been supported by studies in which the distinct and largely independent signaling systems that appear to control cell fate along the

A/P and D/V axes of the neural tube have been dissected in the spinal cord and hindbrain. This review first considers the nature and identity of inductive signals thought to control cell fate along the D/V axis of the neural tube and then discusses the evidence that cells located at constant D/V positions have, at an earlier stage, acquired distinct regional identities in accord with their different A/P levels.

CONTROL OF CELL FATE IN THE VENTRAL NEURAL TUBE

The early development of the neural tube at spinal cord and hindbrain levels is characterized by the appearance of distinct cell types at different D/V positions (Fig. 8–1). In the ventral half of the neural tube, floor plate cells appear at the ventral midline and motoneurons and ventral interneurons are generated in more lateral positions. In the dorsal half of the neural tube, roof plate cells and neural crest cells appear at or near to the dorsal midline and sensory relay interneurons are generated laterally. Each class of neuron is generated bilaterally, with a plane of symmetry that corresponds to the D/V axis of the neural tube.

The signals that initiate this program of cell differentiation act on neural progenitor cells before neural tube closure by imposing an early polarity along the M/L axis of the neural plate. Two distinct groups of non-neural cells appear

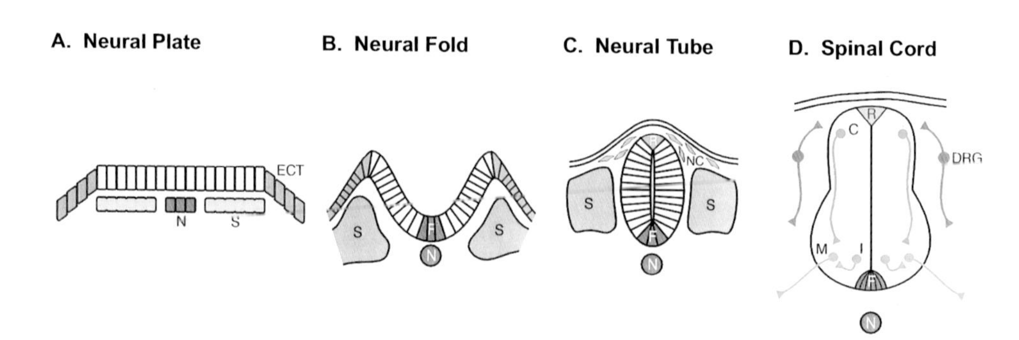

FIGURE 8–1 Stages in the development of the embryonic spinal cord. Establishment of the D/V axis in the developing spinal cord. The four diagrams show successive stages in the development of the spinal cord. (A) The neural plate consists initially of a simple columnar epithelium. Cells at the midline of the neural plate are contacted directly by axial mesoderm cells of the notochord (N). More lateral regions of the neural plate overlie the segmental plate mesoderm, which later gives rise to somites (S), and they are flanked by the epidermal ectoderm (ECT). (B) During neurulation, the neural plate buckles at its midline to form the neural folds and a floor plate (F) forms at the midline. Contact between the midline of the neural plate and the notochord is maintained at this stage. (C) The neural tube is formed when the dorsal tips of the neural folds fuse. Cells in the region of fusion form a specialized group of dorsal midline cells; the roof plate (R) and neural crest cells (NC) emigrate from the dorsal neural tube (D). After neural tube closure, neuroepithelial cells continue to proliferate and eventually differentiate into defined classes of neurons at different dorsoventral positions within the spinal cord. Commissural neurons (C) differentiate dorsally near to the roof plate; motoneurons (M) and interneurons (I) differentiate ventrally near to the floor plate, which by this time is no longer contacted by the notochord. Dorsal root ganglia (DRG) form from postmigratory neural crest cells.

to provide these early patterning signals: axial mesodermal cells of the notochord that underlie the midline of the neural plate and the cells of the epidermal ectoderm that flank its lateral edges. As documented below, the notochord is the source of a ventralizing inductive factor and the epidermal ectoderm is the source of dorsalizing factors. The opposing actions of these two signals appear to be critical in establishing the identity and pattern of cell types generated along the D/V axis of the neural tube.

Notochord-Derived Signals and the Induction of Ventral Cell Fates

The differentiation of each of the cell types generated in the ventral half of the neural tube, notably floor plate cells, motoneurons, and ventral interneurons, depends on signals provided by axial mesodermal cells of the notochord (Placzek, 1995). The requirement for notochord-derived signals has been established in avian embryos by the demonstration that ventral cell types are not generated if the notochord is removed at neural plate stages (Placzek et al., 1990; Hirano et al., 1991; Yamada et al., 1991; van Straaten and Hekking, 1991; Placzek, 1995) (Fig. 8–2). Similarly, ventral cell types are absent in mutant mouse and ze-

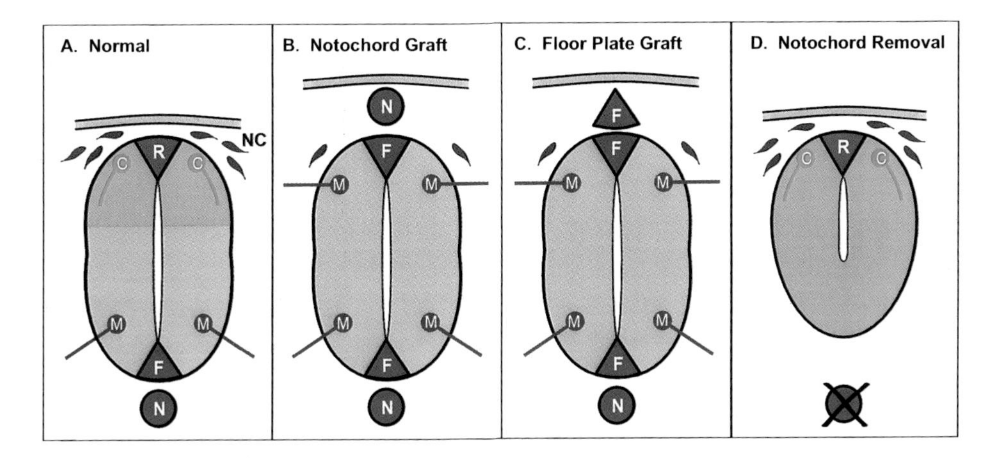

FIGURE 8–2 Influence of the notochord on cell fate along the D/V axis of the neural tube. Diagram summarizing the results obtained from experiments in chick embryos in which a notocord or floor plate is grafted at neural fold stages such that it is later placed above the dorsal midline of the neural tube, or in which the notochord is removed before neural tube closure. (A) Normal organization of the spinal cord showing the ventral midline location of floor plate cells (F) and motor neurons (M) and the dorsal position of roof plate cells (R), neural crest cells (NC), commissural interneurons (C), and a dorsal zone of *pax-3* expression (orange). (B) Dorsal grafts of a notochord result in the induction of a floor plate at the dorsal midline and ectopic dorsal motoneurons. The differentiation of dorsal cell types is suppressed, or, in the case of neural crest cells, significantly reduced. (C) Dorsal grafts of a floor plate also result in the induction of a dorsal midline floor plate and ectopic dorsal motoneurons. Dorsal cell differentiation is also suppressed. (D) Removal of the notochord results in the elimination of the floor plate and motoneurons and the expression of dorsally restricted transcription factors such as *pax-3* in at all D/V levels of the spinal cord. The position of generation of roof plate cells, neural crest cells, and dorsal commissural neurons is, however, not significantly changed.

brafish embryos in which notochord formation is perturbed or prevented (Bovolenta and Dodd, 1991; Weinstein et al., 1994; Ang and Rossant, 1994; Talbot et al., 1995). Notochord-derived signals can also induce ventral cell types in dorsal regions of the neural tube at the expense of dorsal cell differentiation (van Straaten et al., 1988; Yamada et al., 1991; Basler et al., 1993; Monsoro-Burq et al., 1994; Liem et al., 1995).

The differentiation of both floor plate cells and motoneurons can also be induced by the notochord in neural plate explants in vitro (Placzek et al., 1993; Yamada et al., 1993; Tanabe et al., 1995; Marti et al., 1995b). Thus an inductive signal from the notochord is sufficient to initiate the differentiation of these two ventral cell types. In vitro assays have revealed two additional features of this ventral induction process. First, the induction of floor plate differentiation by the notochord requires cell contact whereas the induction of motoneuron differentiation can be achieved in a contact-independent manner by a notochord-derived diffusible factor (Placzek et al., 1990, 1993; Yamada et al., 1993; Tanabe et al., 1995). Second, floor plate cells, although dependent on a notochord-derived inductive signal, subsequently acquire inductive activities indistinguishable from those of the notochord itself (Hatta et al., 1991; Yamada et al., 1991; Placzek et al., 1993). Soon after neural tube closure, the notochord is displaced ventrally and is separated from the ventral neural tube by sclerotomal cells. It is likely, therefore, that the notochord is responsible for the early steps in the patterning of cell types in the ventral neural tube, but that this ventralization process is perpetuated by floor-plate–derived signals.

Sonic Hedgehog as a Mediator of Ventral Inductive Signaling

The leading candidates as mediators of notochord-derived ventralizing inductive signals are members of the vertebrate hedgehog protein family, a class of proteins that has also been implicated in many other tissue-inductive interactions (Johnson and Tabin, 1995). In higher vertebrates, the notochord and floor plate express a single member of this gene family, *Sonic hedgehog (Shh),* over the period that these two midline cell groups exhibit their inductive inference (Krauss et al., 1993; Echelard et al., 1993; Riddle et al., 1993; Roelink et al., 1994; Chang et al., 1994). Misexpression experiments in vivo have shown that *Shh* can elicit ectopic floor plate differentiation (Riddle et al., 1993; Krauss et al., 1993; Echelard et al., 1993; Roelink et al., 1994; Ruiz i Altaba et al., 1995a; Goodrich et al., 1996). In vitro assays have provided evidence that Sonic Hedgehog protein (SHH) mediates both floor plate and motoneuron differentiation (Roelink et al., 1995; Tanabe et al., 1995; Marti et al., 1995b). Moreover, cell lines transfected with *Shh* mimic the ability of the notochord to induce the differentiation of floor plate cells in a contact-dependent manner and the differentiation of motoneurons in a contact-independent manner (Roelink et al., 1994; Tanabe et al., 1995).

The decision of progenitor cells in neural plate explants to differentiate into floor plate cells or motoneurons appears to be influenced by the concentration of SHH to which they are exposed. The concentration threshold of SHH for motor neuron induction ($\sim 10^{-9}$ M or greater) in vitro is about fivefold lower than that required for floor plate differentiation (Roelink et al., 1995). Furthermore, exposure of neural plate explants to high SHH concentrations (10^{-8} M) induces virtually all neural plate cells to differentiate into floor plate cells (Roelink et al., 1995). These observations have raised the possibility that the identity of cell types in the ventral neural tube and the position at which they differentiate are controlled by SHH actions at distinct concentration thresholds.

Biochemical studies on HH proteins have begun to provide insight into the mechanisms by which cells located at different M/L positions within the neural plate could be exposed to different concentrations of SHH. As with other HH proteins, SHH is synthesized as precursor of ~45 kD which is subsequently cleaved to generate a ~20-kD amino terminal product and a ~29-kD carboxy-terminal product (Lee et al., 1994; Porter et al., 1995; Bumcrot et al., 1995). This internal cleavage event is autoproteolytic and is thought to be dependent on a serine-protease–like activity resident in the carboxy-terminal domain of the precursor protein (Lee et al., 1994). Cleavage occurs at a highly conserved Gly-Cys-Phe sequence that is similar to the cleavage site of self-splicing proteins (Koonin, 1995; Porter et al., 1995). The cellular localization of the two HH cleavage products differs: The vast majority of the amino-terminal product (SHH-N) is retained on the cell surface whereas the carboxy-terminal product (SHH-C) is freely diffusible upon secretion (Lee et al., 1994; Bumcrot et al., 1995; Roelink et al., 1995). Importantly, all the activities of SHH, including the ability to induce floor plate and motoneuron differentiation, are mediated by SHH-N (Johnson and Tabin, 1995; Roelink et al., 1995; Marti et al., 1995a; Hynes et al., 1995b). The carboxy-terminal domain of SHH is required to generate SHH-N and also appears to be required for the tethering SHH-N to the cell surface (Porter et al., 1995; Roelink et al., 1995; Bumcrot et al., 1995), through the addition of a cholesterol linkage (Porter et al., 1996).

These biochemical studies suggest a model that could explain how the proteolytic processing of SHH by notochord cells establishes marked differences in the concentration of SHH at varying distances from the notochord (Fig. 8–3). SHH precursor protein synthesized by the notochord appears to be cleaved autoproteolytically to generate both the SHH-N and SHH-C products. Once cleaved, SHH-C is free to diffuse away from the notochord but has no intrinsic inductive activity whereas most of the biologically active SHH-N remains attached to the surface of notochord cells. Thus, cells at the midline of the neural plate that are contacted by the notochord are likely to be exposed to a high local concentration of SHH-N, exceeding the threshold for floor plate differentiation. The low level of SHH-N that is able to diffuse from the notochord could expose cells in more lateral regions of the neural plate to a concentration of SHH-N sufficient to initiate motoneuron differentiation but below the threshold for floor plate induction. Although SHH-N mimics the known inductive activities of the notochord, it remains to be established that the activity of the protein is necessary for the induction of ventral cell types. A requirement for SHH in ventral inductive processes has been suggested by in vitro assays in which antibodies to SHH-N block the notochord-mediated induction of floor plate cells and motoneurons (Marti et al., 1995b; Ericson et al. 1996).

A floor plate induction by surface-associated SHH-N

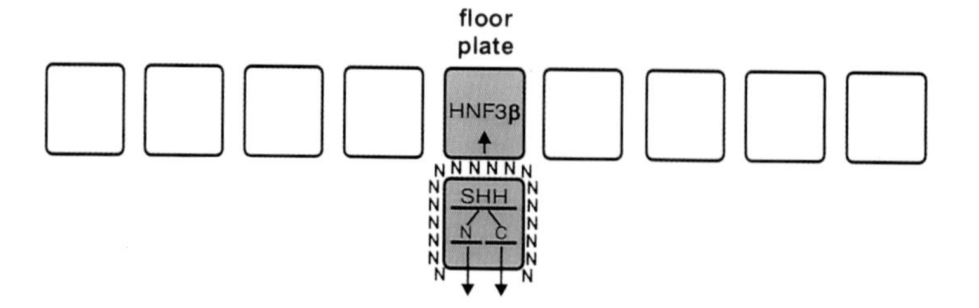

B i motor neuron induction by diffusible SHH-N

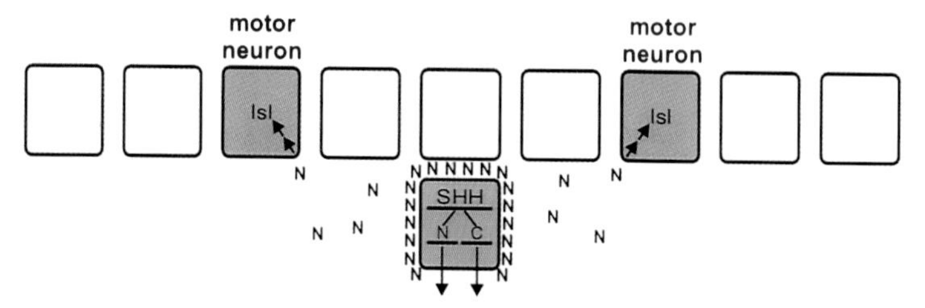

ii motor neuron induction by SHH-N dependent synthesis of a distinct factor

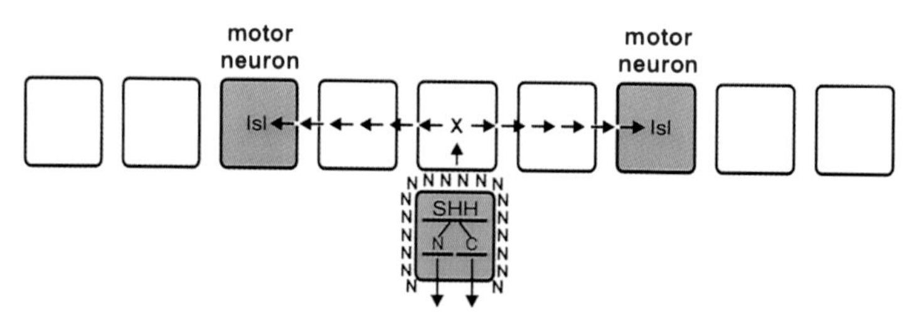

FIGURE 8–3 **Pathways of floor plate and motoneuron differentiation in response to the amino-terminal autoproteolytic product of SHH.** (A) Floor plate induction by the amino-terminal cleavage product of SHH (SHH-N). In this scheme, SHH synthesized by the notochord undergoes autoproteolytic cleavage to generate amino (SHH-N)- and carboxy (SHH-C)-terminal cleavage products. SHH-N is secreted but the vast majority of the protein is retained on the surface of notochord cells. The association of SHH-N with the cell surface generates a high local concentration that is above a threshold for induction of HNF3β and floor plate differentiation in overlying neural plate cells. (B) Two possible pathways by which SHH could induce motoneuron differentiation. (i) This scheme shows that a small proportion of SHH-N diffuses away from the notochord, generating a low concentration that is sufficient to induce motoneuron differentiation in lateral neural plate cells but is below the threshold for floor plate induction. The steps involved in the commitment of neural plate cells to a motoneuron fate are not known. The important element in this scheme is that SHH-N itself exerts both local and long-range inductive actions. (ii) An alternative scheme in which SHH-N induces the synthesis of an intermediary secreted factor (X) that diffuses away from the midline to trigger Isl-1 expression and motoneuron differentiation in lateral neural plate cells. The critical difference between this scheme and that shown in (i) is that SHH-N exerts its entire spectrum of inductive activities through local signaling. For simplicity, the extracellular location of SHH-C is not shown. The differentiation of distinct cell types generated in the region that is interposed between the floor plate and notochord could be induced by a concentration of SHH-N or of X greater than that required to initiate motoneuron differentiation (from Roelink et al., 1995).

The number of relevant threshold responses to SHH remains unclear. The region of the neural tube interposed between floor plate cells and motoneurons gives rise to other distinct cell types (Yaginuma et al., 1990; Yamada et al., 1991; Hynes et al., 1995a; Fan et al., 1996). The induction of these ventral cell types could involve a concentration of SHH-N intermediate between that effective in inducing floor plate cells and motoneurons. SHH-N may therefore operate at several different concentration thresholds to generate distinct cell types in the ventral neural tube, although secondary inductive signals also appear to contribute to the diversification of ventral cell types, as discussed below.

A related issue is whether the differentiation of floor plate cells and motoneurons is triggered directly by SHH-N (Fig. 8–3B) or requires the induction in neural plate cells of secondary secreted factors (Fig. 8–3C). Floor plate differentiation is unlikely to require an intermediary factor. The notochord-mediated induction of the winged helix gene HNF3β, an early marker of floor plate cells (Ruiz i Altaba et al., 1993; Monaghan et al., 1993; Sasaki and Hogan, 1993), is induced in the absence of protein synthesis (Ruiz i Altaba et al., 1995a) and thus is likely to be a direct response to SHH.

Whether SHH is the sole inductive factor involved in the differentiation of motoneurons has not been resolved, in part because progenitor cells exposed to low concentrations of SHH appear to undergo at least one further round of cell division before differentiating into motoneurons and expressing markers such as Isl-1 (Leber et al., 1990; Yamada et al., 1993; Pfaff et al., 1996). Thus, it is possible that SHH secreted by the notochord does not act directly to induce motoneurons but instead acts locally on overlying neural plate cells to induce the expression of a distinct secreted factor, the synthesis of which is independent of floor plate differentiation. This remains a possibility since many of the apparent long-range patterning effects of HH in *Drosophila* are mediated by the induction of intermediary factors, notably the TGF-β–related protein dpp (Perrimon, 1995). Indeed, in some vertebrate tissues the inductive activities of SHH may be mediated by TGFβ-like proteins (Bitgood and McMahon, 1995). In the caudal neural tube, however, this seems less likely since members of the TGF-β family suppress rather than induce motoneuron differentiation (Basler et al., 1993; Pituello et al., 1995; Liem et al., 1995). Moreover, recent experiments using function blocking anti-SHH antibodies show that motor neuron progenators require SHH signaling until late in their final cell division (Ericson et al., 1996) providing strong evidence that SHH acts directly to induce motor neuron differentiation.

The biochemical events that transduce SHH signals in neural plate cells remain poorly characterized. Studies on HH signaling in *Drosophila* (Forbes et al., 1993) have indicated that HH inhibits, directly or indirectly, the activity of the Patched protein, which is likely to function as a transmembrane transporter (Ingham, 1991). The activity of a second transmembrane protein, Smoothed, also appears to be involved in the transduction of HH signals (Hooper, 1994). A cytoplasmic serine theonine kinase, Fused, is also necessary for the transduction of HH signals (Ingham, 1993; Ingham and Hidalgo, 1993). In contrast, the elimination of protein kinase A (PKA) activity mimics HH signaling and activation of PKA may inhibit HH signaling (Perrimon, 1995). These cytoplasmic phosphorylation-dependent steps appear to be transduced in the nucleus through the actions of a zinc finger transcription factor, Cubitus Interruptus (ci) (Orenic et al., 1990; Motzny and Holmgren, 1995). A conserved but paradoxical secondary consequence of HH signaling in *Drosophila* cells is the transcriptional upregulation of the *patched* gene (Forbes et al., 1993; Perrimon, 1995).

In vertebrates, several key components of the HH signal transduction pathway appear to be conserved. A vertebrate *patched* gene is expressed in the neural tube and at other sites of HH signaling and its transcript is induced by SHH (Marigo and Tabin, 1996; Goodrich et al., 1996). The involvement of *patched* in HH signaling in vertebrates may therefore be similar to that in *Drosophila*. Similarly, increasing the level of PKA activity can block the ventralizing activities of SHH in neural plate explants (Hynes et al., 1995a) and in vivo (Hammerschmidt et al., 1996). In addition, blocking PKA function mimics HH signaling (Hammerschmidt et al., 1996). Finally, vertebrate homologues of ci, the *Gli* genes, are expressed in cells that respond to HH signals (Hui et al., 1994). Moreover, mutants in one of these genes, *Gli-3*, result in defects in the development of regions of the forebrain thought to be sensitive to SHH signals (Franz, 1994). Thus, the initial steps of HH signaling in flies and vertebrate are likely to be similar.

The target genes activated in neural plate cells as a direct response to this conserved HH signal transduction pathway remain unknown. The induction of *HNF3β* by the notochord in the presence of protein synthesis inhibitors (Ruiz i Altaba et al., 1995a) provides evidence that this gene is a direct target of SHH signaling in the process of floor plate differentiation. The homeobox genes *nkx2.1* and *nkx2.2* are also induced in ventral neural tube cells by SHH (Barth and Wilson, 1995; Ericson et al., 1995), and *nkx 2.1* function appears to be required for the differentiation of ventral cell types at rostral levels of the neuraxis (Kimura et al., 1996). In contrast, expression of the homeobox genes *msx-1* and *pax-3* is rapidly extinguished in caudal neural plate cells by SHH signaling (Goulding et al., 1993; Liem et al., 1995), raising the possibility that the repression of expression of certain transcription factors by SHH is an essential step in the specification of ventral cell fates. In support of this idea, misexpression of *pax-3* in the ventral neural tube of transgenic mice appears to inhibit floor plate and motoneuron differentiation (Tremblay et al., 1996).

Intrinsic Determinants of Ventral Cell Types

The acquisition of distinct fates by cells in the ventral neural tube appears to involve the expression of cell-specific transcription factors. Studies of early stages of cell specification in the ventral neural tube have provided evidence that transcription factors of the winged helix and LIM homeodomain classes are induced by SHH and have essential functions in the specification of floor plate cell and motoneurons.

Floor plate differentiation

The winged helix transcription factor HNF3β is expressed at high levels by both notochord and floor plate cells (Monaghan et al., 1993; Ruiz i Altaba et al., 1993; Sasaki and Hogan, 1993). Gene targeting studies have shown that mice lacking

HNF3β function fail to form a notochord, indicating that the protein is required for the development of the notochord (Weinstein et al., 1994; Ang and Rossant, 1994). The neural tube of HNF3β mutant mice lacks floor plate cells and motoneurons and exhibits a uniform D/V expression of *pax-3,* a gene that is normally restricted to proliferating cells in the dorsal neural tube (Ang and Rossant, 1994). Since the absence of floor plate cells is expected simply as an indirect consequence of the failure of notochord formation, studies of HNF3β mutant mice have not yet permitted a direct test of a neural requirement of HNF3β in floor plate development. Nevertheless, experiments in which *HNF3β* and a closely related gene, *Pintallavis,* have been misexpressed in *Xenopus* and mouse embryos have shown the generation of ectopic floor plate cells in the dorsal neural tube (Ruiz i Altaba et al., 1993; Sasaki and Hogan, 1994; Hynes et al., 1995a; Ruiz i Altaba et al., 1995b), supporting the idea that these transcription factors are involved in floor plate differentiation. Thus, the expression of HNF3β is likely to be required for the differentiation of floor plate cells as well as the notochord.

Motoneuron differentiation

The first molecular indicator of motoneuron differentiation is the expression of Isl-1 (Karlsson et al., 1990; Ericson et al., 1992), a member of a family of homeodomain-containing transcription factors that possess an amino-terminal pair of zinc-binding LIM domains (Dawid et al., 1995). Isl-1 is expressed initially by all motoneurons, and its expression precedes that of other motoneuron markers, defining an early and common step in motoneuron differentiation (Ericson et al., 1992; Tsuchida et al., 1994; Pfaff et al., 1996).

Analysis of neural tube development under conditions in which Isl-1 expression has been eliminated has shown that this protein is required for the formation of motoneurons (Pfaff et al., 1996). Motoneurons are not generated in *Isl-1* mutant mice or in chick neural tube explants treated with antisense *Isl-1* oligonucleotides although floor plate cells are present and *Shh* is expressed normally. Thus, the absence of motoneurons is not the consequence of a perturbation in the midline-derived signals that trigger motoneuron differentiation; instead, it results from an impairment in the ability of cells to acquire motoneuron properties. In the absence of Isl-1, prospective motoneurons appear not to assume alternative ventral neuronal fates and instead undergo apoptosis (Pfaff et al., 1996).

Although Isl-1 function is necessary for the generation of motoneurons, its exact role in this process is not clear. In part this reflects a lack of information about the precise stage at which cells in the ventral neural tube commit to a motoneuron fate. Lineage analyses have suggested that ventral progenitor cells in the spinal cord are not committed to a motoneuron fate until close to their final cell division (Leber et al., 1990). The fate of motoneurons could be determined immediately before the final division of the progenitors, at a stage similar to that at which specific neuronal fates in the cerebral cortex are thought to be established (McConnell and Kaznowski, 1991). However, Isl-1 is expressed only after prospective motoneurons have undergone their final mitotic division (Ericson et al., 1992; Pfaff et al., 1996). Thus, if motoneuron fate is determined prior to the final division of the progenitor cell, Isl-1 is unlikely to function in this determination process and instead may have a role in the developmental progression of previously committed motoneuron progenitors. If commitment to a motoneuron fate occurs postmitotically, however, Isl-1 could be involved in the initial determination of motoneuron identity, although its widespread expression in other tissues suggests that Isl-1 alone is insufficient to commit cells to a motoneuron fate.

Specification of Interneuron Fates in the Ventral Neural Tube

Several distinct classes of interneurons are generated in the ventral half of the neural tube and can be distinguished by expression of the homeobox genes *En-1, Lim-3, Gsh-4,* and *Lim-2* (Davis and Joyner, 1988; Tsuchida et al., 1994; Li et al., 1994; Zhadanov et al., 1995). These distinct classes of ventral interneurons could be generated independently of SHH signals, as is the case for Lim-2 interneurons (Weinstein et al., 1994) as a response of neural progenitor cells to low concentrations of SHH in a process similar to that implicated in the induction of motoneurons, or in response to secondary inductive signals emanating from ventral cell types that are induced as a primary response to notochord-derived signals.

Evidence that a cascade of inductive signals is involved in the specification of certain ventral interneurons has emerged from analysis of the class of interneurons that express Engrailed-1 (En-1) (Pfaff et al., 1996). En-1 interneurons are first detected immediately dorsal to the motoneuron population and differentiate soon after the first motoneurons. In Isl-1 mutant mice these interneurons are absent from both the spinal cord and hindbrain (Pfaff et al., 1996), indicating that the differentiation of En-1 interneurons is dependent on the generation of motoneurons. Consistent with this idea, the differentiation of En-1$^+$ interneurons can be restored in neural tissue isolated from Isl-1 mutant mice by a signal derived selectively from the ventrolateral region of the neural tube of normal embryos, the region that contains motoneurons. En-1$^+$ interneurons can also be induced in naive chick neural plate explants by ventrolateral neural tube tissue. Taken together, these findings suggest that a signal provided by motoneurons themselves, or by a cell type that is itself dependent on motoneurons, is required for the differentiation of En-1 interneurons (Fig. 8–4).

The dependence of En-1 interneuron differentiation on the generation of motoneurons suggests that the diversification of neuronal subclasses in the ventral neural tube involves serial inductive interactions. Motoneurons appear to be generated as a primary response to SHH, in the sense that their formation is independent of the generation of any other definitive neural cell type, and in turn appear to provide a secondary signal that recruits interneurons. The ability of naive but not SHH-exposed progenitors to generate En-1 interneurons might explain why En-1$^+$ interneurons are generated only in the region of the neural tube dorsal to the motoneuron population. It is nevertheless possible that progenitor cells that have been ventralized by SHH do respond to motoneuron-derived signals, but with the generation of classes of ventral interneurons distinct from the En-1 population. Moreover, the dependence of certain ventral interneurons on motoneuron-derived signals does not exclude the possibility that other classes of ventral interneurons are induced as a primary response to SHH.

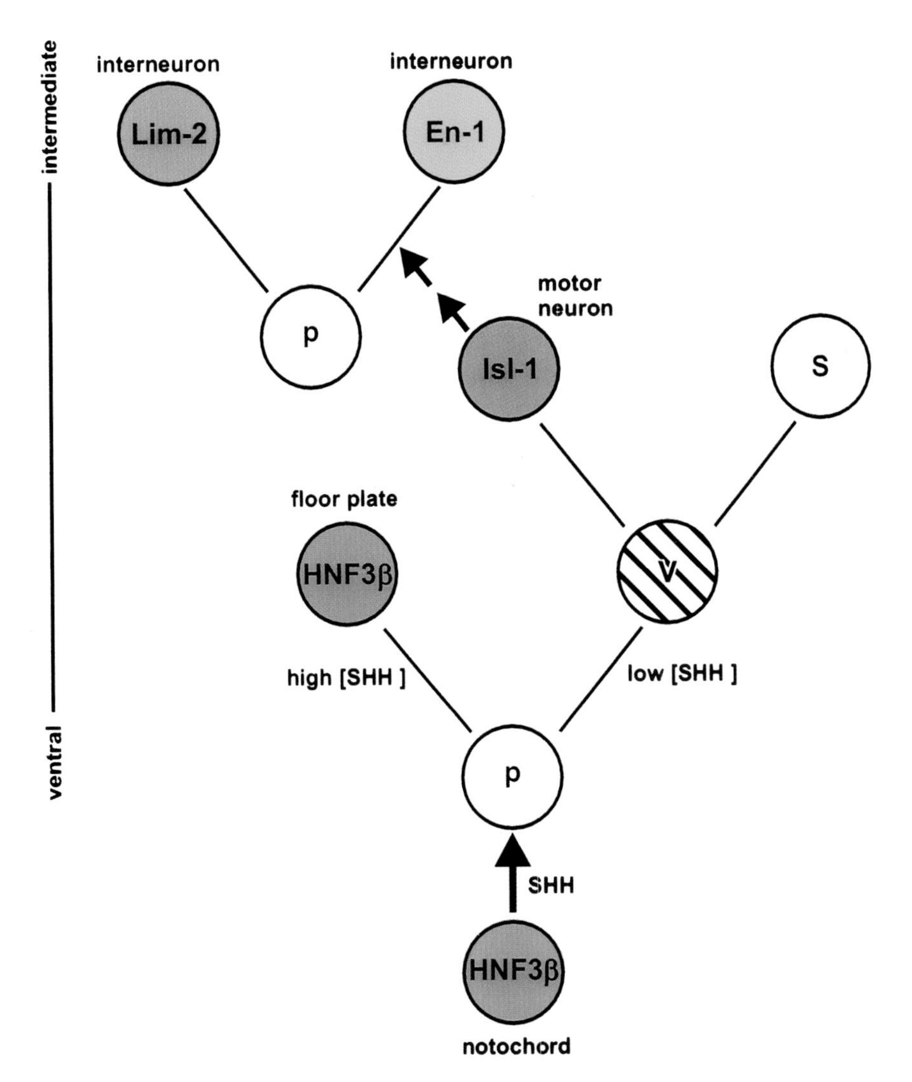

FIGURE 8–4 Sequential inductive interactions control the differentiation of motoneurons and En-1 interneurons. The model outlines the possible steps involved in the generation of floor plate cells, motoneurons, and En-1⁺ interneurons in the ventral neural tube. Floor plate differentiation results from the exposure of naive (*Nkx2.2*⁻, *Msx-1*⁺, and *Pax-3*⁺) neural plate progenitor cells (P) to a high concentration of SHH-N, inducing the expression of HNF3β and consequent floor plate differentiation. Exposure of progenitor cells to a lower concentration of SHH-N results in the generation of one or more ventralized (V) (*Nkx2.2*⁺/⁻, *Msx-1*⁻, and *Pax-3*⁻) progenitor cell types, which retain the capacity for further cell division. The progeny of ventralized cells include motoneurons and other cell types (sibling cell, S). Motoneuron differentiation is accompanied by and requires expression of Isl-1. Motoneurons may then be the source of a secondary inductive signal (double arrows) that acts directly or indirectly on nearby progenitor cells to induce the differentiation of En-1⁺ interneurons. Alternatively, the differentiation of En-1⁺ interneurons could involve a signal from an intervening cell type that is itself dependent on motoneurons (not shown). Naive neural plate progenitor cells give rise to Lim-1⁺/Lim-2⁺ interneurons in the absence of SHH-mediated and motoneuron-dependent signals. The model predicts that ventralized cells do not give rise to En-1⁺ interneurons in response to a motoneuron-derived signal, but they may respond to this signal with the generation of other classes of ventral interneurons. From Pfaff et al. (1996).

CONTROL OF CELL FATE IN THE DORSAL NEURAL TUBE

The cell types that populate the dorsal half of the neural tube—roof plate cells, neural crest cells, and dorsal sensory relay interneurons—derive from progenitor cells located in the lateral half of the neural plate. The differentiation of dorsal cell types in vivo occurs independently of notochord-derived signals (Yamada et al., 1991; Artinger and Bronner-Fraser, 1992; Monsoro-Burq et al., 1994), indicating that dorsal and ventral cell fates are determined by distinct pathways. Neural plate cells could acquire dorsal fates by default, in the sense that naive progenitors are predisposed to differentiate into dorsal cell types unless exposed to a ventralizing signal from the notochord. Alternatively, the generation of dorsal cell types could require exposure of progenitor cells in the neural plate to a distinct dorsalizing inductive signal.

Ectodermal Signals and the Induction of Dorsal Cell Fates

The analysis of cell differentiation in the caudal neural plate and neural tube has suggested that certain of the molecular characteristics of dorsal cell types are acquired by default (Goulding et al., 1993; Liem et al., 1995). At early stages of neural plate formation or under experimental conditions in which neural plate cells have not been exposed to a ventralizing signal, two transcription factors, *pax-3* and *msx-1*, are expressed uniformly along the M/L axis of the neural plate. At later stages in normal embryos these two genes are restricted to cells in the lateral neural plate and dorsal neural tube (Goudling et al., 1993; Liem et al., 1995). The repression of expression of *pax-3* and *msx-1* by cells in the medial region of the neural plate appears to be achieved by a SHH-mediated signal from the notochord (Goulding et al., 1993; Liem et al., 1995). These results show that the expression of certain transcription factors characteristic of cells in the dorsal neural tube appears to be a property inherited from neural plate cells in the absence of additional polarizing signals.

Despite this, definitive dorsal cell fates are not acquired by default and require dorsalizing signals (Liem et al., 1995). Cells of the epidermal ectoderm that flank the neural plate are the source of a contact-dependent dorsalizing signal that induces the differentiation of both roof plate cells and neural crest cells (Moury and Jacobson, 1990; Liem et al., 1995; Sechrist et al., 1995; Dickinson et al., 1995; Selleck and Bronner-Fraser, 1995; Liem et al., 1996). At early neural plate stages, when the epidermal ectoderm and neural plate are initially contiguous, this dorsalizing signal is likely to be transmitted through the plane of the epithelium. As the neural plate folds, however, the basal surface of the epidermal ectoderm contacts an extended region of the dorsal neural epithelium, and thus its range of action could be increased by its transmission across this interface.

BMPs as Dorsalizing Inductive Signals

The early dorsalizing activity of the epidermal ectoderm appears to be attributable to members of the bone morphogenetic protein (BMP) family (Kingsley, 1994) of TGF-β–like proteins (Fig. 8–5). In chick, both *BMP4* and *BMP7* are ex-

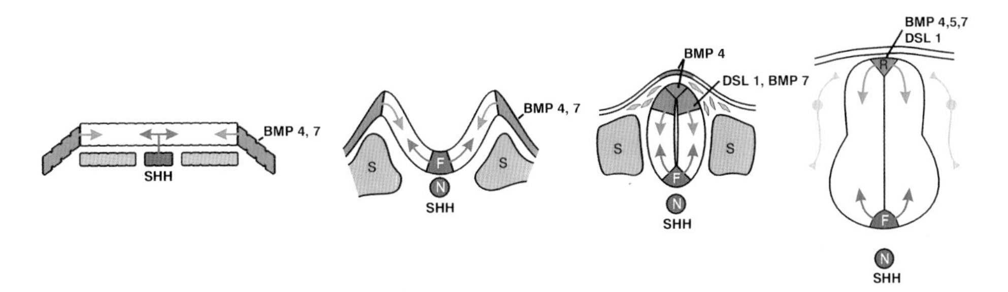

FIGURE 8–5 Transfer of inductive signals from non-neural to neural cell types. The diagram shows the source of ventralizing (SHH, blue) and dorsalizing (BMPs, orange) inductive signals at sequential stages of neural development. BMPs originate in the epidermal ectoderm flanking the lateral edges of the neural plate and SHH is initially expressed in the axial mesoderm. At neural fold stages, SHH begins to be expressed by floor plate cells at the midline of the neural plate and BMPs by cells in the dorsal tips of the neural folds. Soon after neural tube closure, BMPs are expressed in the roof plate and adjacent dorsal region of the neural tube and SHH expression is maintained by floor plate cells. BMP expression is lost from the ectoderm except at the dorsal midline. At the time of neuronal differentiation, BMP expression persists in the roof plate and SHH expression is maintained in the floor plate.

pressed by epidermal ectoderm cells that flank the neural plate and both proteins mimic the ability of the epidermal ectoderm to induce roof plate and neural crest cells (Liem et al., 1995, 1996). Numerous other secreted factors fail to mimic this activity and thus BMPs or related TGF-β–like proteins are currently the only candidates as mediators of ectodermally derived dorsalizing signals. Demonstration of the loss of dorsal cell differentiation after blockade of BMP-mediated signals is, however, necessary to establish that BMPs are required for the specification of dorsal cell fates.

Although the initial source of the dorsalizing signals involved in neural tube patterning appears to be the epidermal ectoderm, four members of the BMP family—*BMP4, BMP5, BMP7,* and *Dsll*—are induced in dorsal midline neural tube cells as a component of the program of roof plate differentiation (Basler et al., 1993; Liem et al., 1995, 1996). Thus, at stages after the separation of the overlying ectoderm, the early short-range dorsalizing signal from the epidermal ectoderm appears to be propagated with the dorsal neural tube through the actions of BMPs expressed by the roof plate. The strategy of using homeogenetic induction to transfer a dorsalizing signal from the epidermal ectoderm to the dorsal midline of the neural tube is similar to that used to propagate ventralizing signals, through the inductive transfer of SHH expression from the notochord to the floor plate (Placzek et al., 1993; Placzek, 1995). Roof-plate– derived BMPs appear to have a role in promoting the differentiation of sensory relay interneurons in the dorsolateral spinal cord (Liem et al., 1996). Thus, BMP-mediated signals provided initially by the epidermal ectoderm and later by the roof plate are likely to induce most and perhaps all definitive dorsal cell types.

The dorsalizing activity of BMPs leaves unresolved the question of how several distinct cell types are generated in response to a qualitatively similar inductive signal. One possibility, raised by analogy with the mechanisms of SHH-mediated signaling ventrally, is that different dorsal cell types are triggered at different BMP concentration thresholds. However, in vitro induction assays indicate that the differentiation of roof plate, neural crest cells, and dorsal sensory relay neuronal fates is achieved at the same BMP concentration threshold (Liem et al., 1997). A second possibility, suggested by the expression of several BMPs in the epidermal ectoderm and dorsal neural tube, is that distinct homodimeric or heterodimeric BMPs (Aono et al., 1995) or other TGF-β family members exhibit qualitatively distinct inductive activities through actions on different BMP receptors. A third possibility is suggested by the observation that roof plate cells and neural crest cells are generated at around the time of neural tube closure, in contrast to dorsal sensory interneurons which differentiate at much later stages (Oppenheim et al., 1988; Liem et al., 1996). Differences in the time at which neural cells are exposed to BMPs could therefore influence the identity of the dorsal cell types that are generated. In support of this latter possibility, in vitro studies have shown that the exposure of newly formed neural plate cells to BMPs derived from the epidermal ectoderm leads to neural crest differentiation whereas naive progenitor cells that have been matured in vitro differentiate into dorsal sensory relay interneurons but not into neural crest cells in response to the same BMP signal (Liem et al., 1997).

Intrinsic Determinants of Dorsal Cell Fates

Several transcription factors have been identified in subsets of cells in the dorsal neural tube but their roles in the specification of dorsal cell fates remain poorly characterized. The paired homeobox genes *pax-3* and *pax-7* and the *msx-1–3* genes are expressed in broad domains in proliferating cells in the dorsal neural tube (Davidson and Hill, 1991; Stuart et al., 1994; Liem et al., 1995; Shimeld et al., 1996). Mutations in the mouse *pax-3* and *pax-7* genes are associated with an impairment in neural crest cell differentiation (Stuart et al., 1994; Mansouri et al., 1996). The mouse *open-brain* mutation also results in a marked impairment in cell differentiation in the dorsal neural tube (Gunther et al., 1994), although the molecular basis of this patterning defect remains to be defined.

Distinct classes of dorsal cells can also be distinguished by the expression of transcription factors. Roof plate cells express the LIM homeodomain protein Lmx-1 (Riddle et al., 1995), neural crest cells express the zinc finger transcription factor slug (Nieto et al., 1994), certain dorsal commissural neurons express the LIM homeodomain proteins LH2A and LH2B (Liem et al., 1997), and dorsal association neurons express Isl-1 (Ericson et al., 1992; Liem et al., 1995). Antisense oligonucleotides directed against *slug* have been shown to perturb the delamination of premigratory neural crest cells in chick embryos (Nieto et al., 1994), providing evidence that this protein is required for appropriate neural crest differentiation. The function of other transcription factors expressed by distinct dorsal cell types has not been examined.

Taken together, these studies on cell differentiation in the dorsal neural tube support the idea that the initial polarization of the neural plate along its M/L axis and the later generation of distinct cell types along the D/V axis of the neural tube are controlled by the coordinate actions of a SHH-mediated ventralizing signal from the notochord and floor plate and a BMP-mediated dorsalizing signal from the epidermal ectoderm and roof plate (Fig. 8–5). SHH is likely to represent a major factor in inducing ventral cell types, conferring the early dorsal restriction in expression of *pax* and *msx* genes and limiting the domain of the neural tube within which the differentiation of definitive dorsal cell types can occur. The maintenance of dorsal cell differentiation in lateral regions of the neural plate in the face of long-range ventralizing signals might depend upon the ability of ectodermally derived BMPs to oppose the actions of SHH-mediated signals. Since SHH can suppress dorsal cell differentiation and BMPs can suppress ventral cell differentiation, the fate of cells in the neural plate appears to depend on whether they are exposed to SHH or to BMPs and on the concentration and timing of exposure to these factors. In medial regions of the neural plate, SHH-mediated signals appear dominant whereas in lateral regions, the influence of BMPs prevails. It remains unclear how cell types that are generated from cells in the intermediate region of the neural plate (Rangini et al., 1991; Lu et al., 1992; Riddle et al., 1995) acquire their distinct fates.

Control of Cell Identity Along the Anteroposterior Axis of the Neural Tube

The early inductive interactions that establish D/V cell fate appear to be conserved in the spinal cord and hindbrain. However, at a constant D/V position there are marked differences in the identity of neural cell types at different A/P levels. Motoneurons represent one prominent group of neurons that exhibit A/P positional variation in subtype identity. In addition, neural crest cells emigrate from the dorsal neural tube in a discontinuous pattern at different A/P levels of the hindbrain (Lumsden et al., 1991). The greatest progress in analyzing how variations in cell identity and fate are established along the A/P axis has been made in the hindbrain (Guthrie, 1996), and this issue is addressed below. We also discuss the evidence that A/P positional differences in motoneuron identity exist in the spinal cord and how they might be established.

Establishment of Segmental Fate in the Hindbrain

At hindbrain levels the neural tube is initially cylindrical but becomes progressively subdivided along its length by constrictions that form a series of eight varicosities, termed rhombomeres (Fig. 8–6) (Graper, 1913; Vaage, 1969; Lumsden, 1990). The segmental pattern of cell types and axonal pathways evident in the chick embryo hindbrain emerges soon after neural tube closure and is virtually complete at the onset of neuronal differentiation. Rhombomeres appear to function as lineally restricted compartments, with little or no mixing of cells across compartment interfaces (Fig. 8–6) (Fraser et al., 1990). Developmental compartments were originally defined in insect larvae and imaginal discs and provide a way of allocating blocks of cells that have distinct properties, permitting each block a degree of autonomy during the period of cell specification (Garcia-Bellido et al., 1973; Lawrence, 1989). In the hindbrain, the formation of

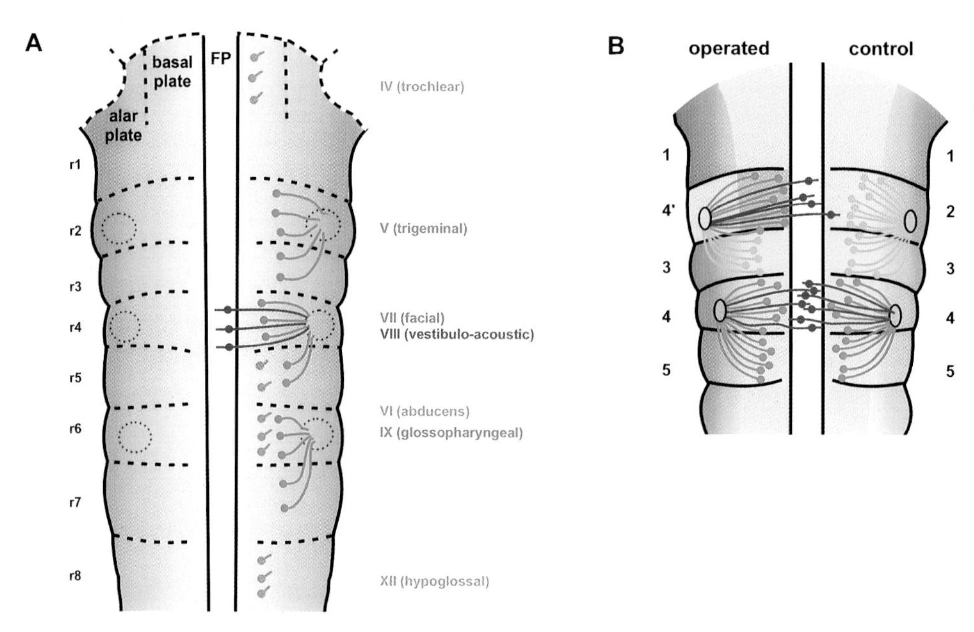

FIGURE 8–6 Compartmental organization of rhombomeres. At stage 8–9, the early expression of transcription factor (Krox20) and Eph kinases delineates distinct rhombomeric subdivisions of the hindbrain in an alternating "pair-rule"-like pattern. At stage 9–10 there are clear restrictions to cell mixing across rhombomere boundaries, which may reflect differences in the adhesive properties of cells in adjacent rhombomeres. At stage 13 and onward, molecular and morphological specializations at interrhombomeric boundaries (b) appear. These include transcription factors (Cook et al., 1995, PLZF), signaling molecules (Mahmood et al., 1995, FGF3), cytoskeletal components (Heyman et al., 1993, vimentin), cell-surface molecules (Lumsden and Keynes, 1989, NCAM-PSA, PNA binding), and extracellular matrix components (Heyman et al., 1993, laminin, chondroitin sulphate proteoglycans—CSPG).

compartments also provides a potential mechanism by which cells in adjacent rhombomeres may interact with each other to establish additional cell states at the boundaries.

The earliest neurons in the hindbrain are laid out in stripes with neuronal differentiation and axonogenesis starting within the confines of alternate, even-numbered rhombomeres, and only appearing later in the odd-numbered rhombomeres. In this way, a "two-segment repeat" pattern is generated which encompasses all of the early forming neuronal systems in the hindbrain. Soon after neural tube closure, two levels of organization can be distinguished in the neuronal pattern of the segmented hindbrain—one involving neurons of the reticular formation and the other involving motoneurons. Eight identified types of reticular neurons are repeated as "segmental homologues" through sequential rhombomeres such that each rhombomere contains a complete set of neurons (Clarke and Lumsden, 1993). Later in development, local variations are superimposed on this basic segmental theme as certain cell types become more numerous in particular rhombomeres, suggesting the differential production and/or selective elimination of some cell types.

Motoneurons also develop in each rhombomere but, by contrast to the reticular neurons, they are organized from the outset into discrete groups in different rhombomeres (Fig. 8–7A) (Lumsden, 1990). The branchial motor nerves emerge from the hindbrain through focal exit points in the lateral wall of even-numbered rhombomeres and grow towards the branchiomeric muscles of the jaws

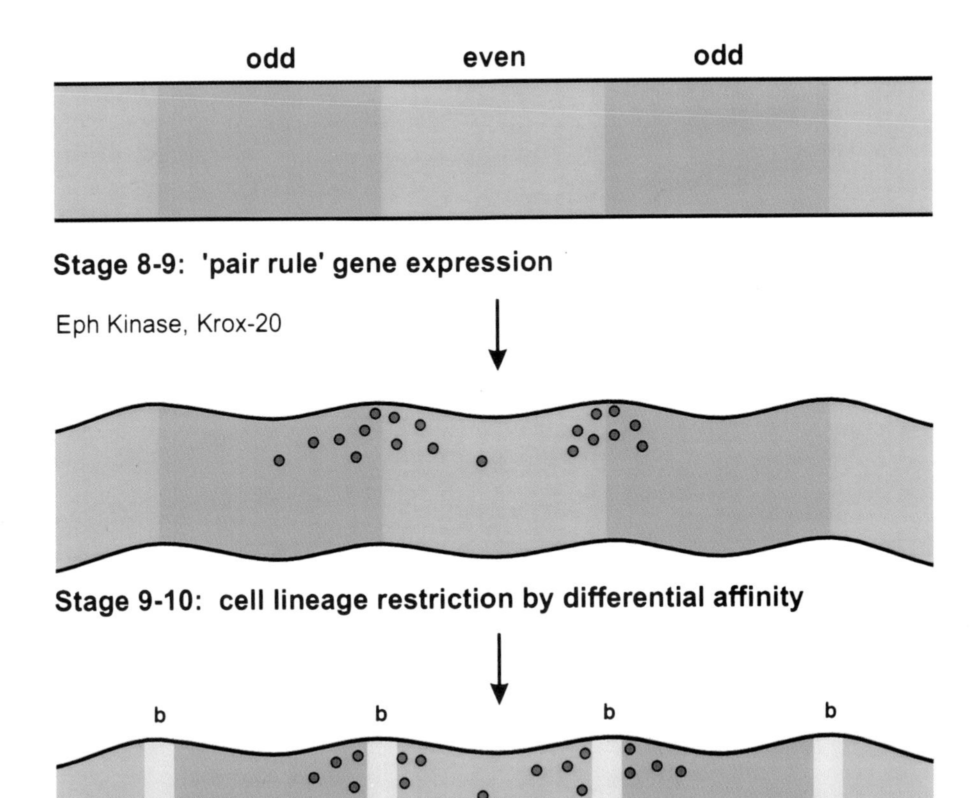

FIGURE 8–7 **Patterning of motoneurons along the A/P axis of the hindbrain.** (A) Diagram showing the distribution of neuronal types in the chick hindbrain at stage 17–20. On the right side of the figure are shown the branchiomotor neurons (green), forming in the basal plate (B) of r2+r3 (Vth nerve, trigeminal), r4+r5 (VIIth nerve, facial), and r6+r7 (IXth nerve, glossopharyngeal). Also shown are the contralaterally migrating efferent neurons (red) of the VIIIth nerve (vestibulo-acoustic), which are located in the floor plate (F) at the r4 level. Somatic motoneurons (orange) form in r1 (IVth nerve, trochlear), r5+r6 (VIth nerve, abducens), and r8 (XIIth nerve, hypoglossal). Cranial-nerve entry–exit points associated with r2, r4, and r6 are shown as dotted circles. (B) Summary showing the result of grafting a right-side r4 in place of the left-side r2, with 180° dorsoventral inversion of the transplant (r4′). The basal plate is shown in yellow; 1–2 days after grafting at stage 9–10, facial motoneurons (dark green cell bodies) and contralateral vestibuulo-acoustic efferent neurons (red cell bodies), both characteristic of r4 but not r2, develop at dorsoventral positions in the graft that are normal for the host. Trigeminal neurons in r2 and r3 are shown in light green. Adapted from Simon et al. (1995).

and pharynx: these nerves extend from neurons contained precisely within two adjacent rhombomeres. Trigeminal neurons (cranial nerve (cn) V), for example, are located in rhombomere (r)2 and r3 and facial motor neurons (cn VII) lie in r4 and r5. Somatic motor nerves to the eye muscles and tongue (cn IV, VI and XII) have a one- or two-segment origin, whereas the efferent nerve to the hair cells of the inner ear (c.n. VIII) develops within a single rhombomere, r4 (Fig. 8–7A) (Simon and Lumsden, 1993).

The segmental disposition of motonuclei in the early hindbrain of the chick embryo bears a close anatomical and functional correlation with target structures with the segmental series of branchial arches that lie beneath the hindbrain. Thus, there is an obvious anatomical relationship between the successive branchiomotor nerves and the muscle plates of successive branchial arches (Lumsden and Keynes, 1989). The early rhythmic activities in the cranial nerves have also revealed the existence of a segmental organization of rhythm-generating neurons associated with the efferent nuclei (Fortin et al., 1995). The rhombomere-specific pattern of efferent neurons is superimposed on the reiterated pattern of reticular cells. Later in development, these segmental origins become obscured as the motonuclei and other rhombomerically derived cell groups condense and migrate to new positions. Thus, in the hindbrain as in the insect, segmentation is a developmental mechanism for specifying the pattern of developing structures, not necessarily for deploying these structures in the adult.

Compartment-like Properties of Rhombomeres

The containment of groups of neuroepithelial cells within rhombomeres has been shown by lineage tracing studies in the chick. The clonal descendants of single marked cells disperse widely within the epithelium, mixing with unmarked cells. Yet, provided the ancestral cell is marked after rhombomeres appear, the spreading clone always remains within a single rhombomere, confined at the interfaces with neighboring rhombomeres (Fig. 8–6) (Fraser et al., 1990). This restriction to cell mingling persists up to at least embryonic day 3 (Birgbauer and Fraser, 1994), suggesting that the critical period for initial precursor specification persists for about 24 hours. Later, young neurons may escape this rhombomeric containment as they differentiate and migrate into the mantle layer, presumably having already acquired specification with respect to their ultimate position. Long-term marking studies using the chick/quail chimera technique have, however, shown that mitotic precursor cells remain confined within rhombomeric domains of the ventricular zone up to late stages (day 10) of embryonic development, when hindbrain neurogenesis is nearing completion (Wingate and Lumsden, 1996).

What cellular mechanism is responsible for segregating cells into compartments? One possibility is that rhombomere boundaries enforce the separation by forming a mechanical barrier to cell dispersal. Later in development, numerous molecules are expressed at rhombomeric boundaries which also become selective conduits for axon growth; however, all of these boundary molecules appear too late to be implicated in the lineage restriction (Fig. 8–6) (Lumsden, 1990;

Heyman et al., 1993, 1995). An alternative possibility is that cells in one rhombomere have a state different from that of the cells in neighboring rhombomeres and that this involves the expression of cell-surface adhesion or repulsion molecules. The interfaces apparent between rhombomeres might therefore be established by the tendency of cells in adjacent rhombomeres to separate from one another. Consistent with this idea, a network of enlarged intercellular spaces exists at the boundary interfaces at the time that rhombomeres become defined by restriction to cell mixing (Heyman et al., 1993). Furthermore, heterotopic grafting experiments (Guthrie and Lumsden, 1991; Guthrie et al., 1993) together with in vitro cell aggregation experiments using vital dye-labeled cells (Wizenmann and Lumsden, 1997), have suggested that rhombomeres partition from each other according to an adhesive differential that obeys a two-segment repeat rule. Thus, cells from even-numbered rhombomeres mix with cells from other evens, and cells from odd-numbered rhombomeres mix with other odds, but odd and even cells segregate when mixed together in cultured aggregates in vitro.

Segmentation of the vertebrate hindbrain bears more than a passing resemblance to segmentation in the fly embryo, where a hierarchy of gap genes, pair-rule genes, and segment polarity genes partition the axis into parasegments, individual anterior and posterior compartments, and then segments. Odd/even pairs of rhombomeres (e.g., r3/r4) may correspond to parasegments, whereas even/odd pairs (r4/r5) may correspond to segments. Although a number of *Drosophila* segmentation genes have vertebrate homologues, these are generally not expressed in a segmental manner in the hindbrain. This may reflect the fact that segmentation in arthropods and vertebrates is convergent: the two organisms having diverged from an unsegmented bilaterian ancestor in which the homeotic genes specified position along the A/P axis without the constraint of compartmentalization (McGinnis and Krumlauf, 1992). Thus, genes involved in the segmentation process in flies need not have the same role in vertebrates.

Several candidate genes that appear to function in the regulatory network that controls hindbrain segmentation have, however, been identified. Notable amongst these is Krox20, a zinc finger transcription factor that is expressed in two sharp stripes in the neural plate that later become r3 and r5 (Wilkinson et al., 1989). Targeted disruption of *Krox20* results in the elimination of these rhombomeres (Schneider-Maunoury et al., 1993) and the fusion of r2-r4-r6 into a single composite region (Schneider-Maunoury, Charnay, Lumsden, in preparation). This phenotype has similarities with that of pair-rule mutants in flies, suggesting that Krox20 may similarly be responsible for generating periodicity along the axis from nonperiodic cues established by upstream genes. Also in a close parallel with fly development, Krox20 exerts downstream control on *Hox* genes. Krox20-binding sites have been found in the enhancer regions of both *Hoxb-2* and *Hoxa-2* that impose r3 and r5 expression on reporter constructs in transgenic mice (Sham et al., 1993; Nonchev et al., 1996).

A second category of genes with a possible role in hindbrain segmentation are those encoding molecules involved in cell–cell interaction. Members of the Eph family of receptor tyrosine kinases, together with certain of their ligands, display rhombomere-restricted expression in alternate rhombomeres, suggesting a function in recognition and the establishment of adhesion differentials between adjacent rhombomeres (Nieto et al., 1992). *Sek-1,* for example, is expressed in r3 and r5 territories in the presegmental hindbrain. Perturbing Sek function in zebrafish and *Xenopus* embryos by overexpressing a presumed dominant negative mutant form of the protein leads to failure to establish sharp interrhombomere boundaries, as judged by irregularly bounded *Krox20* expression domains (Xu et al., 1995).

Hox *Genes Encode Positional Value Along the A/P Axis*

Superimposed on the alternate repeat identity of rhombomeres, the organization of motoneurons in discrete classes in different rhombomeres suggests that each rhombomere also has a unique identity. It is therefore reasonable to consider rhombomeres as units of cell specification and to ask how their individual identity is conferred. The clustered homeobox-containing genes of the *Hox* family are candidates for this role (McGinnis and Krumlauf, 1992), on the basis of their appropriate spatiotemporal expression and their sequence similarity with the *HOM-C* homeotic genes of *Drosophila,* which control parasegment identity in the fly. *Hox* genes are expressed in overlapping, or nested, domains along the axis of the early vertebrate embryo, those at the 3′ ends of the clusters being expressed in the hindbrain, where their anterior expression boundaries coincide with the interfaces between rhombomeres (Wilkinson et al., 1989).

The overlapping distribution of *Hox* transcripts in the hindbrain suggests that their proteins, expressed within the confines of a particular rhombomere, act in a combinatorial manner to set the positional value of individual rhombomeres, and thereby control their identity and phenotypic specializations. The identity of r4, for example, may be conferred by the expression of *Hoxa-1, Hoxb-1, Hoxa-2,* and *Hoxb-2.* The best-characterized of these genes is *Hoxb-1,* the high-level expression of which is confined to r4. Expression is strongly upregulated soon after the rhombomere becomes defined by its boundaries (Marshall et al., 1992). The expression of *Hoxb-1* is stabilized from the developmental stage at which regional identity becomes established. Thus, transplantation of presumptive r4 region, in stage 9 chick embryos, into the more anterior position of r2 has shown that *Hoxb-1* is expressed in the ectopic r4 as strongly as in the normal r4, whereas reciprocal grafts of presumptive r2 placed in the r4 position do not express *Hoxb-1* (Guthrie et al., 1992; Kuratani and Eichele, 1993). In addition, retrograde axonal tracing of branchiomotor nerve nuclei indicates that the phenotypes of the ectopic rhombomeres develop according to their original position (Guthrie et al., 1992; Simon et al., 1995).

Thus, under conditions in which the extent of positional displacement is limited to relevant regions of the hindbrain, *Hox* expression and segmental identity appear to be independent of position in the neuroepithelium from as early as stage 9. After more extensive anterior-to-posterior transpositions, however, a progressive posterior transformation in rhomobmere fate is observed, with the coordinate induction of new *Hox* gene expression (Itasaki et al., 1996). The transplantation of somites adjacent to more anterior regions of the rhombencephalon has the ability to induce similar changes in *Hox* gene expression, suggesting that paraxial mesoderm is a source of the environmental signal responsible for this plasticity. These results demonstrate that under extreme circumstances, the program of rhombomeric *Hox* gene expression can be superseded, but they do not necessarily address the normal mechanism by which expression of *Hox* genes in the hindbrain is established. Other experiments performed on neural-plate–stage embryos before the onset of *Hox* expression in the presumptive hindbrain region have shown that ectopically grafted rhombomeres acquire the profile of *Hox* transcripts and neuroanatomical features characteristic of their new location (Grapin-Botton et al., 1995). Thus, there appears to be a critical period at around the time of neural tube closure (stage 8–9) when *Hox* expression and A/P identity are stably established.

Loss-of-function mutations of rostrally expressed *Hox* genes result in malformations that are consistent with transformation of regional identity. In the *Hoxb-1* null mouse, for example, rhombomere 4 loses certain of its r4-specific characters and takes on those of other rhombomeres within which the gene is not normally expressed (r2) or is expressed only at low levels (r6) (M. Studer, A. Lumsden, unpublished observations). Knockouts of other *Hox* genes have less easily interpretable phenotypes. The *Hoxa-1* mutation, for example, appears to result in deletions both of particular rhombomeres and of specific neuronal nuclei (Mark et al., 1993), neither of which is obviously consistent with a role in conferring specific identity on an existing, repetitive, ground plan.

The *HOM-C* genes are expressed in specific regions of the fly CNS that correspond, in A/P position, to equivalent regions of the vertebrate CNS. Although this might be taken to imply that the nervous systems of flies and vertebrates are themselves homologous (Thor, 1995), this may not be the case. Rather, the conservation of homeotic gene expression domains may illustrate the remarkable conservation of a patterning system that encodes positional values along the A/P axis, independent of the specific structure that forms at any particular positional value.

Hensen's Node and Retinoic Acid Signaling of A/P Axial Position

The expression of *Hox* genes is directed to restricted domains at specific levels along the A/P axis, and for certain *Hox* genes, rhombomere-specific control elements have been identified (Marshall et al., 1992, 1994). If *Hox* genes encode positional value along the A/P axis, what positional signaling mechanism is responsible for activating region-specific promoter elements at appropriate levels of the neuraxis? One candidate mechanism involves retinoid-mediated signals emanating from the node, a key organizing center in the gastrula-stage embryo. The node is a rich source of retinoids (Hogan et al., 1992; Chen et al., 1992), and two lines of evidence support the possibility that retinoic acid (RA) confers positional information on cells at different A/P positions directly through the differential regulation of *Hox* gene expression.

First, both RA deficiency and excess exogenous RA lead to severe neural defects in the hindbrain and branchial arch region. Excess RA causes a dose-dependent A-to-P transformation of cell fate, in which the hindbrain is expanded at the expense of the forebrain and midbrain (Durston et al., 1989). Vitamin A–deficient quail embryos, by contrast, lack posterior rhombomeres (Maden et al., 1996). Second, RA-induced changes in cell fate are associated with changes of *Hox* gene expression patterns in a manner consistent with the principle of colinearity: the correspondence between the location of a *Hox* gene in the cluster and its responsiveness to RA. *Hox* genes at the 3′ end of a cluster respond more rapidly and at a lower RA concentration than do more 5′ *Hox* genes (Simeone et al., 1990; Papalopulu et al., 1991). The changes of *Hox* gene expression following excess RA treatment are followed by stable changes in morphology, including the ordered transformation of anterior rhombomere cell types to those of a more posterior type, suggesting that an RA signal normally regulates the pattern of *Hox* gene expression (Marshall et al., 1992).

RA exerts its effects on development by controlling target gene transcription via two major classes of RA receptors, RARs and RXRs. Both classes of proteins serve as ligand-dependent transcription factors that bind as homo- or heterodimers to RA response elements (RAREs and RXREs) in the promoters of target genes (Chambon, 1994; Marshall et al., 1994; Mangelsdorf et al., 1994). The promoters of at least some of the *Hox* genes contain RAREs which are required for appropriate gene activation (Marshall et al., 1994; Giguere, 1994; Studer et al., 1994). Thus, RA may control directly the expression of *Hox* genes. Since Hensen's node is a rich source of RA in the gastrula-stage embryo, a P-to-A gradient of RA, or an increasing exposure to RA of cells that pass in A-to-P order through the node, has strong candidacy for establishing the nested expression of *Hox* genes along the A/P axis.

Control of Neural Crest Fate by Interrhombomeric Signaling

The fate of cells along the A/P axis of the hindbrain is not established exclusively by the early signals that confer distinct identities to individual rhombomeres. Once formed, signals from certain rhombomeres appear to influence the differentiation and fate of cells in neighboring rhombomeres, thus providing a secondary mechanism for establishing AP positional differences between adjacent cell groups. The periodicity in the generation and/or fate of neural crest cells in the hindbrain appears to be dependent on such a secondary signaling system.

The neural crest at cranial levels is specified morphogenetically prior to its migration from the neural primordium, and emerging neural crest cells appear to transpose their positional information to adjacent mesodermal tissues with which they interact (Noden, 1988). Much of the cranial neural crest derives from the rhombencephalon, and there is a clear relationship between site of emergence of neural crest cells and the segmentation of the neural epithelium (Lumsden et al., 1991). Neural crest cells that emigrate from r1 and r2 contribute to the trigeminal ganglion and the first branchial arch; those from r4 contribute to the facial and vestibulo-acoustic ganglia and the second arch; and those from r6 populate the superior ganglion of the IXth nerve and the third branchial arch.

Separating these three areas of crest production are two axial levels, r3 and r5, which do not contribute to the emergent neural crest once these rhombomeres become delineated. Thus, in the rhombencephalon the neural crest originates from three discontinuous levels and migrates ventrolaterally in three distinct streams.

The discontinuous emigration of the neural crest along the neuraxis is achieved in large part by mechanisms intrinsic to the neurectoderm. Neural crest cell production appears to be continuous along the neuraxis, but most or all neural crest cells are eliminated from r3 and r5, before emigration (Lumsden et al., 1991). In vitro studies have shown that when isolated from their neighbors, odd-numbered rhombomeres, r3 and r5, which are normally depleted will now produce neural crest cells (Graham et al., 1993). When these rhombomeres are cultured in conjunction with an even-numbered neighbor, however, they do not produce neural crest cells. Moreover, surgical manipulations of hindbrain segments in vivo that produce an odd-numbered rhombomere flanked by two host odd-numbered rhombomeres result in the emergence of neural crest cells from both the donor and host odd rhombomeres (Graham et al., 1993). These studies demonstrate that odd-numbered rhombomeres will produce neural crest if they are freed from the influence of even-numbered rhombomeres. Thus, even-numbered rhombomeres appear to exert a repressive effect upon the production of neural crest by odd-numbered rhombomeres, most likely through the induction of cell death.

Analysis of the expression of *msx* homeobox genes in the rhombencephalon has revealed that there is a close relationship between neural crest production and the patterns of *msx* gene expression, as noted in other locations (Davidson and Hill, 1991). *Msx-1* and *msx-2* are expressed in the dorsal aspect of r3 and r5, those rhombomeres depleted of migratory neural crest cells (Graham et al., 1993). Moreover, the expression pattern of *msx-2* in the rhombencephalon is spatially and temporally similar to the pattern of apoptosis, suggesting that *msx* genes play a role in the patterning of the neural crest through their involvement in the selective elimination of specific populations of cells (Graham et al., 1993).

The apoptotic elimination of neural crest cells in r3 and r5 induced by neighboring even-numbered rhombomeres appears to be elicited through the induction of high-level expression of *BMP4* in the neural crest primordium of r3 and r5 (Graham et al., 1994). BMP4 stimulates *msx-2* expression and depletes neural crest cells from r3 and r5 in vitro and is associated with the induction of apoptosis. Thus, it appears that the induction of BMP4 in rhombomeres 3 and 5 by even-numbered rhombomeres is involved in the death of neural crest cells. The sculpting of the neural crest into discrete streams which populate and pattern the branchial arches is therefore achieved by a mechanism that is intrinsic to the neuroepithelium, involving local signals between rhombomeres. Moreover, since BMP4 causes the depletion of neural crest from isolated r3 and r5 but not from r4 or from the neural tube at spinal cord levels (Liem et al., 1995), there appear to be intrinsic A/P differences in the response properties of neural cells, possibly reflecting the deployment of distinct BMP receptors (Graham and Lumsden, 1996).

Integration of Positional Information along the A/P and D/V Axes of the Hindbrain

SHH is produced uniformly along the ventral midline of the hindbrain yet induces different cell fates at distinct A/P levels. SHH therefore appears to be a general ventralizing signal that is interpreted in a specific way by the differential competence of the responsive neural plate tissue. The patterning activities of the notochord and SHH therefore appear to function within the context of a specified A/P regional identity.

Support for this idea has been obtained by transposition of neural tissue along both the A/P and D/V axes of the hindbrain. Rhombomere 4 is characterized by high-level expression of the *Hoxb-1* gene and, at a later stage, by the emergence of a unique cell group adjacent to the floor plate, the contralateral vestibulo-acoustic (CVA) efferent neurons (Fig. 8–7B) (Simon and Lumsden, 1993). Transplantation of presumptive r4 to a more anterior position (in place of r2) at the time of neural tube closure results in maintained *Hoxb-1* expression and the ectopic production of CVA neurons (Fig. 8–7B) (Guthrie et al., 1992; Simon et al., 1995). Furthermore, CVA neurons are produced by dorsal r4 tissue if it is placed close to the inducing tissues of the ventral midline in this ectopic A/P location. Thus, r4-specific ventral cell types are formed irrespective of the DV level of origin of this precursor (Simon et al., 1995).

These results indicate that in the rhombencephalon, cells are first assigned their A/P identity with their D/V identity and ultimate choice of fate awaiting later ventral and dorsal midline-derived signals, presumably SHH and BMPs. The multipotency of precursor cells is restricted first to a repertoire appropriate to A/P position, leaving them in a state of competence to respond to signals that only later determine the specific cell identity appropriate for their D/V position.

Positional Specification of Motoneuron Identity in the Spinal Cord

In the spinal cord, as in the hindbrain, the axons of motoneurons project toward their targets in the periphery, establishing a highly stereotyped pattern of connections. Axial, body wall, and limb muscles and autonomic ganglia are each innervated by a distinct subclasses of motoneurons. The cell bodies of the motoneurons in each of these subclasses occupy discrete rostrocaudal positions in the spinal cord, and their growth cones appear to select specific pathways to reach their cellular targets in the periphery. Manipulations of the neural tube in avian embryos have suggested that the neurons of each motoneuron subclass share properties that are distinct from those of their neighbors (Lance-Jones and Landmesser, 1980a,b; Landmesser, 1992) and that this distinction specifies their organization into distinct motor columns and pools within the spinal cord and the projection pattern of their axons (Lumsden, 1995; Tosney et al., 1995).

Motor column organization

The somatic motoneurons that innervate muscles of the trunk are located in a medial motor column (MMC) that is continuous along the length of the spinal cord (Fig. 8–8A). In contrast, motor neurons that innervate limb muscles form a discontinuous lateral motor column (LMC) present only at brachial and lumbar levels. These two somatic motor columns can be further subdivided according to the positional identity of the muscles they innervate. Motoneurons in the medial subdivision of the MMC (MMC_m) project axons to the axial muscles that lie close to the vertebral column, whereas motoneurons in the more lateral subdivision of the MMC (MMC_l) are found only at thoracic levels and project axons to the muscles of the ventral body wall (Fig. 8–8D) (Gutman et al., 1993). Motoneurons in the medial subdivision of the LMC (LMC_m) project axons to limb muscles that derive from the ventral premuscle mass, whereas neurons in the lateral subdivision of the LMC (LMC_l) project axons to muscles derived from the dorsal premuscle mass. Motoneurons that form the preganglionic motor column of Terni (CT) are found at thoracic and sacral levels and in contrast to somatic motoneurons, innervate neural-crest–derived sympathetic and parasympathetic neurons of the autonomic ganglia (Fig. 8–8).

These columnar subclasses of motoneurons can be distinguished by expression of four LIM homeobox genes: *Islet-1*, *Islet-2*, *Lim-1*, and *Lim-3* (Fig. 8–8B, C) (Tsuchida et al., 1994). At the time that the columnar organization of motoneurons becomes evident, individual cells of each subclass express a specific combi-

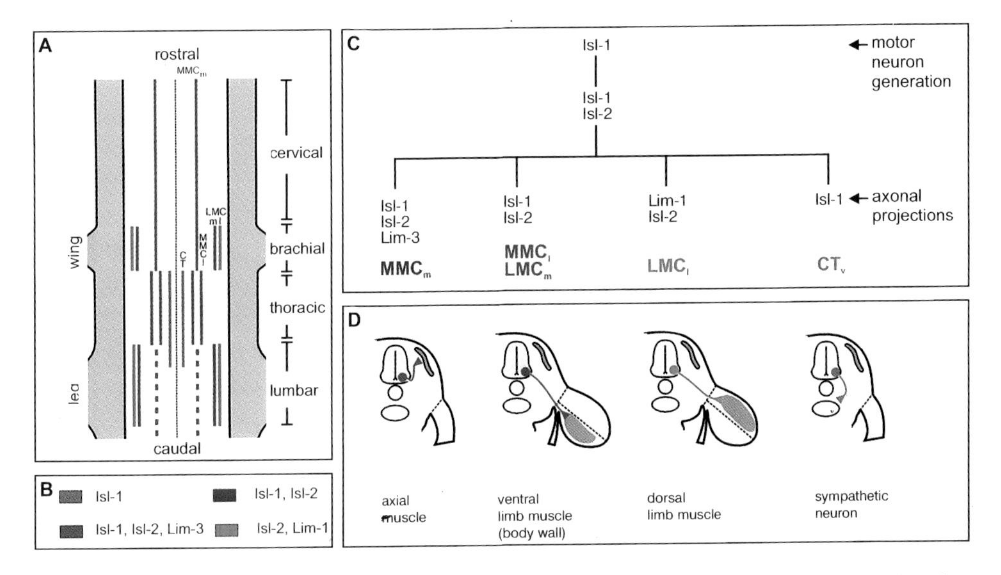

FIGURE 8–8 Schematic representation of the organization of motor columns in the spinal cord and the peripheral targets of motoneurons. (A) Diagram of the position of individual motor columns along the rostrocaudal axis of the spinal cord of a stage 35 chick embryo. The MMC_m is shown in blue (MMC_m), the MMC_l in red (MMC_l), the LMC_m in red (LMC_m), the LMC_l in green (LMC_l), and the CT in brown (CT). Broken blue line signifies the decrease in number of motoneurons in the MMC_m at lumbar levels. (B) Color code of LIM homeodomain protein expression. (C) Temporal sequence of expression of LIM homeodomain proteins by newly differentiating motoneurons. All classes of motoneuron initially express Isl-1 and Isl-2, soon after their birth. Differential expression of LIM homeodomain proteins occurs at around the time of axon extension. (D) Diagram of transverse sections through stage 22–25 chick embryos at different segmental levels, showing the location of motoneurons in the spinal cord and their peripheral targets. Adapted from Tsuchida et al. (1994).

nation of LIM homeobox genes (Fig. 8–8C). Moreover, the retrograde labeling of motoneurons from specific muscle targets has shown that the expression of LIM homeodomain proteins corresponds precisely with the columnar organization of motoneuron subtypes (Tsuchida et al., 1994).

The segregation of LIM homeobox gene expression with the identity of peripheral targets raises the possibility that these genes direct the expression of proteins that control the ability of growth cones to choose between the distinct pathways that they encounter outside the spinal cord—a dorsal path toward the dermomyotome (MMC_m neurons), a ventromedial path toward the sympathetic chain (CT neurons), and a pathway straight out into the limb (LMC neurons), which divides into dorsal (LMC_l) and ventral (LMC_m) branches (Fig. 8–8). LIM homeodomain proteins could, for example, regulate receptors for guidance cues that direct axons selectively along each of these distinct pathways.

The discontinuous organization of motor columns along the rostrocaudal axis of the spinal cord defined by LIM homeodomain protein expression has permitted a preliminary analysis of how different motoneuron subtypes form preferentially at different segmental levels. Experiments in which segments of brachial neural tube at stage 10–11 have been grafted to thoracic levels and vice versa led to a respecification of the columnar organization and identity of motoneurons as assessed by a change in the number and position of motoneurons and in the combination of LIM homeodomain proteins expressed (Ensini et al., 1997). Thus, thoracic regions grafted to brachial levels acquire a LMC whereas brachial regions grafted to thoracic levels lose the LMC and instead acquire a CT. The respecification of cell fate in the spinal cord elicited by these rostrocaudal transpositions appears not to be confined to motoneurons: The entire morphology of the spinal cord is affected and segmentally restricted patterns of *Hox* gene expression are also changed in accordance with the new positions of the graft (Ensini et al., 1997). Thus, the specification of individual motor column identity appears to be established around the time of neural tube closure. These observations suggest that the positional identity of motoneurons along the rostrocaudal axis of the spinal cord is controlled by signaling mechanisms that operate at stages after neural tube closure.

The origin and molecular nature of the signals that control the rostrocaudal positional identity of motoneurons are not known. One possibility is that a long-range signal is transmitted along the A/P axis of the neural tube and defines the fate of cells at different distances from the source of the signal. An alternative, and perhaps more attractive, possibility is that the columnar identity of motoneurons is controlled initially by positionally restricted signals from the paraxial or lateral plate mesoderm that flanks the neural tube. Positionally restricted mesodermal signals have been invoked to trigger the induction of the early limb bud (Crossley et al., 1996) and similar or identical signals could act on neural tube cells to restrict the position at which the LMC is generated.

Motor pool organization

In addition to the allocation of motoneurons to longitudinally organized columns, there is also a highly ordered arrangement of motoneurons within each column. This organization is most clearly evident in the LMC where motoneurons are organized into distinct clusters, or pools, each of which projects to a specific limb muscle target (Landmesser, 1978a,b). Embryonic manipulations in chick have again provided evidence that the neurally derived signals that lead to the acquisition of motor pool identity are fixed prior to the generation of motoneurons and lead eventually to the ability of the axons of motoneurons in discrete pools to navigate muscle-specific pathways in the limb (Lance-Jones and Landmesser, 1980a,b). The positional signals within the neural tube that control the pool-specific identity of LMC motoneurons appear to be established around stage 13, since inversion of the lumbar neural tube along the rostrocaudal axis at stage 13 leads to a respecification of motor pool identity (Matise and Lance-Jones, 1996) whereas such respecification does not occur if inversions are performed at stage 15 (Lance-Jones and Landmesser, 1980a,b).

Intrasegmental identity

Certain lower vertebrates, notably zebrafish, possess primary motor neurons with distinct identities within individual segments of the spinal cord (Fig. 8–9) (Eisen, 1994). Each segment contains a stereotyped set of primary motor neurons, CaP (caudal), MiP (middle), RoP (rostral), and VaP (variable), named on the basis of the position of the motoneuron cell body with respect to the somite border (Fig. 8–9A) (Eisen, 1994). The axon of each primary motoneuron projects to a different domain of the axial musculature via a stereotyped and distinct pathway.

Different primary motoneurons in zebrafish can be distinguished by the combinatorial expression of LIM homeodomain proteins (Inoue et al., 1994; Appel et al., 1995; Tokumoto et al., 1995), suggesting that the same class of transcription factors that establishes the columnar identity of motoneurons in higher vertebrates regulates the diversification of primary motoneurons in lower vertebrates. The transplantation of individual primary motoneurons to a different intrasegmental position soon after their overt differentiation leads to a respecification of the trajectory and target domain of the motor axon (Fig. 8–9) (Eisen, 1991). Such manipulations also alter the combination of LIM homeobox genes expressed by the neuron (Appel et al., 1995), providing support for the idea that the combinatorial expression of LIM homeobox genes is an important determinant of the identity of motoneuron subclasses.

The constancy in primary motoneuron identity with respect to the somite border raises the possibility that local signals emanating from cells at different rostrocaudal positions within the somite control primary motoneuron fate. An involvement of somitic mesoderm in the establishment of primary motoneuron identity has also been suggested by an analysis of the zebrafish *spadetail (spt)* mutation (Ho and Kane, 1990), which perturbs the differentiation of paraxial mesoderm and leads to disruptions in somite development. Primary motorneuron identity, as defined by the LIM homeobox gene expression, is disrupted in *spt* mutants (Tokumoto et al., 1995), although the generalized disruption of mesodermal development in *spt* has precluded an analysis of axonal trajectories.

These studies suggest that highly localized signals specify the segmental arrangement of primary motoneuron identities in the zebrafish spinal cord. Whether a similar intrasegmental organization of motoneurons has been pre-

A

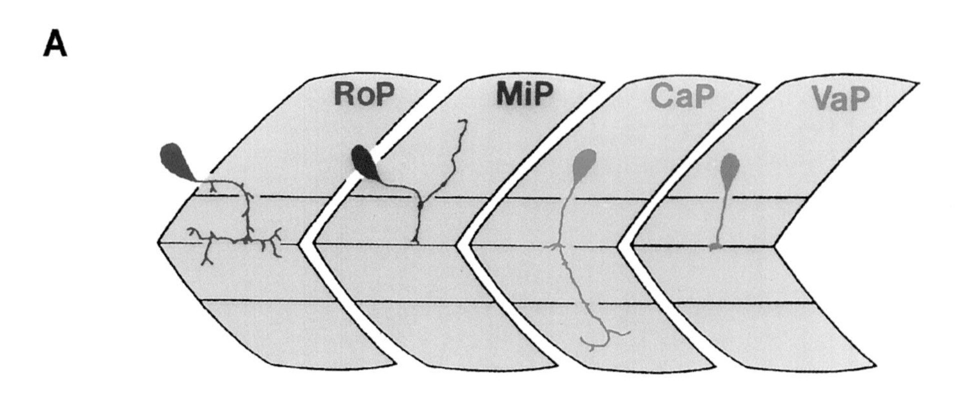

B

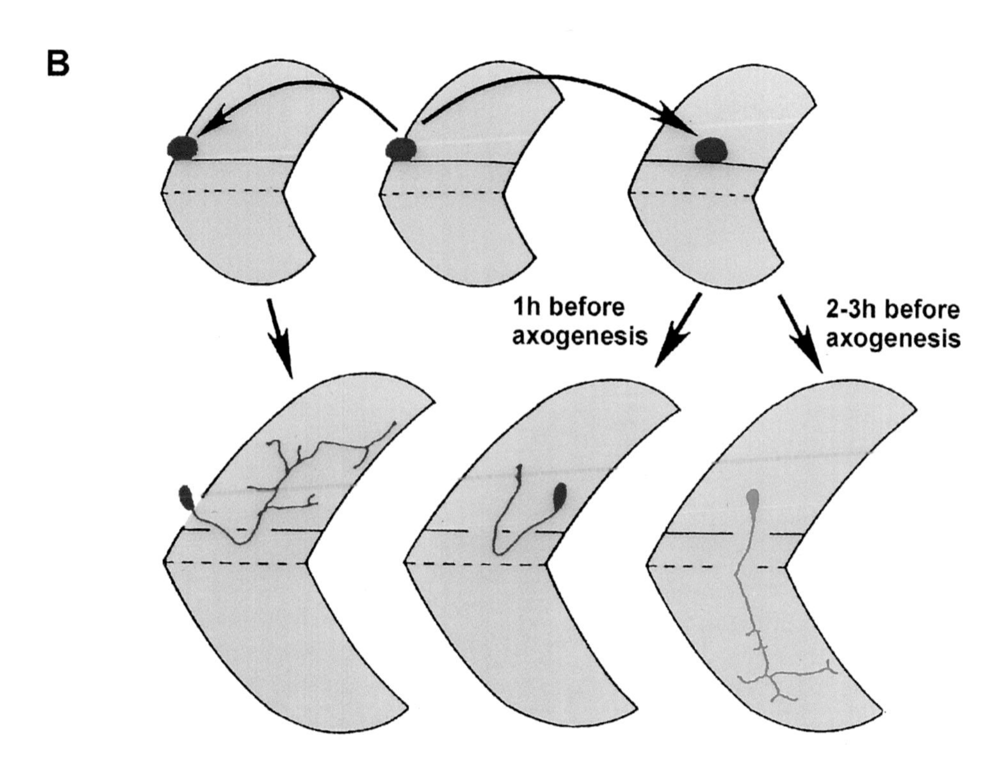

FIGURE 8–9 **Zebrafish motoneurons develop position-specific axonal trajectories.** (A) Location of individual primary motoneurons at different positions within an individual segment of the spinal cord. (B) A single identified motoneuron transplanted from the MiP position of a donor embryo to the same position of a host develops a normal MiP morphology (*left*). Motoneurons develop an MiP axonal trajectory when transplanted from the MiP position to the CaP position about an hour before axogenesis (*middle*) and a CaP morphology when transplanted from the MiP position to the CaP position 2 to 3 hours before axogenesis (*right*). Adapted from Eisen (1994).

served in the spinal cord of higher vertebrates remains unclear. In chick, the autonomic motoneurons in the column of Terni exhibit a distinct intrasegmental arrangement with respect to their peripheral targets (Forehand et al., 1994; Stirling et al., 1995), raising the possibility that local signals might indeed control the intrasegmental identity of certain motoneuron classes in higher vertebrates.

Control of Cell Pattern at Rostral Levels of the Neural Tube

To what extent are the inductive signals and patterning mechanisms employed at caudal levels of the neuraxis conserved at axial levels rostral to the hindbrain? The notochord underlies the neural plate at mesencephalic and caudal diencephalic levels of the neuraxis but does not underlie the prospective telencephalic region (Kingsbury, 1930; Ericson et al., 1995). Similarly, the floor plate extends rostrally into the caudal diencephalon (Puelles and Rubenstein, 1993).

Not surprisingly then, the ability of SHH-mediated signals from the notochord and floor plate to influence the development of the ventral region of the neural tube is not restricted to the spinal cord and hindbrain. Besides inducing motoneurons and floor plate at more rostral levels of the neuraxis, ventral-midline–derived signals are also involved in the development of region-specific neuronal subpopulations. Two such cell groups are the serotonergic neurons of the hindbrain raphe nuclei and the dopaminergic neurons of the midbrain (Fig. 8–10). Both these monoaminergic cell groups develop close to the floor plate and can be induced to differentiate in competent neural plate tissue in response to a notochord- or floor-plate–derived signal and to SHH (Yamada et al., 1991; Wang et al., 1995; Hynes et al., 1995a,b). The response to these midline-derived signals is determined by the position of origin of the responding tissue. Serotonergic neurons can be induced in ectopic regions of the anterior hindbrain by notochord from more posterior axial levels (Yamada et al., 1991) and the induction of dopaminergic neurons in midbrain neuroepithelium can be triggered by floor plate tissue obtained from a distant axial level (Hynes et al., 1995a). Similarly, the response to SHH signaling depends on the rostrocaudal position of origin of the responding tissue (Wang et al., 1995; Ericson et al., 1995; Hynes et al., 1995a,b). These studies indicate that the same ventralizing inductive signal, SHH, is produced uniformly along much of the neuraxis and that the specific fate of ventral neurons depends on the A/P position of the responding neural plate tissue (Fig. 8–10).

The telencephalic subdivision of the forebrain does not contain a floor plate and is never underlain by the notochord. The absence of these midline structures therefore raises the question of how the bilateral organization of the forebrain and the differentiation of its ventral cell types are controlled. It appears that even in this terminal expansion of the CNS a conserved mechanism is used for ventral patterning. *Shh* is transiently expressed in the prechordal mesoderm that is formed anterior to the notochord and is also expressed along the ventral midline of the diencephalon and at later stages at the ventral midline of the telencephalon (Fig. 8–10A) (Echelard et al., 1993; Ericson et al., 1995; Ruiz i Altaba et

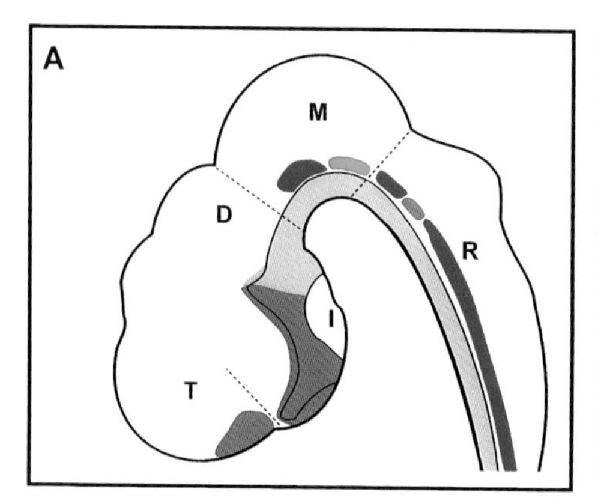

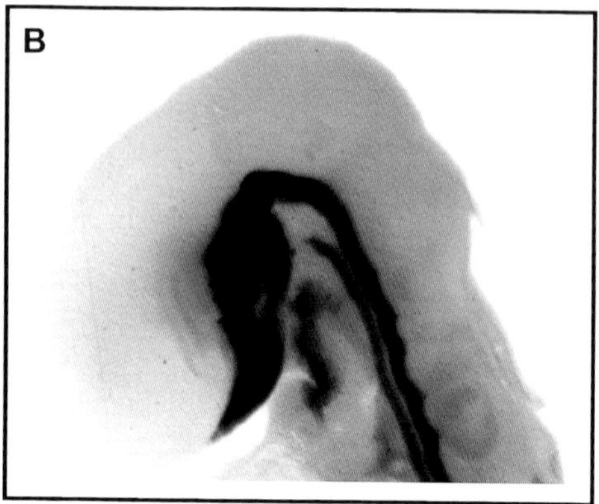

FIGURE 8–10 **SHH-induces distinct ventral cell types at different positions along the rostrocaudal axis of the CNS.** (A) Localization of *Shh* transcripts in the notochord, floor plate, and ventral diencephalon of a stage 16-17 chick embryo. (B) Diagram showing the relationship between *Shh* expression (gray zone) and ventral neuronal classes at different rostrocaudal positions. *Shh* is expressed at the ventral midline of the spinal cord, rhombencephalon (R), and mesencephalon (M) and extends in the rostral diencephalon (D) except at the level of the infundibulum (I). At this stage *Shh* is not expressed in the ventral telencephalon (T). At different midbrain and hindbrain levels, motoneurons (red), serotonergic neurons (brown), and dopaminergic neurons (green) differentiate near the *Shh*-expressing ventral midline cells. In the forebrain, the ventral domain of *Shh* expression is associated with the differentiation of ventral forebrain interneurons (blue) image in A adapted from Ericson et al. (1995); B courtesy of T. Lints.

al., 1995a,b). Moreover, early markers of the ventral telencephalon, notably the *nkx2.1* and *nkx2.2* genes and ventral telencephalic Isl-1^{+}, neurons can be induced by SHH in telencephalic level neural plate explants (Barth and Wilson, 1995; Ericson et al., 1995).

Although *Shh* is expressed in the ventral regions of both the diencephalon and telencephalon, it appears that the diencephalic domain is influential in controlling cell patterning in the ventral forebrain as a whole. *Shh* is expressed in the ventral diencephalon considerably before the appearance of ventral neurons in either forebrain region, and studies in vitro have shown that the midline rostral diencephalic cells can induce ventral forebrain neurons in telencephalic neural plate explants (Ericson et al., 1995). By contrast, the onset of telencephalic expression of *Shh* occurs later in development, coincident with the differentiation of ventral forebrain neurons in that region (Ericson et al., 1995). It is likely, therefore, that SHH inductive signaling extends rostrally from the diencephalon into the telencephalon.

Studies in zebrafish have provided additional evidence that the ventral midline of the diencephalon is a controlling region for patterning the anterior end of the nervous system (Hatta et al., 1994). This midline territory is deleted in the *cyclops* mutant, whose phenotype most obviously involves fusion of the eyes around the anterior pole of the embryo (Hatta et al., 1991). Analysis of the *cyclops* phenotype has revealed that *Shh* is not expressed in prospective telencephalic ventral midline structures, providing additional evidence that the normal expression of *Shh* has consequences that extend beyond patterning the ventral neural tube. In *cyclops* mutants the optic stalk—a region that normally expresses *pax-2*—is diminished, whereas the retina, which normally expresses *Pax-6*, extends throughout the optic territory such that the eyes are fused not by optic stalk tissue but by retina (Hatta et al., 1994).

Shh, or a related gene *twhh*, also appears to have a role in patterning the eye. Ectopic overexpression of *Shh* or *twhh* in zebrafish embryos leads to phenotypes that are reciprocal to those seen in *cyclops*. The domain of *pax-2* expression is extended and encroaches into the territory that would normally express *pax-6* (Macdonald et al., 1995; Ekker et al., 1995). At the cellular level, affected embryos have an enlarged optic stalk and a reduced retina. Although the consequence of interfering with SHH signaling has yet to be analyzed at later stages of eye development, these studies suggest that SHH is responsible for patterning both the ventral forebrain and the optic territories.

The role of BMPs in establishing dorsal cell fates in the midbrain or forebrain has not been determined; however, several BMPs are expressed in the ectoderm that surrounds the diencephalic region of the neural tube (Liem et al., 1995; Lyons et al., 1995), and mice lacking BMP7 exhibit defects in eye development (Dudley et al., 1995; Luo et al., 1995). Thus, the inductive signals involved in dorsal cell differentiation may also be conserved along much of the rostrocaudal axis of the neural tube.

Conclusions

The studies of neural cell differentiation in the spinal cord and hindbrain outlined in this review have provided a preliminary insight into the principles of cellular organization in the developing vertebrate CNS but leave many issues unresolved.

It is now clear that many aspects of cell patterning in the caudal neural tube are highly conserved. The source and identity of signals that establish the early M/L polarity of the neural plate and the later D/V organization of cell types in the spinal cord and hindbrain appear identical, and many of the same genes are induced in neural plate cells in response to these signals. This is perhaps not surprising since the same basic classes of neurons are generated at similar D/V positions in the spinal cord and hindbrain, and these neurons have conserved functions. It is also apparent that the SHH- and BMP-mediated signals that establish cell fate and pattern act along the D/V axis on neural epithelial cells which have received earlier signals that endow them with a stable and heritable indication of their position along the A/P axis of the neural plate (Fig. 8–11). The source, identity, and time of operation of these A/P patterning signals remain unclear, but evidence is emerging that neural inducing factors and retinoids derived from the node region have a role in establishing an initial coarse-grained A/P regional identity within the neural plate. The primary candidates as responders to these early A/P patterning signals, and thus as mediators of A/P positional value, are members of the *Hox* gene family and other transcription factors.

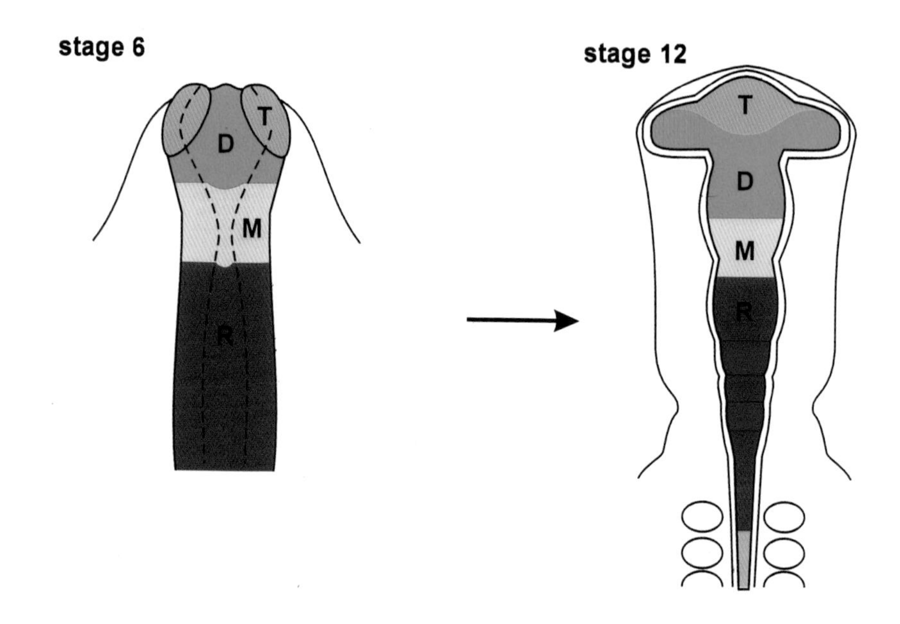

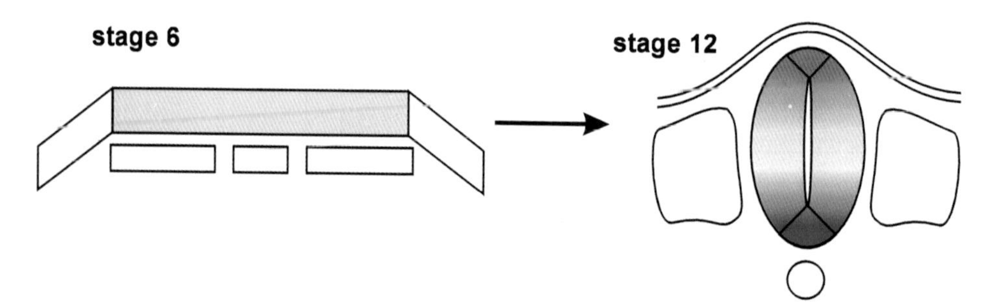

FIGURE 8–11 Sequential restrictions in regional identity along the A/P axis and cell fate along the D/V axis. (A) Diagram shows the early establishment of restrictions in regional identity along the A/P axis of the chick neural plate (stage 6), which anticipates the eventual subdivisions of the neural tube along its A/P axis (stage 12). T, telencephalon; D, diencephalon; M, mesencephalon; R, rhombencephalon. (B) Diagram showing that at the time that early A/P restrictions are apparent in the neural plate, cell fates are not fixed along its M/L axis. D/V cell-fate restrictions established later, at around the time of neural tube closure.

A critical issue remains: To what extent is the identity imposed by these early A/P patterning signals immutable, or is there a later contribution of signals from the paraxial mesoderm that refines or modifies the initial A/P positional identities of neural cells? Transcription of rhombomeres along the A/P axis of the hindbrain has demonstrated a high degree of autonomy of segmental identity, although certain studies suggest that if the extent of displacement is sufficiently great, it may be possible to override intrinsic positional information. At spinal levels, the neural tube appears less rigidly committed to an early A/P positional identity, and neural tube transpositions here result in an apparently complete transformation in the potential identity of motoneurons. Thus, it seems likely that at both spinal and hindbrain levels, the establishment of A/P positional identity involves a combination of early signals emanating from the node region or its axial mesodermal derivatives and later signals from paraxial or lateral plate mesoderm. The relative influence of these early and late signals in establishing A/P fate may, however, differ at the two levels. At hindbrain levels, early node-derived signals may predominate whereas at spinal cord levels, signals from the paraxial or lateral plate mesoderm may be more influential.

One reason that the hindbrain and spinal cord might deploy somewhat different strategies to establish A/P positional identities relates to the informational content intrinsic to the mesodermal cells that surround the neural tube at different A/P levels. The mesoderm adjacent to the hindbrain is unsegmented and appears to be patterned in large part as a passive response to signals derived from the neural tube or its emigrant neural crest cells. The unsegmented disposition of the cranial paraxial mesoderm contrasts strongly with the overtly segmented confirmation of the neural crest derivatives in the branchial arches and their corresponding sensory and motor innervation in the segmented cranial ganglia and hindbrain. Thus, mechanisms for conferring a relatively rigid set of positional values within the neural tube at hindbrain levels may have evolved in compensation for the lack of signaling from the surrounding cranial mesoderm. In this region, segmentally restricted signals impose positional information on the surrounding cranial mesoderm, whereas at spinal cord levels the opposite occurs. Here, the paraxial mesoderm is overtly segmented, and both paraxial and lateral plate mesoderm populations exhibit intrinsic positional identities that can be detected in the A/P level-specific expression of *Hox* genes, in the intrinsic capacity of the somitic sclerotome to shape axial skeletal structures, and possibly in the ability to initiate appendage formation at brachial and lumbar levels. In higher vertebrates, the absence of any overt segmentation of the neural tube at spinal cord levels may therefore be compensated for by the existence of segmentally restricted positional signals that derive from the paraxial or lateral plate mesoderm. Such signals may have a primary role in imposing the positional differences in motoneuron identity detected along the A/P axis of the spinal cord.

What remains less clear is how the fine-grained distinctions in neuronal identity are established within a single segmental level of the spinal cord or hindbrain. For example, how are motoneuron and reticulospinal neurons generated at the same A/P and D/V position in the hindbrain? Does position on the inside–outside (ventricular–pial) axis of the pseudostratified neuroepithelium play any significant role in modulating the response to A/P and D/V signals? Similarly, in the spinal cord how do distinct motor pools acquire their specific identity within the LMC? It seems likely that short-range signals within the neural tube acting subsequent to and independent of mesodermally derived signals will contribute importantly to the later diversification of neuronal cell types. The extent to which progenitor cells within the neural tube acquire lineally restricted programs for cell differentiation also remains unresolved.

Despite these uncertainties, recent studies of cell patterning in the caudal neural tube have drawn a general picture of the role of prospective (nodal) and definitive (paraxial) mesodermal cells in establishing the A/P pattern of the neural plate and neural tube and the specific role of the axial mesoderm and epidermal ectoderm in imposing D/V positional differences. Genetic manipulations in the mouse and zebrafish and cellular studies in avian embryos offer considerable promise for extending these initial insights into the mechanisms of neural cell specification in the vertebrate CNS.

ACKNOWLEDGMENTS

We thank C. Henry, K. MacArthur and I. Schieren for help in preparing the text and figures and J. Dodd, J. Ericson, and S. Pfaff for helpful comments on the manuscript. T.M.J. is an Investigator and A.L. an International Research Scholar of the Howard Hughes Medical Institute.

REFERENCES

Anderson, D., Jan, Y.N. (1997) In: *Molecular and Cellular Approaches to Neural Development*, Cowan, W.M., Jessel, T.M., Zipursky, S.L., eds. Oxford: Oxford University Press.

Alvarado-Mallart, R.M. (1993). Fate and potentialities of the avian mesencephalic/metencephalic neuroepithelium. *J. Neurobiol.* 24:1341–1355.

Ang, S.L., Rossant, J. (1994). HNF-3β is essential for node and notochord formation in mouse development. *Cell* 78:561–574.

Aono, A., Hazama, M., Notoya, K., Taketomi, S., Yamasaki, H., Tsukuda, R., Sasaki, S., Fujisawa, Y. (1995). Potent ectopic bone-inducing activity of bone morphogenetic protein-4/7 heterodimer. *Biochem. Biophys. Res. Comm.* 210:670–677.

Appel, B., Korzh, V., Glasgow, E., Thor, S., Edlund, T., Dawid, I.B., Eisen, J. (1995). Motoneuron fate specification revealed by patterned LIM homeobox gene expression in embryonic zebrafish. *Development* 121:4117–4125.

Artinger, K. B., Bronner-Fraser, M. (1992). Notochord grafts do not suppress formation of neural crest cells or commissural neurons. *Development* 116:877–886.

Barth, K.A., Wilson, S.W. (1995). Zebrafish nkx2.2 is influenced by sonic hedgehog/vertebrate hedgehog-1 and demarcates a neuronal differentiation in the embryonic forebrain. *Development* 121:1755–1768.

Basler, K., Edlund, T., Jessell, T.M., Yamada, T. (1993). Control of cell pattern in the neural tube: regulation of cell differentiation by dorsalin-1, a novel TGF beta family member. *Cell* 73:687–702.

Birgbauer, E., Fraser, S.E. (1994). Violation of cell lineage compartments in the chick hindbrain. *Development* 120:1347–1356.

Bitgood, M.J., McMahon, A.P. (1995). Hedgehog and Bmp genes are coexpressed at many sites of cell-cell interaction in the mouse embryo. *Dev. Biol.* 172:126–138.

Bovolenta, P., Dodd, J. (1991). Perturbation of neuronal differentiation and axon guidance in the spinal cord of mouse embryos lacking a floor plate: analysis of Danforths' short-tail mutation. *Development* 113:625–639.

Bumcrot, D.A., Takada, R., McMahon, A.P. (1995). Proteolytic processing yields two secreted forms of sonic hedgehog. *Mol. Cell. Biol.* 15:2294–2303.

Chambon, P. (1994). The retinoid signaling pathway; molecular and genetic analysis. *Semin. Cell. Biol.*

Chang, D.T., Lopez, A., von Kesseler, D.P., Chiang, C., Simandl, B.K., Renbin, Z., Seldin, M.F., Beachy, P.A., Fallon, J.F. (1994). Products, genetic linkage and limb patterning activity of a murine hedgehog gene. *Development* 120:3339–3353.

Chen, Y.P., Huang, L., Russo, A.F., Solursh, M. (1992). Retinoic acid is enriched in Hensen's node and is developmentally regulated in the early chick embryo. *Proc. Natl. Acad. Sci. U.S.A.* 89:10056–10059.

Clarke, J.D., Lumsden, A. (1993). Segmental repetition of neuronal phenotype sets in the chick embryo hindbrain. *Development* 118:151–162.

Cook, M., Gould, A., Brand, N., Davies, J., Strutt, P., Shaknovich, R., Licht, J., Waxman, S., Chen, Z., Gluecksohn-Waelsch, S., Krumlauf, R., Zelent, A. (1995). Expression of the zinc-finger gene PLZF at rhombomere boundaries in the vertebrate hindbrain. *Proc. Natl. Acad. Sci. U.S.A.* 92:2249–2252.

Cox, W.G., Hemmati-Brivanlou (1995). Caudalization of neural fate by tissue recombination and bFGF. *Development* 121:4349–4358.

Crossley, P.H., Minowada, G., MacArthur, C.A., Martin, G.R. (1996). Roles for FGF8 in the induction, initiation and maintenance of chick limb development. *Cell* 84:127–136.

Davidson, D.R., Hill, R.E. (1991). *Msh*-like genes: a family of homeobox genes with wide-ranging expression during vertebrate development. *Semin. Dev. Biol.* 2:405–412.

Davis, C.A., Joyner, A.L. (1988). Expression patterns of the homeo box-containing genes En-1 and En-2 and the proto-oncogene int-1 diverge during mouse development. *Genes Dev.* 2:1736–1744.

Dawid, I.B., Toyama, R., Taira, M. (1995). LIM domain proteins. *C.R. Acad. Sci.* 318:295–306.

Dickinson, M.E., Selleck, M.A., McMahon, A.P., Bronner, F.M. (1995). Dorsalization of the neural tube by the non-neural ectoderm. *Development* 121:2099–2106.

Doniach, T., Phillips, C.R., Gerhart, J.C. (1992). Planar induction of anteroposterior pattern in the developing central nervous system of Xenopus laevis. *Science* 257:542–545.

Dudley, A.T., Lyons, K.M., Robertson, E.J. (1995). A requirement for bone morphogenetic protein-7 during development of the mammalian kidney and eye. *Genes Dev.* 9:2795–2807.

Durston, A.J., Timmermans, J.P., Hage, W.J., Hendriks, H.F., de Vries, N., Heideveld, M., Nieuwkoop, P.D. (1989). Retinoic acid causes an anteroposterior transformation in the developing central nervous system. *Nature* 340:140–144.

Eagleson, G.W., Harris, W.A. (1990). Mapping of the presumptive brain regions in the neural plate of Xenopus laevis. *J. Neurobiol.* 21:427–440.

Echelard, Y., Epstein, D.J., St. Jacques, B., Shen, L., Mohler, J., McMahon, J.A., McMahon, A.P. (1993). Sonic hedgehog, a member of a family of putative signaling molecules, is implicated in the regulation of CNS polarity. *Cell* 75:1417–1430.

Eisen, J.S. (1991). Determination of primary motoneuron identity in developing zebrafish embryos. *Science* 252:569–572.

Eisen, J.S. (1994). Development of motoneuronal phenotype. *Annu. Rev. Neurosci.* 17:1–30.

Ekker, S.C., Ungar, A.R., Greenstein, P., von Kessler, D.P., Porter, J.A., Moon, R.T., Beachy, P.A. (1995). Patterning activities of vertebrate hedgehog proteins in the developing eye and brain. *Curr. Biol.* 5:944–955.

Ensini, M., Tsuchida, T., Jessell, T.M. (1997). Control of motor neuron subtype identity by signals from paraxial mesoderm. In preparation.

Ericson, J., Thor, S., Edlund, T., Jessell, T.M., Yamada, T. (1992). Early stages of motor neuron differentiation revealed by expression of homeobox gene Islet-1. *Science* 256:1555–1560.

Ericson, J., Muhr, J., Placzek, M., Lints, T., Jessell, T.M., Edlund, T. (1995). Sonic hedgehog induces the differentiation of ventral forebrain neurons: a common signal for ventral patterning within the neural tube. *Cell* 81:747–756.

Ericson, J., Marton, S., Kawakami, A., Roelink, H., and Jessell, T.M. (1996). Two critical periods of sonic hedgehog signaling required for specification of motor neuron identity *Cell*, 87, 661–673.

Fan, C.-M., Kuwana, E., Bulfone, A., Fletcher, C.F., Copeland, N.G., Jenkins, N.A., Crews, S., Martinez, S., Puelles, L., Rubenstein, J., Tessier-Lavigne, M. (1996). Expression patterns of two murine homologs of *Drosophila single-minded* suggest possible roles in embryonic patterning and in the pathogenesis of Down Syndrome. *Mol. Cell. Neurosci.* 7:1–16.

Forbes, A.J., Nakano, Y., Taylor, A.M., Ingham, P.W. (1993). Genetic analysis of hedgehog signalling in the Drosophila embryo. *Dev. Suppl.* 1993:115–124.

Forehand, C.J., Ezerman, E.B., Rubin, E., Glover, J.C. (1994). Segmental patterning of rat and chicken sympathetic preganglionic neurons: correlation between soma position and axon projection pathway. *J. Neurosci.* 14:231–241.

Fortin, G., Kato, F., Lumsden, A., Champagnat, J. (1995). Rhythm generation in the segmented hindbrain of chick embryos. *J. Physiol.* 486:735–744.

Franz, T. (1994). Extra-toes (Xt) homozygous mutant mice demonstrate a role for the Gli-3 gene in the development of the forebrain. *Acta Anat.* 150:38–44.

Fraser, S., Keynes, R., Lumsden, A. (1990). Segmentation in the chick embryo hindbrain is defined by cell lineage restrictions. *Nature* 344:431–435.

Garcia-Bellido, A., Ripoll, P., Morata, G. (1973). Developmental compartmentalisation of the wing disk of Drosophila. *Nature* 245:251–253.

Giguere, V. (1994). Retinoic acid receptors and cellular retinoid binding proteins: complex interplay in retinoid signaling. *Endocr. Rev.* 15:61–79.

Goodrich, L.V., Johnson, R.L., Milenkovic, L., McMahon, J.A., Scott, M.P. (1996). Conservation of the *hedgehog*/patched signaling pathway from flies to mice: induction of a mouse *patched* gene by Hedgehog. *Genes Dev.* 10:301–312.

Goulding, M.D., Lumsden, A., Gruss, P. (1993). Signals from the notochord and floor plate regulate the region-specific expression of two Pax genes in the developing spinal cord. *Development* 117:1001–1016.

Graham, A., Lumsden, A. (1996). Interactions between rhombomeres modulate Krox-20 and follistatin expression in the chick embryo hindbrain. *Development* 122:473–480.

Graham, A., Heyman, I., Lumsden, A. (1993). Even-numbered rhombomeres control the apoptotic elimination of neural crest cells from odd-numbered rhombomeres in the chick hindbrain. *Development* 119:233–245.

Graham, A., Francis, W.P., Brickell, P., Lumsden, A. (1994). The signaling molecule BMP4 mediates apoptosis in the rhombencephalic neural crest. *Nature* 372:684–686.

Graper, L. (1913). Die Rhombomeren und ihre Nervenbeziehungen. *Arch. Mikr. Anat.* 83: 371–426.

Grapin-Botton, A., Bonnin, M.A., McNaughton, L.A., Krumlauf, R., Le Douarin, N.M. (1995). Plasticity of transposed rhombomeres: Hox gene induction is correlated with phenotypic modifications. *Development* 121:2707–2721.

Gunther, T., Struwe, M., Aguzzi, A., Schughart, K. (1994). Open brain, a new mouse mutant with severe neural tube defects, shows altered gene expression patterns in the developing spinal cord. *Development* 120:3119–3130.

Guthrie, S. (1996). Patterning the hindbrain. *Curr. Opin. Neurobiol.* 6:41–48.

Guthrie, S., Lumsden, A. (1991). Formation and regeneration of rhombomere boundaries in the developing chick hindbrain. *Development* 112:221–229.

Guthrie, S., Muchamore, I., Kuroiwa, A., Marshall, H., Krumlauf, R., Lumsden, A. (1992).

Neuroectodermal autonomy of Hox-2.9 expression revealed by rhombomere transpositions. *Nature* 356:157–159.
Guthrie, S., Prince, V., Lumsden, A. (1993). Selective dispersal of avian rhombomere cells in orthotopic and heterotopic grafts. *Development* 118:527–538.
Gutman, C.R., Ajmera, M.K., Hollyday, M. (1993). Organization of motor pools supplying axial muscles in the chicken. *Brain Res.* 609:129–136.
Hammerschmidt, M., Bitgood, M.J., McMahon, A.P. (1996). Protein kinase A is a common negative regulator of Hedgehog signaling in the vertebrate embryo. *Genes Dev.* 10: 647–658.
Harland, R.M. (1997). In: *Molecular and Cellular Approaches to Neural Development*, Cowan, W.M., Jessell, T. M., Zipursky, S.L., eds. Oxford: Oxford University Press.
Hatta, K., Kimmel, C.B., Ho, R.K., Walker, C. (1991). The cyclops mutation blocks specification of the floor plate of the zebrafish central nervous system. *Nature* 350:339–341.
Hatta, K., Puschel, A.W., Kimmel, C.B. (1994). Midline signaling in the primordium of the zebrafish anterior central nervous system. *Proc. Natl. Acad. Sci. U.S.A.* 91:2061– 2065.
Hemmati-Brivanlou, A., Kelly, O.G., Melton, D.A. (1994). Follistatin, an antagonist of activin, is expressed in the Spemann organizer and displays direct neuralizing activity. *Cell* 77:283–295.
Heyman, I., Kent, A., Lumsden, A. (1993). Cellular morphology and extracellular space at rhombomere boundaries in the chick embryo hindbrain. *Dev. Dyn.* 198:241–253.
Heyman, I., Faissner, A., Lumsden, A. (1995). Cell and matrix specialisations of rhombomere bounaries. *Dev. Dyn.* 204:301–315.
Hirano, S., Fuse, S., Sohal, G.S. (1991). The effect of the floor plate on pattern and polarity in the developing central nervous system. *Science* 251:310–313.
Ho, R.K., Kane, D.A. (1990). Cell-autonomous action of zebrafish spt-1 mutation in specific mesodermal precursors. *Nature* 348:728–730.
Hogan, B.L., Thaller, C., Eichele, G. (1992). Evidence that Hensen's node is a site of retinoic acid synthesis. *Nature* 359:237–241.
Hooper, J.E. (1994). Distinct pathways for autonomic and paracrine Wingless signalling in Drosophila embryos. *Nature* 372:461–464.
Hui, C.C., Slusarski, D., Platt, K.A., Holmgren, R., Joyner, A.L. (1994). Expression of three mouse homologs of the Drosophila segment polarity gene cubitus interruptus, Gli, Gli-2, and Gli-3, in ectoderm- and mesoderm-derived tissues suggests multiple roles during postimplantation development. *Dev. Biol.* 162:402–413.
Hynes, M., Poulsen, K., Tessier-Lavigne, M., Rosenthal, A. (1995a). Control of neuronal diversity by the floor plate: contact-mediated induction of midbrain dopaminergic neurons. *Cell* 80:95–101.
Hynes, M., Porter, J.A., Chiang, C., Chang, D., Tessier-Lavigne, M., Beachy, P.A., Rosenthal, A. (1995b). Induction of midbrain dopaminergic neurons by sonic hedgehog. *Neuron* 15:1–20.
Ingham, P.W. (1991). Segment polarity genes and cell patterning within the Drosophila body segment. *Curr. Opin. Genet. Dev.* 1:261–267.
Ingham, P.W. (1993). Localized hedgehog activity controls spatial limits of wingless transcription in the Drosophila embryo. *Nature* 366:560–562.
Ingham, P.W., Hidalgo, A. (1993). Regulation of wingless transcription in the Drosophila embryo. *Development* 117:283–291.
Inoue, A., Takahashi, M., Hatta, K., Hotta, Y., Okamoto, H. (1994). Developmental regulation of islet-1 mRNA expression during neuronal differentiation in embryonic zebrafish. *Dev. Dyn.* 199:1–11.
Itasaki, N., Sharpe, J., Morrison, A., Krumlauf, R. (1996). Reprogramming *Hox* expression in the vertebrate hindbrain: influence of paraxial mesoderm and rhombomere transposition. *Cell* 16:487–500.
Jacobson, C.O. (1964). Motor nuclei, cranial nerve roots, and fibre pattern in the medulla

oblongata after reversal experiments on the neural plate of axolotl larvae. *Zool. Bidr. Uppsala* 36:73–160.

Johnson, R.L., Tabin, C. (1995). The long and short of hedgehog signaling. *Cell* 81:313–316.

Joyner, A.L. (1996). Engrailed, wnt and pax genes regulate midbrain-hindbrain development. *Trends Genet.* 12:15–20.

Karlsson, O., Thor, S., Norberg, T., Ohlsson, H., Edlund, T. (1990). Insulin gene enhancer binding protein Isl-1 is a member of a novel class of proteins containing both a homeo- and a Cys-His domain. *Nature* 344:879–828.

Kengaku, M., Okamoto, H. (1995). bFGF as a possible morphogen for the anteroposterior axis of the central nervous system in *Xenopus. Development* 121:3121–3130.

Kimura, S., Hara, Y., Pineau, T., Fernandez-Salguero, P., Fox, C. H., Ward, J.M., Gonzalez, F.J. (1996). The *T/ebp* null mouse: thyroid-specific enhancer-binding protein is essential for the organogenesis of the thyroid, lung, ventral forebrain, and pituitary. *Genes Dev.* 10:60–69.

Kingsbury, B.F. (1930). The developmental significance of the floor plate of the brain and spinal cord. *J. Comp. Neurol.* 50:177–207.

Kingsley, D. (1994). The TGFβ superfamily: new members, new receptors, and new genetic tests of function in different organisms. *Genes Dev.* 8:133–146.

Knecht, A.K., Good, P.J., Dawid, I.B., Harland, R.M. (1995). Dorsal-ventral patterning and differentiation of noggin-induced neural tissue in the absence of mesoderm. *Development* 121:1927–1935.

Koonin, E.V. (1995). A protein splice-junction motif in hedgehog family proteins. *Trends Biochem. Sci.* 20:141–142.

Krauss, S., Concordet, J.P., Ingham, P.W. (1993). A functionally conserved homolog of the Drosophila segment polarity gene hh is expressed in tissues with polarizing activity in zebrafish embryos. *Cell* 75:1431–1444.

Kuratani, S.C., Eichele, G. (1993). Rhombomere transplantation repatterns the segmental organization of cranial nerves and reveals cell-autonomous expression of a homeodomain protein. *Development* 117:105–117.

Lamb, T.M., Harland, R.M. (1995). Fibroblast growth factor is a direct neural inducer, which combined with noggin generates anterior-posterior neural pattern. *Development* 121: 3627–3636.

Lance-Jones, C., Landmesser, L. (1980a). Motoneurone projection patterns in embryonic chick limbs following partial deletions of the spinal cord. *J. Physiol.* 302:559–580.

Lance-Jones, C., Landmesser, L. (1980b). Motoneurone projection patterns in the chick limb following early partial reversals of the spinal cord. *J. Physiol.* 302:581–602.

Landmesser, L. (1978a). The development of motor projection patterns in the chick hind limb. *J. Physiol.* 284:391–414.

Landmesser, L. (1978b). The distribution of motoneurons supplying chick hind limb muscles. *J. Physiol.* 284:371–389.

Landmesser, L.T. (1992). Growth cone guidance in the avian limb: a search for cellular and molecular mechanisms. In *The Nerve Growth Cone,* Letourneau, P.C., Kater, SB., Macagno, E.R., eds. New York: Raven Press, pp. 373–385.

Lawrence, P.A. (1989). Cell lineage and cell states in the Drosophila embryo. *Ciba Found. Symp.* 144:131–140.

Leber, S.M., Breedlove, S.M., Sanes, J.R. (1990). Lineage, arrangement, and death of clonally related motoneurons in chick spinal cord. *J. Neurosci.* 10:2451–2462.

Lee, J.J., Ekker, S.C., von Kessler, D., Porter, J.A., Sun, B.I., Beachy, P.A. (1994). Autoproteolysis in hedgehog protein biogenesis. *Science* 266:1528–1537.

Li, H., Witte D.P., Branford, W.W., Aronow, B.J., Weinstein, M., Kaur, S., Wert, S., Singh, G., Schreiner, C.M., Whitsett, J.A. (1994). Gsh-4 encodes a LIM-type homeodomain, is expressed in the developing central nervous system and is required for early postnatal survival. *EMBO J* 13:2876–2885.

Liem, K.F., Jr., Tremml, G., Roelink, H., Jessell, T.M. (1995). Dorsal differentiation of neural plate cells induced by BMP-mediated signals from epidermal ectoderm. *Cell* 82: 969–979.

Liem, K., Tremml, G., Jessell T.M. (1997). Roof plate-dependent patterning in the dorsal neural tube: induction of dorsal commissural interneurons by BMP-mediated signals. Submitted.

Lu, S., Bogarad, L.D., Murtha, M.T., Ruddle, F. H. (1992). Expression pattern of a murine homeobox gene, Dbx, displays extreme spatial restriction in embryonic forebrain and spinal cord. *Proc. Natl. Acad. Sci. U.S.A.* 89:8053–8057.

Lumsden, A. (1990). The cellular basis of segmentation in the developing hindbrain. *Trends Neurosci.* 13:329–335.

Lumsden, A. (1995). A "LIM code" for motor neurons? *Curr. Biol.* 5:491–496.

Lumsden, A., Keynes, R. (1989). Segmental patterns of neuronal development in the chick hindbrain. *Nature* 337:424–428.

Lumsden, A., Sprawson, N., Graham, A. (1991). Segmental origin and migration of neural crest cells in the hindbrain region of the chick embryo. *Development* 113:1281–1291.

Luo, G., Hofmann, C., Bronckers, A.L., Sohocki, M., Brdley, A., Karsenty, G. (1995). BMP-7 is an inducer of nephrogenesis, and is also required for eye development and skeletal patterning. *Genes Dev.* 9:2808–2820.

Lyons, K.M., Hogan, B.L., Robertson, E.J. (1995). Colocalization of BMP 7 and BMP 2 RNAs suggests that these factors cooperatively mediate tissue interactions during murine development. *Mech. Dev.* 50:71–83.

Macdonald, R., Barth, K.A., Xu, Q., Holder, N., Mikkola, I., Wilson, S.W. (1995). Midline signaling is required for Pax gene regulation and patterning of the eyes. *Development* 121:3267–3278.

Maden, M., Gale, E., Kostetskii, I., Zilc, M.H. (1996). Vitamin A-deficient quail embryos have only half a hindbrain and other neural defects. *Curr. Biol.* 6:417–426.

Mahmood, R., Kiefer, P., Guthrie, S., Dickson, C., Mason, I. (1995). Multiple roles for FGF-3 during cranial neural development in the chick. *Development* 121:1399–1410.

Mangelsdorf, D.J., Umensono, K., Evans, R.M. (1994). The retinoid receptors. In *The Retinoids: Biology, Chemistry and Medicine,* Sporn, M.B., Roberts, A.B., Goodman, D.S., eds. New York: Raven Press, pp. 319–349.

Mansouri, A., Stoykova, A., Torres, M., Gruss, P. (1996). Dysgenesis of cephalic neural crest derivatives in *Pax7*$^{-/-}$ mutant mice. *Development* 122:831–838.

Marigo, V., Tabin, C. J. (1996). Regulation of *patched* by Sonic hedgehog in the developing neural tube. *Proc. Natl. Acad. Sci. U.S.A. In press.*

Mark, M., Lufkin, T., Vonesch, J.L., Ruberte, E., Olivo, J.C., Dolle, P., Gorry, P., Lumsden, A., Chambon, P. (1993). Two rhombomeres are altered in Hoxa-1 mutant mice. *Development* 119:319–338.

Marshall, H., Nonchev, S., Sham, M.H., Muchamore, I., Lumsden, A., Krumlauf, R. (1992). Retinoic acid alters hindbrain Hox code and induces transformation of rhombomeres 2/3 into a 4/5 identity. *Nature* 360:737–741.

Marshall, H., Studer, M., Popperl, H., Aparicio, S., Kuroiwa, A., Brenner, S., Krumlauf, R. (1994). A conserved retinoic acid response element required for early expression of the homeobox gene Hoxb-1. *Nature* 370:567–571.

Marti, E., Takada, R., Bumcrot, D.A., Sasaki, H., McMahon, A.P. (1995a). Distribution of Sonic hedgehog peptides in the developing chick and mouse embryo. *Development* 121:2537–2547.

Marti, E., Bumcrot, D.A., Takada, R., McMahon, A.P. (1995b). Requirement of 19K form of Sonic hedgehog for induction of distinct ventral cell types. *Nature* 375:322–325.

Matise, M.P., Lance-Jones, C. (1996). A critical period for the specification of motor pools in the chick lumbosacral spinal cord. *Development* 121:659–669.

McConnell, S.K. (1995). Strategies for the generation of neuronal diversity in the developing central nervous system. *J. Neurosci.* 15:6987–6998.

McConnell, S.K., Kaznowski, C.E. (1991). Cell cycle dependence of laminar determination in developing neocortex. *Science* 254:282–285.
McGinnis, W., Krumlauf, R. (1992). Homeobox genes and axial patterning. *Cell* 68:283–302.
Monaghan, A.P., Kaestner, K.H., Grau, E., Schutz, G. (1993). Postimplantation expression patterns indicate a role for the mouse forkhead/HNF-3 alpha, beta and gamma genes in determination of the definitive endoderm, chordamesoderm and neuroectoderm. *Development* 119:567–578.
Monsoro-Burq, A.-H., Bontoux, M., Teillet, M.-A., Le Douarin, N. M. (1994). Heterogeneity in the development of the vertebra. *Proc. Natl. Acad. Sci. U.S.A.* 91:10435–10439.
Motzny, C.K., Holmgren, R. (1995). The Drosophila cubitus interruptus protein and its role in the wingless and hedgehog signal transduction pathways. *Mech. Dev.* 52:137–150.
Moury, J.D., Jacobson, A.G. (1990). The origins of neural crest cells in the axolotl. *Dev. Biol.* 141:243–253.
Nieto, M.A., Gilardi, H.P., Charnay, P., Wilkinson, D.G. (1992). A receptor protein tyrosine kinase implicated in the segmental patterning of the hindbrain and mesoderm. *Development* 116:1137–1150.
Nieto, M.A., Sargent, M.G., Wilkinson, D.G., Cooke, J. (1994). Control of cell behavior during vertebrate development by *slug*, a zinc finger gene. *Science* 264:835–839.
Noden, D.M. (1988). Interactions and fates of avian craniofacial mesenchyme. *Development* 103 Suppl:121–140.
Nonchev, S., Vesque, C., Maconochie, M., Setanidou, T., Ariza-McNaughton, L., Frain, M., Marshall, H., Sham, M.H., Krumlauf, R., Charnay, P. (1996). Segmental expression of Hoxa-2 in the hindbrain is directly regulated by Krox-20. *Development* 122:543–554.
Oppenheim, R. W., Shneiderman, A., Shimizu, I., Yaginuma, H. (1988). Onset and development of intersegmental projections in the chick embryo spinal cord. *J. Comp. Neurol.* 275:159–180.
Orenic, T.V., Slusarski, D.C., Kroll, K.L., Holmgren, R.A. (1990). Cloning and characterization of the segment polarity gene cubitus interruptus Dominant of Drosophila. *Genes Dev.* 4:1053–1067.
Papalopulu, N., Lovell-Badge, R., Krumlauf, R. (1991). The expression of murine Hox-2 genes is dependent on the differentiation pathway and displays a collinear sensitivity to retinoic acid in F9 cells and Xenopus embryos. *Nucleic Acids Res.* 19:5497–5506.
Perrimon, N. (1995). Hedgehog and beyond. *Cell* 80:517–520.
Pfaff, S.L., Mendelsohn, M., Stewart, C.L., Edlund, T., Jessell, T.M. (1996). Requirement for LIM homeobox gene Isl1 in motor neuron generation reveals a motor neuron-dependent step in interneuron differentiation. *Cell* 84:1–20.
Pituello, F., Yamada, G., Gruss, P. (1995). Activin A inhibits Pax-6 expression and perturbs cell differentiation in the developing spinal cord in vitro. *Proc. Natl. Acad. Sci. U.S.A.* 92:6952–6956.
Placzek, M. (1995). The role of the notochord and floor plate in inductive interactions. *Curr. Opin. Genet. Dev.* 5:499–506.
Placzek, M., Tessier-Lavigne, M., Yamada, T., Jessell, T.M., Dodd, J. (1990). Mesodermal control of neural identity: floor plate induction by the notochord. *Science* 250:985–988.
Placzek, M., Jessell, T.M., Dodd, J. (1993). Induction of floor plate differentiation by contact-dependent, homeogenetic signals. *Development* 117:205–218.
Porter, J.A., von, K.D., Ekker, S.C., Young, K.E., Lee, J.J., Moses, K., Beachy, P.A. (1995). The product of hedgehog autoproteolytic cleavage active in local and long-range signaling. *Nature* 374:363–366.
Porter, J.A., Young, K.E. and Beachy, P.A. (1996). Cholesterol modification of hedgehog signaling proteins in animal development. *Science,* 274 255–259.
Puelles, L., Rubenstein, J.L. (1993). Expression patterns of homeobox and other putative

regulatory genes in the embryonic mouse forebrain suggest a neuromeric organization. *Trends Neurosci.* 16:472–479.

Rangini, Z., Ben, Y.A., Shapira, E., Gruenbaum, Y., Fainsod, A. (1991). CHox E, a chicken homeogene of the H2.0 type exhibits dorso-ventral restriction in the proliferating region of the spinal cord. *Mech. Dev.* 35:13–24.

Riddle, R.D., Johnson, R.L., Laufer, E., Tabin, C. (1993). Sonic hedgehog mediates the polarizing activity of the ZPA. *Cell* 75:1401–1416.

Riddle, R.D., Ensini, M., Nelson, C., Tsuchida, T., Jessell, T.M., Tabin, C. (1995). Induction of the LIM homeobox gene *Lmx1* by WNT7a establishes dorsoventral pattern in the vertebrate limb. *Cell* 83:631–640.

Roach, F.C. (1945). Differentiation of the central nervous system after axial reversals of the medullary plate of ambystoma. *J. Exp. Zool.* 99:53–77.

Roelink, H., Augsburger, A., Heemskerk, J., Korzh, V., Norlin, S., Ruiz i Altaba, A., Tanabe, Y., Placzek, M., Edlund, T., Jessell, T.M., Dodd, J. (1994). Floor plate and motor neuron induction by vhh-1, a vertebrate homolog of hedgehog expressed by the notochord. *Cell* 76:761–775.

Roelink, H., Porter, J. A., Chiang, C., Tanabe, Y., Chang, D.T., Beachy, P.A., Jessell, T.M. (1995). Floor plate and motor neuron induction by different concentrations of the amino-terminal cleavage product of sonic hedgehog autoproteolysis. *Cell* 81:445–455.

Rubenstein, Shimamura. (1996). *Neuronal Development,* Cowan, W.M., Jessell, T.M., Zipursky, S.L., eds. Oxford: Oxford University Press.

Ruiz i Altaba, A. (1994). Pattern formation in the vertebrate neural plate [published erratum appears in *Trends Neurosci* 1994 Jul;17(7):312]. *Trends Neurosci.* 17:233–243.

Ruiz i Altaba, A., Prezioso, V.R., Darnell, J.E., Jessell, T.M. (1993). Sequential expression of HNF-3 beta and HNF-3 alpha by embryonic organizing centers: the dorsal lip/node, notochord and floor plate. *Mech. Dev.* 44:91–108.

Ruiz i Altaba, A., Placzek, M., Baldassare, M., Dodd, J., Jessell, T.M. (1995a). Early stages of notochord and floor plate development in the chick embryo defined by normal and induced expression of HNF3β. *Dev. Biol.* 170;299–313.

Ruiz i Altaba, A., Roelink, H., Jessell, T.M. (1995b). Restrictions to floor plate induction by hedgehog and winged-helix genes in the neural tube of frog embryos. *Mol. Cell. Neurosci.* 6:106–121.

Sasai, Y., Lu, B., Steinbeisser, H., De Robertis, E. (1995). Regulation of neural induction by the Chd and Bmp-4 antagonistic patterning signals in Xenopus. *Nature* 376:333–336.

Sasaki, H., Hogan, B.L. (1993). Differential expression of multiple fork head related genes during gastrulation and axial pattern formation in the mouse embryo. *Development* 118:47–59.

Sasaki, H., Hogan, B.L. (1994). HNF-3 beta as a regulator of floor plate development. *Cell* 76:103–115.

Schneider-Maunoury, S., Topilko, P., Seitandou, T., Levi, G., Cohen-Tannoudji, M., Pournin, S., Babinet, C., Charnay, P. (1993). Disruption of Krox-20 results in alteration of rhombomeres 3 and 5 in the developing hindbrain. *Cell* 75:1199–1214.

Schoenwolf, G.C., Smith, J.L. (1990). Mechanisms of neurulation: traditional viewpoints and recent advances. *Development* 109:243–270.

Schoenwolf, G.C., Bortier, H., Vakaet, L. (1989). Fate mapping the avian neural plate with quail/chick chimeras: origin of prospective median wedge cells. *J. Exp. Zool.* 249:271–278.

Sechrist, J., Nieto, A.M., Zamanian, R.T., Bronner-Fraser, M. (1995). Regulative response of the cranial neural tube after neural fold ablation: spatiotemporal nature of neural crest regeneration and up-regulation of *Slug. Development* 121:4103–4115.

Selleck, M., Bronner-Fraser, M. (1995). Origins of the avian neural crest; the role of neural plate-epidermal interactions. *Development* 121:525–538.

Sham, M.H., Vesque, C., Nonchev, S., Marshall, H., Frain, M., Gupta, R.D., Whiting, J.,

Wilkinson, D., Charnay, P., Krumlauf, R. (1993). The zinc finger gene Krox20 regulates HoxB2 (Hox2.8) during hedgehog segmentation. *Cell* 72:183–196.

Shimeld, S.M., McKay, I.J., Sharpe, P.T. (1996). The murine homobox gene Msx-3 shows highly restricted expression in the developing neural tube. *Mech. Dev.* 55:201–210.

Simeone, A., Acampora, D., Arcioni, L., Andrews, P.W., Boncinelli, E., Mavilio, F. (1990). Sequential activation of HOX2 homeobox genes by retinoic acid in human embryonal carcinoma cells. *Nature* 346:763–766.

Simon H., Lumsden, A. (1993). Rhombomere-specific origin of the contralateral vestibulo-acoustic efferent neurons and their migration across the embryonic midline. *Neuron* 11:209–220.

Simon, H., Hornbruch, A., Lumsden, A. (1995). Independent assignment of antero-posterior and dorso-ventral positional values in the developing chick hindbrain. *Curr. Biol.* 5:205–214.

Stirling, R.V., Liestol, K., Summerbell, D., Glover, J.C. (1995). The segmental precision of the motor projection to the intercostal muscles in the developing chicken embryo. A differential labeling study using fluorescent tracers. *Anat. Embryol.* 191:397–406.

Stuart, E.T., Kioussi, C., Gruss, P. (1994). Mammalian Pax genes. *Annu. Rev. Gen.* 28:219–236.

Studer, M., Popperl, H., Marshall, H., Kuroiwa, A., Krumlauf, R. (1994). Role of a conserved retinoic acid response element in rhombomere restriction of Hoxb-1. *Science* 265:1728–1732.

Talbot, W.S., Trevarrow, B., Halpern, M.E., Melby, A.E., Farr, G., Postlethwait, J.H., Jowett, T., Kimmel, C.B., Kimelman, D. (1995). A homeobox gene essential for zebrafish notochord development. *Nature* 378:150–157.

Tanabe, Y., Roelink, H., Jessell, T. (1995). Induction of motor neurons by Sonic hedgehog is independent of floor plate differentiation. *Curr. Biol.* 5:651–658.

Tessier-Lavigne, M.L., Goodman, C. (1996). Mechanisms of axon guidance. In *Neuronal Development,* Cowan, W.M., Jessell, T.M., Zipursky, S.L., eds. Oxford: Oxford University Press.

Thor, S. (1995). The genetics of brain development: conserved programs in flies and mice. *Neuron* 15:975–977.

Tokumoto, M., Gong, Z., Tsubokawa, T., Hew, C.L., Uyemura, K., Hotta, Y., Okamoto, H. (1995). Molecular heterogeneity among primary motoneurons and within myotomes revealed by the differential mRNA expression of novel Islet-1 homologs in embryonic zebrafish. *Dev. Biol.* 171:578–589.

Tosney, K.W., Hotary, K.B., Lance-Jones, C. (1995). Specifying the target identity of motoneurons. *Bioessays* 17:379–382.

Tremblay, P., Pituello, F., Gruss, P. (1996). Inhibition of floor plate differentiation by *pax-3:* evidence from ectopic expression in transgenic mice.

Tsuchida, T., Ensini, M., Morton, S.B., Baldassare, M., Edlund, T., Jessell, T.M., Pfaff, S.L. (1994). Topographic organization of embryonic motor neurons defined by expression of LIM homeobox genes. *Cell* 79:957–970.

Vaage, S. (1969). The segmentation of the primitive neural tube in chick embryos (Gallus domesticus). A morphological, histochemical and autoradiographical investigation. *Erg. Anat. Entwick.* 41:3–87.

van Straaten, H.W., Hekking, J.W. (1991). Development of floor plate, neurons and axonal outgrowth pattern in the early spinal cord of the notochord-deficient chick embryo. *Anat. Embryol.* 184:55–63.

van Straaten, H.W.M., Hekking, J.W.M., Wiertz-Hoessels, E.L., Thors, F., Drukker, J. (1988). Effect of the notochord on the differentiation of a floor plate area in the neural tube of the chick embryo. *Anat. Embryol.* 177:317–324.

Wang, M.Z., Jin, P., Bumcrot, D.A., Marigo, V., McMahon, A.P., Wang, E.A., Woolf, T., Pang, K. (1995). Induction of dopaminergic neuron phenotype in the midbrain by Sonic hedgehog protein. *Nat. Med.* 1:1184–1188.

Weinstein, D.C., Ruiz i Altaba, A., Chen, W.S., Hoodless, P., Prezioso, V.R., Jessell, T.M., Darnell, J.J. (1994). The winged-helix transcription factor HNF-3 beta is required for notochord development in the mouse embryo. *Cell* 78:575–588.

Wilkinson, D.G., Bhatt, S., Chavrier, P., Bravo, R., Charnay, P. (1989). Segment-specific expression of a zinc-finger gene in the developing nervous system of the mouse. *Nature* 337:461–464.

Wingate, R.J.T., Lumsden, A. (1996). Persistence of rhombomeric organization in the post-segmented hindbrain. *Development* 122:2143–2152.

Wizenmann, A., Lumsden, A. (1997). Specific adhesion segregates odd- and even-numbered rhombomeres. *Development. Submitted.*

Xu, Q., Alldus, G., Holder, N., Wilkinson, D.G. (1995). Expression of truncated Sek-1 receptor tyrosine kinase disrupts the segmental restriction of gene expression in the Xenopus and zebrafish hindbrain. *Development* 121:4005–4016.

Yaginuma, H., Shiga, T., Homma, S., Ishihara, R., Oppenheim, R.W. (1990). Identification of early developing axon projections from spinal interneurons in the chick embryo with a neuron specific beta-tubulin antibody: evidence for a new "pioneer" pathway in the spinal cord. *Development* 108:705–716.

Yamada, T., Placzek, M., Tanaka, H., Dodd, J., Jessell, T.M. (1991). Control of cell pattern in the developing nervous system: polarizing activity of the floor plate and notochord. *Cell* 64:635–647.

Yamada, T., Pfaff, S.L., Edlund, T., Jessell, T.M. (1993). Control of cell pattern in the neural tube: motor neuron induction by diffusible factors from notochord and floor plate. *Cell* 73:673–686.

Zhadanov, B.M., Bertuzzi, S., Taira, M., Dawid, I.B., Westphal, H. (1995). Expression pattern of the murine LIM class homeobox gene *lhx3* in subsets of neural and neuroendocrine tissues. *Dev. Dyn.* 202:354–364.

9

The role of *Hox* genes in hindbrain development

MARIO R. CAPECCHI

The mouse hindbrain is a particularly attractive target for molecular genetic analysis. Although it is an enormously complex structure controlling numerous autonomic and voluntary functions, the complexity is generated during development by a rather simple and commonly used developmental paradigm: the generation of complexity by the diversification of repeated units. Early in its development, the mouse hindbrain anlage is transiently subdivided along its rostrocaudal axis into eight metameric segments called rhombomeres (Orr, 1887; Vaage, 1969; Lumsden and Keynes, 1989). Rhombomeres can function as compartments that limit cell movement and thereby create centers that are capable of independent development and diversification through localized gene activity and cell interactions (Fraser et al., 1990). However, it is also apparent that through intercompartmental communication, these units can function collectively to form a scaffold upon which a coherent neural network is built (Glover and Petursdottir, 1991; Clarke and Lumsden, 1993). It is fascinating that these developmental compartments, which play such a critical role in hindbrain development, should have only a fleeting existence. Once the need for more extensive cell migration arises, rhombomere boundaries dissipate and cell movement resumes.

One factor that contributes to mouse hindbrain development's particular amenability to molecular genetic analysis has been the identification of a genetic network that is used to specify cells within rhombomeres. The genes of this network are members of the *Hox* complex, a set of 39 closely related transcription factors that, in a more general context, are used to pattern the entire mammalian embryo. The rostral expression boundaries of a subset of these genes correspond strictly with rhombomere boundaries and each rhombomere can be characterized by the expression of a unique set of *Hox* genes (for review, see McGinnis and Krumlauf, 1992; Krumlauf, 1994). Establishing the individual as well as the collective roles of these *Hox* genes in hindbrain development provides an inviting entry point for unraveling the molecular circuitries responsible for building this neural network. Since the midbrain and the forebrain also appear to be built using a metameric paradigm (Puellus and Rubenstein, 1993), the lessons learned from the analysis of hindbrain development should also contribute to our understanding of how these even more complex structures are built.

In this review I discuss our current understanding of hindbrain development with a particular emphasis on exploring the roles of *Hox* genes in mediating this process. The chapter is divided into three sections: Rhombomeres, *Hox* Genes, and Neural Crest. The first section explores properties of rhombomeres that influence how the neuronal architecture of the hindbrain is formed. The section on *Hox* genes examines their organization, expression patterns, and genetic function. From the genetic analysis of these genes it is hoped that not only will the roles of *Hox* genes in specifying neuronal phenotypes emerge, but that the mutants themselves will provide a dissection of this complex system into workable subunits that will serve as substrates for further analysis. Finally, the positional cues for patterning the head and neck appear to be dependent on the underlying embryonic organization of the central nervous system. The conduit for transferring this positional information is the neural crest, the subject of the final section.

Rhombomeres

What we know about early vertebrate hindbrain organization and development has been derived largely from the analysis of the chick hindbrain. Its comparatively large size and accessibility during embryogenesis have made it a particularly attractive system for experimental manipulation. It is often tacitly assumed that what is observed in the chick hindbrain will be mirrored in the mammal. However, this assumption has not always been found to be true. As an example, the cell bodies associated with the branchiomotor system are distributed in the chick brain with a two-rhombomere periodicity (Lumsden and Keynes, 1989). Early in development their nuclei appear to be confined to rhombomeres 2, 4, and 6 (r2, r4, and r6), and subsequently they are joined by neurons in the adjacent caudal rhombomeres. Thus by stage 17 in the chick, the branchiomotor nerves V, VII, and IX emerge from consecutive pairs of rhombomeres and are in register with their targets in the first three branchial arches. This pattern is not duplicated in the mouse (Fig. 9–1). At a comparable stage of mouse development, nuclei associated with the Vth nerve, the trigeminal nerve, are found in r1, r2, and r3. The mouse VIIth or facial nerve motor pool appears to be similar to that observed in the chick hindbrain, occupying primarily r4 and r5, whereas the IXth (glossopharyngeal) motor pool appears to be confined to r6 (Carpenter et al., 1993). This example is provided not as an indicator that hindbrain development diverges radically between mouse and chick but rather as a cautionary note that development in these two species is likely to be different in detail and that these differences may be as informative as the similarities.

Single-cell marking experiments in the chick hindbrain have shown that rhombomere boundaries function as barriers to cell movement and therefore limit the mixing of cells between neighboring rhombomeres (Fraser et al., 1990). As a consequence, each rhombomere can be considered as a group of cells that

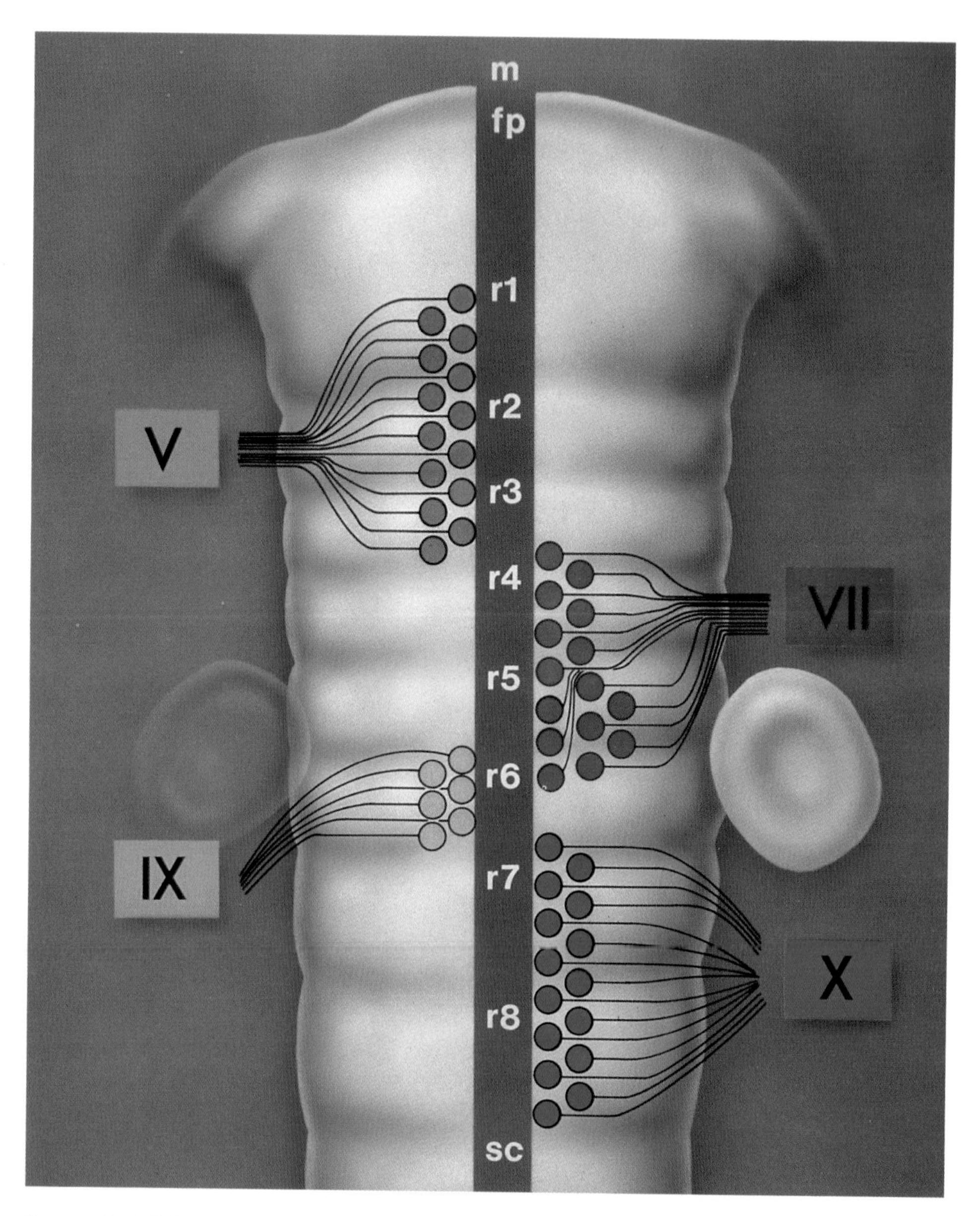

FIGURE 9–1 Schematic representation of the position of efferent neurons in the murine hindbrain. Neurons contributing to cranial nerve V (green circles) are positioned in rhombomeres (r) 1, 2, and 3. The axons from these neurons leave the hindbrain through the lateral wall of r2. Neurons contributing to the cranial nerve VII (red circles) are positioned medially in r4 and r5 as well as more laterally in r5. The medial neurons in r4 and r5 are motoneurons that innervate the facial muscles. The more lateral neurons in r5 represent preganglionic parasympathetic neurons that innervate, for example, the salivary glands. VIIIth nerve efferents also are positioned in r4 and exit the hindbrain at r4. The IXth nerve motoneurons (yellow circles) are positioned in r6 and leave the hindbrain at the same level. Xth nerve motoneurons (blue circles) are positioned medially in r7 and r8 and leave the hindbrain in several rootlets through the walls of both rhombomeres. Rhombomere boundaries are depicted by darker shading. The otocyst is positioned lateral to the hindbrain at the level of r5 and r6. mfp, midline floor plate; sc, spinal cord.

share cell lineages and that function as units able to establish their phenotype based on local information within each compartment. Transplantation of chick rhombomeres to ectopic positions illustrates that cells from even and odd rhombomeres are particularly immiscible (Guthrie and Lumsden, 1991; Guthrie et al., 1993). Juxtaposition of either even with even or odd with odd rhombomeres prohibits the formation of normal rhombomere boundaries between these compartments. As a consequence, extensive cell mixing occurs between these rhombomeres. However, normal rhombomere boundaries are not absolute barriers to cell movement. Approximately 5% of cells from one rhombomere have been observed to cross normal boundaries and settle in adjacent rhombomeres (Birgbauer and Fraser, 1994). It is not clear whether such crossing of boundaries is a directed process or a random one, representing leakage in the system.

Chick rhombomeres acquire specific character early in their development. This is apparent both morphologically and molecularly. For example, in the chick, maturation of neurons is first apparent in rhombomeres 2, 4, and 6 (Lumsden and Keynes, 1989). Cell-surface antigens are expressed in a segmental manner coincident with rhombomeres, suggesting intrinsic cellular differences and possibly reflecting differences in adhesion properties (Layer and Alber, 1990; Trevarrow et al., 1990; Kuratani and Eichele, 1993; Heyman et al., 1995). Each rhombomere also expresses a distinct combination of transcription-factor-encoding genes, as well as of genes encoding growth factors and signal transduction molecules (Wilkinson et al., 1988; 1989a,b; Gilardi-Hebenstreit et al., 1992; Nieto et al., 1992; Becker et al., 1994). Particularly prominent among the genes differentially expressed in rhombomeres are the *Hox* genes. The combination of *Hox* genes expressed within a rhombomere correlates both temporally and spatially with the position of the genes on the chromosomes, such that a 3′ gene in the *Hox* linkage group is expressed prior to and in a more rostral rhombomere than its 5′ neighbor (Murphy et al., 1989; Wilkinson et al., 1989b; Sundin and Eichele, 1990; Hunt et al., 1991; Prince and Lumsden, 1994). Since the hindbrain develops in a rostrocaudal fashion, this correlation suggests that *Hox* genes may be used to generate this progression of development by sequentially activating combinations of selector genes within successive rhombomeres. Successive activation of *Hox* genes within a linkage group could be an intrinsic property of the complex, thereby converting temporal instructions into directed morphological output (i.e., regionalization along an axis).

When is the identity of a rhombomere specified? Rhombomere transplantation experiments suggest that this choice occurs early (Guthrie et al., 1992; Kuratani and Eichele, 1993). Transplantation of prospective rhombomere 4 to the position of r2, prior to overt rhombomere boundary formation, does not alter the specific gene expression pattern normally associated with r4. With continued development the ectopic rhombomeres generate neurons whose morphology and behavior are characteristic of r4, rather than r2. Likewise, transplantation of prospective r2 to the position of r4 does not result in expression of r4-specific markers in the ectopic rhombomere. In these experiments, the fate of cells within these rhombomeres appears to be set prior to the time of transplantation and, importantly, prior to morphological evidence of rhombomere boundary formation. However, not all aspects of rhombomere cell identity appear to be specified at the same time. When Simon et al. (1995) compared transplantation of r4 to the position of r2 with or without a rotation, so as to also alter the position of the ectopic rhombomere with respect to the dorsoventral axis of the embryo, they noted that although the general fate of r4 cells had been set prior to transplantation, the choice of specific r4 cell fates along the dorsoventral axis could be influenced by the local environment (i.e., proximity to the neural floor plate). Thus it appears that at the seven- to ten-somite stage the dorsoventral positional values are still labile, whereas the rostrocaudal values are already fixed.

Although transplantation of presumptive posterior rhombomeres to a more anterior position in the hindbrain does not alter their subsequent fate, transplantation of presumptive anterior rhombomeres to postotic positions in the hindbrain (i.e., posterior to r6) does alter their fate (Grapin-Botton et al., 1995; Itasaki et al., 1996). Posteriorization of the ectopic rhombomeres occurs whether the transplantation is carried out before or after rhombomere boundary formation. The identity adopted by an ectopic rhombomere is dependent both on its source and on the position of the graft. At a given graft site, there is a graded ability in rhombomeres to respond to the new environment, with posterior rhombomeres responding more readily than anterior rhombomeres. However, the response by the same set of rhombomeres is stronger as the position of the graft is moved to more posterior levels, suggesting a stronger signal in more posterior regions. Finally, it appears that even or odd rhombomeres will more readily maintain that identity than switch from one to the other (Itasaki et al., 1996).

Based on the variability of the induced changes of *Hox* gene expression when multiple rhombomeres were transplanted to the same site, Grapin-Botton et al. (1995) suggested that a planar, neurectodermal signal may be responsible for mediating the changes in rhombomere identity. In contrast, Itasaki et al. (1996) suggested that the source of the signal was paraxial mesoderm. Supporting their hypothesis, Itasaki et al. showed that transplantation of posterior somites to more anterior levels resulted in expression by rhombomeres adjacent to the transplant of *Hox* genes that are normally restricted to more posterior regions. The two hypotheses for the source of the posteriorizing signal are not mutually exclusive, since both tissues could be sources of similar or separate signals. However, Itasaki et al. directly looked for the source of such a signal in neurectoderm by transplanting posterior neurectoderm to a more anterior position in the hindbrain. In these experiments, they failed to observe any changes in *Hox* gene expression in the host tissues adjacent to the graft.

These experiments leave little doubt that preprogramming of rhombomeres occurs prior to the formation of rhombomere boundaries; however, such programming is reversible and can be influenced by environmental factors later in development. Thus, rhombomeres do not behave in a strictly autonomous manner but rather appear capable of continually assessing their identity and making appropriate adjustments. The transplantation experiments emphasize developmental plasticity, which is a recurring theme in our own research. As a consequence of mutations in *Hox* genes, we often observe variability in the expressivity of the resulting phenotype. This variability is likely to arise from the use of alternative members of the *Hox* complex to compensate for the mutation. Particularly striking is the variability observed in affected tissues or structures on opposite sides of the same mutant mouse. Such variability cannot be explained in terms of variability in genetic background. Rather it suggests that even in a uniform genetic background there is stochastic variability in the response to the mutation and again emphasizes plasticity during development. The variability in response to a mutation is likely to reflect normal variability in the use of alternative genetic pathways to implement localized adjustments to microenvironmental fluctuations. By using parallel genetic pathways and the ability to make adjustments, the system becomes much more sophisticated and less prone to failure.

The transplantation of rostral rhombomeres to postotic regions of the hindbrain illustrates that rhombomeres can assess their position in the neural tube through signals emanating in part from posterior paraxial mesoderm. Is there evidence for interrhombomere communication? The answer is yes. Neural crest cells emerge from the dorsal margin of each rhombomere. However, the amount of neural crest derived from r1/r2, r4, and r6–r8 is greater than that from r3 and r5. This difference results in part from selective activation of neural crest apoptosis in rhombomeres 3 and 5 (Lumsden et al., 1991). Induction of this pathway in r3 and r5 is dependent on the contiguous presence of even-numbered rhombomeres (Graham et al., 1993). Transplantation of r3 into the position of r4 so that the ectopic rhombomere is sequestered from even-rhombomere neighbors blocks neural crest apoptosis in r3. Similarly in culture, isolated r3s show robust production of neural crest cells, whereas r3s in isolated r2/r3 or r3/r4 pairs do not. Exposure of isolated r3 explants to purified bone morphogenic protein 4 (BMP4) induces neural crest apoptosis. Further support for interrhombomere communication comes from the demonstration that activation or maintenance of *Krox20* expression in r3 and repression of *follistatin* expression in r3 are also dependent on the contiguous presence of even rhombomeres (Graham and Lumsden, 1996).

The genes that are responsible for establishing the segmental pattern in the hindbrain have not yet been identified. Transcription factors such as *Krox*20 or kreisler appear to have important roles in the maintenance of rhombomeres. *Krox*20, which contains zinc finger DNA-binding domains, is expressed in presumptive rhombomeres 3 and 5 (Chavrier et al., 1988; Wilkinson et al., 1989a). Targeted disruption of *Krox20* in mice has shown that formation of presumptive rhombomeres 3 and 5 is initiated, but that following this initiation, these rhombomeres are completely lost (Schneider-Manoury et al., 1993; Swiatek and Gridley, 1993). These findings suggest that *Krox20* is required for the outgrowth or maintenance of these rhombomeres, as opposed to their specification. The *kreisler* mutation was originally identified in mice by defects in hearing and vestibular function. However, these defects appear to be secondary to defects in hindbrain formation (Deol, 1964; Frohman et al., 1993; Cordes and Barsh, 1994; McKay et al., 1994). Rhombomere segmentation in these mutant mice is abnormal from r4 through r6. These defects have been interpreted as a simple deletion of r5 and r6 (McKay et al., 1994). However, such mice would be expected to die at birth from cardiovascular and pulmonary dysfunction (Carpenter et al., 1993; Dollé et al., 1993; Mark et al., 1993) where *kreisler* mutant mice live a normal life span and are fertile. Determining exactly which hindbrain nuclei are missing in *kreisler* mutants requires further analysis. Nevertheless, this gene product is a good candidate for functioning upstream of *Hox* genes, as well as of *Krox*20, in the transcriptional cascade that is used to form the caudal portion of the hindbrain.

The *Hox* genes themselves may be used to establish the segmental pattern in the hindbrain. This would be a departure from the role of the analogous *HomC* genes in *Drosophila*. In the latter organism, *gap, pair-rule,* and *segment-polarity* genes are used to establish the parasegmental pattern in the embryo and the *HomC* genes are then used to establish cell identity within the parasegments. A mutation in *HomC* genes does not change the number of parasegments in the *Drosophila* embryo, but rather the identity of parasegments. However, in the mouse, a mutation in *hoxa-1* does alter the rhombomere pattern (Carpenter et al., 1993; Dollé et al., 1993; Mark et al., 1993). Specifically, the hindbrain is approximately 20% shorter along the rostrocaudal axis. In addition, r5 nuclei that normally send efferent axons through the VIIth nerve are missing, and a series of molecular markers specific to r5 is absent. Based on these observations, *hoxa-1* appears minimally to be required for the maintenance or outgrowth of r5. Defects are also apparent in r4 and r6. Establishing whether *hoxa-1* is used to initiate the formation of the normal rhombomere pattern will require further analysis. The analysis of interactions between the *hoxa-1* and *hoxb-1* mutations should be particularly informative on this issue.

HOX GENES

Hox genes encode helix-turn-helix transcription factors. All metazoa examined share a linked cluster of *Hox* genes (for a review, see Holland and Garcia-Fernandez, 1996). It is presumed that these genes play a fundamental role in controlling axial patterning in all animals (Slack et al., 1993). The 39 mouse *Hox* genes are arranged in four linkage groups designated *Hox A, B, C,* and *D* (Fig. 9–2). Early in vertebrate evolution, an ancestral complex shared with invertebrates was duplicated twice to give rise to the four linkage groups (Kappen et al., 1989). As a consequence, corresponding genes on separate linkage groups, called paralogues, are most closely related to each other. Based on sequence similarities and position on the chromosome, the vertebrate *Hox* genes have been subdivided into 13 paralogous groups (Scott, 1992). The relationship of the mammalian *Hox* genes to the eight *Drosophila Homeotic* genes (*HomC*) is also shown in Figure 9–2. As already mentioned, the physical order of *Hox* genes on the chromosome correlates with the order of the anterior boundaries of expression of each gene along the anteroposterior axis of the embryo (Lewis, 1978; Duboule and Dollé, 1989; Graham et al., 1989). Expression of the 3′ genes extends more rostrally, while expression of those on the 5′ end of the complex is limited to more posterior regions. This correlation holds for the nervous system

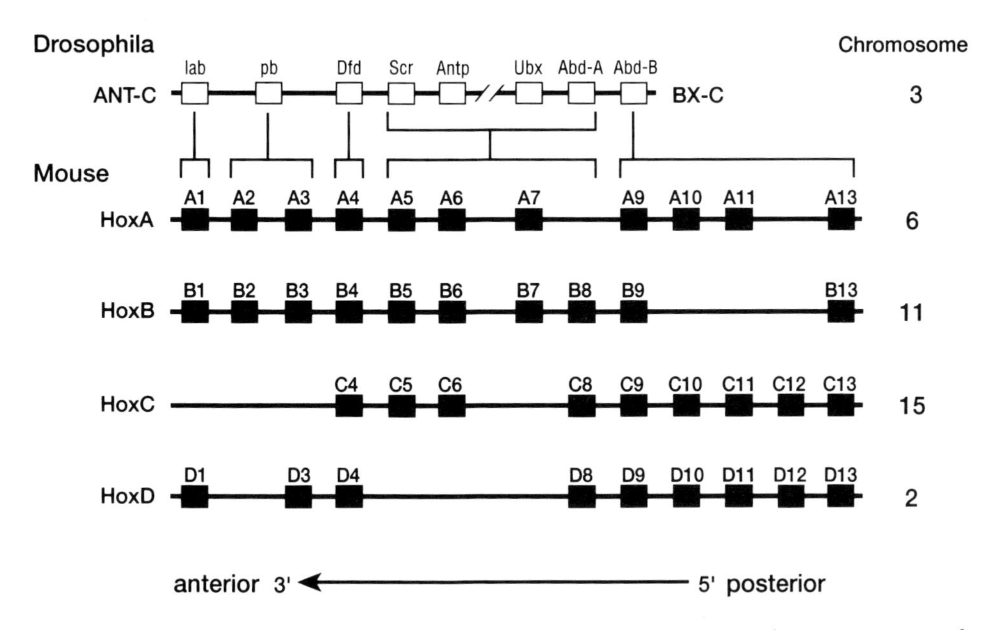

FIGURE 9–2 Alignment of the vertebrate *Hox* linkage groups into paralogous groups and comparison with *Drosophila HomC*. The *Drosophila HomC* is depicted on top, and the four mouse linkage groups HoxA, B, C, and D below. The closed boxes denote the known *Hox* genes. These genes are divided into 13 paralogous groups (Scott, 1992). The relationships, based on DNA and protein sequence similarities, between the eight *Drosophila HomC* genes and the vertebrate *Hox* genes are denoted by solid lines. Thus the *Drosophila labial* gene is most closely related to *hoxa-1* and its paralogues and so on. *Hoxb-13* represents a newly discovered member of the *hoxa-13* paralogous family (Zeltser, Desplan, Heintz, unpublished results). All of the vertebrate *Hox* genes are transcribed from the same DNA strand. The polarity of transcription is indicated at the bottom of the figure.

and for the prevertebral column as well as for subsystems such as the gut, gonadal tissues, and the limbs (Yokouchi et al., 1991, 1995; Roberts et al., 1995; Dollé et al., 1989, 1991; Izpisúa-Belmonte et al., 1991a; Haack and Gruss, 1993). The order of *Hox* genes on the chromosome also correlates with their times of activation with 3′ members being activated before their 5′ neighbors (Izpisúa-Belmonte et al., 1991b). Maturation of the mammalian embryo is directed along an axis. This is true in the hindbrain, in somite-derived structures, and in the subsystems mentioned above. It is therefore tempting to suggest that the ordered activation of *Hox* genes has a causative role in this directed process—that is, that a fundamental, unifying property of *Hox* genes is that they are capable of orchestrating a series of temporal events which are translated into a morphological vector (Duboule, 1994; Davis et al., 1995).

Interestingly, amphioxus, a cephalochordate, has a single linkage group of *Hox* genes with ten or more members (Holland et al., 1994). The anatomy of amphioxus is consistent with expectation for a creature resembling our vertebrate ancestor. It has a dorsal nerve cord, an axial notochord, and bilaterally paired blocks of muscles along the entire length of its body that develop from somites. In amphioxus, the correlation of regionally restricted patterns of *Hox* gene expression with position on the chromosome holds for the nervous system but not for the somites. This observation suggests that the use of *Hox* genes to regionalize the nervous system may be more primitive than the use of this complex to regionalize mesoderm-derived structures (Holland and Garcia-Fernandèz, 1996). I will come back to this point later in the chapter.

It is anticipated that *hoxa-1* through *hoxa-4* and their paralogous family members will be found to function in patterning the hindbrain (Fig. 9–2). This assumption is based on expression patterns. The rostral expression boundaries of *HoxA* genes in the hindbrain are shown in Figure 9–3. *Hox* gene expression commences early in embryogenesis, typically during the formation of the primitive streak (i.e., mouse gestation day 7.5, E7.5). From this posterior position, expression moves rostrally to a specific anterior boundary characteristic for each *Hox* gene, forming a nested set of transcripts extending along the anteroposterior axis. From Figure 9–3, it is evident that *Hox* gene expression respects rhombomere boundaries. However, the anterior limit of expression, for the most anteriorly expressed *Hox* genes, is reached prior to the formation of rhombomere boundaries. What is not evident from Figure 9–3 is that *Hox* gene expression is very dynamic and can change rapidly during development. For example, expression of *hoxa-1* reaches the presumptive boundary between r3 and r4 by 8 days of gestation and then promptly recedes caudally. By E8.5, *hoxa-1* expression is no longer detectable within the hindbrain. Also, *Hox* gene expression is not uniform from one rhombomere to another. Each rhombomere is characterized by the expression of a unique set of highly expressed *Hox* genes. This is referred to as a "*Hox* code" and has been suggested to be responsible for specifying the identity of each rhombomere. In the hindbrain, most paralogous *Hox* genes share the same anterior limit of expression (Hunt et al., 1991). However, notable exceptions exist. For example, *hoxa-2* expression extends into rhombomere 2, whereas the rostral boundary of *hoxb-2* expression is limited to the r2/r3 boundary. The pattern of *hoxb-1* expression is similar to that of *hoxa-1*, except that during the period of recession, *hoxb-1* expression is upregulated in r4, leaving behind a stripe of strong expression restricted to r4. During this second phase of *Hox* gene expression, the anterior boundaries of other *Hox* genes do not appear to change in the hindbrain (i.e., do not recede); however, the expression in selected rhombomeres is upregulated. The pattern of upregulation differs among paralogous members.

A potential upstream regulator of *Hox* gene expression is the retinoid signaling pathway. Exposure of vertebrate embryos to retinoic acid alters *Hox* gene expression and induces neuronal and vertebral malformations (Kessel and Gruss, 1991; Conlon and Rossant, 1992; Kessel, 1992; Maden and Holder, 1992; Marshall et al., 1992; Morriss-Kay, 1993). Retinoic-acid–responsive elements have been identified in close proximity to *hoxa-1*, *hoxb-1*, and *hoxd-4* (Langston and Gudas, 1992; Marshall et al., 1994; Frasch et al., 1995; Ogura and Evans, 1995a,b; Popperl and Featherstone, 1993). Mutations in retinoic acid receptors cause mutant phenotypes that closely resemble those observed in *Hox* gene mutant mice (Lohnes et al., 1993). Finally, exposure of embryonic stem cells and embryos to retinoic acid causes the sequential induction of *Hox* genes, which parallels their physical order on the chromosome (Simeone et al., 1990, 1991; Papalopulu et al., 1991;

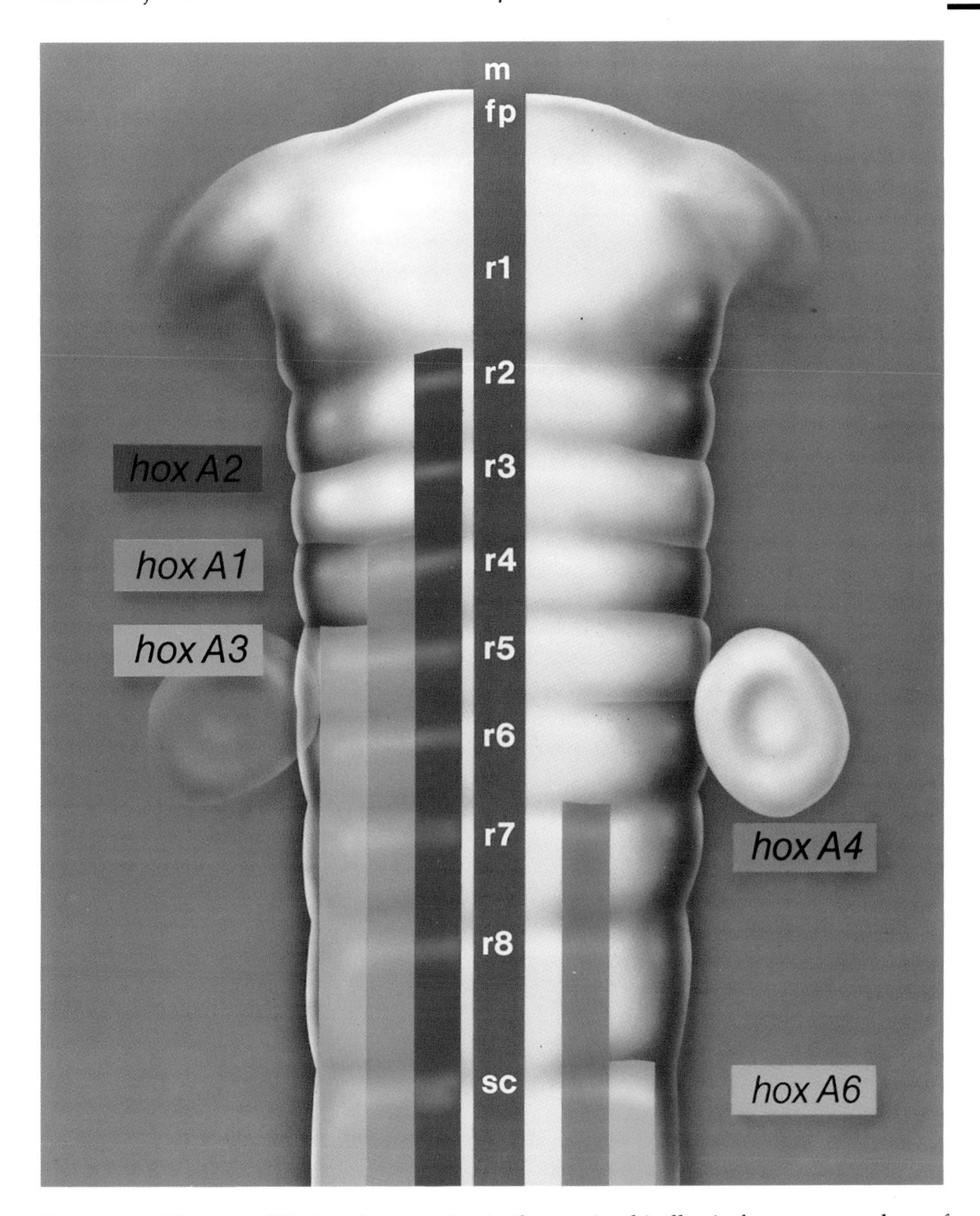

FIGURE 9–3 The rostral limits of expression in the murine hindbrain for some members of the HoxA linkage group. The hindbrain anlage is transiently subdivided into eight segments called rhombomeres (r). The rhombomere boundaries are depicted by darker shading. The otocyst is positioned lateral to the hindbrain adjacent to rhombomeres 5 and 6. The position of major neural crest production in the hindbrain is depicted by red shading. For a more detailed description of *Hox* gene expression, see text. mfp, midline floor plate; sc, spinal cord.

Dekker et al., 1993). The above correlations suggest that the retinoid signaling pathway may be important in activating *Hox* gene expression during normal development. Such activation could occur very early in embryogenesis at the time of patterning of mesoderm in the node. An intriguing possibility is that sequential activation of different *Hox* genes occurs in response to different concentrations of retinoic acid produced in the node (Hogan et al., 1992). An obvious test of this hypothesis is to determine the phenotypic consequence of removing the known retinoic acid response elements, of which there are at least four, from the *Hox* complex. However, it would not be surprising if a parallel system for activating *Hox* genes also existed.

An alternative approach to identifying the upstream regulators of *Hox* genes is to define the *cis*-acting regulatory elements responsible for localizing *Hox* gene expression and then to identify the gene products that interact with these *cis* elements. This is a complex problem. An important character of *Hox* genes is that they are linked in a chromosomal complex. This appears to be universally true in all animals examined. The proper regulation of *Hox* genes may require their presence in the complex for a number of reasons. To achieve appropriate levels of expression may require the sharing of regulatory elements among linked *Hox* genes. The complex may have one or more locus control elements (Grosveld et al., 1987) that regulate multiple *Hox* genes. Local chromosomal environment may be an important contributor to determining the time of gene activation or accessibility to modulating factors. The success of recapitulating appropriate *Hox* gene expression in transgenic mice has been variable (Püschel et al., 1990, 1991; Whiting et al., 1991; Marshall et al., 1992; Sham et al., 1992; Behringer et al., 1993; Gerard et al., 1993; Frasch et al., 1995; Nonchev et al., 1996). In the above cases it was not established whether the transgene is capable of establishing its own expression pattern or whether it is responding to autoregulatory signals from the endogenous genes. Since autoregulation is likely to be shared among paralogous genes, it is experimentally difficult to test this possibility. Most commonly, the transgene recapitulates a subset of the complete endogenous pattern.

Despite the above limitations, transgenic analysis has provided valuable insight into the regulation of *Hox* gene expression and provided a profitable route to defining the location and nature of *cis*-acting elements that mediate expression of *Hox* genes in selected rhombomeres. Using this approach, Krumlauf and his colleagues demonstrated that *Krox20* is required for expression of both *hoxb-2* and *hoxa-2* in rhombomeres 3 and 5 (Sham et al., 1993; Nonchev et al., 1996).

In *Drosophila* there are groups of positive and negative regulators of *HomC* that appear to function by altering chromatin structure. Their function appears to be needed to maintain the proper pattern of *HomC* function once its expression has been activated during embryogenesis. The modifiers include members of the *trithorax* and *Polycomb* group of genes (Kennison, 1993; Struhl and Akam, 1985; Paro, 1990, 1993). Not surprisingly, homologues of these genes are surfacing in mammals. A particularly interesting example is *bmi-1*. Although it was originally identified by Berns and his colleagues in a screen for genes that cooperate during the onset of neoplasia, it was found that loss- and gain-of-function *bmi-1* mutations in the mouse produced, respectively, posterior and anterior homeotic transformations in the vertebral column. The resulting phenotypes resemble homeotic mutations in *Hox* genes (van der Lugt et al., 1994; Alkema et al., 1995). *Bmi-1* has homology to *Posterior sex comb*, a member of the *Drosophila Polycomb group* genes (Brunk et al., 1991). Other members of this gene family, as well as homologues of *trithorax*, have been identified in the mouse. The genetic analysis of these genes will further test the extent of functional homology between vertebrates and invertebrates in the use of these modifiers of *HomC/Hox* gene expression.

The most direct method of assessing *Hox* gene function in the formation of the hindbrain is to analyze loss-of-function mutations. Mice have been generated with targeted gene disruptions in the 12 *Hox* genes most likely to be involved in hindbrain development. Analysis of the phenotypes for a number of these mutants is still in the early phases. Expression of *hoxd-1* has not been detected in the neural tube (Zhang et al., 1994), and *hoxd-1* mutant homozygotes do not exhibit an overt phenotype in the hindbrain or in the cranial ganglia (Puetz, Condie, Capecchi, unpublished results). It has already been noted above that *hoxa-1* mutant homozygotes have defects in hindbrain segmentation (Carpenter et al., 1993; Dollé et al., 1993; Mark et al., 1993). These mutant mice also have ectopic nuclei, extending from r6 to the rostral end of the spinal column, that send their efferent axons to the VIIth cranial nerve (Carpenter et al., 1993). Such neurons are never observed in normal mice and their origin is unknown. Either they represent the specification of new neurons in the caudal region of the hindbrain or, more likely, they are born and specified in a more rostral region of the hindbrain and, as a consequence of defective rhombomere boundaries, migrate to ectopic posterior regions. Finally, *hoxa-1* mutants often exhibit a rostral displacement of VIIth branchiomotor neurons from r4 to r3. Curiously, the displacement can be asymmetric, with one side of the hindbrain having these nuclei in r3 and the contralateral side having them in r4. Again the displacement may be the result of faulty specification or faulty migration. From this description it is apparent that the set of defects in *hoxa-1* mutant mice is complex. The major defect appears to be a failure of outgrowth of the hindbrain from r4 through r5 and extending into r6. This is evident from the reduction in hindbrain length and from the absence of a series of molecular markers that characterize this region (Carpenter et al., 1993; Dollé et al., 1993; Mark et al., 1993). More careful analysis of early time points in development is needed in order to determine whether the absence of r5 reflects a failure in the specification of this rhombomere or in the maintenance and outgrowth of this segment.

Because the early expression patterns of *hoxb-1* and *hoxa-1* are similar, we anticipate that these two genes will be found to function together to form the r4–r6 region of the hindbrain. Further supporting their interaction, *hoxa-1* and *hoxb-1* have been shown to share an autoregulatory loop (Pöpperl et al., 1995). Mice mutant for both genes seem likely to have a more exacerbated phenotype than mice containing either the *hoxa-1* or *hoxb-1* mutation alone, and the phenotype of such double mutants may provide clearer insight into the roles of these genes in hindbrain formation. As an example of strong interactions between two paralogous *Hox* genes, *hoxa-11* mutant mice have a slightly shortened and thickened radius and ulna (Small and Potter, 1993), and *hoxd-11* mutant mice have defects in sculpting of the ends of the radius and ulna (Davis and Capecchi, 1994). However, in mice mutant for both *hoxa-11* and *hoxd-11*, the radius and ulna are almost entirely eliminated (Davis et al., 1995). Thus, not only is the phenotype greatly exacerbated, but these experiments were necessary to show that the two genes function together to specify the outgrowth of the radius and ulna, which was not clear from the analysis of the individual mutants. Furthermore, analysis of early time points shows that the initial bifurcation of the prechondrogenic precursors needed to form the radius and ulna occurs in the double mutant, but that the enormous outgrowth required to generate these long bones does not. Previously, it was not recognized that posterior *Hox* genes would have a major role in controlling the outgrowth of the limb in the proximodistal direction.

The analysis of mice with only the *hoxb-1* mutation is just beginning (Goddard, Rossel, Manley, Capecchi, unpublished results). Mice homozygous for mutations of this gene do not appear to have defects in the segmentation pattern of the hindbrain but rather in the specification of neurons of the hindbrain. In particular, the VIIth nerve motor pool—the neurons that innervate the facial muscles—is absent in *hoxb-1* mutant homozygotes. This absence can be traced to defects early in the development of the hindbrain. Analysis of mutations in other members of the chosen *Hox* set is also underway. For example, *hoxa-2* and *hoxa-3* mutant mice each show defects in the formation of the IXth glossopharyngeal nerve (Gendron-Maguire et al., 1993; Rijli et al., 1993; Manley and Capecchi unpublished results). The defects are similar, showing hypoplasia of the IXth cranial ganglion and sparse connections of the IXth nerve with the hindbrain. Initially, it was believed that *hoxa-3* neural crest defects were restricted to mesenchymal neural-crest–derived structures, but more recent analysis has shown that the IXth cranial nerve is also affected. In summary, the analysis of mice with targeted mutations in *Hox* genes has implicated these genes in specification of neuron identities within rhombomeres, in regulation of cell migration and cell proliferation, and possibly in specification of rhombomeres themselves.

Earlier in this chapter it was suggested that the role of *Hox* genes in patterning the nervous system may be more primitive than their role in patterning mesoderm-derived structures. As a consequence, following the quadruplication of the ancestral *Hox* complex in early vertebrates, the level of redundancy among vertebrate *Hox* genes patterning the nervous system may be particularly high. Analysis of double, triple, and even quadruple mutants may be required to determine the roles of *Hox* genes in the formation of the hindbrain. Moreover, interactions between *Hox* genes are not restricted to paralogous genes. Genes within the same *Hox* linkage group have been shown to interact quantitatively. As an example, *hoxb-5* and *hoxb-6* function together to specify cervical vertebrae (Rancourt et al., 1995), and *hoxd-11, hoxd-12,* and *hoxd-13* have been shown to function synergistically to specify the formation of carpal, metacarpal, and phalangeal bones (Davis and Capecchi, 1996). Even nonparalogous *Hox* genes on separate linkage groups interact (Favier et al., 1996; Davis and Capecchi, 1996). The potential exists for interactions among all 11 of the anterior *Hox* genes postulated to be required for forming the hindbrain. It follows that analysis of mice with multiple mutations may not only be more informative than the analysis of mice with single mutations, but necessary for a full understanding of the regulatory interplay among these genes. In this light, the molecular genetic analysis of the role of *Hox* genes in hindbrain development must be regarded as just beginning.

Other issues of *Hox* gene activity also need further investigation, particularly gain-of-function mutations. Widespread ectopic expression of *Hox* genes usually leads to embryonic lethality (Balling et al., 1989; Kessel et al., 1990; Lufkin et al., 1992; Zhang et al., 1994). The defects induced by a transgene are often found to be anterior to the normal rostral expression boundary of the endogenous gene. Thus, overproduction of the transgene product in more posterior regions appears to be neutral to normal development. In mice this phenomenon is called posterior dominance (Duboule, 1991; Duboule and Morata, 1994). The same phenomenon is also observed in transgenic *Drosophila* that ectopically express *HomC* genes, where it is referred to as posterior phenotypic suppression (Morata et al., 1990; Bachiller et al., 1994). Posterior dominance appears to be a fundamental, but elusive, property of *Hox* genes. Although there are numerous exceptions, defects in mice with *Hox* gene mutations often are restricted to the anterior boundaries of *Hox* gene expression. Posterior dominance would suggest that the identities of cells at a particular axial level are determined by the activity of the most posterior *Hox* gene expressed at that level. Therefore, as a consequence of a mutation in a *Hox* gene, the mutant phenotype would be expected at the axial level at which it is the most posterior *Hox* gene expressed (i.e., at its anterior boundary of expression). What makes the phenomenon elusive are the exceptions. Critical to an understanding of posterior dominance is the determination of whether the phenomenon involves timing and competition between *Hox* genes of relatively equal stature, or an intrinsic, graded dominant phenotype of the more posterior *Hox* genes.

Mice transgenic for the hindbrain affecting *hoxa-1* die in utero and show abnormal development in the anterior region of the embryo (Zhang et al., 1994). The ectopic expression of *hoxa-1* results in strong ectopic expression of *hoxb-1* in the anterior neural tube and in neural crest emigrating from this region. The upregulation of *hoxb-1* appears to result from activation of the autoregulatory loop shared by *hoxa-1* and *hoxb-1* (Pöpperl et al., 1995). Interestingly, activation of *hoxb-1* does not occur in all regions where the *hoxa-1* transgene is expressed. This is consistent with the suggestion by Krumlauf and co-workers that this autoregulatory loop is dependent on cooperative interaction with other transcription factors (i.e., Exd/Pbx) which presumably are not uniformly expressed throughout the embryo.

Neural Crest Cells

Mice mutant for genes, such as *hoxa-1* and *kreisler,* with abnormalities in hindbrain segmentation, also show defects in the formation of more peripheral structures such as the inner ears and cranial ganglia (Lufkin et al., 1991; Chisaka et al., 1992; Deol, 1964; Frohman et al., 1993; McKay et al., 1994). Such observations suggest that patterning of the peripheral structures may be dependent on the integrity of hindbrain segmentation. That is, the positional value of peripheral structures may be derived from positional values in the hindbrain. A natural conduit for passing positional information from the hindbrain to the peripheral structures is the neural crest (LeDouarin, 1982). This follows because (1) neural crest cells contribute to the formation of most of the peripheral structures of the lower head and throat regions (Noden, 1983; Couly et al., 1993), and (2) they migrate from their point of origin on the neural tube, in stereotypic patterns, to specific positions along the rostrocaudal axis of the embryo. For example, r1/r2 neural crest cells migrate into the first branchial arch, those of r4 migrate into the second branchial arch, and so on (Lumsden et al., 1991). Further, it has been suggested that depletion of neural crest populations emanating from r3 and r5 might contribute to maintaining the proper registry between the source and final destination of neural crest (Lumsden et al., 1991). Consistent with the hypothesis that neural crest cells are positionally prepatterned, Noden (1988) demonstrated by transplantation that if neural crest cells from a more rostral position migrate into the second branchial arch, they direct the formation of first-arch–like structures.

The correspondence of *Hox* gene expression in the neural tube with the neural crest emigrating from that position further suggests that positional value could be directly transferred to the branchial arches by continued expression of appropriate combinations of *Hox* genes in the migrating crest cells (Hunt et al., 1991). In broad outline, this unifying hypothesis has merit; in detail it requires revision. *Hox* information characteristic of a rhombomere is not always transferred to the neural crest, and independent specification of *Hox* gene expression in the neural crest is likely in numerous cases (Prince and Lumsden, 1994). Further, neural crest cells of different origin are known to mix. Rhombomeres 3 and 5 are characteristically low in the production of migrating neural crest, but this level is not zero. The neural crest population produced from these rhombomeres, however, migrates into two branchial arches, r3 crest cells into arches 2 and 3 and r5 crest cells into arches 3 and 4 (Sechrist et al., 1993; Birgbauer et al., 1995). This observation breaks down the simple correlation that neural crest originating from one rhombomere migrates into a single branchial arch. Moreover, the patterns of *Hox* gene expression in the branchial arches are very intricate and change with time. These patterns are presumably marking populations of cells whose identity is presently unrecognized. The spectrum of paralogous *Hox* gene expression in the branchial arches is more complex than the spectrum of *Hox* gene expression observed in the hindbrain at the time of neural crest emigration. Thus, some of the *Hox* gene expression must be activated de novo in the branchial arches. Finally, the combination of *Hox* genes expressed by neural crest can be altered by the environment into which they migrate (Saldivar et al., 1996). In such cases, as in the trunk, these cells are not behaving in a cell-autonomous manner but are capable of reacting to their environment.

It has already been noted that mutations in *hoxa-1*, *hoxa-2*, *hoxa-3*, and *hoxb-1* affect the formation of neural-crest–derived cranial ganglia. In addition, *hoxa-3* mutant mice have extensive defects in tissues and cartilages derived from mesenchymal neural crest. This includes formation of the thymus and thyroid, musculature of the throat, and cartilages of the throat and heart tissue (Chisaka and Capecchi, 1991). Neither the amount of neural crest nor its migration pattern appears to be altered by the *hoxa-3* mutation. Therefore, it appears that the loss of *hoxa-3* affects the intrinsic capacity of this neural crest cell population to differentiate and/or to induce proper differentiation of the surrounding pharyngeal arch and pharyngeal pouch (Manley and Capecchi, 1995).

Conclusion

To date very little progress has been made in identifying the targets of *Hox* genes. It is anticipated that since *Hox* genes regulate morphology, their targets will include genes that modulate cell behavior such as proliferation, adhesion, migration, differentiation, and death. However, between *Hox* genes and the genes that regulate cell behavior may be one or more additional layers of transcription factors that further refine the choices of cell identity. It will indeed be a challenge to identify the cascade of genes that intercede between the initiation of regional specification, presumably mediated by *Hox* genes, and the ensuing activities of many groups of precursor cells used to form complex systems such as the hindbrain. However, the molecular and genetic tools are now in hand to proceed with this task. A molecular genetic description of hindbrain development will not only tell us how this machine is made but should also provide deep insight into how it functions. Virtually as a rule in biology, it has been found that the function of a system, whether we are speaking of individual molecules or of complex systems, is derived from its structure. The structure in turn is derived intrinsically from its synthesis. If we understand in detailed molecular terms how the hindbrain is made, we may well be a long way along the road to understanding how it functions.

References

Alkema, M.J., van der Lugt, N.M.T., Bobeldijk, R.C., Berns, A., van Lohuizen, M. (1995). Transformation of axial skeleton due to over expression of *bmi-1* in transgenic mice. *Nature* 374:724–727.

Bachiller, D., Macías, A., Duboule, D., Morata, G. (1994). Conservation of a functional hierarchy between mammalian and insect *Hox/HOM* genes. *EMBO J.* 13:1930–1941.

Balling, R., Mutter, G., Gruss, P., Kessel, M. (1989). Craniofacial abnormalities induced by ectopic expression of the homeobox gene *Hox-1.1* in transgenic mice. *Cell* 58:337–347.

Becker, N., Seitanidou, T., Murphy, P., Mattéi, M.-G., Topilko, P., Nieto, M.A., Wilkinson, D.G., Charnay, P., Gilardi-Hebenstreit, P. (1994). Several receptor tyrosine kinase genes of the *Eph* family are segmentally expressed in the developing hindbrain. *Mech. Dev.* 47:3–17.

Behringer, R.R., Crotty, D.A., Tennyson, V.M., Brinster, R.L., Palmiter, R.D., Wolgemuth, D.J. (1993). Sequences 5′ of the homeobox of the *Hox-1.4* gene direct tissue-specific expression of *lacZ* during mouse development. *Development* 117:823–833.

Birgbauer, E., Fraser, S.E. (1994). Violation of cell lineage compartments in the chick hindbrain. *Development* 120:1347–1356.

Birgbauer, E., Sechrist, J., Bronner-Fraser, M., Fraser, S. (1995). Rhombomeric origin and rostrocaudal reassortment of neural crest cells revealed by intravital microscopy. *Development* 121:935–945.

Brunk, B.P., Martin, E.C., Adler, P.N. (1991). *Drosophila* genes *Posterior Sex Combs* and *Suppressor two of zeste* encode proteins with homology to the murine *bmi-1* oncogene. *Nature* 353:351–353.

Carpenter, E.M., Goddard, J.M., Chisaka, O., Manley, N.R., Capecchi, M.R. (1993). Loss of *Hox-A1* (*Hox-1.6*) function results in the reorganization of the murine hindbrain. *Development* 118:1063–1075.

Chavrier, P., Zerial, M., Lemaire, P., Almendral, J., Bravo, R., Charnay, P. (1988). A gene encoding a protein with zinc fingers is activated during G_0/G_1 transition in cultured cells. *EMBO J.* 7:29–35.

Chisaka, O., Capecchi, M. R. (1991). Regionally restricted developmental defects resulting from targeted disruption of the mouse homeobox gene *hox-1.5*. *Nature* 350:473–479.

Chisaka, O., Musci, T.S., Capecchi, M.R. (1992). Developmental defects of the ear, cranial nerves and hindbrain resulting from targeted disruption of the mouse homeobox gene *Hox-1.6*. *Nature* 355:516–520.

Clarke, J.D.W., Lumsden, A. (1993). Segmental repetition of neuronal phenotypes sets in the chick embryo hindbrain. *Development* 118:151–162.

Conlon, R.A., Rossant, J. (1992). Exogenous retinoic acid rapidly induces anterior ectopic expression of murine *Hox-2* genes *in vivo*. *Development* 116:357–368.

Cordes, S.P., Barsh, G.S. (1994). The mouse segmentation gene *kr* encodes a novel basic domain—leucine zipper transcription factor. *Cell* 79:1025–1034.

Couly, G.F., Coltey, P.M., Le Douarin, N.M. (1993). The triple origin of skull in higher vertebrates: a study in quail-chick chimeras. *Development* 117:409–429.

Davis, A.P., Capecchi, M.R. (1994). Axial homeosis and appendicular skeleton defects in mice with a targeted disruption of *hoxd-11*. *Development* 120:2187–2198.

Davis, A.P., Capecchi, M.R. (1996). A mutational analysis of the 5′Hox D genes: dissection of genetic interactions during limb development in the mouse. *Development* 122: 1175–1185.

Davis, A. P., Witte, D.P., Hsieh-Li, H.M., Potter, S.S., Capecchi, M.R. (1995). Absence of radius and ulna in mice lacking *hoxa-11* and *hoxd-11*. *Nature* 375:791–795.

Dekker, E.J., Pannese, M., Houtzager, E., Boncinelli, E., Durston, A. (1993). Colinearity in the *Xenopus laevis Hox-2* complex. *Mech. Dev.* 40:3–12.

Deol, M.S. (1964). The abnormalities of the inner ear in *kreisler* mice. *J. Embryol. Exp. Morph.* 12:475–490.

Dollé, P., Izpisúa-Belmonte, J.-C., Falkenstein, H., Renucci, A., Duboule, D. (1989). Coordinate expression of the murine *Hox-5* complex homeobox-containing genes during limb pattern formation. *Nature* 342:767–772.

Dollé, P., Izpisúa-Belmonte, J.-C., Brown, J.M., Tickle, C., Duboule, D. (1991). *Hox-4* genes and the morphogenesis of mammalian genitalia. *Genes Dev.* 5:1767–1776.

Dollé, P., Lufkin, T., Krumlauf, R., Mark, M., Duboule, D., Chambon, P. (1993). Local alterations of *Krox-20* and *Hox* gene expression in the hindbrain suggest lack of rhombomeres 4 and 5 in homozygote null *Hoxa-1* (*Hox-1.6*) mutant embryos. *Proc. Natl. Acad. Sci. U.S.A.* 90:7666–7670.

Duboule, D. (1991). Patterning in the vertebral limb. *Curr. Opin. Genet. Dev.* 1:211–216.

Duboule, D. (1994). Temporal colinearity and the phylotypic progression: a basis for the stability of a vertebrate Bauplan and the evolution of morphologies through heterochrony. *Dev. Suppl.:* 135–142.

Duboule, D., Dollé, P. (1989). The structural and functional organization of the murine *Hox* gene family resembles that of *Drosophila* homeotic genes. *EMBO J.* 8:1497–1505.

Duboule, D., Morata, G. (1994). Colinearity and functional hierarchy among genes of the homeotic complexes. *Trends Genet.* 10:358–364.

Favier, B., Rijli, F.M., Fromental-Ramain, C., Fraulob, V., Chambon, P., Dollé, P. (1996). Functional cooperation between the non-paralogous genes *Hoxa-10* and *Hoxd-11* in the developing forelimb and axial skeleton. *Development* 122:449–460.

Frasch, M., Chen, X., Lufkin, T. (1995). Evolutionary-conserved enhancers direct region-specific expression of the murine *Hoxa-1* and *Hoxa-2* loci in both mice and *Drosophila*. *Development* 121:957–974.

Fraser, S., Keynes, R., Lumsden, A. (1990). Segmentation in the chick embryo hindbrain is defined by cell lineage restrictions. *Nature* 344:431–435.

Frohman, M.A., Martin, G.R., Cordes, S.P., Halamek, L.P., Barsh, G.S. (1993). Altered rhombomere-specific gene expression and hyoid bone differentiation in the mouse segmentation mutant, *kreisler (kr)*. *Development* 117:925–936.

Gendron-Maguire, M., Mallo, M., Zhang, M., Gridley, T. (1993). *Hoxa-2* mutant mice exhibit homeotic transformation of skeletal elements derived from cranial neural crest. *Cell* 75:1317–1331.

Gérard, M., Duboule, D., Zákány, J. (1993). Structure and activity of regulatory elements involved in the activation of the *Hoxd-11* gene during late gastrulation. *EMBO J.* 12: 3539–3550.

Gilardi-Hebenstreit, P., Nieto, M.A., Frain, M., Mattei, M.-G., Chestier, A., Wilkinson, D.G., Charnay, P. (1992). An *Eph*-related receptor protein tyrosine kinase gene segmentally expressed in the developing mouse hindbrain. *Oncogene* 7:2499–2506.

Glover, J.C., Petursdottir, G. (1991). Regional specificity of developing reticulospinal, vestibulospinal and vestibulo-ocular projections in the chicken embryo. *J. Neurobiol.* 22:353–376.

Graham, A., Lumsden, A. (1996). Interactions between rhombomeres modulate *Krox-20* and *follistatin* expression in the chick embryo hindbrain. *Development* 122:473–480.

Graham, A., Papalopulu, N., Krumlauf, R. (1989). The murine and *Drosophila* homeobox gene complexes have common features of organization and expression. *Cell* 57:367–378.

Graham, A., Heyman, I., Lumsden, A. (1993). Even-numbered rhombomeres control the apoptopic elimination of neural crest cells from odd-numbered rhombomeres in the chick hindbrain. *Development* 119:233–245.

Grapin-Botton, A., Bonnin, M.-A., Ariza McNaughton, L., Krumlauf, R., Le Douarin, N.M. (1995). Plasticity of transposed rhombomeres: *Hox* gene induction is correlated with phenotypic modifications. *Development* 121:2707–2721.

Grosveld, F., van Assendelft, G.B., Greaves, D.R., Kollias, G. (1987). Position-independent, high-level expression of the human *β-globin* gene in transgenic mice. *Cell* 51:975–985.

Guthrie, S., Lumsden, A. (1991). Formation and regeneration of rhombomere boundaries in the developing chick hindbrain. *Development* 112:221–229.

Guthrie, S., Muchamore, I., Kuroiwa, A., Marshall, H., Krumlauf, R., Lumsden, A. (1992). Neuroectodermal autonomy of *Hox-2.9* expression revealed by rhombomere transpositions. *Nature* 356:157–159.

Guthrie, S., Prince, V., Lumsden, A. (1993). Selective dispersal of avian rhombomere cells in orthotopic and heterotopic grafts. *Development* 118:527–538.

Haack, H., Gruss, P. (1993). The establishment of murine *Hox-1* expression domains during patterning of the limb. *Dev. Biol.* 157:410–422.

Heyman, I., Faissner, A., Lumsden, A. (1995). Cell and matrix specialisations of rhombomere boundaries. *Dev. Dyn.* 204:301–315.

Hogan, B.L.M., Thaller, C., Eichele, G. (1992). Evidence that Hensen's node is a site of retinoic acid synthesis. *Nature* 359:237–241.

Holland, P.W.H., Garcia-Fernàndez, J. (1996). *Hox* genes and chordate evolution. *Dev. Biol.* 173:382–395.

Holland, P.W.H., Garcia-Fernàndez, J., Williams, N.A., Sidow, A. (1994). Gene duplications and the origin of vertebrate development. *Dev. Suppl.*: 125–133.

Hunt, P., Gulisano, M., Cook, M., Sham, M.-H., Faiella, A., Wilkinson, D., Boncinelli, E., Krumlauf, R. (1991). A distinct *Hox* code for the branchial region of the vertebrate head. *Nature* 353:861–864.

Itasaki, N., Sharpe, J., Morrison, A., Krumlauf, R. (1996). Reprogramming *Hox* expression in the vertebrate hindbrain: the influence of paraxial mesoderm and rhombomere transposition. *Neuron*. 16:487–500.

Izpisúa-Belmonte, J.-C., Falkenstein, H., Dollé, P., Renucci, A., Duboule, D. (1991a). Murine genes related to the *Drosophila AbdB* homeotic gene are sequentially expressed during development of the posterior part of the body. *EMBO J*. 10:2279–2289.

Izpisúa-Belmonte, J.-C., Tickle, C., Dollé, P., Wolpert, L., Duboule, D. (1991b). Expression of the homeobox *Hox-4* genes and the specification of position in chick wing development. *Nature* 350:585–589.

Kappen, C., Schughart, K., Ruddle, F.H. (1989). Two steps in the evolution of *Antennapedia*-class vertebrate homeobox genes. *Proc. Natl. Acad. Sci. U.S.A.* 86:5459–5463.

Kennison, J.A. (1993). Transcriptional activation of *Drosophila* homeotic genes from distant regulatory elements. *Trends Genet*. 9:75–79.

Kessel, M. (1992). Respecification of vertebral identities by retinoic acid. *Development* 115: 487–501.

Kessel, M., Gruss, P. (1991). Homeotic transformations of murine vertebrae and concomitant alteration of *Hox* codes induced by retinoic acid. *Cell* 67:89–104.

Kessel, M., Balling, R., Gruss, P. (1990). Variations of cervical vertebrae after expression of a *Hox1.1* transgene in mice. *Cell* 61:301–308.

Krumlauf, R. (1994). *Hox* genes in vertebrate development. *Cell* 78:191–201.

Kuratani, S.C., Eichele, G. (1993). Rhombomere transplantation repatterns the segmental organization of cranial nerves and reveals cell-autonomous expression of a homeodomain protein. *Development* 117:105–117.

Langston, A.W., Gudas, L.J. (1992). Identification of a retinoic acid responsive enhancer 3′ of the murine homeobox gene *Hox-1.6 Mech. Dev.* 38:217–228.

Layer, P. G., Alber, R.A. (1990). Patterning of chick brain vesicles as revealed by peanut agglutinin and cholinesterases. *Development* 109:613–624.

LeDouarin, N.M. (1982). *The Neural Crest*. Cambridge: Cambridge University Press.

Lewis, E.B. (1978). A gene complex controlling segmentation in *Drosophila*. *Nature* 276: 565–570.

Lohnes, D., Kastner, P., Dierich, A., Mark, M., LeMeur, M., Chambon, P. (1993). Function of the retinoic acid receptor γin the mouse. *Cell* 73:643–658.

Lufkin, T., Dierich, A., LeMeur, M., Mark, M., Chambon, P. (1991). Disruption of the *Hox-1.6* homeobox gene results in defects in a region corresponding to its rostral domain of expression. *Cell* 66:1105–1119.

Lufkin, T., Mark, M., Hart, C.P., Dollé, P., LeMeur, M., Chambon, P. (1992). Homeotic transformation of the occipital bones of the skull by ectopic expression of a homeobox gene. *Nature* 359:835–841.

Lumsden, A., Keynes, R. (1989). Segmental patterns of neuronal development in the chick hindbrain. *Nature* 337:424–428.

Lumsden, A., Sprawson, N., Graham, A. (1991). Segmental origin and migration of neural crest cells in the hindbrain region of the chick embryo. *Development* 113:1281–1291.

Maden, M., Holder, N. (1992). Retinoic acid and development of the central nervous system. *Bioessays* 14:431–438.

Manley, N.R., Capecchi, M.R. (1995). The role of *hoxa-3* in mouse thymus and thyroid development. *Development* 121:1989–2003.

Mark, M., Lufkin, T., Vonesch, J.-L., Ruberte, E., Olivo, J.-C., Dollé, P., Gorry, P., Lumsden, A., Chambon, P. (1993). Two rhombomeres are altered in *Hoxa-1* mutant mice. *Development* 119:319–338.

Marshall, H., Nonchev, S., Sham, M.H., Muchamore, I., Lumsden, A., Krumlauf, R. (1992). Retinoic acid alters hindbrain Hox code and induces transformation of rhombomeres 2/3 into a 4/5 identity. *Nature* 360:737–741.

Marshall, H., Studer, M., Pöpperl, H., Aparicio, S., Kuroiwa, A., Brenner, S., Krumlauf, R. (1994). A conserved retinoic acid response element required for early expression of the homeobox gene *Hoxb-1*. *Nature* 370:567–571.

McGinnis, W., Krumlauf, R. (1992). Homeobox genes and axial patterning. *Cell* 68:283–302.

McKay, I.J., Muchamore, I., Krumlauf, R., Maden, M., Lumsden, A., Lewis, J. (1994). The *kreisler* mouse: a hindbrain segmentation mutant that lacks two rhombomeres. *Development* 121:2199–2211.

Morata, G., Macius, A., Urquial, N., Gonzales-Reyes, A. (1990). Homeotic genes. *Semin. Cell Biol.* 1:219–224.

Morriss-Kay, G. (1993). Retinoic acid and craniofacial development: molecules and morphogenesis. *Bioessays* 15:9–15.

Murphy, P., Davidson, D.R., Hill, R.E. (1989). Segment-specific expression of a homeobox-containing gene in the mouse hindbrain. *Nature* 341:156–159.

Nieto, M.A., Gilardi-Hebenstreit, P., Charnay, P., Wilkinson, D.G. (1992). A receptor protein tyrosine kinase implicated in the segmental patterning of the hindbrain and mesoderm. *Development* 116:1137–1150.

Noden, D.M. (1983). The role of the neural crest in patterning of avian cranial skeletal, connective and muscle tissues. *Dev. Biol.* 96:144–165.

Noden, D.M. (1988). Interaction and fates of avian craniofacial mesenchyme. *Dev. Suppl.* 103:121–140.

Nonchev, S., Vesque, C., Maconochie, M., Seitanidou, T., Ariza-McNaughton, L., Frain, M., Marshall, H., Sham, M.H., Krumlauf, R., Charnay, P. (1996). Segmental expression of *Hoxa-2* in the hindbrain is directly regulated by *Krox 20*. *Development* 122:543–554.

Ogura, T., Evans, R.M. (1995a). A retinoic acid-triggered cascade of *Hoxb-1* gene activation. *Proc. Natl. Acad. Sci. U.S.A.* 92:387–391.

Ogura, T., Evans, R.M. (1995b). Evidence for two distinct retinoic acid response pathways for *Hoxb-1* gene regulation. *Proc. Natl. Acad. Sci. U.S.A.* 92:392–396.

Orr, X. (1887). Contribution to the embryology of the lizard. *J. Morphol.* 1:311–372.

Papalopulu, N., Lovell-Badge, R., Krumlauf, R. (1991). The expression of murine Hox-2 genes is dependent on the differentiation pathway and displays a collinear sensitivity to retinoic acid in F9 cells and Xenopus embryos. *Nucleic Acids Res.* 19:5497–5506.

Paro, R. (1990). Imprinting a determined state into the chromatin of *Drosophila*. *Trends Genet.* 6:416–421.

Paro, R. (1993). Mechanisms of heritable gene repression during development of *Drosophila*. *Curr. Opin. Cell Biol.* 5:999–1005.

Pöpperl, H., Featherstone, M.S. (1993). Identification of a retinoic acid response element upstream of the murine *Hox-4.2* gene. *Mol. Cell. Biol.* 13:257–265.

Pöpperl, H., Bienz, M., Studer, M., Chan, S.-K., Aparicio, S., Brenner, S., Mann, R.S., Krumlauf, R. (1995). Segmental expression of *Hoxb-1* is controlled by a highly conserved autoregulatory loop dependent upon *exd/phx*. *Cell* 81:1031–1042.

Prince, V., Lumsden, A. (1994). *Hoxa-2* expression in normal and transposed rhombomeres: independent regulation in the neural tube and neural crest. *Development* 120: 911–923.

Puelles, L., Rubenstein, J.L.R. (1993). Expression patterns of homeobox and other putative regulatory genes in the embryonic mouse forebrain suggest a neuromeric organization. *Trends Neurosci.* 16:472–479.

Püschel, A.W., Balling, R., Gruss, P. (1990). Position-specific activity of the *Hox1.1* promoter in transgenic mice. *Development* 108:435–442.

Püschel, A.W., Balling, R., Gruss, P. (1991). Separate elements cause lineage restriction and specify boundaries of *Hox-1.1* expression. *Development* 112:279–287.

Rancourt, D.E., Tsuzuki, T., Capecchi, M.R. (1995). Genetic interaction between *hoxb-5* and *hoxb-6* is revealed by nonallelic noncomplementation. *Genes Dev.* 9:108–122.

Rijli, F.M., Mark, M., Lakkaraju, S., Dierich, A., Dollé, P., Chambon, P. (1993). A homeotic transformation is generated in the rostral branchial region of the head by disruption of *Hoxa-2*, which acts as a selector gene. *Cell* 75:1333–1349.

Roberts, D.J., Johnson, R.L., Burke, A.C., Nelson, C.E., Morgan, B.A., Tabin, C. (1995).

Sonic hedgehog is an endodermal signal inducing *Bmp-4* and *Hox* genes during induction and regionalization of the chick hindgut. *Development* 121:3163–3174.

Saldivar, J.R., Krull, C.E., Krumlauf, R., Ariza-McNaughton, L., Bronner-Fraser, M. (1996). Rhombomere of origin determines autonomous versus environmentally regulated expression of *Hoxa3* in the avian embryo. *Development* 122:895–904.

Schneider-Maunoury, S., Topilko, P., Seitanidou, T., Levi, G., Cohen-Tannoudji, M., Pournin, S., Babinet, C., Charnay, P. (1993). Distribution of *Krox-20* results in alteration of rhombomeres 3 and 5 in the developing hindbrain. *Cell* 75:1199–1214.

Scott, M.P. (1992). Vertebrate homeobox gene nomenclature. *Cell* 71:551–553.

Sechrist, J., Serbedzija, G.N., Scherson, T., Fraser, S.E., Bronner-Fraser, M. (1993). Segmental migration of the hindbrain neural crest does not arise from its segmental generation. *Development* 118:691–703.

Sham, M.H., Hunt, P., Nonchev, S., Papalopulu, N., Graham, A., Boncinelli, E., Krumlauf, R. (1992). Analysis of the murine *Hox-2.7* gene: conserved alternative transcripts with differential distributions in the nervous system and the potential for shared regulatory regions. *EMBO J.* 11:1825–1836.

Sham, M.H., Vesque, C., Nonchev, S., Marshall, H., Frain, M., Gupta, R.D., Whiting, J., Wilkinson, D., Charnay, P., Krumlauf, R. (1993). The zinc finger gene *Krox-20* regulates *Hox-B2 (Hox2.8)* during hindbrain segmentation. *Cell* 72:183–196.

Simeone, A., Acampora, D., Arcioni, I., Andrews, P.W., Boncinelli, E., Mavilio, F. (1990). Sequential activation of *HOX2* homeobox genes by retinoic acid in human embryonal carcinoma cells. *Nature* 346:763–766.

Simeone, A., Acampora, D., Nigro, V., Faiella, A., D'Esposito, M., Stornaiuolo, A., Mavilio, F., Boncinelli, E. (1991). Differential regulation by retinoic acid of the homeobox genes of the four *HOX* loci in human embryonal carcinoma cells. *Mech. Dev.* 33:215–227.

Simon, H., Hornbruch, A., Lumsden, A. (1995). Independent assignment of antero-posterior and dorso-ventral positional values in the developing chick hindbrain. *Curr. Biol.* 5:205–214.

Slack, J.M.W., Holland, P.W.H., Graham, C.F. (1993). The zootype and the phylotypic stage. *Nature* 361:490–492.

Small, K.M., Potter, S.S. (1993). Homeotic transformation and limb defects in *HoxA-11* mutant mice. *Genes Dev.* 7:2318–2328.

Struhl, G., Akam, M. (1985). Altered distributions of *Ultrabithorax* transcripts in *extra sex combs* mutant embryos of *Drosophila*. *EMBO J.* 4:3259–3264.

Sundin, O.H., Eichele, G. (1990). A homeo domain protein reveals the metameric nature of the developing chick hindbrain. *Genes Dev.* 4:1267–1276.

Swiatek, P.J., Gridley, T. (1993). Perinatal lethality and defects in hindbrain development in mice homozygous for a targeted mutation of the zinc finger gene *Krox20*. *Genes Dev.* 7:2071–2084.

Trevarrow, B., Marks, D.L., Kimmel, C.B. (1990). Organization of hindbrain segments in the zebrafish embryo. *Neuron* 4:669–679.

Vaage, S. (1969). The segmentation of the primitive neural tube in chick embryos *(Gallus domesticus)*. *Adv. Anat. Embryol. Cell Biol.* 41:1–88.

van der Lugt, N.M.T., Domen, J., Linders, K., van Roon, M., Robanus-Maandag, E., te Riele, H., van der Valk, M., Deschamps, J., Sofroniew, M., van Lohuizen, M., Berns, A. (1994). Posterior transformation, neurological abnormalities, and severe hematopoietic defects in mice with a targeted deletion of the *bmi-1* proto-oncogene. *Genes Dev.* 8:757–769.

Whiting, J., Marshall, H., Cook, M., Krumlauf, R., Rigby, P.W., Stott, D., Allemann, R.K. (1991). Multiple spatially specific enhancers are required to reconstruct the pattern of *Hox-2.6* gene expression. *Genes Dev.* 5:2048–2059.

Wilkinson, D.G., Peters, G., Dickson, C., McMahon, A.P. (1988). Expression of the FGF-related proto-oncogene *int-2* during gastrulation and neurulation in the mouse. *EMBO J.* 7:691–695.

Wilkinson, D.G., Bhatt, S., Chavrier, P., Bravo, R., Charnay, P. (1989a). Segment-specific expression of a zinc-finger gene in the developing nervous system of the mouse. *Nature* 337:461–464.

Wilkinson, D.G., Bhatt, S., Cook, M., Boncinelli, E., Krumlauf, R. (1989b). Segmental expression of *Hox-2* homeobox-containing genes in the developing mouse hindbrain. *Nature* 341:405–409.

Yokouchi, Y., Sasaki, H., Kuroiwa, A. (1991). Homeobox gene expression correlated with the bifurcation process of limb cartilage development. *Nature* 353:443–445.

Yokouchi, Y., Nakazato, S., Yamamoto, M., Goto, Y., Kameda, T., Iba, H., Kuroiwa, A. (1995). Misexpression of *Hoxa-13* induces cartilage homeotic transformation and changes cell adhesiveness in chick limb buds. *Genes Dev.* 9:2509–2522.

Zhang, M., Kim, H.-J., Marshall, H., Gendron-Maguire, M., Lucas, D.A., Baron, A., Gudas, L.J., Gridley, T., Krumlauf, R., Grippo, J.F. (1994). Ectopic *Hoxa-1* induces rhombomere transformation in the mouse hindbrain. *Development* 120:2431–2442.

10

Regulation of patterning and differentiation in the embryonic vertebrate forebrain

JOHN L. R. RUBENSTEIN AND KENJI SHIMAMURA

The development of the central nervous system (CNS) follows a relatively simple morphological program. The entire CNS derives from the neural plate, a single layer of neuroepithelial cells. Sequential inductive processes then create regional molecular differences that initiate distinct programs of proliferation, migration, and differentiation in different locations of the CNS. This leads to neurulation (formation of the neural tube), the creation of histogenic domains, and eventually to the wiring of the CNS. Within the CNS, there are four major transverse subdivisions: the spinal cord, hindbrain, midbrain, and forebrain, each with characteristic morphological and histological properties. This chapter deals with cellular and molecular processes that regulate embryonic forebrain development.

The forebrain is the seat of higher conscious and most subconscious functions. In addition, visual and olfactory information is largely processed in forebrain-derived tissues (the retina and olfactory bulbs) (Fig. 10–1). The mature vertebrate forebrain is among the most complex biological ensembles. However, its organization appears to be relatively simple at the neural plate and early neural tube stages, which facilitates molecular and cellular experiments that explore the mechanisms regulating its development. In this chapter, we review studies that address the topological arrangement of the histogenic primordia within the forebrain at the neural plate and embryonic neural tube stages. In particular, we discuss a neuromeric (segmentation) model which postulates that dorsoventral (D/V) and anteroposterior (A/P) patterning mechanisms subdivide the embryonic forebrain into longitudinal and transverse domains. Next, we describe studies that are beginning to define the patterning mechanisms which regulate the expression of genes whose function is linked to the development of specific histogenic anlage. These studies suggest that some common mechanisms may regulate dorsoventral patterning throughout the CNS. Finally, we describe genetic studies that define the function of several classes of transcription factors in regulating forebrain patterning, morphogenesis, and differentiation.

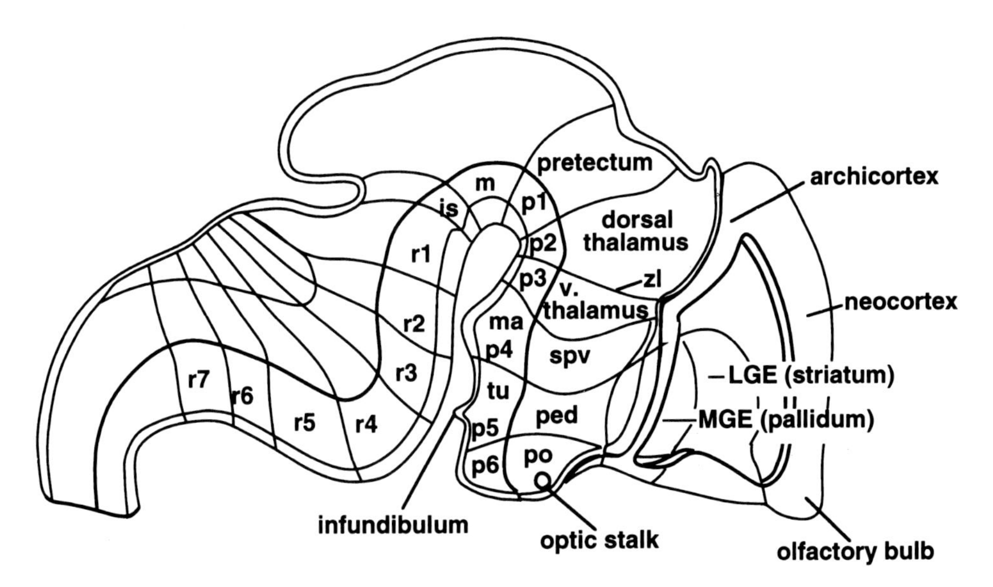

FIGURE 10–1 Schema of the embryonic day 12.5 mouse brain in which the primordia of some forebrain structures are labeled. In addition, longitudinal and transverse subdivisions are indicated. The paired telencephalic vesicles make up the majority of the forebrain mass and can be subdivided into the pallium (roof) and subpallium. The pallium includes the cerebral cortex, which in mammals is further subdivided into various laminar structures: the neocortex, archicortex (hippocampus), and paleocortex (olfactory bulb and olfactory cortex). The subpallium includes components such as the striatum, globus pallidus, septum, and parts of the amygdala (not shown). Ventral to the telencephalon are the eyes and hypothalamic (po, ped, spv, tu, ma) areas. The zona limitans intrathalamic (zl) separates the more anterior ventral thalamus from the dorsal thalamus.

Neuromeric (transverse) components are labeled in their basal plate: r1–r7: rhombomeres; p1–p6 are the theoretical prosomeres. This drawing is a modified version of the Prosomeric model from Bulfone et al. (1993a), Puelles and Rubenstein (1993), Rubenstein et al. (1994), and Puelles (1995). Note that the striatum is one of the derivatives of the lateral ganglionic eminence (LGE) and the pallidum (globus pallidus) is one of the derivatives of the medial ganglionic eminence (MGE). Abbreviations: is: isthmus; m; mesencephalon; ma: mammillary hypothalamus; ped: peduncular area of the hypothalamus, including the anterior hypothalamus; po: preoptic hypothalamus; spv: supraoptic paraventricular area of the hypothalamus; tu: tuberal hypothalamus; v. thalamus: ventral thalamus; zl: zona limitans intrathalamica.

Most experimental studies of forebrain development require an anatomical context in which to interpret the results. Unfortunately, a thorough anatomical understanding of the embryonic forebrain is not yet available. We will begin this chapter by describing a model of embryonic forebrain organization that follows from the same developmental and morphological rules that are known to operate in more posterior regions of the CNS.

Due to the morphological and histological complexity of the adult forebrain, it is extremely difficult to determine the topological relationships of its components. Many of the commonly held assumptions about forebrain topology are based on conclusions made from adult anatomy. (See Puelles and Rubenstein,

1993, and references therein for a discussion of this point.) As with many complex objects, one can better understand forebrain organization if one studies its origins.

Over the last century neuroembryologists have formulated models of forebrain organization that differ from those derived from the adult forebrain. Even among these neuromorphologists, however, differing hypotheses have been advanced (see Puelles et al., 1987). Recent methodological advances are now providing new ways to solve this old problem. Improved vital dyes and transplantation methods are giving fate maps greater resolution (Couly and Le Douarin, 1988; Eagelson et al., 1995). The discovery of molecular markers with regionally restricted patterns of expression within the forebrain are defining domains that are delimited by sharp boundaries (reviewed in Rubenstein et al., 1994; Rubenstein and Puelles, 1994). The results from these studies have revitalized the idea that the forebrain is a neuromeric structure (Bulfone et al., 1993a; Figdor and Stern, 1993). We will briefly review some of the recent studies that have contributed to current formulations of forebrain organization.

GENE EXPRESSION PATTERNS IN THE FOREBRAIN PROVIDE EVIDENCE FOR LONGITUDINAL AND TRANSVERSE SUBDIVISIONS

In the last 10 years, several lines of evidence have suggested that regionalization of the embryonic vertebrate CNS follows a simple organizational principle: The neuroepithelium is subdivided into histogenic primordia by a checkerboard-like grid of domains generated by the intersection of longitudinal columns and transverse segments. Studies of D/V patterning within the spinal cord demonstrate that signals arising from the notochord and non-neural ectoderm generate longitudinal columns of cells with common properties: floor plate, motoneuron columns (basal plate), neuronal columns involved in sensory processing (alar plate), and the roof plate (Yamada et al., 1993; Liem et al., 1995). Segment-like expression of homologues of the *Drosophila* homeotic genes (the *hox* genes) and of the Krox20 gene subdivides the embryonic hindbrain along the A/P axis into transverse domains (Keynes and Krumlauf, 1994). In conjunction with morphological, histological, and cellular data, these studies support the idea that the hindbrain is organized into neuromeric subunits named rhombomeres (Keynes and Lumsden, 1990; Simon et al., 1995).

For almost a century, neuroembryologists have used morphological information to suggest that the brain is segmented (see Vaage, 1969; Puelles, 1995). The basis for this argument is the transient embryonic appearance of repeated transverse constrictions present in the hindbrain, midbrain, and forebrain. Recently, we and others have provided molecular and cellular evidence that the forebrain has a neuromeric organization (Puelles et al., 1987; Salinas and Nusse, 1992; Figdor and Stern, 1993; Bulfone et al., 1993a).

The discovery of gene expression patterns in the forebrain with sharp boundaries near the boundaries of histogenic primordia provided molecular markers for embryonic forebrain subdivisions (Porteus et al., 1991; Price et al., 1991; Robinson et al., 1991; Walther and Gruss, 1991). The boundaries of expression of several genes follow lines that can be conceived of as being either parallel or perpendicular to the longitudinal axis of the forebrain. Based on gene expression patterns and embryological, histological, and morphological information, the Prosomeric model was postulated (Figs. 10–1 and 10–2). This model hypothesizes that the embryonic forebrain is a neuromeric structure subdivided into a grid-like pattern of histogenic domains by longitudinal (columnar) and transverse (segmental) boundaries (Puelles et al., 1987; Bulfone et al., 1993a; Puelles and Rubenstein, 1993; Rubenstein and Puelles, 1994; Rubenstein et al., 1994; Shi-

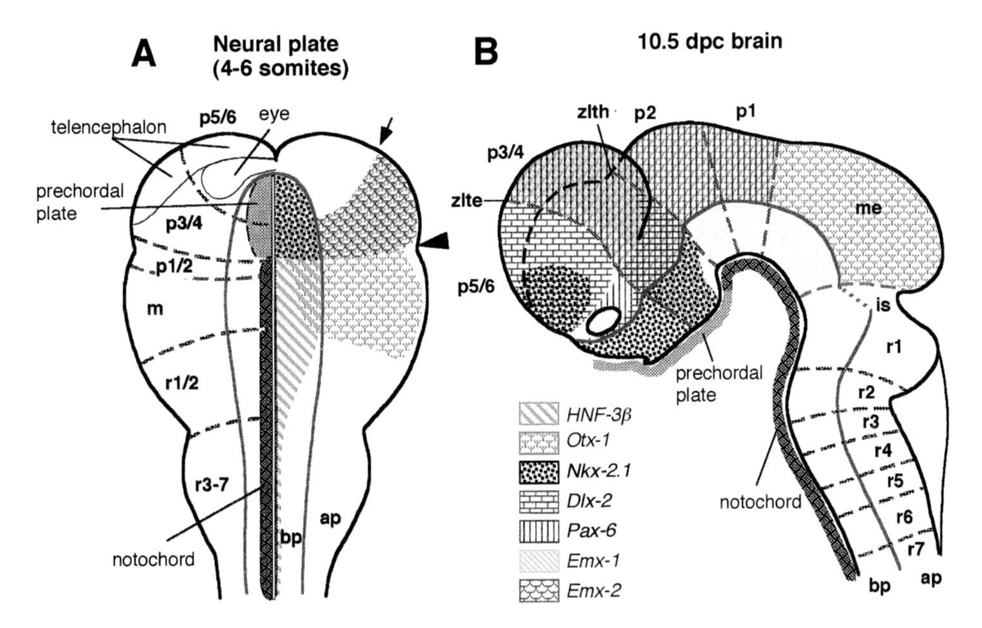

FIGURE 10–2 Schemata showing gene expression patterns and the theoretical organization of the mouse neural plate (4–6 somites) and neural tube (embryonic day 10.5). The expression of HNFβ, Nkx-2.1, Otx-1, and Emx-2 is shown in the neural plate; the expression of Pax-6, Emx-1, Dlx-2, Nkx-2.1, Otx-1, and Emx-2 is shown in the neural tube. The position of the axial mesendoderm (notochord and prechordal plate) is indicated, as are the principal longitudinal columns (basal plate: bp; alar plate: ap). Transverse boundaries are indicated by gray dashed lines that separate theoretical proneuromeric domains in the neural plate and neuromeres in the neural tube. Transverse boundaries in the prosencephalic neural plate are indicated by an arrowhead (a boundary which we hypothesize becomes the zona limitans intrathalamica, zlth) and an arrow (a boundary which we hypothesize becomes the zona limitans intratelencephalica, zlte). The approximate locations of the eye and telencephalic primordia, based on fate-map studies, are shown on the left side of the neural plate schema. Note that Pax-6 is expressed in the neural plate but has been omitted for simplicity (see Shimamura and Rubenstein, submitted; Shimamura et al., in press). It is expressed in a broad domain in the lateral neural plate spanning from about the midbrain/forebrain boundary near the anteriormost edge. The Pax-6 domain appears to consist of two separate regions: a boundary of decreased expression can be seen separating these domains (this boundary is the same as the anterior boundaries of Emx-2 and Otx-1); the anterior domain probably corresponds to the eye primordium and the posterior domain probably corresponds to cortical primordium. Abbreviations: is: isthmus; m: mesencephalon; p1–p6: prosomeres; r1–r7: rhombomeres.

mamura et al., 1995; Puelles, 1995). The longitudinal boundaries segregate columns of cells with similar properties that are specified by D/V patterning mechanisms. The transverse boundaries separate forebrain neuromeres (prosomeres). In the initial formulation, six prosomeres—p1–p6—were postulated. Figdor and Stern (1993) provided cellular evidence for the presence of transverse boundaries in the caudal forebrain or diencephalon. They showed that there are clonal expansion boundaries at the limits of the pretectum, thalamus, and ventral thalamus. Clonal expansion boundaries are found at segment boundaries in the hindbrain (Fraser et al., 1990). While Figdor and Stern's conclusions about the diencephalon are largely consistent with the Prosomeric model (although Figdor and Stern suggest that the pretectum [p1] is subdivided into two segments [D3 and D4]), theories of the organization of the rostral forebrain differ considerably, particularly with regard to the topological relationship of the telencephalon to the rest of the forebrain and the organization of the hypothalamus. In many morphological formulations, the telencephalon is considered the anteriormost brain segment, whereas in the Prosomeric model it is postulated to be a dorsal domain of the most anterior prosomeres (Figs. 10–1 and 10–4).

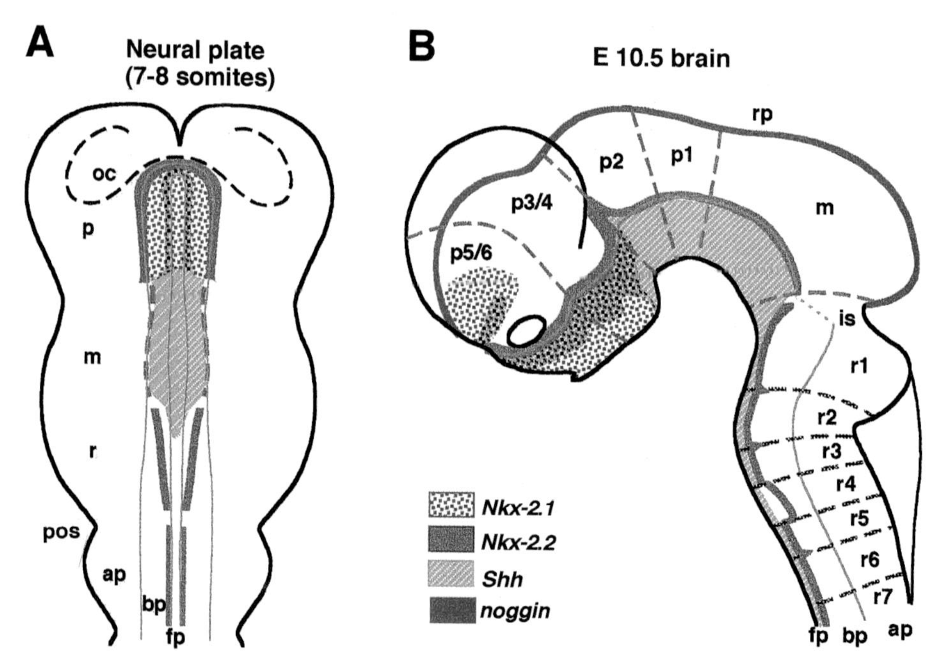

FIGURE 10–3 Schema highlighting gene expression patterns that reflect the longitudinal organization of the mouse neural plate (7–8-somite stage) and neural tube (embryonic day 10.5). The expression of Shh, Nkx-2.1, and Nkx-2.2 is restricted to medial regions of the neural plate (primordia of the floor and basal plates). In the neural tube, expression of these genes is largely restricted to these ventral regions, although a secondary domain of Shh and Nkx-2.1 expression is present in the medial ganglionic eminence of the basal telencephalon. Expression of Noggin is in the roof plate. Abbreviations: ap: alar plate; bp: basal plate; di: diencephalon; fp: floor plate; is: isthmus; me: mesencephalon; MGE: medial ganglionic eminence; oc: optic cup: os: optic stalk; p1, p2: diencephalic prosomeres; po: preotic sulcus; pros: prosencephalon; rh: rhombencephalon.

Controversy regarding the topological organization of the anterior forebrain has arisen in part due to a lack of agreement regarding the trajectory of the longitudinal axis in the forebrain (see Puelles and Rubenstein [1993] and Shimamura et al. [1995]). Recently, this issue has been addressed experimentally using gene expression markers to define the longitudinal organization of the forebrain (Shimamura et al., 1995). In that study, several genes were chosen that are expressed in longitudinal columns in the CNS (ventral markers: *sonic hedgehog* and *nkx-2.2*; dorsal markers: *bf-1* and *noggin*). Following the trajectory of the expression of these ventral and dorsal markers to the front of the brain provided evidence for the longitudinal organization of the forebrain (Figs. 10–3 and 10–4), which is consistent with the Prosomeric model. In this formulation, the telencephalic and

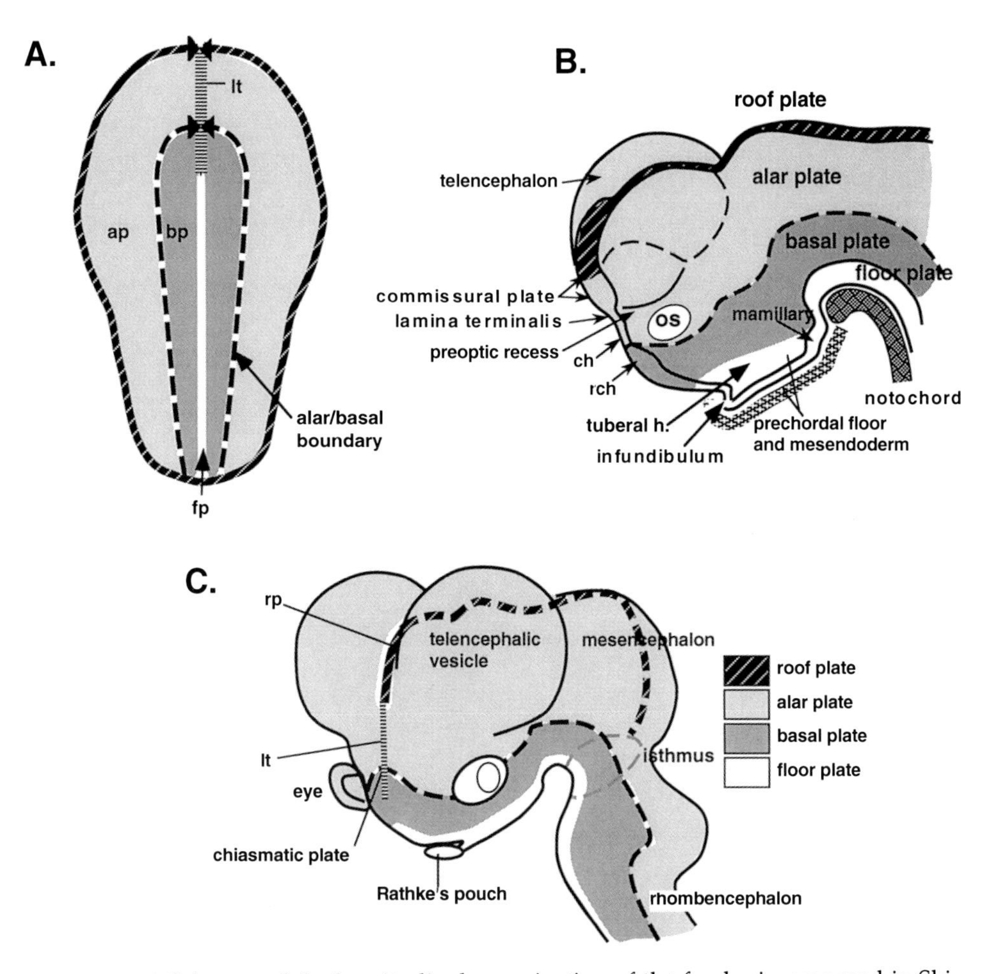

FIGURE 10–4 Schemas of the longitudinal organization of the forebrain proposed in Shimamura et al. (1995). (A) Model of the longitudinal domains in the neural plate, including the primordia of the floor, basal, alar, and roof plates (fp, bp, ap, rp). Note the radial organization of the longitudinal zones at the front of the neural plate. (B) Medial view of the neural tube. (C) Rostrolateral view of the neural tube. Abbreviations: ch: chiasmatic plate; lt: lamina terminalis, rch: retrochiasmatic area.

optic vesicles are alar (dorsal) structures and the tuberal and mammillary hypothalamus are basal (ventral) structures.

Analysis of gene expression patterns in the neural plate are consistent with the Prosomeric model. At this stage, several genes are expressed in the anteromedial neural plate (*nkx-2.1, nkx-2.2, shh*; see Fig. 10–3). This region has been shown by fate maps to give rise to the tuberal hypothalamus (Jacobson, 1959; Couly and Le Douarin, 1988; Eagleson and Harris, 1990; Eagelson et al., 1995). The bf-1 gene is expressed in the anterolateral neural plate, where fate maps have assigned the telencephalic primordium (Fig. 10–5). Note that in the neural tube, *bf-1, nkx-2.2, nkx-2.1,* and *shh* continue to be expressed in regions that are topologically consistent with their expression in the neural plate and consistent with fate map assignments (Shimamura et al., 1995). This result suggests that the expression patterns of these genes are relatively stable markers of regional identity.

While these studies address the D/V organization of the neural plate, they do not define its A/P organization. Recent advances have provided evidence that planar signals from the dorsal mesodermal organizer and vertical signals from

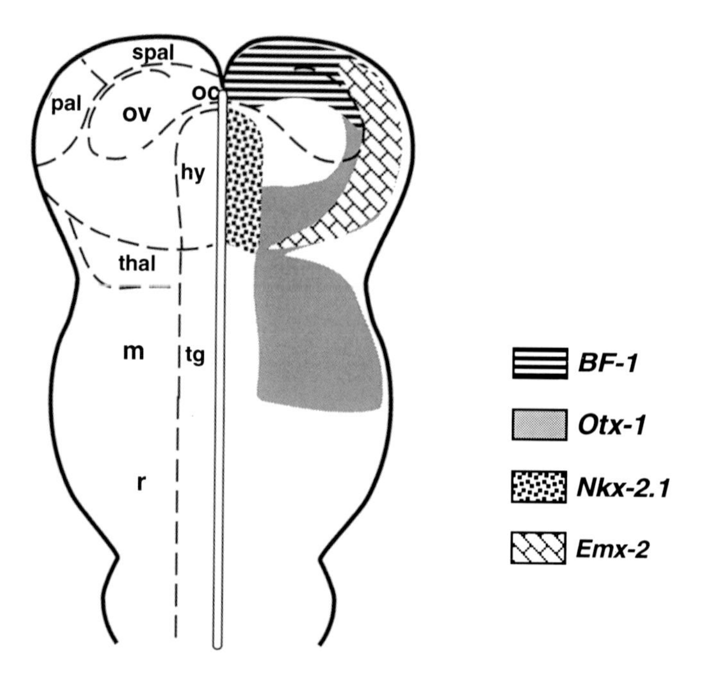

FIGURE 10–5 Schema of the mouse neural plate (7–8-somite stage). On the left side is an approximate fate map; on the right side the expression of BF-1, Otx-1, Nkx-2.1, and Emx-2 is shown. The fate map was drawn using information published by Jacobson (1959), Couly and Le Douarin (1988), Eagleson and Harris (1990), and Eagleson et al. (1995). Note that rostralmost Otx-1 expression is obscured by expression of Emx-2 (see Fig. 10–2A). Abbreviations: hy: tuberal and mammillary hypothalamus; m: mesencephalon; oc: optic chiasm; ov: optic vesicle; pal: pallium (roof of telencephalon or cortex); r: rhombencephalon; spal: subpallium (basal telencephalon, including the basal gangia); tg: tegmentum (basal plate of midbrain); thal: dorsal thalamus.

mesoderm and mesendoderm contribute to the specification of A/P regional differences (reviewed by Doniach, 1995). A two-signal model proposed by Nieuwkoop and by Saxen and Toivonen suggests that the first signal induces the neuroectoderm to assume an anterior neural fate (forebrain and midbrain); candidates for this signal are Noggin, Follistatin, and Chordin (Doniach, 1995). The second signal would be produced in a graded manner to posteriorize the neural plate; candidates for this signal include retinoic acid and basic FGF (Doniach, 1995).

We have recently provided evidence that forebrain regions of the neural plate and neural tube have transverse boundaries of gene expression (Fig. 10–2; Shimamura et al., 1995, 1996). Two transverse expression boundaries are seen for several genes. One approximates the position where the notochord ends and the prechordal plate begins (see arrowhead in Figure 10–2A); at later stages we hypothesize that this boundary becomes the zona limitans intrathalamica (zlth, Fig. 10–2B), the boundary between the dorsal thalamus, p2, and ventral thalamus, p3). In the medial neural plate, *nkx-2.1* expression has a posterior boundary approximately at this position; in the lateral neural plate *emx-2* has a boundary approximately at this position. *otx-1* has decreased expression at this location in both the lateral and medial regions of the neural plate. More rostrally is the anterior expression boundary of several genes: The boundary of *otx-1* spans the width of the neural plate, whereas the expression of *emx-2* is restricted to the lateral neural plate (arrow in Fig. 10–2A); we hypothesize that the anterior limit of *otx-1* expression becomes a boundary which is roughly between the cortical and subcortical domains of the telencephalon; we will call this boundary the zona limitans intratelencephalica (zlte) (Fig. 2B).

Tissue, Cellular, and Molecular Mechanisms that Regulate Ventral Patterning of the Embryonic Forebrain

It is clear that gene expression patterns provide insight into the organization of the embyronic CNS. They also provide clues about the genes that regulate particular developmental processes, thus making these genes useful experimental tools for studying the mechanisms regulating CNS patterning.

One can simplify the analysis of pattern formation in the forebrain to D/V, A/P, and local patterning. Dorsoventral patterning generates longitudinally aligned columns of phenotypically related cells. Anteroposterior patterning generates regional differences in the longitudinal columns (variations on a theme). Local patterning doesn't follow the Cartesian paradigm; it generates region-specific structures such as the eyes and olfactory bulbs. Below, we will discuss approaches to studying each of these types of patterning mechanisms.

Dorsoventral patterning of spinal cord, hindbrain, and midbrain levels of the neural tube (mediolateral patterning within the neural plate) is regulated by the notochord, an axial mesendodermal structure that is located just ventral to the medial neural plate. (Yamada et al., 1993). The notochord ends before the anterior subdivisions of the forebrain; anterior to and continuous with the notochord is the prechordal plate (Sulik et al., 1994). Thus, the prechordal plate is likely to be involved in patterning the medial part of the prosencephalic neural plate.

To directly test this hypothesis, experimental embryological methods have begun to be used with explant cultures of mouse neural-plate–stage embryos. In these experiments, the effects of ablation and transplantation (isotopically and ectopically) of the prechordal plate have been studied (Shimamura and Rubenstein, manuscript submitted). The design of these experiments is shown in Fig. 10–6.

The results from these experiments can be summarized as follows: (1) The prechordal plate can induce medial neural plate molecular patterns (e.g., *nkx-2.1, shh*) in both medial and lateral parts of the neural plate explants. (2) The prechordal plate can repress lateral neural plate molecular properties (*emx-2, pax-6*). (3) The notochord can also induce *nkx-2.1* in the anterior neural plate. (4) The prechordal plate cannot induce anterior properties (nkx-2.1) in more posterior neural plate regions, implying that the neural plate has differing competence along the A/P axis to respond to signals from the axial mesendoderm (Shimamura and Rubenstein, manuscript submitted).

Shh may be the common signal produced by both the prechordal plate and the notochord. Shh is produced by the prechordal plate and notochord (Echelard et al., 1993; Krauss et al., 1993; Roelink et al., 1994) and can induce ventral properties in the spinal cord and midbrain (Roelinke et al., 1995; Marti et al., 1995; Hynes et al., 1995). In addition, Ericson et al. (1995) found that Shh can induce *nkx-2.1* in isolated fragments of the chick anterior neural plate but not in fragments from the posterior neural plate.

A common signaling pathway regulating ventral CNS specification was previously suggested based on the phenotype of zebrafish that are homozygous for the cyclops mutation (Hatta et al., 1994). In these fish there is a deletion of varying amounts of the ventral CNS (the forebrain has the largest deletion, the spinal cord the smallest). In addition, cyclops fish lack *shh* expression in the CNS, which further implicates this molecule in ventralization of the CNS (Krauss et al., 1993; Barth and Wilson, 1995). The width of the ventral nervous system deletion correlates with the width of the *shh* expression domain within the CNS. As discussed earlier, there are additional genes that are expressed along the entire medial neural plate/ventral neural tube including *nkx-2.2* and *hnf3β* (Fig. 10–3; Shimamura et al., 1995), which also suggests a common ventral signaling pathway. Of note in cyclops fish, *nkx-2.2* expression is not detected (Barth and Wilson, 1995). However, if RNA encoding Shh is injected in cyclops fish, *nkx-2.2* expression is rescued (Barth and Wilson, 1995). At this point the function of *nkx-2.2* is not known, although *vnd,* its *Drosophila* homologue, is required for development of a subset of neurons in the ventral nervous system (Jimenez et al., 1995).

The expression of *nkx-2.2* and *shh* is closely related: *nkx-2.2* is in a longitudinal band of cells just dorsal to *Shh*-expressing cells (Fig. 10–3; Shimamura et al., 1995; Barth and Wilson, 1995). This close relationship is consistent with the observation that Shh can induce *nkx-2.2* (Barth and Wilson, 1995; Qui, Shimamura, Rubenstein, unpublished). The fact that little or no *shh* expression is found in

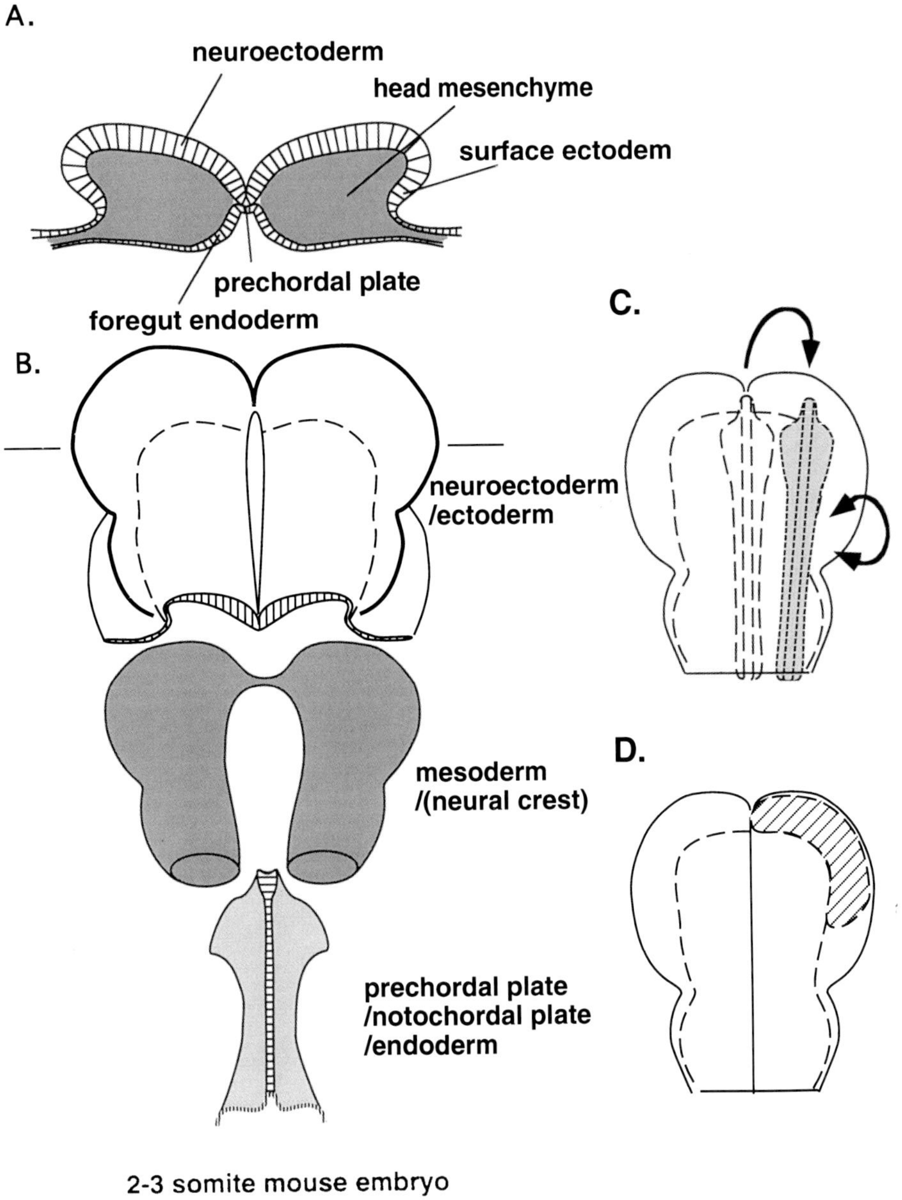

FIGURE 10–6 Schema showing the experimental approach used to identify the location of patterning signals in the prosencephalic neural plate. The ectodermal, mesodermal, and embryonic endodermal (containing prechordal and notochordal plates) layers of neural-plate–stage mouse embryos are enzymatically separated. Then, recombinant tissue explants are generated. The prechordal and notochordal plates (with the embryonic endoderm that is attached to the mesendoderm) are ectopically transplanted at various positions underneath the anterior neural plate. These recombinant explants are cultured for 24 hours and assayed by in situ hybridization for the expression of regional markers. (**A**) Transverse section through the prosencephalic neural plate of a 2–3-somite mouse embryo (at the position of the transverse line in **B**) showing the morphological relationships of the neuroectoderm, surface ectoderm, axial mesendoderm (prechrodal plate), foregut, and head mesenchyme. (**B**) Three tissue layers can be separated using enzymatic digestion. (**C**) The prechordal plate is ectopically transplanted (lateral translocation or rotation) under the neural plate to test its patterning properties. (**D**) The surface ectoderm is excised from the rostral neural plate to test its patterning properties.

nkx-2.2–positive cells suggests that *nkx-2.2* may inhibit the ability of these cells to express *shh* and thereby perhaps restrict the dorsal homeogenetic spread of *shh*-expressing cells. This would in turn restrict the source of *shh* production to a limited ventral region.

The longitudinal band of *nkx-2.2* expression is also distinctive in that this region produces some of the first neurons of the forebrain (Shimamura et al., 1995; Barth and Wilson, 1995). These neurons subsequently project axons, many of which follow the trajectory of the *nkx-2.2* stripe (Shimamura et al., 1995; Barth and Wilson, 1995). Thus, *nkx-2.2*–expressing cells correspond to a specialized longitudinal boundary zone that flanks the source of Shh-producing cells; this domain may provide spatial cues for the generation of early born neurons and the trajectory of their axons.

At present, it is reasonable to speculate that *shh* normally induces the expression of the *nkx* homeobox genes (*nkx-2.1* and *2.2*) in the ventral CNS, and that these genes regulate differentiation of cell types that depend upon their A/P position (e.g., forebrain: hypothalamic, midbrain: substantia nigra, hindbrain, and spinal cord: motor neurons).

TISSUE MECHANISMS THAT REGULATE DORSAL PATTERNING OF THE EMBRYONIC FOREBRAIN

It has been suggested that the parts of the forebrain ventral to the *nkx-2.2* stripe correspond to the basal plate and are specified by signals from the prechordal plate and that regions of the forebrain which are dorsal to the *nkx-2.2* stripe are the alar plate (Shimamura et al., 1995). According to this view the optic and telencephalic vesicles would be alar plate derivatives (Fig. 10–4).

At spinal cord levels, there is evidence that dorsalizing signals are produced in the non-neural ectoderm and that BMP proteins are candidates for the molecules that specify dorsal fate (Liem et al., 1995; Dickenson et al., 1995). To test whether non-neural ectoderm can regulate dorsal specification of the forebrain, the neural plate explant approach was used (Fig. 10–6; Shimamura and Rubenstein, manuscript submitted). The lateral region of the neural plate becomes the dorsal neural tube; thus lateral patterning in the neural plate corresponds to dorsal patterning in the neural tube. To study patterning of the lateral anterior neural plate, the expression of the winged helix transcription factor *bf-1* was analyzed. *bf-1* is expressed in most of the telencephalic vesicles and is essential for their development (Xuan et al., 1995). Expression of *bf-1* is first apparent in the non-neural ectoderm underlying the anterior edge of the neural plate at the three-somite stage. By the eight-somite stage, *bf-1* expression is found in the anterolateral neural plate, the region where fate maps place the telencephalon (Jacobson, 1959; Couly and Le Douarin, 1988; Eagleson et al., 1995). The shape of the *bf-1*–expressing domains in the non-neural ectoderm and the neuroectoderm are almost identical, suggesting that the same signal(s) is inducing expression in both tissues (Shimamura and Rubenstein, manuscript submitted). The neural plate explant system was employed to study the role of the non-neural ectoderm in regulating neural plate *bf-1* expression (Fig. 10–6) (Shimamura and Rubenstein, manuscript submitted). Excision of the non-neural ectoderm prevented induction of *bf-1* expression in the neural plate, implying that specification of the anterolateral neural plate is regulated by factors produced in the adjacent ectoderm. Candidate inducer molecules include *bmps* and *fgfs*. Additional experiments also demonstrated that induction of *bf-1* is restricted to the anterior neural plate, providing evidence that the neural plate has a differing ability along the A/P axis to respond to signals from the non-neural ectoderm.

Theoretical Considerations of A/P Patterning of the Forebrain

The tissues and molecules that generate the A/P pattern in the forebrain are poorly understood. The forebrain is at the end of the neural tube and therefore has topological features that are not present in other areas. In our preferred conception of the organization of the forebrain (the Prosomeric model), the alar plate and basal plate are continuous across the rostral midline in concentric rings (see Shimamura et al., 1995). The midline of the neural plate that directly overlies the axial mesendoderm does not reach to the rostral edge of the neural plate. Posterior to the zona limitans intrathalamica, the ventral midline becomes the floor plate; anterior to the zona limitans we have defined the ventral midline as the prechordal floor region. The zone connecting the end of the prechordal floor region to the rostromedial edge of the neural plate is the lamina terminalis (Fig. 10–4).

As the longitudinal columns arch across the rostralmost neural plate and cross the lamina terminalis, longitudinal and transverse coordinates become the same. This organization implies that in this region, D/V and A/P patterning are equivalent, assuming that there are no additional regionally restricted signals (Figs. 10–4 and 10–7). For instance, the vector of the signals from the anterior end of the prechordal plate is parallel with the A/P axis (see Fig. 10–7A). Likewise, signals produced by the rostralmost non-neural ectoderm have a vector parallel to the A/P axis. The result of this topological organization of patterning signals is the radial arrangement of longitudinal domains at the front of the brain. Such an arrangement would not necessarily generate morphogenetic information for specifying anteroposterior regional differences, unless there are regional differences in the magnitudes or characteristics of the morphogenetic signals at different positions in the axial mesendoderm or non-neural ectoderm. It will be important to discover whether there are morphogenetic signals localized at the front of the neural plate. Anteroposterior patterning information may be generated during gastrulation by planar signals from the node (Ruiz i Altaba, 1994; Doniach, 1995). At later developmental stages A/P patterning signals may be produced by organizers within the neural plate such as the isthmus (see below). Vertical signals may also be regionally localized, as is observed in regions of the axial mesendoderm or mesoderm (Ang et al., 1994). Within the axial mesendoderm, *rpx* and *gsc* expression is restricted to the prechordal plate, whereas

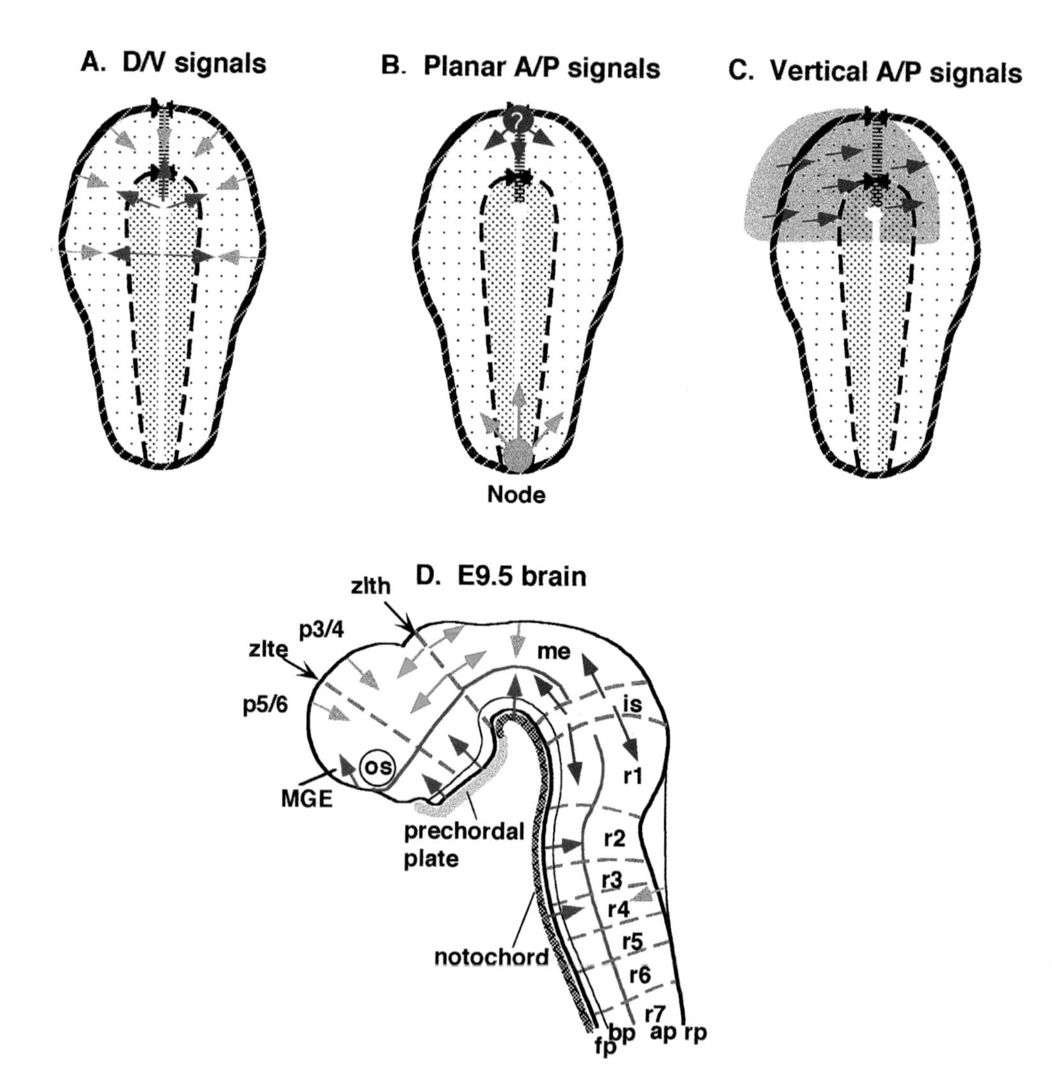

FIGURE 10–7 Schemata of known and theoretical patterning mechanisms in the neural plate (A, B, C) and neural tube (D). (**A**) Dorsoventral (D/V) patterning in the neural plate; medial signals (ventral) originate in axial mesendoderm and later from the midline of the neural plate; lateral signals (dorsal) originate in the non-neural ectoderm flanking the neural plate. (**B**) anteroposterior (A/P) patterning signals can arise within the plane of the ectoderm (planar signals). There is good evidence that the node produces planar signals from the posterior end of the neural plate. We hypothesize that there may be an anterior signaling center. (**C**) A/P patterning may also be regulated by "vertical" signals that are distinct at different A/P positions; these signals arise from tissues underlying the neural plate (e.g., axial mesendoderm and mesoderm). Here, a side view shows signals (purple) being vertically transmitted to the overlying anterior neural plate. (**D**) Once the basic A/P and D/V organization is set up, planar signals arising from transverse domains (e.g., the isthmus) produce morphogenetic effects on the neighboring neuroepithelium and may also restrict the spread of patterning signals. D/V signals, arising from ventral and perhaps dorsal midline structures (e.g., floor and roof plate), refine the longitudinal organization of the brain. At this stage, ventral signaling (Shh expression) spreads into the basal telencephalon, in the region of the medial ganglionic eminence (MGE). Abbreviations: ap: alar plate; bp: basal plate; fp: floor plate; p3/4 and p5/6: preprosomeric 3/4 and 5/6 domains; rp: roof plate; zlte: zona limitans intratelencephalica; zlth: zona limitans intrathalamica.

brachyury expression is restricted to the notochord (Hermesz et al., 1996). Also in this regard, mutation of two homeodomain transcription factors, *otx-2* and *lim-1*, that are expressed in the prechordal mesoderm (head mesoderm anterior to the notochord) results in mouse embryos that lack forebrain and midbrain. These data provide evidence that vertical signals play a role in A/P specification (Acampora et al., 1996; Matsuo et al., 1995; Ang et al., 1996). However, because *otx-2* and *lim-1* are also expressed in the neural ectoderm, one cannot unequivocally assign a central role to the prechordal mesoderm in A/P patterning based on mutations of these genes.

ROLE OF BOUNDARIES IN REGIONAL PATTERNING

A/P patterning within the prosencephalic neural plate produces large transverse domains of gene expression (Fig. 10–2). It is possible that the boundaries between these regions separate distinct histogenic domains and that the boundary zones function both as sources of morphogenetic information and as barriers to restrict the mixing of cells of the different compartments and to restrict the spread of patterning information. There is precedence within the vertebrate CNS for the role of boundaries for each of these functions. Interrhombomeric boundaries restrict clonal expansion (Fraser et al., 1990), intercellular communication via gap junctions (Martinez et al., 1992), and the spread of morphogenetic information (Martinez et al., 1995). As noted earlier, interprosomeric boundaries also restrict clonal expansion (Figdor and Stern, 1993) and the zona limitans (p2/p3) demarcates the limit of the spread of patterning signals from the isthmus (see below; Martinez et al., 1991).

There is also evidence for the role of a transverse zone (the isthmus) as the source of morphogens that regulate A/P patterning (see Fig. 10–7). Transplantation of the isthmus, a specialized region between the midbrain and hindbrain, into the midbrain or caudal diencephalon induces neighboring cells to form an ectopic cerebellum or midbrain, respectively (Martinez et al., 1991; Marin and Puelles, 1994). Transformations are not found anterior of the zona limitans, implying either that this structure is a boundary that prevents the spread of the patterning signal or that anterior of the zona limitans the cells are not competent to respond to the signal(s) from the isthmus, or both.

Wnt-1 and Fgf-8 are two candidate morphogens that are expressed in the isthmus, and both are directly implicated in regulating development of this region. Mutations of *wnt-1* lead to loss of the midbrain and anterior cerebellum (McMahon and Bradley, 1990). Ectopic expression of Wnt-1 in the spinal cord leads to hyperplasia of this structure (Dickenson et al., 1994), suggesting that Wnt-1 is also required for proliferation of the mesencephalon and cerebellum. Ectopic application of Fgf-8 in the anterior mesencephalon/posterior diencephalon induces an ectopic midbrain, (Crossley et al., 1996). It also induces the expression of Wnt-1 and the *engrailed* gene, which are required for development of this region (Joyner, 1996). Collectively, these findings suggest that Fgf-8 is a critical component in the specification of the midbrain.

To date, the isthmus is the only transverse boundary region with proven morphogenetic properties, although it is possible that other or all interneuromeric boundaries share some of these properties. For instance, interrhombomeric boundaries express *fgf-3* (Mahmood et al., 1995) and the zona limitans expresses *shh* (Shimamura et al., 1995; Barth and Wilson, 1995). *shh* expression in the zona limitans begins relatively late (~E9.5 in the mouse), suggesting that it does not have a primary role in setting up the A/P pattern in this region but that it may have a role in generation of more refined patterns in the prosomeres flanking the zona limitans (p2 and p3).

MOLECULAR AND HISTOLOGICAL SUBDIVISIONS IN THE TELENCEPHALON

Gene expression patterns define two principal compartments in the embryonic telencephalon: the pallium (cortex) and the subpallium (lateral ganglionic eminence [LGE] and the medial ganglionic eminence [MGE]) (Fig. 10–1); the fates of each of these regions are approximately known. The pallium gives rises to the neocortex, archicortex, and probably the paleocortex, and expresses a number of genes (*otx-1, emx-1, pax-6, id-2,* and *tbr-1*; Simeone et al., 1992b; Stoykova and Gruss, 1994; Bulfone et al., 1995; Shimamura et al., in press). The LGE is thought to give rise to the striatum (Deacon et al. 1994; Campbell et al., 1995; Sussel, Rubenstein, unpublished) and expresses the *dlx* genes (Porteus et al., 1991; Price et al., 1991; Robinson et al., 1991; Bulfone et al., 1993a; Simeone et al., 1994a). The MGE probably gives rise to the globus pallidus and bed nucleus of stria terminalis among other structures (Song and Harlan, 1994; Sussel, Rubenstein, unpublished) and expresses the *dlx* and *nkx-2.1* genes (Price et al., 1992; Bulfone et al., 1993b; Shimamura et al., 1995).

As described earlier, several genes that are expressed in the embryonic cortex have an anterior boundary in the neural plate that appears to be topologically transverse with respect to the longitudinal axis. This suggests that the pallial/subpallial boundary may be transverse (segmental), and we have named this boundary the zona limitans intratelencephalica (zlte, Fig. 10–2B) (reviewed by Puelles et al., 1987; see Fig. 20; Shimamura et al., 1995, see Fig. 7; Bulfone et al., 1995, see Model 3 in Fig. 8; Shimamura et al., in press). The pallial/subpallial limit may correspond to a neuromeric boundary because the *otx-1* expression boundary extends across the full width of the neural plate and full D/V extent of the neural tube (Fig. 10–2; Shimamura et al., 1995; in press). In addition, Fishell et al. (1993) have shown that ventricular zone cells of the pallium do not cross the cortical/subcortical boundary, providing evidence for a clonal restriction within the ventricular zone between these compartments. Other boundaries with similar properties may also exist within the telencephalon, such as the LGE/MGE limit.

Patterning mechanisms within the telencephalon are just beginning to be examined. We have shown that expression of the *bf-1* gene in the telencephalic anlage is regulated by signals in the non-neural ectoderm (Shimamura and Rubenstein, manuscript submitted). Mutation of *bf-1* leads to hypoplasia of the entire telencephalon with a loss of subpallial structures (Xuan et al., 1995). *bf-1* mutants also fail to express the *dlx-2* gene, whose function is involved in differentiation of the subpallial compartment (Anderson et al., 1995).

The lateral edge of the anterior neural plate becomes the dorsal midline of the telencephalon (roof plate or commissural plate), where Noggin (Shimamura et al., 1995) and fgf-8 are expressed (Heikinheimo et al., 1994; Crossley and Martin, 1995). Both proteins are candidates for dorsal patterning; *fgf-8* is also expressed in the ventral midline of the forebrain, where it may participate in development of the tuberal hypothalamus.

shh may also have a role in telencephalic patterning. *shh* expression appears at ~E9.5 in a small domain adjacent to the MGE (Fig. 10–7D: Shimamura et al., 1995). This domain is separated from the *shh*-positive hypothalamic domain by a thin strip of nonexpressing cells in the region of the optic chiasm. The function of *shh* in the basal telencephalon is unknown, but it is likely that it regulates the patterning of this region, particularly because some of the same genes that are coexpressed with *shh* in the hypothalamus (e.g., *nkx-2.1, islet-1*) are also expressed in the MGE (Shimamura et al., 1995; Ericson et al., 1995).

Patterning of Specialized Forebrain Structures: The Eyes, Pituitary, and Olfactory Bulbs

The previous sections of this chapter have provided evidence favoring a model in which patterning of the forebrain follows a Cartesian coordinate system. However, several forebrain structures, such as the eyes, olfactory bulbs, and pituitary, may not be specified solely by morphogens emanating from D/V and A/P organizers. In our opinion, the eyes, olfactory bulbs, and pituitary form as appendage-like structures within the context of the Cartesian map of the neural plate and tube. Like other patterning mechanisms, the development of these structures requires inductive interactions between neural and non-neural tissues.

The pituitary gland consists of the adenohypophysis and neurohyphophysis; the former is derived from the anterior ridge of the neural plate and the latter is derived from the ventral midline of the forebrain (Couly and Le Douarin, 1988). The anterior ridge folds under the neural plate and gives rise to Rathke's pouch, which comes into contact with the ventral forebrain (tuberal hypothalamic anlage) and leads to the evagination of the infundibular stalk. Thus, Rathke's pouch and the ventral forebrain are thought to have reciprocal inductive interactions that promote each other's development. Mutations of several genes encoding transcription factor (*pit-1, sf-1, brn-2, nkx-2.1, lhx-3*) affect pituitary development, providing an inroad to understanding the molecular control of its histogenesis (Wegner et al., 1993; Shinoda et al., 1995; Nakai et al., 1995; Schonemann et al., 1995; Kimura et al., 1995; Sheng et al., 1996).

The primordia of the eyes are in the anterior neural plate, just caudal to the neural ridge (Jacobson, 1959; Couly and Le Douarin, 1988; Eagleson and Harris,

1990; Eagleson et al., 1995). *pax-6* is expressed in the eye primordia in the neural plate and the lens primordium (Walter and Gruss, 1991; Krauss et al., 1991b; Grindley et al., 1995; Shimamura and Rubenstein, submitted; Shimamura et al., 1996) and is required for eye development (reviewed by Hanson and van Heyningen, 1995; Grindley et al., 1995). *pax-6* mutant mice do not form a lens. It is unclear how the eye fields in the neural plate are initially patterned, but later stages of eye development clearly involve reciprocal interactions between the neuroepithelium and the lens placode (reviewed by Saha et al., 1992).

Cyclopic animals, with a single-midline eye, have a defect in the development of the optic stalk and chiasm (Hatta et al., 1994). Studies of the cyclops mutant of zebrafish have demonstrated a ventral patterning defect of the forebrain that results in the absence of ventral forebrain tissue (Hatta et al., 1994) and the loss of *shh* expression (Krauss et al., 1993; Barth and Wilson, 1995). More recent studies suggest that cyclopia results because of the loss of a midline signal, possibly *shh,* that is required to induce the *pax-2*–expressing optic stalk (MacDonald et al., 1995; Ekker et al., 1995).

The olfactory bulbs are bilateral evaginations from the telencephalic vesicles that receive afferents from the olfactory epithelium. Like the eyes and pituitary, development of these structures requires interactions with localized regions of non-neural ectoderm, in this case the olfactory placodes (see Graziadei and Monti-Graziadei, 1992 and references therein). *pax-6* is expressed in the olfactory placode; mutation of *pax-6* results in the lack of olfactory placode differentiation into the olfactory epithelium, and secondarily, these mice lack olfactory bulbs (Hogan et al., 1986; Grindley et al., 1995). It has been suggested that *pax-6* (*Small eye* mutation) shares a role in development of the olfactory and lens placodes, which secondarily regulate development of the olfactory bulbs and optic vesicles (Hogan et al., 1986), although *pax-6* may also have important functions in the neuroepithelium of the optic and telencephalic vesicles.

ROLE OF TRANSCRIPTION FACTORS IN MORPHOGENESIS AND DIFFERENTIATION OF THE FOREBRAIN

The previous sections of this chapter have focused on the organization of the forebrain neuroepithelium and on the mechanisms that generate the pattern of histogenic primordia. Subsequent steps in forebrain development regulate the differential proliferation of these primordia which contribute to forebrain morphogenesis, followed by cellular differentiation and migration to form the final tissues. The genetic mechanisms that control these later processes are beginning to be elucidated. In the last 5 years, large numbers of candidate regulatory molecules have been identified using cross-species homology or differential expression screens (e.g., Porteus et al., 1992). Mutations in some of them have led to insights into their function and role in histogenesis. Below, we will primarily focus on the large number of transcription factors that have recently been described and the effects that mutations in some of them have on forebrain development.

Homeobox genes are involved in diverse functions including regulating regional specification, proliferation, and differentiation. There are over 25 known homeobox genes that are expressed in the vertebrate forebrain (Table 10–1; Rubenstein and Puelles, 1994). Many of these genes are expressed in regionally restricted patterns whose boundaries approximate the limits of histogenic primordia. For instance, within the telencephalon, expression of the *dlx* genes is limited to the anlage of the basal ganglia.

There are five known vertebrate *dlx* genes (*dlx-1, -2, -3, -5,* and *-6*) that are homologues of the *Drosophila Distal-less (dll)* gene (see references in McGuinness et al. 1996; Cohen et al., 1990); these genes are expressed in the developing forebrain (Fig. 10–2), branchial arches, limbs, and enteric nervous system. In the forebrain, *dlx-1* and -2 genes are expressed largely in the proliferating cells of the ventricular and subventricular zones (Bulfone et al., 1993a,b; Porteus et al., 1994), whereas *dlx-5* and *dlx-6* are expressed in progressively more differentiated cells within the developing basal ganglia (Liu, Rubenstein, unpublished), suggesting that this gene family regulates different stages of differentiation within the forebrain.

Mutation of *dlx-2* results in abnormal differentiation of olfactory bulb interneurons that originate from the basal telencephalon, as well as major morphological transformations of craniofacial bones (Qiu et al., 1995). Mutation of *dlx-1* does not appear to affect basal telencephalon differentiation, but does alter craniofacial morphogenesis (Bulfone, Qiu, Rubenstein, unpublished). However, a mutation in both *dlx-1* and *dlx-2* results in a partial block of differentiation in the striatum and other basal telencephalic regions (Anderson et al., 1995, and unpublished). Thus, it appears that *dlx-1* and *dlx-2* have redundant functions within the basal forebrain and that they are required to enable precursor cells in the lateral ganglionic eminence to differentiate into striatal neurons.

Whereas telencephalic expression of the *dlx* genes is restricted to the subcortical compartment of the embryonic telencephalon (Fig. 10–2; before embryonic day 13 in the mouse), the expression of other homeobox genes is either exclusively (*emx-1, otx-1*) or primarily (*emx-2* and *pax-6*) restricted to the cortical compartment (Fig. 10–2). Mutations in *emx-1, emx-2,* and *pax-6* have been studied. Humans heterozygous for mutations in the *emx-2* gene have schizencephaly, which consists of a cleft in the wall of the cerebral cortex (Brunelli, 1996). Mice homozygous for a mutation in *emx-1* are viable and their cerebral cortex appears normal, except that they lack most of the corpus callosum (Qiu et al., 1996).

Homeodomain-containing proteins often have other conserved functional motifs such as the POU, Paired, LIM, and zinc finger domains. Each of these homeobox subfamilies is quite large (Table 10–1). The *pit-1* gene is the best understood member of this family; it is required for the development of thyrotroph, somatotroph, and lactotroph cells in the adenohypophysis (Wegner et al., 1993). The *pit-1* gene is not expressed in the forebrain, unlike the *brn-1, brn-2, brn-4,* and *oct-6/tst-1/scip* genes (He et al., 1989; Monuki et al., 1989; Suzuki et al., 1990; Hara et al., 1992; Wegner et al., 1993). The expression of these genes has been studied in great detail (Alvarez-Bolado et al., 1995). These authors derived

TABLE 10–1. Homeobox Genes

Gene	Fly homolog	Mutation	Reference
Hox A genes	HOM-C	yes	Krumlauf, 1994
Hox B benes	HOM-C	yes	Krumlauf, 1994
Hox C genes	HOM-C	yes	Krumlauf, 1994
Hox D genes	HOM-C	yes	Krumlauf, 1994
Cdx-1	caudal		Meyer and Gruss, 1993
Chx-10			Liu et al., 1994
Dbx-1 & -2			Lu et al., 1992
Dlx-1 & -2	DII	yes/yes	Porteus et al., 1991; Price et al., 1991; Robinson et al., 1991; Qiu et al., 1995
Dlx-5 & -6	DII		Simeone et al., 1994
Emx-1	Ems	yes	Simeone et al., 1992; Qiu et al., 1995
Emx-2	Ems	yes	Simeone et al., 1992; Brunelli et al., 1996
En-1	En	yes	Joyner and Martin, 1987; Hanks et al., 1995
En-2	En	yes	Joyner et al., 1991; Joyner, 1996
Gbx-1, -2		yes	Murtha et al., 1992; Bulfone et al., 1993a; Chapman and Rathjen, 1995
Gsh-2		yes	Hsieh-Li et al., 1995
Gsh-4		yes	Li et al., 1994
Gtx			Komuro et al., 1993
Msx-1	Msh	yes	Satokata and Maas, 1994; Davidson, 1995
Nkx-1.1	Nk1		Bober et al., 1994
Nkx-2.1	Nk2	yes	Guazzi et al., 1990; Kimura et al., 1995
Nkx-2.2	Nk2		Price et al., 1992
Nkx-5.1	Nk5		Bober et al., 1994
Nkx-5.2	Nk5		Rinkwitz-Brant et al., 1995
Not (zf)			Talbot et al., 1995
Otx-1	Otd		Simeone et al., 1993
Otx-2	Otd	yes	Simeone et al., 1993; Matsuo et al., 1995; Acampora et al., 1996
Otp			Simeone et al., 1994
Pax-3	Prd	yes	Goulding et al., 1991; St-Onge et al., 1995
Pax-6	Prd	yes	Walther and Gruss, 1991; Schmahl et al., 1993
Pax-7	Prd		Jostes et al., 1991; Stoykova et al., 1994
Pbx	Exd		Monica et al., 1991; Rauskolb et al., 1993
Phox-2	Prd		Valarche et al., 1993
Prox-1	Prospero		Oliver et al., 1993
Rpx/Hesx1			Hermesz et al., 1996
Sax-1	Nk1		Schubert et al., 1995
Six1, -2, -3	sine oculis		Oliver et al., 1995
SOHo-1(c)			Deitcher et al., 1994
POU genes			
Brn-1			He et al., 1989; Hara et al., 1992; Wegner et al., 1993
Brn-2	Cf1a	yes	Schonemann et al., 1995; Nakai et al., 1995
Brn-3.0	I-POU		Guerrero et al., 1993
Brn-3.1	I-POU		Wegner et al., 1993
Brn-3.2	I-POU		Turner et al., 1994
Brn-4(RHS2)			LeMoine and Young, 1992; Mathis et al., 1992
Brn-5 (Emb)			Anderson et al., 1993
Cns-1			Bulleit et al., 1994
Oct-1(OTF-1)			Scholer, 1991
Oct-2(OTF-2)			Scholer, 1991

TABLE 10–1. Homeobox Genes (*Continued*)

Gene	*Fly homolog*	*Mutation*	*Reference*
Oct-6(Tst 1)			Suzuki et al., 1990; Monuki et al., 1990
Pit-1(GHF-1)		yes	Ingraham et al., 1988; Wegner et al., 1993
LIM genes			
LH-2	apterous		Xu et al., 1993
Lhx-3	apterous	yes	Zhadanov et al., 1995
Lim-1	apterous	yes	Fujii et al., 1994; Shawlot and Behringer, 1995
Isl-1	apterous	yes	Thor et al., 1991; Pfaff et al., 1996
ZFH genes			
zfh-4	zfh-1		Kostich and Sanes, 1995
deltaEF1	zfh-1		Funahashi et al., 1993

Mouse homeobox genes that are expressed in the embryonic brain. Homologues for many of these genes have been reported in other species, but due to space limitations we apologize that these have not been referenced, unless no mouse homologue is yet known (e.g., Not1, SoHo-1). Column 2 lists the Drosophila homologues of these genes. Column 3 lists whether a mutation of the gene has been described. Abbreviations: c: chick; zf: zebrafish.

a model of forebrain regionalization based in part on the expression of the *brn* genes that differs from the Prosomeric model (Puelles and Rubenstein, 1993; Rubenstein et al., 1994). Although *brn-2* is widely expressed, a loss-of-function mutation results in abnormal differentiation in a limited region of the brain; the endocrine hypothalamic nuclei (paraventricular and supraoptic nuclei) and the posterior pituitary (Nakai et al., 1995; Schonemann et al., 1995).

The *pax* genes encode proteins with a Paired domain, which is found in several *Drosophila* segmentation and tissue-specific regulatory genes (Chalepakis et al., 1993; St-Onge et al., 1995). Mutations of *pax-6* in mice, rats, and humans result abnormal development of the eye (Hill et al., 1991; Glaser et al., 1992; Matsuo et al., 1993; Hanson and van Heyningen, 1995), olfactory epithelium, olfactory bulb, and cerebral cortex (Schmahl et al., 1993). The mutant cerebral cortex has nodular aggregates of small, poorly differentiated cells adjacent to the proliferative zones and in more superficial areas. The cortical plate has no apparent marginal zone, is hypoplastic, and has schizencephalic clefts.

The *pax-3* gene, which also encodes paired and homeodomain motifs (Goulding et al., 1991), is expressed in dorsal regions of the CNS. Mutations of *pax-3* in mice and humans (reviewed in Chalepakis et al., 1993) result in neural tube defects (exencephaly and spina bifida) as well as neural crest defects of the peripheral nervous system and pigmentation and hearing defects (Waardenberg Syndrome) (Tassabehji et al., 1992; Epstein et al., 1993; Tremblay et al., 1995).

Brain defects have been found for two other *pax* genes (*pax-2* and *pax-5*), which encode paired domains but not homeodomains. Humans heterozygous for *pax-2* function have optic nerve colobomas (Sanyanusin et al., 1995), which is consistent with the expression of *pax-2* in the embryonic optic stalks. Mice lacking *pax-5* have abnormal morphogenesis of the inferior colliculus and anterior cerebellum (Urbanek et al., 1994), the same regions that are affected by mutations in Wnt-1 and the *engrailed* genes (Joyner, 1996). *pax-2* is the first gene to be expressed in this region (Rowitch and McMahon, 1995); thus, components of the genetic hierarchy regulating development of regions flanking the isthmus have now been identified.

The functions of homeobox genes in *Drosophila* and yeast are regulated by other proteins. The Polycomb and Trithorax protein families modulate expression from the homeotic genes by repressing and activating their transcription, respectively (Paro, 1995; Simon, 1995). Homologues of the Polycomb and Trithorax genes have been identified in vertebrates and have been shown to have similar functions to their invertebrate relatives (Yu et al., 1995; Alkema et al., 1995). Homeodomain proteins are also regulated by protein–protein interactions. The Extradenticle protein, and its vertebrate homologues (*pbx-1, pbx-2,* and *pbx-3*) (Monica et al., 1991; Rauskolb et al., 1993; Popperi et al., 1995), modulate *hox* protein function via interactions with the *hox* pentapeptide motif (Knoepfler and Kamps, 1995).

The MAD-box gene (*mcm1*) also regulates the yeast a1 and α2 proteins via direct interactions (Vershon and Johnson, 1993). Several MAD-box genes have been reported in the vertebrate brain, although interactions with homeodomain proteins have not been described. (Leifer et al., 1993; Lyons et al., 1995).

There are many other nonhomeodomain transcription factors that are required for forebrain development. Helix-loop-helix (HLH) genes encode proteins that have been implicated in regulating differentiation and cell identity (Table 2). Many of these genes are homologues of *Drosophila* regulatory genes that are expressed in regionally restricted patterns in the embryonic forebrain (e.g., *sim-1* and *sim-2;* Fan et al., 1996). An extensive description of these genes can be found in Chapter 2.

Winged-helix genes encode transcription factors such as *hnf3b,* which is required for ventral patterning of most or all of the CNS (Ang and Rossant, 1994; Weinstein et al., 1994), and the *bf-1* and *bf-2* genes, which are expressed in nonoverlapping regions of the forebrain (Hatini and Lai, 1994). The *bf* gene family is related to the *Drosophila sloppy-paired* and *fork-head* genes, which are involved in segmentation and regional specification (Grossniklaus et al., 1994). As noted earlier in this chapter, mutation of *bf-1* leads to hypoplasia of the telencephalon, particularly of the basal ganglia primordia (Xuan et al., 1995). The authors also provided evidence that there is premature differentiation in the cortical domain.

T-box genes encode DNA-binding proteins that are transcription factors related to the *brachyury* (Kispert, 1995) and the *Drosophila* optomotor blind (Pflugfelder et al., 1992) genes. At least four T-box genes are expressed in the forebrain: *tbx-1, tbx-2, tbx-3* (Bollag et al., 1994), and *tbr-1* (Bulfone et al., 1995). *tbr-1* is expressed only in postmitotic cells in the forebrain, which are largely restricted to the cortical regions of the telencephalon (Bulfone et al., 1995). tbr-1's expression has boundaries within the neocortical primordia before thalamic afferents arrive, suggesting that patterning of regional differences within the neocortex is in part regulated by intrinsic mechanisms. Mutation of *tbr-1* results in the loss of specific classes of neurons in the cortex and olfactory bulb (Bulfone, Rubenstein, unpublished).

The zinc finger motif is found in several classes of transcription factors including the nuclear receptors, Sp1, Gli, and Zfh families. A mutation in a gene

TABLE 10–2.

Gene	*Basic domain*	*Leucine zipper*	*Fly homologue*	*Mutation*	*Reference*
GbHLH1.4 (C)	Yes		daughterless		Helms et al., 1994
Hes-1	Yes		hairy/E(spl)	Yes	Takebayashi et al., 1994; Ishibashi et al., 1994, 1995*
Hes-2	Yes		hairy/E(spl)		Ishibashi et al., 1993
Hes-3	Yes		hairy/E(spl)		Sasai et al., 1992*
Hes-5	Yes		hairy/E(spl)		Takebayashi et al., 1995
Id-1			emc		Duncan et al., 1992; Evans and O'Brien, 1993; Wang et al., 1992
Id-2			emc		Neuman et al., 1993; Zhu et al., 1995*
Id-3			emc		Riechmann and Sablitzky, 1995; Ellmeier and Welth, 1995
Id-4			emc		Riechmann and Sablitzky, 1995
ITF2	Yes				Yoon and Chikaraishi, 1994
Mash-1	Yes		achaete-scute	Yes	Lo et al., 1991;* Guillemot et al., 1993*
Math-1	Yes		atonal		Akazawa et al., 1995
Math-2	Yes		atonal		Shimizu et al., 1995
Mi	Yes	Yes		Yes	Steingrimsson et al., 1994
NeuroD	Yes				Lee et al., 1995
Nex-1	Yes				Bartholoma and Nave, 1994
N-myc	Yes	Yes			Ellmeier et al., 1992
NSCL-1	Yes				Begley et al., 1992
NSCL-2	Yes				Gobel et al., 1992
Sim-1	Yes		single-minded		Fan et al., 1996
Sim-2	Yes		single-minded		Dahmane et al., 1995, Fan et al., 1996
Xash-1 -3 (X)	Yes				Ferreiro et al., 1992; Zimmerman et al., 1993; Allende et al., 1994

encoding an orphan nuclear receptor named *ftx-F1/ad4BP/sf-1* leads to abnormal differentiation of the dorsomedial part of the ventromedial hypothalamic nucleus (Shinoda et al., 1995). The *gli* (*gli-1*, *gli-2*, *gli-3*) and zic genes are related to the *Drosophila* segment polarity gene *cubitus interruptus* (Hui et al., 1994; Aruga et al., 1994). *gli-1* is expressed ventrally, whereas *gli-2*, *gli-3*, and *zic* are expressed dorsally. Mutation of *gli-3* results in a lack of choroid plexus and olfactory bulb, and the cerebral cortex lacks its normal laminar organization (Franzt, 1994).

The discovery of most of the genes described in this section has occurred largely in the last 5 years, suggesting that these are just a small fraction of the transcription factors required for forebrain development. Understanding on the molecular level how these genes regulate development will require elucidation of the cofactors and targets of these transcription factors. Hints regarding this genetic circuitry are likely to come from invertebrate studies; certainly previous work has suggested that genetic pathways regulating early development have been conserved in vertebrates.

SIMILAR GENETIC PATHWAYS REGULATE VERTEBRATE AND INVERTEBRATE CNS DEVELOPMENT

Regional specification along the A/P axis of the abdomen, thorax, and pregnathal head of *Drosophila* embryos is regulated by the *hox* genes (McGinnis and Krumlauf, 1992). The vertebrate *hox* genes appear to have similar roles at analogous axial levels (McGinnis and Krumlauf, 1992; Krumlauf, 1994). Specification of more anterior head segments is controlled by a separate sets of genes in *Drosophila*; they are: empty *spiracle (ems), orthodenticle* (Otd), *sloppy paired–1, –2* (*sp-1,* and *sp-2*), and *buttonhead (btd)* (Cohen and Jurgens, 1991; Wimmer et al., 1993; Jurgens and Hartenstein, 1993; Hirth et al., 1995). There is evidence that a combinatorial code of these genes regulates segmentation and regional specification of the pregnathal head segments (Grossniklaus et al., 1994). Vertebrate homologues of these genes are as follows: *ems: emx-1,* and *emx-2; otd: otx-1,* and *otx-2; sp: bf-1,* and *bf-2; btd: sp1* gene family. As noted above, the *emx-2* and *otx-1* genes are expressed in transverse (segment-like) patterns in the anterior neural plate (Fig. 10–2). *bf-1* is expressed in the rostralmost region of the neural plate which we interpret to be the alar plate of the anterior forebrain (the telencephalon) (Fig. 10–5). *otx-2* expression in the neural plate includes the forebrain and midbrain primordia (Simeone et al., 1992b, 1993; Rubenstein et al., 1994). *emx-1* expression begins after formation of the telencephalic vesicles and is limited to the cerebral cortex primordium (Fig. 10–2; Simeone et al., 1992a,b; Shimamura, Rubenstein, unpublished). Genetic analysis of the function of these vertebrate genes is just beginning to elucidate their roles in forebrain development. To date, the phenotypes of null mutations of *bf-1, otx-2,* and *emx-1* have been reported. Loss of *otx-2* function leads to the apparent absence of the midbrain and forebrain (Acampora et al., 1996; Matsuo et al., 1995; Ang et al., 1996). Loss of *bf-1* function results in hypoplasia of the entire telencephalon and part of the eye (Xuan et al., 1995). Loss of *emx-1* function has a more subtle phenotype; in these mice most of the corpus callosum does not form (Qiu et al., 1996). Although it is premature to make any generalizations about the roles each of these genes plays in the genetic hierarchy of forebrain development, these studies make it clear that these genes are involved in anterior CNS development.

Dorsoventral patterning in the CNS of *Drosophila* and vertebrates also uses homologous genes. For instance, the *ventral nervous system defective (vnd)* gene is required for development of a subset of neurons within a ventral longitudinal column in *Drosophila* (Jimenez et al., 1995). The *vnd* gene encodes a homeodomain protein of the *nk* homeobox gene family (Mellerick and Nirenberg, 1995). Vertebrate *nkx* genes (*nkx-2.1,* and *nkx-2.2*) are expressed in ventral longitudinal columns within the CNS (Fig. 10–3). Mutation of *nkx-2.1* results in abnormal differentiation of the ventral forebrain (Kimura et al., 1995). The *Drosophila* and vertebrate *single-minded* genes are expressed in the ventral CNS; however, the fly gene is expressed in the ventral midline whereas the mouse genes are expressed in stripes just lateral to the midline (Crews et al., 1988; Fan et al., 1996). It is highly probable that similar genetic programs which regulate analogous/

homologous developmental processes in vertebrates and invertebrates will continue to be discovered.

Summary and Future Directions

The forebrain differs from other CNS regions in the diversity of its constituent parts, in its essential role in higher neurological functions, and in the increase of its size, relative to the rest of the CNS, during evolution and ontogeny. Due in part to its complex morphology and the apparent differences in its early development, there have been hypotheses which postulate that its development and organization have features distinct from more posterior CNS domains. Although this point has not been fully clarified, recent studies of the organization of the prosencephalic neural plate and neural tube are providing evidence for a relatively simple underlying structural organization which shares properties with other CNS regions. The Prosomeric model (Fig. 10–1) postulates a neuromeric organization of the forebrain in which longitudinal columns and transverse segments subdivide the neuroepithelium into a checkerboard-like array of histogenic domains. Experimental embryological and genetic studies are beginning to identify the tissues and molecules that generate regional pattern within the forebrain; these investigations are finding some similarities in the patterning mechanisms (e.g., the role of the axial mesendoderm) found in other CNS regions. However, it remains to be determined how forebrain-specific structures (e.g., basal ganglia and cortex) are induced and patterned.

Ongoing and future investigations must focus on several basic subjects. First, studies of the molecular and cellular mechanisms that pattern the prosencephalon need to establish how the primary D/V and A/P organization is set up and how secondary signaling centers are generated that further regionalize the forebrain. Eventually, this work will determine how specific nuclei (e.g., in the thalamus) and domains (e.g., the various subdivisions of the cortex) are induced in the appropriate locations. In addition, it is likely that organizers also modulate cellular proliferation, which secondarily regulates morphogenesis of brain regions.

Once regional specification has established the basic plan of histogenic primordia, differentiation within these domains will proceed. The section on transcription factors describes how mutations in the rapidly growing number of these regulatory genes affect histogenesis of particular forebrain regions. Our understanding of the mechanisms underlying the effect of the mutations is still very superficial. Future studies must determine the molecular details how loss of a regulatory gene product disrupts important processes such as the control of the cell cycle, cell fate specification, cell migration, the production of proteins required for the elaboration, targeting, and function of axons and dendrites.

Work on forebrain development should focus on patterning and histogenesis of the structures that are unique to this region, such as the basal ganglia and cerebral cortex. During evolution of mammals, the cerebral cortex has expanded both in its size and functional importance. In addition, most if not all of the neocortex, which is the cortical region that has grown disproportionately in higher mammals, is apparently a mammalian-specific structure. Future experimental challenges will include understanding how this six-layered structure evolved, how regionalization of cortical subdivisions is achieved, and how novel cortical domains are integrated within existing brain circuitry.

ACKNOWLEDGMENTS

We wish to thank our collaborators Luis Puelles and Salvador Martinez, whose insights have been essential in the development of the Prosomeric model. We also thank the members of the Rubenstein laboratory for making available unpublished data for this chapter, Jill A. R. Helms for critical reading of the manuscript, and Susan Yu for help preparing the references. This work was supported by research grants to J.L.R.R. from The March of Dimes, NARSAD, the John Merck Fund, and the Human Frontiers Science Program (and NIMH RO1 MH49428-01, RO1 MH51561-01A1, and K02 MH01046-01), and to K.S. from the JSPS and Ministry of Education, Science, and Culture of Japan.

REFERENCES

Acampora, D., Mazan, S., Lallemand, Y., Avantaggiato, V., Maury, M., Simeone, A., Brulet, P. (1996). Forebrain and midbrain regions are deleted in *otx2*-mutants due to a defective anterior neuroectoderm specification during gastrulation. *Development* 121:3279–3290.

Akazawa, C., Ishibashi, M., Shimizu, C., Nakanishi, S., Kageyama, R. (1995). A mammalian helix-loop-helix factor structurally related to the product of drosophila proneural gene *atonal* is a positive transcriptional regulator expressed in the developing nervous system. *J. Biol. Chemistry* 270:8730–8738.

Alkema, M.J., van der Lugt, N.M.T., Bobeldijk, R.C., Berns, A., van Lohulzen, M. (1995). Transformation of axial skeleton due to overexpression of *bmi-1* transgenic mice. *Nature* 374:724–727.

Allende, M.L., Weinberg, E.S. (1994). The expression pattern of two zebrafish *achaete-scute* homolog (*ash*) genes is altered in the embryonic brain of the *cyclops* mutant. *Dev. Biol.* 166:509–530.

Alvarez-Bollado, G., Rosenfeld, M.G., Swanson, L.W. (1995). Model of forebrain regionalization based on spatiotemporal pattern of *pou*-III homeobox gene expression, birthdates and morphological features. *J. Comp. Neurol.* 355:237–295.

Anderson, B., Schonemann, M.D., Pearse, R.V. II, Jenne, K. Sugarman, J., Rosenfeld, M.G. (1993). Brn-5 is a divergent POU domain factor highly expressed in layer IV of the neocortex. *J. Biol. Chem.* 31:23390–23398.

Anderson, S., Qiu, M.S., Bulfone, A., Meneses, J., Pedersen, R., Rubenstein, J.L. R. (1995). Functional analysis of *Dlx-1* and *Dlx-2* using gene replacement. *Soc. Neurosci. Abstract* #320.18.

Ang, S.L., Rossant, J. (1994). HNF-3β is essential for node and notochord formation in mouse development. *Cell* 78:561–574.

Ang, S.-L., Conlon, R.A., Jin, O., Rossant, J. (1994). Positive and negative signals from mesoderm regulate the expression of mouse *otx2* in ectoderm explants. *Development* 120:2979–2989.

Ang, S.-L., Jin, O., Rhinn, M., Daigle, N., Stevenson, L., Rossant, J. (1996). A targeted mouse *otx2* mutation leads to severe defeats in gastrulation and formation of axial mesoderm and to deletion of rostral brain. *Development* 122:243–252.

Aruga, J., Yokota, N., Hashimoto, M., Furuichi, T., Fukuda, M., Mikoshiba, K. (1994). A novel zinc finger protein, Zic, is involved in neurogenesis, especially in the cell lineage of cerebellar granule cells. *J. Neurochem.* 63:1880–1890.

Barth, K.A., Wilson, S.W. (1995). Expression of zebrafish *nk2.2* is influenced by *sonic hedgehog*/vertebrate *hedgehog-1* and demarcates a zone of neuronal differentiation in the embryonic forebrain. *Development* 121:1755–1768.

Bartholoma, A., Nave, K.-A. (1994). Nex-1: a novel brain-specific helix-loop-helix protein with autoregulation and sustained expression in mature cortical neurons. *Mech. Dev.* 48:217–228.

Begley, C.G., Lipkowitz, S., Gobel, V., Mahon, K.A., Bertness, V., Green, A.R., Gough, N.M., Kirsch, I.R. (1992). Molecular characterization of *nscl,* a gene encoding a helix-loop-helix-protein expressed in the developing nervous system. *Proc. Natl. Acad. Sci. U.S.A.* 69:38–42.

Bober, E., Baum, C., Braun, T., Arnold, H.-H. (1994). A novel *nk*-related mouse homeobox gene: expression in central and peripheral nervous structures during embryonic development. *Dev. Biol.* 162:288–303.

Bollag, R.J., Siegfried, Z., Cebra-Thomas, J.A., Garvey, N., Davison, E.M., Silver, L.M. (1994). An ancient family of embryonically expressed mouse genes sharing a conserved protein motif with the T locus. *Nat. Genet.* 7:383–389.

Brunelli, S., Faiella, A., Capra, V., Nigro, V., Simeone, A., Cama, A., Boncinelli, E. (1996). Germline mutations in the homeobox gene *emx2* in patients with severe schizencephaly. *Nat. Genet.* 12:94–96.

Bulfone, A., Puelles, L., Porteus, M.H., Frohman, M.A., Martin, G.R., Rubenstein, J.L.R. (1993a). Spatially restricted expression of *Dlx-1, Dlx-2 (Tes-1), gbx-2,* and wnt-3 in the embryonic day 12.5 mouse forebrain defines potential transverse and longitudinal segmental boundaries. *J. Neurosci.* 13:3155–3172.

Bulfone, A., Kim, H.-J., Puelles, L., Porteus, M.H., Grippo, J.F., Rubenstein, J.L.R. (1993b). The mouse *Dlx-2 (Tes-1)* gene is expressed in spatially restricted domains of the forebrain, face and limbs in midgestation mouse embryos. *Mech. Dev.* 40:129–140.

Bulfone, A., Smiga, S.M., Shimamura, K., Puelles, L., Peterson, A., Rubenstein, J.L.R. (1995). T-Brain (Tbr-1): a homolog of *brachyury* whose expression defines molecularly distinct domains within the cerebral cortex. *Neuron* 15:63–78.

Bulleit, R.F., Cui, H., Wang, J., Lin, X. (1994). NMDA receptor activation in differentiating cerebellar cell cultures regulates the expression of a new *pou* gene, Cns-1. *J. Neurosci.* 14:1584–1595.

Campbell, K., Olsson, M., Bjorklund, A. (1995). Regional incorporation and site-specific differentiation of striatal precursors transplanted to the embryonic forebrain ventricle. *Neuron* 15:1259–1273.

Chalepakis, G., Stoykova, A., Wijnholds, J., Tremblay, P., Gruss, P. (1993). Pax: gene regulators in the developing nervous system. *J. Neurobiol.* 24:1367–1384.

Chapman, G., Rathjen, P.D. (1995). Sequence and evolutionary conservation of the murine *gbx-2* homeobox gene. *FEBS Lett.* 364:285–292.

Cohen, S. M., Bronner, G., Kuttner, F., Jürgens, G., Jackle, H. (1990). Distal-less encodes a homeodomain protein required for limb development in Drosophila. *Nature* 338:432–434.

Cohen, S., Jürgens, G. (1991). Drosophila headlines. *Trends Genet.* 7:267–272.

Couly, G., Le Douarin, N.M. (1988). The fate map of the cephalic neural primordium at the presomitic to the 3-somite stage in the avian embryo. *Dev. Suppl.* 103:101–113.

Crews, S.T., Thomas, J.B., Goodman, C.S. (1988). The Drosophila *single-minded* gene encodes a nuclear protein with sequence similarity to the *per* gene product. *Cell* 52:143–151.

Crossley, P.H., Martin, G.R. (1995). The mouse *Fgf8* gene encodes a family of polypeptides and is expressed in regions that direct outgrowth and patterning in the developing embryo. *Development* 121:439–451.

Crossley, P.H., Martinez, S., Martin, G.R. (1996). Midbrain development induced by *Fgf8* in the chick embryo. *Nature* 380:66–68.

Dahmane, N., Charron, G., Lopes, C., Yaspo, M.L., Maunoury, C., Decorte, L., Sinet, P.M., Bloch, B., Delabar, J.M. (1995). Down syndrome-critical region contains a gene homologous to Drosophila *sim* expressed during rat and human central nervous system development. *Proc. Natl. Acad. Sci. U.S.A.* 92:9191–9195.

Davidson, D. (1995). The function and evolution of *msx* genes: pointers and paradoxes. *Trends Genet.* 11:405–411.

Deacon, T.W., Pakzaban, P., Isacson, O. (1994). The lateral ganglionic eminence is the origin of cells committed to striatal phenotypes: neural transplantation and developmental evidence. *Brain Res* 668:211–219.

Deitcher, D.L., Fekete, D.M., Cepko, C.L. (1994). Asymmetric expression of a novel homeobox gene in vertebrate sensory organs. *J. Neurosci.*14:486–498.

Dickinson, M.E., Krumlauf, R., McMahon, A.P. (1994). Evidence for a mitogenic effect of Wnt-1 in the developing mammalian central nervous system. *Development* 120:1453–1471.

Dickinson, M.E., Selleck, M.A.J., McMahon, A.P., Bronner-Fraser, M. (1995). Dorsalization of the neural tube by the non-neural ectoderm. *Development* 121:2099–2106.

Doniach, T. (1995). Basic FGF as an inducer of anteroposterior neural pattern. *Cell* 83: 1067–1070.

Duncan, M., Dicicco-Bloom, E.M., Xiang, X., Benezra, R., Chada, K. (1992). The gene for the helix-loop-helix protein, Id, is specifically expressed in neural precursors. *Dev. Biol.* 154:1–10.

Eagleson, G.W., Harris, W.A. (1990). Mapping of the presumptive brain regions in the neural plate of Xenopus laevis. *J. Neurobiol.* 21:427–440.

Eagleson, G., Ferreiro, B., Harris, W.A. (1995). Fate of the anterior neural ridge and the morphogenesis of the Xenopus forebrain. *J. Neurobiol.* 28:146–158.

Echelard, Y., Epstein, D.J., St-Jacques, B., Shen, L., Mphler, J., McMahon, J.A., McMahon, A.P. (1993). Sonic Hedgehog, member of a family of putative signaling molecules, is implicated in the regulation of CNS polarity. *Cell* 75:1417–1430.

Ellmeier, W., Weith, A. (1995). Expression of the helix-loop-helix gene Id3 during murine embryonic development. *Dev. Dyn.* 203:163–173.

Ellmeier, W., Aguzzi, A., Kleiner, E., Kurzbauer, R., Weith, A. (1992). Mutually exclusive expression of a helix-loop-helix gene and N-myc in human neuroblastomas and in normal development. *EMBO J.* 11:2563–2571.

Epstein, D.J., Vogan, K.J., Trasler, D.G., Gros, P. (1993). A mutation within intron 3 of the *pax-3* gene produces aberrantly spliced mRNA transcripts in the *Splotch (Sp)* mouse mutant. *Proc. Natl. Acad. Sci. U.S.A.* 90:532–536.

Ericson, J., Muhr, J., Placzek, M., Lints, T., Jessell, T.M., Edlund, T. (1995). Sonic Hedgehog induces the differentiation of ventral forebrain neurons: a common signal for ventral patterning within the neural tube. *Cell* 81:747–756.

Evans, S.M., O'Brien, T.X. (1993). Expression of the helix-loop-helix factor Id during mouse embryonic development. *Dev. Biol.* 159:485–499.

Fan, C.M., Kuwana, E., Bulfone, A., Fletcher, C.F., Copeland, N.G., Jenkins, N.A., Martinez, S., Puelles, L., Rubenstein, J.L.R., Tessier-Lavigne, M. (1996). Expression patterns of two murine homologs of *Drosophila single-minded* suggest possible roles in embryonic patterning and in the pathogenesis of Down Syndrome. *Mol. Cell. Neurosci.* 7:1–16.

Ferreiro, B., Skoglund, P., Bailey, A., Dorsky, R., Harris, W.A. (1992). *xash1*, a Xenopus homolog of achaete-scute: a proneural gene in anterior regions of the vertebrate CNS. *Mech. Dev.* 40:25–36.
Figdor, M.C., Stern, C.D. (1993). Segmental organization of embryonic diencephalon. *Nature* 363:630–634.
Fishell, G., Mason, C.A., Hatten, M.E. (1993). Dispersion of neural progenitors within the germinal zones of the forebrain. *Nature* 362:636–638.
Frantz, T. (1994). *Extra-toes (Xt)* homozygous mutant mice demonstrate a role for the *gli-3* gene in the development of the forebrain. *Acta Anat.* 150:38–44.
Fraser, S., Keynes, R., Lumsden, A. (1990). Segmentation in the chick embryo hindbrain is defined by cell lineage restrictions. *Nature* 344:431–435.
Fujii, T., Pichel, J.G., Taira, M., Toyama, R., Dawid, I.B., Westphal, H. (1994). Expression patterns of the murine LIM class homeobox gene *lim1* in the developing brain and excretory system. *Dev. Dyn.* 199:73–83.
Funahashi, J.-I., Sekido, R., Murai, K., Kamachi, Y., Kondoh, H. (1993). δ-crystallin enhancer binding protein δEF1 is a zinc finger-homeodomain protein implicated in postgastrulation embryogenesis. *Development* 119:433–446.
Glaser, T., Walton, D.S., Maas, R.L. (1992). Genomic structure, evolutionary conservation and aniridia mutations in the human *pax6* gene. *Nat. Genet.* 2:232–238.
Gobel, V., Lipkowitz, S., Kozak, C.A., Kirsch, I.R. (1992). *nscl-2*: a basic domain helix-loop-helix gene expressed in early neurogenesis. *Cell Growth Differ.* 3:143–148.
Goulding, M.D., Chalepakis, G., Deutsch, U., Erselius, J.R., Gruss, P. (1991). Pax-3, a novel murine DNA binding protein expressed during early neurogenesis. *EMBO J.* 10: 1135–1147.
Graziadei, P.P.C., Monti-Graziadei, A.G. (1992). The influence of the olfactory placode on the development of the telencephalon in Xenopus laevis. *Neuroscience* 46:617–629.
Grindley, J.C., Davidson, D.R., Hill, R.E. (1995). The role of Pax-6 in eye and nasal development. *Development* 121:1433–1442.
Grossniklaus, U., Cadigan, K.M., Gehring, W.J. (1994). Three maternal coordinate systems cooperate in the patterning of the Drosophila head. *Development* 120:3155–3171.
Giazzi, S., Price, M., De Felice, M., Damante, G., Mattei, M.-G., Di Lauro, R. (1990). Thyroid nuclear factor 1 (TTF-1) contains a homeodomain and displays a novel DNA binding specificity. *EMBO J.* 9:3631–3639.
Gerrero, M.R., McEvilly, R.J., Turner, E., Lin, C.R., O'Connell, S., Jenne, K.J., Hobbs, M.V., Rosenfeld, M.G. (1993). *Brn-3.0*: a *POU* domain protein expressed in the sensory, immune and endocrine systems that selectively functions on non-octamer motifs *Prot. Natl. Acad. Sci. U.S.A.* 90(22):10841–10845.
Hanks, M., Wurst, W., Anson-Cartwright, L., Auerbach, A.B., Joyner, A.L. (1995). Rescue of the *en-1* mutant phenotype by replacement of *en-1* with *en-2*. *Science* 269:679–682.
Hanson, I., Van Heyningen, V. (1995). Pax6: more than meets the eye. *Trends Genet.* 11:268–272.
Hara, Y., Rovescalli, A.C., Kim, Y., Nirenberg, M. (1992). Structure and evolution of four POU domain genes expressed in mouse brain. *Proc. Natl. Acad. Sci. U.S.A.* 89:3280–3284.
Hatini, Tao, W., Lai, E. (1994). Expression of winged helix genes, *bf-1* and *bf-2*, defines adjacent domains within the developing forebrain and retina. *J. Neurobiol.* 25:1293–1309.
Hatta, K., Puschel, A.W., Kimmel, C.B. (1994). Midline signaling in the primordium of the zebrafish anterior CNS. *Proc. Natl. Acad. Sci. U.S.A.* 91:2061–2065.
He, X., Treacy, M.N., Simmons, D.M., Ingraham, H.A., Swanson, L.W., Rosenfeld, M.G. (1989). Expression of a large family of POU-domain regulatory genes in mammalian brain development. *Nature* 340:35–42.
Heikinheimo, M., Lawshe, A., Shackleford, G.M., Wilson, D.B., MacArthur, C.A. (1994). Fgf-8 expression in the postgastrulation mouse suggests roles in the development of the face, limbs, and CNS. *Mech. Dev.* 48:129–138.

Hermesz, E., Mackem, S., Mahon, K.A. (1996). *rpx:* a novel anterior-restricted homeobox gene progressively activated in the prechordal plate, anterior neural plate and Rathke's pouch of the mouse embryo. *Development* 122:41–52.

Hill, R.E., Favor, J., Hogan, B.L.M., Ton, C.C.T., Saunders, G.F., Hanson, I.M., Prosser, J., Jordan, T., Hastle, N.D., van Heyningen, V. (1991). Mouse *Small eye* results from mutations in a paired-like homeobox-containing gene. *Nature* 354:522–525.

Hirth, F., Therianos, S., Loop, T., Gehring, W.J., Reichart, H., Furukubo-Tokunaga, K. (1995). Developmental defects in brain segmentation caused by mutations of the homeobox genes *orthodenticle* and *empty spiracles* in Drosophila. *Neuron* 15:769–778.

Hogan, B.L.M., Horsburgh, G., Cohen, J., Hetherington, C.M., Fisher, G., Lyons, M.F. (1986). *Small eyes (Sey)*: a homozygous lethal mutation on chromosome 2 which affects the differentiation of both lens and nasal placodes in the mouse. *J. Embryol. Exp. Morphol.* 97:95–110.

Hsieh-Li, H.M., Witte, D.P., Szucsik, J.C., Weinstein, M., Li, H., Potter, S.S. (1995). *gsh-2,* a murine homeobox gene expressed in the developing brain. *Mech. Dev.* 50:177–186.

Hui, C.C., Slusarski, D., Platt, K.A., Holmgren, R., Joyner, A.L. (1994). Expression of three mouse homologs of the Drosophila segment polarity gene *cubitus interruptus, gli, gli-2* and *gli-3* in ectoderm- and mesoderm-derived tissues suggests multiple roles during postimplantation development. *Dev. Biol.* 162:402–413.

Hynes, M., Poulsen, K., Tessier-Lavigne, M., Rosenthal, A. (1995). Control of neuronal diversity by the floor plate: contact-mediated induction of midbrain dopaminergic neurons. *Cell* 80:95–101.

Ingraham, H.A., Chen, R., Mangalam, H.J., Elsholtz, H.P., Flynn, S.E., Lin, C.R., Simmons, D.M., Swanson, L., Rosenfeld, M.G. (1988). A tissue-specific transcription factor containing a homeodomain specifies a pituitary phenotype. *Cell* 55:519–529.

Ishibashi, M., Ang, S.L., Shiota, K., Nakanishi, S., Kageyama, R., Guillemot, F. (1995). Targeted disruption of mammalian hairy and Enhancer of split homolog-1 (HES-1) leads to up-regulation of neural helix-loop-helix factors, premature neurogenesis, and severe neural tube defeats. *Genes & Development* 9:3136–3148.

Ishibashi, M., Moriyoshi, K., Sasai, Y., Shiota, K., Nakanishi, S., Kageyama, R. (1994). Persistent expression of helix-loop-helix factor HES-1 prevents mammalian neural differentiation in the central nervous system. *EMBO J.* 13:1799–1805.

Ishibashi, M., Sasai, Y., Nakanishi, S., Kageyama, R. (1993). Molecular characterization of HES-2, a mammalian helix-loop-helix factor structurally related to Drosophila hairy and Enhancer of split. *Eur. J. Biochem.* 215:645–652.

Jacobson, C.-O. (1959). The localization of the presumptive cerebral regions in the neural plate of the axolotl larva. *J. Embryol. Exp. Morphol.* 7:1–21.

Jiménez, F., Martin-Morris, L.E., Velasco, L., Chu, H., Sierra, J., Rosen, D.R., White, K. (1995). *vnd,* a gene required for early neurogenesis of Drosophila, encodes a homeodomain protein. *EMBO J.* 14:3487–3495.

Jostes, B., Walther, C., Gruss, P. (1991). The murine paired box gene, *pax7,* is expressed specifically during the development of the nervous and muscular system. *Mech. Dev.* 33:27–38.

Joyner, A.L. (1996). *engrailed, wnt* and *pax* genes regulate midbrain-hindbrain development. *Trends Genet.,* 12:15–20.

Joyner, A.L., Martin, G.R. (1987). *en-1* and *en-2,* two mouse genes with sequence homology to the drosophila engrailed gene: expression during embryogenesis. *Genes Dev.* 1:29–38.

Joyner, A.L., Herrup, K., Auerbach, B.A., Davis, C.A., Rossant, J. (1991). Subtle cerebellar phenotype in mice homozygous for a targeted deletion of the En-2 homeobox. *Science* 251:1239–1243.

Jürgens, G., Hartenstein, V. (1993). The terminal regions of the body pattern. In *The Development of Drosophila melanogaster,* Bate, Martinez Arias, eds. pp. 687–746. Cold Spring Harbor Laboratory Press.

Keynes, R., Krumlauf, R. (1994). *Hox* genes and regionalization of the nervous system. *Annu. Rev. Neurosci.* 17:109–132.

Keynes, R., Lumsden, A. (1990). Segmentation and the origin of regional diversity in the vertebrate central nervous system. *Neuron* 2:1–9.

Kimura, S., Hara, Y., Pineau, T., Fernandez-Salguero, P., Fox, C.H., Ward, J.M., Gonzalez, F.J. (1995). The *T/ebp* null mouse: thyroid-specific enhancer-binding protein is essential for the organogenesis of the thyroid, lung, ventral forebrain, and pituitary. *Genes Dev.* 10:60–69.

Kispert, A. (1995). The Brachyury protein: a T-domain transcription factor. *Semin. Dev. Biol.* 6:395–403.

Knoepfler, P.S., Kamps, M.P. (1995). The pentapeptide motif of Hox proteins is required for cooperative DNA binding with Pbx1, physically contacts Pbx1, and enhances DNA binding by Pbx1. *Mol. Cell. Biol.* 15:5811–5819.

Komuro, I., Schalling, M., Jahn, L., Bodmer, R., Jenkins, N.A., Copeland, N.G., Izumo, S. (1993). *gtx*: a novel murine homeobox-containing gene, expressed specifically in glial cells of the brain and germ cells of testis, has a transcriptional repressor activity in vitro for a serum-inducible promoter. *EMBO J.* 12:1387–1401.

Kostich, W.A., Sanes, J.R. (1995). Expression of *zfh-4*, a new member of the zinc finger-homeodomain family, in developing brain and muscle. *Dev. Dyn.* 202:145–152.

Krauss, S., Concordet, J.-P., Ingham, P.W. (1993). A functionally conserved homolog of the Drosophila segment polarity gene *hh* is expressed in tissues with polarizing activity in zebrafish embryos. *Cell* 75:1431–1444.

Krumlauf, R. (1994). *Hox* genes in vertebrate development. *Cell* 78:191–201.

Lee, J.E., Hollenberg, S.M., Snider, L., Turner, D.L., Lipnick, N., Weintraub, H. (1995). Conversion of Xenopus ectoderm into neurons by NeuroD, a basic helix-loop-helix protein. *Science* 268:836–844.

Le Moine, C., Young, W.S., III. (1992). *rhs2*, a POU domain-containing gene, and its expression in developing and adult rat. *Proc. Natl. Acad. Sci. U.S.A.* 89:3285–3289.

Leifer, D., Krainc, D., Yu, Y.-T., McDermott, J., Breitbart, R.E., Heng, J., Neve, R.L., Kosofsky, B., Nadal-Ginard, B., Lipton, S.A. (1993). MEF2C, a MDS/MEF2-family transcription factor expressed in a laminar distribution in cerebral cortex. *Proc. Natl. Acad. Sci. U.S.A.* 90:1546–1550.

Li, H., Witte, D.P., Branford, W.W., Aronow, B.J., Weinstein, M., Kaur, S., Wert, S., Singh, G., Schreiner, C.M., Whitsett, J.A., Scott, W.J., Potter, S.S. (1994). *gsh-4* encodes a LIM-type homeodomain, is expressed in the developing central nervous system and is required for early postnatal survival. *EMBO J.* 13:2876–2885.

Liem, K.F., Tremml, G., Roelink, H., Jessell, T.M. (1995). Dorsal differentiation of neural plate cells induced by BMP-mediated signals from epidermal ectoderm. *Cell* 82:969–979.

Liu, I.S.C., Chen, J., Ploder, L., Vidgen, D., Kooy, D.V., Kalnins, V.I., McInnes, R.R. (1994). Developmental expression of a novel murine homeobox gene (*chx10*): evidence for roles in determination of the neuroretina and inner nuclear layer. *Neuron* 13:377–393.

Lu, S., Bogarad, L.D., Murtha, M.T., Ruddle, F.H. (1992). Expression pattern of a murine homeobox gene, *Dbx*, displays extreme spatial restriction in embryonic forebrain and spinal cord. *Proc. Natl. Acad. Sci. U.S.A.* 89:8053–8057.

Lyons, G.E., Micales, B.K., Schwarz, J., Martin, J.F., Olson, E.N. (1995). Expression of *mef2* genes in the mouse central nervous system suggests a role in neuronal maturation. *J. Neurosci.* 15:5727–5738.

MacDonald, R., Barth, K.A., Xu, Q., Holder, N., Mikkola, I., Wilson, S. (1995). Midline signalling is required for *pax* gene regulation and patterning of the eyes. *Development* 121: 3267–3278.

Mahmood, R., Kiefer, P., Guthrie, S., Dickson, C., Mason, I. (1995). Multiple roles for Fgf-3 during cranial neural development in the chicken. *Development* 121:1399–1410.

Marin, R., Puelles, L. (1994). Patterning of the embryonic avian midbrain after experimental inversions: a polarizing activity from the isthmus. *Dev. Biol.* 163:19–37.

Martí, E., Bumcrot, D.A., Takada, R., McMahon, A.P. (1995). Requirement of 19K form of Sonic Hedgehog for induction of distinct ventral cell types in CNS explants. *Nature* 375:322–325.

Martinez, S., Wassef, M., Alvarado-Mallart, R.-M. (1991). Induction of a mesencephalic phenotype in the 2-day-old chick prosencephalon is preceded by the early expression of the homeobox gene. *En. Neuron* 6:971–981.

Martinez, S., Geijo, E., Sanchez-Vives, V., Puelles, L., Gallego, R. (1992). Reduced junctional permeability at interrhombomeric boundaries. *Development* 116:1069–1076.

Martinez, S., Marin, F., Nieto, A., Puelles, L. (1995). Induction of ectopic *engrailed* and fate change in avian rhombomeres: intersegmental boundaries as barriers. *Mech. Dev.* 51: 289–303.

Mathis, J.M., Simmons, D.M., He, X., Swanson, L. W., Rosenfeld, M.G. (1992). Brain 4: a novel mammalian POU domain transcription factor exhibiting restricted brain-specific expression. *EMBO J.* 11:2551–2561.

Matsuo, T., Osumi-Yamashita, N., Noji, S., Ohuchi, H., Koyama, E., Myokai, F., Matsuo, N., Taniguchi, S., Doi, H., Iseki, S., Ninomiya, Y., Fujiwara, M., Watanabe, T., Eto, K. (1993). A mutation in the *pax-6* gene in rat small eye is associated with impaired migration of midbrain crest cells. *Nat. Genet.* 3:299–304.

Matsuo, I., Kuratani, S., Kimura, C., Takeda, N., Aizawa, S. (1995). Mouse Otx2 functions in the formation and patterning of rostral head. *Genes Dev.* 9:1–13.

McGinnis, W., Krumlauf, R. (1992). Homeobox genes and axial patterning. *Cell* 68:283–302.

McGuinness, T., Porteus, M.H., Smiga, S., Bulfone, A., Kingsley, C., Qiu, M., Liu, J.K., Long, J.E., Czernik, A., Xu, D., Rubenstein, J.L.R. (1996). Sequence, organization and transcription of the *Dlx-1* and *Dlx-2* locus. *Genomics* 35:473–485.

McMahon, A.P., Bradley, A. (1990). The Wnt-1 (int-1) proto-oncogene is required for development of a large region of the mouse brain. *Cell* 62:1073–1085.

Mellerick, D.M., Nirenberg, M. (1995). Dorsal-ventral patterning genes restrict *nk-2* homeobox gene expression to the ventral half of the central nervous system of Drosophila embryos. *Dev. Biol.* 171:306–316.

Meyer, B.I., Gruss, P. (1993). Mouse Cdx-1 expression during gastrulation. *Development* 117:191–203.

Monica, K., Galili, N., Nourse, J., Saltman, D., Cleary, M.L. (1991). *pbx2* and *pbx3*, new homeobox genes with extensive homology to the human proto-oncogene *pbx1*. *Mol. Cell. Biol.* 11:6149–6157.

Monuki, E.S., Weinmaster, G., Kuhn, R., Lemke, G. (1989). *scip:* a glial POU domain gene regulated by cyclic AMP. *Neuron* 3:783–793.

Monuki, E.S., Kuhn, R., Weinmaster, G., Trapp, B.D., Lemke, G. (1990). Expression and activity of the POU transcription factor Scip. *Science* 249:1300–1303.

Murtha, M.T., Leckman, J.F., Ruddle, F.H. (1991). Detection of homeobox genes in development and evolution. *Proc. Natl. Acad. Sci. U.S.A.* 88:10711–10715.

Nakai, S., Kawano, H., Yudate, T., Nishi, M., Kuno, J., Nagata, A., Jishage, K.I., Hamada, H., Fujii, H., Kawamura, K., Shiba, K., Noda, T. (1995). The POU domain transcription factor Brn-2 is required for the determination of specific neuronal lineages in the hypothalamus of the mouse. *Genes Dev.* 9:3109–3121.

Neuman, T., Keen, A., Zuber, M.X., Kristjansson, G.I., Gruss, P., Nornes, H.O. (1993). Neuronal expression of regulatory helix-loop-helix factor Id2 gene in mouse. *Dev. Biol.* 160:186–195.

Oliver, G., Sosa-Pineda, B., Geisendorf, S., Spana, E.P., Doe, C.Q., Gruss, P. (1993). *prox 1*, a prospero-related homeobox gene expressed during mouse development. *Mech. Dev.* 44:3–16.

Oliver, G., Mailhos, A., Wehr, R., Copeland, N.G., Jenkins, N.A., Gruss, P. (1995). six3, a murine homolog of the *sine oculis* gene, demarcates the most anterior border of the developing neural plate and is expressed during eye development. *Development* 121: 4045–4055.

Paro, R. (1995). Propagating memory of transcriptional states. *Trends Genet.* 11:295–297.

Pfaff, S.L., Mendelsohn, M., Stewart, C.L., Edlund, T., Jessell, T.M. (1996). Requirement for LIM homeobox gene *isl1* in motor neuron generation reveals a motor neuron dependent step in interneuron differentiation. *Cell* 84:309–320.

Pflugfelder, G.O., Roth, H., Poeck, B., Kerscher, S., Schwarz, H., Jonschker, B., Heisenberg, M. (1992). The *lethal(l) optomotor-blind* gene of *Drosophila melanogaster* is a major organizer of optic lobe development: isolation and characterization of the gene. *Proc. Natl. Acad. Sci. U.S.A.* 89:1199–1203.

Popperi, H., Bienz, M., Studer, M., Chan, S., Aparico, S., Brenner, S., Mann, R., Krumlauf, R. (1995). Segmental expression of *Hoxb-1* is controlled by a highly conserve autoregulatory loop dependent upon *exd/pbx*. *Cell* 81:1031–1042.

Porteus, M.H., Bulfone, A., Ciaranello, R.D., Rubenstein, J.L. R. (1991). Isolation and characterization of a novel cDNA clone encoding a homeodomain that is developmentally regulated in the ventral forebrain. *Neuron* 7:221–229.

Porteus, M.H., Brice, J.E.A., Usdin, T.B., Ciaranello, R.D., Rubenstein, J.L.R. (1992). Isolation and characterization of a library of cDNA clones that are preferentially expressed in the embryonic telencephalon. *Mol. Brain Res.* 12:7–22.

Porteus, M.H., Bulfone, A., Lui, J.-K., Puelles, L., Lo, L.-C., Rubenstein, J.L.R. (1994). Dlx-2, Mash-1 and Map-2 expression and bromodeoxyuridine incorporation define molecularly distinct cell populations in the embryonic mouse forebrain. *J. Neurosci.* 14:6370–6383.

Price, M., Lemaistre, M., Pischetola, M., Di Lauro, R., Duboule, D. (1991). A mouse gene related to *Distal-less* shows a restricted expression in the developing forebrain. *Nature* 351:748–751.

Price, M., Lazzaro, D., Pohl, T., Mattei, M.-G., Rüther, U., Olivo, J.-C., Duboule, D., Di Lauro, R. (1992). Regional expression of the homeobox gene *nkx-2.2* in the developing mammalian forebrain. *Neuron* 8:241–255.

Puelles, L. (1995). A segmental morphological paradigm for understanding vertebrate forebrains: *Brain Behav. Evol.* 46:319–337.

Puelles, L., Amat, J.A., Martinez-de-la-Torre, M. (1987). Segment-related, mosaic neurogenetic pattern in the forebrain and mesencephalon of early chick embryos: I. Topography of AChE-positive neuroblasts up to stage HH18. *J. Comp. Neurol.* 266:247–268.

Puelles, L., Rubenstein, J.L.R. (1993). Expression patterns of homeobox and other putative regulatory genes in the embryonic mouse forebrain suggest a neuromeric organization. *Trends Neurosci.* 16:472–479.

Qiu, M., Bulfone, A., Martinez, S., Meneses, J.J., Shimamura, K., Pedersen, R.A., Rubenstein, J.L.R. (1995). Role of *Dlx-2* in head development and evolution: null mutation of *Dlx-2* results in abnormal morphogenesis of proximal first and second branchial arch derivatives and abnormal differentiation in the forebrain. *Genes Dev.* 9:2523–2538.

Qiu, M., Anderson, S., Chen, S., Meneses, J.J., Hevner, R., Kuwana, E., Pedersen, R.A., Rubenstein, J.L.R. (1996). Mutation of the Emx-1 Homeobox gene disrupts the corpus callosum. *Dev. Biol.* 178:174–178.

Rauskolb, C., Peifer, M., Wieschaus, E. (1993). *extradenticle,* a regulator of homeotic gene activity, is a homolog of the homeobox-containing human proto-oncogene *pbx1*. *Cell* 74:1101–1112.

Riechmann, V., Sablitzky, F. (1995). Mutually exclusive expression of two dominant-negative helix-loop-helix (dnHLH) genes, Id4 and Id3, in the developing brain of the mouse suggests distinct regulatory roles of these dnHLH proteins during cellular proliferation and differentiation of the nervous system. *Cell Growth Differ.* 6:837–843.

Rinkwitz-Brandt, S., Justus, M., Oldenettel, I., Arnold, H.-H., Bober, E. (1995). Distinct temporal expression of mouse *nkx-5.1* and *nkx-5.2* homeobox genes during brain and ear development. *Mech. Dev.* 52:371–381.

Robinson, G.W., Wray, S., Mahon, K.A. (1991). Spatially restricted expression of a member of a new family of murine *Distal-less* homeobox genes in the developing forebrain. *New Biol.* 3:1183–1194.

Roelink, H., Augsburger, A., Heemskerk, J., Korzh, V., Norlin, S., Ruiz i Altaba, A., Tanabe, Y., Placzek, M., Edlund, T., Jessell, T.M., Dodd, J. (1994). Floor plate and motor neuron induction by *vhh-1*, a vertebrate homolog of hedgehog expressed by the notochord. *Cell* 76:761–775.

Reolink, H., Porter, J.A., Chiang, C., Tanabe, Y., Chang, D.T., Beachy, P.A., Jessell, T.M. (1995). Floor plate and motor neuron induction by different concentrations of the amino-terminal cleavage product of Sonic hedgehog autoproteolysis. *Cell* 81:445–455.

Rowitch, D.H., McMahon, A.P. (1995). *pax-2* expression in the murine neural plate precedes and encompasses the expression domains of *wnt-1* and *en-1*. *Mech. Dev.* 52:3–8.

Rubenstein, J.L.R., Martinez, S., Shimamura, K., Puelles, L. (1994). The embryonic vertebrate forebrain: the Prosomeric model. *Science* 266:578–580.

Rubenstein, J.L., Puelles, L. (1994). Homeobox gene expression during development of the vertebrate brain. *Current Topics in Developmental Biology* 29:1–63.

Ruiz i Altaba, A. (1994). Pattern formation in the vertebrate neural plate. *Trends Neurosci.* 17:233–243.

Saha, M., Servetnick, M., Grainger, R.M. (1992). Vertebrate eye development. *Curr. Opin. Genet. Dev.* 2:582–588.

Salinas, P.C., Nusse, R. (1992). Regional expression of the Wnt-3 gene in the developing mouse forebrain in relationship to diencephalic neuromeres. *Mech. Dev.* 39(3):151–160.

Sanyanusin, P., Schimmenti, L.A., McNoe, L.A., Ward, T.A., Pierpont, M.E.M., Sullivan, M.J., Dobyns, W.B., Eccles, M.R. (1995). Mutation of the *pax2* gene in a family with optic nerve colobomas, renal anomalies and vesicoureteral reflux. *Nat. Genet.* 9:358– 364.

Sasai, Y., Kagcyama, R., Tagawa, Y., Shigemoto, R., Nakanishi, S. (1992). Two mammalian helix-loop-helix factors structurally related to Drosophila hairy and Enhancer of split. *Genes & Development* 6:2620–2634.

Satokata, I., Maas, R. (1994). Msx1 deficient mice exhibit cleft palate and abnormalities of craniofacial and tooth development. *Nat. Genet.* 6:348–356.

Scholer, H.R. (1991). Octamania: the POU factors in murine development. *Trends Genet.* 7:323–329.

Schonemann, M.D., Ryan, A.K., McEvilly, R.J., O'Connell, S.M., Arias, C.A., Kalla, K.A., Li, P., Sawchenko, P.E., Rosenfeld, M.G. (1995). Development and survival of the endocrine hypothalamus and posterior pituitary gland requires the neuronal POU domain factor Brn-2. *Genes Dev.* 9:3122–3135.

Schmahl, W., Knoedlseder, M., Favor, J., Davidson, D. (1993). Defects of neuronal migration and pathogenesis of cortical malformations are associated with *Small eye (Sey)* in the mouse, a point mutation at the Pax-6-locus. *Acta Neuropathol.* 86:126–135.

Schubert, F.R., Fainsod, A., Gruenbaum, Y., Gruss, P. (1995). Expression of the novel murine homeobox gene *sax-1* in the developing nervous system. *Mech. Dev.* 51:99–114.

Shawlot, W., Behringer, R.R. (1995). Requirement for Lim1 in head-organizer function. *Nature* 374:425–430.

Sheng, H., Zhadanov, B.M., Fujii, T., Bertuzzi, S., Grinberg, A., Lee, E.J. et al. (1996). The LIM homeobox gene *lhx-3* is essential for the specification and proliferation of pituitary cell lineages. *Science* 272:1004–1007.

Shimamura, K., Rubenstein, J.L.R. Inductive interactions direct the early regionalization of the mouse prosencephalon. Manuscript submitted to Development.

Shimamura, K., Hartigan, D.J., Martinez, S., Puelles, L., Rubenstein, J.L.R. (1995). Longitudinal organization of the anterior neural plate and neural tube. *Development* 121: 3923–3933.

Shimamura, K., Martinez, S., Puelles, L., Rubenstein, J.L.R. (1996). Patterns of gene expression in the neural plate and neural tube subdivide the embryonic forebrain into transverse and longitudinal domains. *Dev. Neurosci. In Press.*

Shimizu, C., Akazawa, C., Nakanishi, S., Kageyama, R. (1995). MATH-2, a mammalian helix-loop-helix factor structurally related to the product of *Drosophila* proneural gene *atonal*, is specifically expressed in the nervous system. *Eur. J. Biochem.* 229:239–248.

Shinoda, K., Lei, H., Yoshii, H., Nomura, M., Nagano, M., Shiba, H., Sasaki, H., Osawa, Y.,

Ninomiya, Y., Niwa, Q., Morohashi, K.I., Li, E. (1995). Developmental defects of the ventromedial hypothalamic nucleus and pituitary gonadotroph in the Ftz-F1 disrupted mice. *Dev. Dyn.* 204:22–29.

Simeone, A., Gulisano, M., Acampora, D., Stornaiuolo, A., Rambaldi, M., Boncinelli, E. (1992a). Two vertebrate homeobox genes related to the Drosophila *empty spiracles* gene are expressed in the embryonic cerebral cortex. *EMBO J.* 11:2541–2550.

Simeone, A., Acampora, D., Gulisano, M., Stornaiuolo, A., Boncinelli, E. (1992b). Nested expression domains of four homeobox genes in developing rostral brain. *Nature* 358: 687–690.

Simeone, A., Acampora, D., Mallamaci, A., Stornaiuolo, A., D'Aprice, M.R., Nigro, V., Boncinelli, E. (1993). A vertebrate gene related to *orthodenticle* contains a homeodomain of the Bicoid class and demarcates anterior neuroectoderm in the gastrulating mouse embryo. *EMBO J.* 12:2735–2747.

Simeone, A., Acampora, D., Pannese, M., D'Esposito, M., Stornaiuolo, A., Gulisano, M., Boncinelli, E. (1994a). Cloning and characterization of two members of the vertebrate *Dlx* gene family. *Proc. Natl. Acad. Sci. U.S.A.* 91:2250–2254.

Simeone, A., D'Apice, M.R., Nigro, V., Casanova, J., Graziani, F., Acampora, D., Avantaggiato, V. (1994b). *Orthopedia,* a novel homeobox-containing gene expressed in the developing CNS of both mouse and *Drosophila. Neuron* 13:83–101.

Simon, H., Hornbruch, A., Lumsden, A. (1995). Independent assignment of antero-posterior and dorso-ventral positional values in the developing chick hindbrain. *Curr. Biol.* 5:205–214.

Simon, J. (1995). Locking in stable states of gene expression: transcriptional control during Drosophila development. *Curr. Opin. Cell Biol.* 7:376–385.

Song, D.D., Harlan, R.E. (1994). The development of enkephalin and substance P neurons in the basal ganglia: insights into neostriatal compartments and the extended amygdala. *Dev. Brain Res.* 83:247–261.

Steingrimsson, E., Moore, K.J., Lamoreux, M.L., Ferre-D'Amare, A.R., Burley, S.K., Zimring, D.C., Skow, L.C., Hodgkinson, C.A., Arnheiter, H., Copeland, N.G., et al. (1994). Molecular basis of mouse *microphthalmia (mi)* mutations helps explain their developmental and phenotypic consequences [see comments]. *Nat. Genet.* 8:256–263.

St-Onge, L., Pituello, F., Gruss, P. (1995). The role of *pax* genes during murine development. *Semin. Dev. Biol.* 6:285–292.

Stoykova, A., Gruss, P. (1994). Roles of *pax* genes in developing and adult brain as suggested by expression patterns. *J. Neurosci.* 14:1395–1412.

Sulik, K., Dehart, D.B., Inagaki, T., Carson, J.L., Vrablic, T., Gesteland, K., Schoenwolf, G.C. (1994). Morphogenesis of the murine node and notochordal plate. *Dev. Dyn.* 201: 260–278.

Suzuki, N., Rohdewohld, H., Neuman, T., Gruss, P., Scholer, H.R. (1990). Oct-6: a POU transcription factor expressed in embryonal stem cells and in the developing brain. *EMBO J.* 9:3723–3732.

Takebayashi, K., Sasai, Y., Sakai, Y., Watanabe, T., Nakanishi, S., Kageyama, R. (1994). Structure, chromosomal locus, and promoter analysis of the gene encoding the mouse helix-loop-helix factor Hes-1. *J. Biol. Chem.* 269:5150–5156.

Takebayashi, K., Akazawa, C., Nakanishi, S., Kageyama, R. (1995). Structure and promoter analysis of the gene encoding the mouse helix-loop-helix factor Hes-5. *J. Biol. Chem.* 270:1342–1349.

Talbot, W.S., Trevarrow, B., Halpern, M.E., Melby, A.E., Farr, G., Postlethwait, J.H., Jowett, T., Kimmel, C.B., Kimelman, D. (1995). A homeobox gene essential for zebra-fish notochord development. *Nature* 378:150–157.

Tao, W., Lai, E. (1992). Telencephalon-restricted expression of *bf-1,* a new member of the *HNF-3/fork head* gene family in the developing rat brain. *Neuron* 8:957–966.

Tassabehji, M., Read, A.P., Newton, V.E., Harris, R., Balling, R., Gruss, P., Strachan, T. (1992). Waardenburg's syndrome patients have mutations in the human homolog of the *pax-3* paired box gene. *Nature* 355:635–638.

Thor, S., Ericson, J., Brannstrom, T., Edlund, T. (1991). The homeodomain LIM protein *Isl-1* is expressed in subsets of neurons and endocrine cells in the adult rat. *Neuron* 7:881–889.

Tremblay, P., Kessel, M., Gruss, P. (1995). A transgenic neuroanatomical marker identifies cranial neural crest deficiencies associated with the *Pax3* mutant *splotch*. *Dev. Biol.* 171: 317–329.

Turner, E.E., Jenne, K.J., Rosenfeld, M.G. (1994). Brn-3.2: a Brn-3-related transcription factor with distinctive central nervous system expression and regulation by retinoic acid. *Neuron* 12:205–218.

Urbanek, P., Wang, Z.Q., Fetka, I., Wagner, E.F., Busslinger, M. (1994). Complete block of early B cell differentiation and altered patterning of the posterior midbrain in mice lacking Pax5/BSAP. *Cell* 79:901–912.

Vaage, S. (1969). The segmentation of the primitive neural tube in chick embryos (Gallus domesticus). *Ergeb. Anat. Entwicklungsgesch.* 41:1–88.

Valarache, I., Tissier-Seta, J.-P., Hirsch, M.-R., Martinez, S., Goridis, C., Brunet, J.-F. (1993). The mouse homeodomain protein Phox2 regulates *Ncam* promoter activity in concert with Cux/CDP and is a putative determinant of neurotransmitter phenotype. *Development* 119:881–896.

Vershon, A.K., Johnson, A.D. (1993). A short disordered protein region mediates interactions between the homeodomain of the yeast α2 protein and the Mcm1 protein. *Cell* 72: 105–112.

Walter, C., Gruss, P. (1991). *pax-6* a murine paried box gene, is expressed in the developing CNS. *Development* 113:1435–1449.

Wang, Y., Benezra, R., Sassoon, D.A. (1992). Id expression during mouse development: a role in morphogenesis. *Dev. Dyn.* 194:222–230.

Wegner, M., Drolet, D.W., Rosenfeld, M.G. (1993). POU-domain proteins: structure and function of developmental regulators. *Curr. Opin. Cell Biol.* 5:488–498.

Weinstein, D.C., Ruiz i Altaba, A., Chen, W.S., Hoodless, P., Prezioso, V.R., Jessell, T.M., Darnell, J.E. Jr. (1994). The winged-helix transcription factor HNF-3β is required for notochord development in the mouse embryo. *Cell* 78:575–588.

Wimmer, E.A., Jackle, H., Pfelfle, C., Cohen, S. M. (1993). A Drosophila homolog of human Sp1 is a head-specific segmentation gene. *Nature* 366:690–694.

Xu, Y., Baldassare, M., Fisher, P., Rathbun, G., Oltz, E.M., Yancopoulos, G.D., Jessell, T.M., Alt, F. W. (1993). LH-2: a LIM/homeodomain gene expressed in developing lymphocytes and neural cells. *Proc. Natl. Acad. Sci. U.S.A.* 90:227–231.

Xuan, S., Baptista, C.A., Balas, G., Tao, W., Soares, V.C., Lai, E. (1995). Winged helix transcription factor BF-1 is essential for the development of the cerebral hemispheres. *Neuron* 14:1141–1152.

Yamada, T., Pfaff, S.L., Edlund, T., Jessell, T.M. (1993). Control of cell pattern in the neural tube: motor neuron induction by diffusible factors from notochord and floor plate. *Cell* 73:673–686.

Yoon, S., Chikaraishi, D.M. (1994). Isolation of two E-box binding factors that interact with the rat tyrosine hydroxylase enhancer. *J. Biol. Chem.* 269:18453–18462.

Yu, B.D., Hess, J.L., Horning, S.E., Brown, G.A.J., Korsmeyer, S.J. (1995). Altered *Hox* expression and segmental identity in MII-mutant mice. *Nature* 378:505–508.

Zhadanov, A.B., Bertuzzi, S., Taira, M., Dawid, I.B., Westphal, H. (1995). Expression pattern of the murine LIM class homeobox gene *lhx3* in subsets of neural and neuroendocrine tissues. *Dev. Dyn.* 202:354–364.

Zhu, W., Dahmen, J., Bulfone, A., Rigolet, M., Hernandez, M.C., Kuo, W.L., Biggs, J., Puelles, L., Rubenstein, J.L.R., Israel, M.A. (1995). Id gene expression during development and molecular cloning of the human Id-1 gene. *Molecular Brain Research* 30: 312–326.

Zimmerman, K., Shih, J., Bars, J., Collazo, A., Anderson, D.J. (1993). Xash-3, a novel Xenopus achaete-scute homolog, provides an early marker of planar neural induction and position along the mediolateral axis of the neural plate. *Development* 119:221–232.

11

Lineage analysis in the vertebrate central nervous system

CONSTANCE L. CEPKO, JEFFREY A. GOLDEN,
FRANCIS G. SZELE AND JOHN C. LIN

The development of the vertebrate nervous system is a problem of such complexity that it is sometimes difficult to know where to start. A description of the lineal relationships among cells in a given area can provide a starting point, and sometimes a great deal more. The need to know lineal relationships has led to the development of new technologies and has pushed some of the classical methods to their limits. The information so gleaned has been worthwhile, providing us with insights into some of the basic strategies employed by progenitors in the generation of myriad cell types.

LINEAGE ANALYSIS DEFINED

Lineage analysis is a description of the genealogy of cells. It is performed by marking a progenitor cell or group of progenitors in such a way that the fate of the progeny can be followed. By marking cells at different ages and analyzing the progeny at different points in development, one can begin to generate ideas as to when and where decisions are being made. One can monitor the developmental paths followed by sibling cells and ask whether their paths vary depending upon the site of origin of the progenitor. A simple analysis of clone sizes over time can allow predictions to be made concerning modes of proliferation. Finally, one can determine the phenotype and location of progeny and analyze these data relative to boundaries or characteristics described by other techniques. The value of lineage data is perhaps best exemplified by the studies of the nematode, *Caenorhabditis elegans,* in which the lineage of each cell is known (Sulston and Horvitz, 1977; Sulston et al., 1983).

Lineage analysis, however, does not test the potency of cells, nor the role of the environment in instructing cells. The data gathered by lineage studies sometimes can rule in or out hypotheses that do model these events. For example, one might propose that the progenitors in a given area of the nervous system are committed to making only glia. If one then finds that many clones contain both glia and neurons, the model is ruled out. Similarly, if one finds that clones do contain only glia, the model is supported, but not proven. To prove the model, the progenitors that give rise only to glia must be exposed to other environments to determine if they are intrinsically unable to give rise to other cell types. The distinction as to what lineage data rule in or out, as opposed to what they merely support, is one that has led to some confusion in the interpretation of lineage studies. Part of this confusion is due to the range of meanings ascribed to the word "lineage." Lineage is sometimes meant to imply a mechanism of fate determination, as in it is "lineage" if a commitment decision is inherited from a progenitor. Lineage as a mechanism is sometimes concluded from studies in which an invariant lineage is observed, as in the above example with glia-only clones. However, an environment that is reproducible can lead to an invariant lineage. *C. elegans* provides examples of invariant lineages that require environmental interactions rather than occur as a result of commitment decisions handed down from mother to daughter (e.g., see Mello et al., 1994).

METHODS OF LINEAGE ANALYSIS

Lineage analyses have been performed with a variety of methods, each with its own advantages and disadvantages. In animals with large, accessible cells, cell marking can be initiated by injection of tracers such as fluorescent dextran (Weisblat et al., 1978). Such molecules are too large to pass between intercellular junctions and thus are passed only to progeny. Labeled cells can be followed by direct observation of the fluorescent tag in whole mounts or in tissue sections. The main advantage of this method is that it is very direct, allowing one to define precisely the location and time of marking. For the same reason, it is fairly clear that most injections are of a single cell, and thus it is straightforward to define all of the descendants as members of the same clone. One disadvantage is that many areas are not accessible for intracellular injection, particularly in early mammalian embryos. Furthermore, the dye eventually is diluted through many cell divisions and the final fate of the progeny of the injected cell cannot always be determined.

A genetic marking method applied very successfully to avian embryos uses chimeras. LeDouarin pioneered this technique by combining chick and quail tissue and following the cells of each species through distinctive nuclear staining patterns (LeDouarin, 1973). LeDouarin and her colleagues have used this method to define most of what is known of the lineal relationships in the avian peripheral nervous system (LeDouarin, 1973) and have more recently extended the approach to the central nervous system (CNS, see below). The advantages of this method are that the origin of the cells being traced is known precisely and, like any genetic method, the marking is stable. The disadvantages of this approach are that relatively large numbers of cells are engrafted, making clonal analysis impossible, and inevitably cells from two different species with somewhat different developmental schedules are intermingled. Fortunately the latter problem has not caused much difficulty for most studies. The method has not been easily extended to other species, although chimeras of two mouse strains which differ in histological properties have been made, and have contributed to our understanding of lineage in some areas.

The most recent method for marking cells is through stable transduction of a histologically detectable gene, such as lacZ or human placental alkaline phosphatase (PLAP). Retroviral vectors and transgenic mice have been used most extensively for this purpose. A retrovirus vector is an infectious virus that transduces a nonviral gene into mitotic cells in vivo or in vitro (Cepko, 1988). These vectors integrate a single copy of the viral genome stably into the host chromosome. Those viral vectors that are useful for lineage analysis have been modified so that they are replication incompetent and thus cannot spread from one cell to another. They are, however, faithfully passed to all daughter cells of the originally infected progenitor cell. One of the advantages of retroviruses as lineage markers is their stability. Another advantage is that they can be delivered to sites in a complex embryo that are otherwise inaccessible to intracellular injection or are difficult to surgically manipulate for transplantation. One further advantage is that clonal analysis is possible. However, there are also disadvantages in the use of retroviral vectors. In situations where sibling cells disperse widely, or where many clones intermingle at an injection site, they present problems of interpretation of clonal boundaries (see below). One can surmount such problems by using libraries of retroviruses that are engineered to encode a large number of distinctive tags. Another disadvantage is that one does not precisely control which cell or cells are infected as the viral inoculum is delivered to an area of mitotic cells and the viruses "choose" which cells to infect. One further disadvantage is that the detection of infected cells is dependent in large part upon the expression of the marker gene from a promoter that may not express in each member of a clone.

Retroviruses have been used for lineage analysis in the mouse, chicken, rat, ferret, and primate CNS, as described below. Recent efforts have resulted in a broadening of the host range of these vectors such that infection of fish and amphibia are now possible (Yee et al., 1994). In addition to providing data on lineal relationships, one other aspect of retroviral vectors that should be noted is that they provide a means for perturbing lineages as they can be engineered to encode a test gene in addition to a histochemical reporter gene. The characteristics of a clone carrying such a gene can be compared to those of clones that carry only the histochemical reporter gene.

As mentioned above, transgenic mice utilizing histochemical reporters such as *lacZ* have also been used to follow lineages. Several approaches are made possible by such transgenic animals and data are beginning to accumulate from these newer methods. One involves the fortuitous integration of *lacZ* on the X chromosome (Tan and Breen, 1993). Clonal marking is then initiated in female mice heterozygous for the insertion as a result of inactivation of one X chromosome early in development. The advantage of this method is that one does nothing invasive nor deliberate to initiate the cell marking. The disadvantage is that one has no control over the timing and marking is only initiated very early in development. A newer method with promise uses a recombinase under control of a particular promoter to initiate marking. *lacZ* is integrated in such a way that it will be expressed only subsequent to activation of the recombinase (Buenzow and Holmgren, 1995). By using different promoters that turn on in different tissues at different times, some control of marking initiation is achieved.

WHAT IS A CLONE?

If clonal analysis is the goal, at some point one must define the sibling relationships of marked cells. The ambiguity in clonal relationships after marking with retroviruses has brought this issue to the fore. There are two types of errors that can be made in evaluating clonal relationships: lumping and splitting errors. A lumping error is made when cells that are members of two or more clones are "lumped," or considered as members of the same clone. A splitting error is made when members of the same clone are interpreted as members of different clones or are "split" into more than one clone. The sensitivity of a conclusion to each type of error is dependent upon the nature of the conclusion. For example, if one wishes to conclude that each clone comprises only a single cell type, then splitting errors must be addressed. If one wishes to conclude that a progenitor is multipotent based upon the inclusion of multiple cell types in a clone, then lumping errors must be addressed.

Examples of lineage analyses conducted in the vertebrate CNS are given below. A synthesis of the data then follows in an attempt to draw parallels and distinctions. We have chosen to use data primarily from studies conducted in vivo and to draw into the picture those data generated by other methods when directly relevant to the interpretation of the lineage data.

RETINA

The retina is an area of the vertebrate CNS that has proven to be especially amenable to studies of development in various species. Its structure is fairly conserved, with six major classes of neuronal cells and one minor class of glial cell (Dowling, 1987). Cell type is easily determined from the morphology of the cells and their location in particular layers. Retinal cells are generated in a sequence from the neurepithelium of the optic vesicle, an outpouching of the ventral diencephalon near the border with the telencephalon. The accessibility that results from this evagination of the neural tube has made it relatively easy to access the mitotic zone for intracellular injection of tracers and delivery of retroviral vectors. Another characteristic of this tissue that has made lineage analyses relatively straightforward is that the generative zone is quite close to the final location of progeny cells. The lack of extensive migration of progeny cells has made the definition of clonal boundaries clear-cut.

Lineage analyses of retina have been performed in rat (Turner and Cepko, 1987), *Xenopus* (Holt et al., 1988; Wetts and Fraser, 1988; Wetts et al., 1989), mouse (Turner et al., 1990), and chick (Fekete et al., 1994). Clones in *Xenopus* were marked with intracellular tracers, either horseradish peroxidase or fluorescent dextran. In the other species, retroviral vectors encoding LacZ, PLAP, or the viral Gag protein (as an immunohistochemically detectable tag) were used. The number of species and variety of techniques applied to retinal lineage analyses have allowed for a satisfying comparison of data and methods.

Clone Size and Mitotic Behavior of Retinal Progenitors

In each species, clone size has been found to vary greatly. In the mouse, injection at embryonic day 14 (E14) led to clones of from one to 234 cells (Turner et al., 1990). Even at the end of development, clones in the rat varied from one to 22 cells (Turner and Cepko, 1987). In the early chick embryo (E3), clones varied from one to several hundred (Fekete et al., 1994). Clones in *Xenopus* initiated in the optic vesicle were generally more clustered in terms of size; one to 16 cells when injected with horseradish peroxidase (HRP) (Holt et al., 1988) and one to 42 when injected with fluorescent dextran (Wetts and Fraser, 1988), presumably reflecting the rapid genesis of the embryonic retina in a 24-hour period in this species. When clones were labeled by injection into the peripheral margin of *Xenopus* retina (Wetts et al., 1989), the area which is capable of generating retina and adjacent pigment epithelium throughout life, the clones exhibited a bimodal distribution, with the smaller class similar to that following optic vesicle injections.

In the retroviral data set, there were many single-cell clones. One aspect of retroviral integration should be noted here regarding this observation. It has been found that the single retroviral DNA molecule generated per infectious virion is integrated during the M phase of the cell cycle (Roe et al., 1993). This results in only one daughter of the first cell cycle being labeled. Thereafter, the integrated viral genome is replicated with the host genome and is passed to all descendants. If the initial integration occurred in a cell that did not go on to divide further, a single-cell clone would occur. If this is the reason why one sees a large number of single-cell clones in the retroviral data set, then one would expect to see the distribution of cell types in single-cell clones reflect the birthdays of different cell types rather than the frequency of a cell type in the data set. Such is the case for the rodent retinal clones (Young, 1985a,b; Turner and Cepko, 1987; Turner et al., 1990), as illustrated by the following examples. (1) Rods, the most abundant cell type, are predominantly late-born, and are frequent in the data set of cells generated after infection at E13–14 in the mouse (83.6% of all cells), but are only 2.0% of single-cell clones. (2) Cones, which are born early, constitute 60% of single-cell clones, but are only 1.5% of all cells in the data set. These data are illuminating with respect to two aspects of retroviral marking. First, they strongly support the above model for single-cell clone generation and support the idea that viral integration occurs near the time of viral infection, as opposed to several days later. This allows single-cell clones to be used for "birthdating data." Second, they suggest that small clones do not result simply from lack of expression of the viral promoter in many members of a clone, or death of most members of a clone. Although both death and failure to express clearly occur, the point here is that they are not the main cause of single-cell clones and, by extension, of small clones in general.

Analysis of clone sizes indicates that there is no set pattern in terms of the mitotic and postmitotic fates of daughter cells. Occasional divisions at E13–14 in the mouse may yield two-cell clones, which are only ten of the 255 clones of >1 cell in the mouse (Turner et al., 1990). In *Xenopus* retina injected with fluorescent dextran, the number of two-cell clones was 2/58 clones of >1 cell (Wetts and Fraser, 1988). Two-cell clones could be the result of a division in which two postmitotic cells are generated. However, given how rare this clone type is, one must consider the possibility that they do not represent such divisions. This brings out another aspect of lineage data; rare clones must be interpreted with caution. For example, two-cell "clones" marked by a retrovirus can result from lumping of two single-cell clones, from death or failure of the virus to express in all siblings, or from a failure to reconstruct a clone properly across sections. However, two-cell clones can be taken as evidence of divisions in which both cells are postmitotic when they are abundant, as in the case of P0 infections in the rat or mouse. In the rat, two-cell clones represent 33.7% (300/891) of clones of >1 cell and in the mouse, 51.2% (144/281) of the clones of >1 cell. Thus it seems that in the early postnatal period, many divisions yield two postmitotic daughters.

To return to the early infections, it must be that many divisions yield two mitotic daughters as retinal growth is closer to exponential than linear in the midembryonic period (Alexiades and Cepko, 1996). However, many divisions must also produce one mitotic and one postmitotic daughter. The latter is suggested by the abundance of one-cell clones (30%) and the data from classical birthdating (Young, 1985a). Further, the progenitors that generate early-born cells (e.g., horizontal cells) must also lead to the generation of late-born cells (e.g., bipolar cells) as both of these cell types can be found in the same clone (Turner et al., 1990). The most likely scenario is thus that some divisions in the embryonic period are asymmetric with respect to mitotic fate in that one daughter can continue to divide which the other is postmitotic, and many divisions are symmetric in the sense that two mitotic daughters are generated. In the early postnatal period, many divisions appear to produce two postmitotic daughters.

Clone Composition Indicates that Retinal Progenitors Are Multipotent

With the exception of all-rod clones in the rodent, the composition of clones of retinal cells in all species is consistent: They have more than one cell type (e.g., Fig. 11–1). Even two-cell clones can comprise utterly distinct cell types. Consider the clones generated by infection of the P0 rat; some comprise one rod photoreceptor and one Müller glial cell (Turner and Cepko, 1987). In fact, no clones of only Müller cells are seen in any of the mapped species. This could be due to the fairly low occurrence of Müller cells. (They comprise only 2.7% of murine retinal cells.) One can also find unlikely pairs of cells in *Xenopus*, such as a ganglion cell and a photoreceptor (Holt et al., 1988). Larger clones in all species are more complex. In *Xenopus*, when injection is into peripheral margin cells, it appears that

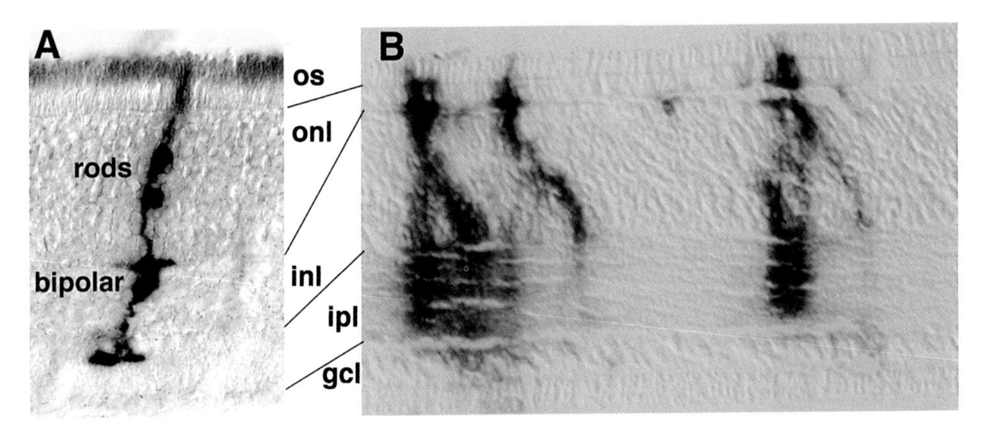

FIGURE 11–1 Clones from infection of the rat (A) and chick (B) retina are shown. In A, a clone from the mature rat retina, infected at P0 with a PLAP-encoding replication-incompetent vector, is shown. This clone contains two rod photoreceptors and one bipolar cell. In B, a fraction of a clone in a chicken retina is shown. The clone was initiated at E3 by infection with a vector similar to that used in A. The clone comprises multiple columns, presumably as a result of mixing of mitotic progenitors in the early retinal neurepithelium. Apparently, after postmitotic progeny begin to be generated, progeny migrate inward radially, with very little dispersion tangentially. onl, outer nuclear layer; inl, inner nuclear layer; ipl, inner plexiform layer; gcl, ganglion cell layer. A is from Cepko et al. (1996) and B is from Fekete et al. (1994).

there are clones with all seven of the major cell classes (Wetts et al., 1989). In the mouse retina (Turner et al., 1990), early clones comprise up to six of the seven classes of cell types. While no clones of all retinal cell types in the rodent have been documented, failure to find them most likely does not reflect the absence of a totipotent progenitor. If one examines the composition of the E13–14–generated murine clones, one finds that the composition of clones overlaps. The simplest explanation for these findings is that large, complex clones arise from totipotent progenitors but do not contain all cell types due to the rarity of certain cell types (horizontal cells comprise 0.3%, Müller cells 2.7%, ganglion cells 2.5%, and cones 2.2%; Young, 1985a). In the chick retina where clones can be initiated very early, it appears that some clones have all retinal cell types, although a detailed analysis of all of the cell types in chick clones has not yet been carried out.

As mentioned above, there are clones consisting only of rods in the rodent retina. Approximately 50% of the rods generated following infection of the rat at P0 can be found in rod-only clones (Turner and Cepko, 1987). The remaining 50% are found in clones with a variety of other cell types, including (in two-cell clones) amacrine cells, bipolar cells, and as mentioned above, Müller cells. Rods are thus frequently born from multipotent progenitors, but there is a possibility that they are also born of restricted progenitors committed to only producing rods. However, since rods comprise over 70% of all cells in the rodent retina, the occurrence of rod-only clones might reflect the influence of the environment on multipotent progenitors, causing them to produce many rods. Lineage analysis cannot determine whether there is a progenitor restricted to rod production, and we and others have turned to other approaches to distinguish among these hypotheses.

The analysis of clones of the retina was greatly aided by the fact that the cells were arranged as linear arrays of sibling cells. Apparently, sibling cells migrate into the developing retina from the ventricular zone (VZ) in a fairly radial manner, with relatively little tangential migration. However, after arriving in the various layers cones and ganglion, horizontal, and amacrine cells apparently can disperse tangentially (Turner et al., 1990; Fekete et al., 1994; Reese et al., 1995). There is also evidence for mixing among early ventricular zone cells of the chick retina in that one can see a lack of coherence of labeled radial arrays of clones generated at E3 (Fig. 1B; Fekete et al., 1994). The cell divisions between E3 and E5 appear to produce siblings that intermingle with nonsiblings; but once postmitotic cells begin to be generated, siblings do not intermingle as extensively with their nonlabeled neighbors. This type of mixing was first seen in clones generated in the chick tectum (Gray et al., 1988) and subsequently in clones in other areas of the click CNS, as discussed further below.

Conclusions

Overall, lineage analysis in the retina has shown the following: (1) Retinal progenitors are multipotent, most likely through the final cell division; (2) the pattern of cell divisions can be symmetric or asymmetric with respect to the mitotic behavior of daughter cells, with no set pattern; and (3) postmitotic retinal cells migrate radially with relatively little tangential migration. The findings have also led to suggestions as to the mechanisms of retinal cell fate determination in which the environment plays a role and in which determination takes place during or shortly after the last cell division (Cepko et al., 1996).

TELENCEPHALON

Several of the structures of the telencephalon of mammals in which lineage analyses have been conducted include the cerebral cortex, hippocampus, and basal ganglia.

The cerebral cortex is a laminated structure, comprising six layers of neurons with characteristic morphologies, patterns of connectivity, and neurotransmitter phenotypes (see Chapter 12). Two major classes of cells are the pyramidal and nonpyramidal neurons, with the pyramidal neurons residing primarily in layers V and II/III, and most nonpyramidal cells in layer IV. The cortex is also divided into cytoarchitectonic areas that are organized orthogonal to the laminae. Each such area subserves a particular function, such as vision or hearing. In the rat cortical neurons originate from the ventricular zone (VZ) of the lateral ventricles between E14 and E21. The cells must travel from their site of genesis through a cell-poor intermediate zone (IZ) to form the cortical plate (CP), a transient structure that later gives rise to the mature cortical laminae. The order of genesis is inside out, with layer VI being formed first and layers II/III appearing last. The hippocampus is a phylogenetically more ancient structure, located along the medial edge of the cerebral cortex. It is also derived from the VZ of the lateral ventricle (Altman and Bayer, 1990) but, although it is also composed of both pyramidal and nonpyramidal neurons, its structure is much simpler. In rodents, the basal ganglia comprise primarily the striatum and globus pallidus, with the principal neurons in the striatum being parcelled into a patch and matrix arrangement. The neurons of the basal ganglia are generated from a VZ adjacent to that of the cerebral cortex, from structures referred to as the ganglionic eminences (Smart, 1976).

The questions addressed by lineage studies of the CNS vary depending upon the structure and function of each area. In the retina, the questions primarily concerned the nature of the final cell types in a clone. Such questions are also of interest for the telencephalon. For example, do cells of different laminae or nuclear groups arise from the same or different progenitors? Do pyramidal and nonpyramidal cells, or patch and matrix cells, arise from the same or different progenitors? Can neurons and glia arise from the same progenitor? Another topic of great interest for the telencephalon concerns the distribution of clones in relation to functional or areal boundaries. Are cells from a single progenitor restricted to a single functional domain in the cerebral cortex such as the somatosensory cortex? Finally, since young cortical neurons in some cases have a long distance to travel when they move from the VZ to the CP, what paths do the sibling cells take? Answers to a number of these questions are now available, although the interpretation of some experimental findings has been difficult due to the extensive migration of telencephalic cells. As we shall see, the answers have in some cases been surprising.

Cortical Clones Can Span Functional Domains Within the Neocortex

Lineage analysis in the cerebral cortex in rodents has been carried out using retroviruses as clonal tags and X-inactivation and murine chimeras. The first experiments in which the LacZ-encoding retrovirus, BAG, was used to infect embryonic rat cortex showed that it would not be a simple matter to analyze lineal relationships in the cortex (Walsh and Cepko, 1988; Luskin et al., 1988; Price & Thurlow, 1988). Unlike the retina, simple, coherent radial arrays of cells emanating from the VZ of the lateral ventricle were not found throughout development. Rather, complex patterns of cell distribution were common. This prevented accurate assignment of clonal boundaries and resulted in an inability to answer most of the above questions. To clarify clonal assignments, we devised an improvement to the retroviral marking method (Walsh and Cepko, 1992). A library of retroviral vectors was constructed in which each member carried the *lacZ* gene along with a short stretch of irrelevant DNA. The irrelevant DNA served as a clonal tag and the library was constructed to have approximately 80 such tags (Walsh et al., 1992). Each tag was identified following recovery by PCR after a cell was identified by virtue of the *lacZ* activity. The libraries have been further improved by inclusion of up to a theoretical maximum of 10^7 tags, comprising oligonucleotides that are sequenced following the PCR (Golden, et al., 1995).

The viral libraries were used to infect E15 rat lateral ventricles and the infected brains were analyzed at E18, E20/21, P3 (Walsh and Cepko, 1993) and at adult stages (Walsh and Cepko 1992). Analysis of E18 brains showed that siblings typically migrated from the VZ in a radial manner. All tags recovered by PCR at E18 were found to identify clones of cells very close to each other, confined to radial columns typically 30–50 μm in diameter (i.e., in the rostrocaudal and mediolateral dimension) and less than 200 μm in the radial dimension. These observations confirmed an earlier study in which BAG was used in mice and in which clonal relationships were inferred using spatial criteria (Austin and Cepko, 1990). However, in both rat and mouse, it appeared that later in embryonic life the cells did not migrate in such a neat and orderly radial fashion. Dispersion of siblings appeared to occur due to migration in multiple, nonradial directions. These nonradial migrations (discussed further below) served to disperse sibling cells over great distances.

The final locations of sibling cells were analyzed in mature brains using both the original BAG viral library (Walsh and Cepko, 1992), as well as a newer and more complex library made in DAP, a vector encoding PLAP, to define clonal relationships (Reid et al., 1995). It was found that sibling cells were frequently distributed in the mediolateral and rostrocaudal dimensions over distances greater than 500 μm. This distance is significant as the known functional domains of the neocortex are for the most part smaller than 500 μm. Approximately 50% of clones initiated at E15 were distributed over >1.5 mm and 50% were dispersed <1.5 mm. These data indicate that most clones do not respect areal or functional boundaries, as predicted by the "radial unit" hypothesis (Rakic, 1988). According to the radial unit hypothesis, the VZ comprises a "protomap" in which progenitors are committed to an areal identity; the progeny from each determined progenitor then reach a particular functional domain by migrating radially along radial glia. In this way, the protomap encoded by VZ cells would be translated into the roughly columnar pattern of the mature cerebral cortex. If this hypothesis were true, each clone of cortical cells would be restricted to a single functional domain. It is important to note here that even if there were radial migration, it would not "prove" that a protomap was encoded by the VZ. The defining feature predicted by a protomap is that clones would respect functional boundaries after the completion of migration.

Dispersion of clones over multiple functional domains was more directly addressed when a histochemical stain was used to examine the barrel field of the somatosensory cortex (Walsh and Cepko, 1992). The definition of the somatosensory cortex histochemically allowed one to localize infected cells with respect to this landmark. Clones that contained cells within the somatosensory cortex did not show siblings restricted to a single barrel, even when the siblings were close together, within approximately 100 μm of each other. These observations support the conclusion of an earlier study in which murine chimeras were made and analyzed for genotype distribution in the somatosensory area (Goldowitz, 1987). Retroviral clones were also found that had cells within the motor and visual cortex, or the somatosensory and auditory cortex, directly demonstrating that clones were not restricted to a functional domain subserving a particular modality (Fig. 11–2).

The clones characterized by the viral library method were quite small on average, with about 40% having only a single PCR-positive cell. As in the retina, this presumably reflects the fact that some viral integrations occur in cells that become postmitotic just after the integration event. However, the finding of many small clones could also reflect a technical aspect of the viral library/PCR method and the frequency of cell death in the cortex. A recent report suggests that cell death within the cortical VZ is substantial (Blaschke et al., 1996), and cortical cell death after cortical plate formation may be presumed to be significant as well. Further, the recovery of the single viral DNA molecule in each labeled cell by PCR is only about 50% efficient. All of these factors mean that the cortical clones defined by PCR are smaller than the clones originally generated and that their spread may be underestimated. It is all the more remarkable that very small clones, comprising just two cells and initiated as late as E17, can span several functional domains and be separated by several millimeters.

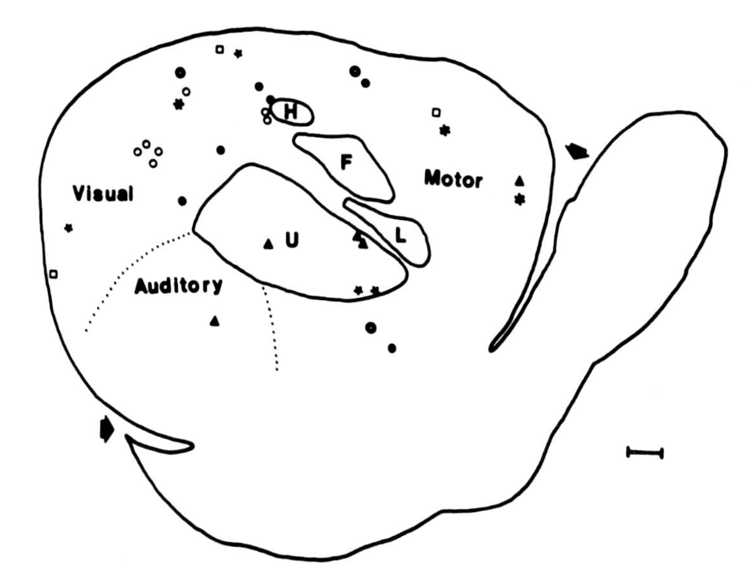

FIGURE 11–2 Clones in the rat cerebral cortex span multiple functional domains. Clones were initiated by infection in the embryonic rat lateral ventricle with the BAG retroviral library. The somatosensory cortex (area demarcated with U [upper lip], F [forelimb], L [lower limb], and H [hindlimb]) was identified using cytochrome oxidase histochemistry and the other areas were localized roughly with reference to this landmark. Each clone is represented by a distinct symbol. Clones from several brains are summarized here. Clones can be seen to cross multiple functional and cytoarchitectonic boundaries. From Walsh and Cepko (1992).

Two Patterns of Dispersion Are Displayed by Cortical Clones

The cortical clones initiated by infections at E15 and defined by PCR formed two patterns of dispersion, clustered and widespread (Walsh and Cepko, 1992; Reid et al., 1995). The clustered clones were composed of one or a few cells located within 1.0 mm of each other. Generally, they were composed of cells of similar morphology (e.g., pyramidal or nonpyramidal) that occupied the same lamina. The widespread clones were composed of cells located >1.5 mm apart, but some were as far as 10 mm apart (Walsh and Cepko, 1993). The widespread clones appeared to be composed of subunits, each of which was similar to a clustered clone in that it usually contained cells of similar morphology, with cells <1.5 mm apart. A similar analysis was carried out following infection at E17; here the frequency of widespread clones dropped to 12.5% (Reid et al., 1995). The spacing of the subunits of the widespread clones along the rostrocaudal axis was examined. Very few clones with subunits spaced 0.5–1.5 mm or 3–4 mm were found. There appeared to be a preferred spacing of subunits with a periodicity of approximately 2–3 mm (Fig. 11–3).

The above data led to the following hypothesis (Reid et al., 1995). There are motile progenitors in the VZ that are infected at E15. The motile progenitors generate mitotic or postmitotic daughters at intervals along the rostrocaudal axis dictated by the length of their cell cycle and their migration rate as they traverse the VZ in a rostrocaudal direction. The mitotic daughters that they generate undergo further divisions in situ, without significant further migration in the VZ. As the daughters of the nonmotile progenitors are frequently of the same type, the nonmotile progenitors might be committed to produce only one daughter type. Alternatively, the similarity of the progeny might be due to environmental

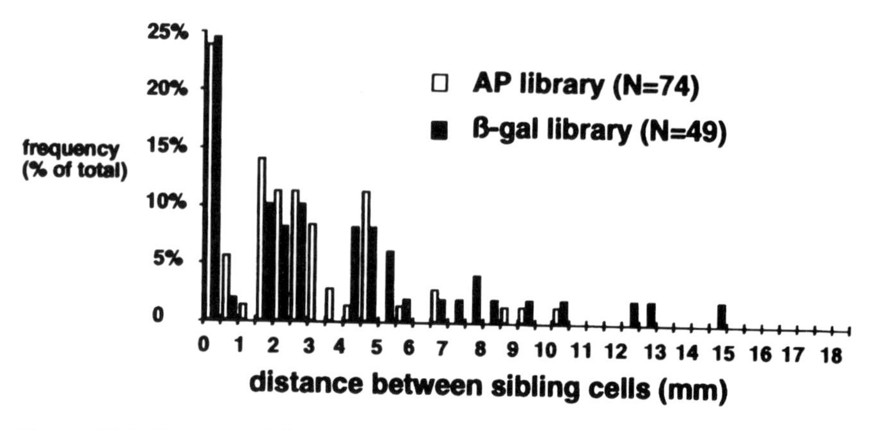

FIGURE 11–3 Rostrocaudal spacing of subunits of clones in the rat cerebral cortex. Clones in the rat cerebral cortex were initiated in the embryonic period by infection with retroviral libraries (Walsh and Cepko, 1992; Reid et al., 1995). Clonal relationships were identified using PCR. The rostrocaudal distance between subunits of clonally related cells was measured and is shown on the X-axis. The frequency of subunits with each distance is shown on the Y-axis. Intersibling spacings of <0.5 mm, 1.5–3.0 mm, and 4–5.5 mm are common, while those of 0.5–1.5 mm are rare. A periodicity of 2–3 mm is suggested. From Reid et al. (1995).

factors. It appears that the nonmotile progenitors do not produce many daughters and that generally daughters are generated over a fairly short time period, perhaps leading to all siblings in the subunit experiencing a fairly similar environment during the cell fate determination process. As transplantation experiments indicate that at least early progenitor cells are not committed to a deep laminar fate, some plasticity must be present (McConnell, 1988; McConnell and Kaznowski, 1991).

Hippocampal Clones Span Functional Domains Within the Hippocampus

Analysis of the distribution of sibling cells with respect to the functional boundaries of the hippocampal formation was carried out using the retroviral vector BAG (Grove et al., 1992). In this study, a library was not used, but the low number of infections per hippocampus and a statistical analysis permitted an estimate of the sibling relationships. Presumptive clones were again quite small, with 50% being single cells and the average being about two neurons (primarily pyramidal neurons) per clone. In this tissue, the siblings were proposed to remain tightly clustered, perhaps because the VZ is close to the mature structure, unlike most regions of the neocortex. The clustered clones, however, cross functional boundaries. The boundaries analyzed here were the CA fields of Ammon's horn, the subiculum, and the prosubiculum. Clones were found to span CA1 and the subiculum, CA3 and CA2, or CA1 and CA2. A model was generated of the frequency at which clones would cross such boundaries if there were no restrictions, using for these calculations the average area occupied by a clone and the areas of the various functional domains. The frequency at which clones crossed borders was found to closely approximate the prediction of the model. Clones that did not cross were thus presumably located within only a single domain probably because clonal dispersion was not extensive, rather than due to some restriction imposed by the cytoarchitectonic boundaries.

Some Telencephalic Clones Span Major Subdivisions of the Telencephalon

With the degree of dispersion summarized above, the question of whether clones spread even more widely to cross the major subdivisions of the telencephalon was raised. To answer this question, the library/PCR method was required to eliminate splitting errors (Walsh and Cepko, 1992, 1993; Reid et al., 1995). Clones were found that spanned the neocortex and evoluntarily more conserved structures, such as the hippocampus (two of 18 adult rat widespread clones), the piriform/entorhinal cortex (six of 18 adult rat widespread clones), and the olfactory bulb (three of 12 P3 rat clones). These clones originated from infections in the embryonic period. Infections into the VZ of the lateral ventricle in the postnatal period also resulted in generation of olfactory bulb neurons, as previously suggested by [^{3}H]thymidine labeling studies (Hinds, 1988a,b). BAG infections targeted to the most anterior portion of the lateral ventricle subventricular zone SVZ led to the production of both granule and periglomerular neurons within the bulb (Luskin, 1993).

Analysis of X-Inactivation Mosaics and Interspecies Chimeras

Patterning in the developing telencephalon has also been examined in mice using X-linked inactivation of the *lacZ* gene (Tan and Breen, 1993). The distribution of blue and white cells in the neocortex should reveal some aspects of patterning that derive from lineage. The type of distribution one might predict depends on when the patterning was initiated relative to when the inactivation occurred and how much mixing of cells occurred in the cortical VZ between inactivation and the patterning events. If the radial unit hypothesis discussed above were correct, one should expect radial columns of either blue or white cells, or columns with some particular ratio of blue to white cells, to predominate in the mature animal. Indeed, radial groups of cells of predominantly one color were seen. Within columns of predominantly blue or white cells, approximately one-third of the cells were of the other color. The columns of predominantly one color were shown to exist only in the medial and dorsolateral regions of the cortex, while labeled and unlabeled cells were completely interspersed in more lateral regious of the cortex (Tan et al., 1995). These dispersion patterns were as predicted by an earlier study of the dispersion of clonally related cells in the cortex of embryonic mice (Austin et al., 1990). In the earlier study, radial migration with some tangential dispersion was predominant in the dorsal and medial cortex but migration into the more lateral regions of the cortex was quite circuitous, presumably due to the curvilinear pattern of radial glia in that region (Gadisseux et al., 1989; Misson et al., 1991). The X-inactivation study of the cortex thus revealed that mixing of clones across the cortex is varied, depending upon the position of the progenitor within the VZ, but it did not reveal how lineage might contribute to the shaping of the final pattern of functional domains. The X-linked inactivation studies did not show stripes of predominantly labeled or unlabeled cells in the hippocampus, which is consistent with the observations of the prior retroviral study (Grove et al., 1992). Similarly, there was extensive mixing in the striatum, which is also consistent with a retroviral lineage study in the rat striatum (see below).

Studies of the neocortex in murine chimeras in which histochemically distinguishable cells at the blastomere stage were mixed have also been carried out (Crandall and Herrup, 1990). Slabs of cells of predominantly one genotype organized in the mediolateral plane, with very little mixing in the rostrocaudal direction, were observed. However, the slabs did not have sharp boundaries, which is consistent with the results from the retroviral and X-inactivation studies.

Lineage Analysis in the Striatum

In the only clonal lineage study of the striatum, two retroviruses, one encoding lacZ (BAG) and one encoding PLAP (DAP), were used to coinfect the rat striatum (Halliday and Cepko, 1992). Radial arrays of cells of only one color were found to result if the animals were infected at E15 and analyzed within 2–3 days of infection. Some of the clones were of a mixed phenotype in that they contained what appeared to be young postmitotic neurons as well as radial glia. Only one radial glial cell appeared in each such clone. The size of the clones indicated that cell divisions had to have occurred such that some divisions produced two mitotic daughters and some produced one mitotic and one postmitotic daughter during this period. As in the cerebral cortex, migration was not simply radial. Starting at about E17, many clones exhibited extensive dispersion in the dorsoventral plane but very little dispersion in the rostrocaudal direction. The dispersion in the dorsoventral plane appeared to occur on a clonal basis; i.e., in some clones, most of all members of the clone appeared to be actively migrating in this direction, while in other clones, few or no members migrated in this direction. Unfortunately, the final phenotype and position of the cells were not analyzed as the retroviral genome became silent in the early postnatal period, thus precluding the analysis. The silencing could be demonstrated by recovery of viral genomes (via PCR) from areas of the striatum in which these were no histochemically labeled cells. The inactivation was quite extensive, in contrast to the findings of inactivation in the cerebral cortex, in which very few inactive genomes were found (Walsh and Cepko, 1992).

There is also lineage data from the striatum that concerns the origin of the SVZ (Halliday and Cepko, 1992). Injection of the viral inoculum was into the ventricle at E15 and presumably virions were only able to access cells lining the ventricle, i.e., VZ cells. Indeed, most clones examined 2–3 days postinfection had labeled VZ cells. However, a fraction of clones at 3–6 days postinfection had cells only within the SVZ. These clones had cells that appeared to be in the M phase of the cell cycle or were tightly apposed as if they had just passed through M. These data suggest that the SVZ-only clones originated from an E15 VZ progenitor. Given at this stage no VZ cells were labeled the VZ progenitor appears to have made only SVZ cells and the SVZ cells remained in the SVZ, making other SVZ cells for at least some portion of the late embryonic period.

Lineage Analysis in the Telencephalon Has Revealed Complex Migration Patterns

The dispersion of telencephalic cells in both rostrocaudal and mediolateral directions has raised a question as to how the cells reach their final locations. As mentioned above, studies of retrovirus-infected clones within a few days after infection in rodents have shown that clones are radially arranged as they exit the VZ. Within the intermediate zone (IZ) the cells become dispersed in the mediolateral plane, the degree of dispersion being dependent on their position of origin in the VZ (Austin and Cepko, 1990). Cells originating in the dorsal and medial areas disperse very little, but cells originating within the lateral and dorsolateral portions of the VZ disperse a great deal. Very little rostrocaudal dispersion is seen during the first few days postinfection in either mouse or rat (Walsh and Cepko, 1993). However, within the next few days, some rostrocaudal dispersion is seen in the more rostral and caudal regions of the mouse cortex. More extensive rostrocaudal dispersion is seen in the rat cortex when brains infected at E15 are examined at E20/21 or P3 with clones having cells dispersed along the rostrocaudal axis and located in the VZ, SVZ, IZ, and CP (Walsh and Cepko, 1993). The rostrocaudal dispersion tends to be greater than the mediolateral dispersion at these ages. Several clones have cells located within the VZ/SVZ throughout the rostrocaudal axis. This subset of clones could reflect the movement of motile progenitors that generate widespread clones as hypothesized above.

A lineage study has been carried out in rhesus monkeys using two types of retroviruses, one encoding a cytoplasmically localized and one encoding a nuclear-localized lacZ (Kornack and Rakic, 1995). The viruses were injected at a time when layers IV, III, and II were being generated. One fetus was analyzed 3 weeks postinfection, while cells were still being generated, and two others were analyzed at 7 weeks postinfection, just after the genesis and migration of the cells in these layers. Migration of cells radially out of the VZ was found, as reported for rodents. However, in the more mature animals, while some cell clusters maintained this arrangement, the cells being aligned radially across several laminae, other clusters of cells were arrayed horizontally within a single lamina. In addition to these two patterns of cells in clusters, individual cells, separated by more than 300 μm from other labeled cells, were also seen. Due to the size of the monkey cerebral cortex, not all sections were analyzed, and since only two viral genomes were used, it was not possible to assess whether the isolated cells were clonally related to members of the horizontal or radial clones. Thus, the distribution patterns seen in this study could represent a subset of the patterns seen in rodents, particularly as subunits of clones in which siblings are tightly grouped and appear to be of similar type.

As the retrovirus studies use fixed tissue for analysis, migration patterns can only be inferred from static pictures of cells taken at intervals postinfection. To address migration patterns directly, several groups have observed the movements of cells within the developing telencephalon. The use of lipophilic dyes and time-lapse videomicroscopy has allowed real-time monitoring of living cells migrating in tissue explants. O'Rourke et al. (1992, 1995) have monitored the migration of young neurons in coronal slices of developing ferret cortex. They found that migration tangential to radial glia occurred in at least 13% of cells. A whole range of angles relative to the orientation of the radial glia, was found; some cells exhibited 90° turns as they appeared to leave radial glial fibers. Tangential migration was observed in all layers of the developing cortex, but was most prominent in the IZ. Tangential movement in the VZ/SVZ has also been inferred by the orientation of TuJ1-labeled young neurons in ferrets (O'Rourke et al., 1995) and mice (Menezes and Luskin, 1994).

Movements within the plane of the VZ up to and greater than 200 μm over a 24-hour period were observed by Fishell et al. (1993) when they used DiI to label cells in the murine VZ. The cells appeared to move randomly rather than in a directed fashion. This is in contrast to the commonly reported substrate-guided migration along radial glia and to the rostral migratory stream of cells moving in the SVZ to the olfactory bulb, which are restricted in orientation and direction

(Lois et al., 1996). Random movement within the VZ followed by radial migration into the cortex could account for some lateral dispersion of clonally related cells in the cortex.

Technical limitations in these experiments restrict the amount of time the labeled live cells can be observed, so the final locations of imaged cells could not be determined. It is notable that no imaged cells have exhibited directed migration in the rostrocaudal direction. Since 73% of the neurons identified by PCR were found in widespread clones (Reid et al., 1995), one might predict that the migration of cells in the rostrocaudal direction should have been observed. However, the fact that the studies reported to date have not shown this type of migration could reflect the type of labeling used and/or the species studied (ferrets and mice vs. rats). Thus, present data from retroviral marking and live imaging suggests that mediolateral migration, either in the IZ or in the upper layers of the developing cortex, contributes to the final widespread dispersion of clonally related cells in the cortex. The compelling spacing of subunits of widespread clones (Reid et al., 1995) could be explained most easily by assuming that motile VZ/SVZ progenitors move rostrocaudally, but direct demonstration of such rostrocaudal migrations within the VZ/SVZ remain to be demonstrated.

Progenitor Cells in the Cortex Produce a Variety of Clone Types

Several retroviral lineage studies have focused on the neuronal and glial phenotypes of sibling cells. Most of the available data have been generated in experiments using a single *lacZ* retrovirus and geometric criteria to assign clonal relationships. As pointed out above, where there is a great deal of migration, splitting errors render interpretation difficult. However, the fact that the PCR analysis has shown that subunits of widespread clones are clustered, permits one to conclude that at least some closely spaced cells are siblings. Parnavelas and colleagues used infection with BAG at E15/E16 to generate material that was analyzed ultrastructurally (Parnaveles et al., 1991; Luskin et al., 1993) or with antisera directed against GABA, glutamate, or calcium-binding proteins (Mione et al., 1994). They report that the majority of clusters (14/15 in the adult rat) comprised cells of either the pyramidal or nonpyramidal type (Parnaveles et al., 1991). They also report that some clusters comprised only neurons (16/27 clusters), only astrocytes (8/27 clusters), some only oligodendrocytes (1/27 clusters), others neurons and astrocytes (1/27 clusters), or in one case oligodendrocytes and a neuron (1/27 clusters) (Luskin et al., 1993). However, a recent report from the same group suggests that approximately 18% (10/51) of clusters are of mixed pyramidal and nonpyramidal phenotype, and 26% (14/54) are of a mixture of neurons and astrocytes (Mione et al., 1995). They also report that clusters are heterogeneous for expression of GABA and glutamate but are homogeneous with respect to calcium-binding protein expression (Mione et al., 1994).

Price and colleagues similarly used BAG to infect the lateral ventricle in E16 rats and geometric criteria to determine whether neurons, astrocytes, and oligodendrocytes were members of individual clusters (Grove et al., 1993). They used morphology, Lucifer Yellow injection of Xgal+ cells, and immunohistochemistry to identify the phenotype of the cells at P14 or P28/29. They report that 62% of the observed clones were of neurons alone, 9% were of gray-matter astrocytes alone, 13% were of white matter cells alone, 6% were of neurons and white matter cells, 4% were of neurons, gray-matter astrocytes, and white-matter cells, and 7% were unidentifiable. They conclude that at E16, most progenitors are restricted to the production of only one cell type but that some progenitors can make both neurons and glial cells. Studies of BAG-infected E16 cortical cells in vitro showed clones of similar composition (Williams et al., 1991). However, there was a skew in the data such that oligodendrocyte clusters were the most abundant, 39–41%, while neurons comprised only 18–31%. Since the in vivo lineage analysis and classical birthdating studies indicate that neurons are the most frequently produced cell type at E16, the in vitro data may reflect an oligodendrocyte-inducing environment.

Goldman and colleagues also used BAG, in combination with the PLAP-encoding virus, DAP, to infect the SVZ of the rat lateral ventricle (Levison et al., 1993). The combination of the two viruses allowed them to address lumping errors. They found that labeled cells formed clusters with very little dispersion in the rostrocaudal direction and only moderate dispersion in the mediolateral direction. Most clusters were composed of either oligodendrocytes or astrocytes. However, 15% of presumed clones consisted of mixed glial cell types, and the few neurons that were present were included in glial clusters. Retroviral infections performed in vitro also showed clusters of either astrocytes of oligodendrocytes (Vaysse and Goldman, 1990; Lubetski et al., 1992). However, a few astrocyte clusters expressed oligodendrocyte markers and vice versa.

The data reported above concerning clones defined using the PCR/library method allows a synthesis of these data and the data from studies in which geometric criteria were used. The fact that subunits of widespread clones, and clustered clones initiated at E15 or E17, are usually uniform in the type of cells that they comprise may indicate that in fact subunits are subclones. The subclones may be the clusters identified by geometric criteria. However, this hypothesis does not necessarily lead to the conclusion that subclones derive from restricted, or committed, progenitors. A more rigorous test of the potential of progenitor cells under a variety of conditions is required for conclusions to be reached about the potential of cells.

Conclusions

Lineage analysis in the telencephalon has shown that clones are often not restricted to known cytoarchitectonic or functional domains. Clones can cross the major subdivisions of the telencephalon, with siblings appearing in, say, the olfactory bulb and neocortex, or the hippocampus and neocortex. Siblings appear to become widely dispersed through migration along a variety of routes and on undefined substrates. Migration of newborn neurons results in dispersion in both mediolateral and rostrocaudal directions, while dispersion of glia appears to be relatively constrained in the rostrocaudal dimension. However, dispersion of neurons is somewhat dependent upon their position of origin in the VZ, with cells originating in the dorsal and medial portions spreading modestly and those originating in the dorsolateral and lateral VZ spreading quite widely. Clones initiated relatively early in neurogenesis can comprise multiple cell types, including both neurons and glia. However, subunits of clones initiated early and clones initiated later can comprise fairly uniform cell types.

Diencephalon

The diencephalon gives rise to the adult thalamus, hypothalamus, and epithalamus. These regions, particularly the thalamus, function as major relays between the telencephalon and more caudal regions of the brain. Early in development, the diencephalon morphologically comprises distinct units known as neuromeres or prosomeres. Classical embryologists have described the diencephalon as arising from four horizontal strips (His, 1893; Herrick, 1910; Khulenbeck, 1973) or neuromeric units (Orr, 1887; Bergquist, 1952; Keyser, 1972) based on the presence of bulges and sulci along the medial walls of the third ventricle. These morphological units have been proposed to be analogous to the rhombomeres (Puelles et al., 1987; Figdor and Stern, 1993; Puelles and Rubenstein, 1993), although this analogy has not been satisfactorily tested (but see Guthrie, 1995; and Chapter 10 in this volume). Examination of the expression patterns of a variety of developmentally regulated genes has shown a correlation with the neuromeric units (Bulfone et al., 1993a,b; Rubenstein et al., 1994; Rubenstein and Puelles, 1994), similar to the correlation noted in the hindbrain. As development proceeds, multiple nuclei or nuclear groups, the functional and anatomical units of the diencephalon, derive from the neuromeres.

Lineage analysis was first conducted in the diencephalon by microinjection of single cells and by DiI and DiO labeling of small cohorts of cells (Figdor and Stern, 1993). More recently, retroviral vectors encoding a library of molecular tags have been used (Arnold-Aldea and Cepko, 1996; Golden and Cepko, 1996). These studies, as well as those investigating lineage in the mesencephalon, rhombencephalon, and spinal cord, have only been conducted in the chick, as technical limitations prevent easy access to mammalian embryos at the early times of development when the cells of these regions are born.

Sibling Cells Respect Neuromeric Boundaries Early, but Cross Neuromeric and Nuclear Boundaries Later in Development

Lineage analysis has addressed two aspects of the distribution of sibling cells in the diencephalon. The first pertains to the crossing of the neuromeric boundaries during development and the second to the distribution of clones with respect to the functional nuclear organization of the diencephalon. Short-term lineal relationships in the diencephalon were analyzed using the method of single-cell microinjection of a fluorescent dye (Figdor and Stern, 1993). Figdor and Stern found that if a cell were injected prior to the formation of a neuromeric boundary (the time of boundary formation varies depending on the particular boundary), its progeny were found within two adjacent neuromeres. However, if a progenitor was labeled after the formation of a boundary, the progeny were never found to cross from one neuromere to another, even if the progenitor was close to the boundary. In fact, progeny were found to accumulate along a boundary as if they were prevented from crossing. Thus, neuromeres behaved as compartments in which movement of clones were confined for up to 48 hours after injection (i.e., to stage 25 in the chick [Hamburger and Hamilton, 1951] which was the latest time point analyzed). However, the analysis of the final patterns of clonal dispersion or of the mature cell types within any one clone was restricted with this technique.

The final patterns of dispersion and cell types generated have been investigated using a library of retroviral vectors for lineage analysis. Degenerate oligonucleotides were cloned into a chick retrovirus encoding PLAP to generate a retroviral library, CHAPOL (Golden et al., 1995). By using PCR to amplify the oligonucleotide encoded by the retroviral genome in an infected cell, and then sequencing the recovered oligonucleotide, clonal relationships were established. CHAPOL was first applied to the study of lineage in the chick diencephalon where, to date, 350 inserts have been recovered with no duplications in independent infections (i.e., a complexity of $>10^5$ inserts with an equal distribution) (Golden and Cepko, 1996). Using CHAPOL, clones were found to spread widely in the rostrocaudal, mediolateral, and dorsoventral directions within the diencephalon (see Fig. 11–4). The average percentage of the diencephalon spanned in the rostrocaudal, dorsoventral, and mediolateral directions was 11% (range 2–54%), 19% (range 1–74%), and 58% (range 0–98%), respectively. The range of distances spanned by a clone was 60–1,140 μm in the rostrocaudal direction, 5–2,940 μm in the dorsoventral direction, and 0–3,550 μm in the mediolateral direction.

Given the dispersion of siblings in the diencephalon, it was of interest to determine if clones crossed neuromeric boundaries. However, detailed maps delineating which mature structures are derived from the developmentally defined neuromeres are not yet available for any species. Thus in order to provide an estimate of the number of clones occupying derivatives of more than one neuromere, the boundaries of neuromeres defined by Figdor and Stern in the stage 40 chick (see Fig. 11–2 in Figdor and Stern, 1993) were used to identify siblings that clearly resided in more than one neuromere. A minimum of 8% (Golden and Cepko, 1996) of clones were found to cross the boundaries of neuromeres. However, 47% of clones were in the region of a proposed border and we were unable to determine if they crossed a border. It is likely that some of these clones did cross a neuromeric border. In the prosomeric model (Rubenstein et al., 1994; Rubenstein and Puelles, 1994) the hypothalamus and thalamus are thought to arise from distinct prosomeres. Since we have observed clones that occupied both the thalamus and hypothalamus, such clones must have crossed prosomeric boundaries (Arnold-Aldea and Cepko, 1996; Golden and Cepko, 1996).

How can one reconcile the apparently disparate behavior of clonally related cells analyzed early and late in development? Several possible explanations can be put forward for the difference in dispersion between the two time points. The

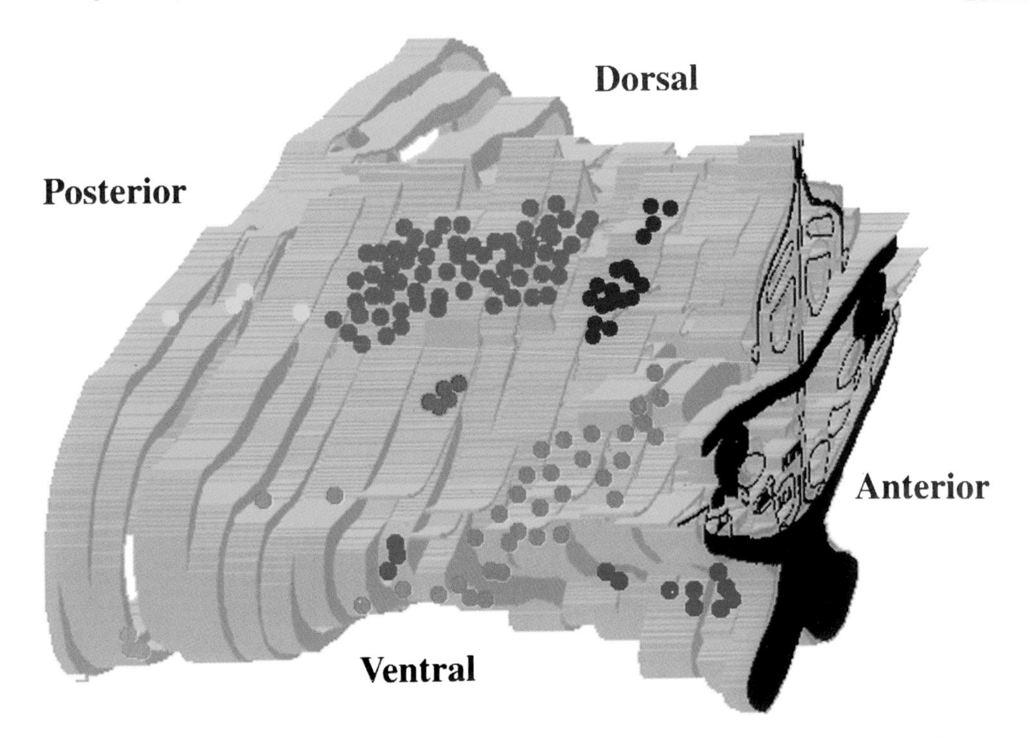

FIGURE 11–4 Three-dimensional reconstruction of clones within the chick diencephalon. Representative clones have been transposed to the surface of a reconstructed diencephalon. Each cell was localized according to its rostrocaudal and dorsoventral position; mediolateral position is not shown. Several clones spread widely in the rostrocaudal plane, while others remain relatively constricted. Dispersion along the dorsoventral axis also varied. The overall extent of dispersion of a clone did not always reflect the number of cells within the clone, as small clones showed similar dispersion patterns to large clones (From Golden and Cepko (1996).

first, and most likely, is that the tracer-labeling studies were analyzed after only 48 hours, whereas in the retroviral study, embryos were harvested 6–16 days beyond the time of infection. Thus it seems likely that progenitor cells give rise to daughters that are initially restricted to within the VZ of a single neuromere, while sometime after the cells leave the VZ, some siblings escape restrictions and move according to other mechanisms or cues. Alternatively, a relatively small subset of clones may cross neuromeric boundaries early in development; such a subset of clones may have been missed by the techniques used in the previous study. This scenario is similar to the recognition of a small percentage of clones that violate rhombomere boundaries early in development (Birgbauer and Fraser, 1994).

The retrovirally marked clones included those in which the sibling cells tended to cluster in one or a small group of nuclei of the diencephalon. These clones are consistent with data derived from labeling with [^{3}H]-thymidine, supporting the idea that subsets of progenitors within the VZ of the third ventricle generate specific regions of the diencephalon. However, other members of some of these clones as well as many other clones were widely dispersed in the diencephalon, indicating that >75% progenitors are not dedicated to producing cells for one nucleus or a small group of nuclei.

We also analyzed retrovirally marked clones in the ventral forebrain, focusing on the hypothalamus, in brains harvested at E8–E10 (Arnold-Aldea and Cepko, 1996). Approximately 96% of the clones were simple radial columns which appeared to respect neuromeric boundaries. However, approximately 4% of the retrovirally marked clones were widely dispersed, with distinctive patterns of dispersion at E8–E10, including bilaterally symmetric clones in the hypothalamus. Another pattern of note is one that is similar to that seen in the widespread clones in the rat cerebral cortex. In these clones, a fairly regular spacing of radial arrays, perhaps like the subunits in the cortical clones, was seen.

Since clones in the diencephalon were found to be widely dispersed, the question arose as to how dispersion occurred. To examine one possible mechanism, the architecture of the radial glia in the diencephalon was explored using DiI labeling (Golden and Cepko, 1996). The labeling of radial glial fibers arising from the third ventricle revealed that fibers extend into the diencephalon from medial to lateral, with a distinct dorsal to ventral slope, that parallels the most common patterns of distribution within clones. It is worth noting that not all clones showed this pattern of dispersion. Several clones showed marked displacement in the dorsal/ventral direction with relatively little dispersion from medial to lateral. This suggests that other mechanisms for clonal dispersion in the dorsal/ventral plane exist in the diencephalon. The nature of the mechanisms used by cells for dorsoventral and rostrocaudal migration are unknown.

Since both symmetric and asymmetric bilaterally distributed clones have been observed (Arnold-Aldea and Cepko, 1996; Golden and Cepko, 1996), some mechanism(s) must exist for cells to cross the midline. At least three possible migration pathways are plausible. A cell could cross through one of the major commissures, which include the anterior, posterior, and supraoptic decussation dorsalis. Analysis at E18 has led to identification of cells within each of these commissures. However, as yet, clonally related cells on each side of the diencephalon and within one of these commissures have not been observed within the infected brains in the current data set. A second pathway to the generation of bilateral clones is for cells to migrate around the anterior, inferior, or posterior limits of the diencephalon. We have not seen this type of migration in the material we have examined, but a more extensive examination of younger brains would be required to exclude these routes. A third possibility is that a population of progenitor cells exists along the midline that is capable of generating siblings that can migrate to populate both sides of the diencephalon. Similar midline cells are present in invertebrates (Crews et al., 1988, 1992; Nambu et al., 1990, 1991) and in vertebrates such as the zebrafish (Hatta et al., 1991), and have been proposed as the source of bilaterally symmetric clones in the chick diencephalon seen at E8 (Arnold-Aldea and Cepko, 1996).

Another pathway for migration suggested by lineage analysis in the diencephalon is from the lateral ventricle. Radial glia, with their cell bodies in the wall of the lateral ventricle and processes that radiate medially into the diencephalon, were observed to have sibling cells, both glial and neuronal, within the body of the diencephalon. These data also indicate that progenitors from the VZ of the lateral ventricles have progeny that populate the diencephalon, as previously proposed by others (Rakic and Sidman, 1969; Altman and Bayer, 1978). No specific nuclei were found to be populated by the clones originating in the lateral ventricles of the chick; however, since only a small number of clones of this type have been found, it is possible that certain nuclei are preferentially populated via this route.

Diencephalic Progenitors Are Multipotent

Progenitors in the diencephalon can be multipotent, producing both neurons and glia (Golden and Cepko, 1996). The frequency of different cell types indicates that most progenitors generate relatively small numbers of neurons and large numbers of glia. The large number of glia in many clones, along with the small numbers of neurons in some of these same clones, is consistent with a multipotent progenitor that divides and first gives rise to one neuron at each cell division. Multiple cell divisions with one neuronal daughter at each division would account for the clones with multiple neurons. One or more mitotic daughters from these same progenitors could continue to proliferate and later produce one or many glial cells. Analysis of single-cell clones (see below) supports this order of genesis, as does [^{3}H]thymidine birthdating (Angevine, 1970).

Twenty-seven percent of all single-cell clones were neurons, indicating that neurons were being born near the time of injection of the retrovirus, at stage 16–17 in development. In contrast, glial cells were rarely born at these stages of development. Despite their high frequency in the total data set, only 6% of the single-cell clones comprised glial cells (other than radial glia). Radial glia comprised 8% of the single-cell clones and one to four radial glial cells were found in 17% of multiple-cell clones. Radial glia were found in clones with neurons and no other glial cells, in clones with other glia but no neurons, and in clones with both neurons and glia. Together these data indicate that radial glia are being born near the time of infection at stage 16–17 in the chick and that the same progenitors that give rise to radial glia also give rise to neurons and other glial cells.

Conclusions

Lineage analysis in the diencephalon has been conducted using single-cell injection and infection with retroviruses. The data indicate the following: (1) Segments in the diencephalon, defined morphologically and by the expression of a variety of developmentally regulated genes, initially form lineage-restricted compartments that are later violated. (2) Progenitor cells in the diencephalon are multipotent. (3) Postmitotic cells in the diencephalon migrate in the rostrocaudal, dorsoventral, and mediolateral planes. Further studies are required to define the timing of nonradial migration, the factors controlling nonradial migration, and the relationship between segments and cell fate determination.

MESENCEPHALON

The mesencephalon is the embryological precursor of the adult midbrain. The midbrain can be divided into ventral and dorsal regions, with the ventral midbrain comprising primarily the substantia nigra and cranial nerve nuclei III and IV. Virtually no information is available on the lineal relationships in this part of the midbrain and thus the remainder of this section addresses the dorsal midbrain, comprising in mammals, the superior and inferior colliculi and in reptiles, fish, amphibians, and birds, principally the optic tectum. Anatomically, the tectum of the chick is organized into 15 laminae (Ramon y Cajal, 1929; LaVail and Cowan, 1971a). The development of the tectum has been studied using a variety of techniques (LaVail and Cowan 1971a,b), including retroviral mediated lineage analysis.

Tectal Clones Disperse in Precise Patterns

Gray et al. (1988) and Galileo et al. (1990) used retroviral vectors encoding the *lacZ* gene, directed to the cytoplasm and/or to the nucleus, to show that clones in the tectum were initially radially oriented. Clones comprised either one coherent column or several parallel, tightly clustered arrays, much as shown in Fig. 11–1 for the retina. Intercalation of infected and uninfected cells within the early VZ, prior to the formation of postmitotic progeny, is the most likely explanation for the multiple columns. The final distribution of siblings in the most frequent clones of the tectum showed a predominately radial distribution with members of a clone populating multiple laminae. Tangential dispersion occurred in several specific locations and in restricted directions. Dorsoventral spread of siblings was seen to begin at E6–E7 (see Fig. 11–5, and Gray and Sanes, 1991). The cells that appeared to leave the radial column(s) of cells moved long distances, up to 3.5 mm in the dorsoventral plane. However, relatively little or no spread in the rostrocaudal direction was found. The tangential migration was predominantly in the lower layers of the tectum, corresponding to the intermediate zone (IZ, layers 13 and 14, see below). These cells eventually differentiated principally into multipolar efferent neurons. In the most superficial layers of the tectum (layer 1), tangential dispersion occurred later in development, beginning at approximately E9. Cells dispersing in the superficial layers appeared to spread in all directions and at least some of these cells differentiated into astrocytes. The dispersion of cells in the more superficial zone of the tectum appeared to involve fewer cells than that in the deeper layers (Gray and Sanes, 1991).

The retinotectal projections form a topographically ordered map upon the surface of the tectum. Although a variety of genes are expressed in gradient fashion across the tectum, no discrete functional domains are known. Thus, the distribution of cells within a clone cannot be analyzed with respect to any known functional boundaries. Clones that extend from the mesencephalon into the diencephalon have also been reported (Golden and Cepko, 1996); this suggests that by E18 in chick development, the primary vesicles are not absolute lineage boundaries.

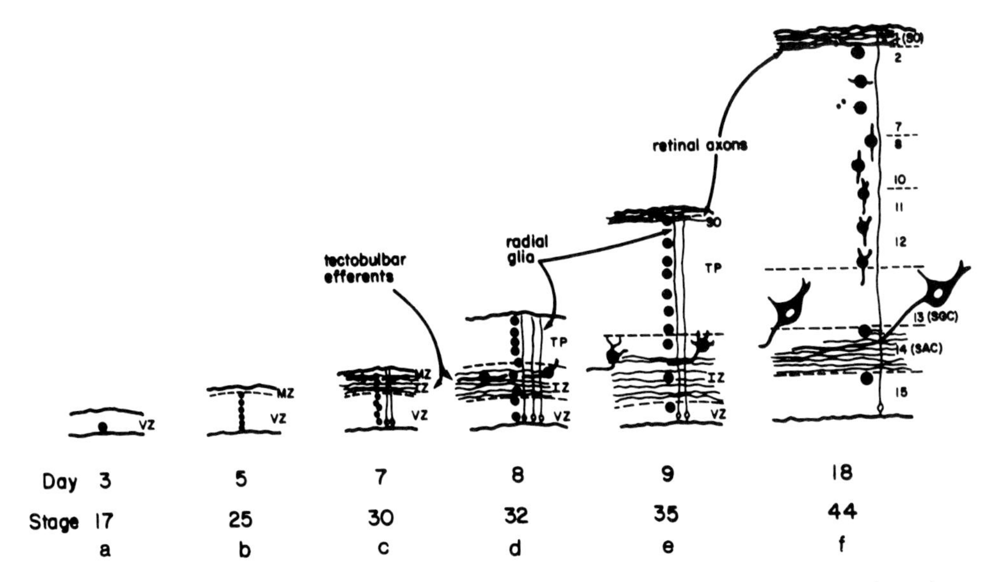

FIGURE 11–5 Migratory paths of clonally related cells in the tectum: summary of results. (a) A cell in the VZ is infected on E3. (b) The clone forms a radial array in the VZ. (c) Multipolar efferent cells migrate tangentially along tectobulbar axons. (d) Most cells migrate along radial glia to form the tectal plate. (e) Presumptive astrocytes migrate tangentially away for the radius. (f) As the tectum matures, cells in the radius differentiate into neurons and astrocytes; the radius is flanked by multipolar efferent cells and clusters of astrocytes derived from tangential migrants. IZ, intermediate zone; MZ, marginal zone; SAC, stratum album centrale; SGC, stratum griseum centrale; SO, stratum opticum; TP, tectal plate; VZ, ventricular zone. From Gray and Sane (1991).

Tectal Progenitors Are Multipotent

The tectum is a highly organized, laminar structure allowing cell type to be determined according to position and morphology. Galileo et al. (1990) used a retrovirus encoding a nuclear-localized β-gal, which allowed double staining with X-gal histochemistry and immunohistochemistry directed against neuronal and glial markers. They concluded that clones contain multiple neuronal types, as well as glia. Radial glia were shown to be members of clones that contain neurons and other glia (Gray and Sanes, 1992). Of interest, they found that approximately one-third of clones generated by infection of the embryo on E3 contained radial glia. However, multiple radial glia were rarely found in a single clone. Injections at E3 showed that 79% of clones with radial glia had a single radial glial cell, 16% had two such cells, and 5% had three; of clones containing radial glia from injections made on E4, 96% had a single radial glial cell.

Conclusions

Lineage analysis has been conducted in the tectum of the chick mesencephalon using retroviral vectors. The data indicate that tectal progenitors are multipotent, as astrocytes, radial glia, and neurons were observed in individual clones. Furthermore, several unusual patterns of clonal dispersion were identified.

Clones initially form simple radial columns, from which there is tangential spread in the dorsoventral plane in specific laminae. The cells that exhibit the most extensive tangential migration appear to become predominantly one cell type in the deeper laminae of the tectum, while a smaller number of cells that exhibit a more modest nonradial migration later in development are in the most superficial layers of the tectum.

CEREBELLUM

The cerebellum comprises the deep cerebellar nuclei, white matter, and the three-layered cerebellar cortex. It is formed from two distinct proliferative zones, a typical VZ and an external granule layer (EGL). The EGL transiently exists on the surface of the cerebellar anlage and has been shown by [^{3}H]thymidine labeling to contain many mitotically active cells. The data discussed here concern the derivatives of the progenitors of the EGL and the VZ.

The Cerebellum Derives from the Mesencephalon and Metencephalon

To define the origin of cerebellar cells, several groups have made chick/quail chimeras using portions of the mesencephalic and metencephalic vesicles (Martinez and Alvarado-Mallart, 1989; Hallonet et al., 1990; Marin and Puelles, 1995). These experiments showed that the cerebellum is derived from the caudal mesencephalon and the rostral metencephalon, including rhombomeres r1 and r2. A longitudinal morphogenetic movement in the dorsal part of the neural tube was seen to create a V-shaped intrusion by the mesencephalon-derived cells into the cerebellar primordium. Thus, the morphologically defined brain vesicles in the early embryo do not represent lineage restriction compartments. This point is confirmed by the observation that some retrovirally labeled clones span the mesencephalic/diencephalic boundary in the chicken (Golden and Cepko, 1996). The origin of different cell types in the cerebellum has also been mapped by the chick/quail chimera technique with the following results: (1) Neurons in the deep cerebellar nuclei and the Purkinje cells are derived from the VZ; (2) the neurons in the molecular layer, the basket and stellate cells, were also found to derive from the VZ (confirmed recently by the retroviral labeling technique, Zhang and Goldman, 1996; see below); (3) the EGL and its derivatives were found to arise exclusively from the metencephalic cerebellum. The significance of this observation will be discussed below in the context of granule cell development.

The EGL Generates Only Granule Cells

Cerebellar granule cells are the most abundant neuronal cell type in the CNS. Their origin from the EGL had been described by Cajal almost a century ago and confirmed by [^{3}H]thymidine labeling experiments in the 1960s (Miale and Sidman, 1961; Hanaway, 1967). The migration of granule cells from the EGL to the internal granule layer (IGL) has been shown to occur along Bergmann glial fibers (Rakic, 1971; Hatten and Mason, 1990). However, the origin of the EGL and other aspects of granule cell development were not definitively established until recently.

Retroviral labeling of the chick cerebellum (Ryder and Cepko, 1994) has revealed that EGL progenitors originate from the entire caudal edge of the cerebellar anlage that forms the rostral part of the rhombic lip. The chick/quail chimera technique has also shown that, unlike other neurons in the cerebellum, the cells of the EGL and the IGL are exclusively of metencephalic origin (Martinez and Alvarado-Mallart, 1989; Hallonet et al., 1990). The migratory pathways of granule cell progenitors in the EGL have been inferred from the chick/quail chimera experiments and from the dispersion of clonally related cells marked with retroviral vectors in the chick cerebellum (Ryder and Cepko, 1994). Once the granule cell progenitors reach the dorsal surface of the cerebellum from the VZ (Fig. 11–6), an extensive course of migration and proliferation ensues within the EGL. The mitotic progenitors initially migrate primarily rostrally in the superficial EGL, while their postmitotic progeny move long distances both medially and laterally in the deep EGL. The final distribution of the granule cell clones was not determined by Ryder and Cepko as the clones were followed only up to E14. Miyake et al. (1993) used a single *lacZ*-bearing virus to infect E13 mouse brain vesicles and analyzed the distribution of granule cells postnatally (up to P52). The authors reported that they did not find any other cell types in the granule cell clusters, in keeping with the idea that the EGL makes only granule cells (see below). Curiously, the granule cell clusters in the mouse always appeared in discrete groups, as opposed to being widely dispersed. The explanation for this apparent discrepancy of migratory behavior of EGL and/or IGL cells in the chick and the mouse is currently unknown. However, it is clear from both studies that granule cell clones do not distribute according to any functionally defined areas in the cerebellum.

Injection of retrovirus was made stereotactically into the rat EGL on P3-4 to address the developmental fates of EGL cells (Fig. 11–7; Zhang and Goldman, 1996). This study has confirmed that EGL cells only make granule neurons and is consonant with the results of a previous experiment in which purified EGL cells from P5–6 mice were isochronically transplanted into the EGL (Gao and Hatten, 1994). However, EGL cells can differentiate into other cerebellar cell types if they are transformed by the viral SV40 large T-antigen before transplantation (Gao and Hatten, 1994), and labeled EGL cells can adopt features of hippocampal granule neurons as shown in an isochronic/heterotopic transplantation experiment (Vicario-Abejon et al., 1995). The EGL cells in the postnatal rodent cerebellum, therefore, give rise exclusively to granule cells in normal development, but the behavior of these cells can be altered either by oncogene transformation or by different environmental cues from another part of the brain.

The Cerebellar VZ Generates all Cerebellar Cell Types Except Granule Cells

In rodents, the neurons of the cerebellar deep nuclei and the cortical Purkinje cells are the first born, with birthdays between E11 and E13. Later, the Golgi type II neurons in the granule cell layer are born. The molecular layer neurons, in-

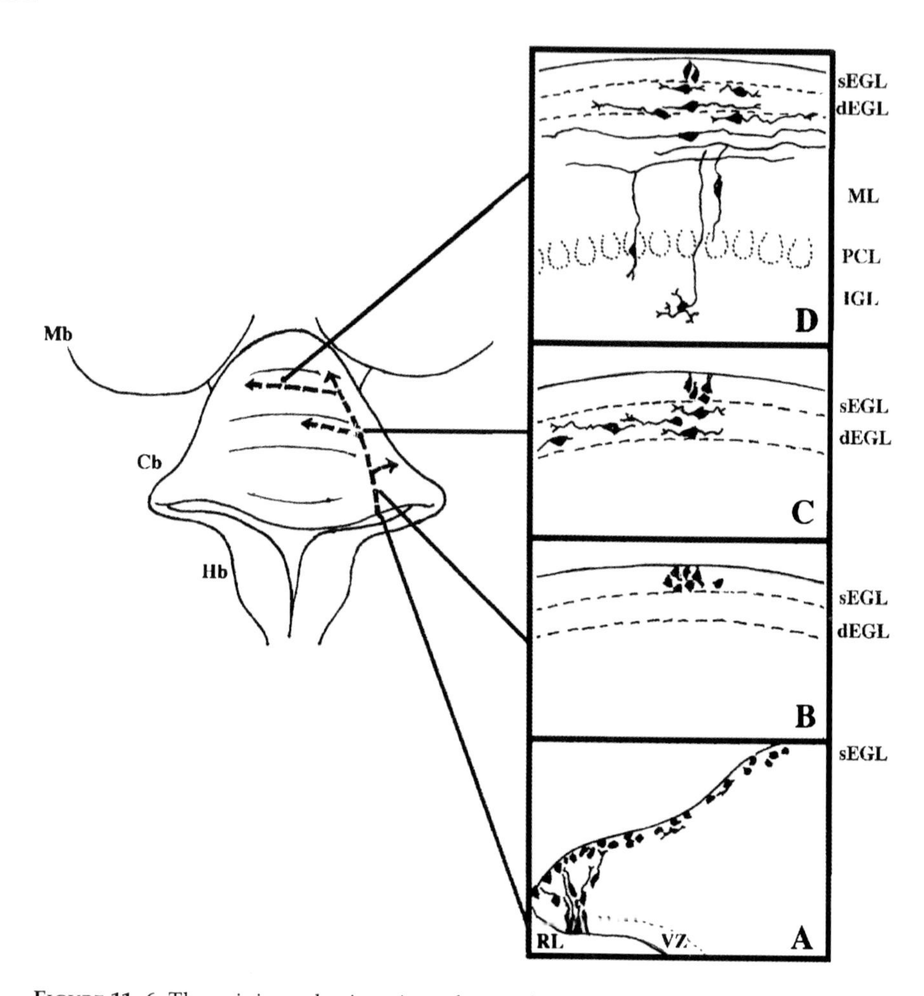

FIGURE 11–6 The origin and migration of granule cells in the cerebellum. On the left side is a dorsal view of an E12 chick cerebellum; the dotted line represents the distribution of a typical granule cell clone marked with a retrovirus. On the right side are drawings of different stages of granule cell development. (A) A sagittal section shows that granule cells originate from a restricted area of the VZ in the posterior cerebellum, the rhombic lip (RL). They migrate to the dorsal surface of the cerebellum, forming a secondary proliferative zone, the EGL. B, C, and D are coronal sections showing later events of granule cell development. (B) Mitotic progenitors migrate rostrally in a narrow strip of the superficial EGL. (C) Postmitotic progenitors migrate medially and laterally in the deep EGL. (D) Granule cells eventually descend into the IGL and assume their mature morphology. Cb, cerebellum; Hb, hindbrain; Mb, midbrain; sEGL, superficial EGL; dEGL, deep EGL; ML, molecular layer; PCL, Purkinje cell layer; IGL, internal granular layer; RL rhombic lip; VZ, ventricular zone. Modified from Ryder and Cepko (1994).

cluding basket cells and stellate cells, become postmitotic mainly in the first postnatal week, and the glial cells are born in the first few postnatal weeks. This stereotypical birthday sequence for the derivatives of cerebellar VZ is similar to those of other regions of the brain (e.g., retina and spinal cord), where the projection neurons are born first, interneurons are born later, and most glial cells are

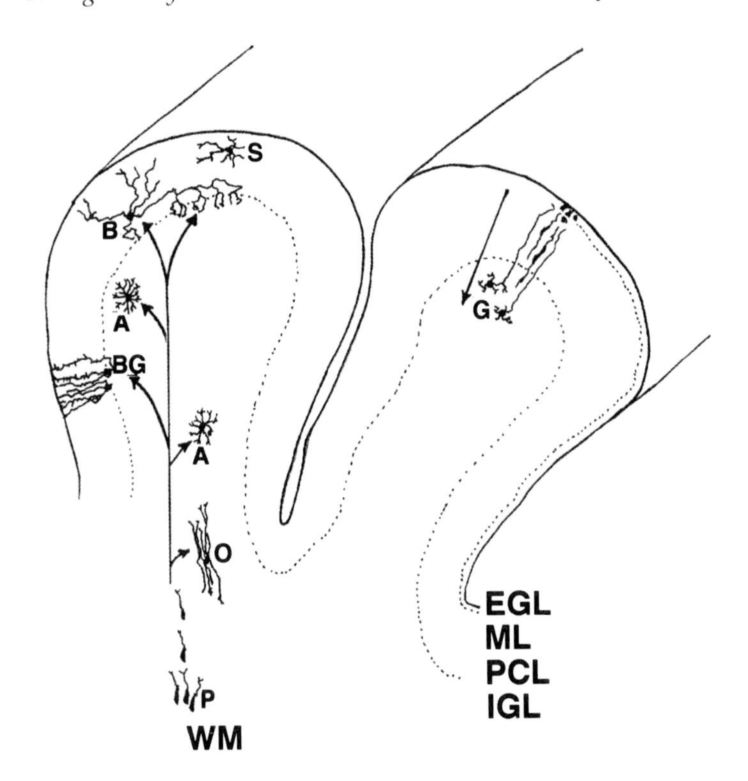

FIGURE 11–7 The dual origins of cerebellar cells. A sagittal section across two folia of the rat cerebellum shows the origins of different cell types in the cerebellum. On the left side, progenitors in the white matter migrate up to the cerebellar cortex and give rise to all the different cell types generated postnatally. On the right side, progenitors in the EGL produce daughters that migrate down into the IGL to become only granule cells. A, astrocyte; B, basket cell; Bg, Bergmann glial cell; O, oligodendrocyte; P, progenitor in WM; S, stellate cell; WM, white matter; G, granule cell; see also legend for Figure 6. Adapted from Zhang and Goldman (1996b).

formed last. However, because the retrovirally labeled cerebellar clones do not form simple radial columns (Miyake et al., 1995; Zhang and Goldman, 1996, MS submitted), lineage analysis of the cerebellar VZ is not straightforward. It is not known at this time whether there exist multipotential progenitors in the cerebellar VZ (as in the retina, for example) or whether different cell types have separate progenitors (as in the EGL).

The study by Miyake et al. (1995) in which a *lacZ*-bearing retrovirus was introduced into the E13 mouse midbrain vesicle revealed labeled Purkinje cells (Miyake et al., 1995). They were single, isolated cells. No labeled deep nuclei neurons were observed, probably reflecting the fact that the viral infection was started very close to end of the birthdays of these neurons. The VZ-derived interneurons were not mentioned in this study, so it is not clear if these cell types were labeled. Labeled glial cells were seen in discrete clusters containing no labeled neurons. Of the 71 glial cell clusters examined, 33 were composed of astrocytes/Bergmann glia, 24 were of oligodendrocytes alone, and ten were of white-matter astrocytes only. This result suggests but does not prove the authors' hypothesis that the cells derived from the VZ have distinct lineages as early as the mouse E13. Potential splitting errors make the conclusions concerning glial lineage only tentative at this time.

Targeting of the postnatal rat cerebellum using stereotaxic injections and retroviruses allowed Zhang and Goldman (1996, MS submitted) to define the origin and the migratory paths of postnatally generated cells. As mentioned above only labeled granule cells were found after injections of virus into the EGL. Molecular layer neurons, Golgi type II neurons, and the entire gamut of glial cells were labeled when the virus was injected into the cerebellar white matter (WM) on P3–4 of the rat. Although the exact clonal relationships could not be established with a single retrovirus, these experiments provided good evidence that all of these cell types can be generated by progenitors in the postnatal WM. While all of these different cell types were labeled if injected on P4–5, mainly oligodendrocytes were labeled when the virus was injected on P14. These observations seem to indicate that the cerebellar VZ progenitors continue to divide and migrate in the WM before they reach the cerebellar cortex, giving rise to different neurons and glial cell types (Fig. 11–7). Zhang and Goldman also observed that the labeled cells in the molecular layer were more widely dispersed than cells in the other layers, confirming the tangential migration of cells in the molecular layer observed in the chick/quail chimera experiments (Hallonet et al., 1990; Otero et al., 1993). This raises the interesting possibility that these cells migrate along the granule cell axons once they reach the molecular layer.

Conclusions

The EGL arises from the VZ in a restricted region of the caudal margin of the cerebellum, forming a proliferative zone separated from the VZ. The progenitors in the EGL produce only neuronal progeny, granule cells. Due to extensive migration within the EGL, clonally related granule cells generally are not confined to any single functional area in the cerebellum. The VZ generates all other neurons and all glia of the cerebellum. As clonal analysis of VZ progeny has not yet been carried out, it is not clear whether the VZ comprises multipotent progenitors or progenitors with restricted fates. In addition, very little is known about the distribution of clonally related VZ cells, particularly those that are generated early.

RHOMBENCEPHALON

The vertebrate hindbrain comprises motor and sensory nuclei of cranial nerves IV–XII, the reticular formation, pontine nuclei, and olivary nuclear complexes and several other nuclear groups (see also Chapters 8 and 9). Because the chick hindbrain is well characterized and accessible, lineage analyses have been carried out in this species.

The Majority of Clones Do Not Cross Rhombomeric Boundaries

The early development of the hindbrain is marked by the appearance of a transient series of bulges, known as rhombomeres. Rhombomeres start to form at stage 9 (E2), and they disappear at about E5. Examination of rhombomeres has revealed that they correspond to repeating units of neuronal differentiation, centers of increased mitotic activity, and domains of regulatory gene expression (Lumsden and Keynes, 1989; Lumsden, 1990; Wilkinson et al., 1989a,b; Wilkinson and Krumlauf, 1990). Neural crest cells are also segmented according to the rhombomeric units, with each segmental stream of neural crest migrating into a corresponding branchial arch (Lumsden et al., 1991). Rhombomere boundaries are marked by concentrations of early axonal tracts and by the expression of a number of cell adhesion molecules and extracellular matrix proteins (Lumsden and Keynes, 1989). When rhombomere r4 is transplanted to a different axial level at or before the appearance of boundaries, it maintains the r4-specific Hox gene expression and neuronal differentiation in the new location (Guthrie et al., 1992). Rhombomeres thus represent segmental units of the hindbrain whose identities are established fairly early during development.

In many respects, rhombomeres resemble the parasegments of the *Drosophila* embryo, which are true compartments in the sense that cells in one parasegment do not mix with those in the neighboring parasegment. Labeling single neuroepithelial cells in the chick hindbrain by intracellular injection of fluorescent dextran, or labeling a small group of cells with the lipophilic dye DiI, was undertaken to test whether rhombomeres are similar compartments. The labeling was carried out either before or after the rhombomeres were formed (stages 6–12), and the labeled clones were examined 24–48 hours later (Fig. 11–8). If labeled before the appearance of rhombomeres, about 25% of the clones crossed a rhombomeric boundary. However, once rhombomeres were formed, clonally related cells usually respected the rhombomeric boundaries, although at least 8% of marked clones did expand across a rhombomeric boundary (Fraser et al., 1990; Birgbauer and Fraser, 1994). These results support the idea that rhombomeres are relative lineage restriction compartments and that lineage may contribute to the establishment of, or be a reflection of, the identity of individual rhombomeres. The relative lineage restriction at the rhombomeric boundaries seems to arise as a consequence of distinct surface adhesion properties of the adjacent rhombomeres (Guthrie and Lumsden, 1991). When an even- and an odd-numbered rhombomere were juxtaposed, in the absence of the original boundary, a new boundary was always formed, whereas no new boundary was formed if two odd- or two even-numbered rhombomeres were put together.

The rhombomeres and the known specific gene expression patterns are set up only to disappear a few stages later. It was thus interesting to ask whether the restrictions to movement between rhombomeres would break down at later stages. Labeling one or a few cells with fluorescent tracers does not allow one to observe clones in mature tissues, because the tracer eventually becomes too dilute. Stable, genetic markers, such as retroviral vectors and the nuclear markers in chick/quail chimeras, have been used to address this question. Hemond and

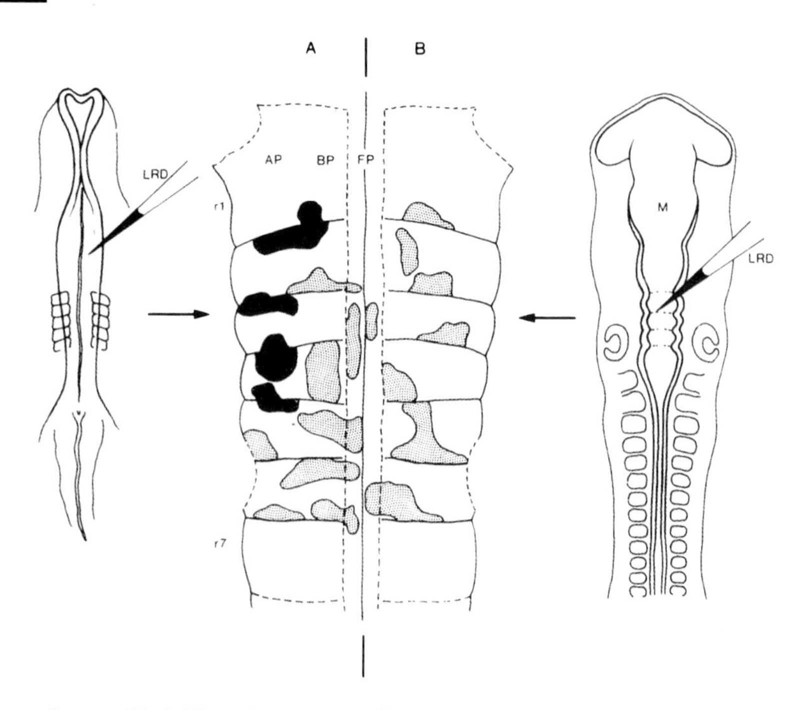

FIGURE 11–8 Rhombomeres are lineage restriction compartments. (A) Injection of a fluorescent tracer before the appearance of rhombomeres frequently gives rise to clones that span rhombomeric boundaries at stage 19. (B) Injections after rhombomere formation produce mostly clones that respect the boundaries. AP, alar plate; BP, basal plate; FP, floor plate; M, midbrain; LRD, lysinated rhodamine dextran. From Fraser et al. (1990).

Glover (1993) used a replication-incompetent retrovirus to infect the neural tube at stages 13–17 when the rhombomeres were formed, and the marked cells were examined at progressively later stages, from stage 29 to stage 35 (E9) (Hemond and Glover, 1993). Since a virus with a single label was used, the authors defined clonal boundaries according to geometric criteria. As a result, only those cells with certain clustering patterns were presumed to be clones. Cells dispersed over a long distance were not analyzed, although they could have been siblings of the more clustered cells, especially considering that the overall infection rate was low (about 75% of the injected brains did not contain any labeled cells). The extent of dispersion of labeled cells at stages 34–35 was so wide that it was not possible to define clones at these stages. Therefore, the question of dispersion of hindbrain clones later in development is currently unresolved.

Examination of the segmental lineage boundaries after the disappearance of rhombomeres has been more successful using chick/quail chimeras (Tan and LeDouarin, 1991; Marin and Puelles, 1995). Tan and LeDouarin (1991) used large grafts extending from the r2/r3 boundary to the r8/spinal cord junction from the chick embryos of stages 10–12. Virtually no cell mixing at the graft/host boundary was noted up to E8, while substantial cell mixing was observed from E9 onward. The longitudinal cell mixing was much more extensive in the ventral hindbrain (up to 125 μg on E9) than in the dorsal hindbrain (10–60 μm on E9).

On E14, the longitudinal migration in the ventral hindbrain, including the pontine nuclei and in the inferior olivary nuclei, was so extensive that both neurons and glia in these nuclei were completely chimeric over their entire volume. The cell mixing in the longitudinal direction was very limited for the motor and sensory nuclei in the dorsal hindbrain. Marin and Puelles (1995) grafted individual rhombomeres (r2 to r6) from stage 11 chick embryos and harvested the chimeras 9–10 days later. Remarkably, the grafted rhombomeres almost always developed at transverse tissue slices in the chimeric hindbrain. The cranial nerve motor nuclei had rhombomeric origins consistent with the two-rhombomere repeat pattern of the early embryo (Lumsden and Keynes, 1989), with only minor contributions to both the trigeminal and the facial nuclei from a third rhombomere. The rhombomeric origins of other cranial sensory nuclei have also been defined with very sparse intersegmental migrations. However, extensive neuronal and glial migrations across several rhombomeres were evident in the pontine nuclei, confirming the observation by Tan and LeDouarin (1991).

It thus appears that there are restrictions to movement that are respected by the majority of neurons in the mature hindbrain. The pontine and inferior olivary nuclei, however, do not appear to follow this rule. In this regard, it is notable that, unlike the other hindbrain nuclei, these ventral structures arise from the rhombic lip and undergo an extensive dorsal-to-ventral migration across the midline during development (Tan and LeDouarin, 1991).

Hindbrain Progenitors Generate Neurons of Similar Phenotypes

As the rhombomere identities appear to become determined fairly early, it was speculated that cell fate choices might be made much earlier in the hindbrain than in the other parts of the brain. To address this possibility, injection of single cells in the chick hindbrain by intracellular tracers was carried out between stages 6 and 12, and neuronal projection phenotypes in the labeled clones were assessed at stages 13–20 (Lumsden et al., 1994). This study revealed that the recognizable neurons of the labeled clones largely comprised just one or two closely related axonal projection phenotypes. Since about 85% of the clones contained neurons of identical or closely related projection phenotypes, and each clone consisted an average of 7.3 neurons, it was concluded that neuronal phenotypes in the chick hindbrain were determined several divisions before these neurons were produced. However, as the authors pointed out, the majority of clones also contained many cells without labeled processes, which could represent mitotically active cells, young glia, or young neurons prior to axonal outgrowth. Therefore, the progenitors in the chick hindbrain could still be multipotent even though the neurons generated during the early period by each progenitor were mostly of a similar phenotype.

To look at the issue of clone composition at later stages of hindbrain development, it appears that retroviral libraries will be required for definitive results. As mentioned previously, Hemond and Glover (1993) infected chick hindbrains at stages 13–17 and were able to estimate the clonal boundaries only before stage 29. In the clusters that were identified at stages 34–35 (E9), when the hindbrain nuclei were fully formed, presumed siblings were found to populate functionally distinct nuclei. Due to the problem of clonal boundary assignment, however, definitive proof that hindbrain progenitors give rise to neurons in different functional nuclei is currently lacking.

Mitotic Behavior of Hindbrain Progenitors

Fluorescent tracer labeling at stages 6–12, with harvests at stages 13–20 (about 48-hour survival period) yielded clones of from one to 64 cells (average of 16 cells per clone) (Fraser et al., 1990; Lumsden et al., 1994). Given the cell cycle of 8 hours at this time of hindbrain development (Guthrie et al., 1991), apparently some divisions produced two mitotic daughters while others gave one postmitotic and one mitotic daughter during the first 48 hours of neurogenesis. In the retroviral analysis of chick hindbrain by Hemond and Glover (1993), infection was initiated at stages 13–17 and harvests were at stages 24–29 (about 50–70 hours later allowing for six to nine cell divisions). The clone size varied from one to 29 cells, with the average being 5.6 cells. This apparently lower number of cells per clone may arise from (1) splitting errors in the clonal assignments, (2) variable expression of the *lacZ* marker gene, and/or (3) cell death at the later stages of development.

Conclusions

Lineage analysis of the chick hindbrain has been done most successfully with intracellular tracer injections and with chick/quail chimeras. The data support the notion that rhombomeres are lineage restriction compartments and that rhombomere fate is determined early. Studies using chick/quail chimeras have shown that lineage restriction is largely maintained later in the mature hindbrain with the exception of the ventral nuclei that are derived from the rhombic lip. It is not known at this time whether the lineally restricted compartments in the hindbrain exist in other vertebrate species. Whether the progenitors in the hindbrain are truly committed to making certain neuronal phenotypes also awaits further experiments.

SPINAL CORD

The spinal cord is a relatively simple and conserved structure, subdivided along its longitudinal axis into cervical, thoracic, lumbar, and sacral regions each of which is connected to a specific region of the body. In cross-section, the spinal cord is composed of white matter (myelinated axons) on the periphery and gray matter (cell bodies) centrally. The gray matter is composed of the dorsal horn, which receives sensory information, a midregion consisting mostly of interneurons, and the ventral horn, which innervates skeletal muscles. Lineage analysis in the spinal cord has addressed whether clones can cross the multiple functional domains and whether motoneurons and other cell types derive from a common progenitor.

Neurons in Spinal Cord Clones Disperse in the Dorsoventral and Mediolateral Planes and Glia Also Disperse Along the Rostrocaudal Axis

Lineage in the chick spinal cord has been examined with retroviruses (Leber and Sanes, 1995; Leber et al., 1990) as well as with chick/quail chimeras (Schoenwolf et al., 1989; Carpenter and Hollyday, 1992; Cameron-Curry and LeDouarin, 1995). Retroviral analysis of the chick spinal cord has yielded information on the dispersion patterns of clonally related cells (Fig. 11–9). Retroviral injections were made at successive stages of development to reveal the initial phases of dispersion. Early (stage 8–9) injection of retrovirus resulted in approximately 70% of clonally related cells dispersing within the VZ (Leber and Sanes, 1995). The pattern of dispersion was suggestive of intercalation between infected and unin-

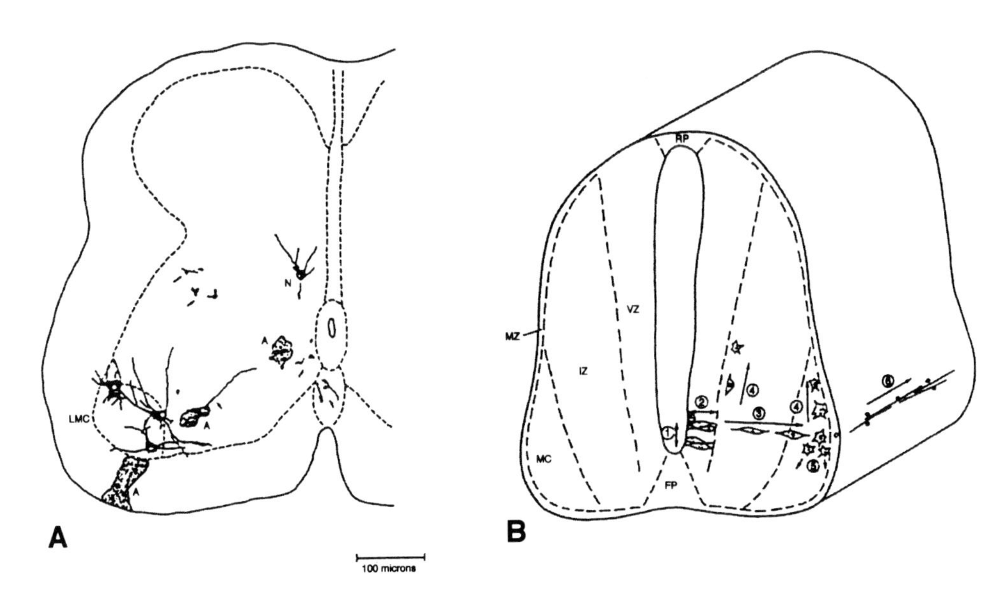

FIGURE 11–9 Retroviral marking of the developing spinal cord. (A) Camera lucida reconstruction of a clone of *lacZ*-marked cells in the chick spinal cord (stage 15 injection, stage 40 sacrifice; Leber et al., 1990). The presence of motoneurons, nonmotoneurons, and astrocytes indicates that the spinal cord contains multipotent progenitor cells. Note also that cells in this group are widely dispersed and include a neuron (N) in a medial position, neurons in the lateral motor column and astrocytes in both the gray and white matter. This pattern shows that clonally related cells in the spinal cord are not confined to single anatomical or functional area. (B) Representation of the major dispersion patterns in the ventral spinal cord (Leber and Sanes, 1995). 1, mixing in the VZ; 2–3, radial migration in the VZ and IZ; 4–5, mediolateral and dorsoventral dispersion of cells; 6, anteroposterior dispersion of glia in the white matter. A, astrocyte; IZ, intermediate zone; FP, floor plate; LMC, lateral motor column; MC, motor column; MZ, marginal zone; N, neuron; RP, roof plate; VZ, ventricular zone. Adapted from Leber et al. (1990—A) and Leber and Sanes (1995—B).

fected cells, as seen in the early chick tectum and chick retina (see above). By stages 15–18, retroviral injections resulted in only 5% of clones with two or more sister cells dispersed within the VZ, indicating that dispersion in the VZ is an early event. Throughout this period, spread in the dorsoventral direction was more common than that in the rostrocaudal direction. However, use of quail/chick chimeras led to detection of prenodal median neuroepithelial cells which were observed to disperse caudally in the VZ (Schoenwolf et al., 1989).

Injections at a fixed period of development (stages 12–14) were followed by sacrifice at various ages to follow other patterns of dispersion (Leber and Sanes, 1995). At early stages (19–21), clustered cells appeared to be migrating in a predominantly radial direction (90% of clones). This was followed at stages 27–29 by approximately 50% of all clones containing cells that dispersed tangentially in the IZ. Some of these cells followed a ventral circumferential route and occasionally crossed the floor plate to the contralateral side. A similar dorsal to ventral migration was described in chick/quail chimeras (Carpenter and Hollyday, 1992). While cells with neuronal morphology did not disperse rostrocaudally, cells with the morphology of immature oligodendrocytes exhibited extensive dispersion rostrocaudally in the white matter. Clonally related glia could spread across several segments and movement could take place in either a rostral or caudal direction. The pattern of dispersion seemed to follow the axonal tracts already present in the white matter at the stages examined.

A very recent study using chick/quail chimeras (Cameron-Curry and Le-Douarin, 1995) and spinal cord grafts produced results very similar to those of the retroviral studies. Extensive mediolateral and dorsoventral dispersion of transplanted E2 cells was seen after both ventral and dorsal grafts. An oligodendrocyte-specific Schwann myelin protein probe (see below) showed rostrocaudal dispersion up to 200–250 μm of young oligodendrocytes in the white matter.

Spinal Cord Clones Span Functional Domains

As a result of the tangential migration, clonally related cells span functionally and anatomically distinct areas of the spinal cord (Leber and Sanes, 1995; Leber et al. 1990). For example, sibling motoneurons were found both in the lateral motor column, which innervates the limb muscles, and in the medial motor column, which innervates the axial muscles. Sibling motoneurons could also be found in both medial and lateral aspects of the lateral motor columns. The majority of clones in the ventral horn contained few if any sibling cells in the dorsal horn and the majority of clones in the dorsal horn were restricted to that sector, presumably reflecting the fact that the early dorsoventral dispersion in the VZ takes place over a rather short period.

Spinal Cord Progenitors Are Multipotent

Retrovirally marked clones were initiated at stages 11–18 in the chicken, when motoneurons (which are among the earliest-born neurons in the cord) were being generated. Clones containing motoneurons were found to contain glial cells, ependymal cells, or other neuronal types in the gray and white matter (Fig. 11–9) (Leber et al., 1990). In fact, the majority (82%) of clones containing motoneurons also contained other cell types. Because the animals were analyzed at stages 38–41, before gliogenesis was complete, the percentage of motoneuron-containing clones that also would contain glia may be even larger in adults. About half (55%) of the clones with motoneurons and other cell types contained interneurons or autonomic preganglion neurons. The morphology of immature glia, as visualized with *lacZ,* makes positive identification difficult. However, it seemed that almost half (42%) of the motoneuron-containing clones contained either white-matter or gray-matter cells that were oligodendrocytes and/or astrocytes. Thus it is clear that, in the spinal cord, progenitor cells do not exclusively give rise to a single cell type.

A number of groups have reported that oligodendrocytes can only be generated from explants of the ventral spinal cord (Warf et al., 1991; Noll and Miller, 1993). The expression of markers of immature oligodendrocytes, DM-20, platelet-derived growth factor receptor (PDGFr), and 2′,3′-cyclic-nucleotide 3′-phosphodiesterase (CNP) in vivo also suggests a ventral origin for oligodendrocyte precursors (Pringle and Richardson, 1993; Timsit et al., 1995; Yu et al., 1994). However, recent transplantation studies of ventral and dorsal spinal cord from quail to chick showed that oligodendrocyte progenitors could be generated in both the ventral and dorsal spinal cord (Cameron-Curry and Le Douarin, 1995). Using a quail-specific SMP probe, it was shown that both the dorsal and ventral halves of the transplanted spinal cord could generate oligodendrocytes. The discrepancy between the in vivo and in vitro studies may stem from the earlier stage of spinal cord development (E2) used in the transplantation study compared with E4 spinal cord used in the in vitro studies. Another difference between the studies is the markers used to define oligodendrocytes. Nevertheless, it appears that both dorsal and ventral E2 chick spinal cord contain progenitors capable of giving rise to oligodendrocytes. The results of previous studies suggest that oligodendrocytes differentiate if presented with inductive signals presumably emanating from notochord or floor plate. What remains uncertain is how progenitors which originate in the dorsal proliferative zones are induced to differentiate into oligodendrocytes. It is possible that cells that migrate into the dorsal spinal cord from the ventral region induce the formation of oligodendrocytes in cells that originate in the dorsal spinal cord. Further work on the mechanisms of induction should clarify these issues.

Conclusions

Spinal cord progenitor cells are multipotent, capable of producing motoneurons, interneurons, and glia. Dispersion, due to mixing of labeled and unlabeled progenitors, can occur early in the VZ. Dispersion due to migration can occur later in the gray matter, primarily in the mediolateral and dorsoventral planes, resulting in a small percentage of clones crossing the midline. Migration of glia in the white matter can occur for considerable distances along the rostrocaudal axis. As a result of dispersion, clonally related cells can be found in distinct functional domains.

LINEAGE AND THE CHOICE OF A NEURONAL OR GLIAL FATE

It is clear that most, if not all neural progenitors are multipotential. This is certainly true in the retina (Turner and Cepko, 1987; Holt et al., 1988; Wetts and Fraser, 1988; Wetts et al., 1989; Turner et al., 1990; Fekete et al., 1994), tectum (Galileo et al., 1990), spinal cord (Leber et al., 1990), and diencephalon (Golden and Cepko, 1996) in that clones frequently comprise both neurons and glia. In the retina the multipotency appears to extend to the final cell division in that clones of only two cells can have, for example, a rod and a Müller glial cell. Similarly, radial glia appear to be generated in the same cell division as neurons in the early tectum (Galileo et al., 1990), striatum (Halliday and Cepko, 1992), and diencephalon (Golden and Cepko, 1996).

The best case for their being a unipotent progenitor in the CNS is the EGL cells of the cerebellum. Lineage analysis has shown that only granule cells are generated by the EGL (Hallonet et al., 1990; Zhang and Goldman, 1996a). More significantly, transplantation experiments do not reveal additional potency (Gao and Hatten, 1994), except that hippocampal granule cells can also be made (Vicario-Abejon et al., 1995). As mentioned above, rods can be found in rod-only clones in the rat retina, but this finding is ambiguous as there are multipotent progenitors that make rods, and rods are very abundant. There is a case to be made that progenitors may be unipotential in the chick hindbrain in that sizable clones with only one defined cell type have been observed (Lumsden et al., 1994). However, since there were also unidentified cells in these clones, and as the clones were not observed after development was complete, it is difficult to conclude that indeed only one cell type was generated. Transplantation of different progenitors to different environments, e.g., isochronic transplantation of cells ventral to dorsal, is in order to test the potency of such progenitors.

There are also reports of unipotent progenitors capable of giving rise only to neurons, only to glia, or only to oligodendrocytes in the cerebral cortex (Price and Thurlow, 1988; Parnevales et al., 1991; Grove et al., 1993; Luskin et al., 1993; Mione et al., 1994). But these lineage studies used either a single, or at most two, retroviral markers and did not take into account splitting errors. Further, extensive cell death in the cortical VZ and SVZ make conclusions of single cell types in cortical clones even more difficult to interpret (Blaschke et al., 1996). In addition to conclusions of restriction concerning glia and neurons in the cortex, conclusions about pyramidal and nonpyramidal cells arising from progenitors committed to making only pyramidal or only nonpyramidal cells have been made (Parnavelas et al., 1991). Committed progenitors giving rise to cells with only GABAergic or only glutaminergic properties were similarly reported (Mione et al., 1994). However, a more recent study using the same method and conducted by the same group, has concluded that 18% of clones contain both pyramidal and nonpyramidal cells, and another 26% of clones contain both astrocytes and neurons (Mione et al., 1995). Nonetheless, it does seem that subunits of clones in the cerebral cortex can have fairly uniform cell types, even rather specific and relatively infrequent neuronal cell types (Reid et al., 1995), and a case can be made for restricted progenitors in the cerebral cortex which coexist with more multipotential progenitors. The confusion around this issue in the cortex highlights the limitations of lineage analysis. When a mixture of different progenitors is present, sorting out the potential of each will require good markers for each type and an understanding of the mechanisms that lead to restriction or commitment.

MIGRATION

Among the features that make the vertebrate nervous system challenging for lineage studies is the fact that neurons are frequently located some distance from their birthplace. Migration, sometimes along circuitous routes, is a prominent feature of the developing nervous system. Cell migration in the nervous system has been investigated using a variety of approaches and a complete review of the literature is beyond the scope of this chapter. Here, we will focus on the insights that have come from lineage analysis and pertinent related studies.

The best-established pathway of migration is along radial glial fibers (called Bergmann glia in the cerebellum; reviewed by Hatten, 1990). A radial glial cell is a bipolar cell with its nucleus in the VZ, a short process attached to the ventricular surface, and a second process that reaches to the pial surface. Cells, particularly young neurons, have been shown to migrate along these fibers en route to their final position. Radial glial fibers have been identified along the entire neuroaxis from the telencephalon to the spinal cord and have been postulated to serve as the primary guiding force for migration of newborn neurons.

Lineage analyses have revealed radial migration in the telencephalon, diencephalon, tectum, cerebellum (granule cells of the cerebellum migrate along Bergmann glia), brainstem, and spinal cord (reviewed in this chapter). These data indicate that radial migration, most likely along radial glia, is frequently the first phase of migration away from the VZ. In addition to radial migration, lineage analyses have revealed nonradial migration in the telencephalon, diencephalon, mesencephalon, cerebellum, and spinal cord.

Migration has been inferred from the position of sibling cells at various times after a progenitor is marked. The results of these studies have raised many questions about how clones disperse. For example, whence do cells migrate? What are the molecular mechanisms underlying guidance along different paths? Is there chemoattraction? What are the factors that tell one cell in a clone to migrate along a particular pathway and a sibling cell to use a different pathway? Answers to most of these questions are presently unknown. However, recent work has begun to address some of these questions.

One seemingly simple question concerns where cells migrate (Fig. 11–10). The emerging answers indicate that migration may occur in different layers and at different times. For example, cell dispersion can occur in the VZ in some areas. Lineage analyses using retroviral vectors have identified multiple clustered columns of labeled cells separated by columns of unlabeled cells in the chick retina (Fig. 11–1B; Fekete et al., 1994), tectum (Gray et al., 1988), diencephalon (Arnold-Aldea and Cepko, 1996; Golden and Cepko, 1996), telencephalon (Szele

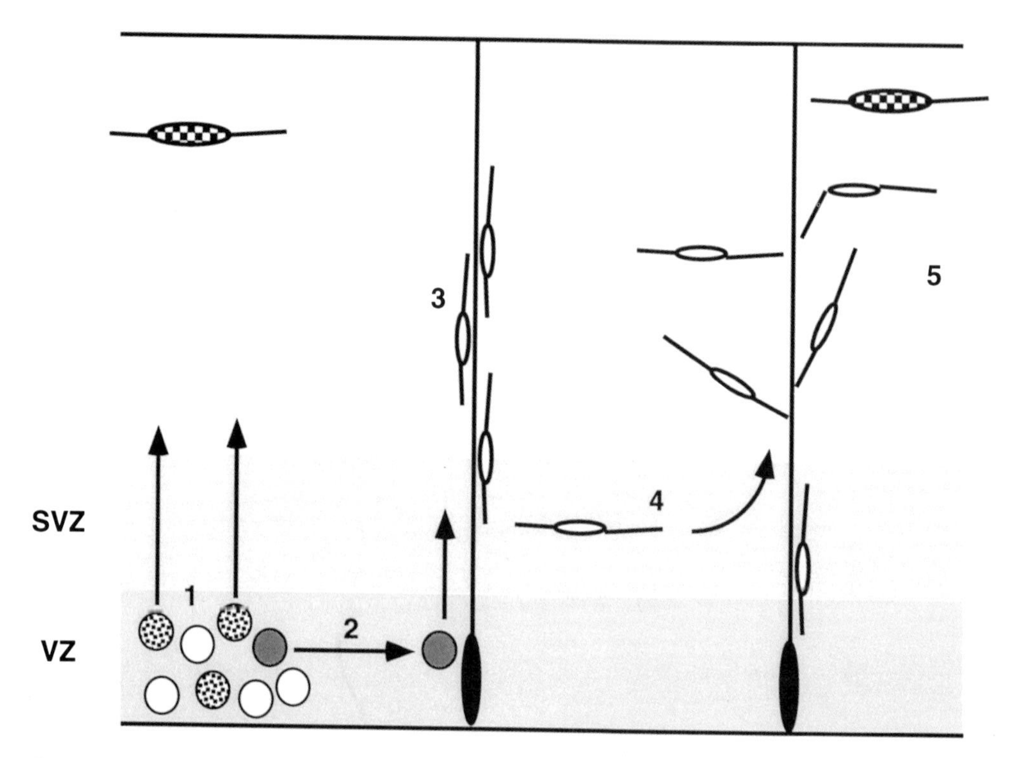

FIGURE 11–10 Modes of cell dispersion. Radial glia are schematized as black cell bodies in the VZ with long processes extending to the pial surface. (1) Mixing of labeled and unlabeled cells in the VZ. (2) Migration in the VZ. (3) Migration along radial glia. (4) Migration in the subventricular zone SVZ. (5) Nonradial migration outside the VZ and SVZ. Two widely dis persed sibling cells are represented by small squares. Wide dispersion could occur by the multiple mechanisms shown. (See text for details.)

and Cepko, unpublished observations), and spinal cord (Leber and Sanes, 1995). Such an arrangement is only observed after injections into areas during the time in which the majority of cells being generated from the VZ cells are also mitotically active VZ cells. It appears as if labeled and unlabeled progenitors become mixed through a simple shuffling or intercalation within the VZ, giving rise to clustered columns of labeled, clonally related cells (Fig. 11–10, model 1). Actual migration has also been hypothesized to occur in the VZ (Fig. 11–10, model 2) (Walsh and Cepko, 1993; Arnold-Aldea and Cepko, 1996), although evidence for VZ migration is at present indirect for the cerebral cortex and is postulated for only a small subset of clones in the chick hypothalamus. Migration within the VZ would be a basis for the wide dispersion observed in some areas of the mature brain (Fig. 11–10).

Migration of cells in the SVZ has been well documented (Fig. 11–10, model 4). The pathway known as the rostral migratory stream lies in the SVZ and has been characterized by a variety of methods. Recently, Lois et al. (1996) have shown that the pathway is surrounded by a meshwork of glial cells. Young neuroblasts appear to migrate as chains of cells moving along each other within the glial meshwork. Production of neurons and migration in this region continues even in adult rodents. Thus, the rostral migratory stream is directed migration in the SVZ, and studies of migration in this region have identified a new structure that may support and direct migration. A less well-defined migration within the SVZ occurs in the embryonic striatum. Cells migrate dorsoventrally in a fairly directed fashion to destinations and along substrates currently unknown.

Lineage analyses using retroviral vectors and imaging of live cells following labelling by DiI have shown that nonradial migration can also occur outside the VZ and SVZ of the cerebral cortex (Fig. 11–10, model 5; Walsh and Cepko, 1988, 1993; Austin and Cepko, 1990, O'Rourke et al., 1992, 1995). Tangential migration outside of the proliferative zones has also been inferred from lineage analyses in the tectum (Gray and Sanes, 1991). In regions of the brain that do not have an obvious IZ and CP, as in the diencephalon and chick telencephalon, nonradial migration has been observed as cells emerge from the VZ and SVZ (Golden, Zitz, and Cepko, and Szele and Cepko, unpublished observations). Again, the cues received by cells that cause them to leave their radial glial supports and migrate in another direction remain unknown.

In addition to identifying where nonradial migration occurs, it is of interest to know what guides or supports cells as they migrate. Some of the examples cited above use glial cells, either radial glial or astroglia in the case of the rostral migratory stream. Axons have been shown to be the structure on which cells migrate in some systems. The gonadotropin releasing hormone (GnRH)-expressing neurons of the hypothalamus take an unusual path to their final position (Lutz et al., 1993; Rugarli et al., 1993). They are born in the olfactory placode and migrate into the hypothalamus along the vomeronasal nerve; in the absence of the vomeronasal nerve these cells fail to migrate.

The evidence that is accumulating indicates that different regions of the nervous system utilize different strategies for cell migration. They vary in the location where migration occurs and presumably in the substrates used for migration. Evidence is also accumulating for the requirement for specific cell–cell and cell–matrix interactions for migration in different regions of the nervous system. Futures studies elucidating the cellular and molecular mechanisms of migration will provide valuable insights into the variety of interesting patterns revealed by lineage analyses.

Lineage and Areal Boundaries

The underlying mechanisms used to pattern the neuraxis and the various structures that derive from it are a subject of intense interest. A longstanding question concerns the role played by lineage in shaping the CNS, particularly whether functionally and anatomically distinct brain regions ("areas") are composed of lineally related cells and whether clones are restricted by areal boundaries. At one extreme is the postulate that areas of the brain are established early in development by lineage mechanisms, such as those operating in the classical compartments of the *Drosophila* wing (Garcia Bellido, et al., 1973). The radial unit hypothesis which states that columns in the cortex are a product of a patterned VZ

(Rakic, 1988) is an example of a model in which lineage as a mechanism plays a major role. At another extreme, distinct areas of the brain could be established independently of lineage, with most of the major patterning occurring as a result of cell–cell interactions. As with most complex systems, the truth undoubtably lies somewhere between these extreme positions, and the relative roles of lineage-based mechanisms and environmental interactions almost certainly vary from region to region.

Lineage analysis has provided support for the role of lineage in shaping some structures. Clones can be found entirely within one functional or anatomical domain, as in the early stages of development of the hindbrain (Fraser et al., 1990) and diencephalon (Figdor and Stern, 1993). Cells arising within a particular rhombomere appear to stay in the dorsal derivatives of that rhombomere late in development, but this is not the case within the ventral derivatives of that rhombomere later in development (Tan and Le Douarin, 1991; Marin and Puelles, 1995). In areas such as these where the majority of clones respect boundaries, and where the clones are large enough and situated such that they have ample opportunity to spread across boundaries, the role of lineage as a mechanism is supported. However, in many other regions of the CNS and at most stages examined, sibling cells in a significant fraction of clones have been found to cross morphologically and functionally distinct areas of the CNS. For example, clonally related cells in the cerebral cortex can occupy functionally distinct domains located great distances from each other (Walsh and Cepko, 1992; Reid et al., 1995). In these areas, it is clear that the simplest models of how lineage mechanisms contribute to patterning cannot pertain to all clones, although lineage-based information may still contribute to cell fate determination in ways that are more subtle than we presently understand.

Lineage studies reveal that neurons and glia behave differently in different parts of the CNS with respect to areal boundaries. In the cerebral cortex, lineally related neurons spread over wide distances, whereas glia tend to be more clustered (Levison et al., 1993). In contrast, glia in the spinal cord white matter disperse once greater distances than neurons in the adjoining gray matter (Leber and Sanes, 1995). In this scenario, different subdivisions of the spinal cord may provide environments more or less favorable to migration.

It may be important to note that when boundaries to clonal dispersion are seen in the developing CNS, they are almost always morphological as well as transient in nature. As mentioned above, in the diencephalon, clonally related cells are restricted to neuromeres at early stages (Figdor and Stern, 1993). However, when analyzed at later stages, clones spread across boundaries (Golden and Cepko, 1996). In the hindbrain as well, the rhombomeres are transient (Vaage, 1969), and the spread of cells across the boundaries occurs in the ventral derivatives of the rhombomeres (Tan and Le Douarin, 1991; Marin and Puelles, 1995).

Even in areas where clones appear to respect boundaries early, there are exceptions. When a large number of clones in the hindbrain were examined, 8% were found to cross rhombomeric boundaries (Birgbauer and Fraser, 1994), and a similar fraction (at least) was found to cross later in the diencephalon (Golden and Cepko, 1996). What does it mean for a small percentage of clones to cross boundaries? Are these "pioneer" clones which serve to establish some subsequent pattern through cell–cell interactions? In the hindbrain, the clones that crossed rhombomeric boundaries did not seem to have any particular type of cell and did not occur in a specific rhombomere or dorsal/ventral location, suggesting that they crossed as a result of a stochastic process. However, it may be significant that the cells that crossed did not do so while in the VZ, but crossed later, after migrating out of the VZ. Similarly, clones that crossed in the diencephalon did so after leaving the VZ, and were otherwise unremarkable, with the exception of a small subset of bilaterally symmetrical clones in the hypothalamus (Arnold-Aldea and Cepko, 1996). Perhaps if the molecular nature of the boundaries was known, it would be possible to assess how clones cross. In some cases the boundaries appear to be physical barriers, such as the axon bundles which subdivide the developing diencephalon. In other cases they may be caused by the differential expression of adhesion or extracellular matrix molecules (Guthrie and Lumsden, 1991), which reflect differential expression of transcription factors (Steindler et al., 1989; Rubenstein and Puelles, 1994; Puelles and Rubenstein, 1993).

It is tempting to speculate that the strategy of patterning the early structures and passing that information on through lineage was retained for some of the more conserved CNS structures. It is possible that early patterning applies only to the cells within the VZ, and is "lost" or overridden by other cues as the cells leave the VZ. The complexity of the brain, as it has evolved, may have necessitated more complex developmental strategies. It is likely that several different types of mechanisms, some lineage-based and others not, are superimposed in different parts of the brain at different times.

The lessons learned from lineage studies to date have provided a wealth of data complementary to that gathered by other approaches. The dissimilarities found in different areas of the CNS point to the need for further studies to address detailed questions concerning lineal relationships. Because some general questions have been answered in some parts of the brain, some may view lineage studies as no longer useful. However, it is important to emphasize that because of its great complexity, probably more remains to be learned of lineal relationships in the CNS than has been discovered so far.

References

Alexiades, M.R., Cepko, C.L. (1996). Quantitative analysis of proliferation and cell cycle length during development of the rat retina. *Dev. Dyn.* 205:293–307.

Altman, J., Bayer, S.A. (1978). Development of the diencephalon in the rat. I. Autoradiographic study of the time of origin and settling patterns of neurons of the hypothalamus. *J. Comp. Neurol.* 182(4 Pt 2):945–971.

Altman, J., Bayer, S.A. (1990). Mosaic organization of the hippocampal neuroepithelium and the multiple germinal sources of dentate granule cells. *J. Comp. Neurol.* 301:325–342.

Angevine, J.B. (1970). Time of neuron origin in the diencephalon of the mouse: an autoradiographic study. *J. Comp. Neurol.* 139:129–188.

Arnold-Aldea, S., Cepko, C. (1996). Dispersion patterns of clonally related cells during development of the hypothalamus. *Dev. Biol.* 173:148–161.

Austin, C.P., Cepko, C.L. (1990). Cellular migration patterns in the developing mouse cerebral cortex. *Development* 110:713–732.

Bergquist, H. (1952). Studies on the cerebral tube in vertebrates. The neuromeres. *Acta Zool.* 33:117–187.

Birgbauer, E., Fraser, S. (1994). Violation of cell lineage restriction compartments in the chick hindbrain. *Development* 120:1347–1356.

Blaschke, A.J., Staley, K., Chun, J. (1996). Widespread programmed cell death in proliferative and postmitotic regions of the fetal cerebral cortex. *Development* 122:1165–1174.

Buenzow, D.E., Holmgren, R. (1995). Expression of the Drosophila gooseberry locus defines a subset of neuroblast lineages in the central nervous system. *Dev. Biol.* 170(2): 338–349.

Bulfone, A., Kim, H.J., Puelles, L., Porteus, M.H., Grippo, J.F., Rubenstein, J.L. (1993a). The mouse *Dlx-2* (*Tes-1*) gene is expressed in spatially restricted domains of the forebrain, face and limbs in midgestation mouse embryos. *Mech. Dev.* 40:129–140.

Bulfone, A., Puelles, L., Porteus, M.H., Frohman, M.A., Martin, G.R., Rubenstein, J.L.R. (1993b). Spatially restricted expression of *Dlx-1, Dlx-2 (Tes-1), gbx-2,* and *wnt-3* in the embryonic day 12.5 mouse forebrain defines potential transverse and longitudinal segmentation boundaries. *J. Neurosci.* 13:3156–3172.

Cameron-Curry, P., Le Douarin, N.M. (1995). Oligodendrocyte precursors originate from both the dorsal and the ventral parts of the spinal cord. *Neuron* 15:1299–1310.

Carpenter, E.M., Hollyday, M. (1992). The location and distribution of neural crest-derived Schwann cells in developing peripheral nerves in the chick forelimb. *Dev. Biol.* 150:144–159.

Cepko, C. (1988). Retrovirus vectors and their applications in neurobiology. *Neuron* 1(5):345–353.

Cepko, C.L., Austin, C.P., Yang, X., Alexiades, M., Ezzeddine, D. (1996). Cell fate determination in the vertebrate retina. *Proc. Natl. Acad. Sci. U.S.A.* 93:589–595.

Crandall, J.E., Herrup, K. (1990). Patterns of cell lineage in the cerebral cortex reveal evidence for developmental boundaries. *Exp. Neurol.* 109:131–139.

Crews, S.T., Thomas, J.B., Goodman, C.S. (1988). The Drosophila *single-minded* gene encodes a nuclear protein with sequence similarity to the *per* gene product. *Cell* 52:143–151.

Crews, S., Franks, R., Hu, S., Matthews, B., Nambu, J. (1992). Drosophila *single-minded* gene and the molecular genetics of CNS midline development. *J. Exp. Zool.* 261:234–244.

Dowling, J.E. (1987). *The retina—an approachable part of the brain.* Cambridge, MA: Harvard University Press.

Fekete, D., Perez-Miguelsanz, J., Ryder, E., Cepko, C.L. (1994). Clonal analysis in the chicken retina reveals tangential dispersion of clonally related cells. *Dev. Biol.* 166: 666–682.

Figdor, M., Stern, C. (1993). Segmental organization of embryonic diencephalon. *Nature* 363:630–634.

Fishell, G., Mason, C.A., Hatten, M. (1993). Dispersion of neural progenitors within the germinal zones of the forebrain. *Nature* 362:636–638.

Fraser, S., Keynes, R., Lumsden, A. (1990). Segmentation in the chick hindbrain is defined by cell lineage restrictions. *Nature* 344:431–435.

Gadisseux, J.F., Evrard, P., Misson, J.P., Caviness, V.S. (1989). Dynamic structure of the radial glial fiber system of the developing murine cerebral wall. An immunocytochemical analysis. *Brain Res. Dev. Brain Res.* 50(1):55–67.

Galileo, D., Gray, G., Owens, G.C., Majors, J., Sanes, J.R. (1990). Neurons and glia arise from a common progenitor in chicken optic tectum: demonstration with two retroviruses and cell type-specific antibodies. *Proc. Natl. Acad. Sci. U.S.A.* 87(1):458–462.

Gao, W.-Q., Hatten, M. (1994). Immortalizing oncogenes subvert the establishment of granule cell identity in developing cerebellum. *Development* 120:1059–1070.

Garcia-Bellido, A., Ripoll, P., Morata, G. (1973). Developmental compartmentalization of the wing disk of Drosophila. *Nature* 241:251–253.

Golden, J.A., Cepko, C.L. (1996). Clones in the chick diencephalon contain multiple cell types and siblings are widely dispersed. *Development* 122:65–78.

Golden, J., Fields-Berry, S., Cepko, C.L. (1995). Construction and characterization of a highly complex retroviral library for lineage analysis. *PNAS* 92:5704–5708.

Goldowitz, D. (1987). Cell partitioning and mixing in the formation of the CNS: analysis of the cortical somatosensory barrels in chimeric mice. *Dev. Brain Res* 35:1–9.

Gray, G., Sanes, J. (1991). Migratory paths and phenotypic choices of clonally related cells in the avian optic tectum. *Neuron* 6:211–225.

Gray, G., Sanes, J. (1992). Lineage of radial glia in the chicken optic tectum. *Development* 114:271–283.

Gray, G., Glover, J., Majors, J., Sanes, J.R. (1988). Radial arrangement of clonally related cells in the chicken optic tectum: lineage analysis with a recombinant retrovirus. *PNAS* 85:7356–7360.

Grove, E.A., Kirkwood, T.B.L., Price, J. (1992). Neuronal precursor cells in the rat hippocampal formation contribute to more than one cytoarchitectonic area. *Neuron* 8:217–229.

Grove, E. A., Williams, B.P., Li, D.-Q., Hajihosseini, M., Friedrich, A., Price J. (1993). Multiple restricted lineages in the embryonic rat cerebral cortex. *Development* 117:553–561.

Guthrie, S. (1995). The status of the neural segment. *Trends Neurosci.* 18:74–79.

Guthrie, S., Lumsden, A. (1991). Formation and regeneration of rhombomere boundaries in the developing chick hindbrain. *Development* 112:221–229.

Guthrie, S., Butcher, M., Lumsden, A. (1991). Patterns of cell division and interkinetic nuclear migration in the chick embryo hindbrain. *J. Neurobiol.* 22:742–754.

Guthrie, S., Muchamore, I., Kuroiwa, A., Marshall, H., Krumlauf, R., Lumsden, A. (1992). Neuroectodermal autonomy of Hox-2.9 expression revealed by rhombomere transpositions. *Nature* 356:157–159.

Halliday, A.L., Cepko, C.L. (1992). Generation and migration of cells in the developing striatum. *Neuron* 9(1):15–26.

Hallonet, M., Teillet, M.-A., LeDouarin, N. (1990). A new approach to the development of the cerebellum provided by the quail-chick marker system. *Development* 108:19–31.

Hamburger, V., Hamilton, H. (1951). A series of normal stages in the development of the chick embryo. *J. Morphol.* 88:49–91.

Hanaway, J. (1967). Formation and differentiation of the external granular layer of the cerebellum. *J. Comp. Neurol.* 131:1–14.

Hatta, K., Kimmel, C.B., Ho, R.K., Walker, C. (1991). The *cyclops* mutation blocks specification of the floor plate of the zebrafish central nervous system. *Nature* 350(6316):339–341.

Hatten, M. (1990). Riding the glial monorail: a common mechanism for glial-guided neuronal migration in different regions of the developing brain. *Trends Neurosci.* 13:179–184.

Hatten, M., Mason, C. (1990). Mechanisms of glial-guided neuronal migration in vitro and in vivo. *Experientia* 46:907–916.

Hemond, S., Glover, J. (1993). Clonal patterns of cell proliferation, migration, and dispersal in the brainstem of the chicken embryo. *J. Neurosci.* 13:1387–1402.

Herrick, C.J. (1910). The morphology of the forebrain in Amphibia and Reptilia. *J. Comp. Neurol.* 20:413–547.

Hinds, J.W. (1988a). Autoradiographic study of histogenesis in the mouse olfactory bulb. I. Time of origin of neurons and neuroglia. *J. Comp. Neurol.* 134:287–304.

Hinds, J.W. (1988b). Autoradiographic study of histogenesis in the mouse olfactory bulb: II. Cell proliferation and migration. *J. Comp. Neurol.* 134:305–322.

His, W. (1893). Vorschlage zur einteilung des gehirms. *Arch. Anat. Entwicklungsgesch* 17: 172–179.

Holt, C.E., Bertsch, T.W., Ellis, H.M., Harris, W.A. (1988). Cellular determination in the Xenopus retina is independent of lineage and birth date. *Neuron* 1(1):15–26.

Keyser, A. (1972). The development of the diencephalon of the Chinese hamster. *Acta Morphol. Neerlando-Scand.* 9(4):379.

Kuhlenbeck, H. (1973). *Overall Morphologic Pattern. The Central Nervous System of Vertebrates.* Basel: Karger, Vol. 3, pt 2.

Kornack, D.R., Rakic, P. (1995). Radial and horizontal deployment of clonally related cells in the primate neocortex: relationship to distinct mitotic lineages. *Neuron* 15:311–321.

LaVail, J., Cowan, W. (1971a). The development of the chick optic tectum. I. Normal morphology and cytoarchitectonic development. *Brain Res.* 28:391–419.

LaVail, S., Cowan, W. (1971b). The development of the chick optic tectum. II. Autoradiographic studies. *Brain Res.* 28:421–441.

Le Douarin, N. (1973). A biological cell labeling technique and its use in experimental embryology. *Dev. Biol.* 30:217–222.

Le Douarin, N. (1982). *The Neural Crest.* Cambridge: Cambridge University Press.

Leber, S., Sanes, J. (1995). Migratory paths of neurons and glia in the embryonic chick spinal cord. *J. Neurosci.* 15:1236–1248.

Leber, S., Breedlove, S., Sanes, J. (1990). Lineage, arrangement, and death of clonally related motoneurons in the chick spinal cord. *J. Neurosci.* 10:2451–2462.

Levison, S.W., Chuang, C., Abramson, B.J., Goldman, J.E. (1993). The migrational patterns and developmental fates of glial precursors in the rat subventricular zone are temporally regulated. *Development* 119:611–622.

Lois, C., Garcia-Verdugo, J.M., Alvarez-Buylla, A. (1996). Chain migration of neuronal precursors. *Science* 271:978–981.

Lubetzki, C., Goujet-Zalc, C., Demerens, C., Danos, O., and Zalc, B. (1992). Clonal segregation of oligodendrocytes and astrocytes during in vitro differentiation of glial progenitor cells. *Glia* 6:289–300.

Lumsden, A. (1990). The cellular basis of segmentation in the developing hindbrain. *Trends Neurosci.* 13:329–335.

Lumsden, A., Keynes, R. (1989). Segmental patterns of neuronal development in the chick hindbrain. *Nature* 337:424–428.

Lumsden, A., Sprawson, N., Graham, A. (1991). Segmental origin and migration of neural crest cells in the hindbrain region of the chick embryos. *Development* 113:1281–1291.

Lumsden, A., Clarke, J., Keynes, R., Fraser, S. (1994). Early phenotypic choices by neuronal precursors, revealed by clonal analysis of the chick embryo hindbrain. *Development* 120:1581–1589.

Luskin, M.B., Pearlman, A.L., Sanes, J.R. (1988). Cell lineage in the cerebral cortex of the mouse studied in vivo and in vitro with a recombinant retrovirus. *Neuron,* 1:635–647.

Luskin, M.B. (1993). Restricted proliferation and migration of postnatally generated neurons derived from the forebrain subventricular zone. *Neuron* 11:173–189.

Luskin, M.B., Parnavelas, J.G., Barfield, J.A. (1993). Neurons, astrocytes, and oligodendrocytes of the rat cerebral cortex originate from separate progenitor cells: an ultrastructural analysis of clonally related cells. *J. Neurosci.* 13:1730–1750.

Lutz, B., Rugarli, E., Eichele, G., Ballabio, A. (1993). X-linked Kallmann syndrome. A neuronal targeting defect in the olfactory system? FEBS *Lett.* 325:128–134.

Marin, F., Puelles, L. (1995). Morphological fate of rhombomeres in quail/chick chimeras: a segmental analysis of hindbrain nuclei. *Eur. J. Neurosci.* 7:1714–1738.

Martinez, S., Alvarado-Mallart, R. (1989). Rostral cerebellum originates from the caudal portion of the so-called "mesencephalic vesicle": a study using chick/quail chimeras. *Eur. J. Neurosci.* 1:549–560.

McConnell, S.K. (1988). Fates of visual cortical neurons in the ferret after isochronic and heterochronic transplantation. *J. Neurosci.* 8:945–974.

McConnell, S.K., Kaznowski, C.E. (1991). Cell cycle dependence of laminar determination in developing neocortex. *Science* 254:282–285.

Mello, C.C., Draper, B.W., Priess, J.R. (1994). The maternal genes *apx-1* and *glp-1* and establishment of dorsal-ventral polarity in the early C. elegans embryo. *Cell* 77:95–106.

Menezes, J.R.L., Luskin, M.B. (1994). Expression of neuron-specific tubulin defines a novel population in the proliferative layers of the developing telencephalon. *J. Neurosci.* 14:5399–5416.

Miale, I., Sidman, R. (1961). An autoradiographic analysis of histogenesis in the mouse cerebellum. *Exp. Neurol.* 4:277–296.

Mione, M.C., Danevic, C., Boardman, P., Harris, B., Parnavelas, J.G. (1994). Lineage analysis reveals neurotransmitter (GABA or glutamate) but not calcium-binding protein homogeneity in clonally related cortical neurons. *J. Neurosci.* 14(1):107–123.

Mione, M.C., Douratsos, L., Parnavelas, J.G. (1995). Birth order and phenotype in lineages of cortical cells. *Soc. Neurosci. Abstract* 21:528.

Misson, J.P., Austin, C.P., Takahashi, T., Cepko, C.L., Caviness, V.S. Jr. (1991). The alignment of migrating neural cells in relation to the murine neopallial radial glial fiber system. *Cerebral Cortex,* 1(3):221–229.

Miyake, T., Fujiwara, T., Fukunaga, T., Takemura, K., Kitamura, T. (1993). Allocation of mouse cerebellar granule cells derived from embryonic ventricular progenitors—a study using a recombinant retrovirus. *Dev. Brain Res.* 74:245–252.

Miyake, T., Fujiwara, T., Fukunaga, T., Takemura, K., Kitamura, T. (1995). Glial cell lineage in vivo in the mouse cerebellum. *Dev. Growth Diff.* 37:273–285.

Nambu, J.R., Franks, R.G., Wharton, K.A., Crews, S.T. (1990). The *single-minded* gene of *Drosophila* is required for the expression of genes important for the development of CNS midline cells. *Cell* 63:63–75.

Nambu, J.R., Lewis, J.O., Wharton, K.A., Crews, S.T. (1991). The Drosophila *single-minded* gene encodes a helix-loop-helix protein that acts as a master regulator of CNS midline development. *Cell* 67:1157–1167.

Noll, E., Miller, R.H. (1993). Oligodendrocyte precursors originate at the ventral ventricular zone dorsal to the ventral midline region in the embryonic rat spinal cord. *Development* 118:563–573.

O'Rourke, N.A., Dailey, M.E., Smith, S.J., McConnell, S.K. (1992). Diverse migratory pathways in the developing cerebral cortex. *Science* 258:299–302.

O'Rourke, N.A., Sullivan, D.P., Kaznowski, C.E., Jacobs, A.A., McConnell, S.K. (1995). Tangential migration of neurons in the developing cerebral cortex. *Development* 121: 2165–2176.

Orr, H. (1887). Contributions to the embryology of the lizard. *J. Morphol.* 1:311–372.

Otero, R., Sotelo, C., Alvarado-Mallart, R.-M. (1993). Chick/quail chimeras with partial cerebellar grafts: an analysis of the origin and migration of cerebellar cells. *J. Comp. Neurol.* 333:597–615.

Parnavelas, K.G., Barfield, J.A., Franke, E., Luskin, M.B. (1991). Separate progenitor cells give rise to pyramidal and nonpyramidal neurons in the rat telencephalon. *Cereb. Cortex* 1:1047–3211.

Price, J., Thurlow, L. (1988). Cell lineage in the rat cerebral cortex: a study using retroviral-mediated gene transfer. *Development* 104(3):473–482.

Pringle, N.P., Richardson, W.D. (1993). A singularity of PDGFa-receptor expression in the dorsoventral axis of the neural tube may define the origin of the oligodendrocyte lineage. *Development* 117:525–533.

Puelles, L., Rubenstein, J. (1993). Expression patterns of homeobox and other putative regulatory genes in the embryonic mouse forebrain suggest a neuromeric organization. *Trends Neurosci* 16(11):472–479.

Puelles, L., Amat, J., Martinez-de-la-Torre, M. (1987). Segment-related mosaic neurogenetic pattern in the forebrain and mesencephalon of early chick embryos: I. Topography of AChE-positive neuroblasts up to stage HH18. *J. Comp. Neurol.* 266:247–268.

Rakic, P. (1971). Neuron-glia relationship during granule cell migration in the developing cerebellar cortex. A Golgi and electron-microscopic study in Macacus rhesus. *J. Comp. Neurol.* 141:283–312.

Rakic, P. (1988). Specification of cerebral cortical areas. *Science* 241:170–176.

Rakic, P., Sidman, R. (1969). Telencephalic origin of pulvinar neurons in the fetal human brain. *Z. Anat. Entwicklungsgeschichte* 129:53–82.

Ramon y Cajal, S. (1929). *Etudes sur la Neurogenesis de Quelques Vertebres,* Guth, L., trans., 1960. Springfield, IL: C. C. Thomas.

Reese, B.E., Harvey, A.R., Tan, S.S. (1995). Radial and tangential dispersion patterns in the mouse retina are cell-class specific. *Proc. Natl. Acad. Sci. U.S.A.* 92:2494–2498.

Reid, C.B., Liang, I., Walsh, C. (1995). Systematic widespread clonal organization in cerebral cortex. *Neuron* 15:299–310.

Roe, T., Reynolds, T.C., Yu, G., Brown, P.O. (1993). Integration of murine leukemia virus DNA depends on mitosis. *EMBO J.* 12(5):2099–2108.

Rubenstein, J.L., Puelles, L. (1994). Homeobox gene expression during development of the vertebrate brain. *Curr. Top. Dev. Biol.* 29:1–63.

Rubenstein, J.L., Martinez, S., Shimamura, K., Puelles, L. (1994). The embryonic vertebrate forebrain: the prosomeric model. *Science* 266(5185):578–580.

Rugarli, E., Lutz, B., Kuratani, S.C., Wawersik, S., Borsani, G., Ballabio, A., Eichele, G. (1993). Expression pattern of the Kallmann syndrome gene in the olfactory system suggests a role in neuronal targeting. *Nat. Genet.* 4:19–26.

Ryder, E.F., Cepko, C.L. (1994). Migration patterns of clonally related granule cells and their progenitors in the developing chick cerebellum. *Neuron* 12(5):1011–1028.

Schoenwolf, G.C., Bortier, H., Vakaet, L. (1989). Fate mapping of the avian neural plate with quail/chick chimeras: origin of prospective median wedge cells. *J. Exp. Zool.* 249: 271–278.

Sidman, R. L. (1961). Histogenesis of mouse retina studied with thymidine-H3. In *The Structure of the Eye,* Smelser, G.K., ed. Academic Press, New York: pp. 487–505.

Smart, I.H.M. (1976). A pilot study of cell production by the ganglionic eminences of the developing mouse brain. *J. Anat.* 121:71–84.

Steindler, D.A., Cooper, N.G.F., Faissner, A., Schachner, M. (1989). Boundaries defined by adhesion molecules during development of the cerebral cortex: the J1/tenascin glycoprotein in the mouse somatosensory cortical barrel field. *Dev. Biol.* 131:243–260.

Sulston, J.E., Horvitz, H.R. (1977). Post-embryonic cell lineages of the nematode Caenorhabditis elegans. *Dev. Biol.* 56:110–156.

Sulston, J.E., Schierenberg, E., White, J.G., Thomson, J.N. (1983). The embryonic cell lineage of the nematode Caenorhabditis elegans. *Dev. Biol.* 100:64–119.

Tan, K., LeDouarin, N. (1991). Development of the nuclei and cell migration in the medulla oblongata—application of the quail-chick chimera system. *Anat. Embryol.* 183:321–343.

Tan, S.-S., Breen, S. (1993). Radial mosaicism and tangential cell dispersion both contribute to mouse neocortical development. *Nature* 362:638–640.

Tan, S.-S., Faulkner-Jones, B., Breen, J., Walsh, M., Bertram, J.F., Reese, B.E. (1995). Cell dispersion patterns in different cortical regions studied with an x-inactivated transgenic marker. *Development* 121:1029–1039.

Timsit, S., Martinez, S., Allinquant, B., Peyron, F., Puelles, L., Zalc, B. (1995). Oligodendrocytes originate in a restricted zone of the embryonic ventral neural tube defined by DM-20 mRNA expression. *J. Neurosci.* 15:1012–1024.

Turner, D.L., Cepko, C.L. (1987). A common progenitor for neurons and glia persists in rat retina late in development. *Nature* 328(6126):131–136.

Turner, D.L., Snyder, E.Y., Cepko, C.L. (1990). Lineage-independent determination of cell type in the embryonic mouse retina. *Neuron* 4:833–845.

Vaage, S. (1969). The segmentation of the primitive neural tube in chick embryos. *Adv. Anat. Embryol. Cell Biol.* 41:1–87.

Vaysse, P.J., Goldman, J.E. (1990). A clonal analysis of glial lineages in neonatal forebrain development in vitro. *Neuron* 5(3):227–235.

Vicario-Abejon, C., Cunningham, M., McKay, R. (1995). Cerebellar precursors transplanted to the neonatal dentate gyrus express features characteristic of hippocampal neurons. *J. Neurosci.* 15(10):6351–6363.

Walsh, C., Cepko, C.L. (1988). Clonally related cortical cells show several migration patterns. *Science* 241:1342–1345.

Walsh, C., Cepko, C.L. (1992). Widespread dispersion of neuronal clones across functional regions of the cerebral cortex. *Science* 255:434–440.

Walsh, C., Cepko, C.L. (1993). Clonal dispersion in proliferative layers of developing cerebral cortex. *Nature* 362(6421):632–635.

Walsh, C., Cepko, C.L., Ryder, E.F., Church, G.M., Tabin, C. (1992). The dispersion of neuronal clones across the cerebral cortex (letter). *Science* 258:317–320.

Warf, B.C., Fok-Seang, J., Miller, R.H. (1991). Evidence for the ventral origin of oligodendrocyte precursors in the rat spinal cord. *J. Neurosci.* 11:2477–2488.

Weisblat, D.A., Sawyer, R.T., Stent, G.S. (1978). Cell lineage analysis by intracellular injection of a tracer enzyme. *Science* 202:1295–1298.

Wetts, R., Fraser, S.E. (1988). Multipotent precursors can give rise to all major cell types of the frog retina. *Science* 239:1142–1145.

Wetts, R., Serbedzija, G.N., Fraser, S.E. (1989). Cell lineage analysis reveals multipotent precursors in the ciliary margin of the frog retina. *Dev. Biol.* 136:254–263.

Wilkinson, D., Krumlauf, R. (1990). Molecular approaches to the segmentation of the hindbrain. *Trends Neurosci.* 13:335–339.

Wilkinson, D.G., Bhatt, S., Cook, M., Boncinelli, E., Krumlauf, R. (1989a). Segmented expression of *Hox-2* homeobox-containing gene in the developing mouse hindbrain. *Nature* 341:405–409.

Wilkinson, D.G., Bhatt, S., Chavrier, P., Bravo, P., Charnay, P. (1989b). Region specific expression of homeobox genes in the embryonic mesoderm and central nervous system. *Nature* 337:461–464.

Williams B.P., Read, J., Price, J. (1991). The generation of neurons and oligodendrocytes from a common precursor cell. *Neuron* 7:685–693.

Yee, J.K., Friedmann, T., Burns, J.C. (1994). Generation of high-titer pseudotyped retroviral vectors with very broad host range. *Methods Cell Biol.* 43 Pt A:99–112.

Young, R. W. (1985a). Cell differentiation in the retina of the mouse. *Anat. Record.* 212:199–205.

Young, R.W. (1985b). Cell proliferation during postnatal development of the retina in the mouse. *Dev. Brain Res.* 21:229–239.

Yu, W.P., Collarini, E.J., Pringle, N.P., Richardson, W.D. (1994). Embryonic expression of myelin genes: evidence for a focal source of oligodendrocyte precursors in the ventricular zone of the neural tube. *Neuron* 12:1353–1362.

Zerlin, M., Levison, S.W., Goldman, J.E. (1995). Early patterns of migration, morphogenesis, and intermediate filament expression of subventricular zone cells in the postnatal rat forebrain. *J. Neurosci.* 15(11):7238–7249.

Zhang, L., Goldman, J.E. (1996a). Generation of cerebellar interneurons from dividing progenitors in white matter. *Neuron* 16:1–20.

Zhang, L., Goldman, J. (1996b). Developmental fates and migratory pathways of dividing progenitors in the postnatal rat cerebelum. J. Comp. Neurol. 370:536–550.

12

Development of the cerebral cortex: mechanisms controlling cell fate, laminar and areal patterning, and axonal connectivity

ANJEN CHENN, JANET E. BRAISTED, SUSAN K. MCCONNELL AND DENNIS D. M. O'LEARY

The precise architecture of the cerebral cortex enables it to process the complex sensory information that forms the basis for our perceptions, conscious thought, and behavior. The mature cortex is organized radially into six distinct layers that are distinguished by the morphology, density, and physiologic properties of the resident neurons. The neurons of different layers project to and receive input from different sites and are functionally interconnected by a stereotyped network of axons and dendrites. The cortex is also organized in a dimension tangential to its surface into distinct areas that subserve different specialized functions such as sensory processing, motor control, and the association between these processes. In setting up the cerebral cortex, the nervous system is faced with a number of challenges including the specification of cortical tissue, the production of specific types of neurons, the migration of neurons from the proliferative zones to their final positions, their elaboration of axons and dendrites, and the establishment and refinement of synaptic connections. Of more unique significance to the proper development of the cerebral cortex is how these fundamental events are orchestrated during development to generate its characteristic laminar and areal framework.

CORTICAL SPECIFICATION

During development, the rostral end of the neural tube develops three vesicles that ultimately give rise to the structures of the forebrain, midbrain, and hindbrain. The dorsolateral aspects of the forebrain evaginate to generate the telencephalic vesicles, from which the cerebral cortex, in addition to the basal ganglia, amygdala, and olfactory bulb, ultimately arises. Although the molecular mechanisms that control these major processes of regionalization are not well understood, progress has been made in identifying genes that are expressed in region-specific patterns within the developing forebrain. It is clear from studies of invertebrate systems such as *Drosophila* that the expression of individual genes or combinations of genes in developing tissues can play important roles in the patterning of the body plan. Furthermore, many of these genes are remarkably conserved through evolution in both structure and function.

What sort of mechanisms might function in patterning the forebrain? The isolation and characterization of regionally expressed genes have provided useful markers and candidates for genes that may control tissue-or region-specific identity. Recent work in a number of labs has unveiled a wide variety of genes that are expressed in different parts of the telencephalon, often correlating with morphologic anatomic boundaries (reviewed in Shimamura et al., 1995). A number of these genes are homologues of *Drosophila* homeobox genes *(orthodentricle [otd]), empty spiracles [ems])* that are essential for proper brain development in flies, suggesting the possibility that they might serve similar roles in vertebrate forebrain patterning. In *Drosophila, ems* plays important roles in the development of the head, and mutations result in the deletion of anterior head structures (Hirth et al., 1995); *otd* mutants are missing anterior brain regions and have defects in ventral/medial neuronal specification (reviewed in Finkelstein and Boncinelli, 1994; and Thor, 1995). Using *engrailed (en)* expression to define segments of the embryonic fly brain, Hirth et al., (1995) showed that mutation of *ems* caused elimination of the second and third brain segments (b2 and b3), while mutation of *otd* eliminated the first brain segment (b1). The regions of the brain affected reflect the domains of expression of *otd* and *ems; otd* is expressed in b1, while *ems* is expressed in b2 and b3.

The expression patterns of the mouse homeobox genes related to the *Drosophila* genes *otd* and *ems* suggest that they may play important roles in the specification of forebrain-specific structures including the cerebral cortex. The mouse homologues *Emx1, Emx2* and *Otx1, Otx2* have nested expression domains in the rostral brain during development (Simeone et al., 1992b; and reviewed in Boncinelli et al., 1993). *Otx2* has the largest domain of expression, encompassing much of the telencephalon, diencephalon, and mesencephalon. *Otx1* is expressed in a subset of the regions in which *Otx2* is expressed, while *Emx2* is expressed in yet more restricted areas (dorsal telencephalon and some of the diencephalon), and *Emx1* is expressed exclusively in the presumptive cerebral cortex and olfactory bulbs. In mice, the onset of expression of these four genes is sequential, with *Otx2* expressed first at E5.5, *Otx1* and *Emx2* (E8–8.5), and then *Emx1*.

The role of *Otx2* in forebrain development has been examined by generating targeted mutations of *Otx2* in mice. The homozygous *Otx2* $-/-$ mice display loss of anterior neural tissues anterior to rhombomere 3 (Acampora et al., 1995; Matsuo et al., 1995; Ang et al., 1996). Molecular markers of forebrain regions such as *Emx2* and *brain factor 1 (BF-1)* are not expressed, nor are midbrain and hindbrain markers such as *HNF3β, netrin1,* and En protein (Ang et al., 1996). It is interesting that structures posterior to regions that express *Otx2* are also deleted, as *Otx2* expression ends at the midbrain–hindbrain boundary (Simeone et al., 1993). *Otx2* −/−mice fail to develop prechordal mesoderm and exhibited deficits in notochord development (Ang et al., 1996). Thus, it is possible that the rostral brain deficits result from the lack of notochord and prechordal mesoderm; the phenotype of *Lim-1* −/−mice (which also lack prechordal mesoderm) is strikingly similar (Shawlot and Behringer, 1995; Ang et al., 1996). Further experiments are necessary to examine whether *Otx2* acts autonomously in the rostral CNS or whether the loss of anterior structures is a consequence of failed inductions from the mesoderm. Even in *Drosophila,* it is not clear whether the anterior brain defects result from intrinsic requirements for *otd* or whether *otd* regulates inductive interactions with surrounding tissues (Thor, 1995). Nevertheless, the phenotype of *Otx2* mutant mice suggests that the function of these genes in head development is remarkably conserved through evolution.

With regard to cortical development, the expression patterns of the *ems* homologues and *BF-1,* a member of the winged-helix transcription factor family, are particularly compelling (Simeone et al., 1992b; Xuan et al., 1995). In mice *Emx2* expression commences in the dorsal forebrain at nine somites (8.25 days). At 9.5 days, *Emx2* is expressed in the dorsal telencephalon and parts of the diencephalon, while *Emx1* is expressed only in the dorsal telencephalon. The onset of *Emx1* expression at E9.5 corresponds to the time when neurogenesis is just beginning (Luskin et al., 1988). *Emx1* is expressed uniformly in the presumptive cortex and in the olfactory bulb. Furthermore, *Emx1* is not expressed in any noncortical structures. *BF-1* expression is restricted to the developing telencephalon, and mice mutant in *BF-1* exhibit marked reductions in telencephalic size (Xuan et al., 1995). Several other genes such as *Nkx-2.1, Nkx-2.2, Shh,* and the *Dlx* genes also exhibit region-specific expression patterns that appear to carve out the forebrain into distinct regions, and some of these genes may prove to be important regulators of the regional identity (Shimamura, et al., 1995). Although these intriguing domains of gene expression suggest a role for these particular genes in determining regional fate, this role remains to be tested.

NEURONAL PRODUCTION

Neuronal Production: Symmetric and Asymmetric Divisions

The cells of the cerebral cortex are generated in the ventricular zone, a pseudostratified epithelium of progenitor cells that lines the lateral ventricles. During development, these progenitors must regulate the production of both sufficient numbers of progenitors and postmitotic neurons. Before the onset of neurogenesis, progenitors must first divide to expand the progenitor cell population. As neuronal production commences, progenitors then must divide to generate young neurons and more progenitors to replenish the precursor population. As neurogenesis proceeds, more and more divisions give rise to neurons, until the rate of neuronal production peaks during the last third of neurogenesis (Miller, 1985; Caviness et al., 1995). Finally, as neurogenesis comes to a close, the progenitors must divide in a terminal fashion, giving rise to neurons, but no progenitors, as the precursor pool is ultimately depleted. It has long been proposed that postmitotic neurons are generated by the asymmetric division of progenitors and that symmetric divisions expand the precursor pool, but these ideas have been inferred only from indirect evidence.

Thymidine birthdating studies and retroviral lineage-tracing experiments have suggested that the cells of the cerebral cortex arise from a series of asymmetric cell divisions of ventricular progenitors. The layers of the cerebral cortex are established in an inside-first, outside-last fashion (Fig. 12–1). Birthdating studies have shown that the cells that differentiate and leave the cell cycle earliest in development give rise to the deepest layers of the cortex while the cells that exit the cell cycle later in development migrate past these early-born neurons to finally reside in the more superficial layers of the cortex (Angevine and Sidman, 1961; Berry and Rogers, 1965; Rakic, 1974). The dilution of thymidine label in superficial layers of the cortex following treatment of young progenitors with [^{3}H] thymidine is consistent with the possibility that progenitors divide asymmetrically to give rise both to postmitotic neurons and progenitors that go

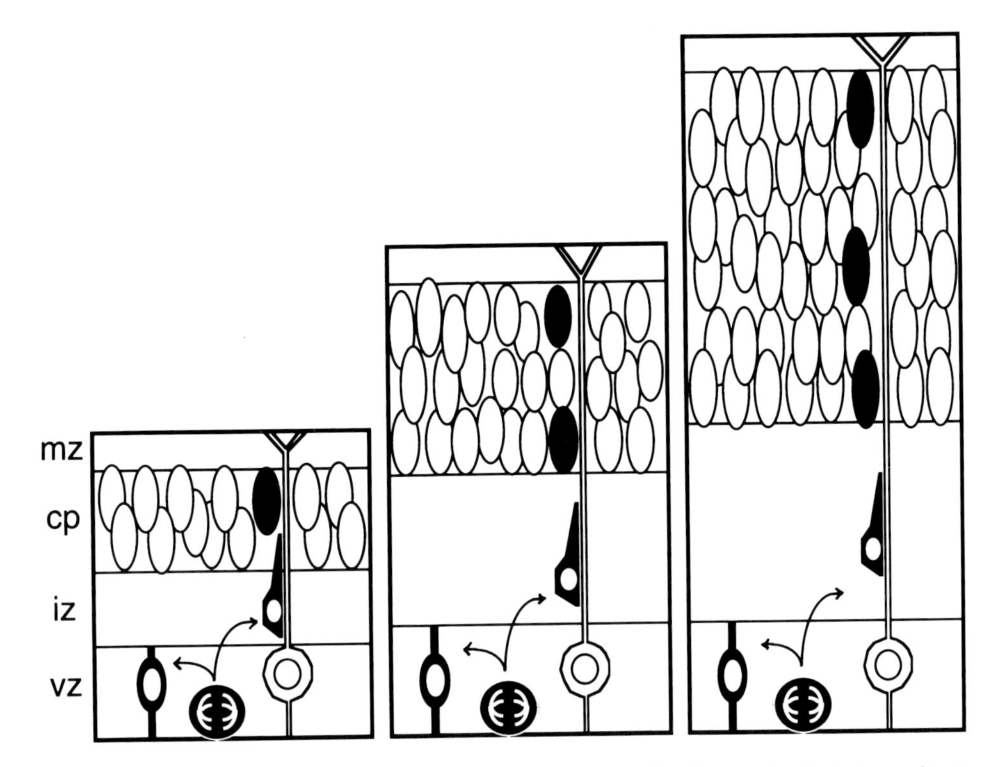

FIGURE 12–1 Cortical neurons are generated in an inside-first, outside-last gradient. Neurons born within the ventricular zone (vz) migrate to populate the deepest layers of the cortical plate (cp). Subsequently generated neurons migrate past the older cells to form the more superficial cortical layers. Cell lineage-tracing experiments indicate that clonally related neurons (black cells) can span several cortical layers, suggesting that precursors can divide asymmetrically within the ventricular zone. (Figure reproduced from McConnell, 1992, with permission from the Society for Neuroscience.)

on to divide again (Rakic, 1974). Progenitor cells marked with a retroviral lineage tracer are able to give rise to clones of progeny that spanned multiple layers of the cortex (Luskin et al., 1988; Price and Thurlow, 1988; Reid et al., 1995). Because the different layers of the cortex are generated at different times during development, the extent of these clones suggests that progenitors may divide asymmetrically to generate a daughter destined for a deep cortical layer and another progenitor that divides again later to give rise to a superficial cortical neuron.

Time-lapse imaging studies in developing ferret cortical slices provide direct evidence that progenitor cells can divide asymmetrically and that cleavage orientation of dividing precursors predicts the fate of the daughters (Chenn and McConnell, 1995). Vertically oriented divisions (cleavage planes perpendicular to the ventricular surface) generate two morphologically and behaviorally identical daughters (Fig. 12–2). Before the onset of neurogenesis, when the progenitor cell population is expanding, nearly all divisions are vertical. These symmetric divisions are thought to be proliferative in nature early in cortical development, and such divisions would serve to expand or maintain the progenitor cell population. Horizontally oriented divisions, in contrast, give rise to basal daughters that behave like young migratory neurons and to apical daughters that remain within the ventricular zone (Fig. 12–2). The basal daughters lose their attachments to the ventricular surface and migrate from the ventricular zone at rates approximating that of migrating neurons. Horizontal divisions are

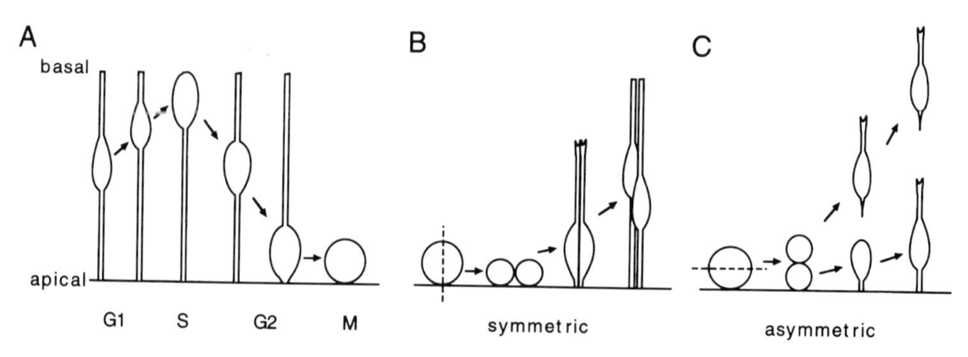

FIGURE 12–2 Interkinetic nuclear migration and models for symmetric and asymmetric division of neuronal precursors. (A) Nuclei of ventricular precursors undergo intracellular migration during the cell cycle. During G1, diploid nuclei rise from the apical ventricular surface. S-phase nuclei reside in the outer (basal) third of the ventricular zone. G2 nuclei migrate apically, and mitosis occurs when the nuclei are located at the ventricular surface. (B) A vertical cleavage (perpendicular to the ventricular surface) generates two daughters that sit side by side, both retaining apical connections. Following mitosis, the nuclei of both daughters move slowly away together from the ventricular lumen and may reenter the cell cycle. (C) A horizontal cleavage (parallel) produces an apical daughter (bottom line of arrows) that retains contact with the apical surface and a basal daughter that loses its apical contact. The apical daughter may be constrained to remain within the ventricular zone, while the basal daughter exhibits behaviors typical of young postmitotic neurons and migrates away. (Reproduced with permission from Chenn and McConnell, 1995, and Cell Press).

seen only rarely prior to the onset of neurogenesis, and their frequency increases with the increasing production of neurons during development. Although the definitive identification of daughters as neurons or progenitors still awaits further study, these imaging studies provide the first direct evidence that ventricular precursors can divide asymmetrically to produce two distinct daughters that behave differently.

How two daughters arising from a common progenitor can adopt different ultimate fates has been a fundamental question in developmental biology. One means by which asymmetry can be established is intrinsic: The two daughters arising from an asymmetric division may differ from each other from the outset by inheriting distinct factors. The asymmetric partitioning of determinants to two daughter cells generates many of the asymmetric divisions of the nematode *Caenorhabditis elegans.* Candidate determinants in *Drosophila* are the prospero and numb proteins, which are asymmetrically localized in neuronal precursors and are required for the correct establishment of cell identity during neurogenesis (reviewed in Doe and Spana, 1995).

During the development of the *Drosophila* nervous system, the prospero and numb proteins are distributed asymmetrically in dividing CNS and PNS neuronal precursors. In the CNS, neuroblasts divide asymmetrically in a stem cell fashion to generate a basal ganglion mother cell (GMC) and another apical neuroblast. The GMCs then divide symmetrically to generate a pair of neurons or glia. Prospero is a homeodomain transcription factor that is asymmetrically localized in the dividing neuroblast and segregated exclusively to the GMC, where it is translocated to the GMC nucleus (Spana and Doe, 1995). Similarly, numb is a novel membrane-associated protein that is asymmetrically localized to the basal cortex of the dividing neuroblast and inherited specifically by the GMC (Spana et al., 1995). Prospero and numb are also localized asymmetrically in peripheral nervous system sensory organ precursors (SOPs), and they are both asymmetrically inherited following cytokinesis (Rhyu et al., 1994). Prospero and numb are required for asymmetric cell division. In the absence of prospero, the GMC fails to develop normally; similarly, loss of numb results in the loss of asymmetry of SOP divisions, causing the daughters of the SOP to differentiate into non-neuronal support cells.

In mammals the first evidence that similar means of generating asymmetry by the unequal distribution of proteins during mitosis was seen in dividing cerebral cortical precursors. In the ferret ventricular zone, Notch 1 immunoreactivity localizes asymmetrically at the basal pole of rounded mitotic precursors (Chenn and McConnell, 1995). Notch1 immunoreactivity is undetectable during interphase but then becomes highly concentrated in a basal cap as the cell enters mitosis (Fig. 12–3). Unlike the distribution of prospero and numb in *Drosophila,* the distribution of Notch1 in a basal crescent on mitotic cells appears to be independent of the orientation of division. Following a vertical cleavage, Notch1 appears to be equally partitioned to the two daughters. In contrast, Notch1 can be asymmetrically inherited following a horizontal cleavage, generating two molecularly distinct daughters. Thus, the amount of Notch1 inherited by the daughters depends on the orientation of cleavage.

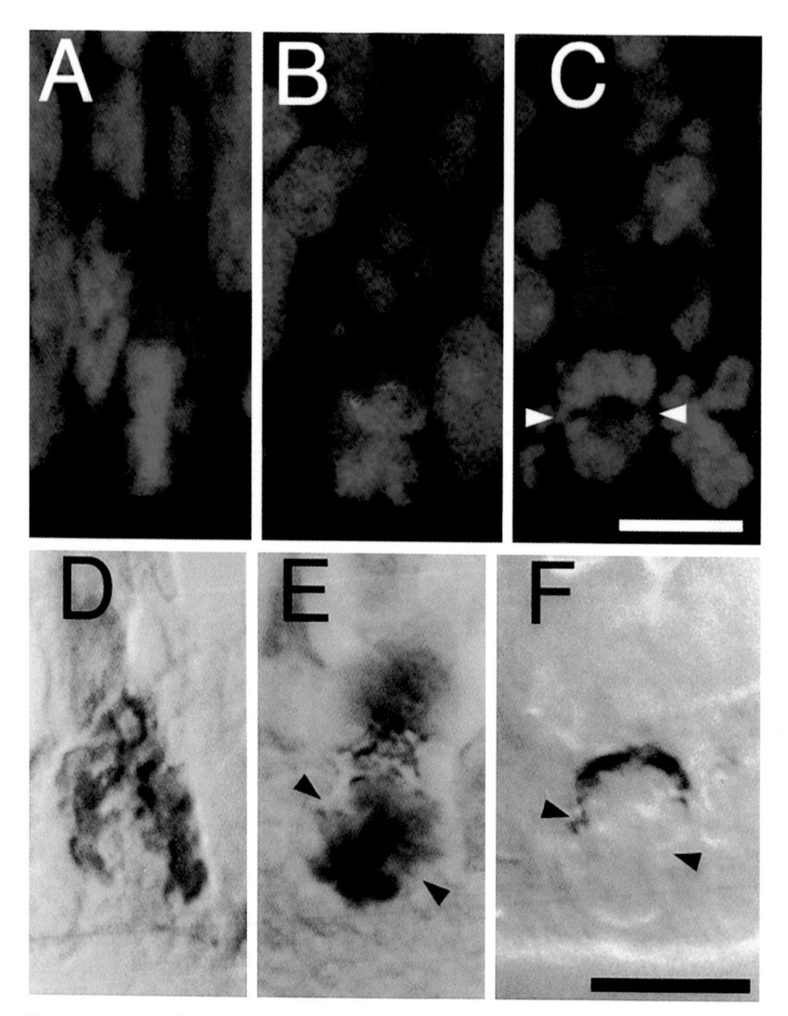

Figure 12–3 Notch1 immunoreactivity is localized asymmetrically to the basal portion of mitotic cells, regardless of their cleavage orientation, in the ferret ventricular zone. (A) Notch1 immunoreactivity (pseudocolored red in A–C) is concentrated at the basal pole of a metaphase cell. The metaphase plate, revealed by propidium iodide staining (green in A–C), is oriented vertically. (B) A crescent of Notch1 immunoreactivity localized basally in a rounded mitotic cell. (C) A basal cap of Notch1 staining in an anaphase cell undergoing a horizontal division (arrowheads mark the predicted cleavage plane). (D) A cell in late anaphase (chromatin revealed by the toluidine blue staining in D–F) localizes Notch1 staining (brown diaminobenzidine [DAB] reaction product in D–F) to its basal pole. Vertical cleavages are likely to distribute Notch1 equally to both daughters. (E) Cells with intermediate cleavage orientations also localize Notch1 asymmetrically. Arrowheads mark the predicted cleavage orientation of this cell in early anaphase. (F) A ventricular cell in early telophase undergoing a horizontal division (arrowheads mark the cleavage furrow). Notch1 staining is associated exclusively with the basal daughter, suggesting that Notch1 is inherited asymmetrically in horizontal divisions. Scale bar in C = 10 μm for A–C; scale bar in F = 10 μm for D–F. (Reproduced with permission from Chenn and McConnell, 1995, and Cell Press).

The possible role of Notch1 in the asymmetric horizontal divisions in the ventricular zone is somewhat puzzling. The expression of Notch in a number of systems appears to correlate with the suppression of neuronal differentiation, whereas in the mammalian ventricular zone, Notch1 appears to be segregated to the basal, or neuronal, daughter following an asymmetric division. The Notch gene of *Drosophila* encodes a large transmembrane protein that functions in many types of local cell interactions. Molecular and genetic evidence suggests that Notch activation by Delta is involved in the decision to produce a neuron or a non-neuronal epidermal cell (reviewed in Artavanis-Tsakonas and Simpson, 1991; Muskavitch, 1994). Loss of Notch or Delta function causes an overproduction of neurons, while the expression of an activated form of Notch results in an overproduction of epithelia at the expense of neurons. Signaling by Delta-Notch homologues in a number of other organisms is also associated with the suppression of the neural fate. Ectopic activity of a *Xenopus* Delta homologue in the embryonic nervous system results in reduced neuronal production, while expression of a truncated nonsignaling Delta caused an overproduction of neurons (Chitnis et al., 1995). Thus, it is surprising that Notch1 protein would segregate to the neuronal daughter of an asymmetric division.

However, Notch also plays a more general role in many developmental processes (reviewed in Artavanis et al., 1995). Notch signaling is involved in a wide variety of cell-fate specification events during *Drosophila* development in addition to neuronal specification, and it is thought that Notch activation may render a cell oblivious to external signals (Artavanis-Tsakonis et al., 1995). Notch activation in the development of the *Drosophila* compound eye causes the R3 and R4 precursors to ignore their normal inductive signals, and differentiate instead into R7 photoreceptors (Fortini et al., 1993). Transient expression of Notch activity appears to interfere with differentiation only temporarily, and once the activity is gone, the cells are able to respond to other cues. It is conceivable that Notch1 activation in the basal daughter of cortical progenitors enables it to ignore other signals within the ventricular zone, including signals to reenter the cell cycle, until the cell has escaped from the ventricular zone.

Recent data have provided evidence that the numb protein interferes with Notch signaling. Genetic evidence in *Drosophila* suggests that *numb* antagonizes Notch signaling, as *numb* mutants have the opposite phenotype as *Notch* mutants (reviewed in Jan and Jan, 1995, see Anderson and Jan chapter 2). In tissue culture cells, numb can antagonize Notch activation, and numb has been shown to physically interact with Notch in these cells (Guo et al., 1996; Spana and Doe, 1996). Recently, a mammalian *numb* homologue has been cloned and characterized. In ventricular cells, numb immunoreactivity is localized in crescents at the apical pole of mitotic precursors, the opposite side of the precursor as Notch1 (Zhong et al., 1996). Like Notch1, numb localization appears to be independent of cleavage orientation.

The intriguing findings that numb interferes with Notch signaling and is located at the opposite pole of dividing precursors provide an attractive hypothesis for how these two proteins might serve to generate asymmetry during mammalian neurogenesis (Fig. 12–4). Numb may antagonize Notch activity by direct interaction during the cell cycle; during mitosis, Notch and numb are segregated to the opposite poles of the cell, where, following a horizontal division, they may be differentially inherited. The basal daughter receives the bulk of the Notch, but none of the numb, and thus Notch activity may go unopposed. In contrast, following a symmetric vertical cleavage, both daughters receive both Notch and numb. Soon after mitosis, in early telophase, numb becomes diffusely localized in the cell cytoplasm (Zhong et al., 1996). This rapid redistribution of numb may allow it to interact quickly with Notch to inhibit Notch function following a symmetric cleavage. These hypotheses about numb and Notch function in mammals await direct experimentation.

Dividing progenitors may utilize the asymmetric localization of proteins in a variety of ways. In *Drosophila* the localization of both numb and prospero are tightly correlated with the direction of cleavage; neuronal progenitors divide so

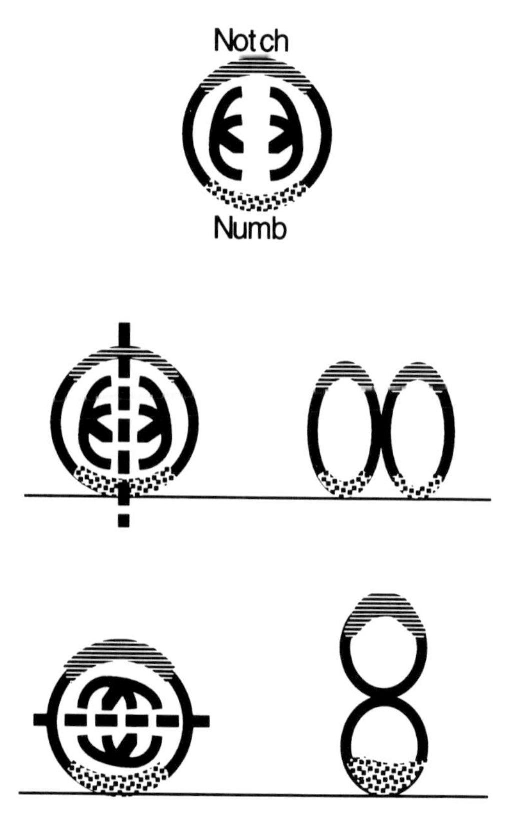

FIGURE 12–4 A model for the inheritance of Notch1 and numb. Notch1 (crosshatched) and numb (stippled) are localized at the opposite poles of mitotic mammalian cortical progenitors. (A) Following a vertical cleavage, both daughters receive both Notch and numb. Soon after mitosis, numb becomes diffusely localized in the cell cytoplasm, possibly allowing it to interact quickly with Notch to inhibit Notch function following a vertical cleavage. (B) In contrast, following a horizontal cleavage, the basal daughter receives the bulk of the Notch, but none of the numb; Notch activity may thus go unopposed in the basal daughter.

that the asymmetrically distributed proteins are inherited exclusively by only one daughter. Mammalian progenitors divide both symmetrically and asymmetrically and may add another degree of regulation by controlling the cleavage angle of progenitors. Vertically dividing progenitors partition Notch1 and numb equally to both daughters to generate molecularly identical daughters, while the basal daughter from a horizontal division exclusively inherits the bulk of Notch1, but no numb, thus distinguishing itself from its sibling.

The regulation of cleavage orientation may be one way that the developing mammalian nervous system controls the rate of neurogenesis. Previous studies have indicated that prior to the onset of neurogenesis, precursors divide to expand the progenitor population. Neuronal production begins slowly, then increases with time and peaks during the last third of neurogenesis (Caviness et al., 1995). Time-lapse imaging of cortical progenitors has shown that the frequency of horizontal cleavages parallels the production of neurons during development (Chenn and McConnell, 1995). At the end of neurogenesis, however, the progenitor pool is depleted as the last neurons are generated. It is possible that these later divisions are symmetric terminal divisions, generating two postmitotic daughters, although further studies need to be performed to address the nature of these late divisions.

Virtually nothing is known about the regulation of cleavage orientation in mammals, but much progress has been made in simpler systems such as the budding yeast *Saccharomyces cerevisiae* and the nematode *C. elegans*. In budding yeast, cortical cues marking the site of previous divisions orient future division axes and the division plane by aligning both the mitotic spindle and a ring of proteins called septins that define the cleavage plane (reviewed in Chant, 1996). Cleavage orientation in the early divisions of *C. elegans* utilizes a microtubule-based mechanism that aligns the mitotic apparatus with the polarity of the cell (reviewed in White and Strome, 1996). This axis of polarity can be either intrinsically determined (the P-cell lineage) or be determined by cell contact (the EMS cell is polarized by contact with the P2 cell) (reviewed in Priess, 1994). Thus, cleavage orientation may be regulated by both intrinsic and extrinsic factors; we do not know whether such mechanisms are conserved in ventricular precursors.

Ventricular cells are clearly polarized with an apical/basal axis. Cleavages in the ventricular zone might be thought of as occurring in two modes: vertical cleavages that align with the intrinsic polarity of the cell and thus apportion out determinants equally to the two daughters, and nonvertical divisions that divide the asymmetry of the cell unequally. The distribution of cleavage angles early in neurogenesis suggests that vertical cleavages may be tightly regulated; the majority of cleavages are within 10°of vertical, while the remainder are evenly distributed from 0 to 80°. It is not known what angle of cleavage is sufficient to generate two distinct daughters; electron micrographs provide an idea of what cleavage angle might be required for the differential inheritance of the apical surface (Hinds and Ruffett, 1971) and suggest that angles as little as 30° from vertical may give rise to an apical daughter that would inherit all of the apical surface and associated junctional signaling apparatus at the expense of its basal sibling. It is not known whether sites of previous divisions are marked on ventricular progenitors, nor whether the cell uses its intrinsic apical/basal polarity to orient the mitotic apparatus. It is also unknown how a progenitor might make the transition from one mode of division to another, and it will be interesting to examine whether extrinsic factors that regulate proliferation or differentiation may also control cleavage orientation.

Neuronal Production: Proliferation and Differentiation Factors

What factors regulate the proliferation of precursors and the production of neurons in the developing ventricular zone? Considerable evidence suggests that diffusible signals can influence the differentiation or proliferation of neural crest cells (Sieber-Blum, 1989), sympathoadrenal precursors (Anderson, 1993), and retinal neuroepithelial cells (Lillien and Cepko, 1992). Recent work has suggested that basic fibroblast growth factor (bFGF) and the neurotrophin NT-3 may play important roles in the proliferation and differentiation of cortical neuronal precursors (Ghosh and Greenberg, 1995). Recently, an intriguing role for the endogenous neurotransmitters GABA and glutamate in regulating proliferation has been suggested as well (LoTurco et al., 1995). Both culture experiments and targeted gene disruptions in vivo have provided insight into possible molecular mechanisms of controlling the proliferation and differentiation of cortical progenitors.

The development of dissociated cortical precursors in vitro can be influenced by the addition of NT-3 or of bFGF (Ghosh and Greenberg, 1995). Application of NT-3 to dissociated embryonic rat cortical progenitors in vitro results in withdrawal from the cell cycle and neuronal differentiation, whereas addition of neutralizing antibodies to NT-3 reduces the amount of neuronal differentiation. In contrast, exposure to bFGF induces these progenitors to proliferate. Interestingly, NT-3 application did not effect neuron number after 7 days in vitro, suggesting that the treatment may accelerate the differentiation process without changing the actual number of neurons produced. Ghosh and Greenberg (1995) point out that other factors probably contribute to neuronal differentiation because animals with targeted disruptions of NT-3 or its receptor TrkC do not have obvious problems with neuronal differentiation.

While bFGF appears to stimulate the proliferation of ventricular cells, membrane-associated factors also appear to be involved. Cortical progenitors that have been dissociated and cultured as single cells differentiate prematurely; however, contact with other cortical cells or astrocytes stimulates the proliferation of multipotent progenitors (Temple and Davis, 1994). Furthermore, membrane homogenates from cortical astrocytes are able to support the proliferation of singly dissociated progenitors. Multipotent precursors isolated from murine embryonic telencephalon can be stimulated by bFGF to proliferate (Kilpatrick and Bartlett, 1995). bFGF is known to be present in the telencephalon during neurogenesis and is associated with cell membranes and the extracellular matrix. It is not known whether bFGF acts directly on the multipotent progenitor cell or indirectly through effects on other cells.

Recent evidence indicates that endogenous neurotransmitters may regulate the proliferation of ventricular progenitors (LoTurco et al., 1995). Ventricular cells express GABA-A and AMPA/kainate receptors type glutamate receptors, and they depolarize in response to GABA and glutamate, respectively. Application of GABA or kainate to cultured slices of developing cortex decreases DNA synthesis, while application of the competitive GABA-A antagonist BMI or the AMPA/kainate receptor antagonist CNQX causes an increase in [^{3}H] thymidine incorporation. Although many questions remain, including the identity of the responding cells, the source of the neurotransmitters, whether the signaling is synaptic, and whether the effects of the neurotransmitters are direct or indirect (reviewed in LaMantia, 1995), the bottom line remains intriguing: the application of GABA or glutamate causes a reduction in DNA synthesis in the ventricular zone.

The possibility that endogenous neurotransmitters could regulate the proliferation of progenitors provides an attractive mechanism by which the developing cortex could "feed back" upon the progenitor population. One could imagine that neurotransmitter release from more mature neurons in the cortical plate could instruct dividing progenitors to differentiate. It would be interesting to examine whether differentiation increases following treatment by GABA or glutamate.

Recent work has provided insight into possible molecular mechanisms that might underlie the proliferation and differentiation of precursors in the telencephalon. The winged-helix transcription factor BF-1 is restricted in its expression to the entire developing telencephalon. Mice with a targeted deletion of BF-1 show marked reductions in the size of the telencephalon. Although it appears that the telencephalon is properly specified and expresses appropriate markers (*Emx2, Emx1,*and *Pax-6*) the proliferation of precursor cells is diminished, and an increased number of Microtubule-Associated Protein-2 (MAP-2)–expressing cells are observed outside the ventricular zone, suggesting that the precursors differentiate prematurely and thus deplete the progenitor cell population (Xuan et al., 1995).

How might BF-1-mediated transcription regulate proliferation? BF-1 may act to enhance the response to mitogens such as bFGF or regulate their production (Xuan et al., 1995). Alternatively, BF-1 may directly regulate the expression of cell-cycle machinery (Xuan et al., 1995). Interestingly, the expression level of BF-1 declines in the proliferative population as the fraction of differentiating cells increases during development. The activity of these winged-helix transcription factors could regulate the growth and differentiation of the cortex by facilitating cell-cycle progression; with less BF-1 present, cell-cycle progression may be slowed and differentiation increased (Xuan et al., 1995).

Neuronal Migration

During the development of the cerebral cortex, young postmitotic neurons migrate long distances from the ventricular zone to attain their final positions in the cortical plate. Derangements in migration may result in improper delivery of neurons to their final locations and malformations of the brain such as focal dysplasias, subcortical heterotopia, improper lamination, and lissencephaly (Sarnat,

1992). The migrational patterns of young neurons to the mature cortex may influence the final regional specificity of cortical areas. It has been suggested that area-specific differences exist early within the ventricular zone and serve as a protomap of the future organization of the cortex; radial migration of neurons to the developing cortical plate may translocate these regional differences outward to the cortex (Rakic, 1988). However, recent studies of neuronal migration suggest that young neurons can employ a variety of migratory pathways as well as radial migration along radial glial fibers.

The migration of young neurons to their final positions in many regions of the developing brain is largely dependent on radial glia. Throughout neurogenesis, radial glia extend long processes from the ventricular to the pial surface to serve as substrates for neuronal migration. Electron micrographic evidence has revealed that migrating neurons form close contacts with radial glial fibers (Rakic, 1978). Studies of cerebellar granule cell migration demonstrate that radial glial fibers of Bergmann glia can serve as substrates for neuronal migration, and cerebellar granule cells can migrate upon these fibers in vitro (Edmondson and Hatten, 1987).

There is now considerable evidence that migrating neurons can also utilize nonradial migratory pathways. Early studies using retroviruses to label clonal progeny of progenitors were hindered by problems in distinguishing between widely dispersed siblings and multiple infection events. These problems were circumvented by infecting progenitors with retroviruses containing a complex library of genetic tags that could be recovered from labeled cells by PCR amplification (Walsh and Cepko, 1988, 1992). Clones of these retrovirally labeled cells are widely dispersed throughout the cerebral cortex, suggesting that neurons are capable of substantial nonradial migration. In addition, transgenic mice in which approximately half of the cells were marked with a transgene suggest that some 30% of neurons appear to spread tangentially (Tan and Breen, 1993; Tan et al., 1995). A mechanism for tangential dispersion has been suggested by time-lapse imaging of migrating neurons in the ferret intermediate zone in which cells employ a variety of migratory pathways, with about 25% moving nonradially. More recently, an extremely widespread pattern of cell dispersion has been revealed after focal injections of DiI into the cortical ventricular and subventricular zones (O'Rourke et al., 1996). During the 1–2 days following injection, labeling cells migrated tangentially through the proliferative and intermediate zones, ending up in regions as distant as the hippocampus and temporal cortex. The migrating cells in this study were postmitotic neurons: They expressed the neuron-specific marker TuJ1 and failed to incorporate bromodeoxyuridine (BrdU) after long periods of cumulative labeling. Finally, time-lapse imaging of the surface of the mouse ventricular zone has provided evidence for a randomly directed, tangential cell movement in this region (Fishell et al., 1993). Although it is not known definitively whether these latter cells are neuronal progenitors, these observations raise the possibility that progenitor cell movement could provide a mechanism for the widespread dispersion of retrovirally labeled clones.

What are the functional consequences of nonradial migration? Transplantation experiments suggest that the identity of cortical areas is specified late in development (O'Leary et al., 1992). When late fetal rat cortex was transplanted to other areas of the newborn rat cortex, the transplanted tissue was able to adopt the phenotype appropriate to the new location. Summarizing a number of heterotopic transplants, the transplanted cortical tissues formed projections appropriate for their new positions and developed the appropriate area-specific architecture and organization. These experiments suggest that cortical area specialization is determined after neuronal migration, and incoming thalamic afferents may serve to pattern the cortex. Consequently, it would appear that a nonradially migrating neuron would adopt the fate of the cells in the location surrounding it and not bring with it area information from its origin. The purpose of nonradial migration thus remains somewhat obscure.

While a strict interpretation of a ventricular protomap model is not supported by the transplant experiments or the findings of nonradial migration, it is possible that an early source of positional information does specify the different cortical areas. An alternative, but related model suggests that early in development, a protomap exists in the ventricular zone to pattern the preplate (O'Leary, 1989). During development, the preplate is the first layer of neurons generated; it is subsequently split into the marginal zone and subplate by the ingrowing neurons of the cortical plate. When the earliest-born neurons migrate from the ventricular zone to generate the preplate, area-specific differences within the ventricular zone could be mapped onto the preplate to direct the ingrowth of thalamic afferents. These thalamic afferents could then provide the local cues to specify region-specific characteristics. Time-lapse studies of cell movements of very early ventricular zone suggest that subplate progenitors migrate strictly radially and thus convey positional information from the ventricular zone (Borngasser and O'Leary, in preparation). Furthermore, molecular evidence suggests early differences between cortical areas. Analysis of a transgenic mouse line generated using regulatory elements from a major histocompatibility complex (MHC) class I gene linked to *lacZ* led to the unexpected finding that the transgene came to be regulated by elements specific to the insertion that restricted β-galactosidase expression to layer IV of the somatosensory cortex (Cohen et al., 1994). Expression of the transgene was detected shortly after the close of neurogenesis on postnatal day 2. When embryonic parietal cortex (presumptive somatosensory cortex) was transplanted to various cortical locations in the postnatal animal, the transplanted tissue maintained its autonomous *lacZ* expression, suggesting that some aspects of area specification occur early in development. Yet, when frontoparietal cortex is transplanted into occipital cortex, the transplanted tissue forms axonal connections typical of its new position (O'Leary and Stanfield, 1989). Thus, other aspects of area identity such as specific long-distance axonal projections can be determined later in development. Although it is clear that neurons utilize diverse modes of migration from the proliferative zone to their final location, the developmental implications of nonradial migration remain to be tested.

NEURONAL SPECIFICATION

Neuronal Specification: Laminar Determination

The cerebral cortex is organized into distinct layers that are distinguished by neuronal density, morphology, connectivity, and physiological properties. Retroviral lineage studies have provided evidence that early cortical progenitor cells are multifated in that they produce clones of progeny that span the cortical layers (Reid et al., 1995). Transplant studies have shown that progenitor cells are multipotent, and laminar identities are determined early in the cell cycle prior to a progenitor's terminal division (McConnell and Kaznowski, 1991). The laminar fate of young progenitors transplanted to older environments correlates with their position in the cell cycle at the time of transplantation (Fig. 12–5). Cells transplanted early in the cell cycle (S phase) produced daughters that migrated to superficial layers of the cortex, a laminar position appropriate for neurons generated in the host. However, if the cells underwent their terminal division in the donor, they migrated to their normal deeper layers (5 and 6) following transplantation. These experiments suggest that there is a critical window of time early in the cell cycle during which progenitors can receive environmental inputs. Indeed, progenitors may forget previous signals and acquire new instructions as they undergo another round of DNA replication in a new environment.

What is the nature of the signals that instruct the transplanted cells where to go? The signals could conceivably be mediated through short-range or long-range diffusible factors or cell–cell contact. One attractive model is that cells in the developing cortical plate feed back upon cells in the ventricular zone, so that once a sufficient number of cells in a particular layer have been generated, they signal the ventricular zone to produce cells of more superficial layers (McConnell, 1992). Anatomical studies of developing subplate show that the growth cones of recently generated subplate neurons travel along the top of the ventricular zone and can provide a potential avenue of communication with ventricular cells (Kim et al., 1991; De Carlos et al., 1992).

Recent experiments have examined the nature of the factors that specify laminar phenotype. Young donor progenitors were cultured under conditions where they maintained their normal anatomical contacts or were dissociated into single cells (Bohner et al., 1995). Progenitors maintained in cultured intact explants during the critical window of time for laminar determination migrated to the deep layers appropriate for the donor following transplantation, demonstrating that normal specification can occur in vitro. In contrast, the dissociated cells cultured at low densities did not acquire a normal deep-layer fate; virtually all migrated to the superficial cortex following transplantation. These cells not only migrated to layer 2/3, a destination appropriate for cells generated in the host, but many also adopted a novel phenotype, migrating to layer 1. These results suggest that an active signaling process is involved in instructing cortical progenitors to generate deep-layer neurons. It is not clear whether migration to the superficial layers is a default pathway followed in the absence of these instructive cues. It is possible that the period during which cell specification can occur

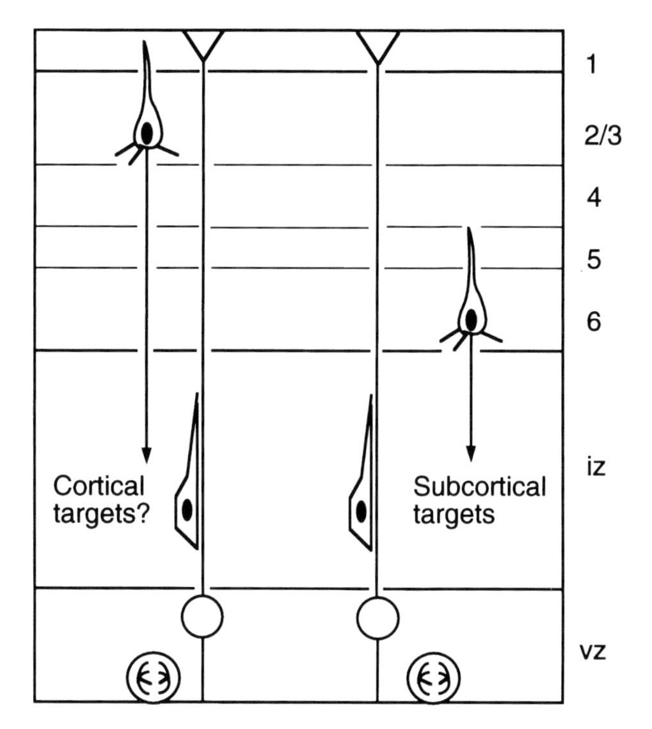

FIGURE 12–5 Outcomes of heterochronic transplantation. Presumptive layer 6 neurons or their progenitors adopt two different outcomes when transplanted into older host brains, in which upper-layer neurons are being generated. (A) Cortical progenitors transplanted during S phase generate daughters that adopt the fate appropriate for neurons being generated in the host environment and migrate to upper layers 2 and 3. (B) Cells transplanted during or after G2 of the cell cycle migrate to their normal deeper layers (5 and 6) and form subcortical projections. These transplants suggest that there is a critical window of time early in the cell cycle during which progenitors can receive environmental inputs. (Reproduced with permission from McConnell, 1992, and Academic Press.)

was extended in culture, allowing the progenitors to receive signals from the host environment instructing them to migrate to superficial layers. Why a population of neurons migrates to layer 1 remains unclear. These cells may simply be confused following the culture period and unable to interpret environmental signals, or they may have receive incomplete instruction.

In order to examine the nature of the cues that specify deep-layer fates, dissociated donor cells were either reaggregated or centrifuged into clusters of cells (Bohner et al., 1995). These groups of cells were maintained in culture for the period of time necessary for laminar specification to occur in vivo, and then transplanted into an older host. This manipulation reconstituted partially the deep-layer specification; over one-half of the transplanted cells migrated to deep laminae, while the remaining cells migrated to layer 2/3. Thus, close association of progenitors can substantially reconstitute laminar specification, suggesting that this process may be mediated through short-range cues or cell contact.

Are older ventricular progenitors equally competent to generate neurons of different layers, or is their potential restricted with development? [^{3}H] thymidine birthdating and retroviral lineage studies suggest that older progenitors normally produce only neurons of upper layers. To examine the competence of late progenitors to form the different layers, dividing cells from more mature ventricular zones (when layer 2/3 neurons are being generated) were transplanted into younger ventricular zones (when neurons of layers 5 and 6 are being generated) (Frantz and McConnell, 1995). Nearly all of the older progenitors maintained their specification and migrated to upper layers, even after one or more rounds of cell division in the new environment. In contrast, young progenitors transplanted isochronically into young ventricular zones populated the appropriate deep layers (Frantz and McConnell, 1995). These transplantation experiments suggest that older progenitors become restricted in their potential as development proceeds and that by the end of neurogenesis, progenitor cells are committed to the production of upper-layer neurons.

How does the developmental potential of progenitors become more restricted during neurogenesis? It is likely that laminar phenotypes reflect early molecular differences between the cortical layers. The transplantation experiments suggest that progenitor cells make decisions concerning their laminar position prior to migration (McConnell and Kaznowski, 1991). In addition, axons of cortical neurons of different layers project to specific laminar targets during development (Katz and Callaway, 1992), and neurons in cocultures make appropriate long-distance connections appropriate for their laminar position (Bolz et al., 1990), also suggestive of molecular differences between layers. The expression patterns of two transcription factors suggest that they may be involved in the establishment of laminar pattern in the developing cortex. The timing and distribution of the homeodomain gene *Otx1* suggests that *Otx1* might play a role in the specification of deep-layer neurons in the cerebral cortex. *Otx1* is expressed in a subset of layer 5 and 6 neurons in the postnatal and adult rat cortex (Frantz et al., 1994b). It is also expressed strongly in the precursors of these deep-layer neurons in the ventricular zone (Frantz et al., 1994b). As neurogenesis proceeds, *Otx1* is downregulated in progenitors that give rise to superficial layers of the cortex, consistent with a role for *Otx1* in deep-layer specification and possibly in the restricted fates of late progenitor cells. Other layer-specific genes include *scip,* a member of the POU-homeodomain gene family, that is expressed in a subset of layer 5 pyramidal neurons beginning at the time that these cells migrate and differentiate. At early times, *SCIP* is also expressed at high levels in neurons of layer 2/3, but it is downregulated by these neurons in adulthood (Frantz et al., 1994a). A cDNA (rCNL3) was recently identified by PCR differential display from the superficial layers of the cortex (2–4) that encodes a putative G-protein–coupled receptor of the cannabanoid receptor family (Song et al., 1994; Levin et al., 1995). rCNL3 appears to be expressed in young differentiated neurons of layers 2–4 of the cortex and is downregulated in the adult cortex. A final gene, *RZR-β,* has been identified by screening of subtracted cDNA libraries generated from superficial and deep cortical layers. *RZR-β* is expressed specifically in cortical layer 4 (Levin et al., 1995).

Neuronal Specification: Neurotransmitter Phenotype

Because of the correlation between the birthdate and final laminar position of a neuron, it is tempting to wonder whether other aspects of neuronal phenotype are determined early. Within the cortex, there are two broad classes of neurons, pyramidal and nonpyramidal. The majority of cortical neurons are pyramidal projection neurons that use excitatory amino acids as neurotransmitters, while the remainder are nonpyramidal interneurons that primarily express GABA (Gotz and Bolz, 1994). The distribution of neurotransmitter type also varies; GABAergic neurons are uniformly distributed among the layers, while the neuropeptide-expressing interneurons are found in particular layers (Gotz and Bolz, 1994).

Do all neurotransmitter subtypes arise from the same precursor? Retroviral lineage-tracing experiments have suggested that glutaminergic pyramidal neurons and GABAergic nonpyramidal neurons arise from distinct progenitors (Parnavelas et al., 1991). Clones consisted of one subtype or the other, and mixed clones were not seen. However, a recent retroviral lineage study employing a library of different genetically tagged retroviruses suggests that while closely clustered sibling cells showed similar laminar position and morphology, their widely dispersed siblings often displayed different phenotypes (Reid et al., 1995). Similarly, when progenitors from embryonic rat cortex were placed into culture, clones marked by a retroviral marker contained both GABAergic and glutamatergic neurons, suggesting that progenitors have the potential to produce both types of neurons (Gotz et al., 1995). Gotz et al. suggest that the discrepancies between the in vivo and in vitro data may suggest that precursors have a greater potential in vitro than is realized in vivo. It is also possible that because earlier retroviral lineage studies assumed that distantly marked cells were unrelated, clone sizes were underestimated. The potential of precursors to express various neurotransmitters has not been tested by transplantation experiments.

The differentiation of neurotransmitter phenotype was compared in slice cultures generated from E16 rat cortex and dissociated progenitors (Gotz and Bolz, 1994). The expression of vasoactive intestinal polypeptide (VIP) appeared to develop normally in slice culture but did not occur in dissociated progenitors, suggesting that close contacts or locally diffusible factors are important for the induction or differentiation of VIP expression. In contrast, the development of glutamate and GABA subtypes occurred similarly in both slice and dissociated cultures (Gotz and Bolz, 1994). Interestingly, the development of glutamate and GABA phenotypes appeared to correlate with the progenitor's position in the cell cycle. Cells dissociated during S phase failed to develop glutamate or GABA immunoreactivity. If maintained in slice culture for 24 hours before dissociation, the progenitors were then able to express these neurotransmitters. These experiments suggest that local environmental influences act to specify neurotransmitter phenotype and that specification to glutamatergic and GABAergic phenotype occurs early in the cell cycle.

Neuronal Specification: Lamina-Specific Connections

The neurons of the mammalian cortex project both within and outside the cortex to create highly organized circuits that enable the processing of sensory information and generation of behavior. This precise ordering of connections between neurons of different layers and neurons within layers is essential for the development of functionally useful neural circuits. A striking feature of adult cortical

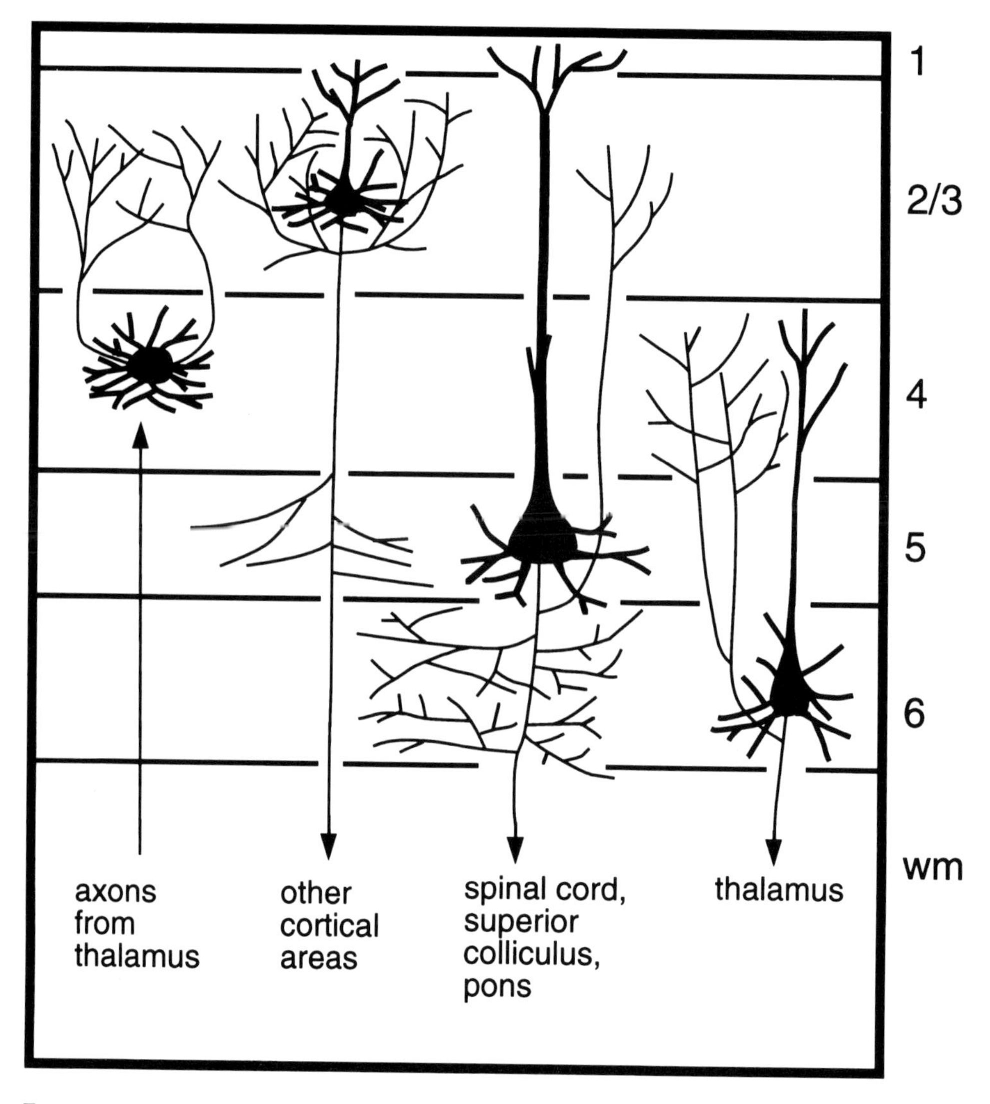

FIGURE 12–6 Local and long-distance axonal targets of the major classes of projection neurons in the primary visual cortex. Abbreviations: wm, white matter (Adapted from Gilbert and Wiesel (1985)).

connectivity is the highly stereotyped, lamina-specific patterns of axonal projections (Figs. 12–6 and 12–7). For example, the long-distance axons of layer 5 descend to targets that include the spinal cord, pons, and superior colliculus, whereas many of the neurons of layer 6 send their long-distance projections to the thalamus. The different layers also show highly specific and characteristic local axon branching patterns.

Recent work has suggested that many aspects of layer-specific axon outgrowth are apparent early during the development of the cortex. Unlike the refinement of retinothalamic or thalamocortical connections that clearly employ

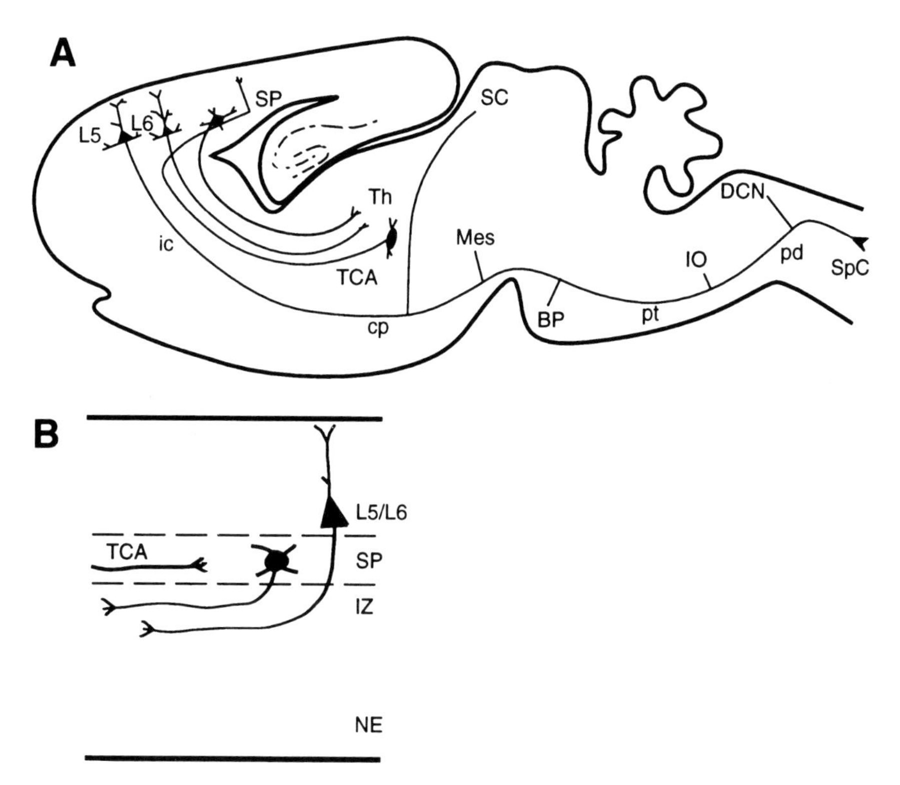

FIGURE 12–7 Cortical efferent and afferent axonal pathways. (A) Schematic of a sagittal section of a neonatal rat brain illustrating the efferent projections of subplate (SP), layer 5 (L5), and layer 6 (L6) neurons, as well as the thalamocortical afferent projection (TCA) of thalamic neurons (Th). SP and L6 axons project through the internal capsule (ic) to the thalamus (Th). L5 axons project through the ic and continue caudally down the neuraxis. As a population, L5 axons project to a number of subcortical targets, including (from rostral to caudal): SC, superior colliculus; Mes, deep mesencephalic nuclei; BP, basilar pons; IO, inferior olive; DCN, dorsal column nuclei; SpC, spinal cord. (B) Schematic depicting the layering of the intracortical pathways of efferent SP, L5, and L6 axons, and of afferent TCA axons. Efferent axons of SP, and later efferent axons of L5 and L6 neurons, extend tangentially through cortex within the intermediate zone (IZ), deep to the SP. TCA axons, which arrive after the extension of SP axons, extend tangentially through cortex to their target area along a path centered on the SP layer, superficial to the path of efferent axons in the IZ. Abbreviations: cp, cerebral peduncle; NE, neuroepithelium; pd, pyramidal decussation; pt, pyramidal tract. Figure adapted from O'Leary and Koester (1993).

activity-dependent mechanisms of axon remodeling, the development of lamina-specific patterns of axonal connections appears to be highly specific even from the outset (Katz, 1991). Intracellular dye injections have been used to examine the development of both horizontal and vertical connections between layers 2/3 and 5 during the development of the cat striate cortex (Callaway and Katz, 1990; Katz, 1991). In the adult cat, axon collaterals from layers 2/3 extend in layer 2/3 and 5 but not in layer 4 (Fig. 12–6). Horizontal connections within layers 2/3 and 5 form periodic clusters of axon branches that link orientation columns. The development of horizontal connections relies on input from patterned visual activity for proper development, and the final pattern of clustered connections results from the refinement of initially exuberant and excessive outgrowth (Callaway and Katz, 1990). Unlike the development of the horizontal connections, vertical connections between the layers develop specifically (Katz, 1991). From the earliest times that axons begin to extend collaterals, the collaterals extend specifically within layers 2/3 and 5, but not layer 4.

The development of vertical projections from layer 4 neurons appears also to follow a similar path. In the adult, layer 4 spiny neurons project extensively to layer 2/3 and within 4, while the projections to layers 5 and 6 are much more sparse (Fig. 12–6) (Gilbert and Wiesel, 1979, 1985). Intracellular fills during development of the cat visual cortex demonstrate that these differences in projection patterns arise early with differential growth and branching of axon collaterals in the superficial layers (Callaway and Katz, 1990). There is no evidence that the differences are caused by collateral pruning; furthermore, the development of vertical specificity of projections does not require patterned activity, as they formed properly in lid-sutured cats (Callaway and Katz, 1991).

The development of specific neuronal projections in slice cultures provides evidence that target selection occurs correctly even without a completely intact pathway to the target. When slices of cortex were cocultured with each other, layer 2/3 neurons projected specifically to layer 2/3 in the adjacent slice (Bolz et al., 1990). In contrast, when cortical slices were cultured adjacent to a thalamic slice, axons that invaded the thalamus arose specifically from neurons in layers 5 and 6. These experiments suggest that neurons in the different cortical layers are not only specified to project to different targets, but the identities of these targets can be discriminated in culture.

Does a neuron's laminar position determine its axonal projection patterns? Several lines of evidence suggest that the proper migration of neurons to their final location may not be necessary for proper axonal targeting. Due to defects in neuronal migration, the cortical layers of the reeler mouse are roughly inverted with layer 1 at the bottom of the cortical plate and layer 6 the most superficial (reviewed in Rakic and Caviness, 1995). Despite the malpositioning of the layers, the neurons of each layer establish normal patterns of axonal connections. In addition, a recently described rat mutant has a doubled cortex; deep to a normal-appearing cerebral cortex lies another complete set of neuronal layers (Lee et al., 1995). This accessory cortex is laminated, and the neurons arrive in this region in an inside-first, outside-last fashion, although the gradient is less strict than in the normal cortex. Retrograde labeling from the ventrobasal nucleus of the thalamus labeled what appeared to be layer 6 neurons in the accessory cortex, while tracer injection into the accessory cortex in one hemisphere retrogradely labeled the equivalent region in the contralateral side. Although these mutant studies suggest that the specific projection patterns of cortical neurons can be dissociated from proper migration, in both cases, the local neighbor relationships of neurons remain relatively intact. It would be interesting to explore the consequences of placing a neuron ectopically in a different layer and ask whether local inductions could alter the cell's normal pattern of connections.

Neuronal Specification: Long-Distance Projection Neurons

The specification of the major classes of long-distance projections from layer 5 neurons, callosal and subcortical, seems to be established early in development. Mature layer 5 cortical neurons send axons either through the corpus callosum to contralateral targets or through the internal capsule to subcortical targets, but not to both targets. Retrograde labeling from these targets in adult and neonatal animals revealed no double-labeled cells (Koester and O'Leary, 1993). From the earliest stages of axon extension, it appears that the populations of callosally and subcortically projecting layer 5 neurons is distinct (Koester and O'Leary, 1993). The isolation of molecular markers distinguishing the two classes of neurons will help determine exactly when these differences arise.

In contrast to this early distinction between projection neuron classes, other aspects of a neuron's projecting phenotype do not become apparent until much later in development. For example, the process of selective axon elimination and stabilization plays a prominent role in generating functionally appropriate layer 5 projections to subcortical targets (reviewed in O'Leary and Koester, 1993). Layer 5 is the exclusive source of cortical projections to the spinal cord and cortical targets in the midbrain and hindbrain (Fig. 12–7). During development layer 5 neurons innervate their targets in the midbrain and hindbrain via collateral projections of corticospinal axons, and as a population, layer 5 neurons in diverse areas of cortex initially develop a similar set of collateral projections (Fig. 12–8). Different combinations of collaterals or primary axon segments are later selectively eliminated, yielding the mature pattern of layer 5 projections functionally appropriate for each area of cortex.

The mechanisms that control the final pattern of layer 5 cortical projections are not well understood, but the process of selective axon elimination and stabilization exhibits considerable plasticity. For example, heterotopic transplantation experiments indicate that the areal location of a piece of developing neocortex influences the development of area-specific layer 5 connections (Stanfield and O'Leary, 1985; O'Leary and Stanfield, 1989). Late fetal visual cortical neurons transplanted to the sensorimotor cortex of newborn rats extend and permanently retain axons to the spinal cord, a subcortical target of sensorimotor cortex. Conversely, sensorimotor cortical neurons transplanted to visual cortex extend and then lose spinal axons but retain a projection to the superior colliculus, a subcortical target of visual cortex (O'Leary and Stanfield, 1989). Thus, the layer 5 projections permanently established by transplanted cortical neurons are appro-

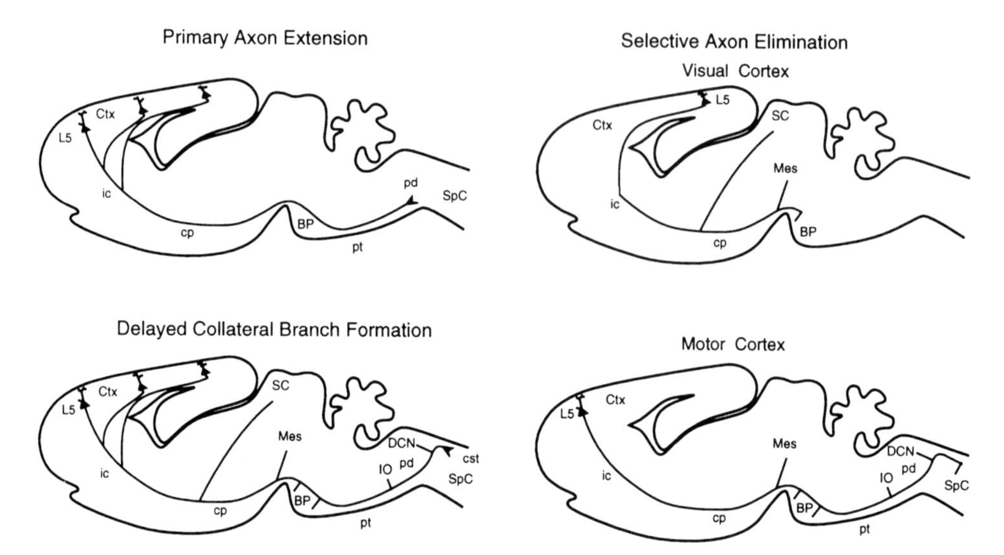

FIGURE 12–8 Development of area-specific subcortical projections of layer 5 (L5) neurons in rat neocortex. This process, illustrated in schematics of a sagittal view, occurs in three phases: (1) Primary Axon Extension: L5 neurons extend a primary axon out of cortex toward the spinal cord (SpC) bypassing their subcortical targets. (2) Delayed Collateral Branch Formation: Subcortical targets are later contacted exclusively by axon collaterals that develop by a delayed extension of collateral branches interstitially along the spinally directed primary axon. As a population, L5 neurons in all areas of the neocortex develop a common set of branches. (3) Selective Axon Elimination: As illustrated for visual and motor cortex, specific collateral branches or segments of the primary axon are selectively eliminated to generate the mature projections functionally appropriate for the area of neocortex in which the L5 neuron is located. Abbreviations: BP, basilar pons; cp, cerebral peduncle; Ctx, neocortex; ic, internal capsule; pd, pyramidal decussation; pt, pyramidal tract; SC, superior colliculus. Figure adapted from O'Leary and Koester (1993).

priate for the cortical area in which the transplanted neurons develop, not where they were born. In conclusion, commitment to a projection neuron type, such as a layer 5 subcortically projecting neuron, seems to occur early in development, whereas other features of projection neuron phenotype, such as the subset of subcortical targets to which a particular layer 5 neuron retains projections, remains modifiable until much later.

AREALIZATION OF THE NEOCORTEX

The adult mammalian neocortex can be parcellated into discrete areas based on their unique connections (both efferents and afferents), distributions of neurotransmitter receptors, and cytoarchitecture (Rakic, 1988). These area-specific characteristics contribute to the unique functional properties of the individual neocortical areas. How are these unique characteristics determined?

The modality of thalamic input to an individual neocortical area largely determines the functional identity of that area. Thalamocortical input also plays a prominent role in the differentiation of the cytoarchitecturally defined, functionally specialized areas of the neocortex from an initially uniform cortical plate (reviewed in Rakic, 1988; O'Leary, 1989). The modality of thalamic input to an individual neocortical area largely determines the functional identity of that area. Thalamocortical input also plays a prominent role in the differentiation of the cytoarchitecturally defined, functionally specialized areas of the neocortex from an initially uniform cortical plate (reviewed in O'Leary, 1989; O'Leary et al., 1994). One example of this concept comes from heterotopic transplantation experiments which test mechanisms involved in the development of "barrels," which are functional and architectural groupings unique to the primary somatosensory cortex of adult rodents and process sensory information from large facial whiskers. Structurally, barrels are distinctive aggregations of layer 4 neurons and extracellular matrix components surrounding a cluster of afferents from the ventroposterior thalamic nucleus (VP), and they are arranged in a pattern that mirrors that of the facial whiskers (Woolsey and Van Der Loos, 1970). During normal development, barrels gradually emerge from initially uniform distributions of layer 4 neurons and VP afferents (Schlaggar and O'Leary, 1993). Interestingly, VP afferents will invade transplants of late embryonic visual cortex placed into somatosensory cortex in newborn rats, segregate into a pattern of barrel-related clusters, and direct the subsequent differentiation of barrel architecture (Schlaggar and O'Leary, 1991). These findings indicate a critical, regulatory role for thalamocortical input in the architectural differentiation of a neocortical area. Therefore, an understanding of the mechanisms that control the development of area-specific thalamocortical projections is crucial to understanding how arealization of the neocortex arises during development.

Thalamocortical projections in the adult are organized such that specific thalamic nuclei project to specific neocortical areas (Hohl-Abrahao and Creutzfeldt, 1991). The development of area-specific thalamocortical projections occurs in two distinct phases. During the first phase, the targeting phase, thalamocortical axons extend through the internal capsule to reach the neocortex and then grow tangentially on an intracortical pathway centered on the subplate layer to reach their specific cortical target area (Ghosh and Shatz, 1992; Miller et al., 1993; Bicknese et al., 1994). After reaching their appropriate target area, thalamocortical pause or wait in the subplate layer for varying amounts of time before entering the second phase, the targeting phase. Although an imperceptible waiting period has been reported in rodents (Catalano et al., 1991), in cats the waiting period has been demonstrated to last as long as 2 weeks (Shatz and Luskin, 1986; Ghosh and Shatz, 1992). During the targeting phase, thalamocortical axons extend collaterals superficially into the overlying cortical plate (Erzurumlu and Jhaveri, 1990; Catalano et al., 1991; Ghosh and Shatz, 1992). The fact that thalamocortical targeting to neocortex occurs in two distinct phases suggests that the cues guiding thalamocortical axons to their specific cortical target areas may be distinct from those that promote the invasion of the cortical plate.

Axons from the principal sensory thalamic nuclei target their appropriate cortical areas in a highly specific manner (Fig. 12–9). An examination of the distribution of retrogradely labeled cells in the thalamus following injections of ret-

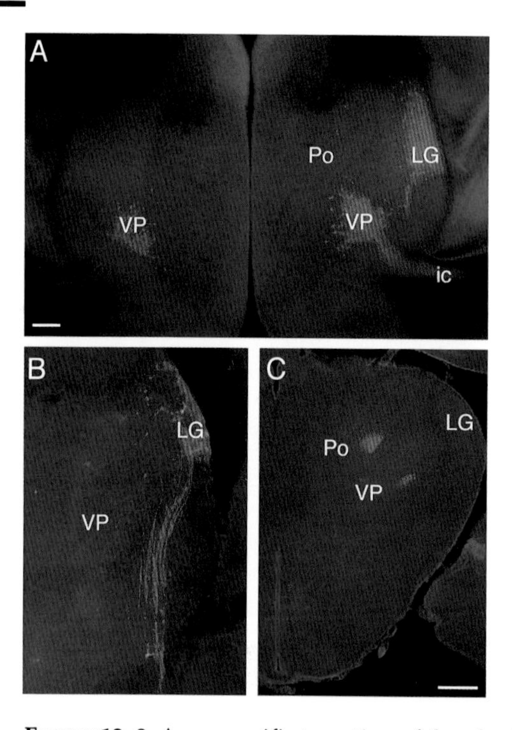

FIGURE 12–9 Area-specific targeting of developing thalamocortical axons. DiI was used as a retrograde tracer in embryonic and postnatal rats to examine the targeting of thalamocortical axons to their correct areas in the rat neocortex. Either large injections of DiI were centered on the subplate, the intracortical pathway of thalamocortical axons, in E17 fixed brains (A, B), or a small injection was restricted to the deep cortical plate without involving the suplate in P1 rats fixed at P3 (C). In A, the midline is in the center, and lateral is to the left and right. In B and C, the midline is to the left, and lateral is to the right. Coronal sections counterstained with bisbenzimide are shown. (A) In the left hemisphere, a single injection made in parietal (sensorimotor) cortex labeled cells in somatosensory thalamic nuclei such as the ventroposterior nucleus (VP). In the right hemisphere, two injections were made, one in parietal cortex and one in occipital (i.e., visual) cortex. Essentially all labeled cells were found in somatosensory thalamic nuclei and in visual thalamic nuclei (e.g., the dorsal lateral geniculate nucleus, LG). (B) A single large injection was made in occipital cortex. The labeled cells were predominantly found in visual thalamic nuclei. (C) A small injection restricted to the cortical plate labeled only cells in somatosensory thalamic nuclei (e.g., VP and the posterior nucleus, Po). The findings from experiments like those shown in A and B indicate that during their initial targeting phase, thalamocortical axons project accurately to their correct target areas. The findings from experiments like those shown in C indicate that during their later invasion phase, thalamocortical afferents extend from the subplate into the cortical plate of their correct target areas. Photos were taken by J.A. De Carlos and B.L. Schlaggar. Abbreviations: ic, internal capsule Scale bar in A = 100 μm in A and 200 μm in B. Scale bar in C = 500 μm.

rograde tracers into various cortical locations in developing rodents found thalamocortical axons to only rarely extend beyond their appropriate cortical target areas or make gross directional errors (Braisted and O'Leary, 1995; DeCarlos et al., 1992). Furthermore, thalamocortical axons only extend collaterals into the cortical plate of appropriate cortical target areas (Crandall and Caviness, 1984; Miller et al., 1993; Braisted and O'Leary, 1995; DeCarlos et al., 1992). The remarkable accuracy in both the targeting and invasion phases of thalamocortical development suggests that an efficient position-dependent marking system is present within the developing neocortex and controls area-specific thalamocortical matching.

What molecules are candidates for playing a role in guiding thalamocortical axons to their appropriate cortical target areas? Antibody and gene expression studies have demonstrated that a number of molecules (including the proto-oncogenes *sis*, *src*, *ras*, and *myc:* Johnston and van der Kooy, 1989), the intermediate filament protein Vimentin (Johnston and van der Kooy, 1989), the novel protein Latexin (Arimatsu, 1994), some POU-domain genes (He et al., 1989; Monuki et al., 1990; Suzuki et al., 1990; Frantz et al., 1994a), and a number of homeobox-containing and other regulatory genes (Tao and Lai, 1992; Xu et al., 1993) are expressed differentially within the developing rodent telencephalon, but none are expressed in a graded or area-specific manner within the developing neocortex. The best candidates to date for a role in arealization of the neocortex are the transcription factor genes *Pax6*, *Emx1*, and *Emx2*. The paired box gene *Pax6* is expressed in a rostral-to-caudal gradient within the neuroepithelium of the dorsal telencephalon (Walther and Gruss, 1991), although it remains to be determined whether this intriguing distribution simply reflects a developmental gradient (i.e., the rostral-to-caudal developmental gradient present across the developing neocortex) or whether this distribution is involved in the arealization of the neocortex. The homeodomain genes *Emx1* and *Emx2* (Simeone et al., 1992a) may not only play important roles in the specification of forebrain-specific structures as discussed above, but may also be involved in arealization of the neocortex. Although *Emx1* is expressed uniformly across the developing neocortex, *Emx2* has a graded expression; but in contrast to *Pax6*, *Emx2* is expressed most prominently caudally in the neocortex. *Emx1* is expressed uniformly in the neocortex, suggesting a nested expression with *Emx2*, similar to *Hox* gene expression in the mammalian hindbrain. Although Emx2, and perhaps Pax6 may be involved in regulating guidance information for thalamocortical axons, since they are DNA-binding proteins rather than membrane-anchored or extracellular matrix molecules, they cannot be directly responsible for the area-specific targeting of thalamocortical axons.

The limbic-system–associated membrane protein (LAMP), a 64–68-kD glycosyl-phosphatidylinositol (GPI)-linked glycoprotein of the immunoglobulin superfamily (Pimenta et al., 1995), has been suggested to influence the development of specific thalamocortical (Barbe and Levitt, 1991, 1992) and corticocortical (Barbe and Levitt, 1995) connections to "limbic" cortex. In normal rats, the lateral dorsal nucleus of the thalamus innervates perirhinal cortex (a limbic cortical region), which expresses LAMP, but does not innervate sensorimotor cortex (a neocortical area), which does not express LAMP. Heterotopic transplants between limbic and neocortical regions of cortex demonstrate that LAMP expression becomes committed by E14. When animals were examined as adults, it was found that lateral dorsal thalamic axons invaded LAMP-positive, but not LAMP-negative, explants, irrespective of either the origin of the donor transplant or the location of the transplant within the host cortex (Barbe and Levitt,

1992). Similar heterotopic transplant experiments demonstrate that LAMP can also influence the organization of corticortical connections in vivo. When E12 or E13 presumptive perirhinal or sensorimotor cortex is transplanted into perirhinal or sensorimotor cortex of neonatal hosts, the development of callosal projections reflects the location of the transplant in the host cortex. In contrast, when the donor cortices are harvested at E14, the callosal projections exhibit phenotypes indicative of both the new host and original donor cortices (Barbe and Levitt, 1995). These studies suggest that during normal development, the invasion of limbic cortex by both limbic thalamic and cortical axons may be regulated by LAMP, and that LAMP is likely to be directly involved in targeting of limbic thalamic and cortical axons to their appropriate cortical target areas. Further evidence that LAMP may be serving as a recognition molecule for the formation of limbic connections comes from antibody perturbation studies; administration of function-blocking anti-LAMP antibodies has been shown to prevent septal cholinergic axons from invading hippocampal explants in vitro (Keller et al., 1989) and, more recently, to result in the misrouting of mossy fiber projections in the hippocampus in vivo (Pimenta et al., 1995).

In contrast to the experiments in limbic cortex, both heterotopic transplant and coculture studies demonstrate that thalamic nuclei which normally project to specific neocortical areas are capable of innervating inappropriate neocortical areas, at least during late stages of thalamocortical development (Chang et al., 1986; Molnar and Blakemore, 1991; Schlaggar and O'Leary, 1991; O'Leary et al., 1992). It is possible that these results are due to a respecification of the molecular phenotype of the cortical explant/transplant. However, by analogy to the commitment of the LAMP phenotype (Barbe and Levitt, 1992), this seems unlikely, since the neocortical tissue was harvested near the end of, or after the cessation of, cortical neurogenesis. So what can we learn from these coculture and transplant studies? Apparently, thalamic axons which normally innervate a specific neocortical area can, if presented with the opportunity, invade inappropriate regions of neocortex. However, these studies do not rule out the possibility that molecular cues that control thalamocortical matching are present within the developing cortex. For instance, the neocortical transplants were placed in newborn host rats, past the age when thalamocortical axons begin to invade the overlying cortical plate. Thus, the thalamocortical axons that have already invaded the host cortical tissue surrounding the transplant have the relatively simple task of reextending and invading the transplanted cortex. Similarly, in the coculture experiments, neurites from embryonic thalamic explants only invaded cortical explants harvested at ages when thalamic axons are already waiting in the subplate beneath their appropriate cortical target area. Therefore, it is possible that in vivo, thalamic axons may be guided to their corresponding cortical target areas by specific cues, but once invasion of the cortical plate commences, general cues may act similarly in all areas of cortex such that thalamic axons invade whatever cortical tissue they are waiting under.

Data from a variety of studies suggest that thalamocortical axons use information available in the subplate layer to select their specific target areas within the neocortex. As mentioned above, thalamocortical axons extend within the subplate layer during the targeting phase and wait in the subplate layer prior to invading the cortical plate (Ghosh and Shatz, 1992); they are in a position to be influenced by potential targeting cues associated with subplate cells. In addition, thalamocortical axons grow past, rather than invade, the cortical plate, overlying regions of the subplate pharmacologically depleted of neurons. This has been shown by Shatz and colleagues (Ghosh et al., 1990; Ghosh and Shatz, 1993), who used the excitotoxin kainic acid to ablate subplate neurons in embryonic cats during the time when axons from the visual thalamic nucleus, the lateral geniculate nucleus (LGN), are waiting in the subplate beneath the visual cortex, but have not yet begun to invade the cortical plate. They found that LG axons do not invade the cortical plate overlying the visual cortex if subplate neurons are absent; instead, they grow past their appropriate target area in an aberrant path. Similar results were obtained when subplate neurons were ablated beneath auditory cortex at the onset of the waiting period for auditory thalamic axons; medical geniculate axons grew past and did not invade auditory cortex (Ghosh and Shatz, 1993). The subplate is also involved in the refinement of visual thalamic connections within their principal target layer 4 during the final stage of thalamocortical development; ablation of subplate neurons underlying visual cortex just prior to ocular segregation of LG axons inhibits the formation of ocular dominance columns (Ghosh and Shatz, 1993).

Investigators have also hypothesized that the axons of subplate neurons may form a scaffold that guides thalamocortical axons as they exit the internal capsule and grow to their appropriate cortical target areas (Blakemore and Molnar, 1990; Molnar and Blakemore, 1995). However, a number of studies have shown that the intracortical pathways of thalamocortical and corticothalamic axons are segregated in both adult (Woodward et al., 1990) and embryonic (De Carlos and O'Leary, 1992; Miller et al., 1993; Bicknese et al., 1994) rats; the intracortical trajectory of subcortically projecting subplate and cortical plate axons lies deep in the intermediate zone (De Carlos and O'Leary, 1992), while the intracortical pathway of thalamocortical axons, which arrive in cortex after subplate axons have exited the cortex, is centered on the subplate layer, superficial to the intermediate zone. The segregation of cortical efferents and afferents makes it unlikely that efferent subplate axons are providing cues that guide thalamocortical axons to their cortical target areas. In addition, it seems unlikely that subplate axons are guiding thalamocortical axons out of the thalamus and into the nascent internal capsule, since thalamocortical and subplate axons both enter the nascent internal capsule at approximately the same time, and pass one another at about the midpoint (Blakemore and Molnar, 1990; De Carlos and O'Leary, 1992; Ghosh and Shatz, 1992). The evidence, therefore, suggests that any subplate-associated cues that may influence the development of area-specific thalamocortical matching are likely to be present in the subplate layer itself.

In contrast, recent experiments have suggested a role for subplate axons in the invasion of appropriate thalamic nuclei by cortical efferent axons. Subcortical projections from the cortex to subcortical target areas such as the thalamus are pioneered by the axons of subplate neurons (McConnell et al., 1989; De Carlos and O'Leary, 1992; McConnell et al., 1994) and have therefore been postulated to play a role in the proper pathfinding and target selection of cortical efferent axons. Although embryonic ablation of subplate neurons results in normal pathfinding of efferent axons out of the visual cortex, through the internal capsule, and into the thalamus, the visual cortical efferent axons often failed to invade their normal thalamic target nucleus, the LG (McConnell et al., 1994).

REFERENCES

Acampora, D., Mazan, S., Lallemand, Y., Avantaggiato, V., Maury, M., Simeone, A., Brulet, P. (1995). Forebrain and midbrain regions are deleted in *Otx2*−/−mutants due to a defective anterior neuroectoderm specification during gastrulation. *Development* 121: 3279–3290.

Ang, S.L., Jin, O., Rhinn, M., Daigle, N., Stevenson, L., Rossant, J. (1996). A targeted mouse *Otx2* mutation leads to severe defects in gastrulation and formation of axial mesoderm and to deletion of rostral brain. *Development* 122:243–252.

Angevine, J.B., Jr., Sidman, R.L. (1961). Autoradiographic study of cell migration during histogenesis of cerebral cortex in the mouse. *Nature* 192:766–768.

Arimatsu, Y. (1994). Latexin: a molecular marker for regional specification in the neocortex. *Neurosci. Res.* 20:131–135.

Artavanis-Tsakonas S., Matsuno, K., Fortini, M.E. (1995). Notch signaling. *Science* 268: 225–323.

Artavanis-Tsakonas, S., Simpson, P. (1991). Choosing a cell fate: a view from the Notch locus. *Trends Genet.* 7:403–408.

Barbe, M.F., Levitt, P. (1991). The early commitment of fetal neurons to the limbic cortex. *J. Neurosci.* 11:519–533.

Barbe, M.F., Levitt, P. (1992). Attraction of specific thalamic input by cerebral grafts depends on the molecular identity of the implant. *Proc. Natl. Acad. Sci. U.S.A.* 89:3706–3710.

Barbe, M.F., Levitt, P. (1995). Age-dependent specification of the corticocortical connections of cerebral grafts. *J. Neurosci.* 15:1819–1834.

Barbe, M.F., Levitt, P. (1991). The early commitment of fetal neurons to the limbic cortex. *J. Neurosci.* 11:519–533.

Berry, M., Rogers, A.W. (1965). The migration of neuroblasts in the developing cerebral cortex. *J. Anat.* 99:691–709.

Bicknese, A.R., Sheppard, A.M., O'Leary, D.D., Pearlman, A.L. (1994). Thalamocortical axons extend along a chondroitin sulfate proteoglycan-enriched pathway coincident with the neocortical subplate and distinct from the efferent path. *J. Neurosci.* 14:3500–3510.

Blakemore, C., Molnar, Z. (1990). Factors involved in the establishment of specific interconnections between thalamus and cerebral cortex. *Cold Spring Harb. Symp. Quant. Biol.* 55:491–504.

Bohner, A.P., Akers, R.M., McConell, S.K. (1997). Induction of deep layer cortical neurons in vitro. *Development,* in press.

Bolz, J., Novak, N., Götz, M., Bonhoeffer, T. (1990). Formation of target-specific neuronal projections in organotypic slice cultures from rat visual cortex. *Nature* 346:359–362.

Boncinelli, E., Gulisano, M., Broccoli, V. (1993). *Emx* and *Otx* homeobox genes in the developing mouse brain. *J. Neurobiol.* 24:1356–1366.

Braisted, J.E., O'Leary, D.D.M. (1995). Axons from the ventrobasal thalamic nucleus pioneer the thalamocortical pathway to rat neocortex. *Soc. Neurosci. Abstr.* 21:798.

Callaway, E.M., Katz, L.C. (1990). Emergence and refinement of clustered horizontal connections in cat striate cortex. *J. Neurosci.* 10:1134–1153.

Callway, E.M., Katz, L.C. (1991). Effects of binocular deprivation on the development of clustered horizontal connections in cat striate cortex. *Proc. Natl. Acad. Sci. U.S.A.* 88: 745–749.

Catalano, S.M., Robertson, R.T., Killackey, H.P. (1991). Early ingrowth of thalamocortical afferents to the neocortex of the prenatal rat. *Proc. Natl. Acad. Sci. U.S.A.* 88:2999–3003.

Caviness, V.S.J., Takahashi, T., Nowakowski, R.S. (1995). Neocortical neuronogenesis: a general developmental and evolutionary model. *Trends Neurosci.* 18:379–383.

Chang, F.L., Steedman, J.G., Lund, R.D. (1986). The lamination and connectivity of embryonic cerebral cortex transplanted into newborn rat cortex. *J. Comp. Neurol.* 244:401–411.

Chant, J. (1996). Septin scaffolds and cleavage planes in Saccharomyces. *Cell* 84:187–190.

Chenn, A., McConnell, S.K. (1995). Cleavage orientation and the asymmetric inheritance of Notch1 immunoreactivity in mammalian neurogenesis. *Cell* 82:631–641.

Chitnis, A., Henrique, d., Lewis, J., Ish-Horowicz, D., Kintner, C. (1995). Primary neurogenesis in Xenopus embryos regulated by a homologue of the Drosophila neurogenic gene Delta. *Nature* 375:761–766.

Cohen, T.M., Babinet, C., Wassef, M. (1994). Early determination of a mouse somatosensory cortex marker. *Nature* 368:460–463.

Crandall, J.E., Caviness, V.S., Jr. (1984). Thalamocortical connections in newborn mice. *J. Comp. Neurol.* 228:542–556.

De Carlos, J.A., O'Leary, D.D.M. (1992). Growth and targeting of subplate axons and establishment of major cortical pathways. *J. Neurosci.* 12:1194–1211.

De Carlos, J.A., Schlaggar, B.L., O'Leary, D.D.M. (1992). Targeting specificity of primary thalamocortical axons in developing rat cortex. *Soc. Neurosci. Abstr.* 18:57.

Doe, C.Q., Spana, E.P. (1995). A collection of cortical crescents: asymmetric protein localization in CNS precursor cells. *Neuron* 15:991–995.

Edmondson, J.C., Hatten, M.E. (1987). Glial-guided granule neuron migration in vitro: a high resolution time-lapse video microscopic study. *J. Neurosci.* 7:1928–1934.

Erzurumlu, R.S., Jhaveri, S. (1990). Thalamic axons confer a blueprint of the sensory periphery onto the developing rat somatosensory cortex. *Dev. Brain Res.* 56:229–234.

Finkelstein, R., Boncinelli, E. (1994). From fly head to mammalian forebrain: the story of *otd* and *otx*. *Trends Genet.* 10:310–315.

Fishell, G., Mason, C.A., Hatten, M.E. (1993). Dispersion of neural progenitors within the germinal zones of the forebrain. *Nature* 362:636–638.

Fortini, M.E., Rebay, I., Caron, L.A., Artavanis, T.S. (1993). An activated Notch receptor blocks cell-fate commitment in the developing Drosophila eye. *Nature* 365:555–557.

Frantz, G.D., McConnell, S.K. (1995). Restriction of late cerebral cortical progenitors to an upper-layer fate. *Neuron* 17:55–61.

Frantz, G.D., Bohner, A.P., Akers, R.M., McConnell, S.K. (1994a). Regulation of the POU domain gene *SCIP* during cerebral cortical development. *J. Neurosci.* 14:472–485.

Frantz, G.D., Weimann, J.M., Levin, M.E., McConnell, S.K. (1994b). Otx1 and Otx2 define layers and regions in developing cerebral cortex and cerebellum. *J. Neurosci.* 14: 5725–5740.

Ghosh, A., Greenberg, M.E. (1995). Distinct roles for bFGF and NT-3 in the regulation of cortical neurogenesis. *Neuron* 15:89–103.

Ghosh, A., Shatz, C.J. (1992). Pathfinding and target selection by developing geniculocortical axons. *J. Neurosci.* 12:39–55.

Ghosh, A., Shatz, C.J. (1993). A role for subplate neurons in the patterning of connections from thalamus to neocortex. *Development* 117:1031–1047.

Ghosh, A., Antonini, A., McConnell, S.K., Shatz, C.J. (1990). Requirement for subplate neurons in the formation of thalamocortical connections. *Nature* 347:179–181.

Gilbert, C.D., Wiessel, T.N. (1979). Morphology and intracortical projections of functionally characterised neurones in the cat visual cortex. *Nature* 280(5718):120–125.

Gilbert, C.D., Wiesel, T.N. (1985). Intrinsic connectivity and receptive field properties in visual cortex. *Vision Res.* 25:365–374.

Gotz, M., Bolz, J. (1994). Differentiation of transmitter phenotypes in rat cerebral cortex. *Eur. J. Neurosci.* 6:18–32.

Gotz, M., Williams, B.P., Bolz, J., Price, J. (1995). The specification of neuronal fate: a common precursor for neurotransmitter subtypes in the rat cerebral cortex in vitro. *Eur. J. Neurosci.* 7:889–898.

Guo, M., Jan, L.Y., Jan, Y.N. (1996). Control of daughter cell fates during asymmetric division: interaction of numb and Notch. *Neuron* 17:27–41.

He, X., Treacy, M.N., Simmons, D.M., Ingraham, H.A., Swanson, L.W., Rosenfeld, M.G. (1989). Expression of a large family of POU-domain regulatory genes in mammalian brain development. *Nature* 340:35–41.

Hinds, J.W., Ruffett, T.L. (1971). Cell proliferation in the neural tube: an electron microscopic and Golgi analysis in the mouse cerebral vesicle. *Z. Zellforsch. Mikrosk. Anat.* 115:226–264.

Hirth, F., Therianos, S., Loop, T., Gehring, W.J., Reichert, H., Furukubo, T.K. (1995). Developmental defects in brain segmentation caused by mutations of the homeobox genes *orthodenticle* and *empty spiracles* in Drosophila. *Neuron* 15:769–778.

Hohl-Abrahao, J.C., Creutzfeldt, O.D. (1991). Topographical mapping of the thalamocortical projections in rodents and comparison with that in primates. *Exp. Brain Res.* 87: 283–294.

Jan, Y.N., Jan, L.Y. (1995). Maggot's hair and bug's eye: role of cell interactions and intrinsic factors in cell fate specification. *Neuron* 14:1–5.

Johnston, J.G., van der Kooy, D. (1989). Protooncogene expression identifies a transient columnar organization of the forebrain within the late embryonic ventricular zone. *Proc. Natl. Acad. Sci. U.S.A.* 86:1066–1070.

Katz, L.C. (1991). Specificity in the development of vertical connections in cat striate cortex. *Eur. J. Neurosci.* 3:1–9.

Katz, L.C., Callaway, E.M. (1992). Development of local circuits in mammalian visual cortex. *Annu. Rev. Neurosci.* 15:31–56.

Keller, F., Rimvall, K., Barbe, M.F., Levitt, P. (1989). A membrane glycoprotein associated with the limbic system mediates the formation of the septo-hippocampal pathway in vitro. *Neuron* 3:551–561.

Kilpatrick, T.J., Bartlett, P.F. (1995). Cloned multipotential precursors from the mouse cerebrum require FGF-2, whereas glial restricted precursors are stimulated with either FGF-2 or EGF. *J. Neurosci.* 15:3653–3661.

Kim, G. J., Shatz, C.J., McConnell, S.K. (1991). Morphology of pioneer and follower growth cones in the developing cerebral cortex. *J. Neurobiol.* 22:629–642.

Koester, S.E., O'Leary, D.D. (1993). Connectional distinction between callosal and subcortically projecting cortical neurons is determined prior to axon extension. *Dev. Biol.* 160:1–14.

Lee, K.S., Lanzino, G., Schottler, F., Collins, J., Kassall, N., Hiramatsu, K., Goto, Y., Hong, S., Berr, S., Caner, H., Yamamoto, H., Omary, R., Okonkwo, D., Jane, J. (1995). A novel rat with a double cortex: 1. General features. *Soc. Neurosci. Abstr.* 21:46.

Levin, M.E., Chenn, A., McConnell, S.K. (1995). Restricted cerebral cortex expression of a candidate G-protein coupled receptor isolated by PCR differential display. *Soc. Neurosci. Abstr.* 21:1044.

Lillien, L., Cepko, C. (1992). Control of proliferation in the retina: temporal changes in responsiveness to FGF and TGFα. *Development* 115:253–266.

Lo Turco, J.J., Owens, D.F., Heath, M.J.S., Davis, M.B.E., Kriegstein, A.R. (1995). GABA and glutamate depolarize cortical progenitor cells and inhibit DNA synthesis. *Neuron* 15:1287–1298.

Luskin, M. B., Pearlman, A.L., Sanes, J.R. (1988). Cell lineage in the cerebral cortex of the mouse studied in vivo and in vitro with a recombinant retrovirus. *Neuron* 1:635–647.

Matsuo, I., Kuratani, S., Kimura, C., Takeda, N., Aizawa, S. (1995). Mouse Otx2 functions in the formation and patterning of rostral head. *Genes Dev.* 9:2646–2658.

McConnell, S.K. (1992). The determination of neuronal identity in the mammalian cere-

bral cortex. In *Determinants of Neuronal Identity,* Shankland, M., Macagno, E., ed. New York: Academic Press, pp. 391–432.
McConnell, S.K., Kaznowski, C.E. (1991). Cell cycle dependence of laminar determination in developing cerebral cortex. *Science* 254:282–285.
McConnell, S.K., Ghosh, A., Shatz, C.J. (1989). Subplate neurons pioneer the first axon pathway from the cerebral cortex. *Science* 245:978–982.
McConnell, S.K., Ghosh, A., Shatz, C.J. (1994). Subplate pioneers and the formation of descending connections from cerebral cortex. *J. Neurosci.* 14:1892–1907.
Miller, M.W. (1985). Congeneration of retrograded labeled corticocortical projection and GABA-immunoreactive local circuit neurons in cerebral cortex. *Dev. Brain Res.* 23:187–192.
Miller, B., Chou, L., Finlay, B.L. (1993). The early development of thalamocortical and corticothalamic projections. *J. Comp. Neurol.* 335:16–41.
Molnar, Z., Blakemore, C. (1991). Lack of regional specificity for connections formed between thalamus and cortex in coculture. *Nature* 351:475–477.
Molnar, Z., Blakemore, C. (1995). How do thalamic axons find their way to the cortex? *Trends Neurosci.* 18:389–397.
Monuki, E.S., Kuhn, R., Weinmaster, G., Trapp, B.D., Lemke, G. (1990). Expression and activity of the POU transcription factor SCIP. *Science* 249:1300–1303.
Muskavitch, M.A.T. (1994). Delta-Notch signaling and Drosophila cell fate choice. *Dev. Biol.* 166:415–430.
O'Leary, D.D.M. (1989). Do cortical areas emerge from a protocortex? *Trends Neurosci.* 12: 400–406.
O'Leary, D.D.M., Koester, S.E. (1993). Development of projection neuron types, axon pathways, and patterned connections of the mammalian cortex. *Neuron* 10:991–1006.
O'Leary, D.D.M., Stanfield, B.B. (1989). Selective elimination of axons extended by developing cortical neurons is dependent on regional locale. Experiments utilizing fetal cortical transplants. *J. Neurosci.* 9:2230–2246.
O'Leary, D.D., Schlaggar, B.L, Stanfield, B.B. (1992). The specification of sensory cortex: lessons from cortical transplantation. *Exp. Neurol.* 115:121–126.
O'Leary, D.D.M., Schlaggar, B.L., Tuttle, R. (1994). Specification of neocortical areas and thalamocortical connections. *Annu. Rev. Neurosci.* 17:419–439.
O'Rourke, N.A., Chenn, A., McConnell, S.K. (1996). Postmitotic neurons migrate tangentially in the cortical ventricular zone. *Development,* in press.
Parnavelas, J.G., Barfield, J.A., Franke, E., Luskin, M.B. (1991). Separate progenitor cells give rise to pyramidal and nonpyramidal neurons in the rat telencephalon. *Cereb. Cortex* 1:463–468.
Pimenta, A.F., Zhukareva, V., Barbe, M.F., Reinoso, B.S., Grimley, C., Henzel, W., Fischer, I., Levitt, P. (1995). The limbic system-associated membrane protein is an Ig superfamily member that mediates selective neuronal growth and axon targeting. *Neuron* 15: 287–297.
Price, J., Thurlow, L. (1988). Cell lineage in the rat cerebral cortex: a study using retroviral-mediated gene transfer. *Development* 104:473–482.
Priess, J.R. (1994). Establishment of initial asymmetry in early *Caenorhabditis elegans* embryos. *Curr. Opin. Genet. Dev.* 4:563–568.
Rakic, P. (1974). Neurons in rhesus monkey visual cortex: systemic relation between time of origin and eventual disposition. *Science* 183:425–427.
Rakic, P. (1978). Neuronal migration and contact guidance in the primate telencephalon. *Postgrad. Med. J.* 54:25–40.
Rakic, P. (1988). Specification of cerebral cortical areas. *Science* 241:170–176.
Rakic, P., Caviness, V.J. (1995). Cortical development: view from neurological mutants two decades later. *Neuron* 14:1101–1104.
Reid, C.B., Liang, I., Walsh, C. (1995). Systematic widespread clonal organization in cerebral cortex. *Neuron* 15:299–310.
Rhyu, M.S., Jan, L.Y., Jan, Y.N. (1994). Asymmetric distribution of numb protein during

division of the sensory organ precursor cell confers distinct fates to daughter cells. *Cell* 76:477–491.

Sarnat, H.B. (1992). *Cerebral dysgenesis: embryology and clinical expression.* New York: Oxford University Press.

Schlaggar, B.L., O'Leary, D.D. (1991). Potential of visual cortex to develop an array of functional units unique to somatosensory cortex. *Science* 252:1556–1560.

Schlaggar, B.L., O'Leary, D.D.M. (1993). Patterning of the barrel field in somatosensory cortex with implications for the specification of neocortical areas. *Perspect. Dev. Neurobiol.* 1:81–91.

Shatz, C.J., Luskin, M.B. (1986). The relationship between the geniculocortical afferents and their cortical target cells during development of the cat's primary visual cortex. *J. Neurosci.* 6:3655–3668.

Shawlot, W., Behringer, R.R. (1995). Requirement for Lim1 in head-organizer function. *Nature* 374:425–430.

Shimamura, K., Hartigan, D.J., Martinez, S., Puelles, L., Rubenstein, J.L. (1995). Longitudinal organization of the anterior neural plate and neural tube. *Development* 121:3923–3933.

Sieber-Blum, M. (1989). Inhibition of the adrenergic phenotype in cultured neural crest cells by norepinephrine uptake inhibitors. *Dev. Biol.* 136(2):372–380.

Simeone, A., Acampora, D., Gulisano, M., Stornaiuolo, A., Boncinelli, E. (1992a). Nested expression domains of four homeobox genes in developing rostral brain. *Nature* 358: 687–690.

Simeone, A., Gulisano, M., Acampora, D., Stornaiuolo, A., Rambaldi, M., Boncinelli, E. (1992b). Two vertebrate homeobox genes related to the Drosophila *empty spiracles* gene are expressed in the embryonic cerebral cortex. *EMBO J.* 11:2541–2550.

Simeone, A., Acampora, D., Mallamaci, A., Stornaiuolo, A., D'Apice, M.R., Nigro, V., Boncinelli, E. (1993). A vertebrate gene related to *orthodenticle* contains a homeodomain of the *bicoid* class and demarcates anterior neuroectoderm in the gastrulating mouse embryo. *EMBO J.* 12:2735–2747.

Song, Z.H., Young, W.S., Brownstein, M.J., Bonner, T.I. (1994). Molecular cloning of a novel candidate G protein-coupled receptor from rat brain. *FEBS Lett.* 351:375–379.

Spana, E.P., Doe, C.Q. (1995). The prospero transcription factor is asymmetrically localized to the cell cortex during neuroblast mitosis in Drosophila. *Development* 121:3187–3195.

Spana, E.P., Doe, C.Q. (1996). Numb antagonizes Notch signaling to specify sibling neuron cell fates. *Neuron* 17:21–26.

Spana, E.P., Kopczynski, C., Goodman, C.S., Doe, C.Q. (1995). Asymmetric localization of Numb autonomously determines sibling neuron identity in the Drosophila CNS. *Development* 121:3489–3494.

Stanfield, B.B., O'Leary, D.D.M. (1985). Fetal occipital cortical neurons transplanted to rostral cortex develop and maintain a pyramidal tract axon. *Nature* 313:135–137.

Suzuki, N., Rohdewohld, H., Neuman, T., Gruss, P., Scholer, H.R. (1990). Oct-6: a POU transcription factor expressed in embryonal stem cells and in the developing brain. *EMBO J.* 9:3723–3732.

Tan, S.S., Breen, S. (1993). Radial mosaicism and tangential cell dispersion both contribute to mouse neocortical development. *Nature* 362:638–640.

Tan, S.S., Taulkner-Jones, B., Breen, S.J., Walsh, M., Bertram, J.F., Reese, B.E. (1995). Cell dispersion patterns in different cortical regions studied with an X-inactivated transgenic marker. *Development* 121:1029–1039.

Tao, W., Lai, E. (1992). Telencephalon-restricted expression of BF-1, a new member of the HNF-3/*fork head* gene family, in the developing rat brain. *Neuron* 8:957–966.

Temple, S., Davis, A.A. (1994). Isolated rat cortical progenitor cells are maintained in division in vitro by membrane-associated factors. *Development* 120:999–1008.

Thor, S. (1995). The genetics of brain development: conserved programs in flies and mice. *Neuron* 15:975–977.

Walsh, C., Cepko, C.L. (1988). Clonally related cortical cells show several migration patterns. *Science* 241:1342–1345.
Walsh, C., Cepko, C.L. (1992). Widespread dispersion of neuronal clones across functional regions of the cerebral cortex. *Science* 255:434–440.
Walther, C., Gruss, P. (1991). *Pax6,* a murine paired box gene, is expressed in the developing CNS. *Development* 113:1435–1449.
White, J., Strome, S. (1996). Cleavage plane specification in C. elegans: how to divide the spoils. *Cell* 84:195–198.
Woodward, W.R., Chiaia, N., Teyler, T J., Leong, L., Coull, B.M. (1990). Organization of cortical afferent and efferent pathways in the white matter of the rat visual system. *Neuroscience* 36:393–401.
Woolsey, T.A., Van Der Loos, H. (1970). The structural organization of layer IV in the somatosensory region (SI) of mouse cerebral cortex. The description of a cortical field composed of discrete cytoarchitectonic units. *Brain Res.* 17:205–242.
Xu, Y., Baldassare, M., Fisher, P., Rathbun, G., Oltz, E.M., Yancopoulos, G.D., Jessell, T.M., Alt, F.W. (1993). *lh-2:* a LIM/homeodomain gene expressed in developing lymphocytes and neural cells. *Proc. Natl. Acad. Sci. U.S.A.* 90:227–231.
Xuan, S., Baptista, C.A., Balas, G., Tao, W., Soares, V.C., Lai, E. (1995). Winged helix transcription factor BF-1 is essential for the development of the cerebral hemispheres. *Neuron* 14:1141–1152.
Zhong, W., Feder, J.N., Jiang, M.-M., Jan, L.Y., Jan, Y.N. (1996). Asymmetric localization of a mammalian Numb homolog during mouse cortical neurogenesis. *Neuron* 17:43– 53.

13

The development of the *Drosophila* visual system

TANYA WOLFF, KATHLEEN A. MARTIN,
GERALD M. RUBIN AND S. LAWRENCE ZIPURSKY

The nervous system comprises numerous cell types arranged and interconnected in a remarkably precise fashion. This precision emerges through the coordinated activities of many genes. To understand the cellular and molecular mechanisms underlying the construction of the nervous system, it is necessary to identify and characterize the individual players and the pathways in which they participate. Because of the array of both classical and molecular genetic approaches in *Drosophila*, it is well suited to such dissection.

Several features of the development of the fly eye make it a particularly favorable system for detailed analysis of neural development. Through the classic studies of Benzer, Ready, and Tomlinson, the cellular dynamics of pattern formation in the developing eye are known at the level of identifiable individual cells. This has provided an important framework for characterizing in detail the effects of mutations. Since the adult eye is easily observed under a dissecting microscope and is not essential for viability, mutations affecting its development can be readily identified. Furthermore, various genetic strategies have been developed which facilitate the identification of genes essential for organismal viability that also participate in controlling eye development. Many of these genes have been cloned and their roles in specific developmental pathways have been studied in detail.

In this review we consider recent progress on eye development. We describe the structure and development of the visual system and the genetic strategies used to identify genes controlling its development. We then discuss the molecular basis of a number of different processes regulating individual steps in visual system development.

THE STRUCTURE OF THE ADULT VISUAL SYSTEM

The *Drosophila* compound eye has approximately 800 hexagonal unit eyes, or ommatidia (Ready et al., 1976) (Fig. 13–1A). Each ommatidium is a precise assembly of 20 cells: eight photoreceptors, four non-neuronal cone cells, two pri-

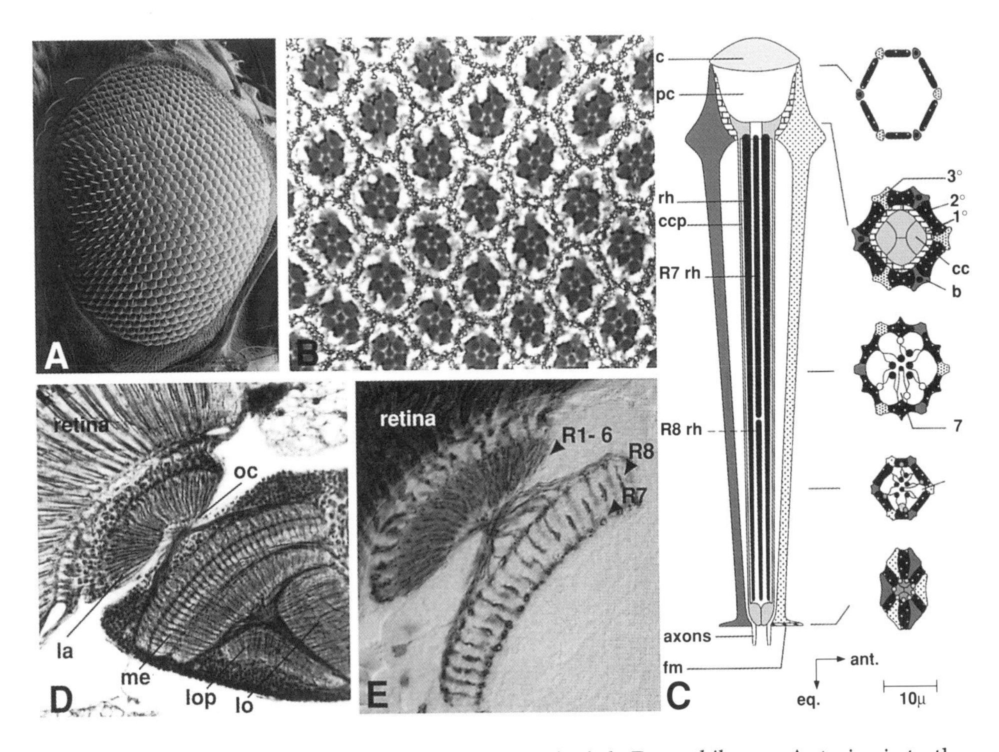

FIGURE 13–1 (A) Scanning electron micrograph of adult *Drosophila* eye. Anterior is to the right. (B) Tangential section through an adult eye. The rhabdomeres, or photosensitive stacks of microvilli elaborated by the photoreceptors, are arranged in an asymmetrical trapezoidal pattern. The rhabdomeres of R1–R6 are large in size and extend the full depth of the ommatidium; those of R7 and R8 are smaller in size. R8's rhabdomere lies below that of R7 and is not visible in this plane of section. The rhabdomeres of R1–R6 form a trapezoid with R3 at the tip; R7's rhabdomere occupies the center of the trapezoid. (See Fig. 13–1C for details.) The cell bodies of the photoreceptors extend radially, behind the rhabdomeres. The two chiral forms of the trapezoid seen in this section fall on opposite sides of the equator; in the dorsal half of the eye the tip of the trapezoid points north and in the ventral half it points south. Anterior is to the right. (C) Schematic of adult ommatidium. The non-neuronal cone cells contribute to two extracellular secretions, the chitinous corneal lens and the nonchitinous pseudocone. The primary pigment cells contribute to the secretion of lens material and all pigment cells optically insulate the ommatidium. c, cornea; pc, pseudocone; rh, rhabdomere; ccp, cone cell process; fm, fenestrated membrane; cc, cone cell; b, bristle; 1–8, photoreceptors R1–R8; 1°, 2°, 3°, primary, secondary, and tertiary pigment cells, respectively. Anterior is to the right. (D and E) Horizontal sections through wild-type adult retina and optic lobes stained with silver (D) or a photoreceptor-specific antibody MAb24B10 (E). Abbreviations in D: lamina (la), first optic chiasm (oc), medulla (me), lobula (lo), lobula plate (lop). (E) R1–R6 cells synapse in the lamina and the R7 and R8 synapse in two distinct layers of the medulla. Anterior is up and medial is to the right. Bar, 20 µm.

mary pigment cells, three secondary pigment cells, one tertiary pigment cell, and one bristle complex containing four cells, two of which die during pupal life (Fig. 13–1C). Each of these cell types can be distinguished by their distinct morphology and position and by the expression of cell-type–specific markers. In this review, we focus on the development of the eight photoreceptor cells (retinula cells or R cells) that form the core of each ommatidium (Fig. 13–1B,C).

The R cells fall into three classes based on spectral sensitivity and synaptic specificity. R1–R6 express Rhodopsin 1 (Rh1) and each R7 expresses one of two different rhodopsins, either Rh3 or Rh4. Microspectrophotometric studies indicate that there are two R8-specific opsins, one of which has recently been identified (S. Brill, personal communication).

The visual processing region of the fly brain, the optic lobe, is composed of four ganglia: the lamina, medulla, lobula, and lobula plate (Fig. 13–1D). Their synaptic organization creates multiple retinotopic maps which underlie the precision of fly vision (reviewed in Meinertzhagen and Hanson, 1993). Unlike the vertebrate retina, the photoreceptor cells of the fly eye project directly to the brain; R1–R6 synapse in the first optic ganglion, the lamina, and R7 and R8 synapse in different layers of the second optic ganglion, the medulla (Fig. 13–1E). Both the lamina and the medulla contain some 800 columnar units which precisely match the number of ommatidial units. There is a direct mapping of visual space from the eye onto the lamina, and an inverted map (along the anterior/posterior axis) created by the first optic chiasm in the medulla.

Perhaps one of the most remarkable examples of specific neuronal connectivity documented in any system is the complex pattern of synapses that R1–R6 neurons elaborate in the lamina (Braitenberg, 1967). The R1–R6 cells within each ommatidium project to six different but neighboring groups of otherwise-identical postsynaptic cells in the lamina, referred to as cartridges. The six R cells which project to a single cartridge have identical optical axes in that they are oriented toward the same point in space. The R7 and R8 cells which share the same optical axis as these six R cells project through the lamina and form synapses at different depths within the same synaptic unit in the medulla neuropil, called a column. The lamina neurons innervated by these R1–R6 cells synapse in the same column of postsynaptic units in the medulla. As a consequence of this precise wiring, the information from the eight R cells looking at the same point in space coalesces in a single medulla column. The other neuropil regions of the optic lobes, the proximal medulla, lobula, and lobula plate, do not receive direct projections from the eye but are innervated by neurons in the optic lobes and the central brain. Although the optic lobe neuronal cell bodies surrounding the neuropils are not distinct, the axonal morphologies are highly invariant and have been described in considerable detail (Fischbach and Dittrich, 1989; Strausfeld, 1976).

THE DEVELOPMENT OF THE COMPOUND EYE

The adult *Drosophila* eye derives from the eye-antennal imaginal disc, an epithelial sac of between six and 23 cells set aside approximately 5 hours into embryogenesis (reviewed in Cohen, 1993). Cells within the eye disc primordium form a columnar epithelium that proliferates throughout larval life (reviewed in Jürgens and Hartenstein, 1993). Pattern formation and differentiation in the eye disc proceed as a wave, beginning at the posterior margin of the eye disc early in the third instar of larval development and advancing anteriorly over a period of about 2 days.

The leading edge of the advancing pattern is marked by the morphogenetic furrow (MF), an apical indentation that runs along the dorsal/ventral axis of the eye disc epithelium (Fig. 13–2A). The MF forms as a consequence of synchronized changes in cell shape associated with early patterning events: The apical surfaces of cells become significantly constricted and cells undergo a dramatic shortening along the apical/basal axis. Ahead of the MF, cells are unpatterned and divide asynchronously (Fig. 13–2B). As the MF progresses anteriorly, it leaves in its wake a developmental gradient of evenly spaced assembling ommatidia such that the least mature units are situated immediately behind the MF and the most mature clusters are located at the posterior edge (Fig. 13–2B,C).

Although early studies on ommatidial assembly argued for a strict lineage relationship between cells in each ommatidium, genetic mosaic studies carried out in the 1970s clearly established that there were no strict cell lineage relationships between cells, arguing that the precise arrangement of cells in the ommatidium was a consequence of cellular interactions (Campos-Ortega and Hofbauer, 1977; Hofbauer and Campos-Ortega, 1976; Lawrence and Green, 1979; Ready et al., 1976; Wolff and Ready, 1991). Similarly, a nonlineage-dependent mechanism of

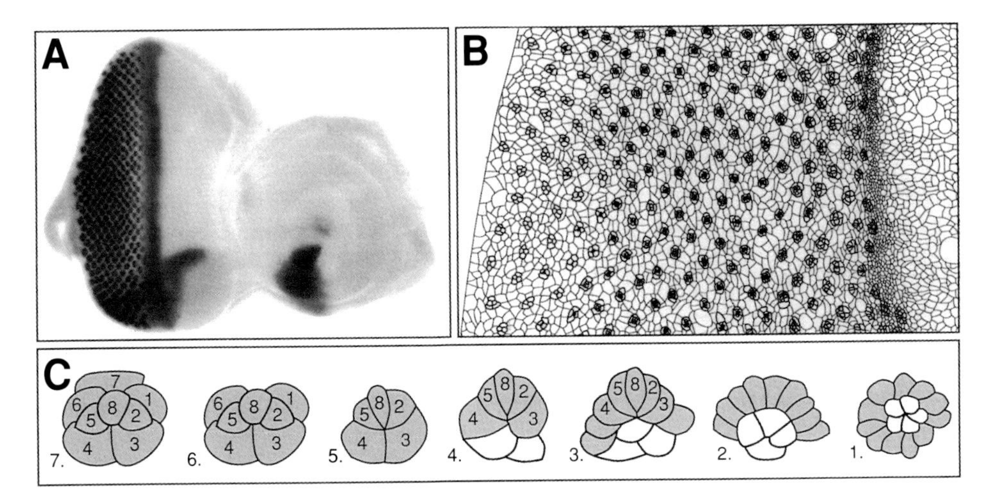

FIGURE 13–2 (A) Third larval instar eye-antennal imaginal disc. The morphogenetic furrow is marked with a *dpp* enhancer trap and is evident as a dark stripe running from dorsal to ventral across the disc. Posterior to the furrow, photoreceptors of developing ommatidial clusters are labeled with Elav, a neuronal antibody. The precursors to the adult ommatidia are distributed in a highly ordered array which prefigures the precision of the adult lattice. (B) Video-lucida drawing of the central portion of a third instar eye disc showing the outlines of cells as seen on the apical surface. The apical surfaces of cells within the furrow are highly constricted. Anterior to the furrow, cells are unpatterned; pattern formation begins in the furrow and continues to the posterior of the eye disc. At the back of the eye disc, ommatidial precursors consist of eight photoreceptors overlaid by four cone cells. Anterior is to the right. (C) Video-lucida tracings of early forms of pattern formation in the third instar eye disc. See text for details.

development has also been shown to operate in vertebrate retinal development (Holt et al., 1988; Turner and Cepko, 1987; Wetts and Fraser, 1988).

Detailed studies of the cellular dynamics of pattern formation (Ready et al., 1976; Tomlinson and Ready, 1987) have laid the groundwork for subsequent molecular and genetic studies of cellular interactions regulating this process. The earliest patterning events in the eye imaginal disc are highly dynamic. In the MF, periodically spaced clusters of cells have been identified based on morphological (Wolff and Ready, 1991; reviewed in Wolff and Ready, 1993) as well as molecular (Baker et al., 1990; Brown et al., 1995; Jarman et al., 1993) criteria. These periodically spaced groups of cells, termed rosettes, consist of an organized collection of 15–20 cells embedded within the MF (Fig. 13–2C-1). Aggregates of disorganized cells, recognized by their more darkly stained membranes in some classical histological preparations (i.e., cobalt sulfide), are also evident in the MF; since their positions approximate the spacing of rosettes, they may represent a "prerosette" stage (not shown).

Early patterning continues as rosettes transform into arcs as some cells dissociate from the group, leaving a curved row of seven to nine elongated cells (Fig. 13–2C-2). The arc then folds in half, beginning at its apex (Fig. 13–2C-3,4). Each group of cells will resolve to the five-cell precluster, in which the identity and position of each cell is highly stereotyped (Ready et al., 1976) (Fig. 13–2C-5). One to two additional cells from the arc are sometimes temporarily associated with the five-cell precluster (Fig. 13–2C-4, unshaded cells) and have been shown to express a subset of R-cell–specific gene products (Bowtell et al., 1989). These cells, termed the mystery cells (Tomlinson and Ready, 1987), are expelled from the cluster and are thought to rejoin the pool of surrounding, undifferentiated cells. Within the precluster there is a precise order of differentiation, with R8 differentiating first, followed by the pair-wise differentiation of R2 and R5, and then R3 and R4. Cells between preclusters undergo a final round of cell division, generating the remaining cells in the ommatidium (see below). R1 and R6 then differentiate and join the cluster (Fig. 13–2C-6), followed by R7 (Fig. 13–2C-7), thus completing the photoreceptor core.

The remaining cells forming each ommatidium are added in stepwise fashion during the end of larval life and the first third of pupal life. The four cone cells are added first, followed by the primary pigment cells, bristles, and finally the secondary and tertiary pigment cells (reviewed in Wolff and Ready, 1993). Two phases of cell death during pupal life remove extra cells within the eye lattice; this normally occurring death can be blocked by artificial expression of cell-death inhibitors (Hay et al., 1994, 1995; Chapter 7, this volume).

THE DEVELOPMENT OF THE OPTIC LOBES

The optic lobes of the adult brain are derived from a contiguous region of posterior head ectoderm in the embryo that lies just posterior to the eye anlage (Green et al., 1993). The optic lobe primordium, containing some 80 cells, invaginates and then pinches off from the surrounding epithelium and forms a vesicle which then fuses with the central brain. The optic lobe primordium remains mitotically quiescent until the latter part of the first larval instar when it separates into two epithelial fields which generate the neurons of the adult optic lobes. The outer proliferation center (OPC) gives rise to the lamina and outer medulla, whereas the inner proliferation center (IPC) gives rise to the lobula complex and the inner medulla (Hofbauer and Campos-Ortega, 1990; White and Kankel, 1978). The development of the lamina and medulla is critically dependent upon retinal innervation (see below).

The larval photosensitive organ, called Bolwig's organ, comprises 12 photoreceptor neurons and is derived from the ventral lip of the invaginating optic lobe primordium (Green et al., 1993). While morphogenetic movements separate the cell bodies of Bolwig's organ from the brain, contact is maintained between the two structures by axonal processes extended by the larval photoreceptors. These axons become ensheathed within a narrow epithelial tube, the optic stalk, which connects the eye disc to the optic lobes (Steller et al., 1987). The synaptic targets of a subset of axons in Bolwig's nerve include three neurons, called the optic lobe pioneers (OLPs) (Tix et al., 1989). The OLPs are located at the base of the optic stalk where their axons fasciculate with Bolwig's nerve. This fascicle projects through the optic lobes and into the central brain. The presence of Bolwig's nerve in the optic stalk prior to R-cell outgrowth led to the proposal that it may function as a pioneer neuron directing subsequent outgrowth of R cells into the optic stalk and into the brain (Steller et al., 1987). Although this is an attractive hypothesis, Bolwig's organ is unlikely to serve this function, since ablation of it by cell-specific expression of diphtheria toxin did not prevent R-cell axons from reaching their targets (Kunes and Steller, 1991). Whether the OLPs function as pioneer neurons for the R-cell projections into the optic lobes has not been critically addressed.

Projections of R-Cell Growth Cones Into the Optic Lobe

Differentiating R cells in the eye imaginal disc extend axonal processes from their basal surface within 2 hours of acquiring a neuronal fate (Tomlinson and Ready, 1987). Axons initially grow basally, piercing the basement membrane of the eye disc epithelium, and enter a region containing glial cells below the differentiating ommatidia. The axons turn and course posteriorly through the optic stalk into the developing optic lobes (Fig. 13–3A,B). The R-cell axons within the optic stalk are arranged in a remarkably precise fashion. Eight axons from each ommatidium form bundles separated by glial cell processes (Meinertzhagen and Hanson, 1993).

Projections of the R cells in the third larval instar form a retinotopic array both in the lamina and the medulla neuropils. R1–R6 elaborate large growth cones in the lamina, forming a discrete plexus bordered by two layers of glial cells (Fig. 13–3A). At this early stage the lamina plexus is largely occupied by R-cell growth cones. The termination of R1–R6 growth cones nestled between two lines of glial cells has led to the speculation that these cells provide a signal for

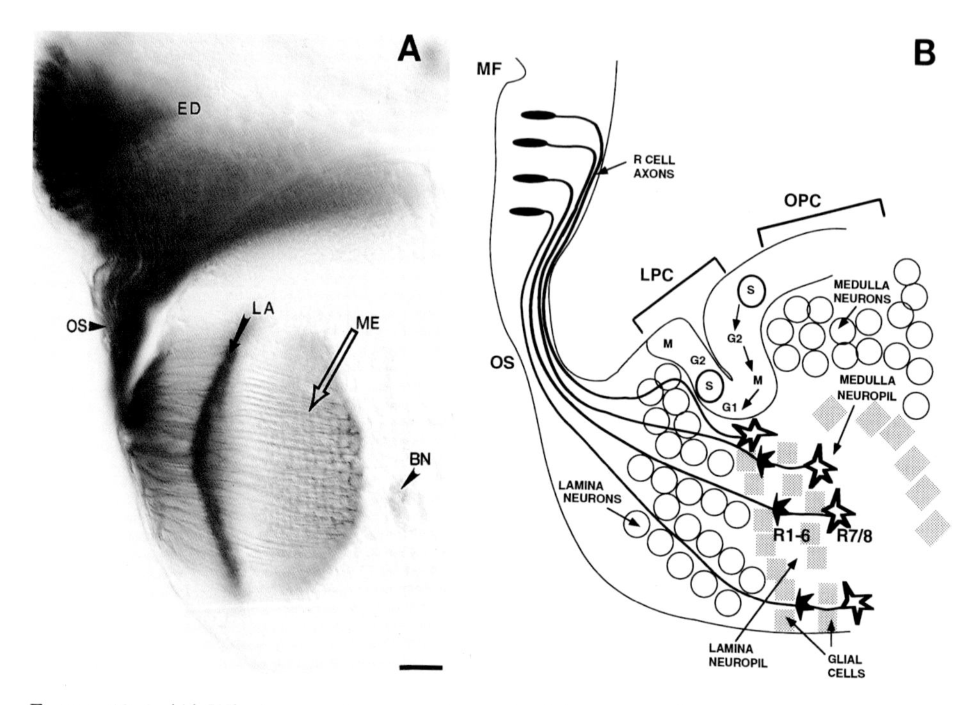

FIGURE 13–3 (A) Whole-mount preparation of wild-type third instar eye–brain complex stained with MAb24B10 to reveal R-cell axons. The growth cones of R1–R6 cells are precisely aligned in the developing lamina neuropil, giving rise to a continuous band of staining. The arrangement of stained R8 growth cones in the medulla neuropil precisely reflects the location of their cell bodies within the eye disc. In this perspective of the optic lobe, lateral is to the left and ventral is down. Abbreviations: eye disc (ED), optic stalk (OS), lamina neuropil (LA), medulla neuropil (ME), terminus of Bolwig's nerve (BN). Bar, 20 μm. (B) A schematic representation of the R-cell projection pattern in a third instar eye brain complex. Note that this perspective is not the same as that shown in A. In this panel, lateral is to the left and posterior is down. Neuroblasts of the outer proliferation center (OPC) give rise to lamina precursor cells (LPC). The LPCs divide twice (S-G1-M, etc.) as they move through the furrow on the surface of the brain. Medulla neurons are generated from the medial edge of the OPC (not shown). R-cell growth cones exit the optic stalk near the LPC and grow into the developing optic lobes. Both the second mitotic division of the LPCs and the aquisition of neuronal fate by their progeny require R-cell innervation (see text). R1–R6 growth cones (filled) terminate between two rows of glial cells (stippled squares) within the lamina neuropil. R8 and R7 growth cones (outlined) pass through the lamina and terminate in the medulla neuropil. The first growth cones from each ommatidium to reach the LPC are likely to be R8s. In more developed posterior regions of the lamina, five lamina neurons of a single cartridge align along a bundle of eight R-cell axons from a single ommatidium.

growth cone termination. As development proceeds, the thin processes of lamina neurons begin to project along R-cell axons through the lamina and into the medulla neuropil. R8 growth cones terminate in the medulla neuropil, which contains both R-cell growth cones and the processes of developing medulla and lobula complex neurons.

Electron micrographic studies indicate that when R-cell axons first enter the developing optic lobe they are not yet complete bundles of eight axons (Meinertzhagen and Hanson, 1993; Tomlinson and Ready, 1987). The precise sequence of R-cell ingrowth into the optic lobes is not known and will require the use of R-cell subclass-specific axonal markers. Based on the ordered differentiation of R cells in the eye disc, sequential outgrowth of different R-cell axons within the same cluster, and the behavior of R1–R6 growth cones in the lamina (Meinertzhagen and Hanson, 1993), it seems likely that R cells enter the developing optic lobe in the order in which they differentiate in the eye disc: R8, followed by R1–R6, and finally by R7.

The emergence of the R1–R6 projection pattern in the lamina has been reconstructed at the EM level and is described in considerable detail by Meinertzhagen and Hanson, (1993). In brief, whereas the R1–R6 neurons from each ommatidium synapse with six different lamina cartridges in the adult, the growth cones of R1–R6 cells from a given ommatidium are initially associated with the same developing cartridge. The R1–R6 growth cones extend an array of filopodial processes throughout the lamina plexus and then undergo a stereotyped pattern of morphological changes, culminating in the divergence of these growth cones to neighboring lamina cartridges. Consequently, each postsynaptic cartridge is innervated by six photoreceptor neurons, all looking at the same point in space.

R Cells Elaborate a Topographic Map

In the third larval instar the terminals of the R-cell axons within the lamina and medulla fields accurately reflect the position of their cell bodies within the eye disc. The retinotopic map created by R-cell growth cones is maintained throughout the complex morphogenetic rotations which occur during the pupal stages (Meinertzhagen and Hanson, 1993). As a result, the direct mapping of R cells to the lamina is maintained, whereas the map within the medulla neuropil is reversed along the anteroposterior axis. This reversal forms the first optic chiasm (Fig. 13–1D).

Little is known about the mechanisms by which the retinotopic array is generated in development. A handful of experiments to address these questions have been reported and interpreted in favor of positional information (Ashley and Katz, 1994; Kunes et al., 1993). For instance, Kunes et al. (1993) traced the projections into the lamina from small patches of ommatidia which develop in eyes carrying mutations that lead to a marked reduction in the number of ommatidia formed (i.e., *Ellipse* and an eye-specific allele of *sine oculis* so^1)). Due to difficulties in independently assessing position along the anteroposterior axis, the specificity of these connections was determined only along the dorsoventral axis. The terminals of the projections were shown to map to the appropriate position along the dorsoventral axis. However, transplantation experiments in cockroaches argue against the importance of positional information in determining the pattern of R-cell projections into the lamina (Nowel, 1981). R cells from transplanted regions of the retina project to the lamina consistent with their position within the host rather than according to their position in the donor tissue.

To critically address the importance of positional information in specifying the formation of a topographic map, it will be necessary to identify and manipulate the spatial expression of molecules imparting such information.

PRODUCTION OF LAMINA NEURONS IS REGULATED BY R-CELL AXONS

The dependence of lamina development on retinal innervation was described over 50 years ago when Power noted that mutants with a reduced number of ommatidia also had reduced optic lobe volume with the greatest reduction in the lamina (Power, 1943). Indeed, lamina neurons were completely absent in *eyeless* mutants. Additional studies by Meyerowitz and Kankel (1978) using genetic mosaic analysis, demonstrated that normal R-cell innervation was necessary for optic lobe development. Related studies in the crustacean *Daphnia* using cell ablation and electron microscopy suggested that direct contact between R-cell axons and lamina precursors was necessary for lamina cell differentiation (Macagno, 1979).

Recently, Steller, Selleck, and colleagues (Nakato et al., 1995; Selleck et al., 1992; Selleck and Steller, 1991) have used bromodeoxyuridine (BrdU) labeling, cell-type–specific markers, and confocal microscopy to analyze the inductive process in the developing third instar optic lobes in more detail. At this developmental stage, three proliferative centers are visible in the optic lobes: the OPC, IPC, and lamina precursor cells (LPC). Cells in the LPC are derived from the posterior edge of the OPC. The LPCs move through a furrow region (Fig. 13–3B), completing two mitotic divisions before differentiating into neurons. R-cell axons are required for LPCs to undergo the second and final mitotic division. Thus, the number of LPCs in S phase is proportional to the number of incoming R-cell fibers. In addition, a separate role for R-cell axons in regulating the differentiation of lamina neurons has been proposed. The *dally* mutant disrupts the two LPC mitotic divisions but it does not prevent LPCs from differentiating into neurons (Nakado et al., 1995). This suggests that the R-cell axons provide at least two cues to developing LPCs, one to induce the final mitotic division and another to direct the differentiation of the postmitotic cell into a neuron. In addition to regulating neuronal development, R-cell innervation is required for differentiation of lamina glial cells (Winberg et al., 1992). The dependence of lamina development on retinal innervation provides a mechanism for matching the number of postsynaptic units formed to the number of ommatidia in the compound eye.

APPROACHES TO GENETIC ANALYSIS OF VISUAL SYSTEM DEVELOPMENT

Development of the *Drosophila* visual system requires the concerted action of thousands of genes. Mutations affecting the visual system have been identified through traditional screens, including behavioral and morphological screens. Initial molecular studies focused on viable mutants whose functions are restricted to eye development, including *glass, rough, sevenless*, and *bride of sevenless*. It has become clear that to fully understand the process of visual system development we need to employ methods that allow us to dissect the role of components that are also used in other tissues and at many developmental stages. Until recently, one limitation to identifying many mutations that inactivate the genes encoding these components has been that they result in lethality. We will briefly review several approaches that have proved successful in identifying the roles of such pleiotropic genes during visual system development.

A majority of the 10,000–15,000 genes present in the *Drosophila* genome are thought to play direct or indirect roles in constructing the visual system. Thaker and Kankel (1992), in an unbiased approach designed to estimate the number of genes required for eye development, produced somatic clones of homozygous mutant cells in the eyes of flies heterozygous for a random selection of X-linked lethals. Of these, one-third produced clones with no phenotype, one-third produced clones with a phenotype, and the final third were either unable to produce clones or produced very small clones, suggesting they were required for cell viability or proliferation. By extrapolation, it is thought that approximately two-thirds of all genes required for viability also play a role in eye development. Some encode components of intra- and intercellular signaling pathways used for communication between cells. It is becoming apparent that most signaling pathways are comprised of proteins that are used during the development of most, if not all, tissues, whose structures and functions have been highly conserved during evolution. Only a very small number of highly tissue-specific components are employed; less than ten genes have been identified that have mutant phenotypes affecting only visual system development. For a nonessential tissue such as the fly eye, mutations in this latter class will produce viable and fertile individuals. Consistent with these genetic data, over 10% of 20,000 enhancer trap lines examined in a series of screens (U. Gaul, L. Higgins, M. Freeman, J. Heilig, G. M. R., unpublished) show expression patterns reflecting gene induction in the visual system during development, but less than five of these lines displayed expression limited to the developing visual system.

Screens Based on the Phenotype of Homozygous Lethal Individuals

The genetic screens carried out by Nüsslein-Volhard and Wieschaus (1980) for embryonic lethal mutations affecting the pattern of the larval cuticle proved to be extremely informative. These screens were based on the generation of homozygous mutant individuals by means of a series of genetic crosses such as those diagrammed in Figure 13–4A. Homozygotes are examined for the phenotype of interest and, if an appropriate phenotype is observed, the mutation is recovered from the heterozygous siblings. Such screens are often of limited use in looking for lethal mutations affecting the development and function of adult structures, since mutant animals often die at early developmental stages. However, in many cases mutant individuals survive long enough to permit the exam-

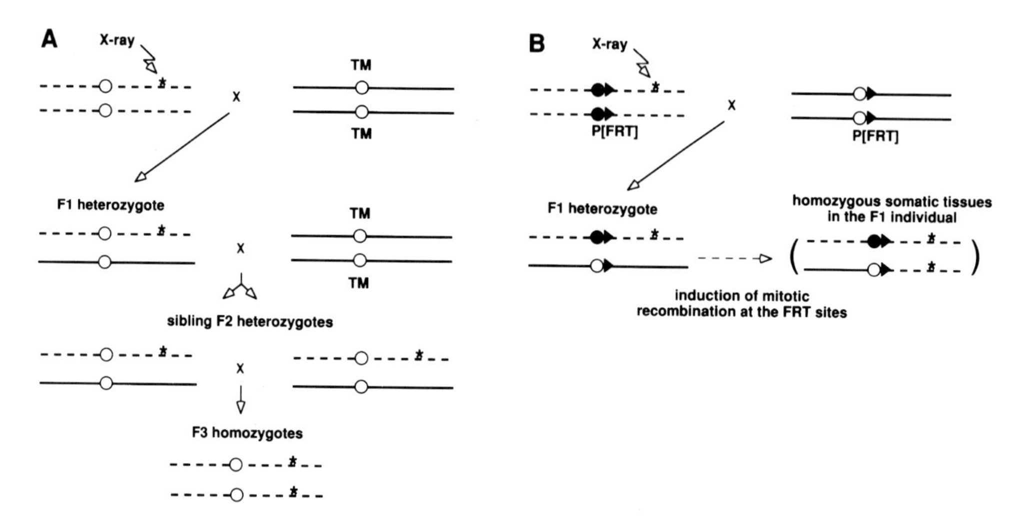

FIGURE 13–4 (A) Diagram of a standard F3 genetic screen. A recessive mutation (*) induced in the germline of the parent is identified in homozygous flies of individual lines that are generated from three generations of crosses. TM, balancer chromosome. (B) Using a strain having a *P[FRT]* element inserted near the centromere of a chromosome arm, induced recessive mutations on that arm can be identified in F1 heterozygous individuals by producing and examining somatic clones of cells that are homozygous for the mutagenized chromosome arm.

ination of phenotypes during larval and pupal stages. This can be the case even if the gene product is required during embryonic development if the mother has placed sufficient wild-type product in the egg during oogenesis to support the early stages of development. In other cases, mutations may disrupt function in the visual system as well as elsewhere but maintain sufficient function for organismal viability.

SCREENS BASED ON THE PHENOTYPE OF HOMOZYGOUS SOMATIC CLONES IN AN OTHERWISE HETEROZYGOUS GENETIC BACKGROUND

Genetic screens that identify mutations by examining mutant phenotypes in homozygous animals are not only biased against mutations in genes with pleiotropic functions but are also laborious, since they require the generation of individual lines of homozygous animals. Although it has been appreciated that lethal mutations affecting adult structures could be identified in mosaics (Garcia-Bellido and Dapena, 1974), the low frequency of induced mosaicism using traditional methods made such an approach impractical for systematic screens. Recently, strains have been specifically engineered to utilize the yeast FRT/FLP site-specific recombination system from yeast to catalyze mitotic recombination (Golic, 1991; Xu and Rubin, 1993). These strains can be used to produce a high frequency of mosaicism for most of the genome and thus provide a powerful way to screen for new mutations that affect development of the visual system.

Instead of isolating mutations in homozygous animals, mutations can be identified by examining clones of mutant tissues within heterozygous animals. Thus, these screens can identify phenotypes in a single generation (F1 screens) as opposed to the three generations required in a standard F2 screen (Fig. 13–4A). Furthermore, unlike standard F2 screens for recessive mutations that are strongly biased against genes in which loss-of-function mutations decrease viability, F1 screens that involve examining mosaic mutant tissues in heterozygous animals allow even lethal mutations to be recovered. Such screens have been used to identify the role of cAMP-dependent protein kinase in morphogenetic furrow progression (Pan and Rubin, 1995) and the role of *phyllopod* in determining the identity of the R1, R6, and R7 photoreceptors (Chang et al., 1995).

Screens Based on Detecting the Phenotype of a Heterozygous Mutant in a Sensitized Background

Screens for second site mutations that modify the phenotype of an existing mutation have proven to be a useful method for discovering additional components of genetic pathways in organisms as diverse as bacteria, yeast, nematodes, and fruitflies. As an example of such a screen applied to the development of the fly visual system, we summarize the work of Karim et al. (1996), who used this approach to identify genes that function downstream of *ras1* during R7 development (see below). Ras1 activation plays a pivotal role in signal transmission from the Sevenless (Sev) receptor tyrosine kinase, and activating mutations in ras1 can bypass the requirement for Sev activation. Since the *Drosophila* compound eye is dispensable for viability and fertility, most deleterious effects associated with the constitutively active Ras1 protein, whose widespread expression would be lethal, can be avoided by expressing it under the control of the *sev*-enhancer/promoter (*sev-Ras1*^{V12}; (Fortini et al., 1992). The *sev-Ras1*^{V12} transgene triggers the presumptive R7 cell to adopt a photoreceptor fate and transforms many of the non-neuronal cone cells into supernumerary R7 cells. The production of ectopic R7 cells by *sev-Ras1*^{V12} disrupts the exterior eye morphology and causes it to become rough in appearance (Fig. 13–5A; Fortini et al., 1992).

Karim et al. (1996) used the *sev-Ras1*^{V12} rough eye phenotype to screen for dominant suppressors and enhancers. The premise of this dominant suppressor/enhancer screen is that, in the sensitized background of *sev-Ras1*^{V12}, a twofold reduction in the dose of a downstream gene (i.e., by mutating one copy of the two present in the diploid genome) will alter the signaling efficiency and thereby visibly modify the rough eye phenotype. This screen was modeled after a successful screen for dominant enhancers of a temperature-sensitive *sev* allele carried out by Simon et al. (1991). Dominant suppressors are expected to carry mutations in genes that act to facilitate Ras signaling, while enhancers are expected to carry mutations in genes that inhibit Ras signaling. An example of a dominant suppressor and a dominant enhancer is shown in Figure 5B and C. Using this assay, Karim et al. (1996) screened approximately 850,000 mutagenized

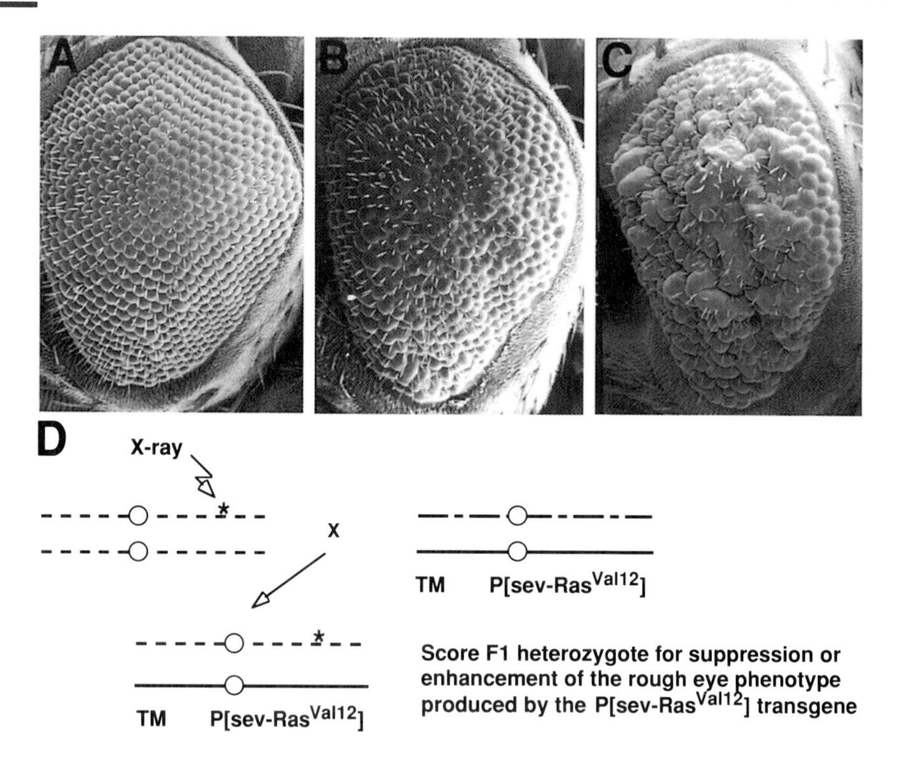

FIGURE 13–5 Scanning electron micrographs of adult eyes showing that the rough eye phenotype produced by an activated Ras1 allele can be suppressed or enhanced by heterozygous mutations in other genes: (A) *sev-Ras1*^{V12}/+. Note that the adult eye appears rough as compared to wild type (see Fig. 13–1A); one can no longer see straight rows of ommatidia and the lenses over many of the ommatidia become fused. (B) *rolled*/+; *sev-Ras1*^{V12}/+. One mutant copy of a positive-acting gene (*rl/MAPK*) reduces signaling efficiency, thereby causing the eye to appear less rough. (C) *yan*/+; *sev-Ras1*^{V12}/+. One mutant copy of a negative-acting gene (*yan*) increases signaling efficiency, thereby enhancing the number of ectopic R7 cells and causing the eye to become extremely rough. (D) Diagram of the genetic crosses used to screen for suppressors or enhancers of activated Ras1. A recessive mutation (*) induced in the germline of the male parent is identified in its progeny that are heterozygous for the mutation and a balancer chromosome (TM) that carries the *sev-Ras1*^{V12} transgene.

progeny and isolated several hundred dominant suppressors and enhancers of *sev-Ras1*^{V12} that defined over 30 genetic loci.

There are many advantages to this type of dominant suppressor/enhancer screen. First, large numbers of mutagenized genomes can be screened, since one is scoring the phenotype of F1 individuals. Second, because many of these dominant suppressors and enhancers are homozygous cell-lethal or homozygous viable with no phenotype, they would not have been identified in screens based on recessive eye phenotypes, either in the whole organism or in homozygous somatic clones. The effectiveness of this approach is demonstrated by the isolation of mutations in several genes of this class encoding proteins such as Raf, MEK, MAPK, as well as in a dozen other critical signaling genes. These genes revealed the outlines of the pathway from Ras1 activation at the membrane to changes in gene expression in the nucleus (Fig. 13–6A,B). Some of these genes appear to be

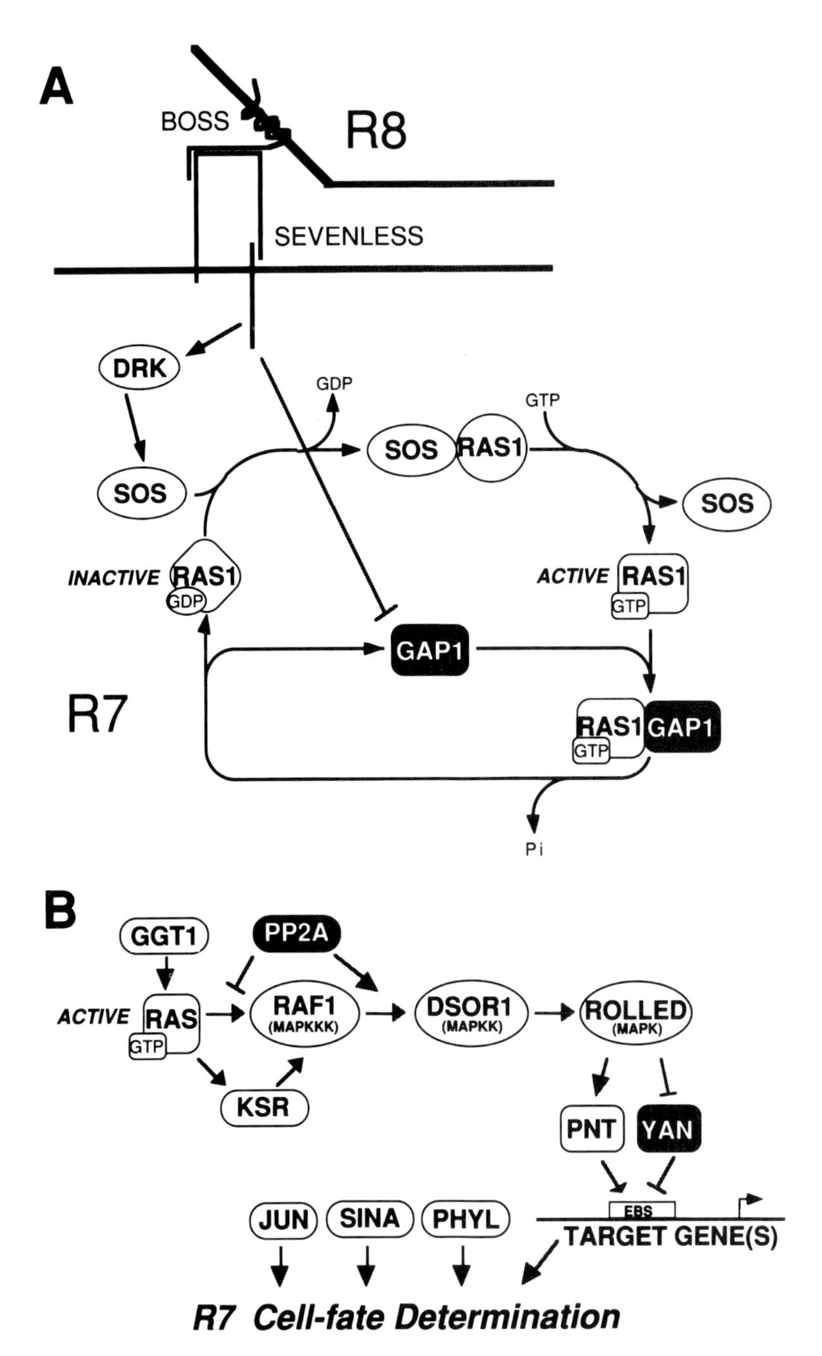

FIGURE 13–6 Summary of the signal transduction pathway initiated by BOSS activation of the SEVENLESS receptor tyrosine kinase. (A) BOSS, a membrane-bound protein expressed on the surface of the R8 cell, binds to the SEV receptor present on the membrane of R7. The activated receptor is thought to bind the adapter protein DRK which in turn binds the guanine nucleotide release factor SOS, bringing it to the membrane where it can activate RAS1. GAP1 is a GTPase activating protein which negatively regulates RAS1 activity. (B) Once activated, RAS1 initiates a kinase cascade consisting of RAF1, DSOR1/MAPKK, and ROLLED/MAPK that leads to the phosphorylation and modification of the activity of the transcription factors YAN and PNT. For additional explanation and references, see Zipursky and Rubin (1994), Wassarman et al. (1995), Karim et al. (1996), and Therrien et al. (1995).

general factors that function in multiple Ras1 signaling events, while others may be cell specific.

Despite these advantages, there are also certain limitations to this type of screen. Even though approximately 850,000 mutagenized progeny were screened, mutations were not isolated in every signaling gene known to play a role in this pathway. As noted earlier, the premise of the screen is that a 50% reduction in the dose of a critical gene might increase or decrease signaling efficiency enough to alter the rough eye phenotype. However, if a particular signaling protein is in vast excess, then a greater-than-50% reduction may be required to visibly alter the rough eye phenotype. In this case only dominant-negative alleles, which are relatively rare, would modify the rough eye sufficiently to be detected. Still other genes might have been missed because they have only a minor effect on the *sev-Ras1*^{V12} rough eye, perhaps because their protein products act downstream of a signal branch point and as such only transmit part of the signal. In this case a null mutation would be expected to suppress *sev-Ras1*^{V12} only weakly.

SCREENS BASED ON PATTERN OF EXPRESSION—ENHANCER TRAP MUTAGENESIS AND IN SITU HYBRIDIZATION TO mRNA

An additional problem confounding the genetic analysis of any developmental process is that a majority of genes, in all multicellular organisms for which the data exist, do not mutate to a phenotype that can be detected under normal conditions in the laboratory. The reasons for this are not entirely clear but probably involve both the difficulties in detecting very subtle abnormalities and redundancy in genetic pathways. Mutations in some of these genes are detected in screens for enhancers and suppressors of other phenotypes. However, the approach that has proven most fruitful is identification of genes based on their expression at a time or place that implicates them as components of a developmental process. Two methods have been used to monitor gene expression: enhancer trap mutagenesis (O'Kane and Gehring, 1987) and in situ hybridization to mRNA in either wholemount or tissue sections (Kopczynski et al., 1996). Once identified, the study of such genes remains problematic since their loss-of-function phenotype provides little information. Several approaches are being tried to surmount this difficulty. Redundancy can be assessed by construction of appropriate double mutant combinations; knowledge of the full coding potential of the genome that is emerging from genome projects will be particularly useful in identifying potentially overlapping functions. Another approach that has had some success is ectopic expression of the gene product, which will, in some fraction of the cases, generate a phenotype that provides insight into the normal function of the gene.

GENETIC ANALYSIS OF VISUAL SYSTEM DEVELOPMENT

Over the past 10 years, considerable progress has been made in uncovering the molecular mechanisms regulating visual system development. These studies have provided insights into a variety of issues of general interest to developmental neurobiologists. In the following section, we briefly review progress in several different areas.

Early Steps in Compound Eye Development

Genetic studies in mouse and human implicated the *pax-6/aniridia* gene as an early acting gene in vertebrate eye development (Hill et al., 1991; Ton et al., 1991; reviewed in Hanson and Van Heyningen, 1995). Quiring et al. (1994) identified a *Drosophila* homologue of these genes and mapped it to the previously identified *eyeless* locus (reviewed in Halder et al., 1995b). *eyeless* encodes a transcription factor containing both a paired box domain and a homeodomain. The sequence of the fly gene shares over 90% sequence identity in both DNA-binding regions with the mouse and human genes. Loss-of-function mutations in the *eyeless* gene disrupt eye development at an early stage, resulting in flies with severely reduced or completely missing eyes. Gehring and his colleagues (Halder et al., 1995a, Quiring et al., 1994) have proposed that the *eyeless* gene is a master regulator of eye development. In support of this view, ectopic expression of the *eyeless* gene in other imaginal discs, such as leg or wing discs, led to the production of ectopic eyes with fully constructed ommatidia containing the normal number and organization of different cell types. Hence, in these ectopic locations *eyeless* can initiate expression of the full complement of target genes required for eye formation. The strong conservation of both sequence and function suggests a common evolutionary origin for the different type of eyes seen in flies and vertebrates. Indeed, expression of mouse *pax-6* in other imaginal discs also induces ectopic fields of ommatidia (Halder et al., 1995a). This challenges the pervasive view among evolutionary biologists that different eyes resulted through a process of convergent evolution.

Since *C. elegans* do not have photoreceptors, the isolation of a *C. elegans pax-6* homologue, *vab-3*, suggests that this family of genes may have initially specified anterior structures more generally and was only later co-opted for eye development (Chisholm and Horvitz, 1995; Zhang and Emmons, 1995). The *vab-3* gene regulates head morphogenesis including the specification of the most anterior lateral hypodermal cell and head neurons. Perhaps the primordial *eyeless/pax-6/aniridia* gene functioned more like other master control genes in the *bithorax* and *Antennapedia* complexes that control regional identity in posterior regions of flies, mice, and human.

Analysis of another early acting eye mutation, *sine oculis (so)*, further suggests that early steps in eye development in *Drosophila* and mammals may be similar. *so* encodes the prototype of a new family of homeobox-containing transcription factors (Cheyette et al., 1994). In *Drosophila so* appears to function downstream from *eyeless* (Halder et al., 1995a; Quiring et al., 1994). The *so* gene is required for the formation of the entire visual system including eye and optic lobe regions of the brain as well as the larval photoreceptor organ. A conserved mouse homologue has recently been isolated and is expressed in the most anterior region of the neural plate including the developing eye and optic tract (Oliver et al., 1995). Using genetic screens in *Drosophila*, it should be possible to identify additional genes that interact with *eyeless* and *sine oculis* to control early steps in eye development. These studies will provide a framework in which to study related genes controlling the development of the vertebrate retina.

Presumably, *eyeless* sets into motion a cascade of gene activity that controls eye development. *eyeless* is expressed throughout the region anterior to the morphogenetic furrow (MF). In response to extracellular signals produced within and just posterior to the MF, cells enter a program of retinal differentiation just anterior to the MF (see below). The competence of anterior tissue to respond to these signals may be conferred by *sine oculis* expression. *so* mutant clones in the eye disc continue to express *eyeless* and to proliferate but cannot initiate retinal patterning events including cell-cycle synchronization, neuronal differentiation, and changes in both cellular morphology and organization (F. Pignoni, S.L.Z., unpublished observations).

LONG-RANGE SIGNALS CONTROL THE PROGRESSION OF THE MORPHOGENETIC FURROW

Recent work has provided insight into the molecular and genetic mechanisms underlying the propagation of pattern in the eye and suggests eye development proceeds as a consequence of a cooperation between events occurring just anterior and posterior to the MF. Two key players known to participate in driving the MF from posterior to anterior across the disc are *hedghog (hh)*, a segment polarity gene, and *decapentaplegic (dpp)*, a transforming growth factor–β homologue (Heberlein et al., 1993; Ma et al., 1993). These two genes play widespread roles in both vertebrate and invertebrate development (see Chapters 1 and 8). Analysis of *hh* and *dpp* function in the fly eye is likely to provide insights into the mechanisms of pattern formation including the identification of other genes in this pathway.

Evidence for the key roles of *hh* and *dpp* in patterning the eye came from analysis of loss-of-function mutants and from ectopic expression studies (Heberlein et al., 1995; Heberlein et al., 1993; Ma et al., 1993). In the third larval instar Hh is expressed, and presumably secreted, by differentiating R cells posterior to the MF; Dpp is expressed in the MF (Fig. 13–7). Disruption of *hh* arrests MF progression and inhibits Dpp expression within the MF. Furthermore, ectopic Hh expression ahead of the MF induces a second MF; cells within this ectopic furrow also express Dpp. This suggests a model for MF progression in which Hh induces Dpp which in turn induces cells just anterior to the MF to form eye tissue. Further support for this model is drawn from the analysis of two factors that appear to negatively regulate *dpp* expression: *patched*, a segment polarity gene, and *pKA-C1*, which encodes the catalytic subunit of cAMP-dependent protein kinase (Fig. 13–7) (Chanut and Heberlein, 1995; Ma and Moses, 1995; Pan and Rubin, 1995; Strutt et al., 1995; Wehrli and Tomlinson, 1995). *pKA-C1* or *patched* loss-of-function mutant clones generated by FRT/FLP-mediated recombination cause the same phenotype as seen with ectopic Hh expression, including ectopic MFs which express Dpp and photoreceptor differentiation.

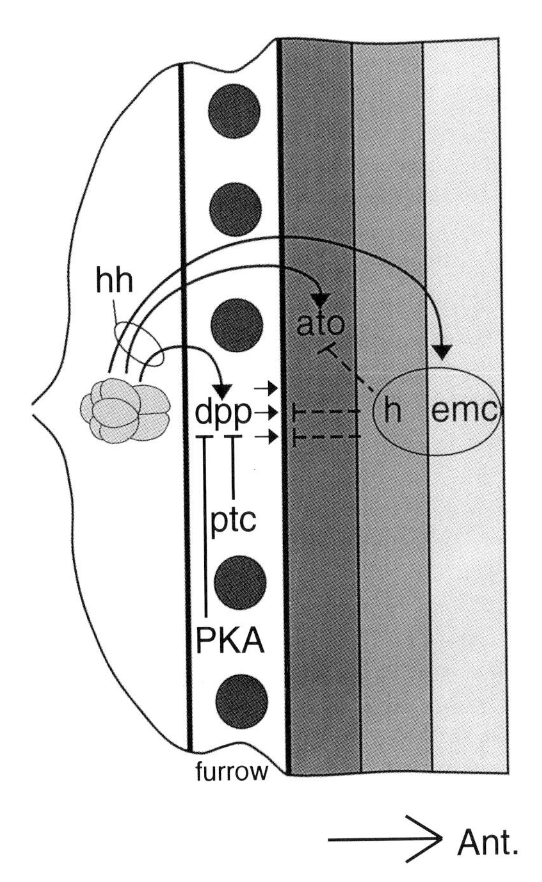

FIGURE 13–7 Schematic of third instar eye disc illustrating the genes involved in furrow progression. See text for details. *hh, hedgehog; dpp, decapentaplegic; ato, atonal; h, hairy; emc, extra macrochaetae; ptc, patched;* PKA, protein kinase A. Anterior is to the right.

MF progression across the eye disc requires continuous cycling of the positive feedback loop between differentiating cells behind the MF and precursor cells ahead of it. Cells induced by Hh/Dpp to differentiate into R cells will later express these same inducers and thereby drive cells anterior to them to form eye tissue. Consequently, genes that influence neuronal differentiation provide a means of regulating MF movement. For example, *atonal (ato),* which encodes a proneural basic helix-loop-helix (HLH) transcription factor, is required for differentiation of R8, the first photoreceptor to be determined (Jarman et al., 1993; see below). In the absence of R8, all future neuronal recruitment is blocked. As a result, Hh is not produced, the MF does not propagate, and *ato* eyes are severely reduced in size, comprising only a few pigment cells.

Whereas *ato* promotes MF movement, *hairy (h)* and *extra macrochaetae (emc)* inhibit it (Brown et al., 1995) (Fig. 13–7). *h* and *emc* are HLH regulatory proteins and are expressed ahead of the MF. *h,* an embryonic segmentation gene, is expressed in a band immediately anterior to the MF; its expression is greatest just anterior to the band of *ato* expression. *emc* is strongly expressed in a stripe anterior to the domain of *h* expression and rapidly decays as *h* expression becomes stronger; it is expressed at low levels within the MF. Whereas loss-of-function mutants in either *h* or *emc* show no, or only a subtle, phenotype on their own, clones of cells doubly mutant for both *h* and *emc* show severe disruptions in patterning. In *emc-h* clones, neuronal development and MF progression are precocious, occurring as much as eight rows in advance of their wild-type counterparts (Brown et al., 1995). This suggests that *emc* and *h* function together to regulate the timing of MF progression and photoreceptor differentiation. The mechanism by which *h* and *emc* regulate these events is unclear, but, since *ato* expression shifts anteriorly in *emc-h* clones, one possibility is that they may act by inhibiting *ato* expression anterior to the MF, either directly or indirectly. If *ato* expression is inhibited, this would prevent proneural determination (see below) and photoreceptor differentiation anterior to the MF in an analogous fashion to the selection of sense-organ precursors, or SOPs, in the peripheral nervous system (see Chapter 2). Alternatively, Emc and H proteins could directly regulate Dpp expression or they could act indirectly by affecting genes involved in interpreting the Dpp signal or inducing Dpp expression.

Together, these studies indicate that events occurring immediately posterior and anterior to the MF are inextricably linked; they are part of a tightly coordinated and continuous developmental program. Differentiating photoreceptors behind the MF produce Hh, which stimulates Dpp as a key determinant of MF movement. The effects of Hh are also conveyed ahead of the MF, where it induces an activator, Ato, and an inhibitor, H, of neural differentiation. *ato* positively regulates neuronal differentiation such that Hh will continue to be produced, and *h* counteracts this effect. This dual regulation in the zone just anterior to the MF may be important in controlling the rate at which the MF progresses across the eye disc. Clearly, once set in motion, this series of genes interacts in a self-regulating cycle to ensure an emergence of pattern from posterior to anterior.

Although considerable progress has been made in studying progression of the MF, little is known about the mechanism by which the MF is initiated at the posterior edge of the eye disc. Prior to photoreceptor differentiation, there is no source of Hh in the eye disc, so initiation must proceed via an *hh*-independent mechanism. *dachsund* is one candidate gene that may be required for this process (Mardon et al., 1994). Furthermore, *wingless*, a member of the *Wnt* gene family (Nusse and Varmus, 1992), is thought to restrict furrow initiation to the posterior margin of the disc (Ma and Moses, 1995; Treisman and Rubin, 1995).

Rux Controls G1 in the Morphogenetic Furrow

Recent studies on the control of cell-cycle progression in the developing eye disc have focused on the mechanism by which cells become synchronized in G1 at the onset of pattern formation within the MF. Anterior to the MF, all cells divide asynchronously in the first mitotic wave (Ready et al., 1976). Using a variety of markers, it has been shown that all cells within the MF are in G1 (Thomas et al., 1994). These cells will either become part of the five-cell precluster and differentiate into neurons or synchronously undergo one final round of cell division in the so-called second mitotic wave. This additional round of division is not required for subsequent cellular differentiation but is required to generate enough cells to fully constitute the cellular composition of each ommatidium (de Nooij and Hariharan, 1995). No further divisions in the eye disc epithelium occur until pupal development, when the cells generating the bristle associated with each ommatidium undergo two divisions (Cagan and Ready, 1989b). Hence, cell-cycle progression in the developing eye is under strict developmental control.

Cell-cycle synchronization occurs in a dorsal/ventral swath of cells just anterior to the MF. This region is marked by the expression of String (Alphey et al., 1992; Thomas et al., 1994), an evolutionarily conserved dual-specificity phosphatase that functions as a mitotic trigger by dephosphorylating and thereby activating CyclinB-Cdc2 (Edgar and O'Farrell, 1989; Kumagai and Dunphy, 1991). Whereas String is normally expressed in cycling cells just prior to the G2/M transition, Hh induces String in all cells just anterior to the MF (Heberlein et al., 1995). All cells in G2 within the String domain are driven through the G2/M transition and into G1. Within the String domain and the MF, cells are prevented from entering S phase by the product of the *rux* locus (Thomas et al., 1994). *rux* encodes a novel protein which prevents cell-cycle progression. The precise mechanism by which it inhibits cell-cycle progression is not yet known. In addition to playing a critical role in controlling G1 progression within the MF, *rux* also prevents cell division in the most posterior, normally quiescent, region of the eye disc.

Loss-of-function *rux* mutations leads to the synchronous entry of all cells within the MF into S phase (Thomas et al., 1994). Although cells assemble into equally spaced clusters, their organization within the clusters and their cellular identities are disrupted. For instance, many ommatidia contain multiple R8 neurons. Hence, the lateral signaling mechanism (see below) that restricts the number of R8 cells in each ommatidium to one is disrupted. It seems likely, then, that at least in some cases cells must be in G1 for normal signaling to take place.

Lateral Signaling Controls R8 Development

The developmental strategy by which a single R8 cell forms within each developing ommatidial unit appears similar to the mechanism by which a single neuroblast or precursors to sensory structures in the peripheral nervous system, so-called sensory organ precursor cells, are selected from a group of developmentally equivalent cells. Our understanding of the mechanisms regulating this process in the SOPs is considerably more advanced than our understanding of R8 selection in the eye. Hence, we only briefly describe this strategy in the context of eye development and refer the reader to Chapter 2.

A common mechanism of cell fate determination regulated by intercellular interactions is the selection of a single cell to assume a specific fate from a group of equivalent cells. Genetic and immunohistological studies argue for the existence of a stepwise process by which a single cell is selected from a lawn of equivalent cells at the anterior edge of the MF to assume an R8 cell fate. The progressive restriction of R8 potential can be visualized by the expression of *ato*. Initially, Ato protein appears as a stripe bordering the anterior edge of the MF and then becomes refined to groups of regularly spaced cells in the MF separated by troughs of nonexpressing cells. Ato expression becomes restricted to groups of two or three precursor cells, termed the R8 equivalence group, as their nuclei rise apically within the epithelium. Expression then resolves to a single cell, the R8 cell (Brown et al., 1995; Jarman et al., 1993; Lee et al., 1996; Cagan, personal communication).

Studies with Notch and Delta provide evidence for a stepwise process of restriction (Cagan and Ready, 1989a; Van Vactor et al., 1991). Notch, a receptor, and Delta, its membrane-tethered ligand, play widespread roles in development (Artavanis-Tsakonas et al., 1995). As a result of incubating either *Notch* or *Delta* temperature-sensitive mutations at the nonpermissive temperatures, many cells within the MF assume an R8 cell fate. In addition, *Notch*^{spl}, a weak viable allele, specifically disrupts the final step of restricting R8 from two or three cells to one. The *scabrous* gene *(sca)* also plays a role in restricting R8 cell fate to a single cell. *sca* encodes a fibrinogen-related secreted factor that can diffuse over several cell diameters. It shows a pattern of expression similar to Ato. While Sca expression occurs more posterior to that of Ato, it is expressed in groups of cells in the MF that overlap Ato expression and quickly becomes restricted to R8 (Baker et al., 1990; Lee et al., 1996; Mlodzik et al., 1990). In loss-of-function *sca* mutations, many ommatidia contain one or two additional R8 cells, suggesting that it functions at a late stage in R8 restriction. Double mutant and morphological studies, however, suggest that *sca* may also regulate earlier steps of cluster formation (Baker and Zitron, 1995). The mechanism by which Sca functions and its relationship to other proteins, which are involved in intercellular communication, namely Notch and Delta, are not known.

The *rough* gene plays a key role in the selection of a single R8 cell. *rough* encodes an eye-specific homeodomain-containing transcription factor (Tomlinson and Ready, 1987). Genetic mosaic studies revealed that *rough* is required in R2 and R5 for normal ommatidial assembly. In *rough* null mutants preclusters are highly disorganized and in nearly all clusters two or three R8s form. The failure to restrict R8 to a single cell is correlated with the persistent expression of Ato in these cells, giving rise to the notion that the *rough* gene negatively regulates *atonal* expression (R. Cagan, personal communication). Further analysis of the role of *rough* and genes involved in cell–cell interaction pathways with which *rough* interacts, such as *Star*, may provide additional insight into the mechanisms regulating R8 selection.

INDUCTION THROUGH DIRECT CELL–CELL CONTACT CONTROLS R7 DEVELOPMENT

The developmental fates of the other cells within the ommatidium are also controlled by cellular interactions. The R7 photoreceptor is the last of the eight photoreceptors to differentiate and its recruitment is the best understood of the dozens of inductive interactions that take place during ommatiditial assembly. The R7 precursor cell—the cell located in the pocket created by the differentiating R8, R1, and R6 cells—appears to have had its developmental potential limited to two alternative cell fates by earlier signaling events; it will develop into an R7 photoreceptor if it receives signals that lead to activation of the Ras signaling pathway, but otherwise will adopt a non-neuronal cone cell fate. The mechanisms that restrict the cell fate options of the R7 precursor, that transmit the inductive signal, and that transduce and interpret this signal in the R7 precursor have been studied intensively (for review, see Zipursky and Rubin, 1994).

Activation of Ras is required in all the photoreceptors for their development (Simon et al., 1991), but it is only for the R7 precursor that we know how this activation occurs. Activation of the transmembrane tyrosine kinase receptor, (Sev), by its ligand, Bride of sevenless or of Boss, provides the primary signal (reviewed in Zipursky and Rubin, 1994), although input from the EGF receptor is also required (Xu and Rubin, 1993). The Sev protein is required only in the presumptive R7 cell itself (Tomlinson et al., 1987), although it is expressed transiently at high levels in at least nine cells (Banerjee et al., 1987; Tomlinson et al., 1987). In contrast, Boss is expressed only in the R8 cell and is required only in that cell for successful development of the R7 cell in the same ommatidium (Krämer et al., 1991; Reinke and Zipursky, 1988). Much of the signal transduction cascade that is initiated by Sev activation has been elucidated by a series of genetic screens for second site mutations that can dominantly modify the phenotypes produced by either the loss- or gain-of-function mutations in known pathway components (see above; Fig. 13–6).

Why does only one of the Sev-expressing cells in the developing ommatidium assume an R7 cell fate? Multiple mechanisms act in concert to restrict induction to the R7 precursor. First, Boss is not diffusible and is expressed only on the surface of the R8 cell (Van Vactor et al., 1991). During normal development, the Sev-expressing cone cell precursors do not contact the Boss-expressing R8 cell. However, when Boss is expressed in all cells in the eye disc the cone cell precursors assume R7 cell fates, implying that during normal development the cone cell precursors fail to assume an R7 cell fate due to the restricted spatial localization of the inductive ligand.

However, the limited presentation of Boss cannot account for the fact that the R1–R6 cells do not become R7 cells. All these cells contact R8, four of them normally express Sev, and Sev expression can be experimentally induced in the other two. Yet these cells do not become R7 cells, indicating that their developmental history has made them somehow unreceptive to the Sev-mediated signal. Consistent with this view, a constitutively active, ligand-independent, form of the Sev receptor (Basler et al., 1990) or activated forms of Ras (Fortini et al., 1992) can cause the cone cells to differentiate as R7-like cells, but they do not appear to be able to induce R7 development in R1–R6 cells.

If the developmental history of the R1–R6 cells—the inputs they have received and the resultant restrictions that have been placed on their developmental potentials—is responsible for their lack of response to Sev, then mutants that disrupt R1–R6 cell fate determination might be expected to permit additional cells to adopt the R7 fate. This prediction was first shown to be true for the *seven-*

up mutation (Mlodzik et al., 1990). The *seven-up* gene encodes a steroid receptor superfamily member that is expressed and required in R1, R3, R4, and R6. In the absence of *seven-up* function, these four cells develop as R7-like cells.

In addition to genes that function to specify the fates of other retinal cells and thereby repress R7 development, such as *rough* and *seven-up,* there are genes that function positively in the R7 precursor to define its response to inductive signals. An example of such a gene is *phyllopod,* which is expressed in and is cell-autonomously required for the development of the R1, R6, and R7 cells (Chang et al., 1995; Dickson et al., 1995). In *phyllopod* mutants, precursors to these three cells develop as cone cells; conversely, ectopic expression of *phyllopod* in the cone cell precursors results in them developing as R7-like cells.

Setting up Ommatidial Polarity

A striking feature of ommatidial structure is its polarity. The organization of cells in the ventral and dorsal halves of the eye are mirror image symmetric along the dorsoventral midline, the so-called equator (Fig. 13–1B). This symmetry is generated in the third larval instar when ommatidial precursors in the dorsal and ventral halves of the eye undergo a 90° rotation. Precursors in the dorsal half of the right eye rotate counterclockwise and those in the ventral half rotate clockwise. Studies on the *nemo* gene, which encodes a serine/threonine protein kinase homologue, indicate that the rotation event is at least a two-step process: in *nemo* mutants, all ommatidial precursors initiate rotation, but arrest halfway, at 45° (Choi and Benzer, 1994).

The rotation signal has not been identified but is thought to be transmitted bidirectionally from the future equator to the dorsal and ventral regions of the disc (Zheng et al., 1995). Frizzled, a cell-surface protein with seven transmembrane domains and some features of G-protein–coupled receptors (Krasnow and Adler, 1994; Park et al., 1994; Vinson et al., 1989), may be involved in mediating this signal (Zheng et al., 1995). Furthermore, studies on *dishevelled,* another polarity mutant, suggest junctional complexes may be involved in transmitting this signal (Theisen et al., 1994).

Coincident with rotation, R3 and R4 undergo morphological movements, altering their contacts with the cluster and breaking the symmetry of the five-cell precluster (Tomlinson, 1985). A correlation between improper fate specification of R3 and R4 and a failure in direction and degree of ommatidial rotation has been noted in several mutants, including *frizzled* (Zheng et al., 1995), *fat facets* (Fischer-Vize and Mosley, 1994), and *argos* (Freeman et al., 1992). This correlation has led to the speculation that R3 and R4 may be involved in guiding the direction of rotation.

Optic Lobe Development

Little is known about the genetic pathways regulating optic lobe development. *so* has been shown to play a crucial role at an early stage of optic lobe development (Cheyette, et al., 1994). It is expressed in a highly specific pattern in the developing embryo just anterior to the dorsal cephalic furrow in a region including, but slightly larger than, the optic lobe primordium. In *so* mutants the optic lobe primordium fails to invaginate and the mutant cells remain at the surface of the embryo. Neither the genetic pathway regulating *so* expression in the optic lobe primoridium nor the targets of *so* lying downstream in the pathway have been identified. Molecular and genetic studies of subsequent steps in optic lobe development have focused on the control of cellular proliferation and the establishment of precise patterns of neuronal connectivity.

Cell–Cell Interactions Control Neuroblast Proliferation

Cells in the optic lobe primordium complete their final embryonic division just prior to invagination. These cells, the optic lobe neuroblasts, remain quiescent, presumably in the G0 stage of the cell cycle, until late first instar (Ebens et al., 1993). They then begin to divide symmetrically to produce more stem cells and later asymmetrically to produce ganglion mother cells and postmitotic neurons. In larvae mutant for the *anachronism* (*ana*) locus, neuroblasts begin dividing 8 hours earlier than in wild type and by 30 hours posthatching *ana* optic lobes have three times as many neuroblasts in S phase (Ebens et al., 1993). *ana* is required for the maintenance but not the establishment of a quiescent state. *ana* encodes a novel glycoprotein secreted by glial cells which surround the larval neuroblasts. As a consequence of the precocious onset of optic lobe neuroblast proliferation, the transition from symmetric to asymmetric cell division occurs earlier, resulting in the precocious generation of optic lobe neurons. The timing of eye development is not advanced; R cells innervate a developmentally more advanced target region, resulting in marked defects in their projection patterns.

trol mutants have the opposite effect on optic lobe neuroblasts; in *trol* larvae the quiescent neuroblasts fail to reactivate the proliferation program, resulting in very reduced optic lobes (Datta, 1995). Interestingly, *ana*/*trol* double mutants show the precocious proliferation phenotype characteristic of *ana*. Hence, *trol* cells are not defective in proliferation per se but rather in their ability to reactivate stem cells from the quiescent state. The study of the developmental control of neuroblast stem cell proliferation may provide important insights into the mechanisms controlling stem cell development in other systems.

A Genetic Approach to Identifying Components Regulating the Formation of R-Cell Connectivity

Recently, a genetic approach to studying R-cell neuronal connectivity has been undertaken (Garrity et al., 1996; Martin et al., 1995). R-cell axonal projections into the optic lobes can be visualized both in the adult and during development using either MAb24B10, which recognizes a photoreceptor-cell–specific cell adhesion molecule or a reporter gene in which a photoreceptor-cell–specific responsive promoter has been fused to a cytoplasmic version of *E. coli* β-galactosidase (pGMR*lacZ*) (Moses and Rubin, 1991). These markers have proven to be of sufficient specificity and reliability to use as reagents to screen R-cell projection patterns in mutagenized larval eye–brain complexes (Martin et al., 1995). Over 200 mutations disrupting the R-cell connectivity patterns from some 7,000 mutagenized lines screened have been identified on the X, second, and third chromosomes (Garrity et al., 1996; Martin et al., 1995).

Since R-cell connectivity is a rather late event in the formation of the visual system, mutations altering connectivity may not do so directly, but only indirectly by disrupting earlier steps in eye development or in the formation of the target region. Although some genes may encode proteins with multiple roles, including guidance and targeting, as well as other aspects of visual system development, demonstrating that one role is independent of another, is problematic. Accordingly, an important step in the analysis of R-cell connectivity mutants is an assessment of pattern formation and differentiation in the eye as well as optic lobe development prior to innervation. Upon examination with various markers for cell fate determination and pattern formation in both the eye and optic lobe, the vast majority of mutations were found to disrupt aspects of visual system development prior to the formation of the initial projection patterns (see Martin et al., 1995).

The detailed knowledge of pattern formation in the eye provides strict criteria for assessing whether R-cell fate determination and patterning occur normally in mutants. Assessment of optic lobe development has proven more problematic. Because R-cell innervation induces early steps in lamina patterning (Selleck and Steller, 1991) one can only meaningfully assess the target region prior to innervation. At this early stage of target development there are few instructive markers. Thus, the development of the target prior to innervation can be assessed only superficially. This leaves open the possibility that any mutation which functions within the target region may disrupt the patterning of the target prior to R-cell innervation. Hence, only those genes required in R cells that do not affect patterning events in the eye can be considered strong candidates for encoding molecules with specific connectivity functions, including those which function within the growth cone to mediate interactions with the environment. Genetic mosaic analysis facilitates identification of this class of genes.

As discussed earlier, genetically mosaic eyes can be produced easily. To assess whether a connectivity mutation is required in the eye, flies are produced which contain a mutant patch of retinal tissue that innervates an otherwise normal optic lobe. Larvae are X-irradiated after the separation of the eye and optic lobe primordia. Due to the low frequency of mitotic recombination it is highly unlikely that mitotic recombination will generate mutant optic lobe tissue in the same individual in which a mutant eye clone is formed. The projections from the mutant clone can be assessed in adult paraffin sections stained with silver or in cryostat sections stained with MAb24B10.

Mutations in three genes—*limbo* (Martin et al., 1995), *quo vadis* (B. Poeck, S.L.Z., unpublished), and *dreadlocks (dock)* (Garrity et al., 1996)—affect pathfinding and/or neuronal targeting and show a strong retinal requirement and normal R-cell fate determination (Fig. 13–8A–C). Hence, these genes are candidates for encoding growth cone components regulating guidance. Below we describe the *dock* gene, the first of this group to be molecularly characterized.

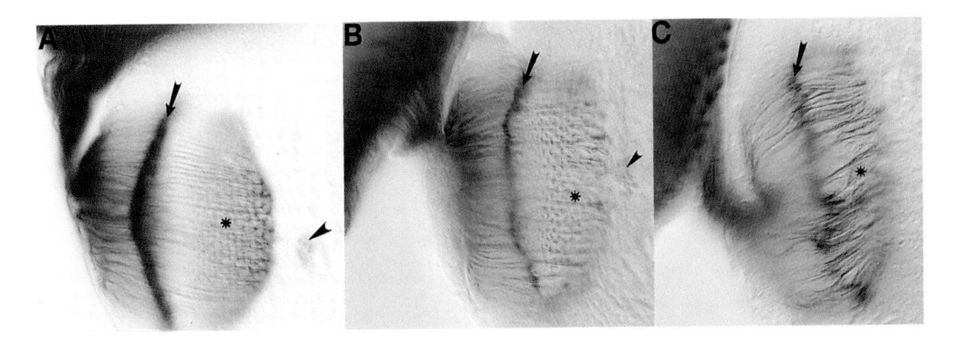

FIGURE 13–8 R-cell projection patterns in connectivity mutants. Third instar optic lobes stained with photoreceptor-specific antibody MAb24B10 in Wild-type (A), *limbo* (B), and *dreadlocks* (C). (B) In *limbo* the lamina neuropil is uneven and slightly reduced and the medulla is hyperinnervated. (C) In *dreadlocks* the lamina is severely disorganized and the medulla is hyperinnervated. The organization of R8 growth cones within the medulla is also severely disrupted. Lateral is to the left and ventral is down. Symbols: lamina neuropil, arrow; medulla neuropil, asterisk; terminus of Bolwig's nerve, arrowhead (only in focus in A and B). Bar, 20 μm.

DOCK IS AN SH2/SH3 ADAPTER PROTEIN CONTROLLING GROWTH CONE SIGNALING

In *dock*, R-cell axons project from the basal region of the disc epithelium in a manner indistinguishable from wild type. EM analysis revealed that the organization of R-cell axons in eight axon-containing fascicles is largely normal. As *dock* axons bundles exit the optic stalk they show defects in pathfinding (Fig. 13–8C). Bundles exiting the stalk frequently cross one another. Occasionally, fibers from positions in the eye disc that normally innervate targets distinct from one another coalesce to form larger bundles. Crossing of fibers is frequently seen both in the lamina and medulla projections. In addition to pathfinding defects, there are marked abnormalities in targeting seen in developing preparations with R1–R6 axons frequently projecting through the lamina into the medulla. Sections of genetically mosaic individuals stained with a marker for R1–R6 axons showed striking defects in which large numbers of fibers project through their normal target region in the lamina and terminate at variable positions in the medulla neuropil. The R-cell projection defects were rescued by expression of the *dock* transgene specifically in photoreceptor neurons.

The *dock* gene encodes an adapter protein comprising three N-terminal SH3 domains and single C-terminal SH2 domain. It has the same domain structure as the human protein Nck (Lehmann et al., 1990) and shares some 60% identity within the SH2 and SH3 domains. Since Dock and Nck share the critical amino acids determining SH2- and SH3-binding specificities, it is likely that these proteins will interact with similar proteins. Adapter proteins link tyrosine kinase signaling pathways to downstream effectors (see earlier section on Sev pathway). Dock does not interact genetically with either the EGF receptor or Sev receptor tyrosine kinases during eye development as does the adapter protein

Drk. Dock protein is expressed on axonal processes and growth cones in developing R cells as well as in neurons in the developing optic lobes.

dock mutants show massive defects in more central regions of the visual system that are neither innervated by R cells nor dependent on retinal innervation for their development. In addition, *dock* flies have poor motor control and reduced viability. These defects are rescued in their entirety by expression in postmitotic neurons, arguing that Dock does not function in signaling events required for cell proliferation or for cell fate determination. This raises the intriguing possibility that Dock's function is specialized for signaling events in the growth cone of many different types of neurons.

DOCK MAY COUPLE TYROSINE KINASE SIGNALING PATHWAYS TO GROWTH CONE MOTILITY

Based on the structure and subcellular location of Dock, it was proposed that Dock links tyrosine kinase signaling pathways to changes in growth cone motility. *dock* mutations affect pathfinding and target recognition. This may reflect a requirement for *dock* early in R-cell guidance which then leads indirectly to defects in later stages. Alternatively, Dock may transmit signals regulating multiple guidance steps to a common set of intracellular signaling pathways controlling actin cytoskeletal structure in the growth cone. This latter scheme would be analogous to the function of the Grb2 adaptor which couples different signals to a common pathway of cellular proliferation.

Previous studies in several systems have implicated tyrosine kinases in regulating growth cone motility. In *Drosophila,* the Derailed receptor tyrosine kinase RTK related to the human c-Ryk proto-oncogene controls the guidance of a small population of neurons in the embryonic nervous system (Callahan et al., 1995). In addition, mutations in both receptor tyrosine phosphatases and nonreceptor tyrosine kinases give rise to guidance defects (Desai et al., 1996; Gertler et al., 1989; Krueger et al., 1996). In vertebrates, the Eph family of tyrosine kinase receptors has also been shown to regulate neuronal connectivity (Henkemeyer et al., 1996). Interestingly, recent biochemical studies have implicated the mammalian Nck protein in signal transduction pathways activated by the Eph family of RTKs (T. Pawson, personal communication).

Whereas the SH2 domain of Dock likely binds to specific tyrosine residues phosphorylated in response to extracellular signals, the SH3 domains are likely to interact with various effectors linking tyrosine kinase signaling to changes in growth cone motility. Based on studies on the control of actin cytoskeleton in fibroblasts, it seems likely that these changes will be mediated by Rho-family members of the ras-related GTPases, CDC42, Rac, and Rho (Nobes and Hall, 1995). Recent studies in mammalian Nck suggest ways Dock may control cytoskeletal changes through its interaction with Cdc42. mPAK3 (Bagrodia et al., 1995), a serine/threonine kinase, binds to both human Nck and the GTP-bound form of CDC42. Perhaps phosphorylation of tyrosine kinase receptors in the growth cone nucleates the formation of a complex including, in addition to the receptor, Dock, mPAK3, and GTP-CDC42. Biochemical studies have also shown that Nck binds to WASP (Rivero-Lezcano et al., 1995), the product of the human Wiskott-Aldrich syndrome gene. WASP also binds to CDC42 and colocalizes with actin in cells (Symons et al., 1996). Furthermore, Wiskott-Aldrich syndrome mutant cells appear defective in regulating the structure of the actin cytoskeleton (Ochs et al., 1980). Alternatively, since Dock may bind to phosphorylated FAK (focal adhesion kinase), as does Nck (Schlaepfer et al., 1994), Dock might function as an adapter to couple cytoskeletal changes initiated by integrin binding to the extracellular matrix.

Formation of Optic Chiasms Requires the Homophilic Cell Adhesion Molecule IrreC

Topographic maps between ganglia are formed in the optic lobe. For instance, the five lamina neurons L1–L5 follow the R7/R8 axons associated with each lamina cartridge through the optic chiasm and into a topographically appropriate medulla column. Mutations at the *irregular chiasm C (irreC)* locus disrupt the formation of both the first and second optic chiasm (between the medulla and lobula/lobula plate; see Fig. 13–1D). Mutant alleles of this gene were identified in histological screens pioneered by Martin Heisenberg and his colleagues (Heinsenberg and Bohl, 1979) in which sections of adult brains were analyzed for gross structural abnormalities. *irreC* encodes a homophilic cell adhesion molecule of 140 kD with five immunoglobulin repeats (Ramos et al., 1993). Its closest vertebrate relative is DM-GRASP/SC1/BEN, which is expressed on the surface of chick retinal ganglion cell axons (Polterberg and Mack, 1994). IrreC is expressed in a dynamic pattern with expression at high levels at early stages of axonal outgrowth and then falls to very low levels (Schneider et al., 1995). Later as synapses form during pupal development IrreC immunoreactivity is seen in specific layers of the developing medulla and lobula neuropils.

Careful phenotypic analysis of loss-of-function alleles of *irreC* and the effects of ectopic expression of the IrreC protein have provided strong evidence for the importance of selective axon adhesion in regulating the formation of precise neuronal arrays forming between ganglia. These studies are best documented for the lamina neurons. During normal development the lamina neurons from posterior cartridges project to developing columns in the proximal medulla. Normally these axons course over the anterior surface of the developing medulla neuropil and then dive down along the preexisting R7/R8 bundle into the retinotopically appropriate column. As a consequence of rotation of the lamina and medulla neuropil relative to one another, these fibers will form a chiasm in the adult. In *irreC* mutants, fibers from the posterior edge of the lamina frequently fasciculate with adjacent posterior medulla neurons called C and T neurons. These neurons project along the posterior surface of the medulla and then to their appropriate column within the medulla neuropil. As a consequence of this misrouting the posterior lamina fibers no longer course through the chiasm but instead project around the posterior edge of the medulla. The C and T neurons express the cell adhesion molecule Fasciclin II as do the lamina neurons. Perhaps the loss of IrreC on lamina neurons facilitates Fasciclin II–mediated adhesion to C and T axonal surfaces. More anteriorly located lamina neurons are not misrouted because they lack proximity to the C and T processes.

Misexpression of IrreC using the *elav* promoter results in expression on all axons, thus disrupting the transient expression of IrreC on different fiber populations. Surprisingly, this leads to defects in axon fasciculation similar to those observed in *irreC* loss-of-function mutants. For instance, axons from posterior lamina neurons fasciculate with C and T neurons. Presumably this results from IrreC-mediated adhesive interactions between these neurons (i.e., during normal development, C and T neurons do not express appreciable levels of IrreC at this stage of development). The studies on IrreC are noteworthy in that they provide some of the best evidence for the involvement of specific cell adhesion molecules in regulating patterns of neuronal pathway selection and fasciculation in any system.

CONCLUDING REMARKS

Over the past 10 years considerable progress in understanding the molecular pathways regulating the formation of the fly visual system have been made. Early studies focused on describing the structure of the adult visual system and its development and laid the groundwork for future genetic and molecular approaches. Studies in the late 1980s and early 1990s led to the identification of key regulators of various steps in visual system development. These became starting points for detailed analysis of developmental pathways and led to the design of approaches to their molecular dissection. As further progress is made on the molecular characterization and eventual sequencing of the *Drosophila* genome, the previously often burdensome task of cloning and molecularly characterizing mutant loci will be eased considerably. This will provide researchers with access to many different developmental genes and genetic and molecular tools to dissect the pathways regulating other key pathways in visual system development. Given the remarkable findings that many and perhaps most developmental genes have been highly conserved during evolution, we believe the fly visual system will continue to provide starting points for the genetic dissection of additional pathways regulating nervous development in all organisms.

ACKNOWLEDGMENTS

We would like to thank Ulrike Heberlein, Jessica Treisman, and Volker Hartenstein for extensive discussions and editorial comments, David Wassarman and Henry Chang for critical comments on the manuscript, and Karen Ronan for help in preparing the manuscript.

References

Alphey, L., Jiminez, J., White-Cooper, H., Dawson, I., Nurse, P., Glover, D.M. (1992). *twine*, a cdc25 homolog that functions in the male and female germline of Drosophila. *Cell* 69:977–988.

Artavanis-Tsakonas, S., Matsuno, K., Fortini, M.E. (1995). Notch signalling. *Science* 268: 225–232.

Ashley, J.A., Katz, F.N. (1994). Competition and position-dependent targeting in the development of the Drosophila R7 visual projections. *Development* 120:1537–1547.

Bagrodia, S., Taylor, S.J., Creasy, C.L., Chernoff, J., Cerione, R.A. (1995). Identification of a mouse $p21^{cdc42/rac}$ activated kinase. *J. Biol. Chem.* 270:22731–22737.

Baker, N.E., Zitron, A.E. (1995). Drosophila eye development: *Notch* and *Delta* amplify a neurogenic pattern conferred on the morphogenetic furrow by *scabrous*. *Mech. Dev.* 49: 173–189.

Baker, N.E., Mlodzik, M., Rubin, G.M. (1990). Spacing differentiation in the developing Drosophila eye: a fibrinogen-related lateral inhibitor encoded by scabrous. *Science* 250: 1370–1377.

Banerjee, U., Renfranz, P.J., Pollack, J.A., Benzer, S. (1987). Molecular characterization and expression of *sevenless*, a gene involved in neuronal pattern formation in the Drosophila eye. *Cell* 49:281–291.

Basler, K., Yen, D., Tomlinson, A., Hafen, E. (1990). Reprogramming cell fate in the developing *Drosophila* retina: transformation of R7 cells by ectopic expression of *rough*. *Genes Dev.* 4:728–739.

Bowtell, D.D.L., Kimmel, B.E., Simon, M.A., Rubin, G.M. (1989). Regulation of the complex pattern of *sevenless* expression in the developing Drosophila eye. *Proc. Natl. Acad. Sci. U.S.A.* 86:6511–6515.

Braitenberg, V. (1967). Patterns of projection in the visual system of the fly. I. Retina-lamina projections. *Exp. Brain Res.* 3:271–298.

Brown, N.L., Sattler, C.A., Paddock, S.W., Carroll, S.B. (1995). *hairy* and *emc* negatively regulate morphogenetic furrow progression in the Drosophila eye. *Cell* 80:879–887.

Cagan, R.L., Ready, D.F. (1989a). The emergence of order in the Drosophila pupal retina. *Dev. Biol.* 136:346–362.

Cagan, R.L., Ready, D.F. (1989b). *Notch* is required for successive cell decisions in the developing Drosophila retina. *Genes Dev.* 3:1099–1112.

Callahan, C.A., Muralidhar, M.G., Lundgren, S.E., Scully, A.L., Thomas, J.B. (1995). Control of neuronal pathway selection by a Drosophila receptor protein-tyrosine kinase family member. *Nature* 376:171–174.

Campos-Ortega, J., Hofbauer, A. (1977). Cell clones and pattern formation: on the lineage of photoreceptor cells in the compound eye of Drosophila. *Wilhelm Roux's Arch. Dev. Biol.* 181:227–245.

Chang, H.C., Solomon, N., Wassarman, D.A., Karim, F.D., Therrien, M., Rubin, G.M., Wolff, T. (1995). *Phyllopod* functions in the fate determination of a subset of photoreceptors in Drosophila. *Cell* 80:463–472.

Chanut, F., Heberlein, U. (1995). Role of the morphogenetic furrow in establishing polarity in the Drosophila eye. *Development* 121:4085–4094.

Cheyette, B.N.R., Green, P.J., Martin, K., Garren, H., Hartenstein, V., Zipursky, S.L. (1994). The Drosophila *sine oculis* locus encodes a homeodomain-containing protein required for the development of the entire visual system. *Neuron* 12:977–996.

Chisholm, A.D., Horvitz, H.R. (1995). Patterning of the *Caenorhabditis elegans* head region by the *pax-6* family member *vab-3*. *Nature* 377:52–59.

Choi, K., Benzer, S. (1994). Migration of glia along photoreceptor axons in the developing Drosophila eye. *Neuron* 12:423–431.

Cohen, S.M. (1993). Imaginal disc development. In *The Development of Drosophila*

Melanogaster, Martinez Arias, A., Bate, M. eds. Cold Spring Harbor, NY: Cold Spring Harbor Laboratory Press, pp. 747–841.

Datta, S. (1995). Control of proliferation activation in quiescent neuroblasts of the *Drosophila* central nervous system. *Development* 121:1173–1182.

de Nooij, J.C., Hariharan, I. (1995). Uncoupling cell fate determination from patterned cell division in the Drosophila eye. *Science* 270:983–985.

Desai, C.J., Gindhart, J.G., Jr., Goldstein, L.S.B., Zinn, K. (1996). Receptor tyrosine phosophatases are required for motor axon guidance in the Drosophila embryo. *Cell* 84:599–609.

Dickson, B., Dominguez, M., van der Straten, A., Hafen, E. (1995). Control of Drosophila photoreceptor cell fates by *phyllopod,* a novel nuclear protein acting downstream of the Raf kinase. *Cell* 80:463–472.

Ebens, A.J., Garren, H., Cheyette, B.N.R., Zipursky, S.L. (1993). The *Drosophila anachronism* locus: a glycoprotein secreted by glia inhibits neuroblast proliferation. *Cell* 74: 15–27.

Edgar, B.A., O'Farrell, P.H. (1989). Genetic control of cell division patterns in the Drosophila embryo. *Cell* 57:177–187.

Fischbach, K.F., Dittrich, A.P.M. (1989). The optic lobe of Drosophila melanogaster. I. A Golgi analysis of wild-type structure. *Cell Tissue Res.* 258:441–475.

Fischer-Vize, J.A., Mosley, K.L. (1994). *marbles* mutants: uncoupling cell determination and nuclear migration in the developing Drosophila eye. *Development* 120:2609–2618.

Fortini, M.E., Simon, M.A., Rubin, G.M. (1992). Signalling by the Sevenless protein tyrosine kinase is mimicked by Ras1 activation. *Nature* 355:559–561.

Freeman, M., Klämbt, C., Goodman, C.S., Rubin, G.M. (1992). The *argos* gene encodes a diffusible factor that regulates cell fate decisions in the Drosophila eye. *Cell* 69:963–975.

Garcia-Bellido, A., Dapena, J. (1974). Induction, detection and characterization of cell differentiation mutants in *Drosophila. Mol. Gen. Genet.* 128:117–130.

Garrity, P.A., Rao, Y., Salecker, I., Zipursky, S.L. (1996). Drosophila photoreceptor axon guidance and targeting require the *dreadlocks* SH2/SH3 adapter protein. *Cell* 85:639–650.

Gertler, F.B., Bennett, R.L., Clark, M.J., Hoffmann, F.M. (1989). Drosophila *abl* tyrosine kinase in embryonic CNS axons: a role in axonogenesis is revealed through dosage-sensitive interactions with *disabled. Cell* 58:103–113.

Golic, K.G. (1991). Site-specific recombination between homologous chromosomes in Drosophila. *Science* 252:958–961.

Green, P., Hartenstein, A.Y., Hartenstein, V. (1993). The embryonic development of the *Drosophila* visual system. *Cell Tissue Res.* 273:583–598.

Halder, G., Callaerts, P., Gehring, W.J. (1995a). Induction of ectopic eyes by targeted expression of the *eyeless* gene in *Drosophila. Science* 267:1788–1765.

Halder, G., Callaerts, P., Gehring, W.J. (1995b). New perspectives on eye evolution. *Curr. Opin. Genet. Dev.* 5:602–609.

Hanson, I., Van Heyningen, V. (1995). Pax6: more than meets the eye. *Trends Genet.* 11:268–272.

Hay, B.A., Wolff, T., Rubin, G.M. (1994). Expression of baculovirus P35 prevents cell death in Drosophila. *Development* 120:2121–2129.

Hay, B.A., Wassarman, D.A., Rubin, G.M. (1995). Drosophila homologs of baculovirus inhibitor of apoptosis proteins function to block cell death. *Cell* 83:1253–1262.

Heberlein, U., Wolff, T., Rubin, G.M. (1993). *Star* is required for neuronal differentiation in the Drosophila retina and displays dosage-sensitive interactions with Ras1. *Developmental Biol.* 160:51–63.

Heberlein, U., Singh, C.M., Luk, A.Y., Donohue, T.J. (1995). Growth and differentiation in the Drosophila eye coordinated by *hedgehog. Nature* 373:709–711.

Heisenberg, M., Bohl, K. (1979). Isolation of anatomical brain mutants of Drosophila by histological means. *Z. Naturforsch.* 34:143–147.

Henkemeyer, M., Orioli, D., Henderson, J.T., Saxton, T.M., Roder, J., Pawson, T., Klein, R.

(1996). Nuk controls pathfinding of commissural axons in the mammalian CNS independently of its tyrosine kinase domain. *Cell* 86:35–46.

Hill, R.E., Favor, J., Hogan, B.L.M., Ton, C.C.T., Saunders, G.F., Hanson, I.M., Prosser, J., Jordan, T., Hastie, N.D., Van Heyningen, V. (1991). Mouse *Small eye* results from mutations in a paired-like homeobox-containing gene. *Nature* 354:522–525.

Hofbauer, A., Campos-Ortega, J. (1976). Cell clones and pattern formation: genetic eye mosaics in Drosophila melanogaster. *Wilhelm Roux's Arch.* 179:2700–2704.

Hofbauer, A., Campos-Ortega, J.A. (1990). Proliferation pattern and early differentiation of the optic lobes in Drosophila melanogaster. *Roux's Arch. Dev. Biol.* 198:264–274.

Holt, C.W., Bertsch, T.W., Ellis, H.M., Harris, W.A. (1988). Cellular determination in the *Xenopus* retina is independent of lineage and birthdate. *Neuron* 1:15–26.

Jarman, A.P., Grau, Y., Jan, L.Y., Jan, Y.N. (1993). *atonal* is a proneural gene that directs chorodotonal organ formation in the Drosophila peripheral nervous system. *Cell* 73: 1307–1321.

Jürgens, G., Hartenstein, V. (1993). The terminal regions of the body pattern. In *The Development of Drosophila Melanogaster,* Martinez Arias, A., Bate, M., eds. Cold Spring Harbor, NY: Cold Spring Harbor Laboratory Press, pp. 687–746.

Karim, F.D., Chang, H.C., Therrien, M., Wassarman, D.A., Laverty, T., Rubin, G.M. (1996). A screen for genes that function downstream of Ras1 during Drosophila eye development. *Genetics* 143:315–329.

Kopczynski, C.C., Davis, G.W., Goodman, C.S. (1996). A neural tetraspanin, encoded by *late bloomer,* that facilitates synapse formation. *Science* 271:1867–1870.

Krämer, H., Cagan, R.L., Zipursky, S.L. (1991). Interaction of bride of sevenless membrane-bound ligand and the sevenless tyrosine-kinase receptor. *Nature* 352:207–212.

Krasnow, R.E., Adler, P.N. (1994). A single Frizzled protein has a dual function in tissue polarity. *Development* 120:1883–1893.

Krueger, N.X., Van Vactor, D., Wan, H.I., Gelbart, W.M., Goodman, C.S., Saito, H. (1996). The transmembrane tyrosine phosphatase DLAR controls motor axon guidance in Drosophila. *Cell* 84:611–622.

Kumagai, A., Dunphy, W.G. (1991). The Cdc25 protein controls tyrosine dephosphorylation of the Cdc2 protein in a cell-free system. *Cell* 64:903–914.

Kunes, S., Steller, H. (1991). Ablation of Drosophila photoreceptor cells by conditional expression of a toxin gene. *Genes Dev.* 5:970–983.

Kunes, S., Wilson, C., Steller, H. (1993). Independent guidance of retinal axons in the developing visual system of Drosophila. *J. Neurosci.* 12:752–767.

Lawrence, P.A., Green, S.M. (1979). Cell lineage in the developing retina of Drosophila. *Dev. Biol.* 71:142–152.

Lee, E.C., Hu, X., Yu, S.Y., Baker, N.E. (1996). The *scabrous* gene encodes a secreted glycoprotein dimer and regulates proneural development in Drosophila eyes. *Mol. Cell Biol.* 16:1179–1188.

Lehmann, J.M., Riethmuller, G., Johnson, J.P. (1990). Nck, a melanoma cDNA encoding a cytoplasmic protein consisting of the Src homology units SH2 and SH3. *Nucleic Acids Res.* 18:1048.

Ma, C., Moses, K. (1995). *Wingless* and *patched* are negative regulators of the morphogenetic furrow and can affect tissue polarity in the developing Drosophila compound eye. *Development* 121:2279–2289.

Ma, C., Zhou, Y., Beachy, P.A., Moses, K. (1993). The segment polarity gene *hedgehog* is required for progression of the morphogenetic furrow in the developing Drosophila eye. *Cell* 75:927–938.

Macagno, E.R. (1979). Cellular interactions and pattern formation in the development of the visual system of *Daphnia magna* (crustacea, branchiopoda). *Dev. Biol.* 73:206–238.

Mardon, G., Solomon, N., Rubin, G.M. (1994). *dachshund* encodes a nuclear protein required for normal eye and leg development in Drosophila. *Development* 120:3473–3486.

Martin, K.A., Poeck, B., Roth, H., Ebens, A.J., Conley Ballard, L., Zipursky, S.L. (1995).

Mutations disrupting neuronal connectivity in the Drosophila visual system. *Neuron* 14:229–240.

Meinertzhagen, I., Hanson, I. (1993). The development of the optic lobe. In *The Development of Drosophila Melanogaster,* Martinez Arias, A., Bates, M., eds. Cold Spring Harbor, NY: Cold Spring Harbor Laboratory Press, pp. 1363–1490.

Meyerowitz, E., Kankel, D. (1978). A genetic analysis of visual system development in *Drosophila melanogaster. Dev. Biol.* 62:112–142.

Mlodzik, M., Baker, N.E., Rubin, G.M. (1990). Isolation and expression of *scabrous,* a gene regulating neurogenesis in Drosophila. *Genes Dev.* 4:1848–1861.

Moses, K., Rubin, G.M. (1991). *glass* encodes a site-specific DNA-binding protein that is regulated in response to positional signals in the developing Drosophila eye. *Genes Dev.* 5:583–593.

Nakato, H., Futch, T.A., Selleck, S.B. (1995). The *division abnormally delayed (dally)* gene: a putative integral membrane proteoglycan required for cell division patterning during postembryonic development of the nervous system in *Drosophila. Development* 121: 3687–3702.

Nobes, C.D., Hall, A. (1995). Rho, Rac, and Cdc42 GTPases regulate the assembly of multimolecular focal complexes associated with actin stress fibers, lamellipodia and filopodia. *Cell* 81:53–62.

Nowel, M.S. (1981). Formation of the retina-lamina projection of the cockroach: no evidence for neuronal specificity. *J. Embryol. Exp. Morphol.* 62:241–258.

Nusse, R., Varmus, H.E. (1992). Wnt genes. *Cell* 69:1073–1087.

Nüsslein-Volhard, C., Wieschaus, E. (1980). Mutations affecting segment number and polarity in Drosophila. *Nature* 287:795–801.

O'Kane, C.J., Gehring, W.J. (1987). Detection in situ of genomic regulatory elements in Drosophila. *Proc. Natl. Acad. Sci. U.S.A.* 84:9123–9127.

Ochs, H.D., Slichter, H.J., Harker, L.A., Von, B. W., Clark, R.A., Wedgwood, R.J. (1980). The Wiskott-Aldrich syndrome: studies of lymphocytes, granulocytes and platelets. *Blood* 55:243–252.

Oliver, G., Mailhos, A., Wehr, R., Copeland, N., Jenkins, N.A., Gruss, P. (1995). *Six3,* a murine homolog of the *sine oculis* gene, demarcates the most anterior border of the developing neural plate and is expressed during eye development. *Development* 121: 4045–4055.

Pan, D., Rubin, G.M. (1995). Detection in situ of genomic regulatory elements in Drosophila. *Cell* 80:543–552.

Park, W.J., Liu, J., Adler, P.N. (1994). *frizzled* gene expression and development of tissue polarity in the Drosophila wing. *Dev. Genet.* 15:383–389.

Polterberg, G.E., Mack, T.G.A. (1994). Cell adhesion molecules SC1/DMGRASP is expressed on growing axons of retina ganglion cells and is involved in mediating their extension on axons. *Dev. Biol.* 165:670–687.

Power, M.E. (1943). The effect of reduction in numbers of ommatidia upon the brain of *Drosophila melanogaster. J. Exp. Zool.* 34:33–71.

Quiring, R., Walldorf, U., Kloter, U., Gehring, W.J. (1994). Homology of the *eyeless* gene of Drosophila to the *small eye* gene in mice and *aniridia* in humans. *Science* 265:785–789.

Ramos, R.G.P., Igloi, G.L., Lichte, B., Baumann, U., Maier, D., Schneider, T., Brandstätter, J. H., Fröhlich, A., Fischbach, K.-F. (1993). The *irregular chiasm C-roughest* locus of *Drosophila,* which affects axonal projections and programmed cell death, encodes a novel immunoglobulin-like protein. *Genes Dev.* 7:2533–2547.

Ready, D.F., Hanson, T.E., Benzer, S. (1976). Development of the Drosophila retina, a neurocrystalline lattice. *Dev. Biol.* 53:217–240.

Reinke, R., Zipursky, S.L. (1988). Cell-cell interaction in the Drosophila retina: the *bride-of-sevenless* gene is required in photoreceptor cell R8 for R7 cell development. *Cell* 55: 321–330.

Rivero-Lezcano, O.M., Marcilla, A., Sameshima, J.H., Robbins, K.C. (1995). Wiskott-

Aldrich syndrome protein physically associates with Nck through Src homology 3 domains. *Mol. Cell Biol.* 15:5725–5731.

Schlaepfer, D.D., Hanks, S.K., Hunter, T., van der Geer, P. (1994). Integrin-mediated signal transduction linked to Ras pathway by Grb2 binding to focal adhesion kinase. *Nature* 372:786–791.

Schneider, T., Reiter, C., Eule, E., Bader, B., Lichte, B., Nie, Z., Schimansky, T., Ramos, R.G. P., Fischbach, K.-F. (1995). Restricted expression of the IrreC-Rst protein is required for normal axonal projections of columnar visual neurons. *Neuron* 15:259–271.

Selleck, S.B., Steller, H. (1991). The influence of retinal innervation on neurogenesis in the first optic ganglion of Drosophila. *Neuron* 6:83–99.

Selleck, S.B., Gonzalez, C., Glover, D.M., White, K. (1992). Regulation of the G1-S transition in post-embryonic neuronal precursors by axon ingrowth. *Nature* 355:253–255.

Simon, M.A., Bowtell, D.D.L., Dodson, G.S., Laverty, T.R., Rubin, G.M. (1991). Ras1 and a putative guanine nucleotide exchange factor perform crucial steps in signaling by the Sevenless protein tyrosine kinase. *Cell* 67:701–716.

Steller, H., Fischbach, K.F., Rubin, G.M. (1987). *disconnected:* a locus required for neuronal pathway formation in the visual system of Drosophila. *Cell* 50:1139–1153.

Strausfeld, N.J. (1976). *Atlas of an Insect Brain.* New York: Springer-Verlag.

Strutt, D.I., Wiersdorff, V., Mlodzik, M. (1995). Regulation of furrow progression in the Drosophila eye by cAMP-dependent protein kinase A. *Nature* 373:705–709.

Symons, M., Derry, J.M.J., Karlak, B., Jian, S., Lemahieu, V., McCormick, F., Francke, U., Abo, A. (1996). Wiskott-Aldrich syndrome protein, a novel effector for the GTPase CDC42Hs, is implicated in polymerization. *Cell* 84:723–734.

Thaker, H.M., Kankel, D.R. (1992). Mosaic analysis gives an estimate of the extent of genomic involvement in the development of the visual system in Drosophila melanogaster. *Genetics* 131:883–894.

Theisen, H., Purcell, J., Bennett, M., Kansagara, D., Syed, A., Marsh, J.L. (1994). *dishevelled* is required during *wingless* signaling to establish both cell polarity and cell identity. *Development* 120:347–360.

Thomas, B.J., Gunning, D.A., Cho, J., Zipursky, S.L. (1994). Cell cycle progression in the developing Drosophila eye: *roughex* encodes a novel protein required for the establishment of G1. *Cell* 77:1003–1014.

Tix, S., Minden, J.S., Technau, G.M. (1989). Pre-existing neuronal pathways in the developing optic lobes of Drosophila. *Development* 105:739–746.

Tomlinson, A. (1985). The cellular dynamics of pattern formation in the eye of Drosophila. *J. Embryol. Exp. Morphol.* 89:313–331.

Tomlinson, A., Ready, D.F. (1987). Neuronal differentiation in the *Drosophila* ommatidium. *Dev. Biol.* 120:366–376.

Tomlinson, A., Bowtell, D.D., Hafen, E., Rubin, G.M. (1987). Localization of the Sevenless protein, a putative receptor for positional information, in the eye imaginal disc of Drosophila. *Cell* 51:143–150.

Ton, C.C.T., Hirvonen, H., Miwa, H., Weil, M.M., Monaghan, P., Jordan, T., Van Heyningen, V., Hastie, N.D., Meijers-Heijboer, H., Drechsler, M. (1991). Positional cloning and characterization of a paired box- and homeobox-containing gene from the *aniridia* region. *Cell* 67:1059–1074.

Treisman, J.E., Rubin, G.M. (1995). *wingless* inhibits morphogenetic furrow movement in the Drosophila eye disc. *Development* 121:3519–3527.

Turner, D.L., Cepko, C.L. (1987). A common progenitor for neurons and glia persists in rat retina late in development. *Nature* 238:131–136.

Van Vactor, D.L., Jr., Cagan, R.L., Krämer, H., Zipursky, S.L. (1991). Induction in the developing compound eye of Drosophila: multiple mechanisms restrict the R7 induction to a single retinal precursor cell. *Cell* 67:1145–1155.

Vinson, C.R., Conover, S., Adler, P.N. (1989). A Drosophila tissue polarity locus encodes a protein containing seven potential transmembrane domains. *Nature* 338:263–264.

Wehrli, M., Tomlinson, A. (1995). Epithelial planar polarity in the developing Drosophila eye. *Development* 121:2451–2459.

Wetts, R., Fraser, S.E. (1988). Multipotent precursors can give rise to all major cell types of the frog retina. *Science* 239:1142–1145.

White, K., Kankel, D.R. (1978). Patterns of cell division and cell movement in the formation of the imaginal nervous system in Drosophila melanogaster. *Dev. Biol.* 65:296–321.

Winberg, M.L., Perez, S.E., Steller, H. (1992). Generation and early differentiation of glial cells in the first optic ganglion of *Drosophila melanogaster. Development* 115:903–911.

Wolff, T., Ready, D.F. (1991). The beginning of pattern formation in the Drosophila compound eye: the morphogenetic furrow and the second mitotic wave. *Development* 113: 841–850.

Wolff, T., Ready, D.F. (1993). Pattern formation in the Drosophila retina. In *The Development of Drosophila Melanogaster,* Martinez Arias, A., Bate, M., eds. Cold Spring Harbor, NY: Cold Spring Harbor Laboratory Press, pp. 1277–1325.

Xu, T., Rubin, G.M. (1993). Analysis of genetic mosaics in developing and adult Drosophila tissues. *Development* 117:1223–1237.

Zhang, Y., Emmons, S.W. (1995). Specification of sense-organ identity by a *Caenorhabditis elegans pax-6* homolog. *Nature* 377:55–59.

Zheng, L., Zhang, J., Carthew, R.W. (1995). *frizzled* regulates mirror-symmetric pattern formation in the Drosophila eye. *Development* 121:3045–3055.

Zipursky, S.L., Rubin, G.M. (1994). Determination of neuronal cell fate: lessons from the R7 neuron of Drosophila. *Annu. Rev. Neurosci.* 17:373–397.

14

Neurotrophins and visual system plasticity

CARLA J. SHATZ

The functioning of the nervous system depends on the precision and complexity of its underlying circuitry. The construction of these circuits occurs in several steps. The basic framework of connectivity is established initially via mechanisms that involve recognition, by neuronal growth cones, of fixed and diffusible guidance signals arrayed along forming pathways and within specific targets (reviewed in Tessier-Lavigne & Goodman, 1996). However, once axons arrive at their targets, they must sort to form the precise sets of connections present in the adult. This sorting process is thought to involve at least two sequential steps. First, axons use positional information to seek their general addresses within the target. In the vertebrate retinotectal system, candidates for providing positional cues include the Eph family of receptor tyrosine kinases and their ligands (Cheng et al., 1995; Drescher, et al., 1995; Tessier-Lavigne, 1995). Next, synaptic connections are refined into the adult precision by a process that requires synaptic transmission and the coordinated activity of groups of pre- and postsynaptic cells (reviewed in Katz and Shatz, 1996), and this results in the gradual structural remodeling of the presynaptic terminal arbor.

While a great deal is now known about the requirement for patterned neural activity in this process, much less is understood about the underlying cellular and molecular mechanisms. In particular, little is known about the signaling thought to occur between pre- and postsynaptic cells that drives the process of synaptic remodeling. This review will consider recent evidence suggesting that neurotrophins may function in the activity-dependent remodeling of connections during the development of the mammalian visual cortex. In the first section, evidence of a requirement for a retrograde signal from the postsynaptic to the presynaptic neuron will be considered. Then I will review evidence that neurotrophins may be good candidates for this signal.

THE GENICULOCORTICAL SYNAPSE AND THE FORMATION OF OCULAR DOMINANCE COLUMNS

In the adult visual system of higher mammals, the thalamic inputs to the cortex arise from the lateral geniculate nucleus (LGN) and terminate on the neurons of cortical layer 4. The axon arbors of LGN neurons are strictly segregated within layer 4 according to eye preference into a series of alternating patches. These eye-specific patches in layer 4 form the structural basis for the functionally defined system of ocular dominance columns that include all cortical layers. However, the ocular dominance columns in layer 4 are not present at the outset (Rakic, 1977; LeVay et al., 1978, 1980; reviewed in Shatz, 1990). LGN axons representing both eyes initially share common territory within layer 4, and then remodel their axon arbors, gradually clustering branches into eye-specific patches. This anatomical rearrangement of the presynaptic axon is accompanied functionally by a corresponding change in the synaptic physiology of layer 4 neurons, the majority of which are initially driven binocularly, but which finally come to respond to visual stimulation through one eye only.

A REQUIREMENT FOR ACTIVITY-DRIVEN SYNAPTIC COMPETITION

Over the past 30 years, many lines of evidence have accrued that point to a requirement for an activity-driven synaptic competition between LGN axons for common postsynaptic layer 4 neurons during the formation of ocular dominance columns (reviewed in Shatz 1990; Goodman and Shatz, 1994). The first indication of competitive interactions came from the pioneering studies of Hubel and Wiesel, who examined the effects of visual deprivation on the functional organization of the primary visual cortex (Wiesel and Hubel, 1963a,b; Hubel and Wiesel, 1970; Hubel et al., 1977). If one eye is deprived of vision by eyelid suture for several weeks during a critical period in neonatal life, the majority of visual cortical neurons (>90%) are driven only by the open eye—there is a dramatic physiological shift in the ocular dominance of cortical neurons away from an even distribution in which the cortex is shared about equally by inputs from the two eyes. Anatomically, the consequence of early eye closure is that within layer 4, the patches of input from LGN axons representing the open eye are far larger than normal, while those representing the closed eye are relegated to very small regions (note, however, that the periodicity of a right plus a left eye column is unchanged, and remains at about 1 mm in width) (Hubel et al., 1977; Shatz and Stryker, 1978). However, if both eyes are sutured closed, then the columns do form (Stryker and Harris, 1986; Mower et al., 1985; Wiesel and Hubel, 1965), although the process takes much longer and eventually the responses of cortical neurons to visual stimulation are seriously degraded. These observations argue for competition rather than simply lack of use because unequal use of the two eyes can bias the outcome of ocular dominance column formation in favor of the more active eye, while balanced use produces columns of equal size.

These use-dependent effects on the functional and structural organization of the visual cortex are a consequence of alterations in the level or patterning (or both) of neural activity within the visual pathways. Perhaps the most graphic demonstration of this fact comes from experiments in which the inputs from both eyes are completely silenced during the period in which ocular dominance columns normally form in the visual cortex by means of intraocular injections of tetrodotoxin (TTX, a blocker of voltage-sensitive sodium channels). Under these circumstances, the eye-specific patches within layer 4 do not form (Stryker and Harris, 1986), and LGN axons remain widely branched, failing to restrict their terminal arbors (Antonini and Stryker, 1993a,b).

The experiments described above indicate that neural activity is required for the structural remodeling of LGN axons to produce the final pattern of connectivity within layer 4 of visual cortex. A major question is, how can activity instruct in the remodeling of axons according to eye of origin? Many lines of experiments (reviewed in Shatz, 1990; Goodhill and Lowel, 1995) have indicated that correlated neural activity arising from each eye, coupled with the ability of the postsynaptic cell to detect that activity, is necessary for the adult patterning of connections to emerge in the visual system. For example, Stryker and Strickland (1984) found that if TTX was injected into both eyes to silence the visually driven firing of ganglion cells, and then implanted stimulating electrodes were used to deliver shocks to the optic nerve, the segregation of LGN axons into the system of ocular dominance columns in cortical layer 4, as assessed physiologically, could occur only if the stimulation of the two nerves was asychronous. Synchronous stimulation of the optic nerves prevented segregation, as indicated by the fact that cortical layer 4 neurons remained binocularly driven.

Not only is there a requirement for correlations in the pattern of presynaptic activity, but several other experiments also argue for a mechanism that must operate to reinforce synapses that are coactive (Fregnac et al., 1988; Miller et al., 1989a; Shatz, 1990; Lowel and Singer, 1992). To give one clear case, Fregnac et al. (1988) demonstrated that the ocular dominance or orientation preference of cortical neurons can be altered during the critical period by pairing postsynaptic depolarization of cortical neurons with visual stimulation. For example, a binocular neuron responding initially more strongly to the left eye could be made to shift its response preference toward the right eye by pairing depolarization produced by potassium ejection from the recording electrode in the cortex with right eye stimulation; this shift could be an enduring one, lasting for hours.

Cellular Correlates of Activity-Dependent Synaptic Remodeling

This feature of synaptic transmission—an increase in the strength of presynaptic inputs whose firing is correlated with that of the postsynaptic cell—is characteristic of Hebb synapses (1949). Such synapses could operate during cortical development to strengthen all coactive inputs to layer 4 cortical neurons from LGN axons receiving inputs from one eye (e.g., "cells that fire together wire together"). At the cellular level, these enduring changes in synaptic strength are hypothesized to involve the mechanism of long-term potentiation (LTP), in which specific patterns of high-frequency stimulation, or pairing of presynaptic stimulation and postsynaptic depolarization, are known to result in increases in the strength of synaptic transmission lasting for hours to days (Madison et al., 1991; Malenka and Nicoll, 1993; Malenka, 1994). In the developing mammalian visual system, there is good evidence for the presence of LTP, not only at synapses within the cortex (Komatsu et al., 1988; Artola et al., 1990; Kirkwood et al., 1995), but also specifically at the geniculocortical synapse (Crair and Malenka, 1995). However, while LTP is certainly an attractive cellular mechanism for the activity-dependent strengthening of synaptic connections that must underlay the segregation of LGN axons into ocular dominance columns, it should be noted that at this point it is only a cellular correlate. Direct experiments are still needed to make a causal link between LTP and the segregation of LGN axons into ocular dominance columns in layer 4.

For the ocular dominance columns to form, there is a clear need not only for the strengthening and growth of synapses from LGN axons representing one eye, but also for the weakening and elimination of those representing the other eye. Again there is an attractive cellular mechanism that could operate in the cortex to accomplish synaptic weakening: long-term depression (LTD). Various forms of LTD have been described in hippocampal and neocortex (Stanton and Sejnowski, 1989; Dudek and Bear, 1993; Bear and Malenka, 1993; Malenka, 1994; Bear and Abraham, 1996). In both structures, low-frequency stimulation of presynaptic inputs results in weakening of synaptic transmission (homosynaptic LTD); in hippocampus, lack of presynaptic activity when the postsynaptic neuron is active can also result in weakening of the inactive input (heterosynaptic LTD). Both forms of LTD could contribute to the elimination of synaptic inputs from LGN axons representing one eye in regions of layer 4 ultimately belonging to a column from the other eye. At present, though, as is the case with LTP, the linkage between LTD and the formation of ocular dominance columns remains one of correlation rather than causation.

ACTIVITY IN THE POSTSYNAPTIC CELL IS REQUIRED FOR THE FORMATION OF OCULAR DOMINANCE COLUMNS

In thinking about possible cellular mechanisms of ocular dominance column development, a major issue is whether the activity of the postsynaptic neurons—the layer 4 cortical neurons—is necessary. Accumulated evidence argues very strongly that this is the case (reviewed in Shatz, 1990). A key experiment (Reiter and Stryker, 1988; Hata and Stryker, 1994) was to infuse muscimol, a GABA-A agonist that silences all cortical neurons, during the period in which monocular eye closure is known to alter the formation of the ocular dominance columns both anatomically and physiologically. The consequences of blocking all postsynaptic activity both on the eye preference of cortical neurons and on the pattern of ocular dominance columns in layer 4 were assessed. Recall that monocular deprivation usually causes a physiological shift in the responses of cortical neurons in favor of the nondeprived eye and a larger-than-normal occupancy of cortical layer 4 by LGN axons representing the nondeprived eye at the expense of deprived eye axons (Shatz and Stryker, 1978; LeVay et al., 1980). However, within the muscimol-infused region of cortex, just the reverse occurred: There was a physiological dominance of the *deprived* eye inputs and a larger-than-normal anatomical occupancy of layer 4 by LGN axons representing the *deprived* eye! Despite these surprising results, this experiment demonstrates quite clearly that the activity of the postsynaptic cell is intimately involved in the process of remodeling of LGN synapses.

Why would the deprived eye be at an advantage in the silenced region of cortex? One possibility is that the very large mismatch in activity between the presynaptic inputs from LGN axons representing the nondeprived eye and the (silent) postsynaptic cells has set the scene for synaptic weakening rather than strengthening of the most active inputs. Similar observations have been made when (N-methyl-D-aspartate glutamate) (NMDA) receptors were blocked by minipump infusions of 2-amino-5-phosphono-pentanoic acid (AP5). Again there was a paradoxical physiological shift in favor of the deprived eye within the infused region of cortex (Bear et al., 1990); unfortunately, the status of the pattern of ocular dominance columns in layer 4 was not assessed in those experiments. Because the visually driven response of most cortical neurons is mediated by NMDA-receptor–gated currents (Miller et al., 1989b; Fox et al., 1989), it is likely that blockade of NMDA receptors was similar in its postsynaptic effect to that of the muscimol infusions—the postsynaptic cell was silenced, or at least prevented from achieving sufficient depolarization to permit activity-dependent strengthening of the most active presynaptic inputs. It would be extremely useful to know how much spontaneously evoked postsynaptic activity remains in the cortex under both infusion conditions.

Neurotrophins as Candidates for a Retrograde Signal in Synaptic Plasticity

The experiments considered above indicate that the pattern of synaptic connections underlying the formation of ocular dominance columns in the developing mammalian visual cortex is controlled by patterns of neural activity in the geniculocortical afferents in conjunction with cellular mechanisms that enable the detection of these patterns and the consequent modulation of synaptic strength. The fact that the postsynaptic cell is involved, coupled with the fact that the ultimate outcome is a structural remodeling of the presynaptic terminal arbors of LGN axons, implies that there must be a retrograde message released from the postsynaptic cell that acts on the presynaptic terminals. Moreover, there must be some constraints on how this retrograde signal works. Such a signal must somehow not only be regulated by neuronal activity but also be effective in causing *selective* stabilization and growth only at those presynaptic

synaptic terminals whose activity is well correlated with that of the postsynaptic neuron.

In the past few years, members of the family of nerve growth factors known as the neurotrophins have become attractive candidates as retrograde signals in activity-dependent synaptic remodeling in the mammalian nervous system (reviewed in Lo, 1995; Thoenen, 1995; Bonhoeffer, 1996; Ghosh, 1996). This growth factor family consists of four members—NGF, BDNF, NT-3, and NT-4—and their high-affinity receptors: members of the Trk family of tyrosine kinases (Davies, 1994; Kaplan and Stephens, 1994; Snider, 1994; Lindholm et al., 1994; Bothwell, 1995; Lewin and Barde, 1996). Traditionally, these growth factors and their receptors are known to play essential roles during early development in controlling the programmed cell death, the survival, and the differentiation of neurons. Yet, it has remained a puzzle why, in the adult cerebral cortex and hippocampus, levels of neurotrophin and receptor expression remain high, and why levels of certain neurotrophins are very rapidly modulated by neural activity, induced, for example, by seizure or LTP (Gall and Isackson, 1989; Ernfors et al., 1991; Isackson et al., 1991; Patterson et al., 1992; Castren et al., 1992). One possibility is that the neurotrophins are playing a role in the synaptic plasticity known to persist into adulthood in these cortical structures.

Several lines of recent evidence suggest there is this additional and novel role for neurotrophins in modulating the strength of synaptic transmission. For example, at developing neuromuscular synapses in vitro, NT-3 and BDNF (but not NGF) application resulted in the potentiation of both evoked and spontaneous synaptic transmission; this potentiation was dependent on the presence of the neurotrophin and disappeared upon washout (Lohof et al., 1993). In hippocampal slices, Schuman and colleagues have shown that a robust and long-lasting form of potentiation of synaptic transmission can be induced in CA1 by low-frequency stimulation of the Schaeffer collaterals in the presence of NT-3 or BDNF (but not NGF; Kang and Schuman, 1995a,b); this potentiation occurs within 15–20 minutes of neurotrophin application, and once established, does not require the ongoing presence of the neurotrophin. The enhancement can be blocked by application of K252a, a generic tyrosine kinase inhibitor, suggesting that it may operate through the activation of Trks (see also Levine et al., 1995). Of note, the neurotrophin-induced enhancement of synaptic transmission does not occlude LTP induced by high-frequency stimulation, suggesting that the two forms of synaptic enhancement may use partially independent signaling pathways. In another complementary set of experiments, Korte et al. (1995) demonstrated that transgenic mice lacking BDNF have diminished hippocampal LTP. These observations taken together support the hypothesis that neurotrophins can modulate the strength of synaptic transmission within minutes of application. However, many fascinating questions remain: Is this effect pre- or postsynaptic or both? Do the effects of exogenous application of neurotrophins require concurrent presynaptic and/or postsynaptic activity, or can simple application without stimulation also induce synaptic strengthening? The major question remains: Do neurotrophins indeed play an endogenous role in regulating synaptic strength in vivo?

Strong hints that neurotrophins can indeed act as activity-dependent signals in vivo comes from a recent study by Ibanez and colleagues on the adult rat neuromuscular junction (Funakoshi et al., 1995). Blockade of neuromuscular transmission with α-bungarotoxin decreased NT-4 mRNA levels in the muscle, while electrical stimulation of denervated muscle caused an increase. Exogenous addition of NT-4 protein resulted in extensive sprouting of the intact muscle nerve, suggesting that NT-4 release from muscle normally controls the pattern and extent of muscle innervation. However, again many crucial questions remain to be addressed: Is NT-4 released from the muscle with muscle activity? Is NT-4 the endogenous ligand for TrkB, which is known to be expressed by adult motoneurons, or is it somehow acting indirectly to influence nerve branching? And, does the administration of NT-4 actually increase the strength of neuromuscular synaptic transmission?

A Role for Neurotrophins in Activity-Dependent Visual System Development?

Experiments in the mammalian visual system lend further support to an in vivo role for neurotrophins in activity-dependent synaptic plasticity. The visual system is an excellent model in which to examine this question in the sense that neural activity is known to be required for formation of the adult pattern of ocular dominance within the primary visual cortex (see above). However, as considered below, it is a complex system in which the sites of expression and action of neurotrophins are just beginning to be understood.

The first indication that neurotrophins might be involved in activity-dependent synaptic remodeling during visual system development came from studies of Maffei and colleagues on the effects of administration of NGF on the eye preference of cortical neurons following monocular visual deprivation (MD), in rodents and cats. Daily intraventricular administration of NGF during the critical period for the effects of MD prevented the physiologically assessed shift in ocular dominance in favor of the nondeprived eye (Maffei et al., 1992; Carmignoto et al., 1993) (a result similar to that obtained with muscimol or z-aminophosphovaleric acid APV infusions). NGF administration also prevented the well-known anatomical shrinkage of the cell bodies of LGN neurons receiving inputs from the deprived eye (Berardi et al., 1993; Domenici et al., 1993). In addition, administration of a monoclonal antibody to NGF could extend the period during which monocular eye closure alters the ocular dominance of cortical neurons (Domenici et al., 1994). Since a similar extension of the critical period has been obtained with dark-rearing, a manipulation that alters levels of neural activity (Cynader and Mitchell, 1980), the authors interpreted their results to indicate that neural activity normally regulates NGF levels, which in turn regulate the pattern of branching of geniculocortical axons. Moreover, when NGF is available in excess, axons from the deprived eye are protected against the effects of disuse, and when NGF availability is blocked by NGF antibody administration, the normal pace of development is slowed significantly.

While there is no doubt from these studies that NGF can alter the outcome of the effects of molecular visual deprivation on the physiology of cortical neurons, the interpretation that NGF is acting directly on LGN axons is highly questionable. First, the high-affinity receptors for NGF—TrkA—are known to be present only on cholinergic axons originating from the basal forebrain, and these axons are well known to participate in regulating activity-dependent plasticity in visual cortex (Kasamatsu and Pettigrew, 1976; Bear and Singer, 1986). In contrast, recent studies in cats and ferrets indicate that there are high levels of TrkB and TrkC in LGN and cortex (but undetectable levels of TrkA) and that LGN neurons are immunoreactive for TrkB during the appropriate period of development (Allendoerfer et al., 1994; Cabelli et al., 1994). Second, since NGF and NGF antibodies were administered by intraventricular injection (rather than local application confined to visual cortex), it is very likely that the basal forebrain cholinergic axons were indeed affected directly. Finally the extremely high levels of NGF administered (>2 mg/day) make it possible that NGF acted as a competitive antagonist of an endogenous ligand of TrkB or TrkC (cross-talk between Trks is known to occur at high levels of neurotrophins; Lewin and Barde, 1996). Taken together, these considerations make it highly unlikely that NGF acts directly on LGN axons in vivo.

The results of two more recent experiments favor the interpretation that a ligand of TrkB—either BDNF or NT-4 (or both)—is involved in the activity-dependent control of LGN axon branching during development. Cabelli et al. (1995) examined the effect of intracortical infusions of neurotrophins on the segregation of geniculocortical axons into ocular dominance columns within layer 4 of the cat's visual cortex. This experiment was motivated by the hypothesis that the well-known activity-dependent competition between LGN axons representing the two eyes for territory in cortical layer 4 could involve a competition for a neurotrophin that is somehow available in limited amounts from the postsynaptic cell in an activity-dependent manner. Infusion of BDNF or NT-4, but not NT-3 or NGF, during a 2-week period in which the LGN axons normally segregate to form ocular dominance columns in layer 4 prevented the segregation as assessed anatomically by the method of transneuronal autoradiographic labeling following an intraocular injection of tritiated amino acids. In these experiments, the neurotrophins were infused directly into visual cortex via a minipump, and only the infused region was affected; ocular dominance columns formed normally in adjacent stretches of noninfused cortex, indicating a local site of action. The most parsimonious interpretation of this experiment is that since LGN neurons are known to express TrkB at the relevant times in development, BDNF or NT-4 may be acted directly on the LGN axon terminals in layer 4. The excess of exogenously supplied neurotrophins available to LGN terminals may have removed competitive interactions that normally drive axonal remodeling, leading to unregulated growth of LGN axon terminals even in the presence of normal patterns of neural activity. (Exogenously applied BDNF does promote increased branching of individual retinal ganglion cell axon terminals in *Xenopus* optic tectum; Cohen-Cory and Fraser, 1995.) It is now essential to know exactly how the branching patterns of individual LGN axons have been altered by the neurotrophin infusions into cat visual cortex.

A second experiment lends further support to a hypothesized role for TrkB ligands in activity-dependent synaptic remodeling in visual cortex. Riddle et al. (1995) assessed the consequences of adding exogenous neurotrophins for the effects of monocular eye closure in the ferret during the critical period. Eye closure, in addition to preventing LGN axon terminals representing the deprived eye from gaining their fair share of territory within cortical layer 4, also is known to cause a corresponding shrinkage of cell bodies within the LGN (Wiesel and Hubel, 1963a). However, the shrinkage of LGN neuron cell bodies receiving input from the deprived eye could be prevented when latex microspheres saturated with NT-4, but not BDNF, NT-3, or NGF, were neatly injected into layer 4. This observation is consistent with the suggestion that LGN axons normally compete for an endogenous ligand of TrkB available within layer 4 and that availability of ligand from the postsynaptic cell, or the ability of the presynaptic terminals to respond to it, is regulated by neural activity. Addition of excess neurotrophin to layer 4 would permit LGN axons from the deprived eye to obtain neurotrophins independent of their level of neural activity and thereby protect them from the effects of monocular deprivation. In this regard, it would be fascinating to know if, within the treated area of layer 4, the ocular dominance columns formed normally with equal widths in the two eyes even following monocular deprivation, or if instead, no columns formed at all—as might be predicted from the results of Cabelli et al. (1995).

The results of all of the experiments considered thus far, based on the exogenous administration of neurotrophins, suggest a role for BDNF and/or NT-4 in activity-dependent synaptic remodeling during visual system development. Yet, there remain many crucial missing links. First, it is currently not known if any of the exogenously supplied neurotrophins indeed act directly on LGN axons. While it is reasonable to suppose that they do, in view of the fact that LGN neurons express TrkB, it is possible that there were also indirect effects: At least in ferret and cat the neurons of cortical layers 2/3 and 5 in particular are strongly immunoreactive for TrkB (Cabelli et al., 1994), and in organotypic slice cultures of ferret visual cortex the addition of BDNF can alter the branching pattern of the basal dendrites of layer 4 neurons, while NT-4 can affect the dendrites of layer 5/6 neurons (McAllister et al., 1995). Second, if ligands of TrkB are indeed involved here, they should be expressed by layer 4 cortical neurons at the relevant developmental times, and their levels of expression ought to be regulated by neural activity in a highly local fashion. To date, BDNF mRNA has been detected within neurons of the primary visual cortex in cats, ferrets, and rats (Lein et al., 1995; Ernfors et al., 1990), and neural activity is known to regulate the global levels of BDNF expression in rat visual cortex (Castren et al., 1992; Schoups et al., 1995). However, while BDNF is expressed in layers 2/3 and 5 and 6, it has not been detected in significant amounts in layer 4 neurons, at least at the mRNA level. And virtually nothing is known about NT-4 expression patterns in visual cortical neurons.

Even more important to establish the validity of any neurotrophic hypothesis of activity-dependent synaptic competition in the visual system would be to demonstrate a requirement for an *endogenous* neurotrophic factor. One approach might be to examine the effects of genetically altering neurotrophins or receptors in mice. Although there are now a large variety of transgenic and "knockout" mice available (Snider, 1994), unfortunately the mouse visual system does not have ocular dominance columns, and consequently the ability to compare the effects of such genetic manipulations on both structural and physiological organization of synaptic connectivity is limited. Thus, there is a great need to explore other possible examples of activity-dependent synaptic remodeling during development—perhaps in the somatosensory system—in this species.

Another nongenetic approach in the visual system of higher mammals is to show that competitive antagonists, or antibodies against specific neurotrophins, prevent synaptic remodeling. Preliminary results of minipump infusions of the Trk-IgG's (Shelton et al., 1995) (constructed from the extracellular domain of each member of the Trk family linked to the Fc tail of a human IgG) which compete with native Trk receptors for binding neurotrophins indicate that infusion of excess TrkB-IgG protein, but not TrkA-IgG or TrkC-IgG, prevents the segregation of LGN axons into ocular dominance columns (Cabelli et al., 1996). In this regard, similar experiments are needed to demonstrate a role for an endogenous neurotrophin in hippocampal synaptic plasticity as well. Very recently, experiments by Figurov et al. (1995) in hippocampal slices demonstrated that infusion of TrkB-IgG substantially attenuates LTP produced by tetanus (but not by pairing of presynaptic low-frequency stimulation with postsynaptic depolarization), again implicating an endogenous ligand of TrkB in hippocampal synaptic plasticity in vitro.

PERMISSIVE VS. INSTRUCTIVE ROLES FOR NEUROTROPHINS IN VISUAL SYSTEM DEVELOPMENT

The results considered above suggest a role for neurotrophins—in particular, ligands of TrkB—in activity-dependent competition the visual system. But exactly what role? Could neurotrophins act as a retrograde signal, being released from the postsynaptic neuron in an activity-dependent fashion and acting rapidly and locally on co-active presynaptic inputs to enhance synaptic transmission, as is required by models of visual cortical plasticity considered earlier in this chapter? It seems unlikely that neurotrophins themselves could carry this kind of instructive activity-dependent signal in which the precise timing of activity in the pre- and postsynaptic neurons controls whether or not synapses are strengthened or weakened. The requirement for synchrony in firing operates on a time scale of tens to hundreds of milliseconds, whereas once neurotrophin binds to Trk receptors, signaling is likely to be sustained over much longer periods, degrading the process of coincidence detection. (A more attractive candidate for a retrograde signal in this regard is nitrous oxide [NO]; Barinaga, 1991; Schuman, 1995). And there is also the problem of explaining, in the context of a simple model of neurotrophin action, how cortical infusion of muscimol coupled with monocular visual deprivation leads to a *weakening* of the more active inputs from the nondeprived eye (see above; Reiter and Stryker, 1988).

However, it is conceivable that neurotrophins could act in a permissive fashion to set the scene for activity-dependent competition. For instance, neurotrophin levels might modulate the threshold for synaptic plasticity in visual cortex (Bienenstock et al., 1982; Bear, 1995; Deisseroth et al., 1995). In a recent study, Bear and colleagues (Kirkwood et al., 1996) demonstrated that the threshold could be modulated by altered visual experience: dark-rearing diminished the extent of LTD and enhanced the degree of LTP that could be elicited over a range of stimulation frequencies in slices of rat visual cortex, effectively sliding the threshold in favor of synaptic enhancement. Could it be that high levels of neurotrophin and consequent Trk activation also slide the threshold toward enhancement of active inputs, while low levels of neurotrophin might shift the threshold toward synaptic weakening of active inputs? Preliminary results from Bear and Huber (1996) support this idea as well: incubation of cortical slices in BDNF for 2–5 hours diminishes the amplitude of LTD produced by low-frequency stimulation.

A permissive role for neurotrophins also could help explain the muscimol infusion results, assuming that levels of neurotrophins are very low in cortex in which postsynaptic activity has been blocked by muscimol. Thus, neurotrophins, acting in concert with patterned neuronal activity, would be required for synaptic modification. Recently, Barres and colleagues (Meyer-Franke et al., 1995) have discovered that the effects of growth factors on the survival of retinal ganglion cells in vitro can be enhanced greatly by depolarizing the cells with potassium chloride, implying a convergence of intracellular signaling pathways for neural activity and growth factors controlling neuronal survival. An implication of this observation is that similar interactions between growth factors and neural activity might occur in the process of synaptic remodeling in the visual cortex.

Obviously, many questions remain to be answered regarding a role for neurotrophins in controlling activity-dependent synaptic remodeling in visual system development, and I have only highlighted some of the more pressing ones here. The idea that the neurotrophins and their receptors may function not only in the regulation of cell death and survival early in development but also in controlling the strengthening or weakening of synapses and the consequent structural remodeling of axons during subsequent development is an attractive one for many reasons, including the fact that a great deal is already known about the molecular control of growth and survival by this family of molecules. It has been over 20 years since Hubel and Wiesel first described the effects of abnormal visual experience on the functional and structural organization of the mammalian visual system. It would be exciting indeed if we were on the verge of a satisfying molecular explanation of these phenomena.

ACKNOWLEDGMENTS

I thank Ed Lein for his critical and helpful reading of the manuscript. Experiments cited here from this laboratory were supported by the National Eye Institute. C.J.S. is an Investigator of the Howard Hughes Medical Institute.

REFERENCES

Allendoerfer, K.L., Cabelli, R.J., Escandon, E., Kaplan, D.R., Nikolics, K., Shatz, C.J. (1994). Regulation of neurotrophin receptors during the maturation of the mammalian visual system. *J. Neurosci.* 14:1795–1811.

Antonini, A., Stryker, M.P. (1993a). Rapid remodeling of axonal arbors in the visual cortex. *Science* 260:1819–1821.

Antonini, A., Stryker, M. (1993b). Development of individual geniculocortical arbors in cat striate cortex and effects of binocular impulse blockade. *J. Neurosci.* 13(8):3549–3573.

Artola, A., Brocher, S., Singer, W. (1990). Different voltage-dependent thresholds for inducing long-term depression and long-term potentiation in slices of rat visual cortex. *Nature* 347:69–72.

Barinaga, M. (1991). Is nitric oxide the 'retrograde messenger'? *Science* 254:1296–1297.

Bear, M.F. (1995). Mechanism for a sliding synaptic modification threshold. *Neuron* 15:1–4.

Bear, M., Abraham W. (1996). Long-term depression in hippocampus. *Annu. Rev. Neurosci.* 19:437–462.

Bear, M.F., Huber, K.M. (1996). Brain derived neurotrophic factor (BDNF) modulates synaptic plasticity in rat visual cortex. *Soc. Neurosci. Abstr.* 22:1729.

Bear, M.F., Malenka, R.C. (1993). Synaptic plasticity: LTP and LTD. *Curr. Opin. Neurobiol.* 4:389–399.

Bear, M.F., Singer, W. (1986). Modulation of visual cortical plasticity by acetycholine and noradrenaline. *Nature* 320:172–176.

Bear, M.F., Kleinschmidt, A., Gu Q., Singer, W. (1990). Disruption of experience-dependent synaptic modifications in striate cortex by infusion of an NMDA receptor antagonist. *J. Neurosci.* 10:909–925.

Berardi, N., Domenici, L., Parisi V., Pizzorusso, T., Cellerino, A., Maffei, L. (1993). Monocular deprivation effects in the rat visual cortex and lateral geniculate nucleus are prevented by nerve growth factor (NGF). I. Visual cortex. *Proc. R. Soc. Lond. Biol.* 251:17–23.

Bienenstock, E.L., Cooper, L.N., Munro, P.W. (1982). Theory for the development of neuron selectivity: orientation specificity and binocular interaction in visual cortex. *J. Neurosci.* 2:32–48.

Bonhoeffer, T. (1996). Neurotrophins and activity-dependent development of the neocortex. *Curr. Opin. Neurobiol.* 6:119–126.

Bothwell, M. (1995). Functional interactions of neurotrophins and neurotrophin receptors. *Annu. Rev. Neurosci.* 18:223–253.

Cabelli, R.J., Radeke, M.J., Wright, A., Allendoerfer, K.L., Feinstein, S.C., Shatz, C.J. (1994). Developmental patterns of localization of full-length and truncated TrkB proteins in the mammalian visual system. *Soc. Neurosci. Abstr.* 20:37.

Cabelli, R.J., Hohn, A., Shatz, C.J. (1995). Inhibition of ocular dominance column formation by infusion of NT-4/5 or BDNF. *Science* 267:1662–1666.

Cabelli, R.J., Shelton, D., Shatz, C.J. (1996). An endogenous ligand of trkb is required for ocular dominance segregation. *Soc. Neurosci. Abstr.* 22:276.

Carmignoto, G., Canella, R., Candeo, P., Comelli, M.C., Maffei, L. (1993). Effects of nerve growth factor on neuronal plasticity of the kitten visual cortex. *J. Physiology* 464:343–360.

Castren, E., Zafra, F., Thoenen, H., Lindholm, D. (1992). Light regulates expression of brain-derived neurotrophic factor mRNA in rat visual cortex. *Proc. Natl. Acad. Sci. U.S.A.* 89:9444–9448.

Cheng, H.-J., Nakamoto, M., Bergemann, A., Flanagan, J.G. (1995). Complementary gradients in expression and binding of ELF-1 and Mek4 in development of the topographic retinotectal projection map. *Cell* 82:371–381.

Cohen-Cory, S., Fraser, S. (1995). Effects of brain-derived neurotrophic factor on optic axon branching and remodeling in vivo. *Nature* 378:192–196.

Crair, M., Malenka, R. (1995). A critical period for long-term potentiation at thalamocortical synapses. *Nature* 375:325–328.

Cynader, M., Mitchell, D.E. (1980). Prolonged sensitivity to monocular deprivation in dark-reared cats. *J. Neurophysiol.* 43(4):1026–1039.

Davies, A.M. (1994). The role of neurotrophins in the developing nervous system. *J. Neurobiol.* 25:2334–1349.

Deisseroth, K., Bito, H., Schulman, H., Tsien, R.W. (1995). A molecular mechanism for metaplasticity. *Curr. Biol.* 5(12):1334–1338.

Domenici, L., Cellerino, A., Maffei, L. (1993). Monocular deprivation effects in the rat visual cortex and lateral geniculate nucleus are prevented by nerve growth factor (NGF). II. Lateral geniculate nucleus. *Proc. R. Soc. Lond. Biol.* 251:25–31.

Domenici, L., Cellerine, A., Berardi, N., Cattaneo, A., Maffei, L. (1994). Antibodies to nerve growth factor (NGF) prolong the sensitive period for monocular deprivation in the rat. *Neuroreport* 5:2041–2044.

Drescher, U., Kremoser, C., Handwerker, C., Loschinger, J., Noda M., Bonhoeffer, F. (1995). In vitro guidance of retinal ganglion cell axons by RAGS, a 25 kDa tectal protein related to ligands for Eph receptor tyrosine kinases. *Cell* 82:359–370.

Dudek, S.M., Bear, M.F. (1993). Bidirectional long-term modification of synaptic effectiveness in the adult and immature hippocampus. *J. Neurosci.* 13:2910–2918.

Ernfors, P., Wetmore, C., Olson, L., Persson, H. (1990). Identification of cells in rat brain and peripheral tissues expressing mRNA for members of the nerve growth factor family. *Neuron* 5:511–526.

Ernfors, P., Bengzon, J., Kokaia, Z., Persson, H., Lindvall, O. (1991). Increased levels of messenger RNAs for neurotrophic factors in the brain during kindling epileptogenesis. *Neuron* 7:165–176.

Figurov, A., Pozzo Miller, L., Olafsson, P., Wang, T., Lu, B. (1995). Neurotrophins regulate synaptic responses to tetanic stimulation and LTP in the hippocampus. *Nature* 381: 706–709.

Fox, K., Sato, H., Daw, N.W. (1989). The location and function of NMDA receptors in cat and kitten visual cortex. *J. Neurosci.* 9:2443–2454.

Fregnac, Y., Shulz, D., Thorpe, S., Bienenstock, E. (1988). A cellular analogue of visual cortical plasticity. *Nature* 333:367–370.

Funakoshi H., Belluardo, N., Arenas, E., Yamamoto, Y., Casabona, A., Persson, H., Ibanez, C. (1995). Muscle-derived neurotrophin-4 as an activity-dependent trophic signal for adult motor neurons. *Science* 268:1495–1499.

Gall, C.M., Isackson, P.J. (1989). Limbic seizures increase neuronal production of messenger RNA for nerve growth factor. *Science* 245:758–761.

Ghosh, A. (1996). Cortical development: with an eye on neurotrophins. *Curr. Biol.* 6(2): 130–133.

Goodhill, G., Lowel, S. (1995). Theory meets experiment: correlated neural activity helps determine ocular dominance column periodicity. *Trends Neurosci.* 18(10):437–439.

Goodman, C.S., Shatz, C.J. (1993). Developmental mechanisms that generate precise patterns of neuronal connectivity. *Cell* 72(Suppl):77–98.

Hata, Y., Stryker, M. (1994). Control of thalamocortical afferent rearrangement by postsynaptic activity in developing visual cortex. *Science* 265:1732–1735.

Hebb, D.O. (1949). *The Organization of Behavior.* New York: John Wiley.

Hubel, D.H., Wiesel, T.N. (1970). The period of susceptibility to the physiological effects of unilateral eye closure in kittens. *J. Physiol.* 206:419–436.

Hubel, D.H., Wiesel, T.N., LeVay, S. (1977). Plasticity of ocular dominance columns in the monkey striate cortex. *Philos. Trans. R. Soc. Lond. Biol.* 278:377–409.

Isackson, P.J., Huntsman, M.M., Muray, K.D., Gall, C.M. (1991). BDNF mRNA expression is increased in adult rat forebrain after limbic seizures: temporal pattern of induction distinct from NGF. *Neuron* 6:937–948.

Kang, H., Schuman, E.M. (1995a). Long-lasting neurotrophin-induced enhancement of synaptic transmission in the adult hippocampus. *Science* 267:1658–1662.

Kang, H.J., Schuman, E.M. (1995b). Neurotrophin-induced modulation of synaptic transmission in the adult hippocampus. *J. Physiol.* 89:11–22.

Kaplan, D., Stephens, R. (1994). Neurotrophin signal transduction by the Trk receptor. *J. Neurobiol.* 25(11):1404–1417.

Kasamatsu, T., Pettigrew, J.D. (1976). Depletion of brain catecholamines: failure of ocular dominance shift after monocular occlusion in kittens. *Science* 194:206–209.

Katz, L.C., Shatz, C.J. (1996). Synaptic activity and the construction of cortizel circuits. *Science* 274:1133–1138.

Kirkwood, A., Lee, H.-K., Bear, M. (1995). Co-regulation of long-term potentiation and experience-dependent synaptic plasticity in visual cortex by age and experience. *Nature* 375:328–331.

Kirkwood, A., Rioult, M.E., Bear, M.F. (1996). Experience-dependent modification of synaptic plasticity in visual cortex. *Nature* 381:526–528.

Komatsu, Y., Fujii, K., Maeda, J., Sakaguchi, H., Toyama, K. (1988). Long term potentiation of synaptic transmission in kitten visual cortex. *J. Neurophysiol.* 59:124–141.

Korte, M., Carroll, P., Wolf, E., Brem, G., Thoenen, H., Bonhoeffer, T. (1995). Hippocampal long-term potentiation is impaired in mice lacking brain-derived neurotrophic factor. *Proc. Natl. Acad. Sci. U.S.A.* 92:8856–8860.

Lein, E., Hohn, A., Shatz, C.J. (1995). Reciprocal laminar localization and developmental regulation of BDNF and NT-3 during visual cortex development. *Soc. Neurosci. Abstr.* 21:1795.

LeVay, S., Stryker, M.P., Shatz, C.J. (1978). Ocular dominance columns and their development in layer IV of the cat's visual cortex. *J. Comp. Neurol.* 179:223–244.

LeVay, S., Wiesel, T.N., Hubel, D.H. (1980). The development of ocular dominance columns in normal and visually deprived monkeys. *J. Comp. Neurol.* 191:1–51.

Levine, E.S., Dreyfus C.F., Black, I.B., Plummer, M.R. (1995). Brain-derived neurotrophic factor rapidly enhances synaptic transmission in hippocampal neurons via postsynaptic tyrosine kinase receptors. *Proc. Natl. Acad. Sci. U.S.A.* 92:8074–8077.

Lewin, G., Barde, Y.-A. (1996). Physiology of the neurotrophins. *Annu. Rev. Neurosci.* 19: 289–317.

Lindholm, D., Castren, E., Berzaghi, M.P., Blochl, A., Thoenen, H. (1994). Activity-dependent and hormonal regulation of neurotrophin mRNA levels in the brain—implications for neuronal plasticity. *J. Neurobiol.* 25:1362–1372.

Lo, D. (1995). Neurotrophic factors and synaptic plasticity. *Neuron* 15:979–981.

Lohof, A. M., Ip, N.Y., Poo, M.-M. (1993). Potentiation of developing neuromuscular synapses by the neurotrophins NT-3 and BDNF. *Nature* 363:350–353.

Lowel, S., Singer, W. (1992). Selecti on of intrinsic horizontal connections in the visual cortex by correlated neuronal activity. *Science* 225:209–212.

Maffei, L., Berardi, N., Domenici, L., Parisi, V., Pizzorusso, T. (1992). Nerve growth factor (NGF) prevents the shift in ocular dominance distribution of visual cortical neurons in monocularly deprived rats. *J. Neurosci.* 12:4651–4662.

Malenka, R. (1994). Synaptic plasticity in the hippocampus: LTP and LTD. *Cell* 78:535–538.

Malenka, R.C., Nicoll, R.A. (1993). NMDA-receptor-dependent synaptic plasticity: multiple forms and mechanisms. *Trends Neurosci.* 16(12):521–527.

Madison, D.V., Malenka, R., Nicoll, R.A. (1991). Mechanisms underlying long-term potentiation of synaptic transmission. *Annu. Rev. Neurosci.* 14:379–397.

McAllister, A.K., Lo, D.C., Katz, L.C. (1995). Neurotrophins regulate dendritic growth in developing visual cortex. *Neuron* 4:791–803.

Meyer-Franke, A., Kaplan, M.R., Pfrieger, F.W., Barres, B.A. (1995). Characterization of the signaling interactions that promote the growth and survival of retinal ganglion cells in culture. *Neuron* 15:805–819.

Miller, K.D., Keller, J.B., Stryker, M.P. (1989a). Ocular dominance column development: analysis and stimulation. *Science* 245:605–615.

Miller, K.D., Chapman, B., Stryker, M.P. (1989b). Visual responses in adult cat visual cortex depend on N-methyl-D-aspartate receptors. *Proc. Natl. Acad. Sci. U.S.A.* 856:5183–5187.

Mower, G.D., Caplan, C.J., Christen, W.G., Duffy, F.H. (1985). Dark rearing prolongs physiological but not anatomical plasticity of the cat visual cortex. *J. Comp. Neurol.* 235:448–466.

Patterson, S.L., Grover, L.M., Schwartzkroin, P.A., Bothwell, M. (1992). Neurotrophin expression in rat hippocampa slices; a stimulus paradigm inducing LTP in CA1 evokes increases in BDNF and NT-3 mRNAs. *Neuron* 9:1081–1088.

Rakic, P. (1977). Prenatal development of the visual system in the rhesus monkey. *Philos. Trans. R. Soc. Lond. Biol.* 278:245–260.

Reiter, H.O., Stryker, M.P. (1988). Neural plasticity without postsynaptic action potentials: less-active inputs become dominant when kitten visual cortical cells are pharmacologically inhibited. *Proc. Natl. Acad. Sci. U.S.A.* 85:3623–3627.

Riddle, D.R., Lo, D.C., Katz, L.C. (1995). NT-4 mediated rescue of lateral geniculate neurons from effects of monocular deprivation. *Nature* 378:189–191.

Schoups, A.A., Elliott R.C., Friedman, W.J., Black, I.B. (1995). NGF and BDNF are differentially modulated by visual experience in the developing geniculocortical pathway. *Dev. Brain Res.* 86:326–334.

Schuman, E. (1995). Nitric oxide signalling, long-term potentiation and long-term depression. *Nitric Oxide in the Nervous System.* London; San Diego, Academic Press. pp. 125–147.

Shatz, C.J. (1990). Impulse activity and the patterning of connections during CNS development. *Neuron* 5:745–756.

Shatz, C.J., Stryker, M.P. (1978). Ocular dominance in layer IV of the cat's visual cortex and the effects of monocular deprivation. *J. Physiol.* 281:267–283.

Shelton, D.L., Sutherland, J., Gripp, J., Camerato, T., Armanini, M.P., Philips, H.S., Carroll, K., Spencer, S.D., Levinson, A.D. (1995). Human trks: molecular cloning, tissue distribution, and expression of extracellular domain immunoadhesins. *J. Neurosci.* 15(1): 477–491.

Snider, W.D. (1994). Functions of the neurotrophins during nervous system development: what the knockouts are teaching us. *Cell* 77:627–638.

Stanton, P.K., Sejnowski, T.J. (1989). Associative long term depression in the hippocampus: induction of synaptic plasticity by Hebbian covariance. *Nature* 339:215–218.

Stryker, M.P., Harris, W. (1986). Binocular impulse blockade prevents the formation of ocular dominance columns in cat visual cortex. *J. Neurosci.* 6:2117–2133.

Stryker, M.P., Strickland, S.L. (1984). Physiological segregation of ocular dominance columns depends on the pattern of afferent electrical activity. *Invest. Ophthalmol. Vis. Sci.* (Suppl.)25:278.

Tessier-Lavigne, M. (1995). Eph receptor tyrosine kinases, axon repulsion, and the development of topographic maps. *Cell* 82:345–348.

Tessier-Lavigne, M., Goodman, C.S. (1996). The molecular biology of axon guidance. *Science* 274:1123–1133.

Thoenen, H. (1995). Neurotrophins and neuronal plasticity. *Science* 270:593–598.
Wiesel, T.N., Hubel, D.H. (1963a). Effects of visual deprivation on morphology and physiology of cells in the cat's lateral geniculate body. *J. Neurophysiol.* 26:978–993.
Wiesel, T.N., Hubel, D.H. (1963b). Single cell responses in striate cortex of kittens deprived of vision in one eye. *J. Neurophysiol.* 26:1003–1017.
Wiesel, T., Hubel, D. (1965). Comparison of the effects of unilateral and bilateral eye closure on cortical unit responses in kittens. *J. Neurophysiol.* 28:1060–1072.

15

Linking layers and connecting columns

The development of local circuits in visual cortex

LAWRENCE C. KATZ

Within the six layers of the mammalian neocortex, over 90% of the synapses originate from other cortical neurons. In the visual cortex, there is a growing consensus that many cortical properties—orientation and direction selectivity, end-stop inhibition, and interactions from beyond the classical receptive field—arise in whole or in part from the local circuits created by richly interconnected neuronal assemblies (reviewed in Gilbert, 1992; Martin, 1988). While there has been considerable progress in deciphering the patterns of anatomical connectivity and functional interactions in the adult brain, relating specific circuits to specific functional properties remains a challenge. One potentially powerful strategy employs correlative developmental analyses to relate the emergence of particular neuronal assemblies to emergence of a distinct visual response property, such as orientation. On a more general level, uncovering the mechanisms by which such assemblies form may provide insight into the capabilities of the adult cortex and into the forces which drive its assembly. In this review I will discuss recent experimental approaches that attempt to decipher these mechanisms and relate these findings to the emergence of functional architecture, especially orientation selectivity, in the visual cortex.

The local circuits that comprise the neocortex can be separated into two broad categories: vertical and horizontal. Historically, cortical microanatomy has emphasized vertical (or radial) circuits: connections that link neurons in different layers with one another. These connections form the basis of the cortical column—specifically, interconnecting groups of neurons that lie roughly in vertical register with one another. The other dimension of cortical connectivity is across the tangential extent of a single layer, and these connections are termed horizontal, or lateral, connections. These connections span many millimeters and extend well beyond the region of a single cortical column.

The distinction between vertical and horizontal connections is somewhat artificial, as the same neuron can provide connections that participate in both systems. For example, pyramidal cells in layers 2 and 3 form numerous vertical connections within the same column in layers 2/3 and 5 but are also the source of the rich system of horizontal connections that spans many millimeters in those layers (Gilbert and Wiesel, 1979, 1983; Martin and Whitteridge, 1984). However, as discussed in more detail below, from a developmental standpoint the formation of vertical and horizontal connections follows fundamentally distinct rules: Vertical connections develop through highly specific patterns of outgrowth, probably relying on layer-specific molecular cues, while horizontal connections undergo significant structural and functional remodeling, under the influence of neuronal activity, to achieve their adult patterns.

LAMINAR SPECIFICITY OF VERTICAL CONNECTIONS

Given the prominence accorded to remodeling and synaptic rearrangements during the development of connections throughout the mammalian brain, especially the cerebral cortex, it is surprising that the connections between cortical layers are completely specific from the very earliest times in development. In contrast to the development of laminar specificity in the retinogeniculate system, for example (see Chapter 14), which relies on elimination of some portion of "inappropriately" situated synaptic connections, the formation of circuits connecting the six cortical layers involves selective elaboration of projections to the appropriate layers rather than any elimination of inappropriate connections.

Laminar specificity in early development was already apparent from early Golgi studies by Lund (Lund et al., 1977), who first suggested that the diversity of specific patterns of dendritic and axonal arbors of layer 6 cells in monkeys emerged by specific outgrowth, rather than by selection of different patterns from a common earlier pattern. Both this older and more recent work (Wiser and Callaway, 1996) suggest that the laminar patterns of monkey layer 6 axons are related to distinct functional streams—the magnocellular and parvocellular—present in the macaque visual system. These distinct streams originate in the retina from retinal ganglion cells with distinct patterns of activity, yet functional specificity is presumably achieved without resorting to sorting based on differential activity patterns, even though such differences are clearly present.

Although suggestive, Golgi studies suffer from numerous limitations, including incomplete staining of axonal arbors. In fact, several of the pyramidal cell types described in layer 6 based on Golgi staining are not observed using intracelluar staining in adult brain slices (Wiser and Callaway, 1996), leaving open the possibility that they represent some earlier, immature forms that do undergo rearrangements. More direct evidence for laminar specificity comes from several studies in which the development of laminar patterns of arborizations were followed using intracellular staining from the earliest outgrowth of axons until adulthood. In the case of pyramidal cells in layer 2/3—which, in the adult cat, ferret, and primate project exclusively to layers 2/3 and 5, bypassing layer 4—collateral branches are never found in inappropriate target layers (Burkhalter et al., 1993; Katz, 1991). Sprouting of collaterals is confined to those portions of the axons that transit the appropriate layers, and the number of primary collaterals appears to increase rapidly, but not to undergo any detectable elimination. Similarly, layer 4 stellate cells, which project primarily to layer 2/3 undergo a progressive and conservative elaboration of their interlaminar projection with no evidence for "inappropriate" connections to erroneous layers.

Cues for Laminar Specificity

What drives these patterns of laminar specificity? In the visual cortex, much of the observed specificity in circuits (such as ocular dominance columns) is attributed to the effects of spontaneous or evoked neuronal activity on refining less accurate initial connections (reviewed in Shatz, 1990). However, the available evidence suggests that patterns of neuronal activity do not drive the formation of layer-specific termination patterns. Binocular deprivation during the time that either layer 2/3 or layer 4 cells are forming their layer-specific patterns has no effect on their emergence or specificity (Katz, 1991), in contrast to the marked effects of such manipulations on the specificity of horizontal connections (see below). While these results are suggestive, the definitive experiments (such as blocking all spontaneous or evoked retinal activity, or blocking all cortical activity) have not yet been done.

If we accept the suggestion that laminar specificity is activity independent, the identity of the cues used by the developing system to determine such specificity remains to be determined. It is likely that laminar specificity can be achieved either by a cell-surface or locally diffusible factor. There is suggestive evidence for the presence of membrane-bound molecules that might regulate the laminar specificity of thalamic connections to layer 4 (Hubener et al., 1995), as well as for molecules that allow the construction of layer-specific connections between brain areas (reviewed in Bolz et al., 1993; Molnar and Blakemore, 1995). Developing cortical neurons clearly acquire their laminar fate very early in development, even before they have migrated into their final laminar position (see Chapter 12), and, once in position, they exhibit layer-specific patterns of transcription factor expression.

Nevertheless, the identity of putative guidance factors in the cortex remains completely unknown. In the case of layer 2/3 pyramidal cells, which form connections in layer 2/3 and 5, but not in layer 4, the cues might consist of repulsive signals originating from layer 4. Such repulsive cues must be specific for layer 2/3 cells, however, as layer 4 cells are able to grow axons perfectly well. Thus, generally repulsive substrates (such as the myelin-associated inhibitory protein, IN-1; Schwab et al., 1993) are not likely to be involved. Others, such as repulsive axon guidance signal RAGS (Drescher et al., 1995) or other members of the Eph tyrosine kinase family, or molecules of the semaphorin/collapsin family (Goodman, 1994; Luo et al., 1993), which have differential effects on specific neuronal populations, are more likely candidates. In the mammalian superior colliculus, topographic specificity is at least partly controlled by the branching specificity of retinal axons (Roskies and O'Leary, 1994); in an in vitro stripe assay system branching is apparently controlled by inhibitory molecules linked to postsynaptic membranes by a phosphotidylinositol (PI) linkage. Thus far, layer-specific cell-surface molecules have not been found, although the finding that certain homeobox transcription factors are found in specific layers and are developmentally regulated (Frantz et al., 1994), strongly implies the existence of such markers.

It is also possible that locally diffusible factors should provide the signals to initiate sprouting along certain segments of the axon. In this regard, it is interesting that the signal regulating interstitial sprouting of corticopontine axons (which originate from layer 5 pyramidal cells) appears to be a locally diffusible factor rather than a cell-surface cue (Sato et al., 1994). One attractive class of signals is the neurotrophins, as the neurotrophin family of growth factors causes sprouting of neurites in tissue culture. In the developing ferret cortex, neurotrophins and their cognate receptors, the Trk family of tyrosine kinase receptors, are distributed in specific layers (Allendoerfer et al., 1994); moreover, we have recently shown that laminar patterns of dendritic growth can be regulated by specific neurotrophins (McAllister et al., 1995). The same specificity might account for the differential sprouting and growth of layer-specific axon collaterals as well.

DEVELOPMENT OF HORIZONTAL PROJECTIONS AND CONNECTIONS

In addition to the synaptic linkages between the six cortical layers, pyramidal cells in the neocortex form elaborate, long-range horizontal projections within certain cortical layers. These projections are especially prominent in layers 2/3 and layer 5. The pyramidal cells in layer 2/3 of visual cortex, for example, form lateral connections that span more than 5 mm in cats, ferrets, and primates, including humans (see, e.g., Burkhalter and Bernardo, 1989; Gilbert and Wiesel, 1979, 1983; Martin and Whitteridge, 1984; Rockland and Lund, 1983). In the striate cortex, these lateral projections form several periodic clusters of synaptic boutons, spaced at about 1-mm intervals. Numerous experiments from several laboratories have shown that these clusters link groups of neurons with similar functional properties (Gilbert and Wiesel, 1989; Livingstone and Hubel, 1984; Malach et al., 1993). This is especially obvious in the cat and ferret visual cortex, in which clusters link groups of neurons sharing similar orientation preference. Because these lateral connections extend well beyond a given neuron's classical receptive field, these orientation-column–specific links have been implicated in a wide range of orientation-specific psychophysical and electrophysiological phenomena involving interactions from outside the classical receptive field (reviewed in Gilbert, 1994; Singer, 1995); these connections also seem involved adult cortical plasticity in response to peripheral lesions (see, e.g., Chino et al., 1992; Das and Gilbert, 1995a,b; Kaas, 1991; Merzenich and Sameshima, 1993).

From a developmental standpoint, there are suggestions from modeling studies that lateral interactions amongst cortical neurons are required for the formation of the normal pattern of orientation domains in adult cortex. Indeed, the fact that this system of lateral interactions is so closely linked to the pattern of orientation columns has led to suggestions that these lateral links may be required to actually establish some aspects of orientation tuning (Miller, 1994). In the following sections, I will review some of the basic aspects of the structural and functional development of this system and then consider whether this system is involved in setting up the system of orientation columns.

Normal Development of Horizontal Connections

How is the orientation-column–specific pattern of synaptic connections established during the course of development? In contrast to the vertical connections made by these same neurons (layer 2/3 pyramids), the adult patterns of horizontal projections are not present in young animals but instead develop via an activity-dependent process involving the selective growth and elimination of axon collaterals. Anatomical experiments using either retrograde or anterograde tracers or intracellular injections have consistently revealed a similar pattern of events in cats, ferrets, monkeys, and humans. Initially, pyramidal cells extend long, unbranched axon collaterals that arborize over an extensive lateral extent of the developing cortex, seemingly without regard to the possible pattern of orientation columns (Fig. 15–1). Subsequently, certain of these unbranched axons retract, while other branches begin to branch repeatedly, forming first crude and later more refined clusters that arborize preferentially in iso-orientation columns (reviewed in Katz and Callaway, 1992; see also Callaway and Katz, 1990; Kennedy et al., 1994; Lubke and Albus, 1992). In cats, crude clusters appear by about postnatal day 8, approximately the time of eye opening, and refinement is complete by about P30. In ferrets, a more altrical species, recent results demonstrate that clusters emerge at approximately P26, a week or so prior to eye opening, and reach adult levels of refinement by about P43 (Durack and Katz, 1996). Thus, the appearance of the initial pattern of clusters does not seem to require visual experience, but in carnivores refinement takes place subsequent to eye opening. In macaque monkeys, refined clusters are clearly evident prenatally (Yoshioka, 1994). Thus, in primates the system of clustered connections, like the organization of ocular dominance columns (Horton and Hocking, 1996), is present by birth and does not require visual experience to emerge. Taken together with results from the ferret, in which crude clusters are present before eye opening (Durack and Katz, 1996), and the requirements for activity for the development of this system (see below), this suggests that the activity cues required to generate clusters can be gleaned from endogenous activity patterns present prenatally as well as exogenous patterns generated by postnatal visual experience.

Although this general scheme—elaboration of unbranched arbors, formation of crude clusters, and refinement to adult patterns—has been observed by several investigators in several species, some differences between laboratories in the details of this process remain. Earlier suggestions that the development of these connections involved a significant change in the tangential extent of connections (Luhmann et al., 1991) have been discredited by numerous subsequent studies (Callaway and Katz, 1990; Durack and Katz, 1996; Galuske and Singer, 1996; Kennedy et al., 1994; Lubke and Albus, 1992). Lubke and Albus (1992) observed an extremely rapid transition—a few days—between unclustered pat-

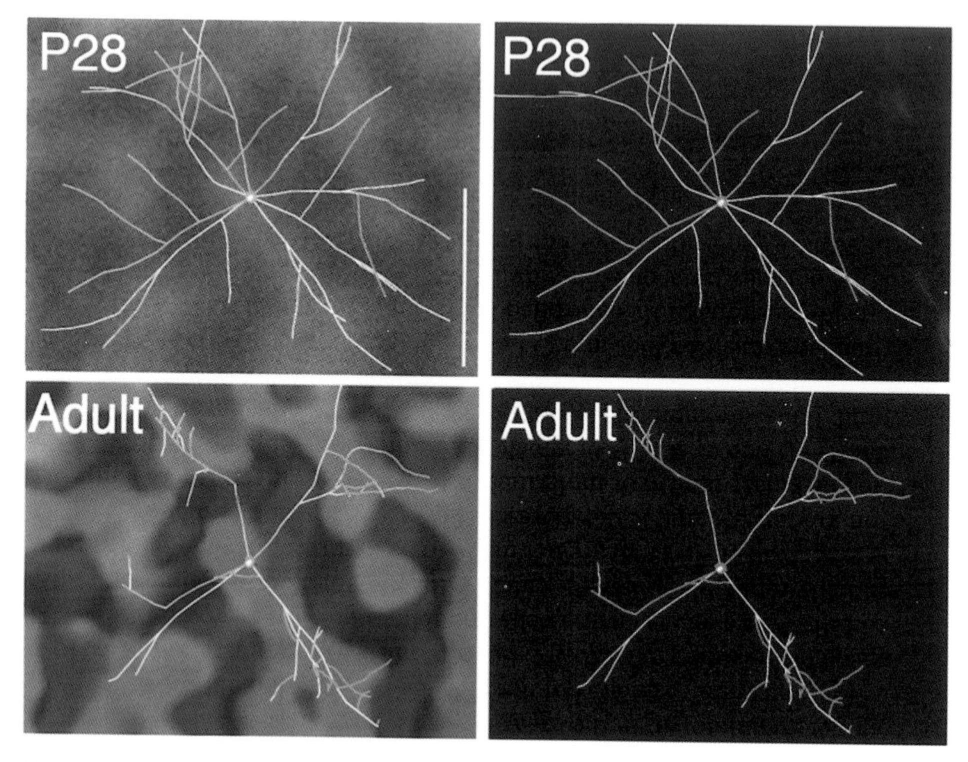

FIGURE 15–1 Morphological development of clustered horizontal projections in ferret visual cortex. Tangential views of the patterns of axonal collaterals (dendrites are not shown) are indicated on the right; the relationship of axon collaterals to orientation columns is shown on the left. By postnatal day (P) 28, which is prior to eye opening, lateral projections have reached their maximal lateral extent but have a random relationship to the emerging system of orientation columns. Through selective addition of collaterals in iso-orientation columns, and elimination of long, unbranched collaterals from inappropriate regions, the adult pattern of clustered projections emerged. Scale bar: 1 mm.

terns and the adult pattern and claimed that no period of "refinement" existed. However, recent evidence in ferrets (Durack and Katz, 1996), using more modern anatomical methods, makes it clear that for some period of time at least, horizontal axons form clusters that are less specific than those seen in the adult.

RELATIONSHIP BETWEEN ANATOMICAL AND FUNCTIONAL REARRANGEMENTS IN THE EMERGENCE OF CLUSTERED CONNECTIONS

While the anatomical rearrangements that accompany cluster formation are obvious using any of a number of anterograde, retrograde, or intracellular tracing methods, the functional significance of these rearrangements is an open question. Conspicuous anatomical rearrangements occur in many regions of the developing nervous system, and in most cases these are assumed to reflect changes in the distribution of functional synaptic connections. However, in most developing systems, it is extremely difficult to determine the relationship between patterns of anatomical projections and patterns of functional synaptic connections. While in adult brains, synaptic terminals or boutons are readily discernible using either light or electron microscopy, or immunohistochemistry for synaptic vesicle proteins, during early development morphologically defined synapses are often difficult to observe. In the developing cortex, developing axon collaterals do not form obvious boutons, staining with synaptic-vesicle–specific proteins reveals a broad distribution of vesicles and their associated proteins, and in the electron microscope, the signatures of mature synapses, such as clusters of vesicles and defined pre- and postsynaptic specializations, are absent (Aghajanian and Bloom, 1967; Cragg, 1972; Miller, 1981). In the specific case of the development of clustered horizontal connections, two functional approaches—scanning laser photostimulation and optical imaging with voltage-sensitive dyes—strongly suggest that the changes in functional connectivity are not entirely predictable from changes in anatomical projection patterns.

Assessing the functional changes that accompany the retraction and elaboration of horizontal axonal collaterals requires detecting changes in the strength, sign, and number of functional synaptic inputs either made or received by a given pyramidal cell. In the case of layer 2/3 pyramidal cells, we have used scanning laser photostimulation to map the pattern of functional inputs converging onto a single cell at various times during development. Photostimulation is based on the highly localized, laser-induced release of "caged" glutamate, combined with whole-cell patch clamp recordings in brain slices. By stimulating many thousands of different locations over several square millimeters of a cortical brain slice, it is possible to "map" the pattern of functional synaptic connections converting on a single, identified cell (Callaway and Katz, 1993; Dalva and Katz, 1994; Katz and Dalva, 1994). This can be done at a variety of postnatal ages in the developing ferret visual cortex, and, knowing the anatomical state of horizontal connections at the same ages (Durack and Katz, 1996), the correlation between structural and functional rearrangements can be determined.

From this sort of analysis, it is clear that there are significant differences between the anatomical and functional state of horizontal connections, especially at early stages in their development (Fig. 15–2) (Dalva and Katz, 1994). During the time that neurons are extending long, unbranched collaterals, photostimulation results demonstrate that neurons only receive inputs from cells located in the immediate vicinity of their cell body (within ~300 μm). This suggests that synapses are only formed along the proximal portions of the developing pyramidal cell axons, and few if any functional synapses originate from more distal sites. Thus, while axons initially cover an extensive area, for about a week they are incapable of sustaining activity over a similarly extensive area, and what activity they are able to generate remains confined to the approximate dimensions of a single cortical column.

Subsequently, during the period of formation of crude clusters, very large numbers of synapses are added to the proximal regions of the axon arbor, and, with the onset of axonal branching, functional synapses begin appearing in more distal regions. At this stage, we infer that functional interactions between neurons in different columns become possible, but most of a given neuron's intrinsic

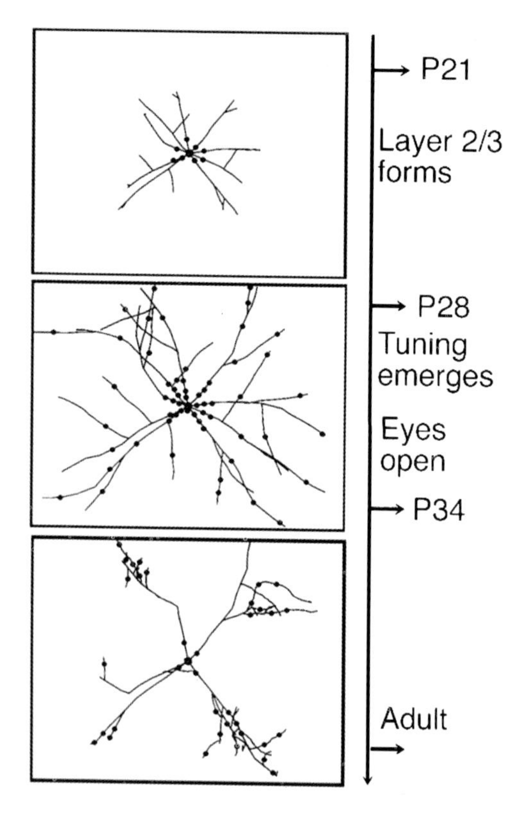

FIGURE 15–2 Functional connections and anatomical projections in the development of horizontal interactions, relative to major developmental milestones. The anatomical state of projections is shown in the drawings, based on intracellular dye injections; the functional state of synapses, as determined by photostimulation, is indicated by the black circles. At early ages long, unbranched axons appear to be devoid of functional synaptic connections and incapable of mediating long-range synaptic interactions. During the stage of "crude" clusters, neurons form a higher density of synapses along proximal regions of their axonal arbor. Subsequently many of these are lost, but numerous additional connections are added along the secondary and tertiary branches that comprise the clusters.

inputs arise from within the cell's "home" column. At this stage, lateral inputs, although numerous, are not located in the discrete clusters characteristic of the adult. As mature clusters form, the balance between local and longer-range inputs changes: With the onset of specific secondary and tertiary branching, the number of functional synapses originating from within the "home" column declines markedly, while those from more lateral regions are now organized into discrete clusters.

Thus, despite the presence of axon collaterals, at early ages these are incapable of transmitting synaptic activity over significant lateral distances. The ability to interact with more distant neurons begins to emerge only when axons begin branching, which in itself may serve as a cue for synaptogenesis. The observation that early horizontal axons are only weakly functional is buttressed by observations using microstimulation and real-time voltage-sensitive dye recordings. In these experiments (Nelson and Katz, 1995) the patterns of activity evoked by local stimulation were followed during the development of lateral connections. In agreement with the results from photostimulation, voltage-sensitive dye recordings (which detect primarily dendritic depolarization) show that activation is restricted to a narrow columnar region during the time that long, unbranched axons are present (from P21 to P28 in ferrets). Once branching begins and crude clusters emerge, the ability of these collaterals to evoke activity at more distant sites begins to increase, and by P43, when anatomical clusters are clearly evident, lateral activation detected optically has also reached adult levels.

Neuronal Activity and the Emergence of Clustered Connections

In contrast to the seemingly activity-independent mechanisms that guide the emergence of vertical connections, the emergence of the adult pattern of clustered connections requires neuronal activity. Prolonged binocular deprivation produced by eyelid suture results in clusters that are clearly abnormal: although neurons lose some unbranched axons and add numerous secondary branches, these secondary branches are distributed over much larger cortical regions than normal (even larger than the "crude" clusters present in young animals) (Callaway and Katz, 1991). Thus, on a morphological level at least, axonal arbors appear to undergo the same sequence of developmental events—loss of unbranched collaterals and addition of secondary branches—but they do so with much degraded specificity. This implies that some cue necessary for the proper organization of clusters has been degraded or eliminated by lid suture. The most likely cue is activation of orientation-selective cells, as lid suture attenuates all but the highest and lowest spatial frequencies.

The observations (1) that horizontal connections link isofunctional cortical regions and (2) that lid suture altered the organization of adult columns led us to suggest that the mechanism guiding the normal formation of clusters was based on the correlated activity of pre- and postsynaptic neurons in orientation columns responding to the same preferred orientation. In a series of elegant experiments, Lowel and Singer convincingly demonstrated that the pattern of horizontal connections could be predictably modified by alterations in the patterns of cortical correlations (Löwel and Singer, 1992). To do this, they rendered young cats strabismic, which greatly decreases the correlations in activity between the two eyes. It is well documented that this manipulation leads to increased segregation of ocular dominance columns and makes neurons throughout the thickness of the cortex monocular: neurons in each column respond almost exclusively to stimulation of one eye or the other, but not to both. Using combined 2-deoxyglucose labeling (to reveal the pattern of ocular dominance columns) and localized injections of a fluorescent retrograde tracer, they demonstrated that in normal animals, clusters had no special relationship to ocular dominance columns. However, in the strabismic animals, clusters were only found in ocular dominance columns with the same eye preference as that at the injection site. In normal animals, in which numerous binocular neurons are present, the only important correlation required for forming clusters is that neurons at different locations have similar orientation tuning. However, in the strabismic animals, even neurons with similar orientation tuning will not be correlated unless they are also in the same ocular dominance column.

The standard explanation for correlation-dependent patterns of connectivity is some variant of the coactivation rule of Hebb. In this model, certain connections (i.e., those present in the appropriate columns) would be selected and strengthened, presumably via a long-term potentiation LTP-like mechanism, while those in "inappropriate" orientation columns would be weakened (perhaps via long-term depression (LTD) and presumably lost. Although appealing, and widely used to explain this and similar phenomena (see, e.g., Bear and Kirkwood, 1993; Bear and Malenka, 1994; Kirkwood and Bear, 1994; Singer, 1995), several aspects of the development of clustered horizontal connections do not readily fit into such a "selectionist" model. As discussed above, both photostimulation and optical recording demonstrate that at the time the purported selection is occurring, remote horizontal connections are weak or absent. Thus, there is not a large, preexisting repertoire of synaptic connections from which to select on the basis of activity patterns. The major events in cluster development seem to involve the local *addition,* accompanied by axon branching, of large numbers of active synapses. Hence it seems more likely that the cues correlated activity is providing are not necessarily related to strengthening or weakening individual connections, but rather are cues required to initiate formation of new synapses and axon branches.

A possible mechanism might involve the local release of factors that are known to be involved in synaptogenesis and branch addition in cortex such as the neurotrophins (Cabelli et al., 1995; McAllister et al., 1995; Riddle et al., 1995). One need only posit that neurons in distant, coordinately activated columns release neurotrophins, and that only axon collaterals whose parent cell body has been also activated are capable of responding to the synaptogenic effects of the neurotrophins (Fig. 15–3). This model does not require that developing collaterals have functional synapses along their length, but only that the collaterals would be capable of responding (perhaps via the Trk family of neurotrophin receptors) to the presence of a synaptogenic factor. This "constructionist" view of the role of activity (Purves, 1988) seems to more plausibly account for the development of the system of clustered connections than mechanisms based on large-scale regressive events.

At this point, the source of the spontaneous activity driving the formation of clusters prior to eye opening is unknown, but retinal waves are especially attractive candidates (see Chapter 14). Waves are present in the retina of developing ferrets from birth until about P30 (Wong et al., 1993), which encompasses most of the time during which horizontal connections are first forming. Moreover, one can imagine that the "wavefront" of a retinal wave could produce an oriented line of activity in the retina; this could serve as an initial cue for constructing sys-

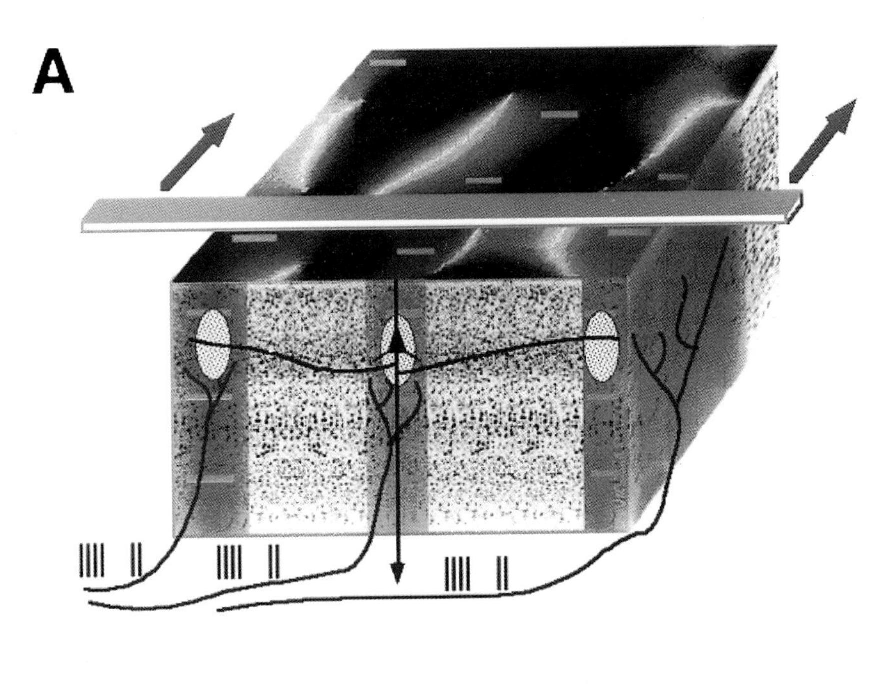

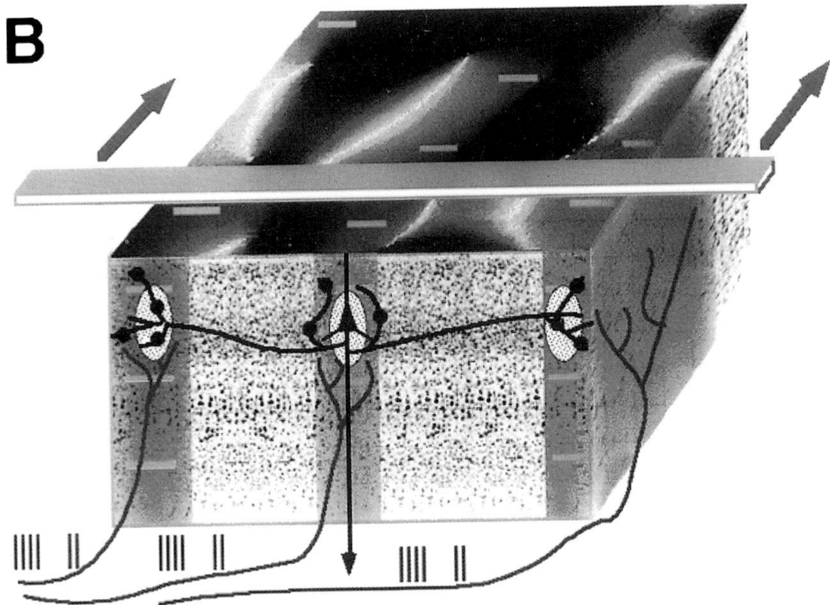

FIGURE 15–3 Development of clustered connections by directed growth. (**A**) In this hypothetical scenario, correlated activity (either spontaneous, or induced by a visual stimulous, such as the oriented bar shown above the cortex) from thalamic afferents produces local zones of high levels of secreted growth factors, such as neurotrophins. Only active collaterals can respond to these factors, leading, in **B,** to the local addition of branches and synaptic connections (indicated by the black circles). Note that this model does not require selection of a subset of synapses from a cohort of already-present functional synapses; as such, it differs significantly from correlation-based models that require synapses to be strengthened or weakened.

tems of connections related to orientation in the cortex. Although it is clear that both horizontal connections and orientation selectivity are influenced by activity, it will be necessary to specifically interfere with the pattern of waves in the retina to test whether their temporal or spatial pattern can actually provide cues required to construct these systems.

CLUSTERED CONNECTIONS AND THE EMERGENCE OF ORIENTATION TUNING

Since the discovery of orientation-tuned neurons, considerable experimental effort has been devoted to determining the circuitry that gives rise to this fundamental property of visual cortex. If we accept that patterned activity, probably related to orientation selectivity, is required to drive the formation of the adult pattern of clusters, then it is interesting to consider whether clusters form before, after, or along with the establishment of orientation columns. This is important, since if clusters developed well prior to orientation columns, it would be possible for them to serve as a scaffold for constructing the adult pattern of columns. Several lines of evidence from the anatomical and functional studies described above make it clear that there is an intimate relationship between the emergence of orientation tuning and the development of clustered horizontal connections.

This issue has been studied most carefully in ferrets. Because ferrets are born much earlier in their development than cats, it is possible to record stable neuronal responses even prior to the time of natural eye opening, a feat which is most difficult in cats. Chapman and Stryker (1993) carefully examined the emergence of orientation selectivity in the ferret and found that until approximately P32 only about 25% of cells met adult criteria for orientation selectivity. Thereafter, the proportion of orientation-tuned cells rapidly increased, reaching adult levels of 75% orientation-tuned cells by about P43. It appears, therefore, that orientation columns in ferret first emerge some time around P32, a conclusion buttressed by optical recordings of intrinsic signals. In such recordings, orientation columns can first be discerned at about this age; during the following week the orientation signals become stronger but the size and position of the nascent columns do not change (Bonhoeffer and Grinvald, 1996; Chapman et al., 1996). In Figure 15–4, the development of orientation tuning, based on extracellular single-unit recordings in vivo, is shown along with several measures of the anatomical and functional development of lateral connections. The proportion of orientation-tuned cells begins to increase at the same time that clusters begin to refine and when these collaterals become capable of transmitting excitation laterally, as assessed by voltage-sensitive dyes.

This comparison permits several conclusions. First, it is clear that neither functional horizontal connections nor their organization into clusters precedes the development of orientation tuning or columns. Thus, it is unlikely that some preexisting pattern of clusters serves to organize the pattern of orientation columns. On the flip side, the development of clusters does not substantially lag the emergence of orientation tuning and columns, implying that the system of

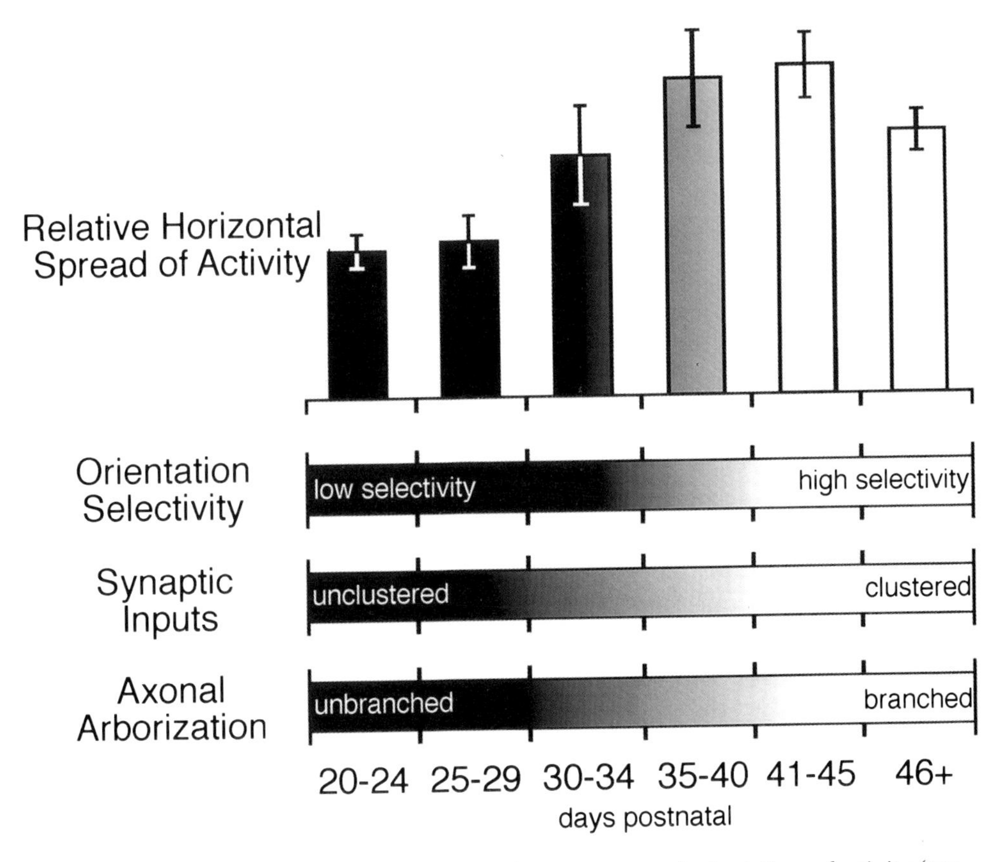

Figure 15–4 Temporal relationships between the emergence of orientation selectivity (top bar) and the state of horizontal connections as determined from voltage-sensitive dye recordings (top histogram), photostimulation (middle bar), and intracellular staining (bottom bar). Graduated shading indicates periods during which a parameter is changing. Long-range synaptic interactions, as determined using either voltage-sensitive dyes or photostimulation, develop only after axons begin branching; orientation–selectivity and functional long-range interactions develop in concert with one another. From Nelson and Katz (1995), with permission.

orientation columns does not develop first and then subsequently instruct the pattern of lateral connections. Instead, the development of these two systems, at this point, seems inextricably linked. It may be that incremental changes in the degree of orientation tuning incrementally improve the specificity of clusters, which in turn incrementally improves orientation tuning, leading to a feed-forward cycle of development.

There is as yet little direct evidence for a role of lateral connections in the development of orientation columns. Silencing cortical activity via tetrodotoxin (TTX), or attenuating it via binocular deprivation, both delay the developmental emergence of oriented cells, and both manipulations have effects on horizontal corrections as well (Callaway and Katz, 1991; Chapman and Stryker, 1993). However, this argument is again correlational. Attempts to disrupt the pattern of orientation columns by disrupting horizontal connections (using a piece of

Teflon inserted into the cortex) failed to uncover any deficit, suggesting that lateral interactions were not required to form orientation columns (Lowel and Singer, 1990). However, in these experiments the status of orientation columns was assessed using metabolic labeling with 2-deoxyglucose, which allows examining the organization of orientation domains responding to a single stimulus. In preliminary experiments, we have found that interrupting lateral interactions in layer 2/3 between P32 and 39 locally disrupts the contiguity of the pattern of orientation columns but not the degree of orientation tuning within individual columns (Weliky and Katz, 1995). Interruptions of these connections at P43, after they have reached their mature configuration, have no effect. Although suggestive, resolving the issue of the relationship between orientation column development and cluster development will require more direct experimentation to determine whether orientation columns can develop in the absence of lateral interactions, or, conversely, whether clustered connections can emerge even in the absence of obvious orientation columns.

CONCLUSIONS

Our current views of the development of cortical circuits are dominated by descriptive and phenomenological findings. While it is intriguing that vertical circuits result from highly specific patterns of outgrowth, the next stage of understanding can only be achieved by determining the layer-specific molecular cues that guide this specificity. It is possible that these cues are shared with other portions of the developing brain, but it is also possible that some entirely new classes of layer-specific molecules are waiting to be discovered. Advances in molecular techniques—such as differential display and subtractive hybridization—should allow the identification of numerous candidate molecules. Unfortunately, at this point many of the assays for the roles of different molecules are rather cumbersome and crude, relying on subtle differences in the degree of branching of neurites in various conditions. It also seems important to develop more reliable, direct assays to test the candidate molecules that arise from molecular screens. Given the emerging sophistication of transgenic and knockout mice, it also seems worthwhile to examine the developmental specificity of connections in a rodent system, which is likely to be more experimentally tractable than the carnivore and primate systems currently used.

For horizontal connections, two large issues remain outstanding: do these connections play a role in establishing the pattern of orientation columns, and how do patterns of activity alter the growth, branching, and retraction of these connections? It seems important to go beyond simply blocking activity with TTX or attenuating it with lid suture. More direct manipulations of the patterns of activity converging on the cortex, rather than simply manipulating the levels of activity, should provide more refined insights into the role of neuronal activity in constructing these circuits. In this case, examining these issues in mice is more problematic, as rodents do not seem to have clear patterns of orientation columns, and their horizontal connections, although present (Burkhalter and

Charles, 1990), are not nearly as well developed as in carnivores or primates. Despite this impediment, reexamining these systems in rodents may pay rich dividends in allowing more sophisticated manipulations of activity patterns and the molecules involved in transducing these patterns, such as the neurotrophins. Despite the fact that mice do not have defined ocular dominance columns, the development and plasticity of ocular dominance in mice share many characteristics with the same processes in cats and monkeys (Gordon and Stryker, 1996); the development of orientation selectivity is likely to be similar as well.

Acknowledgments

Work in my laboratory was supported by National Institute of Health grant EY07690 and by a Scholar's Award from the McKnight Foundation for Neuroscience. I am grateful to M. Dalva, D. Nelson, and M. Weliky for their many contributions to the work described here.

References

Aghajanian, G.K., Bloom, F.E. (1967). The formation of synaptic junctions in developing rat brain: a quantitative electron microscopic study. *Brain Res.* 6:716–727.

Allendoerfer, K.L., Cabelli, R.J., Escandon, E., Kaplan, D.R., Nikolics, K., Shatz, C.J. (1994). Regulation of neurotrophin receptors during the maturation of the mammalian visual system. *J. Neurosci.* 14:1795–1811.

Bear, M.F., Kirkwood, A. (1993). Neocortical long-term potentiation. *Curr. Opin. Neurobiol.* 3:197–202.

Bear, M.F., Malenka, R.C. (1994). Synaptic plasticity: LTP and LTD. *Curr. Opin. Neurobiol.* 4:389–399.

Bolz, J., Gotz, M., Hubener, M., Novak, N. (1993). Reconstructing cortical connections in a dish. *Trends Neurosci.* 16:310–316.

Bonhoeffer, T., Grinvald, A. (1996). Optical imaging based on intrinsic signals. In *Brain mapping: the methods,* Toga, A. W., Mazziotta, J. C., eds. San Diego: Academic Press, pp. 55–97.

Burkhalter, A., Bernardo, K.L. (1989). Organization of corticocortical connections in human visual cortex. *Proc. Natl. Acad. Sci. U.S.A.* 86:1071–1075.

Burkhalter, A., Charles, V. (1990). Organization of local axon collaterals of efferent projection neurons in rat visual cortex. *J. Comp. Neurol.* 302:920–934.

Burkhalter, A., Bernardo, K.L., Charles, V. (1993). Development of local circuits in human visual cortex. *J. Neurosci.* 13:1916–1931.

Cabelli, R.J., Hohn, A., Shatz, C.J. (1995). Inhibition of ocular dominance column formation by infusion of NT-4/5 or BDNF. *Science* 267:1662–1666.

Callaway, E.C., Katz, L.C. (1990). Emergence and refinement of clustered horizontal connections in cat striate cortex. *J. Neurosci.* 10:1134–1153.

Callaway, E.C., Katz, L.C. (1991). Effects of binocular deprivation on the development of clustered horizontal connections in cat striate cortex. *Proc. Natl. Acad. Sci. U.S.A.* 88: 745–749.

Callaway, E.C., Katz, L.C. (1993). Photostimulation using caged glutamate reveals functional circuitry in living brain slices. *Proc. Natl. Acad. Sci. U.S.A.* 90:7661–7665.

Chapman, B., Stryker, M.P. (1993). Development of orientation selectivity in ferret visual cortex and effects of deprivation. *J. Neurosci.* 13:5251–5262.

Chapman, B., Stryker, M.P., Bonhoeffer, T. (1996). Development of orientation maps in ferret primary visual cortex. *J. Neurosci.* 16:6443–6453.
Chino, Y.M., Kaas, J.H., Smith, E.L., Langston, A.L., Cheng, H. (1992). Rapid reorganization of cortical maps in adult cats following restricted deafferentation in retina. *Vision Res.* 5:789–796.
Cragg, B.G. (1972). The development of synapses in cat visual cortex. *Invest. Ophthalmol.* 11:377–385.
Dalva, M.B., Katz, L.C. (1994). Rearrangements of synaptic connections in visual cortex revealed by laser photostimulation. *Science* 265:255–258.
Das, A., Gilbert, C.D. (1995a). Long-range horizontal connections and their role in cortical reorganization revealed by optical recordings of cat primary visual cortex. *Nature* 375:780–784.
Das, A., Gilbert, C.D. (1995b). Receptive field expansion in adult visual cortex is linked to dynamic changes in strength of cortical connections. *J. Neurophysiol.* 74:779–792.
Drescher, U., Kremoser, C., Handwerker, C., Loschinger, J., Noda, M., Bonhoeffer, F. (1995). In vitro guidance of retinal ganglion cell axons by RAGS, a 25 kDa tectal protein related to ligands for Eph receptor tyrosine kinases. *Cell* 82:359–370.
Durack, J.C., Katz, L.C. (1996). Development of horizontal projections in layer 2/3 of ferret visual cortex. *Cereb. Cortex* 6:178–183.
Frantz, G.D., Weimann, J.M., Levin, M.E., McConnell, S.K. (1994). Otx1 and Otx2 define layers and regions in developing cerebral cortex and cerebellum. *J. Neurosci.* 14:5725–5740.
Galuske, R.A. W., Singer, W. (1996). The origin and topography of long-range intrinsic projections in cat visual cortex: a developmental study. *Cereb. Cortex* 6:417–430.
Gilbert, C.D. (1992). Horizontal integration and cortical dynamics. *Neuron* 9:1–13.
Gilbert, C.D. (1994). Learning. Neuronal dynamics and perceptual learning. *Curr. Biol.* 4:627–629.
Gilbert, C.D., Wiesel, T.N. (1979). Morphology and intracortical projections of functionally characterized neurons in the cat visual cortex. *Nature* 280:120–125.
Gilbert, C.D., Wiesel, T.N. (1983). Clustered intrinsic connections in cat visual cortex. *J. Neurosci.* 3:1116–1133.
Gilbert, C.D., Wiesel, T.N. (1989). Columnar specificity of intrinsic horizontal and corticocortical connections in cat visual cortex. *J. Neurosci.* 9:2432–2442.
Goodman, C.S. (1994). The likeness of being: phylogenetically conserved molecular mechanisms of growth cone guidance. *Cell* 78:353–356.
Gordon, J.A., Stryker, M.P. (1996). Experience-dependent plasticity of binocular responses in the primary visual cortex of the mouse. *J. Neurosci.* 16:3274–3286.
Horton, J.C., Hocking, D.R. (1996). An adult-like pattern of ocular dominance columns in striate cortex of newborn monkeys prior to visual experience. *J. Neurosci.* 16:1791–1807.
Hubener, M., Gotz, M., Klostermann, S., Bolz, J. (1995). Guidance of thalamocortical axons by growth-promoting molecules in developing rat cerebral cortex. *Eur. J. Neurosci.* 7:1963–1972.
Kaas, J.H. (1991). Plasticity of sensory and motor maps in adult mammals. *Annu. Rev. Neurosci.* 14:137–167.
Katz, L.C. (1991). Specificity in the development of vertical connections in cat striate cortex. *Eur. J. Neurosci.* 3:1–9.
Katz, L.C., Callaway, E.M. (1992). Development of local circuits in mammalian visual cortex. *Annu. Rev. Neurosci.* 15:31–56.
Katz, L.C., Dalva, M.B. (1994). Scanning laser photostimulation: a new approach for analyzing brain circuits. *J. Neurosci. Methods 54:205–218.*
Kennedy, H., Salin, P., Bullier, J., Horsburgh, G. (1994). Topography of developing thalamic and cortical pathways in the visual system of the cat. *J. Comp. Neurol.* 348:298–319.
Kirkwood, A., Bear, M.F. (1994). Hebbian synapses in visual cortex. *J. Neurosci.* 14:1634–1645.

Livingstone, M.S., Hubel, D.H. (1994). Specificity of intrinsic connections in primate primary visual cortex. *J. Neurosci.* 4:2830–2835.

Lowel, S., Singer, W. (1990). Tangential intracortical pathways and the development of iso-orientation bands in cat striate cortex. *Brain Res. Dev.* 56:99–116.

Löwel, S., Singer, W. (1992). Selection of intrinsic horizontal connections in the visual cortex by correlated neuronal activity. *Science* 255:209–212.

Lubke, J., Albus, K. (1992). Rapid rearrangement of intrinsic tangential connections in the striate cortex of normal and dark-reared kittens: lack of exuberance beyond the second postnatal week. *J. Comp. Neurol.* 323:42–58.

Luhmann, H.J., Singer, W., Martinez-Millan, L. (1991). Horizontal interactions in cat striate cortex: I. anatomical substrate and postnatal development. *Eur. J. Neurosci.* 2:344–357.

Lund, J.S., Boothe, R.G., Lund, R.D. (1977). Development of neurons in the visual cortex of the monkey (Macaca nemestrina): a Golgi study from fetal day 127 to postnatal maturity. *J. Comp. Neurol.* 176:149–188.

Luo, Y., Raible, D., Raper, J.A. (1993). Collapsin: a protein in brain that induces the collapse and paralysis of neuronal growth cones. *Cell* 75:217–227.

Malach, R., Amir, Y., Harel, M., Grinvald, A. (1993). Relationship between intrinsic connections and functional architecture revealed by optical imaging and in vivo targeted biocytin injections in primate striate cortex. *Proc. Natl. Acad. Sci. U.S.A.* 90:10469–10473.

Martin, K.A.C. (1988). From single cells to simple circuits in the cerebral cortex. *Q. J. Exp. Physiol.* 73:637–702.

Martin, K.A.C., Whitteridge, D. (1984). Form, function and intracortical projections of spiny neurons in striate visual cortex of the cat. *J. Physiol.* 353:463–504.

McAllister, A.K., Lo, D.C., Katz, L.C. (1995). Neurotrophins regulate dendritic growth in developing visual cortex. *Neuron* 15:791–803.

Merzenich, M.M., Sameshima, K. (1993). Cortical plasticity and memory. *Curr. Opin. Neurobiol.* 3:187–196.

Miller, K.D. (1994). A model for the development of simple cell receptive fields and the ordered arrangement of orientation columns through activity-dependent competition between ON- and OFF center inputs. *J. Neurosci.* 14:409–441.

Miller, M.W. (1981). Maturation of rat visual cortex. II. A combined Golgi-electron microscope study of pyramidal neurons. *J. Comp. Neurol.* 203:555–573.

Molnar, Z., Blakemore, C. (1995). How do thalamic axons find their way to the cortex? [Review]. *Trends Neurosci.* 18:389–397.

Nelson, D.A., Katz, L.C. (1995). Emergence of functional circuits in ferret visual cortex visualized by optical imaging. *Neuron* 15:23–34.

Purves, D. (1988). *Body and brain: a trophic theory of neural connections.* Cambridge, Mass.: Harvard University Press.

Riddle, D.R. Lo, D.C., Katz, L.C. (1995). NT-4-mediated rescue of lateral geniculate neurons from effects of monocular deprivation. *Nature* 378:189–191.

Rockland, K.S., Lund, J.S. (1983). Intrinsic laminar lattice connections in primate visual cortex. *J. Comp. Neurol.* 216:303–318.

Roskies, A.L., O'Leary, D.D. (1994). Control of topographic retinal axon branching by inhibitory membrane-bound molecules. *Science* 265:799–803.

Sato, M., Lopez-Mascaraque, L., Heffner, C.D., O'Leary, D.D. (1994). Action of a diffusible target-derived chemoattractant on cortical axon branch induction and directed growth. *Neuron* 13:791–803.

Schwab, M.E., Kapfhammer, J.P., Bandtlow, C.E. (1993). Inhibitors of neurite growth. *Annu. Rev. Neurosci.* 16:565–595.

Shatz, C.J. (1990). Impulse activity and the patterning of connections during CNS development. *Neuron* 5:745–756.

Singer, W. (1995). Development and plasticity of cortical processing architectures. *Science* 270:758–764.

Weliky, M., Katz, L.C. (1995). Disruption of horizontal connections in developing visual cortex induces orientation map discontinuities. *Soc. Neurosci. Abstr.* 25:509.6.
Wiser, A.K., Callaway, E.M. (1996). Contributions of individual layer 6 pyramidal neurons to local circuity in Macaque primary visual cortex. *J. Neurosci.* 16:2724–2739.
Wong, R.O., Meister, M., Shatz, C.J. (1993). Transient period of correlated bursting activity during development of the mammalian retina. *Neuron* 11:923–938.
Yoshioka, T. (1994). Development of clustered lateral connections in macaque visual cortex. *Soc. Neurosci. Abstr.* 24:457.3.

Index